1995

Species Diversity in Ecological Communities

Species Diversity in Ecological Communities

HISTORICAL AND GEOGRAPHICAL PERSPECTIVES

EDITED BY
Robert E. Ricklefs
and
Dolph Schluter

THE UNIVERSITY OF CHICAGO PRESS • CHICAGO AND LONDON

Robert E. Ricklefs is professor of biology at the University of Pennsylvania. Dolph Schluter is associate professor of zoology at the University of British Columbia.

The University of Chicago Press, Chicago 60637
The University of Chicago Press, Ltd., London
© 1993 by The University of Chicago
All rights reserved. Published 1993
Printed in the United States of America
02 01 00 99 98 97 96 95 94 93   1 2 3 4 5

ISBN: 0-226-71822-0 (cloth)
        0-226-71823-9 (paper)

Library of Congress Cataloging-in-Publication Data

Species diversity in ecologcal communities : historical and
    geographical perspectives / edited by Robert E. Ricklefs and Dolph
    Schluter.
        p.   cm.
      Includes bibliographical references (p. 365) and index.
      1. Species diversity.   2. Biotic communities.   3. Biogeography.
    I. Ricklefs, Robert E.   II. Schluter, Dolph.
    QH541.15.S64S63 1993
    574.5′24—dc20   93-16747

Contents

Preface

This book grew out of our conviction that the discipline of community ecology would benefit from a broadening of its paradigms. Ecological studies of the past thirty years have presumed that interactions among populations within small areas are the fundamental forces regulating community structure. However, this paradigm failed to solve one of the monumental problems of biology: the origin and maintenance of global patterns of biodiversity. A tenet common to the ecological theories of diversity, which were based largely on interspecific competition and enemy-victim interactions, was that coexistence of species would depend predictably on local environmental conditions. Yet, disparities were often found in the numbers of species present in similar environments in different parts of the globe, hinting that larger-scale processes were also at work and might even dominate local ones. Until recently, community ecologists largely ignored such vexing disparities. Indeed, from the 1960s to the 1980s ecology largely spurned its sister disciplines of systematics, biogeography, and paleontology and the insights they provided into the larger-scale processes and historical events that have influenced species richness. By the late 1980s, however, the intellectual climate of ecological study was changing. The failure of the local-process paradigm to achieve a consensus on the causes of diversity opened the door for fresh approaches.

In our casual discussions at several meetings, we discovered a common interest in a new approach to community ecology. Robert Ricklefs has had a long-standing interest in species diversity and had incorporated historical thinking in understanding island biogeography in a series of papers with George W. Cox on the concept of "taxon cycles." Dolph Schluter had worked on the ecology of bird communities on islands and continents, investigating the evolution of assemblages of Darwin's finches in the Galápagos Archipelago and of other finch taxa in mainland Kenya. In the mid-1980s both of us turned our attention to global patterns of species diversity, in particular to the degree of convergence (or lack thereof) in species richness between ecologically similar but geographically separate localities. This common interest led us to organize a symposium, The Historical and Geographical Determinants of Community Diversity, which was held at the annual meeting of the Ecological Society of America in Snowbird, Utah, in August 1990.

We had already decided prior to the meeting that the topic required broader coverage than a symposium afforded and that an edited volume containing a variety of viewpoints would make an important contribution to the changing discipline. We solicited contributions from a large number of colleagues who had active interests in species diversity and well-developed perspectives. We were pleased and somewhat horrified by the enthusiastic response that followed, a response that accounts for the size of this book. Although several chapter authors contributed to the symposium, this book should not be viewed as conference proceedings. We selected authors on the basis of two criteria: we desired a broad and even coverage of approaches to the problem of species diversity, ranging from the local, contemporary perspectives to the global and historical points of view; we also chose authors with diverse, strongly held opinions. The authors of this volume have admirably and gratifyingly met our expectations.

One hopes that a new book will be all things to all people. Our hopes for *Species Diversity in Ecological Communities* are that the book will provide a summary statement of the discipline for researchers actively concerned with diversity issues; that it will serve as a basis for discussions and seminars for graduate students in ecology; that it will help shift the collective opinion of ecologists to a more balanced view of the factors responsible for global patterns of biodiversity; that it will help bring about a reunion between community ecology and evolutionary biology, systematics, biogeography, and paleontology.

To whatever extent this book realizes these ambitions, its success will have resulted from the efforts of many individuals. By all accounts the authors of this volume have been enthusiastic, conscientious, and prompt. Certainly they have made the task of the editors easy, exciting, and pleasurable, and they contributed substantially to the editing process by providing insightful reviews of other chapters. Susan Abrams, the book's editor at the University of Chicago Press, has been a constant source of encouragement, wisdom, and friendship. The editorial and production skills of many individuals have gone into this project, and we give special thanks to Jean Eckenfels and Norma Roche.

In the end, we hope that this book will stimulate discussion and new inquiry into one of the most fascinating and challenging phenomena in nature—global patterns of species diversity.

June 1993

Robert E. Ricklefs   Dolph Schluter
Philadelphia   Vancouver

1

Species Diversity
An Introduction to the Problem

Dolph Schluter and Robert E. Ricklefs

Early in this century, species diversity was regarded primarily as a historical phenomenon, a reflection of the accumulation of species over time, and thus a subject outside the realm of the developing discipline of ecology (Wallace 1876; Willis 1922; Fischer 1960). By the early 1960s, the rise of population biology had changed all that (Kingsland 1985), and diversity had come to be perceived as the outcome of ecological interactions, particularly competition, that were resolved quickly within small areas and habitats (MacArthur 1972). For the past three decades, biologists have struggled with ecological concepts of the maintenance of diversity, and their confidence in their accomplishment has waxed and waned.

Recently, ecologists have begun to deal seriously with ecological processes across large temporal and spatial dimensions, recognizing connections between the local habitat and global biogeography and between the moment and the long history of life on earth (Ricklefs 1987). Many ecologists have embraced a more balanced view that patterns of diversity are caused by a variety of ecological and evolutionary processes, historical events, and geographical circumstances. Now we must develop new concepts, data, and analyses that match the scales of the processes responsible for diversity phenomena. Charting this new terrain has been the motivation for this book, which builds upon other recent syntheses of community ecology (Strong et al. 1984; Diamond and Case 1986; Kikkawa and Anderson 1986; Gee and Giller 1987; Wiens 1989a). This volume, however, concentrates on diversity as its general theme and highlights regional processes and historical events.

These considerations suggest an expanded program for the ecological study of community diversity. We emphasize the importance of research in the following areas: (1) continued assessment of hypotheses based on local interactions among species, including the degree to which local assemblages exchange individuals and species with one another; (2) continued testing of the prediction of community convergence through comparative analysis; (3) evaluation of the relationship between local and regional diversity; (4) specification of the temporal and spatial scales of ecological and evolutionary processes; and (5) exploration of the role of unique historical and geographical circumstances in the development of communities through time. These endeavors will require a broadening of the concept of the ecological community and the incorporation of novel kinds of data into community analysis. To a varying degree, the chapters in this book explore this agenda; we provide a more detailed assessment of what is needed in our concluding chapter at the end of this volume.

This book has three purposes. The first is to review modern ideas about how local and "mesoscale" processes influence the number of coexisting species. The purely local perspective is addressed by the chapters in part I ("Local Patterns and Processes"), which summarize theories relating diversity to local processes, including the relationship between diversity and habitat productivity. Mesoscale models address spatially structured landscapes of habitat patches or graded ecological conditions (part II, "Coexistence at the Mesoscale"). Mesoscale approaches retain the flavor of population thinking in that they model systems in which the equilibrial coexistence of species depends on the local structure of the system. They also expand the geographical scale of ecological concern by recognizing that the properties of a local site depend to a large degree on processes external to it (emigration and immigration). The chapters in part II explore the implications of population interactions in subdivided landscapes for diversity within landscapes, including the relationship between species and area.

The second purpose of this book is to present case studies of various regions and taxa. These provide a fresh look at the phenomena of biodiversity and emphasize the importance of historical, geographical, and phylogenetic information. Part III ("Regional Perspectives") includes chapters that explore patterns of diversity within biogeographical regions or within particular taxa, emphasizing the empirical basis of diversity phenomena, assessing the principle of community convergence, and evaluating various explanations of diversity patterns. Diversity does not correspond perfectly to conditions of the local physical environment; similar habitats on different continents often support strikingly dissimilar numbers of species (Orians and Paine 1983; Schluter and Ricklefs, chap. 21). Such disparities force us to entertain the possibility that patterns of diversity are influenced by historical and geographical circumstances as well as by local ecological factors. Island biotas have taught us that diversity is a reflection of the size and geographical position of an island as well as the variety of its habitats (MacArthur and Wilson 1967). Continental biotas similarly may reflect geography and age (e.g., Cody 1975). Part IV ("Historical and Phylogenetic Perspectives") sets out historical and phylogenetic approaches to studying the development of ecological communities, using information from systematics, biogeography, and paleontology to reconstruct the history of diversity within regions.

Our inability to explain global patterns of biodiversity through local environmental circumstances is a starting point in the search for more powerful explanatory tools. Hence our third purpose is to suggest the beginnings of a program of research into biodiversity on expanded spatial and temporal scales. We wish to provide a new framework for the study of diversity that emphasizes the use of comparative, geographical, and historical data to investigate the development of biological communities. Broader concepts demand new kinds of data and new methods of analysis and interpretation. At a time when ecological study has become increasingly reductionist and experimental, we emphasize the merits of comparative information—a revitalized natural history—and the idea that patterns of diversity may derive, at least in part, from ecological and evolutionary processes expressed in unique settings. There is a strong parallel here between studies of diversity and studies of phenotypic evolution. Evolutionists accept heredity and selection as general principles, but also recognize that different forms of life arise through unique, unpredictable, and unrepeatable historical sequences. Just as morphology and behavior are in part phylogenetically constrained, so do patterns of diversity have a historical, evolutionary component.

THE PHENOMENA OF SPECIES DIVERSITY

General Patterns

The most fundamental data of diversity are the numbers of species in different places. Ecologists have discovered relationships between these data and latitude, climate, biological productivity, habitat heterogeneity, habitat complexity, disturbance, and the sizes and distances of islands. These patterns have been reviewed so often (for example, Fischer 1960, 1961; Pianka 1966; Stehli, Douglas, and Newell 1969; Brown and Gibson 1983; Stevens 1989; Ricklefs 1990) that we will mention them here only in broad outline. Several of these relationships have suggested mechanisms that might regulate diversity; a general and comprehensive theory of diversity must account for all of them.

Within most groups of organisms, the average number of species in a sampling area of a given size reaches its maximum in tropical latitudes and decreases both northward and southward toward the poles. In many cases, the latitudinal gradient in diversity is very steep. Tropical forests, for example, may support ten times as many species of trees as forests with similar biomass in temperate regions (Latham and Ricklefs 1993). The ability to predict this pervasive pattern has been regarded as a crucial test of diversity theories, although so many factors vary in parallel with latitude that latitudinal patterns cannot by themselves support any one theory over another (Connell and Orias 1964; Pianka 1966; Ricklefs 1977).

Occasionally, the general trend in diversity is reversed, as it is for shorebirds, parasitoid wasps, and freshwater zooplankton, of which more species occur at high and moderate latitudes than in the tropics. These counterexamples may reflect the latitudinal distribution of particular habitat types (tundra does not exist in the tropics), the history of the evolution of a taxon (for example, the Betulaceae—birches and poplars—almost certainly diversified in the north temperate region, to which they are largely restricted at present), or ecological circumstances peculiar to a particular group. For example, Janzen (1981) has suggested that the extreme host specialization of parasitoids has mitigated against the diversification of these taxa in the tropics, where their hosts tend to be rare.

Species richness appears to be related generally to climate (Terborgh 1973). In particular, conditions that favor biological production—warm temperatures and abundant precipitation in terrestrial ecosystems—are often associated with high diversity (Richerson and Lum 1980; Currie and Paquin 1987; Currie 1991; Wright, Currie, and Maurer, chap. 6). Ecologists recently have recognized peaks of diversity at intermediate levels of biological production (Whittaker and Niering 1975; Tilman 1982). This pattern has led to the proposition that low habitat fertility reduces diversity through nutrient stress and that high fertility removes the limitations imposed by nutrient stress, resulting in simplified communities as the outcome of competitive exclusion (see Tilman and Pacala, chap. 2; Rosenzweig and Abramsky, chap. 5).

Disturbance may also influence species richness. Landscapes typically include a mosaic of patches of disturbance of different extent and intensity such that each patch exists in some stage of succession, out of equilibrium with the prevailing climate in its species composition (Watt 1947). Thus, the entire mosaic, which is part of a larger regional equilibrium, contains more species than any individual patch. This observation is the basis of the "intermediate disturbance hypothesis" (Connell 1975), in which disturbance is regarded as a stress that precludes species at high levels and fails to prevent competitive exclusion by a few superior competitors at low levels (Paine 1966; Connell 1978; Huston 1979; Keough 1984), resulting in the greatest species diversity at an intermediate level of disturbance.

Most of these ideas have their origins in empirically perceived patterns. Other processes implicated in the regulation of diversity have occurred to ecologists as a result of more theoretical considerations. These include density-dependent predation (Paine 1966; Janzen 1970; Clark and Clark 1984), and temporal and spatial stochasticity (Sale 1977, 1978; Hubbell 1979; Levins 1979; Chesson and Warner 1981; Thiery 1982; Chesson 1986, 1991; Clarke 1988). Theories based on these processes obtain little empirical support from geographical patterns of diversity because ecologists lack a general understanding of the geographical distribution of these factors. That ecologists generally have a higher regard for empirically based hypotheses emphasizes our difficulty in distinguishing between phenomena as inspirations for hypotheses and as tests of the mechanisms proposed to explain them.

Deviations from the nearly pervasive relationship between diversity and habitat productivity have directed attention to other patterns. One of the most striking of these is the connection between diversity and habitat complexity. Salt marshes are extremely productive but harbor few species of plants and animals; deserts occupy the other end of the productivity gradient but may support a diverse flora and fauna. MacArthur and MacArthur (1961) formalized the relationship between the diversity of bird species and the complexity of vegetation, noting that the number of bird species in a habitat varies in direct proportion to the number of layers of vegetation (basically herb, shrub, and tree). Similar relationships between diversity and habitat complexity have been found in other groups (e.g., Pianka 1967).

Regional or landscape complexity has also been implicated in patterns of diversity. Ecologists have long been aware of the greater diversity of mountainous regions compared with flatlands (Simpson 1964b; Cook 1969). This pattern may arise because of the increased numbers of species distributed allopatrically on isolated mountains or in isolated valleys, the greater variety of habitats included within sampling areas of virtually any size, and the increased numbers of species coexisting within habitats.

Finally, the influence of geography on diversity has long been apparent in island settings, where the number of species tends to increase with island size and to decrease with distance from sources of colonists (Hamilton et al. 1963; Connor and McCoy 1979; Abbott 1980). MacArthur and Wilson's (1963, 1967) theory portraying diversity on islands as a balance between colonization and extinction emphasized the influence of both local and regional processes on the diversity of local communities (fig. 1.1). It is ironic that island biogeography gained its major impetus as a field of study at the same time that ecologists began to seek means of looking at local community diversity that assigned a smaller role to the influence of external processes.

Definition of the Community

Communities usually are defined by spatial, functional, or taxonomic association or by dynamic interactions within the food web (Shimwell 1971; MacArthur 1972). Within local associations of potentially interacting taxa, species may be compartmentalized into smaller ecological units, often called "guilds," within which interactions are strong but between which interactions are weak (Root 1967; Yodzis 1982; Pimm 1982; Cohen 1989). Food web analysis has suggested that the number of feeding links per species is more or less independent of diversity, suggesting that the fine-scale structure of the food web is conserved (Briand 1983b; Cohen 1989; Sugihara, Schoenly, and Trombla 1989; Martinez 1992; cf. Pimm and Lawton 1980). Ecologists have not determined the degree to which such detailed characterization of diversity phenomenology is necessary to understanding the generation and maintenance of geographical patterns of diversity. The most useful description of diversity and its organization undoubtedly relates to the particular processes that determine diversity. However, without a knowledge of these processes, the measurement of diversity instead must depend on hypotheses concerning its regulation. The common practice of enumerating species does not take into account their varied ecological roles and presumes implicitly that communities are organized in a more or less homogeneous fashion. More detailed appraisals of food webs and niche relationships relate local interactions to the regulation of diversity and to the organization of community structure.

For the most part, communities have a spatially open structure, in which the species that coexist in a given locality may extend more or less independently into other communities (Gleason 1926; Whittaker 1953, 1972). Community boundaries associated with physical discontinuities may be apparent on a local scale, for example, in delimiting the biota of small ponds, the flora of serpentine barrens, the fauna of pitcher plants, or the herbivores of a particular plant. However, even within communities that are locally discrete, species replace one another geographically. Clearly, the strength of interactions weakens at increasing temporal, spatial, or ecological distance; a community is not limitless in extent. It is equally apparent, however, that functional concepts of community embody a hierarchy of ecological interactions at different

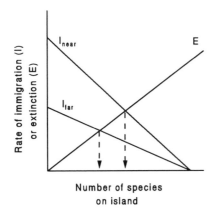
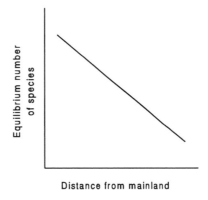

Figure 1.1 MacArthur and Wilson's (1963, 1967) model relating species diversity on islands to distance from the nearest mainland source of colonists. Equilibrium diversity reflects the balance between rate of extinction and rate of immigration, the latter being higher for islands close to the mainland.

scales of time, space, and both ecological and evolutionary distance (Allen and Starr 1982; O'Neill et al. 1986).

The landscape includes a multitude of partially overlapping distributions of species (Whittaker 1967; Terborgh 1971, 1985). The distributional boundaries of each species reflect its relationship with the environment, including physiological responses to physical characteristics and various interactions with other species (including predators, prey, pathogens, and hosts, as well as pollinators, dispersal agents, and other mutualists). Every point in time and space, or any sample within a larger temporal and spatial frame, includes a certain diversity of species, higher taxa, and ecological types. Thus, the phenomenology of community diversity includes patterns of variation with respect to the size of the sample (e.g., species-area curves), ecological conditions (e.g., latitudinal gradients in diversity), and similarities between samples (e.g., species turnover along gradients and comparisons between continents).

The Measurement of Diversity

We can portray diversity most simply as species richness, that is, number of species. But because richness increases in direct relation to number of individuals, area, and variety of habitats sampled, differences in sampling methods may introduce statistical and ecological bias into comparisons of diversity among different localities. As more individuals are included, the probability of discovering rare species increases. The relationship between diversity and sample size depends on the probability distribution of taxa among abundance classes. These distributions have been described by a variety of mathematical relationships, each of which predicts a somewhat different relationship between diversity and sample size and suggests a different approach to normalizing diversity with respect to the number of individuals sampled (Pielou 1969, 1977; Whittaker 1972; Magurran 1988). Normalization procedures transform counts of species to a rate (d) at which species (S) are added with expansion of the size of the sample (N). Thus, d becomes a measure of diversity.

In comparative studies, either sampling should be consistent with respect to number of individuals, or species counts should be normalized with respect to sample size (Simberloff 1979). Ecological variation over the temporal and spatial dimensions of the sample may augment diversity because of the increased number of areas, habitats, or seasons included. However, samples of large ecological dimension usually include large numbers of individuals as well. Consequently, different sample-related causes of variation in diversity may confound one another in comparative studies (Southwood 1961; Southwood, Moran, and Kennedy 1982; Karban and Ricklefs 1983; see Underwood and Petraitis, chap. 4).

A related problem involves the ecological importance of common and rare species. A simple count gives equal weight to all taxa, whether they occur repeatedly in a sample or are represented by a single individual. Ecologists have devised diversity indices that weight the contributions of species according to their abundance, usually discounting rare species to some degree (Hurlbert 1971; Whittaker 1972; Pielou 1977). However, because the abundances of species within samples tend to exhibit regular patterns of distribution (Whittaker 1972; May 1975a), sample size, species richness, and various indices of species diversity are generally interrelated. Therefore, species richness, preferably normalized with respect to sample size, is a suitable measure for most broad-scale comparisons of diversity.

Comparisons of diversity at taxonomic levels higher than species (i.e., genus, family, order) may provide clues to the origin of diversity patterns to the extent that taxonomic level reveals the historical development of biotas. In addition, ecological diversity reveals the degree to which species fill niche space and may therefore provide insight into the regulation of local diversity. Taxonomic diversity may or may not be closely related to ecological diversity (Cousins 1991). The latter may be estimated directly, and independently of taxonomic diversity, by calculating diversity indices for the variety of ecological roles, diet components, and microhabitat uses exhibited by an assemblage of species (e.g., Root 1967; Pianka 1973). Alternatively, one may apply multivariate ordination techniques to define spaces whose axes are constructed from morphological or ecological information, or from morphological measurements as surrogates for ecological information (e.g., Karr and James 1975; Gatz 1979; Ricklefs and Travis 1980; Miles and Ricklefs 1984). The volume of such a space estimates the ecological diversity of the sample. Taxonomic species richness and ecological or morphological diversity correspond closely in studies of birds and mammals (Ricklefs and Miles 1993), although this may not be the case for other taxa (e.g., Lawton, Lewinsohn, and Compton, chap. 16).

Alpha, Beta, and Gamma Diversity

Whittaker (1972) recognized that the total diversity within a large area, which he called gamma diversity, could be partitioned into two components: local (alpha) diversity and turnover of species between habitats or localities (beta diversity). This partitioning of diversity is central to understanding the contribution of large-scale processes to local diversity because it elucidates the connection between local and regional species richness. Whittaker related total diversity to its components by the expression.

gamma diversity = alpha diversity $\times$ beta diversity.

Because alpha and gamma diversity in this formula are expressed in units of number of species, beta diversity is a dimensionless number, which Whittaker (1972) referred to as "full changes in, or turnovers of, species composition." Thus, two identical samples have a beta diversity of 1, while two unique samples with the same alpha diversity have a beta diversity of 2.

Cody (chap. 13) has suggested that in addition to turnover of species between habitats, species also replace one another over distance within a habitat type, a phenomenon he refers to as the gamma component of regional diversity. Cody's gamma diversity may reflect the allopatric distribution of closely related taxa; it may also express changes in physical environment within a given habitat type.

Whittaker's beta diversity includes the separate contributions that the habitat breadths of species and the variety of habitats in the region make to regional diversity. Alternatively, beta diversity may be thought of as an average property of species within a region and may be distinguished from contributions to regional diversity made by habitat variety and other dimensions of the total sample. Accordingly, beta diversity would be the inverse of the average number of localities or habitats occupied by each species within the larger region. In this case, the contribution of beta diversity to regional diversity would be measured by multiplying it by the number of localities or habitats in the regional sample, that is, by the dimension of the sample. Thus,

gamma diversity (number of species) = alpha diversity (number of species in a locality or habitat) × beta diversity (inverse of the specific dimension, e.g., 1/mean number of habitats or localities occupied by a species) × sample dimension (total number of habitats or localities).

In this way, we can differentiate between the separate contributions of species properties (habitat breadth) and regional heterogeneity (number of habitats). For example, in nine habitats in Trinidad, Cox and Ricklefs (1977) tabulated a total of 108 species of passerine birds. The average diversity per habitat (alpha diversity) was 28.2 species, and the average beta diversity among species was 0.43 habitats^{-1}. Thus, within rounding errors,

$$108 \text{ species} = 28.2 \text{ species} \times 0.43 \text{ habitats}^{-1} \times 9 \text{ habitats.}$$

In the same nine habitat types on the small island of St. Kitts in the northern Lesser Antilles, regional (island) diversity was partitioned as follows:

$$20 \text{ species} = 11.9 \text{ species} \times 0.19 \text{ habitats}^{-1} \times 9 \text{ habitats.}$$

Thus, the difference in regional diversity between Trinidad and St. Kitts reflects differences in both alpha diversity and beta diversity (habitat breadth), with the size of the habitat sample held constant.

In this example, habitat was discrete and arbitrarily defined. In practice, habitat can be quantified as a continuous variable (for example, vegetation height, elevation, temperature, precipitation, or distance per se); beta diversity thus would be the inverse of the average dispersion of habitat values of a species. Beta diversity would be multiplied by the range of habitat values within the region to obtain a dimensionless number that scales local diversity to regional diversity.

In principle, alpha diversity also can be partitioned into niche space components within the local sample. Each local assemblage of species occupies a certain niche volume, within which its total diversity equals the product of the average number of species occupying each point in niche space, the inverse of the average volume of niche space occupied by a species, and the total volume of niche space occupied by the entire assemblage. MacArthur (1972) captured the essence of this relationship when he described the number of species in the community (S) by the expression

$$S = \frac{R}{U}(1 + ca)$$

where R is the total niche volume, U the average volume occupied by a species, a the average niche overlap between neighboring species, and c the number of neighbors in the niche space. The ratio R/U is the number of niches of size U that can be fit into volume R without overlap; $1 + ca$ is related to the average number of species occupying each point in niche space.

Irrespective of how each component of diversity is labeled, the partitioning of diversity into point coexistence, the inverse of the number of points occupied by a species, and the number of points within the entire sample provides a framework for conceptualizing the structures of communities in terms of both species and environmental characteristics. Up to this point, studies of diversity have not produced samples of sufficient detail to resolve this hierarchy of structure in ecological systems. Comparisons, including most of the examples in this volume, have relied on data accumulated in diverse, usually descriptive studies undertaken for unrelated purposes. However, much of what follows in this book will clarify the need for systematic sampling of diversity over large gradients of ecology and geography.

A HISTORY OF IDEAS ABOUT DIVERSITY

Diversity as a Historical Phenomenon

Early explanations for geographical patterns of diversity primarily were based on geographical and evolutionary history. Wallace (1876) attributed the higher diversity of the tropics to the uninterrupted persistence of tropical climates there. Toward the poles, diversity was periodically set back through phases of extinction brought about by severe climate changes associated with glaciation. In short, the tropics had simply had more time to accumulate species.

The theme of history was given its fullest expression in J. C. Willis's influential 1922 book, *Age and Area*. Willis was a plant systematist familiar with the geography and ecology of the floras of several areas, particularly India and Ceylon; his meticulous studies revealed to him a pervasive relationship between area of geographical distribution and local abundance. Willis suggested that species arise by macromutation (saltation). As a result, newly formed species usually are not well suited to their environments and remain rare at the site of their origin until they have adapted by Darwinian evolution to local conditions. During the phase of adaptation, the species becomes more abundant and spreads, occupying progressively larger and larger areas, and gives rise to new species by macromutations within its population (fig. 1.2). Moreover, the larger the area over which this process occurs, the greater the population of each species, and the greater the probability that additional species will arise by mutation. Willis believed that species accumulate over time and that the diversity of a region is therefore directly related to its age and its area.

Regardless of the particulars of Willis's theory, which

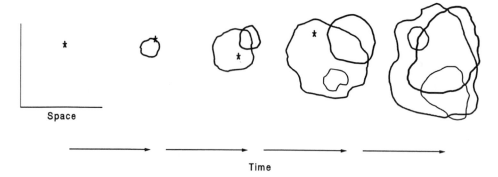

Space

Time

Figure 1.2 A representation of Willis's (1922) model for the accumulation of species within areas over time. Species are produced by macromutation and then spread as they adapt to local conditions, subsequently giving rise to new species by saltation.

seems quaint to us now, it represented thinking about diversity through much of the early part of this century (Fischer 1960; Ross 1972a). Ecology textbooks published during the 1940s and 1950s did not deal with diversity at all; instead, they were preoccupied with the relationship of individual organisms to their physical environments, the dynamics of populations, and the functioning of ecosystems. Diversity apparently was considered outside the realm of ecology and, at any rate, as a subject not amenable to scientific investigation.

The Rise of Community Ecology

By 1960, ecologists had claimed patterns in species diversity as a problem for investigation and had embarked on three decades of intense research into community ecology. The early development of population biology has been chronicled lucidly by Kingsland (1985), and the consequences of this development for the study of species diversity have been touched on by Ricklefs (1987). Briefly, during the 1920s, interest in projecting the growth of human populations led to the development of mathematical models of population dynamics (Pearl and Reed 1920). These models were the basis from which Lotka, Volterra, Gause, and Nicholson developed the mathematics of species interactions during the late 1920s and early 1930s. One of the major results of this work was the principle of competitive exclusion, which held that species closely matched ecologically would compete strongly for resources and therefore could not coexist.

This prediction was quickly put to the test by laboratory experiments in which related species were placed together in containers with limited amounts and varieties of resources (Gause 1934). Almost always, one species outcompeted and excluded the other, usually within a few tens of generations (Miller 1967). It was almost thirty years before ecologists would conduct field experiments on a large scale to investigate the efficacy of competition and other interactions in natural populations (Connell 1961a, 1961b, 1983; Schoener 1983; Sih et al. 1985).

David Lack (1944) was the first modern ecologist to point out that closely related species living together in nature might coexist by partitioning resources between them. Others such as Grinnell (1917) and Elton (1927) had noted such differences in ecological roles but had not

placed their observations in the context of community theory.

Hutchinson (1957, 1959) extended Lack's insights to the partitioning of resources within the community as a whole and conceptualized the ecological positions of species as occurring in a multidimensional niche space. The axes of this space were the various biotic and abiotic factors in the environment along which species could partition resources. MacArthur related this formal concept of the niche to the problem of species diversity. Noting the phenomenon of resource partitioning among ecologically similar species (warblers, *Dendroica*: MacArthur 1958) and that the number of species varied in direct relation to the complexity of the environment (number and length of resource axes: MacArthur and MacArthur 1961), he suggested that competitive interactions between species limited the packing of species in niche space and thus placed an upper limit on the number of species in a community (MacArthur 1964, 1965). By then, "community" was defined in the functional sense as an assemblage of interacting species.

Throughout this development of community theory, MacArthur and others recognized that while local diversity might be constrained by ecological interactions, the species richness of a large geographical region was influenced by historical processes and events (MacArthur 1969). Thus, certain regions might contain comparatively many or comparatively few species due to their isolation from sources of colonists or to extinctions during the Pleistocene. But local habitats could nonetheless become saturated ecologically and achieve a predictable local level of diversity. Regional diversity could vary while local diversity remained constant if the turnover of species between habitats and localities (beta diversity: Cody 1975) increased in direct relation to regional diversity.

By 1967, MacArthur and Levins had formalized the concept of limiting similarity, that is, the maximum degree to which to species could overlap in niche space and continue to coexist locally (MacArthur and Levins 1967; May and MacArthur 1972). Soon theoreticians had begun to explore model communities consisting of matrices of interaction coefficients between every pair of species in the local assemblage (MacArthur 1972; Vandermeer 1972). May (1975b) and others began to examine the dynamical and equilibrium properties of these systems during the 1970s (see Case 1990; Drake 1990b). Among the conclusions of this body of work were that (1) diverse communities were inherently less stable than depauperate ones; (2) less diverse communities could be invaded more readily than more diverse ones; (3) stability (i.e., the rate of return to equilibrium) increased as the strength of interaction coefficients increased; and (4) environmental stochasticity generally reduced the probability of species coexistence (see Tilman and Pacala, chap. 2).

Regardless of how one views the applicability of these theoretical studies, they strengthened the notion that the equilibrium properties of communities, including the number of coexisting species, were determined largely by ecological interactions—primarily competition—between species within the context of the local physical environment. Food web theory (Pimm 1982; Cohen 1989; Yodzis, chap. 3), and its conclusions about food chain length and level of connectance between trophic categories of species, arose from the same tradition of investigating the dynamical properties of model systems.

Community theory developed in tandem with several other lines of investigation. Model systems were devised to measure the interaction terms of the community matrix directly and, particularly, to search for nonlinearities and higher-order interactions between species (Vandermeer 1969; Wilbur 1972; Neill 1974). Ecologists attempted to characterize the community matrix by field measurements of ecological overlap (Pianka 1973; Cody 1974; Schoener 1974) or, as a proxy for ecology, by morphological similarity (Hespenheide 1973; Ricklefs and Travis 1980; Grant and Schluter 1984; Schoener 1984). Other researchers tried to determine whether distributions of closely related species were strongly consistent with competition theory (e.g., altitudinal distributions of birds: Terborgh 1971, 1985; Diamond 1975; Noon 1981; Schluter 1982; Schluter and Grant 1982). Finally, this period saw the growth of a strong tradition of field experimentation in community ecology.

If competition and other interactions are in fact responsible for community properties, then one should be able to demonstrate resource partitioning and the existence of species interactions in natural settings. By and large, ecologists have not been disappointed. No two species are exactly alike; resource partitioning can be demonstrated within most groups of organisms. Experimental removals of competitors and predators have revealed strong interactions among species in a large proportion of cases (Schoener 1983; Connell 1983; Sih et al. 1985). Local assemblages of species appear to be highly interactive and to partition resources in such a way as to reduce competition. However, some types of organisms are not well represented in these experiments (e.g., phytophagous insects, pelagic fishes), and it remains to be seen whether such conclusions also apply to them.

During the past fifteen years, the theory of community diversity has been expanded in significant new directions. Abrams (1977) showed that the condition of limiting similarity does not hold under all consumer-resource dynamics in model systems. This and other results led to the realization that community theory makes few general predictions as to what we should see in nature. The theory is nonetheless useful in suggesting what *might* occur, and specific models can be tailored for application to specific cases. Tilman (1982, 1988) developed resource-based models of the interactions between consumers that predicted negative, rather than positive, relationships between the productivity of a habitat and the number of species that could coexist. Discussion of these ideas may be found in this volume (see Tilman and Pacala, chap. 2; Rosenzweig and Abramsky, chap. 5). Other theories suggested that more than one species could exist on a single resource in the presence of certain types of environmental variation (Armstrong and McGehee 1980; Fagerström 1988), thereby decoupling to some extent the predicted relationship between resource diversity and species diversity (see, however, Chesson 1991).

Diversity theory and research developed from the premise that one could understand the structure of a community by understanding processes that occurred over ecological scales of time within small areas of habitat. Mesoscale models incorporating spatial subdivision or continuous spatial variation in the environment (see part II of this volume, "Coexistence at the Mesoscale") extended the scope of the ecological community and produced some important results. These included the realization that the number of species occurring in a locality is greatly affected by external as well as internal processes and that local communities need not be locally stable in order to persist.

Challenges to Diversity Theory

Competition-based theories of community structure predict that species should distribute themselves nonrandomly in niche space, either through evolutionary character displacement (Lack 1947; Grant 1972, 1986) or as a result of the competitive exclusion of close neighbors, depending on the relative rates of evolutionary divergence and exclusion. A spatial corollary to this prediction suggests that ecologically similar species should be nonrandomly distributed geographically; this was investigated primarily by examining the distributions of taxa among independently colonized islands (Diamond 1975). In the late 1970s these predictions came under statistical scrutiny (Strong, Szyska, and Simberloff 1979; Connor and Simberloff 1979; Simberloff and Boecklen 1981). Detailed studies of Darwin's finches in the Galápagos archipelago have since essentially confirmed Lack's conclusions with regard to character displacement (Schluter and Grant 1984; Schluter, Price, and Grant 1985). In many people's minds, however, the statistical controversy over nonrandom spacing called into question the basic premises of community theory, and community ecology appeared to shift direction considerably during the ensuing decade. In particular, community ecologists have focused more closely on experimental studies of the nature of population interactions than on the consequences of those interactions for the coexistence of species within communities.

The statistical challenge to competition-based diversity theory was a healthy and necessary development of the discipline of community ecology, but it left many basic issues concerning community diversity unresolved. Competition can influence diversity when the probability of competitive exclusion increases in direct relation to the number of species in the community (MacArthur and Wilson 1963). If competition is to limit diversity, its impact must swamp the effects of other processes. Demonstrating evolutionary consequences of competition, such as character divergence, and even the existence of resource partitioning between competing species provides neither necessary nor sufficient support for the hypothesis that interactions between species regulate the actual number of species present.

Furthermore, the inability to demonstrate nonrandom niche occupancy or spatial distribution is not sufficient evidence to reject competition-based diversity theory; species form open associations without fixed spatial boundaries, and the evolutionary effects of competition need not express themselves within a particular locality. Evolution integrates selection throughout a species' entire range, which includes many associations of species having only partially overlapping membership. Thus, the effects of contrasting selective pressures in localities among which individuals disperse may maintain the ecological position of a species out of evolutionary equilibrium with local interactions. Showing that the dispersion of species in ecological space does not differ from random placement with respect to the positions of other species (Ricklefs and Travis 1980) does not disprove the efficacy of competition in regulating diversity.

Testing the Predictions of Convergence and Local Saturation of Species Diversity

In its simplest manifestation, the theory of local diversity states that community diversity is the outcome of competition and predation among its members, and that this outcome depends upon the physical conditions of the environment, such as productivity, disturbance, and structural complexity of the habitat. It leads to three general predictions: First, local diversity should be correlated with features of the environment, especially the diversity of resources. Second, if competition is a strong force, diversity in local sites should be near saturation; that is, diversity within any habitat type should reach a ceiling that is independent of the number of species in the regional pool (saturation: Terborgh and Faaborg 1980). Third, independently assembled communities in similar habitats on different continents should contain similar numbers of species.

Cody (1966a) noted similarities in bird species richness in comparisons of North American and South American grassland communities. However, Recher (1969) was the first to formally attempt to test convergence of community diversity; he compared the relationship between bird species diversity and diversity in foliage height in eastern North America and in Australia, and found them to be similar (fig. 1.3). Since then, other tests have produced mixed results (Cody 1975; Orians and Paine 1983); many counterexamples to the principle of convergence—what Ricklefs and Latham (chap. 20) call "diversity anomalies" (similar habitats supporting different numbers of species

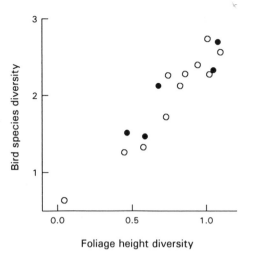

Figure 1.3 The relationship between bird species diversity and foliage height diversity in North America (open symbols) and Australia. (After Recher 1969.)

in different parts of the world)—also have been noted. Schluter (1986) developed a formal statistical treatment of convergence, which we apply to a variety of data sets (Schluter and Ricklefs, chap. 21). Most of these show strong local habitat effects on diversity that are consistent across regions (convergence). But the data also reveal substantial differences between ecologically similar sites in different regions, which suggests that historical and geographical circumstances peculiar to specific regions leave their imprint on local community characteristics.

Habitat saturation was first investigated by Terborgh and Faaborg (1980), who found that the local diversity of birds on West Indian islands increased with total island-wide diversity of bird species but leveled off above a certain limit (saturation). Using a data set obtained from different West Indian islands, Cox and Ricklefs (1977) observed that local diversity varies consistently with habitat but also increases monotonically with total island diversity (Ricklefs 1987). Both alpha and beta diversity increase as approximately the square root of total island (regional, or gamma) diversity (fig. 1.4). Cornell (1985b) tested the principle of community saturation in cynipid gall wasps. He considered individual species of oaks in California as habitat islands, and related the average local diversity of wasps on each species of oak to the total number of species recorded over the whole range of the oak species. Cornell found no evidence for saturation.

The Resurgence of Historical/Geographical Thinking in Community Ecology

The comparative study of communities was an important undertaking in the 1960s and 1970s, when several investigators tested whether communities in different parts of the world had converged in structure and diversity (e.g., Cody 1974, 1975; Pianka 1975; Karr 1976; Mares 1976; Pearson 1977; Cody and Mooney 1978; Shmida 1981). Several authors noted differences in diversity among regions (incomplete convergence), which they ascribed to

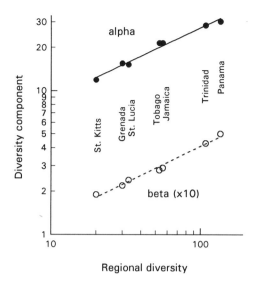

Figure 1.4 Relationship of alpha diversity (average number of species per habitat) and beta diversity (1/average number of habitats per species) to regional (island) diversity in birds of the Caribbean area. Based on standardized censuses of nine matched habitats at each locality. (Data from Cox and Ricklefs 1977; Wunderle 1985.)

historical or other factors. Karr (1976) explicitly cited historical factors to explain differences in diversity between African and Central American bird communities. Pearson (1977) concluded, "From theorized geological history alone one can predict the rank order of the number of bird species on the six [tropical dry forest] plots. The importance of species numbers to much of community structure theory and the biotic interactions associated with co-occurring species makes historical factors a vital part of understanding these communities."

Although community theory based on local processes has assumed prominence in the ecological literature of the past three decades, historical explanations of diversity surfaced frequently in literature peripheral to ecology. Harking back to A. R. Wallace, Fischer (1960) believed that, "Biotic diversity is a product of evolution and is therefore dependent upon the length of time through which a given biota has developed in an uninterrupted fashion," adding that, "the low-level tropics represent that part of the globe which has been least affected by climate fluctuations." Whittaker (1972) stated that although "alpha diversities of birds, and gamma diversities of islands, appear to reach saturation or steady-state levels . . . for terrestrial plants and insects increase of species diversity, with elaboration of niche hyperspace and division of the habitat hyperspace, is a self-augmenting evolutionary process without any evident limit." In Ross's (1972a) view, published in the same issue of *Taxon*, ". . . the input side of the species diversity of a community is a combined function of (1) the geologic longevity of its ecological conditions and (2) the number of times the community area undergoes geographic disjunctions and reconnections." To these authors, history and especially evolutionary change were crucial terms in the diversity

equation, but because local and regional diversity were not clearly differentiated, these insights did not touch upon the issues confronting community ecologists at the time.

Systematists, biogeographers, and paleontologists emphasized the large-scale spatial and temporal components of diversity while overlooking MacArthur's distinction between local and regional diversity. Ecologists concerned with the local community were at the same time beginning to explore the effects of larger-scale processes on local species coexistence. These efforts were based on the spatially partitioned models referred to earlier and on the recognition that local populations may be maintained by immigration from elsewhere (supply-side ecology: Roughgarden, Gaines, and Pacala, 1987; source-sink models: Shmida and Wilson 1985; Pulliam 1988; Stevens 1989; see also Holt, chap. 7).

This same period witnessed the development of equilibrium models of diversity on islands by MacArthur and Wilson (1963, 1967). In these models, the number of species reflected the outcome of opposing processes of extinction (a local process) and colonization (a regional process), but these ideas were never applied to mainland diversity. Whittaker (1972) emphasized that island diversity was not equivalent to local (alpha) diversity. Indeed, few parallels were drawn between models of island (regional) diversity and those of local community (alpha) diversity, the two being made effectively independent by the phenomenon of habitat specialization (beta diversity).

Recently, the isolation of the local community as an ecological unit has begun to break down in the minds of ecologists, who now acknowledge that communities comprise a variety of processes that occur over different spatial and temporal dimensions. Therefore, processes external to the local community—on whatever scale the local community may be defined—may influence the number of species it contains. These larger-scale patterns may be attributed, first, to deterministic or stochastic processes within a regional ecological landscape (see part II, "Coexistence at the Mesoscale"), and second, to unique historical and geographical circumstances and events. Landscape effects include coexistence within spatially subdivided habitats and habitat mosaics in which local interactions that tend to reduce diversity are balanced by the influx of individuals and species from other patches of the same habitat and from other habitats. Migration influences community equilibria within ecological dimensions of time and space. The properties of local communities (i.e., those occupying patches within landscapes) assume equilibrium statistical distributions. Over the landscape as a whole, diversity depends on the spatial configurations of habitats and the dispersal properties of populations. Within larger regions, processes that affect the production of species over evolutionary time come into play.

Unique patterns of diversity may be created by generalizable population processes operating within unique settings; each outcome reflects a singular course of community development. Although historical considerations may account for a substantial part of the patterns of diversity, we cannot use them as a basis for theory, and they can be elucidated solely by historical reconstruction.

THE CURRENT ISSUES

Community ecology is entering a period of spirited integration and synthesis. Much of this resurgence will focus on the temporal and spatial dimensions of processes that influence species diversity. At least seven types of processes contribute to patterns of diversity.

1. Local ecological interactions within small areas of uniform habitat are the population processes of classic diversity theory. Both competition and predation tend to reduce diversity through the elimination of taxa (local extinction), although predation may promote coexistence of prey species under some circumstances (e.g., Paine 1974; Witman 1987). These processes may constrain diversity locally but do not adequately address the regional production of species. Unless local communities are ecologically saturated, local ecological interactions are necessary, but are not sufficient, to explain patterns of diversity.

2. The movement of individuals between patches of the same kind of habitat underscores the importance of regional (external) processes and the ephemeral nature of local populations, and hence the dynamic nature of the local community. A statistical equilibrium is achieved over a number of small habitat patches, which we can characterize by a mean and variance in the number of patches occupied by a population and by persistence times and sizes of populations, both in individual patches and in the area as a whole. This level of process includes both the steady-state and stochastic components of regional extinction (MacArthur and Wilson 1967; MacArthur 1969; Holt, chap. 7).

3. The dispersal of individuals between habitats, variously referred to as mass effects, rescue effects, and source-sink dynamics, reflects the mosaic nature of the ecological landscape and the interdependence of local and regional diversity due to migration between habitats (Shmida and Wilson 1985; Pulliam 1988; Auerbach and Shmida 1987; Stevens 1989). Accordingly, species can be maintained in habitats that are unproductive by immigration from more productive ones. Thus, local diversity may reflect the variety and size of habitat patches within the larger region.

4. The spread of taxa within regions according to their habitat of origin and their subsequent ecological diversification may be responsible in part for prevalent relationships between habitat and diversity (see Schluter and Ricklefs, chap. 21). Taxa originate and diversify within certain habitat types and require evolutionary change to expand into other habitats (see Farrell and Mitter, chap. 23; Ricklefs and Latham, chap. 20). Thus, the relation-

ship between diversity and habitat may depend upon the histories and sizes of habitats as well as upon ecological conditions within particular habitat types. These considerations recognize that community development, including extinction of taxa, has a long evolutionary history constrained by ecological conservatism of taxa within clades (Ricklefs and Cox 1972; Bambach 1985; Vermeij 1987; Jablonski 1987; Jablonski and Bottjer 1990a, 1991b; Farrell, Mitter, and Futuyma 1992; Ricklefs and Latham 1992).

5. Allopatric production of species within regions depends on the particular geographical configurations of habitats, which of course differ in their influence according to the dispersal abilities and other properties of taxa. Regions with different spatial arrangements of habitats and barriers to dispersal, as well as different climates, may vary markedly in rates of species production and consequent regional diversity (Fedorov 1966; Cracraft 1985). While it is often cited in discussions of diversity, speciation has not been studied systematically from ecological and biogeographical perspectives (Rosenzweig 1975; Otte and Endler 1989).

6. The exchange of taxa between regions often depends on unique events and geographical configurations, such as those that occur when barriers between major land masses or ocean basins break down or when habitats are displaced by global climate changes and glaciation. Prominent examples include the completion of the land bridge between North and South America approximately three million years ago, which resulted in extensive migration of animals, particularly from north to south (Stehli and Webb 1985). The intermittent opening of the Bering land connection has also resulted in the exchange of taxa between Asia and North America on a large scale (Stucky 1990). Because biotal exchange may elevate diversity within regions of mixing (Vermeij 1991b), instances of exchange may provide insights into the regulation of local diversity.

7. Finally, many types of unique events may lead to episodes of extinction that reduce diversity for periods long enough to require cladogenesis and biotic exchange for its recovery. Pleistocene cooling and glaciation caused massive extinctions and displacements of communities, particularly in Europe (Latham and Ricklefs, chap. 26). Mass extinctions caused by bolide impacts or other global catastrophies have occasionally reset the diversity clock for many taxa (Flessa 1986; McLaren and Goodfellow 1990, Kauffman and Fagerstrom, chapter 27).

We hope that the 1990s will be a decade of synthesis for ecological, evolutionary, biogeographical, systematic, and paleontological studies. The issue of species diversity provides a common ground for all these disciplines, from which a richer understanding of the natural world will develop.

PART ONE

Local Patterns and Processes

Of the various ideas presented in this book, one is paramount: species diversity in any locality is a balance between two opposing sets of forces. On one hand, local abiotic processes, interactions between species, and chance tend to reduce diversity. On the other hand, immigration from other communities outside the locality tends to augment diversity. Part I of this book is concerned with the first set of forces: local processes that constrain local species diversity.

Numerous lines of evidence demonstrate that local forces affect the number of coexisting species, although their mechanisms are incompletely understood. Researchers have found changes in species diversity following experimental nutrient enrichment, predator removal, predator introduction, and climate alteration. The commonest outcome of such experiments is local extinction, leading to a reduction in diversity. However, it is conceivable that such perturbations, while causing the loss of species unable to accommodate changes in their environment, also create new ecological opportunities that may be filled by other species over long periods by adaptive radiation and invasion of the locality.

Tilman and Pacala open with a review of theoretical ideas on processes affecting the stable coexistence of competitors and the maintenance of local diversity. Simple theory predicts that species richness can be no greater than the number of limiting resources, a prediction difficult to reconcile with the facts. For example, hundreds of species of phytoplankton may inhabit a lake in which only three or four nutrients are limiting. Tilman and Pacala identify many restrictive assumptions of the simple theory (no spatial or temporal variation, equilibrium is attained, simple trophic structure, no limiting physical factors, no neighborhood effects, and simple life histories). They then explore the theoretical consequences for diversity of relaxing these assumptions. Their survey has special relevance for plants, but most of their ideas apply to animals as well. Tilman and Pacala apply some of these ideas to understanding diversity patterns in nature, including the pervasive latitudinal gradient in species diversity. This empirical theme also is picked up by later chapters.

A major issue addressed by the chapters in Part One is the relationship observed between species diversity and productivity. On a global scale, the relationship is apparently positive, as demonstrated by Wright, Currie, and Maurer. These authors suggest that the pattern results from the greater extinction probability (or lower incidence) of species at the low population sizes sustained by low productivity. Tilman and Pacala note that this explanation may not apply to the latitudinal gradient in terrestrial plant species diversity (which is also a productivity gradient), because the overall density of individual plants in a given plant formation is not greater in the tropics than in temperate regions.

When localities within a region are compared, the relationship between productivity and diversity is usually hump-shaped, as demonstrated by Tilman and Pacala and by Rosenzweig and Abramsky. Tilman and Pacala suggest that the relationship results from the reduced heterogeneity of resources at the productivity extremes. For example, for plants, the productivity gradient extends from habitats with low nutrient supply and high light intensity to habitats with high nutrient supply and low light intensity. Low diversity results at the low productivity extreme because all species are limited by soil nutrients, and the best competitor dominates. At the high productivity extreme all species are light-limited, and the best competitor for light dominates. In between these extremes, some species are light-limited while others are nutrient-limited, and coexistence results.

Rosenzweig and Abramsky survey other explanations for the hump-shaped pattern and suggest an interesting alternative of their own. They argue that position along the productivity gradient may be a evolutionarily conservative trait—a single clade cannot exploit the full range of productivities. They further suggest that limited use of the productivity gradient by a single clade is enforced by competition between clades, each of which is superior at a different point along the continuum (other phylogenetic constraints besides competition could also be operating). This explanation may not apply to hump-shaped relationships that include all clades of a potential guild, such as all Borneo trees, but the idea is worth investigating. It is not known whether clades exploit a limited range of productivities; such a pattern would be most likely when productivity reflected other environmental conditions such as temperature or soil type.

Tilman and Pacala mention a novel explanation for the hump-shaped pattern, attributed to J. S. Denslow, namely, that diversity is highest at intermediate productivities simply because intermediate productivity is the most widespread environmental condition. Diversity is lower at the productivity extremes because such conditions are present in only a small area of habitat. These ideas present great challenges to the empirical biologists who will devise tests for them. They should also motivate further attempts to describe the relationship between diversity and productivity, since most existing information is based on indirect measures of productivity.

Yodzis reviews studies of the environmental causes of "trophospecies" diversity, the number of trophic categories connected in a community food web. Thus he emphasizes functional categories within communities, rather than species number per se, paying particular attention to the possible causes of food chain length. Of special note is the weak or negative relationship observed between trophodiversity and environmental productivity; this contrasts with the clearly positive relationship seen for species diversity within trophic categories (Wright, Currie, and Maurer).

Ecologists have yet to determine which environmental factors constrain local diversity; this problem will remain an active area for further research. Progress will come in part from experimental approaches, a theme taken up by Underwood and Petraitis. Their primary concern is how (and whether) one can compare the results of experiments carried out in different locations. They raise practical issues, such as the confounding effects of the use of different plot sizes by different researchers and the different ranges of factor effects tested (e.g., predator densities). They also identify the statistical problems of comparing studies in which the strength of effect is measured using only standard sums of squares in ANOVA. One solution to the last problem may be to present not only typical ANOVA statistics from such experiments, but also the "response surface" of estimated effects, which would allow researchers to compare results over the overlapping ranges of experimental effects.

The effects of local factors on the diversity of species coexisting in small areas of homogenous habitat will continue to be an important area of ecological research. The chapters in Part One develop a strong theoretical basis for devising comparative and experimental analyses of the relationship between diversity and the local environment.

The Maintenance of Species Richness in Plant Communities

David Tilman and Stephen Pacala

In 1959, G. E. Hutchinson posed one of the major conceptual dilemmas of ecology: How is it that numerous species are able to persist in the same habitat? Although many natural habitats are species-rich, classical models predicted that an environment should be dominated by one or a few species. Hutchinson offered a broad outline of potential solutions to this paradox of diversity. He pointed out that the classical models assumed that interactions went to equilibrium, that food webs were composed of just two trophic levels, and that habitats were spatially homogeneous. He asserted that each of these assumptions was violated in nature, and that these violations might allow numerous species to coexist. This idea spawned theoretical and experimental studies which have now offered numerous alternative solutions to the paradox of diversity. In this chapter we summarize and critique these alternative theories and briefly mention relevant observational and experimental data. Our goals are to point out the many alternative mechanisms that can maintain biodiversity and to suggest ways to test for these. Although most theories apply to organisms on any trophic level, we focus on the maintenance of species-rich terrestrial plant communities.

The purpose of ecological theory is to provide simplified explanations of patterns in nature and to explore the logical implications of alternative simplifying assumptions. Theories of the maintenance of local species richness can be classified as those that assume (1) limitation by resources versus limitation by resources and physical factors in combination; (2) a spatially homogeneous habitat versus a spatially heterogeneous habitat; (3) long-term equilibrium versus nonequilibrium conditions; (4) a habitat in which colonization is unimportant versus a habitat in which patches are linked by colonization; (5) a simple trophic structure, such as in a consumer-resource interaction, versus multiple trophic levels, as in a resource-consumer-predator food web; (6) individual organisms that influence and experience the entire habitat versus each organism influencing and experiencing only its particular neighborhood; and (7) species with interspecific trade-offs versus functionally identical species.

The paradox of diversity posed by Hutchinson dealt with the simplest of these situations, i.e., a spatially homogeneous habitat, at equilibrium, with a simple trophic structure, no limiting physical factor, no neighborhood effects, no habitat patchiness, and simple life histories. In the simple model that results from these assumptions, the number of species that can coexist at equilibrium can be no greater than the number of limiting resources (e.g., MacArthur and Levins 1964; Levin 1970, 1979; Armstrong and McGehee 1980). Simply put, this occurs because there are only as many equations defining equilibrial biomasses for species as there are limiting resources. Eliminating the unlikely case of functionally identical species (species with identical R^* values, i.e., species that reduce the concentrations of resources, at equilibrium, to identical levels), this means that the number of coexisting species must be less than or equal to the number of limiting resources. A critical parameter of such models of resource competition is R^*, which is the resource concentration at which the resource-dependent growth rate of a species exactly balances its total loss rate from herbivory, senescence, mortality, and all other sources of biomass or nutrient loss (Tilman 1982). When several species compete for a single limiting resource, the one species with the lowest R^* is predicted to displace all others.

Although this model places an upper bound on species richness in such simple habitats, the actual number of species that could coexist depends on the traits of the species. With two limiting resources, for instance, two consumer species can stably coexist only if one is a better competitor for one resource but a poorer competitor for the other, and if each species consumes relatively more of the resource that more limits its growth (e.g., Tilman 1982). Under these circumstances, each species inhibits itself more than it inhibits the other species, which is the condition required by the Lotka-Volterra model for stable equilibrial coexistence of two species. Similar trade-offs are required for N species to coexist on N resources. These trade-offs provide the basis for testing the relevance of this model to local diversity patterns. Once the limiting resources have been experimentally identified, it can be determined whether the coexisting species have the requisite differences in competitive abilities and uptake characteristics, and whether changes in resource supply rates cause the predicted changes in the relative abundance of species (see Tilman 1982).

Levins (1979) discussed models similar to the simple models of resource competition described above, but in which fitness is also constrained by various physical factors. A physical factor differs from a resource in that a physical factor, such as temperature or pH, is experienced by an organism but is not consumed by it. Levins (1979) showed that the number of species that could stably coexist in a homogeneous, equilibrial habitat was less than or equal to the number of limiting resources plus the number of constraining physical factors.

Although a model that assumes that organisms live in a spatially homogeneous, equilibrial, two-trophic-level habitat may explain the abundances of the dominant species (reviewed in Tilman 1982), it rarely explains local biodiversity. A lake may have three or four limiting nutrients and two or three constraining physical factors, and yet contain hundreds of phytoplanktonic algal species. A tropical forest or a prairie can also contain hundreds of species of vascular plants, and yet only have a few limiting resources and constraining physical factors. These failures of this model imply that one or more of its assumptions are too simple to explain most biodiversity patterns. In the sections below, we discuss the effect on biodiversity of making this model more complex by relaxing its simplifying assumptions, one assumption at a time. In so doing, we summarize a vast literature. We present the conclusions of the mathematics, and offer intuitive or graphic explanations for these conclusions, but do not present the underlying mathematical details.

The more complex, and thus more realistic, models presented below are only a partial explanation for diversity patterns, because these models deal only with the forces that can allow numerous species to coexist and do not consider additional countervailing forces that influence extinction rates. These countervailing forces, which should apply to any models of species persistence, are briefly discussed later.

PROCESSES ALLOWING PERSISTENCE OF NUMEROUS SPECIES

Spatial Heterogeneity

Let us assume that organisms live in an equilibrial habitat, that there are two trophic levels (a resource level and a consumer level), that organisms have simple life histories and that local abundances are not limited by dispersal, but that the habitat is spatially heterogeneous. Spatial homogeneity would mean that all individuals, of all species, experience the same resource supply rates, the same resource concentrations, and the same intensities of physical factors no matter where the individuals might live in the habitat. Spatial heterogeneity is measured by the extent of plant-to-plant differences, both in resource supply rates and concentrations and in the intensities of physical factors experienced by individual organisms. Because individual organisms can "average" resource supply rates and physical factors through time and through the region in which they forage, the relevant components of spatial heterogeneity are the individual-to-individual differences in these averages.

A Single Limiting Resource. If a habitat has a single limiting resource, and if organisms are only interacting via mutual exploitation of this resource, spatial heterogeneity in its supply rate should have no effect on equilibrial species diversity. At all points throughout the habitat, the one species with the lowest R^* for the limiting resource would grow to a sufficient density at equilibrium to reduce the resource concentration down to its R^*, at which level no other species could survive (e.g., Hsu, Hubbell, and Waltman 1977). Spatial heterogeneity in resource supply rates would lead to comparable spatial heterogeneity in the equilibrial density of the one species with the lowest R^*, and all other species would be driven to extinction.

One Resource and One Physical Factor. A physical factor can modify the resource requirements (R^* values) of species through its effect on resource-dependent growth rates and/or loss rates. If species have trade-offs in the dependence of their growth and loss on a physical factor, then spatial heterogeneity in that physical factor could allow numerous species to coexist on a single limiting resource. This is easily illustrated graphically (fig. 2.1). Interspecific trade-offs with respect to a physical factor, such as temperature, could give each species its optimal competitive ability (i.e., lowest R^*) at a different intensity of the physical factor. If this is so, then individual-to-individual spatial heterogeneity in the physical factor, or a geographical gradient in the intensity of the physical factor, could allow an almost unlimited number of species to coexist on a single limiting resource. For instance, Habitat 1 has spatial heterogeneity in a physical factor (fig. 2.1A). This heterogeneity would allow species B, C, and D to coexist because each is a superior competitor (has lower R^*) in a portion of Habitat 1. Habitat 1 could be invaded by other species if they had appropriate trade-offs (fig. 2.1B). As long as each species has a temperature at which it has an R^* value lower than that of any other species (as illustrated for species W, X, Y, and Z), there is no simple limit to the number of such species that could coexist on a single limiting resource in a spatially heterogeneous habitat.

Are these trade-offs likely? Such differentiation has been reported for temperature-dependent growth of freshwater algae (Jitts et al. 1964; Tilman, Mattson, and Langer 1981) and seems likely for pH-dependent growth and temperature-dependent growth (e.g., Barbour, Burk, and Pitts 1987, 326–355) of terrestrial plants. Temperature differentiation may be partially the result of the "optimal temperature effect" that has been observed in all enzymatic reactions, and which is caused by the joint effects of temperature on an enzyme's conformation (shape) and on the kinetic energy of its substrate.

Two or More Limiting Resources. If species have trade-offs in their abilities to compete for two or more limiting resources, then spatial heterogeneity in the supply rates of the resources could allow numerous species to coexist (Tilman 1982). Trade-offs in competitive abilities mean that each species has its greatest competitive ability at a particular ratio of supply rates of the limiting resources. To see this, consider the resource-dependent growth isoclines of figure 2.2A. Species A, B, C, and D have trade-offs in their abilities to compete for R_1 versus R_2. Species A is the superior competitor for R_1 but the inferior competitor for R_2. Species D is the superior competitor for R_2 but the inferior competitor for R_1. Species B and C are

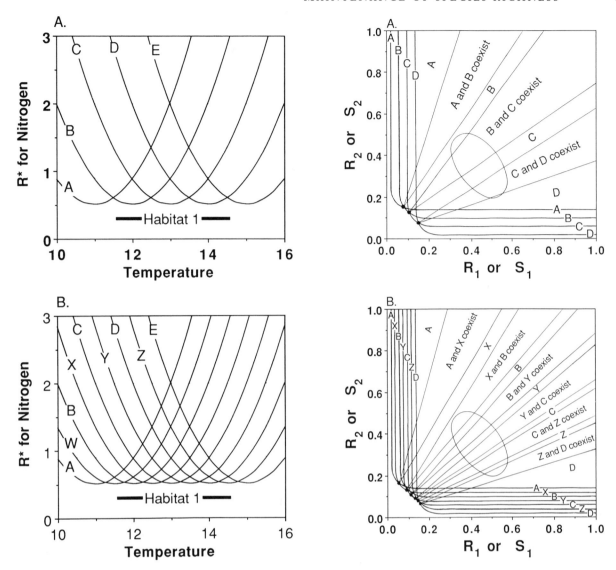

Figure 2.1 (*A*) The temperature dependence of nitrogen competition among species A, B, C, D, and E. In a homogeneous habitat with a given temperature, the species with the lowest R^* for nitrogen at that temperature should completely displace all others; thus, species A wins at 10° C and B wins at 12° C. Habitat 1 is spatially heterogeneous, with different points in the habitat having different mean temperatures. The range in temperatures within habitat 1 is shown by the thick line. This range allows species B, C, and D to coexist because each has regions in the habitat in which it is the superior nitrogen competitor. (*B*) If there are interspecific trade-offs in the ability to compete at different temperatures, species W, X, Y, and Z could invade habitat 1 and persist with species B, C, and D.

Figure 2.2 (*A*) The zero net growth isoclines of four species (labeled A–D) for two limiting resources, R_1 and R_2, and the consumption vectors associated with these, define habitats in which either a single species is competitively dominant or in which a pair of species can stably coexist (see Tilman 1982 for details). The oval represents the individual-to-individual spatial heterogeneity in these resources. Because this oval overlaps the regions in which they can exist, species B, C, and D should persist in the heterogeneous habitat represented by this oval, assuming that their abundances are not limited by dispersal. (*B*) Species X, Y, and Z have trade-offs that make them intermediate between some of the pairs of species shown in part A. This leads to regions in which species X, Y, and Z can persist with A, B, C, or D, and allows six species to coexist in the heterogeneous habitat represented by the oval. As long as new species have such trade-offs, there is no limit to the number that can coexist in a heterogeneous habitat.

intermediate in their competitive abilities for R_1 and R_2. If these species forage optimally for R_1 and R_2, then each will consume resources in the proportion in which it is equally limited by them (Tilman 1982). This leads to stable two-species equilibrium points and to regions of the resource supply plane in which each species is dominant, or in which a pair of species stably coexist at equilibrium (fig. 2.2A).

Spatial heterogeneity in such a habitat can be quantitatively described by a closed curve that encompasses the

range in average resource supply rates experienced by different individuals living in different locations in the habitat. For instance, the oval in figure 2.2A might encompass the spatial heterogeneity in R_1 and R_2 supply rates in a habitat. This amount of heterogeneity would allow species B, C, and D to coexist, but does not overlap the resource supply points for which species A is able to survive in competition. This representation provides an upper bound for the effects of resource spatial heterogeneity on diversity because it assumes that propagules of all species can always find all appropriate resource patches.

The habitat represented by the oval in figure 2.2A could be invaded by a species that was intermediate between species A and B (such as species X of figure 2.2B), or intermediate between species C and D (such as species Z), and by ones intermediate between these, etc. As long as the invading species has the appropriate trade-off, this model provides no upper limit to the number that could coexist in a spatially heterogeneous habitat with two or more limiting resources (Tilman 1982; fig. 2.2B). Such trade-offs seem likely because they could be the direct and unavoidable result of different patterns of allocation. To be a superior competitor for a resource, a plant must allocate more of its materials or energy to the structures or processes involved in obtaining, conserving, and efficiently using that resource, which means that the plant has relatively less to allocate to the acquisition and efficient use of another limiting resource. Given plants with similar allocation to reproduction, such allocation-based trade-offs constrain each plant to being a superior competitor for a particular ratio of essential resources, and mean that species with different allocation patterns should be superior competitors in different habitats (Chapin 1980; Tilman 1988).

Temporal Heterogeneity: Nonequilibrium Conditions

Suppose now that we substitute temporal variability in resource supply rates and/or physical factors for spatial heterogeneity. Trade-offs among species like those discussed above will ensure that different species will be the dominant competitors at different times (e.g., Levins 1979; Armstrong and McGehee 1980; Chesson 1986; Ives 1988; Cohen and Levin 1991; Hanski and Cambefort 1991). It seems reasonable that species should be able to coexist if each species can "disperse," through time, from a period in which it is the dominant competitor to the next such period. Communities maintained in this way would be composed of temporal fugitives.

Chesson (1983, 1984, 1985, 1986), Shmida and Ellner (1984), and Comins and Noble (1985) have derived mathematical conditions that allow the coexistence of temporal fugitives. The most important of these is the stochastic boundedness criterion, which states that a temporal fugitive persists if it has a geometric average finite population growth rate that allows it to increase when rare. This mathematical condition has intuitive appeal because it implies that extinctions will not occur if each species, when rare, increases in abundance more during years good for the species than it decreases during bad years, and if the good years occur at sufficiently close intervals.

Although the stochastic boundedness criterion is itself simple, the types of population dynamics and temporal variation that satisfy it can be complex. Chesson and Case (1986), Warner and Chesson (1985), and Shmida and Ellner (1984) discuss two biologically reasonable cases in which temporal heterogeneity facilitates coexistence. First, consider a group of annual species with dormant seed that compete for a single limiting resource. Temporal environmental variation that differentially affects species' seed production or germination rates can facilitate coexistence, whereas variation that affects only seed survivorship will not. Second, consider a group of perennial species that compete only during the regeneration that follows the death of an adult plant. Coexistence is facilitated by environmental fluctuations that differentially affect fecundity or competitive ability during regeneration, but not by fluctuations that affect only adult death rates.

The similarity between the effects of spatial and temporal heterogeneity on diversity is more than a convenient analogy. Chesson (1985) has shown that the stochastic boundedness criterion often converges on the corresponding criterion for a spatially heterogeneous environment as longevity (and thus the ability to disperse through time) increases, but the effectiveness of temporal heterogeneity in maintaining diversity is generally less than or equal to the effectiveness of spatial heterogeneity.

While Chesson and co-workers focused either on abstract, general models or on simple phenomenological models of competition, Armstrong and McGehee (1976, 1980) and Levins (1979) examined the mathematical effects of resource fluctuations on coexistence in a spatially homogeneous habitat with two trophic levels. If all species had simple, linear growth responses to resource concentrations, they found that fluctuations had no qualitative effect on coexistence: the number of coexisting species was still never greater than the number of limiting resources. However, they found that an unlimited number of species could persist indefinitely if the species had the appropriate nonlinear dependencies of per capita growth rates on resource availability.

An essential feature of nonlinear growth responses is that the long-term average growth response of a species to fluctuations does not equal its growth response to the long-term average resource availability (Levins 1979). Depending on the shape of its nonlinearity, a species could either benefit from resource fluctuations or be inhibited by them. For instance, a species could be an exploiter of resource pulses by growing rapidly after a resource pulse. This could allow it to maintain itself in a fluctuating habitat that, on average, had insufficient resources for its survival. Other species (such as ones with a classical Michaelis-Menton saturating resource-dependent growth response) could be inhibited by fluctuations and require a higher average resource level to survive in a fluctuating environment. These would do better in more constant environments.

These different responses to resource fluctuations mean that co-occurring species, in essence, experience different average environments when growing in a fluctuating environment. Such differences could allow many more species to persist in a fluctuating environment than there were resources if the species had the appropriate trade-offs in their long-term responses to the fluctuations (Levins 1979; Armstrong and McGehee 1980). The long-term interactions among the species would be stable persistence if each species, when abundant, more rapidly depleted the resource situation that favored it, and thus tended to inhibit itself more than it inhibited other species. Indeed, the long-term stable persistence of species caused by nonlinear exploitation of pulses can be modeled as an "equilibrial" process (e.g., Armstrong and McGehee 1980; Levins 1979) using resource-dependent isoclines in which each frequency or magnitude of resource fluctuation can function as a distinct "resource" or "limiting factor" (depending on the shape or magnitude of the nonlinearity). One way for an organism to be an exploiter of pulses, for instance, is for it to have an excess capacity to acquire its limiting resource, such that it rapidly consumes any resource pulse and then stores this for future use. Such luxury consumption and storage should have costs, because the organism doing this would have to allocate more to uptake and storage than would an organism that specialized on longer-term average resource availability through efficient use and conservation of the resource. Thus luxury consumers with high maximal uptake rates may be at a competitive disadvantage in a more stable environment. Although it seems plausible that the necessary trade-offs may be the unavoidable result of different patterns of allocation to resource uptake, storage, efficient use, or conservation, this has not yet been formally analyzed mathematically.

In theory, nonlinear resource-dependent growth curves and resource fluctuations could explain the persistence of an unlimited number of species (Armstrong and McGehee 1980). Levins (1979) showed that the number of species must be less than the number of resources plus distinct nonlinearities, which also can give unbounded diversity because each species can have a distinctly nonlinear resource-dependent growth function. Sommer (1984, 1985) and Grover (1988, 1989) have shown that resource fluctuations can allow many more algal species to coexist in controlled laboratory situations than there are limiting nutrients. Annual plants and understory herbs seem ripe candidates for species that may persist in habitats by exploiting temporal variance in resource supply. Tests of this hypothesis will require determination of the effects of the magnitudes and temporal patterns of resource fluctuations on diversity and on long-term growth rates of species.

Competition and Colonization in Spatial Habitats

Another major trade-off that can influence species diversity is that between the competitive ability of a species and its ability to colonize an open site. The rich literature on this topic illustrates numerous alternative conceptual approaches to modeling the competition-colonization trade-off. Skellam (1951), Levins and Culver (1971), Horn and MacArthur (1972), Levin and Paine (1974), Armstrong (1976), Werner and Platt (1976), Platt and Weis (1977), Pickett and White (1985), Tilman (1990) and many others demonstrated that this trade-off can, in theory, determine the pattern of succession or of species richness. The underlying mathematics can be relatively complex because it must be assumed, in some manner, that individual organisms are discrete, i.e., that each individual occupies a different point in space. Metapopulation models (e.g., Levins and Culver 1971; Hastings 1980; Gilpin and Hanski 1991) provide analytically tractable approximations. All of these models assume that the species that is the best

competitor for any given point in space is kept from displacing all other species at all other points in space because it is a poor colonist. This allows a species that is a better colonist, but a poorer competitor, to persist with the superior competitor in a habitat in which periodic openings occur. In theory, this trade-off can allow an unlimited number of species that all compete for the same limiting resource to persist (Tilman 1993). Another way to think of these models is to regard open sites as a second resource (Tilman 1982). If each species performs best at a different point in the successional sequence that follows the creation of an open site, then periodic, localized disturbances or the random death of individual organisms could allow numerous species to stably persist in a two-trophic-level habitat that is spatially homogeneous in the supply rates of limiting resources.

There can be strong allocation-based interspecific trade-offs between colonization and competition. Werner and Platt (1976) showed that goldenrods that were better competitors because they produced larger seeds, were also poorer colonists because they produced fewer, more poorly dispersed seeds. *Schizachyrium scoparium* and *Andropogon gerardi* are the best nitrogen competitors at Cedar Creek Natural History Area by virtue of their 70%–85% allocation to root, but are extremely poor colonists because of their low allocation to seed or rhizome (Tilman 1990; Tilman and Wedin 1991a, 1991b). They require 11–17 years to initially colonize an abandoned field, and another 20–40 years to spread across a field and dominate it. This provides a large window of time during which inferior nitrogen competitors, such as *Agrostis*, *Agropypon*, and *Poa*, can colonize and dominate open sites. Recruitment limitation may thus be as important for terrestrial plant communities as it is for intertidal habitats (e.g., Roughgarden, Gaines, and Pacala 1987; Roughgarden, Gaines, and Possingham 1988).

Many variations on this idea have been reviewed and synthesized by Petraitis, Latham, and Niesenbaum (1989). They showed that there must be trade-offs in the ways that species respond to open sites if numerous species are to persist on one or a few resources. In addition to the competition-colonization trade-off mentioned above, other trade-offs that lead to a multispecies successional sequence in a recently opened patch can allow many species to persist in a patchily disturbed habitat (Petraitis, Latham, and Niesenbaum 1989).

Trophic Complexity: Three Trophic Levels

All the models discussed so far assume that a system consists of consumers and their resources. As such, the only constraint acting on a consumer species is the availability of its limiting resources. The addition of a predator species that preys on consumer species would create a three-trophic-level system, i.e., a system with resources, consumers, and predators. Predators provide additional constraints on consumer species that could allow more consumer species to coexist than there are limiting resources.

One simple way that a third trophic level could allow many species to persist is through the effects of predators, pathogens, or diseases on the R^*s of the consumers. Densities of predators, pathogens, or diseases might vary randomly in space. Because the R^* of a species is influenced by the loss (mortality) rate it experiences, spatial variation in predator densities could cause different species to have the lowest R^* values in different regions. This could potentially allow more species to persist than there were limiting resources. In this case, the effect of predators would be much like the effect of spatial variation in a physical factor (see fig. 2.1B).

Janzen (1970) and Connell (1971) suggested that local frequency-dependent seed or seedling predation could allow the persistence of numerous species of tropical trees. Although the significance of this process for tropical diversity has not yet been resolved (e.g., Hubbell 1979, 1980), the concentration of species-specific seed and seedling predators and pathogens that may occur around an individual adult tropical tree could prevent recruitment of individuals of that species in that locality, and thus promote diversity.

The dynamics of a three-trophic-level interaction can also have major effects on multispecies coexistence. Levin, Stewart, and Chao (1977) developed this idea mathematically and tested it using a sugar, bacteria, and a bacterial virus. They showed that several different strains of bacteria could coexist on a single resource in a homogeneous habitat if the bacterial strains that were more susceptible to viral attack were better competitors for the sugar, and if the bacterial strains that were resistant to the virus were poorer resource competitors.

This approach was extended and generalized to demonstrate that, in theory, an unlimited number of consumer and predator species could coexist on a single resource in a homogeneous habitat if the better competitors were more susceptible to predation (Tilman 1982). Such trade-offs would be likely if the morphological structures or behaviors that allowed an organism to be a superior competitor necessarily increased its risk of predation. For instance, a plant could minimize its chance of herbivory by allocating to secondary compounds or to thick, tough leaves (Coley, Bryant, and Chapin 1985; Gulmon and Mooney 1986; Bazzaz et al. 1987). However, energy or material allocated to herbivore defense cannot simultaneously be allocated to the acquisition of a limiting resource. Thus, it is plausible that a superior competitor might necessarily be more susceptible to herbivory, although there are some plant traits that may reduce herbivory and increase competitive ability (Coley, Bryant, and Chapin 1985). Numerous studies have demonstrated the existence of trade-offs between competitive ability and susceptibility to predation (e.g., Paine 1966; Lubchenco 1978; Sih et al. 1985; Menge and Sutherland 1987) and have shown that changes in predation intensity may influence species richness.

Neighborhood Effects

Individual plants are sedentary and take resources only from their immediate vicinity. This "neighborhood" perspective (Pacala and Silander 1985; Pacala 1986a, 1986b; Goldberg 1987) brings two additional entries to our list of factors that facilitate plant diversity. First, plant spatial distributions are typically intraspecifically clumped and are sometimes interspecifically segregated. Because of local resource use, even fine-scale intraspecific clumping and interspecific spatial segregation serve to increase the strength of intraspecific competition relative to interspecific competition. General models, such as the Lotka-Volterra model, predict that coexistence becomes more likely as between-species competition becomes weaker relative to within-species competition.

A variety of factors lead to clumped and segregated spatial distributions. Propagules produced by a plant tend to be clustered around it. Finite dispersal thus leads to intraspecific aggregation. Also, the process of interspecific competition itself may cause spatial distributions to fragment into pockets of individuals with relatively weak interspecific interactions. Theoretical models of neighborhood competition imply that the feedback among dispersal, competition, and spatial distribution could allow the long-term coexistence of more species than there are resources (Pacala 1986a).

In an influential review, Connell (1983) reported that only approximately one-third of experimental field studies found statistically significant evidence of competition among plant species. However, Kelly, Tripler, and Pacala (1993) reanalyzed those and recent studies and found competition in more than two-thirds of the studies that used plots centered on an individual plant or on a small cluster of target individuals. However, in plots established without reference to the locations of individual plants, only approximately one-fourth of the density manipulations showed interspecific competition. This suggests that differences in spatial distributions may cause a nearly threefold difference in the detected frequency of competition. Its implication is that natural plant diversity is facilitated by clumping and spatial segregation at spatial scales smaller than the sizes of the quadrats commonly employed by experimenters.

A second way in which sedentary habit and spatially local resource use could augment diversity has been less thoroughly explored than the effects of spatial distribution. Individual plants are known to create spatial and temporal heterogeneity in resource supply rates (Wedin and Tilman 1990). The mechanisms responsible include interspecific differences in litter chemistry (Gosz, Likens, and Borman 1973; Harmon 1986; Aber and Melillo 1982; Melillo, Aber, and Muratore 1982; Nadelhoffer, Aber, and Melillo 1983; McClaughtery et al. 1985), resource uptake (Horn 1971; Helvey and Patric 1965; Swift et al. 1975; Nadelhoffer, Aber, and Melillo 1983, 1984; Pastor et al. 1984), through-fall chemistry (Zinke 1962; Gersper and Holowaychuk 1970, 1971; Parker 1983; Crozier and Boerner 1986), and kinds and numbers of natural enemies. It is plausible that plants could coexist because of the spatial and temporal heterogeneity that they create. However, there are few theoretical or empirical studies of this possibility. Clearly, there is a need for additional work on the role of "biotic" heterogeneity in maintaining plant diversity.

Competitively Identical Species

An alternative explanation for high diversity is that many species may be competitively identical (Hubbell 1979;

Hubbell and Foster 1986). This could occur if the selective forces in a given environment favored a general suite of traits, leading to convergence on these traits and thus to species that were effectively competitively identical. Comparably, if speciation were caused by sympatric reproductive isolation, such as occurs with changes in ploidy, in the timing of flowering, or in the pollination vector attracted to flowers, the species complex thus formed might contain species that were effectively competitively identical because of their common morphology and physiology. Although it seems unlikely that there would ever be species that are exactly identical in their competitive ability, this may not be necessary. As species become increasingly similar, the rate of competitive displacement slows. If two species were competitively identical, competitive displacement would not occur.

However, identical species do not stably coexist. Rather, their relative abundances should drift in a process that leads to the random walk to extinction of all but one species (Hubbell and Foster 1986). Thus, this theory predicts that, in the absence of colonization and speciation, a single species should eventually exclude all others. However, the rate of random walk to extinction of species depends on the population sizes of the species, on the magnitude of their population fluctuations, and on their patterns of spatial distribution. Hubbell and Foster (1986) have suggested that the population sizes of tropical tree species may have been sufficiently great, and their population fluctuations sufficiently small, to allow rates of speciation of tropical tree species to approximately balance extinction rates.

Although there are aspects of the structure and diversity of plant communities that contradict the hypothesis that the plants in any given community are competitively identical (Tilman 1982, 1988), there may well be guilds of functionally identical plant species (Hubbell and Foster 1986). The differences among the guilds could reflect trade-offs in the responses of the plants to their environmental constraints. This would cause the guilds to be separated along environmental gradients. The similarity within a guild could reflect either convergent evolution in response to a particular set of environmental constraints or a common evolutionary history. This approach is an interesting counterpoint to the approach that explains diversity solely in terms of interspecific trade-offs. It may be difficult to distinguish between guilds with effectively identical species and those with minute trade-offs along an environmental gradient. One difference is that an underlying trade-off would give each species a slight advantage in a particular microenvironment and thus would add some stabilizing force to their interaction. This stabilizing force would tend to counteract some environmental randomness, and thus might allow the persistence of more species in the face of random walks than in the case of identical species.

Summary of Processes allowing Persistence. Models that assume that resource supply rates and physical factors are spatially homogeneous, that each organism is spread uniformly throughout an environment, that resources do not fluctuate, that localized mortality does not occur, and that higher trophic levels are unimportant, predict the stable coexistence of no more species than there are limiting resources and physical factors (Levins 1979; Armstrong and McGehee 1980). These models are unable to explain the diversity of most terrestrial and aquatic plant communities, almost all of which have many more species than limiting resources and physical factors (Tilman 1982). However, models that violate a single one of these simplifying assumptions, and that assume that organisms have trade-offs in their abilities to respond to their constraints, predict the stable persistence of a potentially unlimited number of species. Models that are made even more complex by violating more than one of these simplifying assumptions can also predict the potential existence of an unlimited number of species. The models presented above are a small subset of all the potential models that could explain the stable persistence of high-diversity communities in a locality. Almost any model that assumes some sort of environmental complexity, and that assumes allocation-based trade-offs in the abilities of organisms to respond to their constraints, has the potential to predict the existence of many more species than there are limiting resources and physical factors. Thus, the question becomes not, "Why are there so many species?", but, "Why are there a particular number of species, and not many, many more?"

PROCESSES THAT LIMIT DIVERSITY

Small Population Size

As we have seen, all the models that can potentially explain the long-term stable persistence of a large number of species have a similar structure: All assume that there are one or more factors that constrain the fitness of individual organisms, and that organisms have unavoidable trade-offs in their abilities to respond to those constraints. In their simplest form, all these models tacitly assume that a habitat is infinitely large. This comes from the mathematical assumption that a species will persist even if there is a time when it is represented mathematically by some fraction of a single individual. On the surface, this is a justifiable assumption, because models are solved for the population densities of species, i.e., the number of individuals per unit area, and it is necessary to multiply these numbers by the total area of the habitat to determine actual densities of species in the habitat. This, though, points out that species diversity is area dependent. A species cannot survive if it is ever represented by less than one individual. For many species, the minimum viable population size would be much greater than one individual.

The finite size of habitats, and thus of population densities, should place a limit on species diversity, reducing the infinite potential diversity predicted by trade-off models to some finite level. An important component of this finite limit is the random walk to extinction of rare species (May 1973). In a finite environment containing competing species, the appearance of each new species would reduce the average number of individuals per species. This would increase the chance that some species would fall below its minimum population size and undergo a random walk to extinction. Extinction would occur when either deterministic or stochastic processes reduced the density of a species below its minimum threshold. As Levinton (1979) showed, such processes limit diversity and can lead to an equilibrium in which speciation is balanced by extinction.

Limiting Similarity

The density of a species is influenced by the number of other species, especially similar species, with which it is competing. Thus, the more species there are with a particular suite of resource requirements, the lower will be the density of each species, and the greater will be its chance of extinction in a finite environment. This is the essential basis of classical limiting similarity theory (e.g., MacArthur 1968, 1970; May 1973; May and MacArthur 1972). Although simple models of plant competition for essential resources along a resource ratio gradient do not have any analytical limit to similarity, it seems likely that there will be a limit to similarity once the models are modified to include competition in a finite environment, neighborhood competition, or dispersal. Limits to similarity are also limits to diversity.

Habitat Fluctuations and Extinctions

As May (1973) pointed out, greater fluctuations in population densities should increase the probability of random walk to extinction for rare species, and thus decrease diversity. All else being equal, more species should coexist in a stable environment than in one in which population densities fluctuate. Thus, environmental fluctuations may have opposite effects. On one hand, they may increase the rate of stochastic extinction; on the other hand, they may facilitate coexistence (Armstrong and McGehee 1980). The effects of fluctuations on extinction rates are consistent with Wallace (1878), who suggested that the tropics might be more diverse than temperate regions because tropical habitats may be more stable and thus have lower extinction rates. Environmental fluctuations do influence diversity. A severe drought that led to about a 50% decrease in living plant biomass caused the local extinction of about 40% of the plant species in four grassland/prairie fields (Tilman and El Haddi 1992). The probability of a species being lost increased with its pre-drought rarity, which supports May's (1973) conclusion that environmental fluctuations increase extinction rates of rare species and thus limit diversity.

PATTERNS IN DIVERSITY

Although the best way to determine the relative importance of the processes that influence species coexistence and extinction is through direct experimental tests, considerable insights can be provided by comparative studies. We discuss, below, four major diversity patterns and the insights that these provide.

Spatial Heterogeneity and Organism Size

Robert May (1978, 1986), who summarized and extended previous data compilations by Van Valen (1973a),

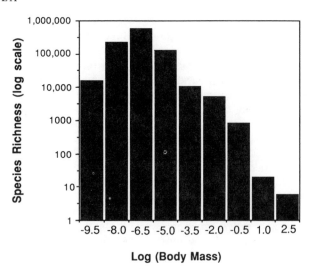

Figure 2.3 Preliminary data gathered by May (1978, 1986) on the relationship between the total number of animal species on earth and their body size. May expressed size as length, L. To prepare this figure, May's estimates were converted to body mass using the assumption that mass is proportional to L^3. A simple regression of log (species richness) on log(body, mass), using the seven data points on the declining portion of this figure, led to an estimated slope of -0.56. All these numbers are extremely preliminary, and should serve, we hope, to encourage a more rigorous collection and synthesis of such data.

Janzen (1977), and others, has highlighted an intriguing, but not yet well-quantified, relationship between animal body size and biodiversity (fig. 2.3). The vast majority of the earth's animals, it seems, have relatively small body sizes; there are many fewer species with large body sizes. May (1986) offered several speculative explanations for this pattern, including the possibility that it might be caused by the fractal nature of spatial heterogeneity. As Morse et al. (1985) pointed out, a habitat may be more heterogeneous to a small organism than to a large one. This is comparable to the classical problem of measuring the shoreline of an island. The length of the shoreline depends on the length of the line segment (the ruler) with which the shore is measured. A larger organism may experience less heterogeneity in the same way that a longer ruler measures less shoreline. May suggested that the quantitative relationship between organism size and habitat heterogeneity, which may be represented by the fractal dimension of the habitat, might help to explain the large number of small-sized animal species.

May's ideas, and the data needed to document the pattern he discussed, deserve much further attention. The apparently lower diversity among the very smallest animals (fig. 2.3) needs to be explored. A potential problem with the spatial heterogeneity hypothesis as an explanation for the large number of small animals comes from a consideration of optimal foraging by small organisms. Most small animals are insects, many of which are herbivores or parasitoids. Small organisms experience a highly heterogeneous environment with high fitness costs for traveling from one site to another. In combination with interspecific differences in plant secondary chemistry, this could favor insect herbivores that specialize on a single plant species and spend much of their life cycle on a single plant individual. Such specialization is the expected evolutionary outcome of optimal foraging, and gives rise to "switching resources" (Tilman 1982). Similar host specialization is expected, and found, in insect parasitoids (Hassell 1978). However, specialization leads to strong directional selection favoring the best specialist on each resource. When this occurs, spatial heterogeneity in resource availability does not allow any more small animal species to coexist than there are distinct resources (Tilman 1982, 259–265). If this view is correct, the fractal dimension of plant spatial heterogeneity would not explain the higher diversity of smaller animals because their diversity would be independent of spatial heterogeneity. However, if each plant species is viewed as providing four to six distinct resources to herbivores (leaves, stems, flowers, roots, etc.), the existence of 500,000 terrestrial plant species is clearly capable of completely explaining the existence of many more insect herbivores than have yet been described. The number of parasitoid species could be similarly explained.

The question, then, is why there are so few species of large-bodied animals. Based on allometric relationships of metabolic costs (Peters 1983), a 10-fold increase in the size of organisms (as represented by their length, L) should lead to about a 180-fold decrease in their population density. Assuming that the mass of an individual, M, is proportional to L^3, this implies that $D = c_1 M^{-0.75}$, where M is the mass of an organism, D is its density (number/area), and c_1 is a constant. Morse et al. (1985) and May (1986) combined this with the effect of the fractal nature of habitats to suggest that $D = cM^{-1}$. Marquet et al. (1990) sampled the number of individuals in species of different sizes within rocky intertidal communities. Where D has units of individuals/m^2 and M has units of g/individual, they found that $D = 37M^{-0.77}$. Thus, a 10-fold increase in body mass in their intertidal community led to a 5.9-fold decrease in population density.

This lower abundance of larger organisms should increase their chance of random extinction. Higher extinction rates would allow fewer large species to persist than small species, and could account for the pattern reported in figure 2.3. May (1986) reported that S, the number of species with a given body size, was approximately proportional to L^{-2}, where L is body length. If body mass, M, is proportional to L^3, then $L = c_2 M^{1/3}$. When combined with S being proportional to L^{-2}, this implies that $S = c_3 M^{-2/3}$, where c_3 is a constant. Thus the data in May (1986) suggest that a 10-fold increase in body mass is associated with roughly a 4.6-fold decrease in species richness. This number is close to the 5.9-fold decrease in population density that Marquet et al. (1990) found to be associated with a 10-fold increase in body mass and with Peters's (1983) allometric relationship. These results support our hypothesis that the lower diversity of larger animals could be caused by higher extinction rates associated with their lower population density. Clearly, these are currently ballpark numbers. A rigorous evaluation of this hypothesis will require more detailed information on the relationship between species richness and body size, between body size and population density, and between ex-

tinction rates and population densities. Moreover, underlying theory is needed to explain these relationships, especially that between body size and population density.

Although smaller and more numerous organisms should experience lower extinction rates, they also may experience greater population fluctuations for several reasons. First, on average, smaller organisms have greater maximal population growth rates, which may give them a greater tendency to oscillate (e.g., May 1977). Second, smaller organisms also have shorter lifespans and a lower ability to store resources, and thus live in a more turbulent environment than do larger organisms. As already discussed, greater population fluctuations should increase the chance of extinction. The balance between these forces should determine the shape of the curve shown in figure 2.3.

Productivity Gradients and Biodiversity

Many features of plant communities, including their diversity, change along productivity gradients (e.g., Vitousek 1982; Tilman 1982). More productive areas generally have higher rates of supply of limiting soil resources (e.,g., Pastor et al. 1984; Nadelhoffer et al. 1990), which leads to greater plant biomass, and thus greater interception of incident light. Thus a productivity gradient is a gradient from habitats with low rates of supply of a limiting soil resource and high light penetration to habitats with high rates of supply of soil resources and low light penetration. To understand how such a gradient influences diversity, it is necessary to know how spatial heterogeneity in nutrients and light changes along the gradient, and how the resource requirements of the competing species are distributed along the gradient.

Because height is a major determinant of competitive ability for light, plants that dominate increasingly productive habitats tend to be taller and thus more massive (Horn 1971; Givnish 1982). Analysis of data in Whittaker (1975) and Gorham (1979) suggests that height increases in a sigmoid manner with productivity, that the mass of an individual plant increases as more than the cube of its height, and that plant density decreases as more than the square of height. These relationships suggest that the average temperate forest tree occupies an area that is roughly 4 times larger than that of a boreal tree, and that the average boreal tree occupies an area roughly 1,000 times larger than that of a tundra plant. These changes in plant size influence the effective spatial heterogeneity of a habitat. Larger plants forage over a larger region and thus may average over some of the point-to-point spatial heterogeneity in resource supply rates. Thus, all else being equal, more productive habitats may have less spatial heterogeneity in resource supply rates because they have larger organisms.

Some effects of these changes on diversity are illustrated in figure 2.4. In very nutrient-poor habitats, all species are limited by the soil nutrient, and the best competitor for this should be competitively dominant, independent of the extent of spatial heterogeneity. As the nutrient supply rate increases, the region of spatial heterogeneity (shown as a circle) should move diagonally, with increased nutrients leading to decreased light. How-

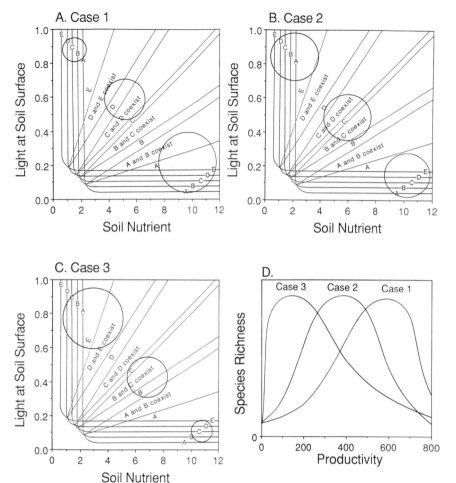

Figure 2.4 (*A–C*) Three possible patterns of changes in plant-to-plant resource heterogeneity in response to changes in productivity (cases 1, 2, and 3). As in figure 2.2, each circle represents a given heterogeneous habitat, and the number of species (A–E) that can coexist in that habitat is shown by the number whose regions are overlapped by the circle. (*D*) The qualitative effects of the three different productivity-dependent patterns of spatial heterogeneity (cases 1, 2, and 3) on species richness.

ever, the effect of increased productivity on the plant-to-plant resource heterogeneity (i.e., on the size of the circle) is less clear. The amount of point-to-point spatial heterogeneity in soil nutrient supply rates should increase as the average supply rate of the nutrient increases. But the larger plants that dominate more productive habitats would forage over a larger area and thus potentially reduce or eliminate the effect of point-to-point heterogeneity on plant-to-plant heterogeneity.

There are three qualitatively different possibilities. In case 1, the increased heterogeneity associated with an increased mean rate of nutrient supply is assumed to outweigh the effects of the increased size of individual plants (fig. 2.4A). Thus, in this case, plant-to-plant heterogeneity (the size of the circle) increases with productivity, and species richness is a unimodal function of productivity, with the mode skewed toward productive habitats (fig. 2.4D). The declining portion of the species diversity curve is caused by most species becoming limited by the same resource, light, in highly productive habitats, with the one species that is the best competitor for light displacing the others. Case 2 assumes that the increased plant size associated with increased productivity exactly compensates for increased point-to-point heterogeneity, making plant-to-plant heterogeneity (the size of the circle) independent of productivity (fig. 2.4B). This gives a unimodal productivity-diversity curve that has its peak at intermediate values (fig. 2.4D). If soil heterogeneity is of biotic origin, its spatial scale should approximate that of individual plants, thus also leading to case 2. Case 3 assumes that the increased size of plants in productive habitats more than overcomes the increased point-to-point heterogeneity, giving a decrease in effective heterogeneity as nutrient supply increases (fig. 2.4C). This leads to a unimodal curve, with peak diversity in nutrient-poor habitats (fig. 2.4D).

Abrams (1988) reviewed Tilman (1982) and suggested that species richness could be either a unimodal function of productivity (as predicted in Tilman 1982 and as shown in Fig. 2.4D) or could be a monotonically increasing function of productivity. However, his prediction of diversity being an increasing function of productivity was based on an ecologically untenable assumption for plants. He assumed that increases in the supply rate of one or more resources need not lead eventually to limitation by some other resource. Because plants require essential resources, the addition of any resource necessarily makes another resource more limiting (Tilman 1982). Even if all nutrient resources were added in the appropriate proportions, plants would still become limited by light as productivity was increased. Light limitation is the unavoidable fate of plants living in highly productive habitats. As an increasingly large proportion of the plant species in a

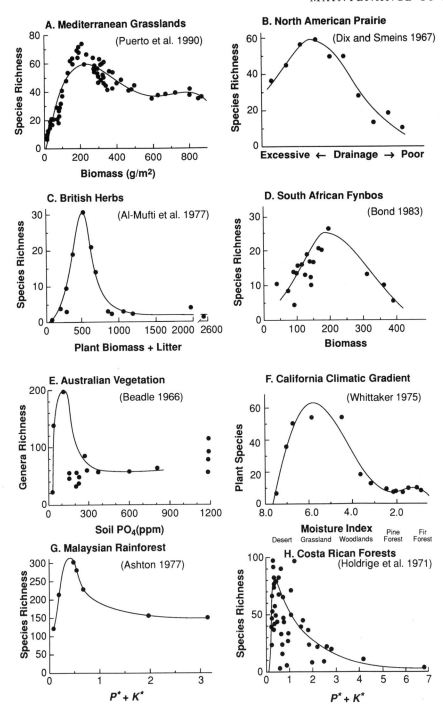

Figure 2.5 The observed relationships between species richness and measures of habitat productivity for a wide variety of plant communities. (A–F): Redrawn from Puerto et al. (1990), Dix and Smeins (1967), Al-Mufti et al. (1977), Bond (1983), Beadle (1966), and Whittaker (1975), respectively. Curves shown for (*A*), (*B*), and (*F*) were fit using polynomial regressions. (*G*) and (*H*) use data from Ashton (1977) and Holdrige et al. (1971), respectively, but were graphed in this manner, after data analysis, by Tilman (1982). P^* and K^* are normalized concentrations of soil phosphorus and potassium, which were summed to give an index of soil fertility.

community become limited by light, species richness is predicted to decline, which contradicts Abram's (1988) assertion.

Another factor that would restrain the diversity of productive plant communities is the sharp decrease in the number of individual plants per unit area as productivity increases (e.g., Gorham 1979). This results from the large increase in plant body size caused by light competition in productive habitats. Tundra, for instance, can have 10^4 more individual plants per unit area than deciduous forest (Gorham 1979). Thus, more productive habitats have, in general, much lower population densities, which should increase the chance of random walks to extinction for rare species, and thus restrain diversity.

The available observational evidence (fig. 2.5; Beadle 1966; Dix and Smeins 1967; Holdridge et al. 1971; Whittaker 1975; 100, Ashton 1977; Al-Mufti et al. 1977; Huston 1980; Bond 1983; Puerto et al. 1990; and see Tilman 1982) supports the hypothesis that plant diversity is a unimodal function of productivity or of other measures of nutrient supply rates. We know of no cases in which diversity is a simple increasing function of productivity or nutrient supply. The most diverse tropical forests (fig. 2.5G, H), grasslands (fig. 2.5A, B, and C) and shrublands (fig. 2.5D) occur in relatively unproductive habitats. Put another way, the most productive tropical forests, grasslands, and shrublands are not the most diverse. Some data suggest a diversity peak skewed to low-nutrient habitats, and others suggest a peak in intermediate habitats. Nutrient addition experiments commonly show declines in diversity, which could be interpreted as meaning that the fertilized habitats were moved beyond the diversity peak. However, fertilization should not increase the effective spatial heterogeneity of a habitat (see fig. 2.4B; case 2), and thus may not be a good method for determining the effect of natural variations in productivity on diversity. There have not yet been studies in which soil heterogeneity was varied experimentally and the effects of this on species diversity determined. There is also insufficient data on the spatial pattern of soil heterogeneity in relation to the size of plants. A more thorough knowledge of the spatial patterning of soil resources, of changes in this pattern along productivity gradients, and of changes in organism size along productivity gradients will be needed to develop the spatial heterogeneity hypothesis beyond the qualitative features illustrated in figure 2.4.

Denslow (1980) extended species packing arguments to suggest that there should be more species evolved to deal with the most common habitat conditions, and fewer species that are specialized on less common habitat conditions. Because habitats of "intermediate" productivity are most common, and less productive or more productive habitats are rarer, she suggested that diversity should peak in the intermediate habitats. This, then, gives a diversity curve like those of figure 2.4D just because some habitats are more common, support more individuals, and thus have a lower chance of random walks to extinction for species specialized on them.

Disturbance Gradients and Diversity

Levin and Paine (1974) showed that habitats characterized by a colonization-competition trade-off should have maximal diversity at intermediate rates of disturbance, given that there is a fixed pool of potential species. In these models, intermediate disturbance rates generate the greatest effective spatial heterogeneity because they create the full range of successional habitats, with some habitats being newly disturbed, some being of intermediate time since disturbance, and others being late successional. In contrast, at high disturbance rates, almost all areas are newly disturbed, and at low disturbance rates almost all are late successional. Indeed, this prediction is qualitatively quite similar to that of peak diversity at intermediate productivity. In both cases, the extremes (very unproductive habitats or undisturbed habitats, and highly productive habitats or highly disturbed habitats) are predicted to be dominated by a single species. The greatest effective heterogeneity occurs in intermediate habitats, and this heterogeneity allows numerous species to coexist. Several studies support this perspective (Connell 1978; Lubchenco 1978; Petraitis, Latham, and Niesenbaum 1989).

Latitudinal Gradients in Species Diversity

If increased productivity, such as would result from increasing the supply rate of a limiting resource, leads to increased population densities, this should also lead to increased species diversity because of a decreased probability of random walks to extinction. This is the conceptual basis for the "species-energy" theory of diversity (Wright 1983), which is a variant on island biogeographic theory of diversity. Species-energy theory assumes that the productivity of a habitat determines the number of individuals that can survive per unit area. This is multiplied by the area of the habitat to give the total number of individuals that can live in the habitat. The probability of extinction is assumed to decrease as the total number of individuals increases. The probability of immigration, or of speciation, is assumed to remain as it is in island biogeographic theory. In practice, the theory is applied by using a measure of total habitat productivity (productivity per unit area × habitat area), rather than the total number of individuals, in the construction of species-area curves. When actual evapotranspiration is used as an estimate of primary productivity, diversity patterns on islands of vastly different sizes and latitudes seem to be reasonably described by the theory (Wright 1983).

This approach offers a potential explanation for latitudinal gradients in species diversity. However, more rigorous tests of the underlying assumptions of the model are needed. For instance, the model assumes that the total number of individual organisms in a habitat is directly dependent on total productivity. This may be a good approximation for birds, and perhaps for mammals, but is incorrect for terrestrial plants. The $-3/2$ thinning relationship (Yoda et al. 1963) and Gorham's application of this to monospecific stands show that increases in productivity lead to major decreases in the densities (number per area) of individual plants. In Gorham's (1979) comparison of natural monospecific stands of 29 different vascular plant species, a tenfold increase in plant standing crop was associated with a hundredfold decrease in the number of plant stems per m^2. The equation Gorham fit to his data can be rewritten as $D = (1.36 \times 10^8)S^{-2.04}$, where D is plant density, in stems/m^2, and S is above-ground standing crop, in g/m^2. Thus, small increases in standing crop lead to sharp decreases in plant density. In general, tropical habitats are much more productive and have greater standing crops than temperate habitats do (Whittaker 1975). Thus, they have many fewer adult plants per unit area. According to the logic underlying the species-energy theory, this should make the forests of a tropical island less diverse than those of a comparably sized temperate island, but the opposite is true (e.g., Wright 1983). Indeed, when Gorham's (1979; Gorham, personal communication) relationships are applied to the island data of Wright (1983) to calculate the estimated number of individual plants per island, temperate and high-latitude islands have orders of magnitude more individual plants, but many *fewer* species, than more equatorial islands. This directly contradicts the underlying logic of the species-energy theory as applied to plants. Similarly, mainland boreal and tundra habitats are much less productive than tropical habitats and have immensely more plant stems per unit area than do tropical forests. Because the geographical extent of the North American tundra is greater than that of the Central American tropics, species-energy theory predicts that plant species richness should be orders of magnitude greater in tundra than in the tropics. However, tundra is much less species-rich than tropical forests are. Thus, the underlying logic of the species-energy theory fails to explain latitudinal gradients in plant species diversity. Its apparent ability to explain such gradients, in a statistical sense, when species richness is graphed against total energy (productivity per unit area × area) may just be another way of representing the well-known effects of area (which is correlated with spatial heterogeneity and probability of colonization) and latitude on diversity. Although it may describe plant diversity relationships, the species-energy approach does not seem to explain them because its underlying logic is not supported. Even if total energy provides a better empirical fit than latitude and area, our analysis suggests that some explanation other than that underlying species-energy theory must be sought to explain this fit.

There are several alternative explanations for latitudinal gradients. First, the species-energy theory may be correct in predicting that the latitudinal gradient is mainly caused by higher extinction rates toward the poles, but it may be ascribing the wrong reasons to such extinctions. Clearly, higher extinction rates cannot be caused by lower population densities of plants toward the poles. Although plant population densities are higher toward the poles, weather is much more variable, which may increase the probability of extinction of rare species. This hypothesis, originally suggested by Wallace (1878), is consistent with May's (1973) model, in which the probability of random walks to extinction increases with increasing population fluctuations. Glaciation is another major disturbance event that occurs on a time scale relevant to speciation and extinction. Glaciers cause the local extinction of almost all species. Species can repopulate an area after glacial recession only by long-distance migration, which is extremely slow (e.g., Davis 1986). Moreover, habitat compression caused by glacial expansion may lead to high rates of global extinction of temperate and boreal species.

Thus, it may be that temperate and boreal habitats are much less diverse than possible, and are greatly undersaturated with species (Cornell and Lawton 1992) because of past extinctions and slow recolonization after glacial recession. Other hypotheses, such as longer periods between disturbances for evolution in the tropics, may also explain this latitudinal trend. Latitudinal gradients in biodiversity are one of the most striking and repeated patterns on earth, and yet are still not understood. Clearly, much more observational and experimental work is needed, as are theoretical treatments that look at the joint effects of the forces that cause species loss and those that allow persistence. However, the work presented here suggests that the answer to this mystery may lie not so much in understanding the causes of high diversity in tropical habitats, but in explaining why temperate and boreal habitats are so species-poor.

CONCLUSIONS

There are a myriad of forces that can allow the local persistence of species. In most cases, all that is required are two or more environmental factors that constrain fitness and unavoidable trade-offs in the ways organisms respond to these constraints. However, the actual constraints and trade-offs that explain local diversity have not yet been identified in most habitats.

The number of species that exist in a given habitat, and the number that exist upon the earth, represent the total effects of the various forces that allow species to evolve and persist and the various forces that cause extinction. The next major step in understanding the causes of patterns in species diversity will come from linking models of species persistence with models of species extinction. This approach can provide us with new predictions of patterns in species diversity that can be tested against existing observational and experimental data. Moreover, we need more detailed descriptions of natural diversity patterns, field studies of the causes of species loss, and experiments that determine the effects of environmental constraints on diversity. Wise conservation of the remaining biotic riches of the earth will require significant increases in our understanding of the fundamental forces that have led to the appearance and persistence of species-rich communities.

3

Environment and Trophodiversity

Peter Yodzis

A community is, more or less literally, a set of organisms that live together. For a generation of American ecologists, under the inspiration of Robert MacArthur, "community" was implicitly understood to mean "competitive community": a set of species that (lives together and) competes for some common requisite of life.

Several developments raised the question to what extent these competitive communities really are sensible units of study. For instance, it is well documented that predators can profoundly affect the structure of competitive communities (as reviewed by Yodzis 1986). This means that in an important sense a competitive community is not necessarily a closed system. To understand it fully, we have to expand our notion of the system to include a higher "trophic level." If that is the case, then perhaps we also need to include other prey of those predators, and animals that consume the predators, and so on. Indeed, it is thinkable that we will not arrive at a sensible unit of study until we have included *all* the organisms that live together in some location. Surely we need at least to explore this viewpoint, without necessarily asserting that it is the only acceptable one.

During the past decade or so, a group of researchers has been trying to articulate a "whole community" viewpoint through the study of community food webs. In this chapter I review what has been learned about the relationship between local environment and community diversity in a food web setting. I will use the term *community* in the remainder of this chapter to mean the set of *all* organisms living in some area. I will use the term *guild* when I want to refer to any functionally defined subset of a community (such as, for example, a "competitive community").

A food web is a set of "species" that live together, and a specification of which species eat which other species. Such a specification can be summarized conveniently in a picture such as fig. 3.1, in which each species is represented by a circle (called a *vertex* in the jargon of graph theory), and an arrow is drawn from each prey species to each of its predator species.

There is a trivial point to be made here, which is so frequently overlooked, and so important, that I am going to make it anyway. The point is that, like community studies, food web studies too can suffer from a failure to include interactions with other species of importance. For instance, many food web pictures, along the lines of figure 3.1, are *sink food webs* (Cohen 1978), which start with a set of predators (perhaps only one), and include their prey,

and the prey of their prey, and so on. Others might start with a set of prey (perhaps only one, or all the members of one guild), and include all their predators, and all their predators' predators, and so on—these are *source food webs* (Cohen 1978). These sorts of food webs will in general not be fully inclusive of the species in a community.

The food webs that one really wants to understand are community food webs. A *community food web* is a community, as I have defined it, together with the associated food web, as I have defined: it is the set of all species in some area, together with a specification of which species eat which other species. Most of the ideas of current food web theory were formulated with community food webs in mind, and I will discuss only those here: henceforth "food web" will mean community food web.

Most published food webs have obvious omissions; the rest probably have subtle omissions of species. In principle, we could never be sure that we had not missed some species in any attempt to document a community food web. When we say "all" species in this context, we mean (certainly at our current level of empirical sophistication) "enough for our purposes"; that nothing *essential* is left out.

In this chapter I review efforts to relate local diversity to local environment in a food web setting. Since food web structure is not yet a part of the standard knowledge of every ecologist, I first sketch some basic definitions. Next I outline the theoretical ideas that have been offered in explanation of food web structure (including those that have not yet been shown to be relevant to the issue of how environment influences that structure), then discuss the influence of environment on diversity, and finally point out some connections of this theme with other themes in this volume.

BASIC DEFINITIONS

The food web view is a trophic view of nature; direct interspecific interactions that are not trophic are ignored. (This does not mean that, say, exploitative competition is ignored, for that is an indirect interaction [Yodzis 1989a, 132–133, 192].) In the conventions of food web theory, a *predator* is a species that consumes some other species, and a *prey* is a species that is consumed by some other species. A predator that is not also a prey is a *top* species; a prey that is not also a predator is a *basal* species; all others are *intermediate* species. For instance, in figure 3.1, butterfish (number 12) is a top species, flagellates and dia-

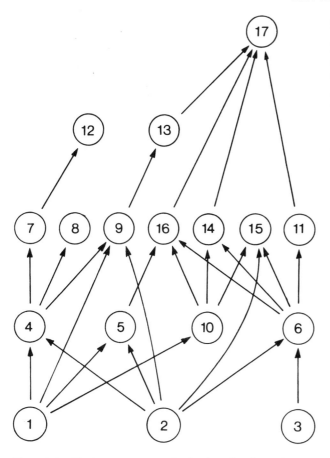

Figure 3.1   The Narragansett Bay food web. 1, flagellates, diatoms; 2, particulate detritus; 3, macroalgae, eelgrass; 4, *Acartia*, other copepods; 5, sponges, clams; 6, benthic macrofauna; 7, ctenophores; 8, meroplankton, fish larvae; 9, Pacific menhaden; 10, bivalves; 11, crabs, lobsters; 12, butterfish; 13, striped bass, bluefish, mackerel; 14, demersal species; 15, starfish; 16, flounder; 17, humans. (After Kremer and Nixon 1978).

and of prey constitutes a *trophospecies,* and these are the basic entities of a food web (the circles in figure 3.1). This lumping of biospecies into trophospecies is central to the contemporary food web enterprise; without this artifice it would be impossible to come anywhere near compiling legitimate community food webs. Even if it were possible to compile food webs composed of biospecies, they would be too complicated to work with—unless one started out by lumping together all the species with identical predators and prey. Henceforth, I will use the unqualified term "species" to mean trophospecies.

The notion of trophospecies makes food web theory practicable, but it highlights the perilous nature of a purely trophic point of view. Many taxonomically oriented researchers will see cherished and, to their minds, essential functional distinctions vanish into a formless aggregation. Biologists schooled in classical niche theory might at first find the notion of trophospecies somewhat nonsensical: how can (bio)species coexist if they have exactly the same prey? The answer might be that they partition along some other niche dimensions such as prey size or microhabitat utilization. Thus most of the processes of classical niche theory occur *within* these sets called trophospecies.

Another fundamental food web concept is the trophic link. If species A feeds on species B, then we say there is a *link* from species B to species A. In a depiction like figure 3.1, a link is represented by an arrow from species B's vertex to species A's. *Connectance* is a measure of the density of links within a food web; this concept is defined and discussed in more detail below.

A *food chain* of *length* $k - 1$ is a sequence of species $s_1, s_2, \ldots, s_k$ such that s_i feeds on s_{i-1} for all $i = 2, \ldots, k$. For instance, in figure 3.1 the sequence of species: $\{1,5,16\}$ forms a food chain of length 2. A *maximal food chain* is a food chain that starts with a basal species and ends with a top species. In figure 3.1, the food chain $\{1,5,16\}$ is not a maximal food chain; $\{1,5,16,17\}$ is.

Many ecologists think of whole communities in terms of "trophic levels" such as those shown in figure 3.2: there are primary producers, which occupy a first "trophic level"; herbivores feed on primary production (a second "trophic level"); secondary consumers ("trophic level" 3) eat herbivores; tertiary consumers feed upon the secondary consumers, and so on. The entire community is reduced to one immense food chain. There is even a tendency in the contemporary literature to apply the term "food web" to this concept. It is important to understand that this concept is very different from that of a food web—even if we agree to aggregate trophospecies into some kind of supercategories that we call "trophic levels."

Suppose we tried to do that with the food web of figure 3.1. Species 4, for instance, would be at the second "trophic level." But where would we put species 9? It feeds on primary production, which ought to put it at "trophic level" 2, but it also feeds on species 4, which is at "trophic level" 2, which would put species 9 at "trophic level" 3.

As you may have guessed from my use of "scare quotes" for the term "trophic level," it is a problematic concept. There are at least six different ways that one might define the term (Yodzis 1989a, 209), each of use in

toms (number 1) are basal species, and benthic macrofauna (number 6) are intermediate species.

There is an obvious way to alleviate the difficulty of including "everything" when one tries to compile a community food web. Consider, for example, the Narragansett Bay food web of figure 3.1. There are probably hundreds of species of flagellates in this community; to list all of them would be an immense task in itself, though only a small part of compiling the entire food web. The compilers of figure 3.1 simply lumped all of them together into the category, "flagellates," on the basis that, *from a trophic point of view* (which is the viewpoint of the food web), there is no significant difference between one flagellate species and another, and in this way we can be sure we have not left out any flagellates. Indeed, from a trophic point of view, it is not even worth distinguishing between flagellates and diatoms (at least, that is the position adopted by the compilers of figure 3.1), so they are all lumped together into one category.

Species that have exactly the same prey and exactly the same predators are not trophically distinct. The set of *all* the biospecies that share some particular set of predators

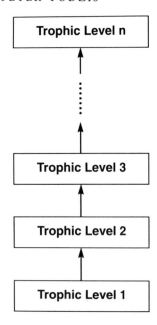

Figure 3.2   A food chain conceptualized as a series of trophic levels.

a different context. Rather than biasing our thinking by applying the term "trophic level" to any one of these, we would do better to invent at least six different new terms. At the very least, an author who uses the term "trophic level" is obliged to state his definition.

So long as one recognizes that it is an approximation and formulates it carefully (and does not call it a "food web"!), a food chain representation such as figure 3.2 is a thinkable first approach to a whole-community viewpoint, and this approach has produced some interesting ideas (review: Fretwell 1987). While my own bias is that this tradition eventually will have to be expanded to a true food web viewpoint, I should acknowledge that there is some empirical evidence in support of the food chain approximation (Persson et al. 1992).

Other reviews of food web data and theory, more comprehensive than this one but less thorough with respect to the themes of the present volume, are by Pimm (1982), Yodzis (1989a, ch. 8), Lawton (1989), and Cohen, Briand, and Newman (1990).

CONCEPTUAL DIFFICULTIES

Nobody, least of all food web researchers themselves, would deny that the food web data currently available, which were generally gathered by investigators whose main interests were in particular groups within the community rather than in the community as a whole, are seriously flawed. Also, and to some extent the cause for these flaws, there are fundamental difficulties with the food web concept itself. I will begin by explicating these difficulties.

The many imperfections in our current data base certainly limit our ability to arrive at valid and useful generalizations from food web analysis. But they do not limit this ability to zero. As I hope to demonstrate in this chapter (see also the reviews cited above), some very promising

insights are beginning to emerge from such analyses—sufficiently promising to have instigated the daunting enterprise of producing food web data of far higher quality (for instance, Winemiller 1990; Martinez 1991; Polis 1991; Hall and Raffaelli 1991). An article on our theme written in ten years will undoubtedly make this one look very naive—but at our current level of sophistication, we can already see enough structure to be confident that articles will still be written on our theme in ten years.

Community Boundary

I have glibly defined communities and community food webs in terms of all the species living together "in some area," as though it were obvious what is an appropriate "area" to use for this purpose. As discussed in the introductory essay by Ricklefs and Schluter (chap. 1), very often it is by no means simple to explicate an appropriate boundary for a community. This problem is difficult enough for a single population; it is still more severe for a community. Even supposing we can define a meaningful boundary for each population, these will generally be different for different populations in a community, sometimes very much so.

Aggregation into Trophospecies

The trophospecies concept would be relatively uncontroversial if we obtained our food web data by starting out with food webs based on biospecies, then systematically lumped together into trophospecies those that are trophically identical. In practice, it usually has been done, at least in part, the other way around: first trophospecies are chosen, then feeding relationships are established. The data we have were collected over a long period of time by many different researchers. There has been no consistent protocol for aggregation, neither among different studies nor even necessarily within a given study. Figure 3.1 is typical in this regard, with basal species very highly aggregated, top species distinguished almost to biospecies, and intermediate species quite mixed in their degree of aggregation.

We need to explicate which food web properties are relatively robust with respect to the degree of aggregation used, and work along these lines has begun. Sugihara, Schoenly, and Trombla found that, while some food web properties are not robust, others are (including chain length, ratio of the total number of predator species to the total number of prey species, fraction of species that are top, intermediate, and basal, and the rigid circuit property [Sugihara 1982, 1984]). However, Martinez (1991), working with a very finely resolved food web (182 recognized taxa, comprising 93 trophospecies) and using a much broader range of degrees of aggregation, found number of links per species, upper connectance, the distribution of chain lengths, and average chain length linking each species to basals to be very sensitive to aggregation; connectance and the proportions of top, intermediate, and basal species and numbers of links between them to be less sensitive; but only directed connectance (the ratio of the number of trophic links to the square of the number of trophospecies) to be robust.

In response to the past decade's work on food web theory, a new generation of researchers (Winemiller 1990; Martinez 1991; Polis 1991; Hall and Raffaeli 1991) is taking up the challenge of producing a body of food web data that comes far closer to the ideal sketched at the beginning of this section than does our current base of data. This very exciting development will undoubtedly lead to new insights.

Weak Links

A food web is a topological object, not a metric one. That is to say, it is framed in terms of the existence or nonexistence of a feeding relationship between each pair of trophospecies. But there is an obvious metric (quantitative) aspect of feeding that is left out in such a formulation. Each trophic interaction has a strength that we could measure in terms of the magnitude of the flow of material, or a fraction of the predator's total ingestion, or some such number. What about very weak trophic links? Should they be included as food web links? Most researchers seem inclined to the view that very weak links are irrelevant, and that it is not an accurate representation of the system to include them with (inevitably in a food web framework) status equal to that of very strong links. But how weak is "very weak"? Decisions of this kind are made in the reporting of all food webs in the empirical literature. Even if one takes the extreme view that every link should be counted, no matter how weak, then one is confronted with the question of whether one has missed some links in one's sampling procedures. Again, there have been no protocols, and there might be a great deal of variation in how different empirical researchers have handled this issue.

Paine (1980) distinguishes three approaches to linkage strength: the topological (food web) approach, in which only the presence or absence of links is considered; the flow approach, in which each link has a quantitative strength attached to it, determined by material flows; and the functional approach, in which linkage strengths are assigned on the basis of perturbation experiments. Weak links, in the flow sense, pose the difficulties just sketched for the topological viewpoint. Because the topology one arrives at will depend upon the "cutoff" level chosen for link strength, one could assert that the topological viewpoint has no existence logically independent of the flow viewpoint. As emphasized by Paine, the functional approach can give link strengths dramatically different from those of a flow approach. In principle, both the topological viewpoint and the functional viewpoint can be derived from the flow viewpoint, but these procedures—especially the latter—are fraught with difficulties. Responses observed in perturbation experiments can be highly sensitive to the overall pattern of flow strengths in a fixed food web topology (Yodzis 1988). This renders it difficult if not impossible to establish clear connections between the functional and flow approaches.

Winemiller (1990) has amassed a very interesting data set and has used it to address the question how the treatment of weak links affects perceived food web structure. His data are from tropical aquatic systems. They are resolved to the biospecies level for all fishes and include a quantitative measure of the strength of each link: the pro-

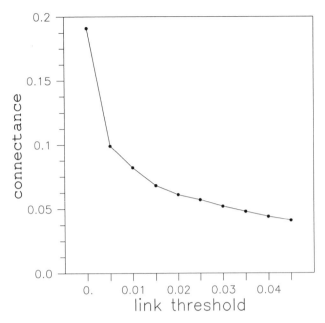

Figure 3.3   The relationship between connectance and link threshold for the wet season food web of the Caño Maraca estero. Link threshold is the lower limit on the strength of links to be included in the food web. (After Winemiller 1990.)

portion by volume of the total predator gut content represented by each prey species. This enables Winemiller to compute how several food web statistics (connectance, mean number of predators per node, and mean number of prey per node) vary as the cutoff level of link strength is changed. Figure 3.3 shows a typical example: as the cutoff level increases from zero, these quantities change at first quite rapidly with respect to cutoff level, then level off to slower variation. This suggests the encouraging view that above a certain small value, the cutoff level chosen is not critical. However, more work on this vital issue is needed, particularly since the cutoff level in figure 3.3 is based on dietary proportions, rather than flows.

Some of the structure that has been observed in the food web data could even be an artifact of the inevitable omission of the weakest links. I discuss this possibility at the end of the section on food web theories.

Seasonal and Other Temporal Variation

Food webs as construed in current research are static objects, although it is quite common for the food web in a given locality to change more or less regularly with the seasons, or to change its structure in response to temporal variation on other relatively long time scales. In common with many other fields of inquiry in theoretical ecology, right now we can do little more than plead that the static food web represents some sort of time average, or perhaps the configuration at a key point in the annual cycle. However, Schoenly and Cohen (1991), summarizing what is currently known of temporal food web variation, find that in those cases where we do have time-specific data, while constructing a cumulative web over all times does introduce systematic biases in certain structural characteristics, those biases are not large.

Life Stages and "Unstructured" Food Webs

In the currently dominant food web paradigm, age/stage structure is ignored; this despite the fact that the feeding habits of many animals change drastically with age. Particularly in aquatic systems, where prey are often selected more on the basis of size than of species, the addition of age/stage categories can yield a much higher connectance in terms of species, and can lead to loops of the form, species A eats species B which eats species C which eats species A. Loops of this kind are rare if attention is restricted to adults, as is done in the conventional data (though no less eminent a naturalist than William Shakespeare observed, "A man may fish with the worm that hath eat of a king, and eat of the fish that hath fed of that worm").

An extreme form of stage-dependent feeding, in which prey selection is entirely by size and, in terms of species, "everything eats everything," was modeled by Isaacs (1972, 1973, 1976), who called it an "unstructured food web." Isaacs felt that such models were appropriate for marine systems. He also suggested a way to test for this kind of unstructure. It is well known that concentrations of trace contaminants can increase as the substances move through a food chain. Similarly, one would expect to see this pattern of bioaccumulation in food webs that are structured in the conventional way. Isaacs showed (1973, 1976) that any such effect would be much attenuated in an unstructured food web. Empirical studies in marine systems have tended to find the pattern of bioaccumulation expected of structured food webs (for instance, Mearns et al. 1981; Rau et al. 1983).

Despite The Bard's pithy observation, these results on marine bioaccumulation lend considerable credibility to the conventional viewpoint. While no one would deny the basic truth of stage-dependent feeding in, say, marine systems, perhaps the feeding activities of adult stages are dominant in some sense (such as share, among life stages, of energy ingested over a lifetime).

FOOD WEBS AND COMMUNITY DIVERSITY

This volume is organized around the issue of community diversity. I think food web research has given us some new perspectives on that issue.

The first new perspective that I identify makes a virtue of necessity. Usually when ecologists think of "diversity," we think of the diversity of biospecies. The food web viewpoint encourages us to see not biospecies, but trophospecies when we look at a community. From this standpoint, it is natural to think of trophospecies diversity rather than biospecies diversity. From a food web viewpoint, then, diversity is diversity of feeding habits. Of course, there is a relationship between this kind of diversity and biospecies diversity, but it has not yet been clearly articulated. The point I want to make is that it is not necessarily a priority to articulate such a relationship. Perhaps it would be more interesting to broaden our whole notion of diversity in the direction suggested by the food web viewpoint. Indeed, having gotten that far, we might

be alert to still other facets of diversity. So, strictly speaking, I will be discussing something that one might call *trophodiversity* (diversity of trophospecies) in this chapter.

The other new perspective on diversity that we can gain from a food web viewpoint is a realization that trophodiversity has at least three dimensions. The first of these is diversity *within* trophospecies. By this I mean the number of biospecies (or other meaningful categories) contained in each trophospecies in a food web. Some trophospecies are taxocenes (for instance, species 4, 6, 7, 8 in figure 3.1), which "traditionally" might have been treated as competitive guilds. Others consist of one or a very few biospecies. Still others are something in between. So intratrophospecific diversity might not be a very clean concept. However, due to the rather arbitrary nature of our current food web data set, it is hard to say at this point what the nature of trophospecies would be in a data set based on adherence to consistent protocols.

The other two dimensions have to do with the diversity *of* trophospecies. But this has at least two aspects that need to be distinguished: *horizontal diversity* and *vertical diversity*. Horizontal diversity has to do, in some sense, with diversity within "trophic levels." This might involve any of the various different definitions one could suggest for "trophic level" (Yodzis 1989a, 209), or it might be framed in terms of the number of basal or of herbivorous species (Briand 1983b), or in terms of dominant cliques (which I define and discuss in the section on environmental variability).

Vertical diversity has to do with food chain lengths. Here again there are numerous formulations that might be of interest: the length of the longest food chain, the average of the lengths of all maximal food chains, and the average of the lengths of the longest chains linking each top species to a basal one come to mind.

Also of interest is the relationship between horizontal and vertical diversity: some food webs are wide and short, others are narrow and tall. To some extent the *shape,* in this sense, of food webs is related to habitat type (see below).

FOOD WEB THEORY

As of this writing, we have no theory of food webs that is generally accepted as correct. What we do have is a pretty good idea what some of the important theoretical issues are, and a number of different theoretical approaches, no one of which can plausibly be regarded as a complete theory in itself. Taken together in some way, these approaches might contain the essentials of a theory for food webs. In this section I briefly sketch these approaches, and in the following sections I summarize both the data and the (mostly tentative) theoretical insights bearing on the matters at issue in this volume.

Energy Flow

The notion that energy flow plays a role in structuring food webs is one of the oldest ideas in food web theory (Elton 1927; Lindeman 1942; Hutchinson 1959). Clearly, energy does propagate "upward" through food chains, with less energy available the higher one moves through a chain. One would think this ought to have implications for food web structure, some of them quite obvious (such as limitations on the lengths of food chains), but strictly speaking this need not necessarily be so. For instance, if some other process were to limit food chain lengths to something shorter than allowed by energetic constraints, then energetic constraints would become less relevant. The idea that energy flow does play a role has become somewhat unfashionable, but has been defended on empirical grounds by Yodzis (1981, 1984a, 1984b; see also the section on productivity and productive space below) and Schoener (1989).

Nutrient Cycling

Similarly, nobody denies that nutrients cycle in ecosystems, but it is not as clear whether this has implications for food web structure. The view that nutrient cycling, together with local stability (discussed below) does indeed affect food web structure has been championed by DeAngelis (1980, 1992). DeAngelis's work shows that the return time of an equilibrium can be influenced at least as much by nutrient cycling times as by food web structure per se (see the section on environmental variability below).

Body Size

As a rough generality, larger animals tend to eat smaller ones. Thus body size might well constrain food web structure. For example, Elton (1927) and Hutchinson (1959) proposed that food chain lengths might be determined by predator/prey body size ratios together with structural constraints on large animals. This idea has been formulated within the context of "unstructured food webs" (see above) by Platt and Denman (1977, 1978) and Silvert and Platt (1978, 1980). Warren and Lawton (1987) and Cohen et al. (1993a) have explored the notion that body size constraints might provide the biology behind the cascade model (see below).

Local Stability

One can associate population dynamical models with food webs and ask whether there are associations between food web structure and the behavior of those dynamical models. Researchers who adopt this approach usually have in mind associations between aspects of food web structure and stability of equilibria of the models. "Stability" is sometimes understood in the sense of the presence of local stability, and sometimes in the sense of the "resilience" (inverse of the return time) of a locally stable equilibrium. Associations of this sort have been found (DeAngelis 1975; Pimm and Lawton 1977; Lawton and Pimm 1978; Pimm and Lawton 1978; DeAngelis 1980; Pimm 1982; Pimm and Rice 1987), but not everybody agrees that resilience in this sense has much to do with the persistence of systems in nature. And as with energy flow (or any of these ideas), the implications of local stability for food web structure may be rendered less relevant by other processes that constrain structure before local sta-

bility becomes an issue. I discuss the relationship between resilience and food chain length below in the section on environmental variability.

A variant of the use of local stability as an organizing principle is the idea of "life at the frontier of stability," proposed by Cohen and Newman (1988). These authors suggest that, for reasons unspecified, communities evolve in such a way as to become less stable, and tend to be found just at the boundary in parameter space at which they would become unstable.

Assembly

Several authors have explored the idea that food web structure is an outcome of the assembly of communities by sequentially arriving species. One can formulate different versions of this common theme, depending upon which constraints one takes into account. In the food web context, assembly has been taken to be constrained by energy flow (Yodzis 1981, 1984b), by a weak version of trophic speciality (Sugihara 1982; 1984), and by Lotka-Volterra dynamics (Post and Pimm 1983; Drake 1985, 1988, 1990b, 1991, 1993). As I discuss below, this view of food web structure is particularly rich in ramifications for the exploration of historical and geographical influences on food web structure.

Drake's version of assembly, developed by using Lotka-Volterra dynamical models in extensive computer simulations of the assembly process, has implications for the "local stability" viewpoint. In Drake's simulations, as new species colonize a community, local stability indeed tends to decrease. But before it can decrease to zero, a community emerges that cannot be invaded by any more new species. The frontier of stability is never reached, and the equilibrium's return time has a lower bound, imposed by whatever it is that confers the invasion resistance property on Drake's end states.

The Cascade Model

The cascade model (Cohen and Newman 1985; Cohen, Newman, and Briand 1985; Cohen, Briand, and Newman 1986; Newman and Cohen 1986; Cohen, Briand, and Newman 1990) is based on two assumptions: (1) all species in the community are ordered, such that each species can (but does not necessarily) eat any species that stands below it in the ordering; (2) any species will eat any one of those below it in the ordering with a fixed probability that is inversely proportional to S, the total number of species in the community. The second assumption is empirically motivated; it is equivalent, when the number of species is large, to the well-known observation that the product SC, where S is species richness and C is connectance, is roughly constant (Rejmánek and Stary 1979; Yodzis 1980; Briand 1983b; however, this generalization seems to be faltering in some of the newer work: Schoener 1989; Winemiller 1989; Martinez 1992). If L is the total number of links in a food web, connectance $C = L/S(S - 1)$. So another way of stating this condition is that the number of links per species is roughly constant. The basis for the ordering in assumption (1) is left open by Cohen and his co-workers, but body size is one obvious possibility (Warren and Lawton 1987; Cohen et al. 1993a).

Operational Issues

I mentioned earlier the conceptual problems associated with the treatment of all feeding relationships on an equal footing in food webs, no matter how strong or weak the associated flows of material. Either by choice or by necessity (because we may fail to observe very weak trophic links even if we intend to include them), an empirical food web filters the data by ignoring links below a certain strength. Recently, theorists have begun to ask whether some of the structure we observe might not be deduced as a consequence of such a filter.

Cohen and Newman (1988) showed that a cutoff of interaction strengths below a certain level, together with the notion of "life at the frontier of stability" (discussed above), implies the hyperbolic law, SC = constant, which, as mentioned above, is one of the oldest and best-known empirical food web generalities. Kenny and Loehle (1991) have derived the same relationship from a cutoff of link strength below a certain level, together with an assumed strength-rank distribution for links, where link strength is measured in terms of biomass or carbon transfer.

ENVIRONMENT AND FOOD WEB STRUCTURE

In this section I discuss the influence of local processes on food web diversity, both in the theoretical framework just sketched and in the food web data currently available.

Productivity and Productive Space

If energy flow plays a role in structuring food webs, then there ought to be systematic differences between food webs from regions of low primary productivity and food webs from regions of high primary productivity. It was suggested already by Hutchinson (1959) that one such manifestation ought to involve vertical diversity: in some sense food chains should be longer in more productive environments, because after a given sequence of energy transfers along a chain, more energy input at the bottom of the chain will mean more energy output at the top, perhaps enough output to support another consumer— which would make the food chain longer, if an appropriate consumer were available to colonize.

Pimm (1982) surveyed data from the International Biological Program (IBP) documenting both primary production and number of "trophic levels," and found no relationship between the two. Briand and Cohen (1987) computed the average value, over all maximal food chains, of food chain length for each of 113 food webs drawn from the literature, and found no tendency for this quantity to be larger in webs from environments of high productivity than in webs from environments of low productivity. However, Persson et al. (1992), in a study of 11 temperate lakes of low to intermediate productivity, found that while the least productive lakes lacked secondary carnivores (piscivores), these were present in the more productive lakes. Thus, while energy strongly limits food chain lengths when productivity is low, there seems to be no gross overall tendency for food chains to be longer in more productive environments.

However, there are more subtle effects of productivity on food chain lengths. The first of these is very much a food web effect. After all, the energetic reasoning that relates food chain length to productivity is perfectly straightforward if we are thinking of an isolated food *chain,* but it becomes more complicated in a food *web.* In that setting, a top species receives energy from basal species through many different food chains, generally of different lengths. If these were all of equal importance energetically, then we could conclude that, generally, "food chains should be longer in more productive environments." However, the very logic behind this notion—that longer food chains dissipate more energy—suggests that among the several chains linking a top species to basals, it is the shorter that are more important energetically, and which therefore ought to be most influenced by production. Indeed, the existence of *arbitrarily long* food chains is perfectly consistent with the energy flow hypothesis, so long as the species which are high in those chains are *also* connected to basal species through sufficiently short chains. (Notice that, for combinatorial reasons, the average over all food chain lengths in a food web, as used by Briand and Cohen [1987], is weighted in favor of the longer chains.)

Therefore, if energy flow influences food web structure, the *shortest* food chains linking top species to basal species should be longer in more productive environments; the lengths of the longer chains linking top species to basal species need not increase with productivity.

Table 3.1 shows the frequency distribution of the lengths of the shortest food chain linking each top species to a basal species in 30 food webs whose environments Briand and Cohen (1987) were able to classify as low productivity (22 webs) or high productivity (8 webs) and which do not contain "man" as a trophic category (since the presence of human exploitation may alter energetic relationships in ways that are difficult to anticipate). About twice the fraction of top species in low-productivity environments, as compared with those in high-productivity environments, have a shortest chain length of 1; while about twice the fraction of top species in highly productive environments, compared with those in less productive environments, have a shortest chain length of 3. Both of these trends indicate longer shortest food chains in more productive environments. However, at these sample sizes, the statistical significance of this pattern is low (P = .22 in a chi-squared contingency table analysis).

As mentioned above, the *longest* food chains linking top species to basals need *not* be constrained by energy. Indeed, as we see in table 3.2, which gives the frequency distributions for these chains in the same 30 food webs as table 3.1, the longest food chains may tend to be *shorter* in more productive environments.

This result is strongly biased by the marine systems. Of the 30 food webs in question, 20% are marine, with all but one of those in low-productivity environments. But of the 6 webs that contain at least one food chain with a

Table 3.1. Shortest Food Chain Lengths for Top Predators in Different Environments

Chain Length	Low-productivity environments	High-productivity environments
1	28 (27%)	3 (12%)
2	65 (62%)	16 (67%)
3	12 (11%)	5 (21%)

Note: Total number of top predators (percentages of the set of all top predators in parentheses) whose shortest food chain to some basal species has the specified length, for food webs in low-productivity (22 webs) and high-productivity (8 webs) environments.

Table 3.2 Longest Food Chain Lengths for Top Predators in Different Environments

Chain Length	Low-productivity environments	High-productivity environments
1	15 (14%)	1 (4%)
2	35 (33%)	8 (33%)
3	25 (24%)	6 (25%)
4	16 (15%)	4 (17%)
5	8 (8%)	4 (17%)
6	1 (1%)	0
7	1 (1%)	0
8	2 (2%)	1 (4%)
9	0	0
10	2 (2%)	0

Note: Total number of top predators (percentages of the set of all top predators in parentheses) whose longest food chain to some basal species has the specified length, for food webs in low-productivity (22 webs) and high-productivity (8-webs) environments.

length greater than 5, 66% are marine. In the context of marine systems, it is widely believed that there is a size-efficiency basis for food chain lengths: smaller organisms are more efficient at nutrient absorption through their surfaces, so in marine environments basal species tend to be smaller where productivity is lower, and size-structured food chains can be longer if they start at a smaller size (Ryther 1969). In such a situation, the longest food chains linking top species to basals can be quite long in regions of low productivity—even though at the same time the *shortest* chains linking those same top species to basals seem severely constrained by energy (table 3.1).

Just how severely constrained the shortest food chains are can be appreciated from the following. The distributions in table 3.1 seem to cut off very sharply above the length 3, while those in table 3.2 appear to drop off more gradually as length increases. We can check the sharp cut-off of table 3.1 in a larger sample. Of the 113 food webs considered by Briand and Cohen (1987), only 30 could be classified as low or high productivity, and those 30 food webs gave us the 129 top species used for tables 3.1 and 3.2. But if we do not make the low-high productivity distinction, we can look at the distribution of shortest chain lengths for all 507 top species in the complete set of 113 webs. We find a startling regularity. It turns out that of these 507 top species, *none* has a shortest chain length longer than 4, and *only two* have a shortest chain length longer than 3. One of these two, the heron in web number 6, a California tidal flat, cannot be considered to be supported by the tidal flat community, as the heron ranges

very widely, both geographically and in habitat type, in its feeding (J. R. Krebs, personal communication). The other, "raptors" in web number 47, a South Florida swamp, may be similarly wide-ranging, but the specification here is too vague to enable further investigation.

Returning again to the *longest* chains linking top species to basals: in systems with relatively clear and discrete boundaries, such as lakes and streams, the *size* of the spatial area containing the community may play a role, through what Schoener (1989) calls the *productive space hypothesis*. Top species tend to have relatively large body sizes, and therefore they have to contend both with relatively large individual energy requirements and with relatively diffuse sources of energy. Of course energy becomes diffuse by the time it reaches the end of a long food chain; and even though, as indicated in the preceding paragraph, these animals tend also to exploit the richer energy sources available through shorter food chains, they must compete for those sources with other creatures that are specifically adapted for consuming them. Because of these circumstances, top species tend to have to feed over a larger spatial area. In extreme cases this area might even span several "communities" as defined in most biologists' understanding, as in the case of the heron discussed in the preceding paragraph.

Therefore, one would expect that smaller communities with clear and discrete boundaries might be missing some of these larger animals (if we leave out "tourist" species like the heron, as food web empiricists usually do), resulting in the *longest* food chains being shorter when the community's spatial area is smaller. Schoener (1989) has ranked the lentic and lotic food webs in a modified form of Briand and Cohen's collection of 113 webs by the length of the very longest food chain in each web, and has compared this ranking with spatial extent. There is a lack of exact measurements of the sizes of these communities, but the general trends seem as expected. The food webs ranked the lowest occur in very small bodies of water, while a fair proportion of the largest lakes and rivers are high in Schoener's ranking.

The significance of such "excluded top predators" for overall food web structure has thus far been explored only piecemeal. In the first instance, of course, we expect communities in discrete habitats of small spatial extent to be less diverse due to the exclusion of these species. But what of the effect on the rest of the community of the release from predation of their putative prey species? According to Paine's (1980 and earlier works cited therein) notion of keystone predators, these effects could be dramatic (particularly in terms of reducing diversity). Work by Paine and others on substrate-limited systems has tended to show a pattern of diversity enhancement by top predators, but on the whole, the influence of top predators is complex and diverse (reviews: Yodzis 1986; Carpenter 1988).

Environmental Variability

Another local process that affects food web structure is environmental variability. This is an extremely difficult process to treat theoretically, and I do not know of a fully satisfactory account. While the food web data are quite clear as to the effects of environmental variability, theory has been less successful in this area.

First the theory: The only attempt that I am aware of is the well-known work of Pimm and Lawton (1977), who adopted a certain version of the local stability viewpoint and applied it to vertical diversity. They investigated the stability of equilibria associated with food chains of differing lengths. While their food chains were never actually unstable, there was a tendency for the systems with longer food chains to require longer times to return to equilibrium after a disturbance. May (1973) had suggested that in a fluctuating environment, systems whose return times are too long may fail to persist—roughly speaking, because the restoring force of the equilibrium's local dynamics would not be strong enough, over a long time period, to overcome the disruptive force of random disturbances. Therefore, Pimm and Lawton concluded that food chains should be shorter in more variable environments.

The only claim I have seen of empirical evidence in support of this conclusion is the suggestion by Pimm (1982) and Pimm and Kitching (1987) of support by their finding that in tree hole communities, species feeding higher in the food chains recolonized more slowly after an experimental disturbance. But surely the dynamics of recolonization, in which the very *dimension* of phase space itself changes, is completely different from the "sufficiently small" perturbations of local stability theory (Yodzis 1989b; Lawton and Warren 1989).

On the other hand, there is empirical evidence against Pimm and Lawton's claim. Briand and Cohen (1987) classified food web environments as having high or low environmental variability, calling an environment "fluctuating" if the original report documenting its food web indicated temporal variations (be they periodic or random) of substantial magnitude in temperature, salinity, water availability, or any other major physical parameter. They found no significant difference between fluctuating and constant environments in mean maximal food chain length, particularly after correcting for the effect of habitat dimensionality (see below). As I pointed out earlier, the metric used by Briand and Cohen is not an appropriate one for testing the energy flow hypothesis because of its emphasis on the longer food chains, but for this same reason, it ought to provide a good test for the putative destabilizing effects of long food chains.

So empirical data fail to support the idea that environmental variability strongly influences vertical diversity. Somewhat oddly, environmental variability does have an effect on food web structure that is quite clear in the data, but has yet to be addressed by theory. I am not quite certain whether the structure involved in this effect can be viewed as a component of diversity, so I will describe it and let the reader decide.

If every species consumed every other species, then the total number of food web links would be $S(S - 1)$ (where S = species richness = total number of species). If L is the actual number of links in a food web, the web's *connectance* (sometimes called "lower connectance") is defined as $C = L/S(S - 1)$. It is a measure of the density of trophic links: the number L/S of trophic links per species, divided by the largest possible number $(S - 1)$ of trophic links per species.

Another notion of connectance was suggested by Yodzis (1980). Let K be the number of species pairs that either are connected by a trophic link or have a prey in common. Define the *upper connectance* $C_u = K/S(S - 1)$. Because species that have a prey in common are potential competitors (through *direct* acts of interference, as distinct from the *indirect* effect of exploitative competition already built into their sharing of a resource), this is a measure both of link density and of the density of potential direct interactions.

Now, the point is that, while the density of links, as measured, say, by C, is not sensitive to environmental variability, the density of potential direct interactions is. In particular, Briand (1983b) showed that, empirically, for a given species richness S, upper connectance C_u tends to be larger in constant environments than in fluctuating environments. This difference has to do not with the number of trophic links, but with the way in which those links are organized in the food web, resulting in more potential competitive interactions—more trophic niche overlap—in constant than in fluctuating environments.

There is another way of looking at this circumstance, which is, perhaps, more closely akin to something that one might want to call horizontal diversity. I need first to define some terms.

A *clique* is a set of species such that every pair of species in the set has at least one prey species in common. A *dominant clique* is a clique that is "maximal" in the sense that it is contained in no other clique. For instance, in the Narragansett Bay food web of figure 3.1, the dominant cliques are the following sets of species: {7,8,9}, {4,5,9,10}, {4,5,6,9,15}, {11,14,15,16}, {12}, {13}, {17}. The first of these dominant cliques {7,8,9} consists of the predators of species 4, the copepods. The second dominant clique {4,5,9,10} is a guild of microherbivores. Members of the third dominant clique are detritivores, and members of the fourth dominant clique are consumers of benthic macrofauna. Each of the remaining three dominant cliques is the unique predator of at least one species.

Dominant cliques can be thought of as an unambiguous mathematical definition (within the context of the food web viewpoint) of Root's (1967) concept of guilds (Yodzis 1980, 1982). Since food webs are a purely trophic framework, these are *trophic guilds*.

Figures 3.4 and 3.5 show the number and average size of dominant cliques in Briand's (1983b) collection of 40 food webs, distinguishing between food webs from fluctuating and from constant environments. Especially in fluctuating environments, the average size of dominant cliques is remarkably constant with respect to the total number of trophospecies (trophospecies richness). To some extent, this is a reflection of the relative constancy of the number of links per species noted above, but, like upper connectance, it also is an outcome of the way that those linkages are organized. Variations in overall food web diversity, in the sense of trophospecies richness, are

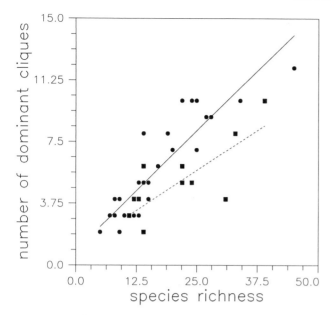

Figure 3.4   The relationship between number of dominant cliques and species richness in 40 food webs. Circles, food webs from fluctuating environments; solid line, linear regression for those points; squares, food webs from constant environments; dashed line, linear regression for those points. (After Yodzis 1982.)

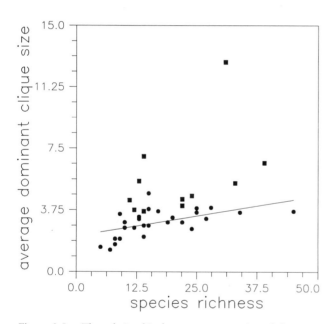

Figure 3.5   The relationship between average size of dominant cliques and species richness in 40 food webs. Circles, food webs from fluctuating environments; solid line, linear regression for those points; squares, food webs from constant environments. (After Yodzis 1982.)

almost entirely due to variations in the number of trophic guilds, as opposed to variations in their size, particularly in fluctuating environments.

One is tempted to offer the generalization that overall food web diversity is controlled primarily by the number of possible functional groupings among species that can live in a given environment, with the feasible sizes of such groupings constrained rather narrowly by forces not yet understood. There may well be merit in such a view, but in food web studies, because of the peculiar nature of our current data base (discussed above), one must always beware of possible artifacts of those peculiarities. In particular, the lack of consistency in how biospecies are aggregated into trophospecies makes it very dangerous to formulate generalizations about trophospecies richness. Perhaps more trophospecies-rich food webs contain more trophic guilds simply because their compilers have differentiated trophic categories more finely.

Comparisons in which species richness is factored out are far less observer dependent, thus more trustworthy. As we see in figures 3.4 and 3.5, for a given species richness, food webs in constant environments have fewer, but larger, dominant cliques (Yodzis 1982). This suggests that in constant environments there are fewer broad categories of trophic type—fewer trophic guilds—than in fluctuating environments (*less* overall trophic diversity in a certain sense), but that trophic guilds tend to be larger in constant than in fluctuating environments (*more* horizontal diversity in another sense).

Another noteworthy, if somewhat subtle, difference between food webs from constant and fluctuating environments can be seen in figures 3.4 and 3.5. Food web structure seems to be more tightly constrained in fluctuating than in constant environments. Notice, for example, how tight the linear regression for fluctuating environments is in figure 3.5, while the points from constant environments show more variability in dominant clique size. This contrast shows up in other contexts as well. Yodzis (1981) found that the structure of food webs from fluctuating environments could be accounted for on the basis of a simple process of energetically constrained assembly far more easily than could the structure of food webs from constant environments; while Briand and Cohen (1984) found the proportions of top, intermediate, and basal species; and Cohen and Briand (1984) the total number of links and the numbers of links categorized as top-intermediate, top-basal, and so on; to be more variable in food webs from constant environments than in webs from fluctuating environments.

Environmental Dimensionality

Briand (1983a) stressed the importance of another aspect of environment, its dimensionality, in the diversity of food webs. He called an environment *two-dimensional* if it is essentially flat, like a grassland, tundra, lake bottom, or rocky intertidal, and *three-dimensional* if it is clearly solid in its geometry, like a pelagic water column or a forest canopy. He also recognized that some environments are of *mixed* dimensionality, with both a two-dimensional and a three-dimensional aspect. For instance, many aquatic communities include both pelagic and benthic components.

Briand pointed out that environmental dimensionality affects both vertical and horizontal diversity as well as the relationship between them, which we might call the *shape* of a food web: webs from two-dimensional environments tend to be shorter and wider, webs from three-dimensional environments to be taller and narrower. Part of this tendency was confirmed by Briand and Cohen (1987), who found that average maximal food chains for food webs from three-dimensional environments were longer than those from two-dimensional environments. Indeed, these authors concluded that dimensionality is the major determinant of average maximal food chain length (which, it will be recalled, emphasizes the longer chains linking each top species to basals). This is another trend that is not addressed by any of the current theory.

Habitat Type

Briand has produced several studies that explore the influences of specific habitat types on food webs. The first of these was an avowedly somewhat informal attempt to see whether food webs might tend to segregate with respect to habitat type in a suitable parameter space (Briand 1983b). Briand came up with figure 3.6, which plots 40 food webs in a space whose axes are SC_u (the product of species richness and upper connectance) and the percentage of species that are herbivores (that is, which feed exclusively on basals). As already noted, the first of these variables has to do both with the density of feeding links and with how they are organized in terms of trophic overlap, and is related to diversity within and among trophic guilds. The second variable can be viewed as a rough measure of overall horizontal food web diversity. As is evident from inspection of figure 3.6, food webs from intertidal, forest, estuarine, pelagic, and mixed terrestrial habitats appear as distinct groups in this space. There is a tantaliz-

ing suggestion here that food webs tend to be more similar within than between habitat types, regardless of geographical location and taxonomic composition.

Briand has continued these studies, using discriminant and cluster analyses. Within the current standard collection of 113 food webs, most of the habitat types are not represented frequently enough to apply these techniques. However, freshwater systems are relatively well represented, and Briand (1985) has analyzed a data set consisting of 21 such webs (12 lentic and 9 lotic), together with 20 food webs from other habitat types.

In a principal component analysis of this data set, three principal components were found to be operationally significant. The loadings on these factors are listed in table 3.3, and the ordination of food webs using factors F1 and F2 is depicted in figure 3.7. The first component, F1, is a shape factor: it discriminates food webs that are long and thin (high scores) from those that are short and wide (low scores). The second component discriminates on the basis of linkage complexity and dietary generalization. Communities with high upper connectance and a large fraction of omnivores score high on this factor, while those with relatively few links and pronounced food partitioning score low.

The most striking feature of figure 3.7 is the segregation between stream communities and the other habitat types. This is highly significant ($P < .0001$). Stream food webs tend to be highly connected, wide, and short. Lake and river food webs tend to be thinner and longer, with less trophic complexity (as measured by upper connectance). Lake food webs appear roughly similar to terrestrial and marine systems ($P = .22$ in a simple discriminant analysis), and quite indistinguishable from the two rivers (Thames and Cam) surveyed. However, the lakes are grouped rather closely, except for Lake Texoma and Lake Rybinsk, whose food webs are longer, thinner, and more complex in terms of upper connectance. Interestingly, these two outliers are the only two reservoir lakes included in the survey.

Briand (1985) suggested that the shortness of stream food webs could be explained by the energetic poverty of

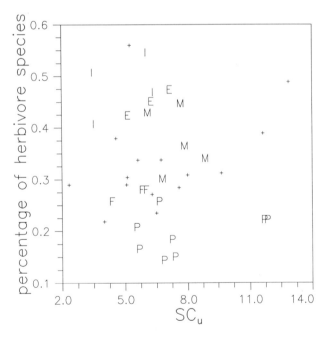

Figure 3.6 Segregation of 40 food webs from different habitat types as a function of SC_u (S = species richness, C_u = upper connectance) and percentage of herbivore species. Habitat types: I, intertidal; F, forest; E, estuarine; P, pelagic; M, mixed terrestrial; +, other types that do not segregate well (rivers and marshes). (After Briand 1983b.)

Table 3.3 Principal Component Analysis of 41 Food Webs

Variable	Factor loadings		
	F1	F2	F3
Average food chain length	0.95***	0.12	−0.11
Maximal food chain length	0.89***	0.22	0.04
Fraction of species that are basal	−0.04	−0.18	0.86***
Fraction of species that are top	−0.69***	0.05	−0.25
Fraction of species that are herbivores	−0.74***	0.07	−0.09
Fraction of intermediate links	0.97***	−0.11	−0.13
Fraction of top links	−0.73***	−0.04	−0.45**
Linkage complexity (SC_u)	0.01	0.82***	−0.07
Fraction of species that are omnivores	0.26	0.46**	0.46**
Fraction of prey species with only one predator	0.06	−0.71***	0.07

Source: Briand 1985.
Note: Asterisks indicate significant correlations (two-tailed test): * for $P < 0.05$, ** for $P < 0.01$, *** for $P < 0.001$.

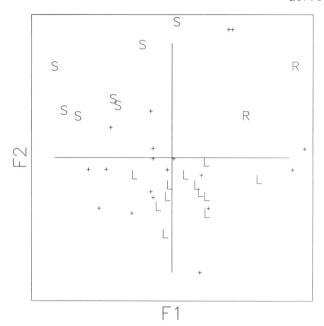

Figure 3.7   Ordination of 41 food webs in the space of the principal factors F1 and F2 of Table 3.3 Habitat types: S, stream; L, lake or river (except for reservoirs); R, reservoir; +, other types. (After Briand 1985.)

the detrital base. On the basis of this hypothesis one would also predict that the food webs of river communities, based largely on autochthonous plant production, would have longer food chains than upstream communities, as seems to be the case in Briand's data set (his table 3, not reproduced here). Briand also proposed that the relatively high upper connectance of the streams could be linked to their very constant ambient temperature (while granting that they do experience marked fluctuations in the supply of allochthonous organic matter).

Another principal component analysis by Briand (1983a), with a larger sample of 62 food webs, produced several distinctions in discriminant analyses, some by now familiar to us from other studies I have discussed, some novel:

1. fluctuating (low SC_u) versus constant (high SC_u) environments; $P = .0001$

2. pelagic (higher average maximal food chain length) versus sublittoral benthic (shorter); $P = .004$

3. two-dimensional (lower average maximal food chain length and larger percentage of herbivores) versus three-dimensional (longer, thinner) environments; $P = .009$

4. pelagic three-dimensional aquatic (fewer specialist predators) versus three-dimensional terrestrial (more specialists) habitats; $P = .015$

5. intertidal (low SC_u) versus sublittoral (high SC_u) habitats; $P = .015$

Schoener (1989) rediscovered some of these trends, and added a few more, in correlation and ANOVA analyses of a modified version of 98 food webs from the current Briand-Cohen collection. With respect to the issue of diversity, in which I am including various measures of food

web "shape," his results on the distribution of species among basal, top, and intermediate categories are worth mentioning. He finds that the fraction of basal species is highest for estuarine and lotic systems, and lowest for marine pelagic systems. The latter have mostly intermediate species (another indication of their tallness), whereas estuarine and lotic webs have few intermediate species (and are short in terms of longest food chains). The fraction of top species is lowest for marine pelagic webs and highest for lotic webs.

Schoener also uses comparisons among habitat types to further evaluate his productive space hypothesis (discussed in the section on productivity). Looking at the longest food chain in each web, he finds a median (taken over the set of 98 food webs) value of five links in marine pelagic webs, four links in terrestrial, marine benthic, intermediate estuarine, and lentic webs, and three links in marine estuarine and lotic webs. The marine, lentic, and lotic systems are ordered as one would expect from applying the productive space hypothesis.

Of course, many of these contrasts between habitat types reflect the general influences of productivity, productive space, variability, and dimensionality of the local environment. For instance, Briand suggests that food chains may tend to be longer in pelagic than in sublittoral benthic food webs (point 2 above) because of a lower assimilation efficiency in benthic systems, where energy transfer is largely detritus-based, and that the lower SC_u of intertidal relative to sublittoral systems (point 5) may be due to a greater degree of environmental fluctuation (point 1).

Or is it the other way around? One would like to separate cleanly the specific influences of habitat type from the general influences of productivity, variability, dimensionality, and possibly other universal variables, but unfortunately at this point we simply do not have enough empirical food webs to do so. There are already worrisome problems of sample size in some of the less finely differentiated comparisons in the current food web literature. If nothing else, the past decade's work on a flawed data base of empirical food webs has highlighted the need for more, and especially better, food web data.

ASSEMBLY, HISTORY, AND LANDSCAPES

This chapter has focused on local processes and food web diversity, but these topics are connected with other themes in this volume. As I mentioned when I surveyed food web theories, some researchers have explored the viewpoint that food web structure is the outcome of a process of assembly. From this standpoint it is very natural to consider the influence of historical and regional factors.

Some work along these lines has been carried out by Drake (1988, 1990b), who used model systems to analyze the assembly of communities in a patchy environment from a central pool of species. His assembly algorithm used Lotka-Volterra dynamical models to judge whether a species could invade and to find the result of a successful invasion. The environment of each local patch was identical to every other one, so any heterogeneity that emerged was generated entirely by random differences in the order in which species colonized. As in his studies of local assembly mentioned above (Drake 1985), each patch attained an invasion-resistant end state, so that the whole landscape reached a persistent end configuration.

Drake found that these theoretical landscapes are extremely patchy. The exact nature of the local end state attained in each patch depends on its history of randomly ordered colonizations. As a test for the role of history, Drake attempted to reconstruct his end states by using a species pool consisting only of those species present in a given end state, but otherwise using the same assembly algorithm. In several hundred such reassembly attempts, he was able to reconstruct only one end state, and only a single history involving that particular set of species reached that particular end state.

However, there is structure amid all this randomness. At a regional level, examining the regional distribution of species in his landscapes, Drake found distributions somewhat akin to those generated by Hanski's (1982) core-satellite hypothesis (i.e., most species are either quite common or quite rare within the landscape as a whole). At a local level, while the number of trophospecies in food webs varies by a factor of approximately 2 (J. A. Drake, personal communication), they are similar to each other and to observed webs in terms of food web structural metrics: prey/predator ratio, proportion of herbivores, mean food chain length, proportion of specialist predators, connectance, and upper connectance. Vertical trophodiversity is relatively constant, while overall trophodiversity (in the sense of trophospecies richness) and, presumably, horizontal diversity, are quite variable within such a landscape.

In his studies of the communities inhabiting *Nepenthes* pitcher plants in Malaysia, Beaver (1985) found a similar pattern: quite variable species richness, but little variation in connectance or ratio of the number of predators to the number of prey. He did, however, observe some variation in vertical diversity, though it is difficult to compare his results with Drake's simulations, as he used a different metric. Drake (1991, 1993), working with laboratory microcosms, has begun to dissect the historical factors responsible for these regional landscapes. He finds the timing of events, particularly relative to the invasions of certain key species, to be crucial. However, his microcosms are not yet complex enough to enable an analysis of the role of food web properties in the assembly of landscapes.

4

Structure of Intertidal Assemblages in Different Locations: How Can Local Processes Be Compared?

A. J. Underwood and Peter S. Petraitis

Considerable effort and ingenuity has gone into experimental analyses of the structure and dynamics of assemblages of species on intertidal rocky shores. Much of this work has been preoccupied with local processes, largely because these are the ones most amenable to manipulative experimentation and because intertidal ecologists tended to assume that small-scale variation was maintained by local processes. This focus is appropriate when explanations are sought for patterns that are apparent at relatively small spatial scales. Thus, the most often-cited studies have all been done on very small parts of a shore or on relatively small parts of several shores. The choice of scale to examine in experiments has been predicated on previous observations of patterns of abundances that varied over small spatial scales. For example, Paine (1974) observed major differences in the cover of sessile species and in the diversity of species over a few meters from high to middle levels on a shore. Underwood, Denley, and Moran (1983) described processes of predation, competition, and variable recruitment over scales of tens to hundreds of meters, leading to small-scale experimentation in several patches over similar scales.

There is a widespread view that sufficient small-scale experimentation will lead to more general or large-scale understanding of the outcome of ecological processes. The usual procedure is to group or average a number of different studies with a view to extracting generality from the collection of individual experiments, as done explicitly, for example, in reviews of competition by Connell (1983) and Schoener (1983) and of predation by Sih et al. (1985). Menge and Farrell (1989) hoped that a general understanding of the structure of assemblages would emerge from a collection of studies.

Nowhere has there been an attempt to evaluate the degree to which the quantitative outcomes of experiments of various designs and power, on populations with different histories and different spatial and temporal variations of abundances, are comparable or even commensurable. To date, no one has provided evidence that accumulation and synthesis of numerous small-scale studies is a valid method for interpreting general or larger-scale patterns.

A major problem with such approaches, which is not a problem for the interpretation of the individual studies, is

that comparisons from one place to another are made very difficult unless each area has been studied at similar scales. Given well-designed experiments, it is relatively easy to demonstrate that various processes, such as competition, predation, and physical disturbance, alter the structure of intertidal assemblages (Connell 1972, 1975; Menge and Sutherland 1976, 1987; Sousa 1979a, 1979b, 1980; Underwood 1984). Well-designed experiments have not always been used (see Hulbert 1984; Underwood 1986a; Underwood and Denley 1984). Ecologists have, however, failed to answer some of the more interesting questions about the relative importance of processes and the difficulties of making regional comparisons.

Before it is possible to make quantitative comparisons of the magnitude, rate, or importance of some ecological process from one habitat or biogeographical region to another, it is important to ensure that the information being compared is commensurable. Only the most superficial and probably qualitative comparison of different regions will be possible unless data are gathered within each region in comparable ways, at comparable intensity, and at similar spatial and temporal scales. Assuming that the relevance of some process affecting diversity can best be ascertained from quantitative sampling and controlled, replicated experimentation, we wish to explore what might be the difficulties in using such data to compare studies from different times and places. Specifically, we attempt to determine what sorts of information might be required from studies in one or a few places in one geographical area so that the identity, magnitudes, rates, and relative effects of different processes could be reliably compared with those found in studies on similar species in other places.

We focus on the further difficulties imposed by differences in the designs of experiments and in the procedures used to manipulate intertidal assemblages. There are other problems that we do not discuss. We believe that any of these problems will be similar in other habitats and for other types of animals and plants. Comparisons of ecologies of faunas and floras from one place to another are obviously beset with historical differences that are not amenable to unraveling or explanation by current experimentation. Such comparisons are also made difficult by the endemism of faunas and floras that result in there being quite different complexes of species in different areas. There are also differences in the philosophies and biases that underlie different studies (Underwood and Fairweather 1986). Intertidal systems are only illustrative of a general problem. This examination of the difficulties should provide guidance that will increase the potential comparability of future studies.

What is a Local Assemblage?

An Assemblage

Any attempt to study the fauna of a region (however locally defined) will founder unless the unit for study is clearly defined. It is a pity that the term "community" has come to mean different things to different researchers (Simberloff 1980; Underwood 1986b). To avoid implications of Clementsian (1936) "super-organisms," we should no longer use the term, because it comes with a collection of meanings that are mostly counterproductive (see Underwood 1986b). Here, we use the term "assemblage" to indicate co-occurring species in a given habitat at a specific time and place.

The Species

One of the difficulties that besets any attempt to compare patterns and processes is that different studies have usually identified the unit of study using different criteria. For example, although there is a considerable literature purporting to be about communities, the majority of such studies is, in fact, on taxocoenes or taxonomic assemblages—i.e., groups of closely related species (Pianka 1978; Feinsinger and Colwell 1978; Jaksic 1981; Underwood 1986b). Whether these are appropriate or cohesive units of study is open to question. The patterns of distribution and abundance in a guild of species (i.e., the species in some area that use similar resources in a similar manner: Jaksic 1981) are more likely to be indicative of the processes operating that affect resources in a particular habitat. There are, for example, cases where investigation of a guild (Root 1967) would be considerably more useful than any amount of study on only one taxonomic component of the assemblage in an area.

In intertidal studies, it has commonly been the case that the unit for study involves quite different types of organisms, from many different phyla, but that are in the same guild. These widely different types of organisms are often studied simultaneously because they all require space. In intertidal assemblages, space on the shore is an absolute prerequisite for all the sessile species (algae, barnacles, etc.).

In addition, however, other species that are not in exactly the same guild may use similar resources. Different types of organisms in an assemblage might have similar problems with physical environmental processes or with predators (see the experimental studies on processes operating on algae, crustaceans, and mollusks by Dayton [1971] and by Underwood, Denley, and Moran [1983], and their derivatives [e.g., Dungan 1986]). Most mobile species of grazers and predators require space over which

to forage. They may therefore compete with sessile species for such space. Thus, when free space is in short supply, grazing limpets cannot feed on microalgae (Underwood and Jernakoff 1981). Some limpets are territorially aggressive in defending space, as a result of which they can maintain a supply of food (Branch 1971, 1976; Stimson 1970, 1973).

The identification of complete groups of species that share a common resource is a difficult problem. It is, perhaps, easier in some habitats, such as rocky intertidal ones, than elsewhere, where identification of the nature of resources, particularly resources shared by many species, may be much more difficult.

Even in studies on rocky shores, the obvious relationship between space and food has not assured that the whole assemblage is used as the appropriate unit for study. Very small, cryptic, or very mobile species are routinely ignored, even though they are known to be important. For example, microscopic algae are the major source of food for many intertidal grazers (Castenholz 1961; Underwood 1984), but there have been remarkably few studies on them (exceptions are those by Castenholz [1963] and MacLulich [1986a, 1986b]). Amphipods and other small crustaceans are difficult to sample because of their small size and great activity, so they are often ignored (but there are exceptions: see Nicotri 1977; Underwood and Verstegen 1988).

Other "difficult" organisms are sponges and ascidians that do not have identifiable individuals (although there are increasingly sophisticated techniques for studying their demography: Hughes 1984; Jackson 1986). Where attempted, analyses of their ecology can be informative (e.g., Connell and Keough 1985; McGuinness 1984b). Other organisms that tend to be ignored are those that visit intertidal areas only during high tide (but there have been some experimental studies on fishes; see Paine and Palmer 1978). Still other species are difficult because of taxonomic uncertainty (e.g., many species of algae tend to be lumped under collective groupings because of the great difficulty of distinguishing them in the field; e.g., Littler and Littler 1984). Paine (1974) and others have discussed some of the difficulties and choices that must be made in defining an assemblage to study.

The net result of all of this uncertainty is twofold. First, there is no evidence that anyone really studies a "community" in the strictest sense of the word (the organisms living and coevolving in a particularly defined habitat). At best, a fairly arbitrary chosen subset becomes the focus for attention. Second, comparisons among habitats or regions are made very difficult when decisions about what sort of subset of species and the species themselves differ from one habitat or geographical region to another. Sometimes there are constraints on the unit of study caused by the sheer number of species in the assemblage. For very good reasons, Menge, Ashkenas, and Matson (1983) divided the fauna on Panamanian seashores into "functional groups," largely as a response to the great richness of the fauna, which would have required enormous numbers of experiments to examine on a species-by-species basis. Nevertheless, it is not possible to know how the ecological relationships of different species in a "group" may differ. In other habitats on rocky shores, quite closely related species can have very different interactions and therefore very different effects on other species in the assemblage (e.g., Underwood and Fairweather 1986). Functional groupings are only useful if the similarity of "function" of all the species in each group is known and not simply assumed.

Obviously, dealing with individual species, as opposed to considering them in functional groups or guilds, creates a taxonomic scale of difference from one locality and study to another. So far, however, decisions about how to deal with many species tend to be based on the requirements of logistics, the status of taxonomy in an area, and simple choice by different researchers. There is no obvious agreed theoretical base from which to make these decisions, and the consequences of making them differently in different places will frustrate geographical or other comparisons for some time to come. This is an important area that needs more research.

How Local Is Local: Spatial Scales

The definition of "local" is largely subjective. There is no easy method available for determining whether the processes being investigated in a study of intertidal organisms are really localized. The choice of sampling unit (for example, the size of a sampling quadrat or experimental plot) may bias perceptions about the "localness" of a process. For example, the per capita mortality of a species of prey, when measured in small quadrats, may show very large variation among quadrats. Rates of mortality are then considered to be relatively unpredictable. Such variability may be due to the foraging behavior of predators that traverse many meters to search for prey, but feed intensively in a small patch of prey once they have been located. In contrast, when estimates of mortality are made in larger quadrats, these average out much of the smaller-scale variation, making predation appear to be more predictable. In this latter case, the effects of predation would appear not to be localized—they would seem similar from one part of a shore to another.

Similarly, small-scale differences in topography may affect the original settlement or establishment of populations of sessile or semi-sessile species (such as barnacles [Connell 1961a; Wethey 1984], mussels [Petraitis 1991], and snails [Fretter and Manly 1977]), leading to considerable spatial variability at a very localized scale. When measured in appropriately small sampling units, those local processes affecting settlement that are driven by local features of topography will be evident. If densities are recorded in larger quadrats, the small-scale variability in settlement and colonization of a shore will not be seen, and no very localized processes will need to be invoked as explanations of variations in density from one place to another.

Comparisons from one place to another are also made difficult by the different spatial scales used by different investigators. For example, Menge (1976) studied six geographical locations along the coast of New England. The locations were chosen to represent different degrees of wave action. At each location, he chose only one midshore site. Ignoring the possible confounding of differences in wave action and in geography (there was only one, different, site for each level of wave action), Menge's study provided data that could be used to calculate spatial variances at a scale of tens to hundreds of kilometers. The coastline of New South Wales is worth comparing with that of New England because of the similarity of some of the experimental procedures used to study it. The coast of New South Wales, however, has so far mostly been described with spatial variances estimable at a scale of tens to hundreds of meters (from site to site at the one general locality: Underwood, Denley, and Moran 1983). An exception is the study of recruitment and subsequent survival of barnacles (*Tesseropora rosea*) by Caffey (1985), who demonstrated considerable spatial and temporal variance in these processes. Any attempt to compare the processes operating in intertidal assemblages in New England and New South Wales is frustrated by the lack of local replication (at a scale of hundreds of meters) in the studies in New England and by the lack of larger-scale comparisons (data from different locations along the coast) in New South Wales.

Several intertidal studies have been done at several sites on one shore, or on several shores but at only one site on each shore (e.g., Dayton 1971). Still others are done at only one site (Sousa 1979a, 1979b, 1980; Paine 1974), making it impossible to separate (or unconfound) the components of variation from one study to another that are due to geographical differences (the ones being examined in a comparative review) from those that are due to intrinsic spatial variation (i.e., from place to place within the same geographical region).

For example, McGuinness (1984b) demonstrated the great problems of interpreting experimental studies of intertidal boulder fields done in more than one place. His results about the relative importance of "intermediate disturbance" in boulder fields differed from those of Sousa (1979a, 1979b, 1980). McGuinness examined four boulder fields (at high or low levels on two different shores) in New South Wales; Sousa examined one boulder field in California. It is therefore impossible to determine whether their results differed because of biogeographical differences or local spatial variation.

Problems with Temporal Scales

Another problem that frustrates biogeographical comparisons of local assemblages is the time scale over which many studies are done. For example, studies are often done at a single time of year, or in only one season, making comparisons from one place to another potentially confounded by the time of year, unless studies are done at similar times, or in the same season. Care must be taken to ensure that differences in the magnitudes and rates of relevant processes and in the structure of assemblages from one time to another do not confound spatial comparisons. The only way that such spatial comparisons could be made valid is to ensure that each study in the various different localities has been done at several times, preferably randomly chosen (in the statistical sense of random: see Winer 1971, Underwood 1981, for examples). The variation from time to time will then have been averaged over the processes investigated in each place. This ensures that comparisons from place to place are not confounded by any potential differences from time to time.

Yet another problem is that studies are often short. For example, even though a long-term (say three- or four-year) experimental study may be available from each of two geographical areas, these can only be compared if it is known that the two periods studied are, in fact, equivalent. If one study (even one that has taken many years) was done during a period of exceptional weather (such as during an El Niño) and the second study was done in a different area during different prevailing environmental conditions (for example, between El Niños), the two studies differ in two ways. Under these circumstances, it is not possible to identify which (if any) components of difference are attributable to locality and which are due to prevailing weather during each study. Alternatively, a lack of differences between the two areas does not necessarily indicate a lack of geographical differences. Instead, the intrinsic differences due to geography and different locality may have been canceled by concomitant differences between the two prevailing patterns of weather. This sort of confounding is routinely considered and often controlled for in small-scale comparisons (although there are exceptions; see Hurlbert 1984), but has usually been ignored in attempts to make larger spatial comparisons.

No matter how diligent experimenters are, there remains the problem of how typical or representative a study of limited duration in any area may be. For example, local populations of the intertidal barnacle *Chamaesipho tasmanica* in New South Wales varied considerably in recruitment from year to year from 1972 to 1976. Populations nevertheless persisted because, on average, every site studied received sufficient recruits to replace animals dying (for whatever cause). During the period 1980–1983, however, there was no recruitment to any of the populations. Mortality due to predation was of major importance, and many local populations went extinct (Underwood, unpublished data). Interpretations of processes operating in the assemblage would have been quite different if data from only one of these two periods were available. It has not yet been determined which pattern may be "typical" of the system—a task that may take longer than the time for which any type of research program could be funded, or during which researchers could maintain enthusiasm (or life!).

Most detailed studies that unravel mechanisms and processes maintaining structure in intertidal assemblages do not have sufficiently long time scales to determine how often certain phenomena occur, and how representative any one picture of the system may be. This point can be illustrated by considering some detailed experiments on the fauna and flora of boulder fields in California that have revealed a number of patterns and processes (Sousa 1979a, 1979b, 1980; Dean and Connell 1987a, 1987b, 1987c). Some of these studies had been repeated and sufficient experimental evidence was available from one boulder field to make the authors reasonably confident that their interpretations of phenomena were robust and

reliable. Patterns of disturbance in the boulder field (due to rocks being rolled around and turned over) dictated much of the spatial and temporal variation in assemblages on boulders of different sizes. A large storm then hit the area, disrupting assemblages. After this, there was a massive arrival of larvae of a tube worm (*Phragmatopoma*), which fused the boulders together, making them a relatively undisturbable habitat (Connell, personal communication). Thus, sudden events at a time scale different from that of the experiments changed the whole system. Comparisons with other localities before or after the storm could have given quite different pictures of similarity or difference, but it would not have been clear which pattern (i.e., before or after a storm) typified the coastlines being compared.

It is not yet possible to determine in many published accounts from only one or a few places, or from only one or a few years, whether studies in numerous localities would average the various possible temporal patterns. This is theoretically possible, using the logic that if sufficient places are examined, they may show the entire possible array that would be seen temporally (e.g., Watt 1955). Unless it can be verified that this is indeed the case, comparisons between short-term studies, even if in many places, may still confound spatial and temporal differences.

Connell (1983) has pointed out the difficulties of comparing the intensity or outcome of competitive interactions in different field studies that are done in only one place or at only one time. In intertidal habitats, the experiments on competition between two gastropods by Underwood (1978, 1984) are still one of the very few examples of a competitive interaction that has been examined in identical experiments at more than one place and at several different times. The snail *Nerita atramentosa* routinely outcompeted the limpet *Cellana tramoserica* for food, but the intensity of interspecific effects varied among experiments. The limpets always showed considerable intraspecific competition leading to decreased survival, although the rates of mortality varied among experiments. The experiments have been repeated several times independently (Underwood 1976, 1978, 1984; Creese and Underwood 1982; Fletcher 1984). As a result, the differences in experimental results that have been obtained in different seasons and at different heights on the shore can now be interpreted. The relationships between availability of food, density of grazers, and the ultimate survival of the two species are reliably predictable (Underwood 1985, 1992). Without the repeated experimentation, no such coherent interpretation would be possible. Other processes, such as predation and the effects of physical disturbances, are just as difficult to interpret when there has been no repetition of the experimental study under different conditions or in different places.

Which Processes Are Local?

Assuming that a realistic definition has been made of the spatial and temporal scales that are appropriate for a study, there are still problems to overcome about the types of processes that should be considered local. Much early work on the nature of intertidal assemblages assumed that rates of birth were relatively constant and that only processes affecting rates of death were important determinants of local variation (e.g., Underwood and Denley 1984; Underwood and Fairweather 1989). This preoccupation with death has led to predation, competition, and reductions in density or diversity of species due to physical processes of disturbance being considered the most important aspects of the local dynamics of intertidal assemblages.

Most of these studies had very restricted spatial scales of investigation—usually a matter of a few square meters on rocky shores. For many of the processes that influence the abundances and distribution of intertidal organisms, particularly the sessile fauna and the plants, such scales are probably appropriate. Even in studies of interactions, such as competition for food, among mobile invertebrate species, few intertidal species regularly move more than a few meters while foraging (Branch 1984; Chapman and Underwood 1992). Considerable structure can be created within a few meters of refuges by mobile grazers and predators, again, many of which do not move far to feed (e.g., Garrity and Levings 1981; Moran 1985; Fairweather 1988; Fairweather, Underwood, and Moran 1984). Even the most wide-ranging predatory invertebrates do not operate over many tens of meters (Paine 1974; Dayton 1971; Menge 1972, 1979). Thus, small spatial scales seem appropriate for many of the processes known to be important in intertidal habitats.

There are, of course, other processes that operate over larger spatial scales. Drift algae supply food for abalone and, in some places, for urchins (reviewed in Chapman and Underwood 1992). The distances over which the algae are moved are not known, but are likely to exceed the few meters studied in many intertidal and subtidal experiments (Dayton and Tegner 1984; Kingsford and Choat 1986). It is well known that grazing fishes feed over very large areas (Choat 1982). Some predatory animals also forage over large distances. For example, the cling-fish *Sicyases sanguineus* is swept onto a shore by waves, clings on with a specially modified pair of fins, and ravages a small area, removing many species as prey (Paine and Palmer 1978). The fish then departs and is presumably swept back onto the shore elsewhere. The scale over which each fish affects a shore is not currently known.

Useful insights have been gained into the structure and dynamics of assemblages by defining the unit for study as a small area. The story will, however, remain a partial one for the obvious reason that the arrival, frequency of foraging, dynamics, etc., of predators are at scales not amenable to investigation at the small scale usually chosen.

More regional processes cannot easily be investigated experimentally because their dynamics are too difficult to sample over the relevant spatial scales (given the usual budgetary constraints). These processes are usually ignored and assumed to be constant. Often, such larger processes also have marked annual periodicity which would require long runs of data before any patterns became obvious. As an example relevant to intertidal assemblages, consider the processes operating on the different stages of the life-cycle of many species that have widespread dispersal by planktonic larval stages. Propagules may move

over distances of hundreds of miles (Scheltema 1971). As a result, the dynamics and starting densities of whole assemblages and the interactions that occur within any localized site are affected by the coastal and oceanic processes that dictate the dispersal, survival, and settlement of the larvae (Butman 1987). This point has often been forgotten in extrapolations and generalizations about the structure of rocky shore assemblages (reviewed in Underwood and Denley 1984; see also Underwood and Fairweather 1989). More recent theoretical and empirical investigations have, however, considered the importance of processes affecting larval input (Roughgarden, Gaines, and Iwasa, 1984; Roughgarden, Gaines, and Pacala 1987). In fact, the whole notion has been given a new ecological "buzzword" ("supply-side ecology") to demonstrate how pleased we are collectively with its reappearance in our thinking (Lewin 1986; Young 1987; Underwood and Fairweather 1989).

Finally, there are three other larger-scale processes operating that may limit the interpretations of locally defined studies. The first is the weather. Storms, wave action, general patterns of circulation, etc., are all subject to very large, often global, influences. As an example of the ways that such processes can affect the distribution and abundance of shallow-water marine species, consider Dayton and Tegner's (1984) documentation of changes in coastal kelp forests in California. Intrusions of southerly warm water devastated the kelp forests in the usually cold water of the Californian coast. Such weather-driven changes in marine climate overrode the smaller-scale patchy processes that had been the focus of most studies in these habitats (e.g., Dayton et al. 1984).

Weather can also cause unpredictable, and presumably long-term, changes that are not amenable to formal analysis after the event unless studies are in progress when the effects of the unusual weather appear. An example is the series of effects of excessively cold weather in Britain during the 1960s, summarized by Crisp (1964). The abundances of numerous species were reduced by the onset of cold weather, causing some restrictions in geographical distributions. The changes in sizes of populations also set the scene for long-term changes in abundance of some species. Only because a number of small-scale localities were being, or had previously been, examined was it possible to hazard some understanding of the effects of the extreme pattern of weather. Such smaller-scale studies on their own, without knowledge of the effects of larger changes due to cold weather and without the integrated comparative information that became available because of the large effects of cold weather, would each have been at a loss to understand what was happening.

Second, there are biogeographic considerations, sometimes involving the global machinations of humans. The invasion of western Britain by the intertidal barnacle *Elminius modestus* affected the structure and dynamics of local assemblages (Crisp 1958; Crisp and Southward 1959). Another example is the more recent spread of the seaweed *Sargassum muticum* in Britain. This species has also caused changes in the structure of local assemblages, which could not have been interpreted without recourse to larger-scale mapping studies on the distribution and spread of the alga (Boalch and Potts 1979; Critchley, Farnham, and Morrell 1983). There have been several well-documented cases of changes from one location to another in the sizes and morphology of intertidal snails (*Littorina obtusata*), which were only interpreted when the pattern of introduction of a predatory crab (*Carcinus maenas*) was understood (Hadlock 1980). Determining the pattern of spread of the crab was a study at a much larger spatial, and over a much longer temporal, scale. Similarly, Bertness (1984) has investigated the role of, and changes in intertidal assemblages due to, the introduced snail *Littorina littorea* in the northeastern United States.

Finally, no matter how well defined the spatial and temporal scales are, the genetic continuity of the populations being studied may require quite different spatial considerations. It is not always clear how a "local" population should be defined, given the large spatial scales of dispersal that many species undergo. There is evidence that many widespread species with long-term planktonic dispersal are genetically homogeneous over long distances (Berger 1973; Gooch 1975; reviews by Burton 1983; Hedgecock 1986). This tends to reinforce the view that any local population defined by some combination of temporal and spatial considerations will necessarily be genetically homogeneous. In contrast, there has been some recent evidence that very small-scale localized interactions can alter the genetic composition of species (Murphy 1976; Hedgecock 1986). In other cases, genetic differences within populations are maintained through time, despite large-scale patterns of dispersal (Levinton and Suchanek 1978; Todd, Havenhand, and Thorpe 1988; Koehn, Turano, and Mitton 1973; Gosling and McGrath 1990).

Where species do not have widespread pelagic dispersal, there are often differences in their morphology that affect their interactions in assemblages. Whelks (*Nucella lapillus*) with thicker shells are less susceptible to predation by intertidal crabs in sheltered locations than are the thinner-shelled morphs found on open, wave-exposed coasts (Kitching, Muntz, and Ebling 1966; Kitching and Lockwood 1974). In contrast, the thinner-shelled forms seem better able to withstand the effects of wave shock on open coasts. The differences are maintained by direct development of the young, which isolates the two difference types of populations, although there may also be chromosomal differences between populations (Staiger 1957). There are many complications in this example. Whelks grow differently when transplanted into different areas, but also according to their place of origin (Burrows and Hughes 1990). Palmer (1990) has also shown that the presence of predatory crabs may influence the morphology of growing whelks. Identifying genetically homogeneous or other locally (morphologically) distinct populations of intertidal animals is no easy task. Thus, the very nature of the relevant unit of study (the "local" population being investigated in some locality) is itself a major problem for widely dispersed populations, or for mixtures of very finely divided groups of species.

As a result of this sort of mixing of populations in any one place, some of which may be genetically similar to other populations elsewhere and some of which may be

much more localized, there will be a different understanding of the degree to which any local interaction, pattern of use of a resource, or source of mortality may be indicative of the process for the whole population. For species with small-scale or no dispersal, localized studies will be quite informative about the relative importance of processes to whole populations that are themselves localized (e.g., Ayre 1982; McKillup 1983). These kinds of studies will, however, provide very little information about processes operating in similar assemblages elsewhere that have different genetically structured populations (McKillup 1983). In contrast, in widespread populations, local studies in representative areas of habitat will provide adequate information on the average effects of processes, but will provide no precise data about what will happen elsewhere within the range of the species. Again, but for different reasons, such studies will provide little general information about widespread assemblages. This time, however, the reason will be that the relative importance of different processes cannot be known for the whole population when it is spread over large distances and the studies are small in scale.

COMPARISONS OF PROCESSES

Relative Importance of a Process

Before we consider comparisons of the importance of any local process from one region to another, it is worth considering how to assess the relative importance of a process operating in one locality.

The procedure recommended by Welden and Slauson (1986) is a theoretically useful definition of, and operational procedure for estimating, the relative importance of some process as an influence on abundance, size, rate of mortality, or other ecological variable of interest (although its serious shortcomings are identified later). They suggested that, in experimental analyses, the relative importance of a process or factor could best be determined by the use of analyses of variance. These statistical procedures are well suited to partitioning the variation in biological variables to identifiable sources or causes of differences (e.g., Fisher 1932). In field experiments, Welden and Slauson's procedure consists of determining how much variability is associated with the process of interest and how much is attributable to all other processes operating on the organisms. Using a framework of multiple regression, Welden and Slauson's measure of relative importance is the coefficient of multiple correlation. They distinguished between the importance and the intensity of a process by defining the intensity of a process as the linear regression coefficient.

As an example, consider the relative importance of predation on the rate of mortality of some species. The effect of predation on the number or diversity of species could equally well be used as an example. Although it is clear that predators must increase the rate of mortality of their prey, the question of relevance is, does this matter? If predators consumed only a tiny percentage of the prey before new juveniles recruit into the prey's population, the abundance of prey would be only slightly altered by predation. Alternatively, predators may consume only sick individuals that would have quickly died of diseases if there were no predators (in which case predation is not at all important). This is not a trivial point, because many experiments test for the effects of predation in situations in which all other processes are held constant or are removed. For example, intertidal barnacles are often consumed by predatory whelks. Barnacles at low levels on the shore may, however, be smothered by algae (Barnes 1955; Denley and Underwood 1979) or may be killed by competitors (e.g., Connell 1961a; Dayton 1971; Menge 1976). Predators may only be eating animals that would die of some other cause before much more time elapsed— in which case predation makes no difference to the abundance of the barnacles, to their pattern of occupancy of space on the shore, or to the overall structure of the assemblage. Even though mortality of barnacles may be largely unaffected by predators, observations may, and experiments will, allow estimation of the presence or magnitude of predation (whether predators occur; how many prey they eat per capita per unit time). But as Welden and Slauson (1986) pointed out and as illustrated above, these data will not indicate whether predation is an important source of mortality.

The following example illustrates some of the problems of assessing the relative importance of different ecological processes. A typical experimental analysis of predation in the field consists of plots from which predators are removed and plots in which predators can forage naturally (e.g., Connell 1970; Dayton 1971; Menge 1976; Fairweather, Underwood, and Moran 1984). Treatments are assigned at random to randomly located plots. After some predefined period of time (i.e., the relevant temporal scale, however that is to be determined), the numbers of prey in each plot are recorded. These data provide estimates of the rate of, or percentage of, mortality and the surviving density. If the plots are monitored frequently, the rates of consumption per capita of predator can also be determined (Connell 1970; Menge 1978a, 1978b; Fairweather and Underwood 1983). In the simplest such experiments, in only one single area, the data are amenable to a one-factor analysis of variance. Four different circumstances are modeled in table 4.1. In the first two situations, predators are effective and decrease the percentage of surviving prey to less than half of that in areas where there are no predators (table 4.1A). In the second two experiments, predators are far less active or effective.

For each case, two different situations are considered. In the first situation, the variation among replicates in percentage survival of prey (i.e., the residual or error sum of squares) is quite small (table 4.1, experiments 1 and 3). In the other, it is large (table 4.1, experiments 2 and 4). This variation is entirely attributable to the differences from plot to plot in the activities of predators and to the mortality in different plots due to other causes (weather, disturbance, competition, diseases, senility, parasites). Where such sources of mortality are relatively uniform or homogeneous in their combined effects, the variance among replicate plots will be small (as in table 4.1A, experiment 1, and B, experiment 3). Where the impact of different sources of mortality (including predation) varies widely from one place to another, this variance will be larger (as in table 4.1A, experiment 2, and B, experiment 4).

In the first cases (Table 4.1A), predation was intense. It caused 83% of the deaths of prey (of the 60% mortality in the presence of predators, five-sixths was due to predation). In the second set of experiments, predation accounted for only 67% of the mortality of prey (of the 30% of prey dying, one-third died of non-predatory

Table 4.1 Analyses of the Relative Importance of Predation

(A) Numerous or voracious predators		
Plots	+predators	−predators
Mean percentage of prey alive at end of experiment	40	90
	Experiment 1	Experiment 2
Variance among replicate plots	500	2500
Source of variation Df	Sums of squares	Sums of squares
Predators + vs − 1	13000	15000
Among replicates 18	9000	45000
Total 19	22000	60000
Importance of predation:	=13000/22000=59%	=15000/60000=25%
(B) Sparse or less voracious predators		
Plots	+predators	−predators
Mean percentage of prey alive at end of experiment	70	90
	Experiment 3	Experiment 4
Variance among replicate plots	500	2500
Source of variation Df	Sums of squares	Sums of squares
Predators: + vs − 1	2500	4500
Among replicates 18	9000	45000
Total 19	11500	49500
Importance of predation:	=2500/11500=22%	=4500/49500=9%

Note: Proportion of variability in experimental data that is attributable to the effects of predators. In each experiment, the numbers of surviving prey were recorded in 10 replicate plots with and 10 replicate plots without predators, all with the same starting density of prey.

causes). These measures conform to Welden and Slauson's (1986) sensible and readily interpretable measure of intensity of a process.

In these experiments, the percentage of total variation that is attributable to predators (table 4.1) indicates the relative importance of predation, as defined by Welden and Slauson (1986). Where predators were most active and variance among plots was small (table 4.1A, experiment 1) this was 59%. Welden and Slauson would interpret this to mean that the activities of predators contribute 59% to the variation in rates of mortality (percentage of prey killed during the experiment) from one part of the sampled habitat to another.

Under the same intensity of predation in a habitat with greater intrinsic variation from plot to plot, the relative importance of predation is smaller (table 4.1, experiment 2). This difference occurs because all sources of mortality (including, perhaps, predation) are more variable in their influences on the rates of mortality and the resulting percentage survival of prey. In terms of explaining the differences from plot to plot, predation is not very important. The presence or absence of predators does not explain much of the difference (variance) among plots. Note that the experiments in table 4.1 all have the same design. Estimates of importance using variance components (see below) would produce the same interpretations. Relative importance of predation would appear smaller in table 4.1B than in table 4.1A.

So far, so good. Welden and Slauson (1986) did, however, assume that their operational definition of importance was independent of (in their term, "decoupled" from) the intensity of a process. This is not the case. Consider the examples in table 4.1B. Here, the intensity of predation is smaller (predators remove far fewer prey, resulting in greater percentage survival of prey). The estimates of relative importance have also decreased (which is an inevitable consequence of analysis of variance). Thus, intensity and Welden and Slauson's definition of importance of a process are always correlated in these analyses.

More Than One Process Operating Simultaneously

Comparisons among habitats of a single process and its effects on an intertidal population or assemblage have limited value and cannot address issues of the relative importance of one process versus others. Most species exist in a "web of complex interactions" (Darwin 1859). Understanding patterns of abundance or diversity of species will usually necessitate the development of descriptive models that incorporate a range of processes that act simultaneously and that may interact in complex ways that vary from place to place and from time to time (see the examples reviewed in Underwood and Denley 1984). Comparative studies from one location or region to another must therefore incorporate procedures for interpreting the roles of more than one process.

Two or more processes can be compared efficiently by orthogonal contrasts. For example, the differences in mean numbers of animals (or rates of mortality, rates of growth, width of distribution along a gradient, etc.) can be analyzed in experimental populations subjected to different rates of predation and, say, rates of physical distur-

bance. Menge (1978a, 1978b) demonstrated the possibilities with complex analyses of rates of feeding by predatory whelks under different physical conditions. Underwood (1984) unraveled the effects of increasing density of two species of intertidal snails on their survival under different environmental conditions. In the first case, the effects of predation and physical stresses, and in the second case, the effects of competition, seasonality and tidal height, could be analyzed separately. Also, interactions between combinations of two or more processes (i.e., situations where the effect of one process is influenced by the other; they do not operate independently) could also be identified.

From such analyses, it is theoretically possible to estimate the relative importance of several coincidentally acting processes, or their combinations. Two statistical procedures have been used. First is the procedure recommended by Welden and Slauson (1986) using estimates of the percentage contribution of the sum of squares attributable to some process to the total variability (i.e., the total sums of squares) in the variable being examined. The total variability can be partitioned to estimate how much is attributable to different sources of variation. These sources of variation include the planned variations due to processes such as competition and predation which occur because these processes have been manipulated in the experiment. Other sources of variation are those that estimate differences from one replicate to another. These represent "noise" in the assemblage (i.e., they are presumably the result of processes operating at smaller and larger spatial and temporal scales than those currently being investigated). These spatial and temporal differences among replicates form the basis for investigating how the outcome of some process varies from place to place and from time to time.

For example, Levinton (1985) recorded that intraspecific competition among snails, *Hydrobia totteni*, explained (or was associated with) about 67% of the variation in the growth rates of the snails. In contrast, only some 7% of the variation in the abundance of algal foods could be explained by the differences in experimental densities of snails. Levinton therefore concluded that interference competition, the direct effect of increased density, was more important than the available supply of food as a determinant of the growth rate of the snails.

A second and related procedure for estimating the relative importance of several processes uses the variance components from an analysis of multiple sources of variation to derive measures of the relative importance of the different factors. Calculation of the relative magnitudes of effects of a set of treatments for different sorts of analyses is fairly straightforward (e.g., Underwood 1981; Winer 1971). As an example, Caffey (1985) used the relative size of variances in an attempt to determine the relative importance of temporal and spatial variation in the abundance of juvenile barnacles along a coastline.

There are, again, problems in the use of these procedures. Consider an experiment involving two different processes. The first, predation, is investigated by setting up experimental plots from which predators are excluded or removed. These plots are then contrasted with control

Table 4.2   Analyses of the Effects of Predation at Five Experimental Densities of Prey

(A) Mean Densities of Surviving Prey at Different Initial Experimental Densities of a Non-prey Competitor

Density of competitor	+Predators	−Predators
100	100	120
110	90	105
120	80	90
130	70	75
140	60	60

(B) Analysis of Variance of Above Means

Source of varation	Df	Mean square	% of total sums of squares	% magnitude of treatment effects
		Experiment 1 [a]		
Predators:+ vs −	1	800.0	6.7	6.3
Among densities	4	2393.8	80.1	78.1
Interaction	4	143.8	4.8	3.1
Residual	20	50.0	8.4	12.5
		Experiment 2 [b]		
Predators:+ vs −	1	1550.0	6.5	6.3
Among densities	4	4737.5	79.1	78.1
Interaction	4	237.5	4.0	3.1
Residual	50	50.0	10.4	12.5
		Experiment 3 [c]		
Predators:+ vs −	1	1700.0	5.2	4.6
Among densities	4	4887.5	59.6	56.7
Interaction	4	387.5	4.7	2.3
Residual	50	200.0	30.5	36.4

[a]$n=3$ replicates of each treatment; variance in each treatment=50
[b]$n=6$ replicates of each treatment; variance in each treatment=50
[c]$n=6$ replicates of each treatment; variance in each treatment=200

plots to which predators have normal access. The densities of surviving prey are counted at the end of the experiment in both types of plots. Simultaneously, the experiment can be used to investigate another factor, for example, the effects of different densities of a competitor that is not eaten by the predator. Predation may be less important in areas in which competitive interactions are intense. Such an experiment is analyzed in table 4.2. In the first set of situations (table 4.2B (i)), there are three replicates of each of the ten experimental treatments, and the variance among replicates in each treatment is set at 50. The percentage of the total sum of squares attributable to predation or to the (five) different densities of competitors can be seen in the table as 6.7% and 80.1%, respectively. The percentage of relative magnitudes of effects of treatments is also shown as 6.3% and 78.1%, respectively.

The percentage of sums of squares (i.e., the relative importance according to Welden and Slauson [1986]) will vary according to the number of replicates used in the experiment. This is illustrated in table 4.2B, by comparison of experiments 1 and 2. In the second case, there were twice as many replicates as in the first. The percentages of total sums of squares attributable to predation and to differences among the densities were slightly different in the two cases. These differences may or may not be trivial, depending on the degree to which the amount of replication changes. As a result, the percentage relative magnitude of effects of treatments is a potentially more reliable measure of relative importance. This measure is com-

pletely unaffected by the number of replicates used in each experiment. For both procedures, the relative importance of predation or of differences in density of the competitor depends on the amount of residual variation. When the amount of variation that occurs naturally from one replicate plot to another, but that is not due either to predation or to the starting densities of the prey, is large (200 in table 4.2B, experiment 3), the relative importance of predation and starting density by either of the two statistical measures is small.

If differences in residual variance are ignored, the relative importance of different processes could be estimated by the ratio of the relevant magnitudes of treatment effects. This is not Welden and Slauson's (1986) definition of relative importance. In all three examples in table 4.2, the effects of different densities of competitors were 12.4 times greater, and therefore more important, than the effects of predators (78.1% / 6.3% = 12.40 and 56.7% / 4.6% = 12.33).

There is, however, a serious problem with knowing how to deal with the interaction components in such analyses. When interactions are large (not as illustrated in table 4.2), there are two consequences for interpretation of the experiments. First, this serves as statistical evidence that the two (or more) processes being investigated are not independent of one another. In other words, the effect of predation depends on the density of the prey. Simultaneously, any differences from place to place that are due to the densities of the competitor are also affected by the presence or absence of predators. Under such circum-

Table 4.3   Analyses of the Effects of Predation at Three Experimental Densities of Prey

Source of varation	Df	Mean square	% of total sums of squares	% magnitude of treatment effects
		Experiment 1 [a]		
Predators:+ vs −	1	500.0	5.5	4.9
Among densities	2	3800.0	83.5	82.0
Interaction	2	200.0	4.4	3.3
Residual	12	50.0	6.6	9.8
		Experiment 2 [b]		
Predators:+ vs −	1	950.0	5.2	4.9
Among densities	2	7550.0	82.7	82.0
Interaction	2	350.0	3.8	3.3
Residual	30	50.0	8.2	9.8
		Experiment 3 [a]		
Predators:+ vs −	1	1100.0	4.7	3.8
Among densities	2	7700.0	65.5	63.3
Interaction	2	500.0	4.3	2.5
Residual	30	200.0	25.5	30.4

Note: See table 4.2A for densities.
[a] $n=3$ replicates of each treatment; variance in each treatment=50
[b] $n=6$ replicates of each treatment; variance in each treatment=50
[c] $n=6$ replicates of each treatment; variance in each treatment=200

stances, there is no meaning, either in the sense of testing hypotheses about processes (see Underwood 1981; Winer 1971) or in being able to understand the apparent relative importance of different processes, because of the presence of interaction. In other words, it is virtually impossible to put any meaning on the relative importance of predation in isolation from the relative importance of different densities of prey when these two processes interact. At the same time, it would be meaningless to use the interaction term to calculate the relative importance of the two processes combined. There is no clear way to proceed in interactive systems and their analyses.

For the purpose of comparing the processes in different regions, on different coastlines, in different localities, etc., there is another major problem with these statistical manipulations of the data. The problem is that the data collected in experiments in the various regions, localities, etc., must be of identical design. This can be illustrated by comparing the outcome of the experiments analyzed in table 4.3 with those discussed above in table 4.2. In the analyses in table 4.3, only three experimental densities of competitors were considered (the first, third, and fifth indicated in table 4.2A). This experiment is an equally legitimate method for attempting to determine the intensities of predation, differences among densities of competitors (over the same range of densities) and any interaction between these factors. There is no compelling reason why three as opposed to five different densities might, or might not, have been chosen by experimenters in two different regions. Nevertheless, in each of the three cases illustrated in table 4.3, the relative importance of predation is now apparently smaller (although not much changed in this example) than in the previous experiments described in table 4.2. This is an inevitable result of the fact that the variances among replicates are not altered in any way by the size of the experiment (i.e., the number of experimental treatments and the number of replicates) because they are in an intrinsic property of the biology of the system. When there are fewer, but very different densities of com-

petitors, the amount of variability in the data that is attributable to those densities will be larger. The only circumstances under which this will not be true are those in which there is no effect (i.e., no intensity) of any process attributable to different densities. Even the ratios of the relevant variance components will change with different-sized experiments. The only legitimate contrasts that can be made are those of experiments of identical size. This point has also been made by Sih et al. (1985).

It appears that alternative methods for determining the importance of different factors are needed before it is reasonable to expect that the relative importance of different ecological processes can be compared from one locality to another.

Comparisons from One Place to Another

An alternative approach to the use of statistical estimation of relative importance through analyses of variance has been used by Foster (1990). He suggested that the crucial issue was that of generality of a process. In other words, he wished to investigate how general a process such as competition or predation was in terms of its outcome. He studied twenty different rock platforms randomly sampled along the coast of California. On each shore, he sampled the distributions and abundances of major components of the assemblage that occupied space on the rocks. He had a number of hypotheses in mind about processes that should be operating. For example, he considered that competition between some species of algae (which had already been demonstrated in previous experiments [Foster 1982]), should, if it were a general process, lead to distinct, predictable distributions of the algae being demonstrable on a large proportion of the shores. It turned out that in his large sample of sites, very few of the patterns of distribution of potentially competing species of seaweeds could actually be explained by the previously identified competitive interaction. The relative importance of competition to the distribution of these algae was therefore small, as estimated by the proportion of sites in which the process might be able to explain any of the distribution or abundance or other patterns in the assemblage. Such studies would be very useful, not only for determination of such matters as the relative importance of different processes but also, as discussed below, to ensure that comparisons from one region or locality to another are unconfounded. Foster's (1990) procedure of using a large number of representative sites along a coastline will also allow valid comparisons from one region or coastline to another.

It is also important in biogeographical comparisons to remember that there is variation among locations in the processes causing similar patterns. As an example, consider Connell's (1961a) studies in Scotland and on the northwestern coast of the United States. His study in Scotland revealed the importance of inter- and intraspecific competition for space and of predation by whelks as determinants of the patterns of distribution of two temperate species of barnacles. The work also indicated the subsequently forgotten importance of larval settlement in the development of these patterns (Grosberg 1982; Underwood and Denley 1984). In contrast, similar studies on similar species on the northwestern coast of the United States demonstrated that a different mixture of processes of predation were determinants of vertical distribution of barnacles there, with competition having no importance (Connell 1970). Thus, there were very large differences in the most important processes operating in different temperate locations. It would therefore be very unwise and remarkably premature to compare such temperate rocky habitats with apparently similar areas in tropical waters, unless several independent sites had been studied in each region (see also Underwood and Fairweather 1986).

A necessary procedure would be to do such experimental studies using random block designs, each "block" being a locality or region or site along a coastline or in a biogeographical region. The experiments would then be constructed in an equivalent, orthogonal manner in each of the blocks. If sufficient blocks (or sites) were investigated, the relative importance of processes could be estimated using the proportion of blocks in which each process occurred, or in which each occurred with some minimal intensity (as in Foster 1990). The average conditions prevailing on a coastline (or in a biogeographical region) could also be compared with those prevailing on a different coastline (or in a different region), using the blocks as replicate sites. This has not yet been done in any intertidal study.

Some researchers have examined processes at a number of sites (e.g., Dayton 1971; Menge 1976; Underwood, Denley and Moran 1983). In all cases, however, these were chosen to represent different conditions or different habitats. For example, Dayton's (1971) and Menge's (1976) sites were chosen to represent different conditions of exposure to wave action. These were valid and proper choices for the purposes of the original studies. It is not, however, possible to use these sites as though they were randomly chosen (or otherwise representative) pieces of coastline to contrast one coastline with another. Each site is, in fact, an unreplicated situation in terms of the physical habitat. There was only one site at each level of wave exposure. Any comparison using single sites is therefore fraught with difficulties (Hurlbert 1984). There is also no way to be sure that the gradients of wave exposure from one region to another are identical. It would, therefore, be very difficult to make meaningful quantitative comparisons of the processes operating by contrasting such sets of data.

Valid comparisons require a set of representative sites along each coast (region, area) to be compared. Two methods could be used to choose such sites. First, sites could be chosen to randomize such features as latitude, wave exposure, aspect, type of rock, etc. Experiments done with replicates of all treatments in such a set of sites (the blocks) would probably have large variances for comparisons among sites. Intensities of predation, competition, recruitment, disturbance, and other processes would all be very variable among such a choice of sites. Estimates of importance would be properly representative of the processes operating across all sites within a region. The variability among sites would, however, make it difficult to interpret the processes operating along a coastline and to detect any differences in the intensities, rates, and relative contributions of ecological processes between one coastline and another.

A second procedure would be to stratify the habitats before attempting any comparative analysis. For example, experiments could be done only on wave-exposed, midshore habitats on shores composed of hard rocks. Several such sites on each coastline could usefully be compared. To attempt this sort of study for several different types of habitat (sheltered and exposed shores, hard and soft rocks, different heights on the shore, etc.) with adequate replication of each type of habitat would be a daunting task. It would also be one that would be difficult to fund because of the person-power needed to gather the experimental data. Nevertheless, such replication of sites would be mandatory if several different types of habitats were to be compared across biogeographical regions. Clearly, the former procedure of randomizing numerous physical features of habitat would, under most circumstances, be preferable because it would remove the difficulties of choosing only one or a limited subset of habitats.

However replicated, representative sites are chosen, care must be taken to ensure that the experimental design includes sufficient replication at appropriate levels so that the interesting statistical tests will have sufficient degrees of freedom to be powerful. For an experimental analysis of predation, the ideal design would replicate treatments and controls at a number of sites in each of several locations (table 4.4). The various locations would provide generality for any experimental results. Locations might be different headlands or rock platforms, etc. Sites would be randomly chosen within each location to ensure that comparisons among locations (and any assessments of interactions of predation and location) are statistically valid. Sites would be some tens to hundreds of meters apart in each location. Within each site, there would be several replicated plots with and several without predators. Note, however, that the F-ratios for some of the appropriate tests do not involve the residual (i.e., among replicates) mean square (Underwood 1981; Winer 1971).

Table 4.4 Analyses of the Effects of Predation at Several Sites within a Number of Locations

Source of variation		General DF	Df for this example	F-ratio versus
Predators: + vs −	P	$a-1$	1	$P \times L$
Among locations	L	$b-1$	3	$S(L)$
Among sites within locations	$S(L)$	$b(c-1)$	8	R
Predation × Location	$P \times L$	$(a-1)(b-1)$	3	$P \times S(L)$
Predation × sites within locations	$P \times S(L)$	$(a-1)b(c-1)$	8	R
Residual	R	$abc(n-1)$	96	—

Note: Locations were chosen at random to represent a particular habitat (e.g., exposed shores). Experiments were replicated at several randomly-chosen sites within each location. The design included $a=2$ levels of predation (plus or minus predators) replicated in $n=5$ plots in each of $c=3$ replicate sites in each of $b=4$ locations.

Thus, it would be more profitable to increase the number of sites or locations than to increase the number of replicate plots per site. Ideally, cost-benefit analyses (e.g., Underwood 1981) would be used to calculate the optimal allocation of effort to the different levels of replication in the experiment (i.e., to determine the best number of replicate locations, sites, and plots).

The experimenter must also consider how the locations are chosen. If they are chosen to stratify the habitats (for example, there need to be locations on exposed and sheltered shorelines to investigate differences between these habitats), then replicate locations of each type of habitat are necessary, resulting in a more complex design. If locations are randomly chosen as representative parts of a given habitat, significance tests for differences between the treatments would use the Treatment × Location interaction term (table 4.4). The degrees of freedom available and, to a great extent, the power of the test are determined by the number of locations in the experiments.

CONCLUSIONS

In light of the foregoing consideration of some of the problems of making comparisons from one region or coastline to another, it is clear that a number of different factors must be considered before such comparisons are realistic. First, there are obvious requirements that the organisms or assemblages chosen for study should be directly comparable. Although there are often very large taxonomic differences in the suite of species present in one part of the world as compared with another, there are, nevertheless, usually functioning similar, or apparently similar, species in each assemblage. For example, the dominant users of space in intertidal rocky habitats tend to be barnacles and mussels in most parts of the world. Even though the species may differ, their patterns of life history, growth, and many of the interactions that they are involved in tend to have striking similarities (e.g., very similar sets of competitive interactions among limpets have been described in New South Wales, Australia, by Creese and Underwood [1982], and in Costa Rica by Ortega [1985]). Comparative experiments by Dayton (1971), Underwood, Denley, and Moran (1983), and Dungan (1986) all described a similar set of interactions involving algae, barnacles, and grazers, indicating many similarities of interactive processes in different regions of the world. It is probably not too difficult to ensure that the organ-

isms or set of organisms contrasted from place to place are strictly comparable.

It is also essential that the chosen habitats be directly comparable. Otherwise, attempted comparisons from one coastline to another will be confounded by local differences in habitats. There are, for example, considerable differences in such processes as rates of predation, competitive interactions, etc., from one shore to another. Such differences are often dependent on the degree of wave exposure (see the comparative work in Menge 1976, 1978a, 1978b). If a sheltered shore on one coastline were to be contrasted with an exposed shore on another coastline, any differences would probably be indicative of differences in wave exposure rather than differences between the two biogeographical regions.

The processes being contrasted must also be comparable. For example, predation by large generalist predators, such as the sea star *Pisaster*, and predation by smaller specialist predators, such as whelks, are very different (Paine 1974; Connell 1970). The larger, more generalist predators consume a wide variety of prey, clearing patches of the surface of rocks, making the resource available for new colonists. In contrast, the smaller whelks tend to eat their prey one at a time, and can only a clear a patch of prey after a prolonged period of feeding activity, often only where there are large numbers of whelks. If predation by similar types of predators were contrasted, differences from one region to another would be revealed. If, however, one type of predator in one region were chosen to be contrasted with a different type of predator in another region, there would be considerable difficulty in determining how much of any observed difference between two coastlines was due to differences in biogeographical regions and how much was due simply to the fact that different types of predators were being examined.

Another consideration is that the size of experiments (the numbers of treatments and replicates) will also influence perceptions of how important a process might be. Ideally, experimental studies should be planned so that experiments of similar design and size are done in each location. Otherwise, differences in the apparent importance of local processes will appear as though they were differences from one region to another, rather than being due to the different sizes of the experiments.

Statistically valid comparisons require sufficient replication of observations or experiments in each of the regions being compared. This may be very difficult to achieve over the range of a biogeographical province. Nevertheless, without some attention to the variations that occur from place to place within a region, comparisons among regions will always be difficult, if not impossible, to interpret. Choosing spatial and temporal scales to ensure sufficient replication is a difficult part of experimental design. One strategy that may assist in this is to design experimental studies investigating the intensities and rates of different processes with the same spatial scales as those of the original observations about the patterns being explained. For example, if the patterns of distribution of a species along a coastline suggest that a gradient of wave exposure is an important process, replication of shores with different intensities of wave exposure must be included in the experimental analyses. If, in contrast, patterns are being described at the scale of a few meters on a single shore, then scales of replication for the experimental studies need to be similar and there is no necessity for the experiments to be repeated over many shores. This is not a trivial point. There has been argumentation in the literature over the structure of assemblages on which some observations were made at scales of kilometers, but for which the experimental studies on the processes affecting patterns of species diversity were made at scales of a few meters. There is very little commensurability between these two sorts of investigations and there is no clear evidence that we should expect similar processes to be operating at each of these scales. For example, observations about the structure of assemblages and the local diversity of species of fish on small tropical patch reefs can readily be explained by processes including colonization and subsequent survival of individual fish (Sale and Dybdahl 1975). Subsequent tests of such explanations at much larger scales (hundreds of thousands of meters) are not relevant and cannot be contradictory (although this has been claimed: Gladfelter, Ogden, and Gladfelter 1980). These subsequent studies were at a completely different scale and undoubtedly involved many different processes (Allen and Starr 1982).

Unless due attention is paid to these factors in future experimental studies in any locality, only glimpses of the similarities and differences among localities can emerge and a complete interpretation will be unlikely to be realistic. Future studies of patterns and processes on intertidal rocky shores should work toward choosing appropriately and sufficiently replicated sets of data so that such comparative endeavors become practicable.

5

How Are Diversity and Productivity Related?

Michael L. Rosenzweig and Zvika Abramsky

THE PATTERN

The relationship of primary productivity and species diversity on a regional scale (10^6 km²) is not simple. But within such regions, and perhaps even larger ones, a pattern is emerging: as productivity rises, first diversity increases, then it declines. Some ecologists already accept this pattern as an empirical phenomenon. At least one standard ecology textbook reports it in detail (Begon, Harper, and Townsend 1990).

Evidence for the hump-shaped regional pattern is accumulating (see also Wright, Currie, and Maurer, chap. 6). It was first proposed for plants (Whittaker and Niering 1975). Tilman (1982) developed a theory that predicts the pattern, and he suggested that various data from plant communities fit it. (See Tilman and Pacala, chap. 2, for some of these examples. But note that they rest on controversial surrogate variables of unknown ability to stand in for productivity.) Perhaps the least controversial plant example comes from Mediterranean vascular plants (Shmida, Evenari, and Noy-Meir 1986; Shmida, unpublished data; Aronson and Shmida 1990) (fig. 5.1).

The humped-shaped pattern exists in Middle Eastern desert rodents (fig. 5.2), both in the psammophilic assemblage and in the lithophilic assemblage (Abramsky and Rosenzweig 1984). Owen (1988) documented it for Texas carnivores (fig. 5.3).

Productivity is the rate at which energy flows through an ecosystem (e.g., kj/m²/yr). But in all the cases we cite, ecologists used an index of productivity rather than measuring it directly. In arid and semiarid landscapes, precipitation supplies the index. Actual evapotranspiration works for terrestrial systems with a wide variety of temperature and precipitation regimes (Rosenzweig 1968). Without such indices, we wouldn't have enough data to investigate any regional trends in productivity.

On the ocean's floor, depth itself (which determines the availability of light for photosynthesis) governs productivity (e.g., Smith 1978). Rex (1981) showed that many bottom-dwelling marine taxa show the hump-shaped pattern: Gastropoda, Polychaeta, Protobrancha, Cumacea, invertebrate megafauna, and fishes. Figure 5.4 shows the cumacean pattern. In figure 5.5, we see a group not mentioned in Rex's review, brachiopods in Antarctic waters (this pattern comes from data in Foster 1974).

Rodents of the Southwestern United States

Among the most curious cases of the hump-shaped pattern is that of rodents of the southwestern United States.

Two groups studied it, and each discerned a piece of the whole relationship (fig. 5.6). Brown (1975) discovered that diversity grows with productivity—at least in parts of the Southwest. And Owen (1988) demonstrated that it *declines*—at least in Texas. Brown's data for the Great Basin actually show two data points in decline, but with only those two points, he conservatively interpreted the decline as an island effect. Now that we know that rodent diversity declines as productivity increases all across Texas, we need no longer resort to the island effect to explain the deviation from the overall pattern, because no deviation exists.

Desert rodent species diversity increases with precipitation (the surrogate variable for productivity in deserts: Rosenzweig 1968) until rainfall reaches about 350 mm per year (which happens in southeastern Arizona). Then rodent diversity begins its long decline across the whole state of Texas. Rodent species diversity drops by two-thirds as productivities rise to the levels found in eastern Texas along the Louisiana border.

The pattern in the United States resembles the one in the Negev desert (Abramsky, Brand, and Rosenzweig 1985). Also, the increase phase matches the one that Meserve and Glanz (1978) found along a 1000-km latitudinal transect in arid biomes of Chile (figure 5.6A).

Tropical Vertebrates

Who would have guessed that higher productivities depress tropical bird and tropical mammal diversities? Yet we know two studies that suggest they do. One (fig. 5.7) is straightforward: Australian tropical mammal diversity rises to a peak from the dry-wet tropics to the wet tropical highlands. It then declines substantially in the wet lowlands (Rosenzweig and Braithwaite, unpublished data).

The second study requires a bit of explanation. Every birdwatcher, every fan of public television, and certainly every ecologist knows that bird species fairly ooze from the lowland tropics. Nobody has ever found another place that comes close. Censuses of birds in a lowland floodplain forest in the Amazon basin of Peru reveal 319 species in a 97-ha study site, and point diversities of over 160 species (Terborgh et al. 1990). Such point diversities exceed the highest of those of any uniform habitat in North America by at least 300% (Terborgh et al. 1990). Yet we are about to claim that in one sense, compared with the less productive higher elevations, the lowlands are depauperate!

Remember, productivity is not the only influence on diversity. Perhaps the best established influence of all is area.

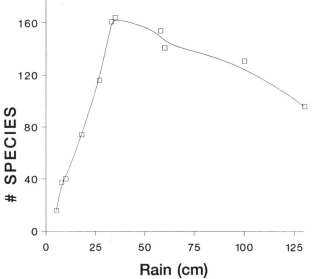

Figure 5.1 Plant species diversity in 0.1-ha plots peaks in regions with intermediate rainfall in Israel, Turkey, and Spain. (Data from Shmida 1985; Shmida, unpublished data; Aronson and Shmida 1990.)

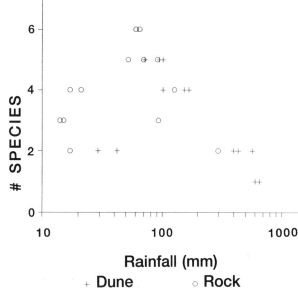

Figure 5.2 Diversity and productivity in Middle Eastern rodent assemblages. Rainfall is a good index of productivity in such arid and semiarid habitats. (After Abramsky and Rosenzweig 1984.)

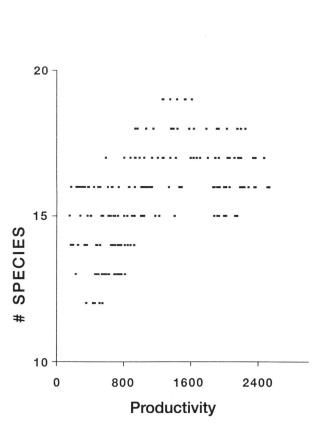

Figure 5.3 Diversity and productivity in Texas carnivores. The humped shape is evident, but the peak is displaced to a productivity of about 1400g/m²/yr—an order of magnitude higher than that for the rodents' peaks in figures 5.2 and 5.6. Productivity is estimated by a function of actual evapotranspiration. (After Owen 1988.)

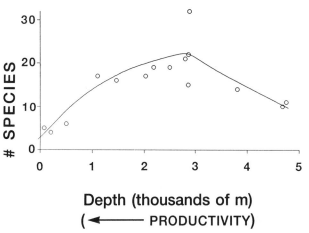

Figure 5.4 Atlantic cumaceans reach peak diversities at intermediate depths. Productivity correlates inversely with depth. These animals live on the bottom and were sampled with an epibenthic sled in a transect from Gay Head to Bermuda. Only samples with at least 100 individuals appear in the graph. This is one of a number of cases cited by Rex (1981) showing benthic marine invertebrate diversity peaking over intermediate depths. (Data from Jones and Sanders 1972.)

In any biogeographical province, larger areas harbor more species. We ecologists have been proving area's importance for decades (reviews include Connor and McCoy 1979; McGuinness 1984a). So we must not ignore area when it comes time to compare the diversities of tropical uplands and lowlands. We must factor out its effect before we decide what productivity is doing. Rahbek (C. Rahbek, personal communication) has done this for neotropical birds (fig. 5.8). (As you read the following summary of his findings, keep in mind that elevation

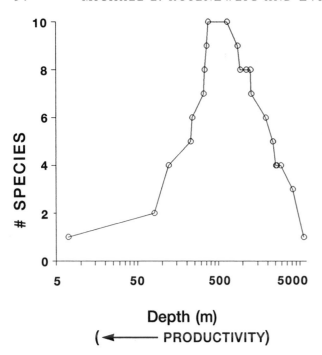

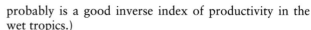

Figure 5.5 Antarctic brachiopods share the diversity pattern of many other marine invertebrate taxa like Cumacea (fig. 5.4). Productivity correlates inversely with depth. Diversity is the maximum number of species known at a given depth in Antarctic waters, rather than the number in a collection. Depths we report in the figure are each the uppermost or lowermost depth known for at least one of the species. (Data from Foster 1974.)

probably is a good inverse index of productivity in the wet tropics.)

Rahbek found that, in the Neotropics, lowland tropical area far exceeds that of any other elevation. But, over similar-sized areas, the more productive lowlands have fewer bird species than subtropical elevations do. For example, a lowland area of 10^5 km^2 has about 526 species, but at subtropical elevations, an area of that size has 855 species.

The temperate-elevation bird species-area curve is like the lowlands'; a 10^5-km^2 region at temperate elevation has 461 species. The high-elevation curve is the lowest of the four; its 10^5-km^2 region has only 176 species. Thus, from high elevations down to subtropical elevations, the more productivity, the more bird diversity. But the even greater productivity at low elevations does not add to diversity; it decreases diversity. We see so many more birds in the lowlands merely because the lowlands are so extensive.

The literature points out that ecologists often confound area and productivity (Wright 1983; Turner, Gatehouse, and Corey 1987). But some have tried to separate them (Abramsky and Rosenzweig 1984; Owen 1988; Turner, Lennon, and Lawrenson 1988). Rahbek's work teaches us how astonishing and rewarding it is to disentangle them.

Other studies also point out mid-elevation peaks in diversity along tropical transects. These include ferns (Tryon 1989) and bryophytes (Gradstein and Pocs 1989), although not angiosperms (Stocker and Unwin 1989) nor amphibians and reptiles (Scott 1976). But each of these cases ought to be reconsidered in the light of Rahbek's method.

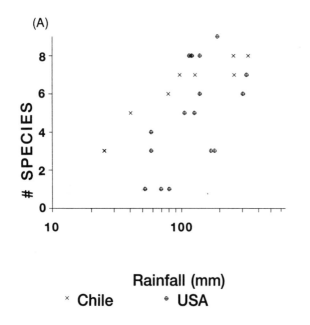

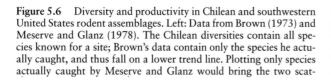

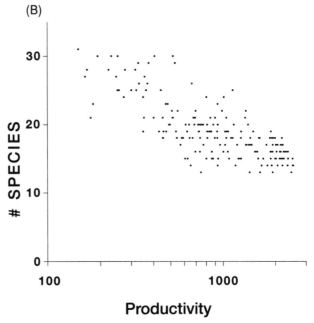

Productivity

Figure 5.6 Diversity and productivity in Chilean and southwestern United States rodent assemblages. Left: Data from Brown (1973) and Meserve and Glanz (1978). The Chilean diversities contain all species known for a site; Brown's data contain only the species he actually caught, and thus fall on a lower trend line. Plotting only species actually caught by Meserve and Glanz would bring the two scat-

tergrams in line, but would diminish the accuracy of the Chilean trend. Right: Owen's (1988) graph of rodent assemblages in Texas. Owen's area of lowest productivity (150 g/m^2/yr.) corresponds to a rainfall of 205 mm; thus his data begin on the right end of Brown's. Owen's data come from much larger areas, which accounts for their much higher numbers of species.

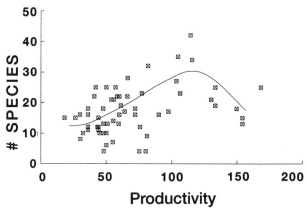

Figure 5.7 Tropical mammal diversity in a large variety of wet and wet-dry tropical habitats in Australia. Diversity declines over the highest productivities. Productivity is estimated as a soil fertility index multiplied by a climatic index; the latter takes temperature and rainfall into account. The curve merely approximates the trend. (Data from Rosenzweig and Braithwaite, unpublished.)

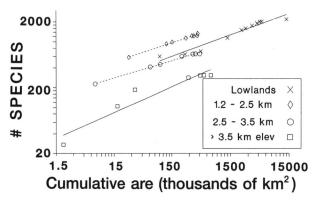

Figure 5.8 For a fixed area, diversity of neotropical birds is highest at subtropical elevations, not in tropical lowlands. (Data from Rahbek, unpublished.)

Fossil Invertebrates

Ziegler (1965) noted the good correlation of sediment type and ocean depth: the finer the sediment, the deeper the water in which it was deposited. He and other paleobiologists began to use that correlation to describe ancient environments and their communities. Although paleobiologists agree that sediment type does not parallel ocean depth precisely (Shabica and Boucot 1976; Johnson and Potter 1976; Hurst 1976; Watkins 1979), they also agree that the correlation is good enough to rank the depth of different deposits (e.g., Mikulic and Watkins 1981). So, sediment type becomes an index of relative productivity.

We have estimates of invertebrate diversity over depth gradients for a number of periods in the Ordovician and the Silurian. All show the hump-shaped pattern. Even Llandovery time of the Lower Silurian, for which we could find no summary data on diversity, shows the pattern. Ziegler, Cocks, and Bambach (1968) note that its *Lingula* community had both the least diversity and the shallowest environment. The "most diverse" community, the *Clorinda,* was in deeper water. Later, Cocks and Rickards (1969) identified a "Marginal *Clorinda*" community, seaward of the *Clorinda,* with very few species, fewer even than the *Lingula.* Seaward of that were the graptolitic mudstones with, essentially, no shelly species at all.

In Figures 5.9, 5.10, and 5.11 we plot three fossil transects. Were any of these to stand alone, it might not convince you. But the pattern occurs repeatedly. The combined Silurian and Ordovician investigations represent some 75 million years of fossil history. And each graph incorporates a very large amount of data. For example, each point of the Ludlow series (Watkins 1979) of six communities (fig. 5.10) is the mean diversity per 50 individuals per sample. There are 200 samples, and no Ludlow point comprises fewer than 11. The Ordovician graph (fig. 5.11) synthesizes "about 2000 samples collected through about 5 km of strata representing about 200,000 individual identifications" (Lockley 1983).

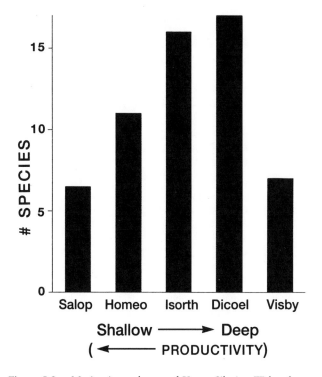

Figure 5.9 Marine invertebrates of Upper Silurian Wales show peak diversity at intermediate ocean depths. Most species are brachiopods. Diversities are averaged over an unspecified number of fossil collections, each of which had 100 to 200 specimens. Communities: *Salopina; Homoeospira/Sphaerirhynchia; Isorthis; Dicoelosia; Visbyella.* (After Hancock, Hurst, and Fürsich 1974.)

What Leads to the Pattern?

Ecologists have a new generalization to contemplate. Within regions about the size of small to medium-sized nations, species diversity is often—perhaps usually?—a unimodal function of productivity (or some well-accepted index of it like rainfall or nutrient supply). We do not wish to see how many other cases of this pattern we can collect. We want to set the stage for the explanation of the pattern and its mechanism(s).

Theoreticians and empiricists have long thought that

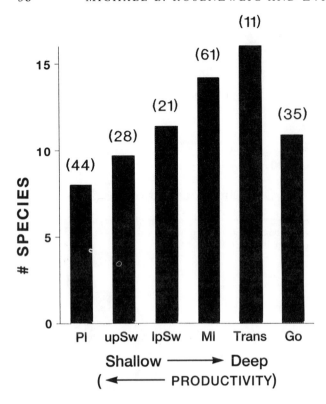

Figure 5.10   Marine invertebrates of brachiopod-dominated Lower Silurian Wales also show peak diversity at intermediate ocean depths. Each diversity is the average of the diversities of the number of collections appearing in parentheses over its bar. The diversity of each collection was calculated by rarifying it to 50 individuals following the method of Sanders (1968). Communities: *Protochonetes ludloviensis,* upper phase of *Sphaerirhynchia wilsoni,* lower phase of *Sphaerirhynchia wilsoni, Mesopholidostrophia laevigata,* "transition fauna," *Glassia obovata.* (Data from Watkins 1979.)

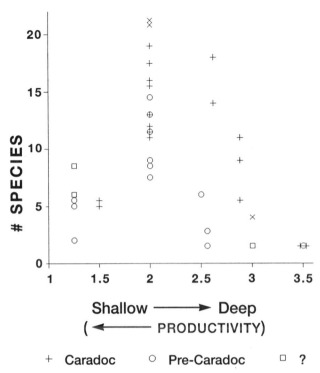

Figure 5.11   Marine invertebrates of brachiopod-dominated Welsh Ordovician communities also show the hump-shaped pattern. Caradoc communities had more species at intermediate depths than did Pre-Caradoc communities (of Llandielo, Llandvirn, or Arenig times), but the pattern remained. Only the brachiopods were reported for three communities (×); we assumed this constituted 90% of the fauna (probably still an underestimate, but that will not affect the overall pattern). So we added two species each to the two such Caradoc communities (both with depth equal to 2), and half a species to the one (of Caradoc or Pre-Caradoc age) (with depth equal to 3). The three points symbolized with a square also may be either Caradoc or Pre-Caradoc. The depth variable itself is relative; we constructed it from Lockley's figure 6 by assigning its top a value of 0, its bottom a value of 4, and interpolating the approximate midpoints of the generic depth ranges he reports. (Data from Lockley 1983; thanks to Martin Lockley for advice on producing this figure from his review.)

there must be a causal relationship between diversity and productivity (e.g., Connell and Orias 1964; Leigh 1965; MacArthur and Pianka 1966; Pianka 1966; Rosenzweig 1971; Brown 1973; Tappan and Loeblich 1973). But they have disagreed about the mechanism that underlies it and even about the direction of the correlation (Valentine 1976; Elseth and Baumgardner 1981). Some—typically those studying aquatic systems or enriched plant systems—found fewer species at higher productivities. Others—especially those studying terrestrial vertebrates—found the opposite.

The fact that diversity is a hump-shaped function of productivity explains how it is that careful scientists could reach opposite conclusions about the direction of the relationship. Brown (1973, 1975), for example, was looking at the increase phase, over the poor end of the productivity spectrum (see also Bramlette 1965; Tappan 1968; Meserve and Glanz 1978). Students of eutrophication (e.g., Sawyer 1966; Whiteside and Harmsworth 1967; Lipps 1970; Hessler and Jumars 1974) usually focused on the rich end (the decrease phase).

Knowing what the pattern is, we should begin to progress in explaining it. Actually, the increase phase of the relationship troubles no one. It is felt to be on a firm theo-

retical basis: A poor environment supplies too meager a resource base for its would-be rarest species, and they become extinct.

Preston (1962) developed the theory behind this idea. He showed that the abundances of species in a region fall into a unimodal distribution called the log normal. Abundances equal "individuals per unit area." So, if the total area of a region is small, then the rarest species of the log normal distribution will have insufficient abundance to survive. He used the metaphor of the veil-line to describe this effect: a small area veils the existence of all the species whose total abundance falls below a critical minimum. The critical minimum is the veil-line.

Wright and his associates (see chap. 6) have been extending Preston's theory to the variable of productivity; they call their idea "species-energy theory." Using it, one can see that an unproductive region will have species that are relatively rare for their abundance rank. The region's veil-line should occur at a fairly high abundance rank, and all lower-ranking species will be veiled (i.e., nonexistent).

Maybe we should suspect such a universally accepted explanation. But we have no reason to. Thus, to us, the decrease phase presents the real puzzle: Why, past a certain point, does enhanced productivity tend to reduce the number of species? In the next section, we discuss hypotheses to account for the decrease phase.

HYPOTHESES TO EXPLAIN THE DECREASE PHASE

The literature presents nine different hypotheses to explain the decrease phase. For convenience, we have organized them into three arbitrary groups. The first group includes hypotheses with an explicit dynamical model relating productivity to diversity. The second group's hypotheses lack such a model at present. Each of the third group's hypotheses maintains that diversity merely correlates with productivity; it identifies the true cause as some other variable, i.e., time, space, disturbance rate, or covariance of different species' population densities.

We treat each hypothesis under three subheadings: First we state it briefly. Then we explain why it might be true. Then we evaluate it. Please understand that this format requires us to play devil's advocate on behalf of a few hypotheses; that is, we begin by arguing for them as best we can, although we do not believe them. It seems only fair for each to have its day in court. But in the evaluations, we state our real opinions and the reasons for them. Although losers do emerge from the list of hypotheses, you will soon see that we are not sure whether there are any winners.

Hypotheses with Mechanism Included

Environmental Heterogeneity

HYPOTHESIS: Under conditions of extreme low productivity, there is not much habitat or resource heterogeneity; the landscape is uniformly barren. The average spot will not sustain any species. As productivity rises, the average variety of micronutrient combinations in fertile sites increases, or some fertile spots have more light with sparser nutrients while others have less light with richer nutrients. In either case, plant diversity increases and plant physiognomies diversify, allowing animal diversity to increase also.

Past a certain point, more productivity has the opposite effect. It reduces heterogeneity of micronutrient combinations and habitats. Productivity tends to be spread more evenly within and between years, reducing the variety of viable temporal specializations. Diversity declines.

Notice that this hypothesis explains both the increase and decrease phases of the productivity pattern. It does not require low productivity to cause rarity and thus higher extinction rates. So it predicts the entire hump-shaped pattern with no help from any other hypothesis.

REASONING: The variety of habitats in space and time underlies much of the specialization that supports diversity (Rosenzweig 1987a). Most ecologists would agree that relatively barren areas offer only a few kinds of habitable times and places: the mean habitat is inhospitable, although some unusually favorable times or places cross the line and support niches. Moderately productive areas have excellent mean habitats. But their variance also encompasses a wealth of different sorts of exploitable niche opportunities. Very productive areas also have excellent mean habitats. But their variance rarely presents significantly different challenges to life. A productive patch that falls to half its mean productivity is still very productive. Similarly, a patch 0.5 km away with half the productivity is still very productive.

Tilman (1982), using micronutrient combinations as the measure of habitat specialization, argues persuasively for this hypothesis. More recently, (Tilman 1987; Tilman and Pacala, chap. 2) he has been exploring a similar model. Newman (1973) noted that as nutrients increase, light becomes more and more of a problem for competing plants. This sets up a gradient along which Tilman imagines each plant species to have a specialty. Regions of poor productivity will include little of the whole gradient and, therefore, few of the plant species that specialize along it. Regions of high productivity will also support few of the species because they too include little of the gradient; most of their sites will cause intense competition for light.

Abrams (1988) developed this model formally. He discovered that it could also lead to a monotonic rise in diversity as productivity grows. But Tilman and Pacala (chap. 2) doubt that the assumptions necessary for monotonicity can ever be satisfied.

Once the plant pattern appears, a similar animal pattern will evolve. Innumerable studies recognize the importance of plant diversity (in physiognomy as well as species) to the maintenance of animal diversity.

EVALUATION: Much evidence favors this hypothesis. For example, United States rodents have habitat specializations that suggest it. As productivity rises to the point where cover approaches 100%, the aspects of habitat heterogeneity that support mammalian diversity in places of intermediate productivity seem to decline. Shrubs diminish in favor of grasses. Patches of open ground, which seem to be crucial for the existence of bipedal rodents, become smaller (Rosenzweig 1977b; Lemen and Rosenzweig 1978; Brown, Reichman, and Davidson 1979; Kotler 1985). Both Rosenzweig (1973) and Whitford et al. (1978) showed that experimentally simplifying the structure of the plant community reduces rodent diversity.

Israel's sand-dwelling rodents also seem to fit. Areas of intermediate productivity have patchy ground cover exploitable both by species usually living in denser cover and by those often found where there is little cover (Rosenzweig, Abramsky, and Brand 1984; Abramsky, Brand, and Rosenzweig 1985; Rosenzweig and Abramsky 1985; Rosenzweig and Abramsky 1986).

Tilman (1987) reports the results of systematic experiments to study the effects of nutrient enrichment in various successional stages from new field to woods. Increasing the productivity drives diversity down. In three years, more than 60% of the species disappeared from plots that received high nitrogen treatments. He believes "that nutrient addition makes plots more homogeneous spatially, forcing more species to compete for the same limiting resource."

Certainly, the natural history survey data taken in the area of Tilman's experiments support his hypothesis (Inouye et al. 1987). The older a field, the more nitrogen and the fewer species it has.

The literature repeatedly tells us that nutrient enrichment will depress plant species diversity. Huston (1979) traced such experiments back to 1882. He cites numerous other examples in plants.

Goldberg and Miller (1990) bring us up to this decade. Their experiments also support Tilman's interpretation. Adding water greatly increased productivity, but did not change diversity. Water amounts do not usually vary much over the small spatial scale of such experiments. On the other hand, adding nitrogen causes only a small increase in productivity, but greatly depressed the diversity. Nitrogen, like other chemical properties of soil, often varies greatly at small spatial scales; adding a uniformly applied extra amount masks such variation.

The heterogeneity hypothesis does leave a nagging question: Is it tautology? Habitat and resource heterogeneity partitioning are evolved responses of organisms. We know no a priori reason to expect that life will subdivide any particular variance into more niches than it will a smaller variance. Species discriminate habitats because natural selection forces them to. More species mean more selection for finer habitat discrimination. Remember what Janzen (1967) took for granted in the title of his classic paper: Mountain passes are higher in the tropics. The same degree of physical change places more of a restriction on tropical species than on temperate ones. Physical environmental heterogeneity is continuous and, in principle, it should be infinitely subdivisible by life.

That is why birds in Puerto Rican rainforest recognize only two foliage levels while birds in Panamanian rainforest recognize four (MacArthur, Recher, and Cody 1966). Both forests present the same physiognomic complexity, but the one in Puerto Rico houses a depauperate island fauna. Under no pressure to subdivide the forest as finely, the birds of Puerto Rico don't.

This point is worth a second example. One of the world's richest floras grows on the impoverished soil of southwestern Australia. Another grows on the similarly nutrient-poor soil of the fynbos in South Africa. To a human being, especially an untrained one, the plant cover in these places looks monotonous. The botanist finds few growth forms (Adamson 1927)—many heathlike plants in Australia, for instance. And most of the plants come from fairly closely related species. Nevertheless, these plants have broken up their temporal world into a succession of flowering times; a continuum of narrow absolute range has evolved into a riot of distinct habitats.

If species are forced to recognize more habitats when there are more species, then we chase our tails to say that the reason there are so many species is that there are so many habitats. To make a statement about the intrinsic effect of habitat variance, we need an independent model. It must start with a habitat continuum. It must show that speciation and extinction rates vary with productivity so that we wind up with more species (and thus more recognized habitats) at intermediate productivity levels. We cannot start with an effect, notice that the effect correlates

closely with its cause, and conclude that therefore, the effect is the cause.

Dynamical Instability

HYPOTHESIS: More productivity reduces dynamical stability. The loss of dynamical stability increases extinction rates, reducing diversity.

REASONING: One of us (MLR) invented this hypothesis as a theory of eutrophication (Rosenzweig 1971). Higher productivity reduces the negative feedback of competition within species, and increases the positive feedback that predatory control produces (Rosenzweig, 1977a). There is a net loss of dynamical stability at higher productivities.

Wollkind (1976) extended the theory to three-level food chains. Riebesell (1974) extended it to competition between species.

EVALUATION: This hypothesis breaks down when natural selection has enough time to do its job. Rosenzweig and Schaffer (1978) showed that natural selection tends to restore the dynamical stability of enriched systems. That is how some extremely productive biomes like coral reefs and tropical forests can be extraordinarily diverse.

Yet, many have reported that diversity does decline after nutrient enrichment. In addition to the plant examples that we mentioned in connection with the previous hypothesis, there are a number of aquatic examples (e.g., Swingle 1946; Yount 1956; Schindler 1990). We cannot easily explain all these cases as the result of increasing competition for light, or homogenization of habitat. So perhaps the hypothesis of dynamical instability does apply to some recently enriched environments, especially aquatic ones like eutrophic lakes. At least the stability hypothesis makes a field-testable prediction that could someday force us to accept it: Enrichment should be accompanied by an increase in oscillatory dynamics and a decline in return times (Pimm 1982).

Changes in Ratio of Predators to Victims

HYPOTHESIS: As productivity increases, predators absorb much more than a proportional share, and reduce the diversity of consumers.

REASONING: Predators can absorb most or all of any increase in standing crop caused by higher productivities (Rosenzweig 1971, 1972). Moreover, Oksanen et al. (1981) determined theoretically that increased productivity adds controlling trophic levels in a food chain. A new higher trophic level should considerably reduce the standing crops of the species beneath it. Smaller standing crops should produce higher extinction rates and diminished diversity.

EVALUATION: Predator-victim diversity ratios are almost constant for a wide variety of species diversities (Mithen and Lawton 1986; Pimm, Lawton, and Cohen 1991). That fact contradicts the prediction of the hypothesis.

Even if we did not know the predator-victim ratios, we would distrust the hypothesis. The theory on which it is based depends on the presence of weak interaction coefficients among predator individuals. While zooplankton may not exhibit strong intraspecific interactions, vertebrates must. So may many other life forms. Individuals that form dominance hierarchies or steal one another's territories don't fit the theory. But such life forms (e.g., Texas carnivores) do exhibit the hump-shaped pattern.

Furthermore, a lot of work has demonstrated that predators can add to the sustainable diversity of their victims (the keystone predator effect). Much of this work consists of incontrovertible field experiments done in all sorts of biomes on various taxa (e.g., Summerhayes 1941; Paine 1966; Hay 1985). More recently, theoretical studies have explained how this can happen through apparent competition or competition for predator-free space (Holt 1977, 1984; Jeffries and Lawton 1984).

The predator-victim ratio hypothesis fails.

Hypotheses with Unspecified Mechanism

Intertaxon Competition

HYPOTHESIS: Once a critical productivity is reached, a multispecies taxon (e.g., a class or an order) cannot absorb any more. All further productivity increases go to a competing taxon. In fact, the competing taxon takes even more than the increase. It actually reduces the productivity going to the first taxon. So the first taxon declines in diversity.

Notice that this hypothesis claims that there is only one relationship between diversity and productivity: The more productivity, the more diversity. The twist is that, according to this hypothesis, higher general productivities mean lower productivities to some taxa. (The predator-victim ratio hypothesis shares this twist, but otherwise it is too weak to consider further.)

REASONING: Intertaxonomic competition exists. For instance, granivorous rodents do compete with ants (Brown, Davidson, and Reichman 1979; Davidson, Inouye, and Brown 1984). Brown and Davidson (1977) detected a likely suppression of rodent diversity by ants in their east-west productivity transect. Possibly, each type (*bauplan?*) of organism competes best at a restricted set of productivities and is largely defeated at others.

EVALUATION: Competitive abilities among species often do differ along a richness gradient (Rosenzweig 1987a, 1991; Keddy 1990). Why shouldn't the same be true among higher taxa? Also, taxa differ in the productivities at which their diversities peak, just as we expect from this hypothesis. Compare Texas rodents to Texas carnivores, for example. Rodents peak west of El Paso in Arizona; carnivores peak in East Texas at far higher productivities. Or compare the ocean depths at which the various taxa cited by Rex (1981) reach their peak diversities. Each is different from the others.

Nevertheless, this hypothesis needs much more work. It needs a coherent theoretical treatment, which would multiply its predictions and make it easier to test. One untested prediction: The broader the taxonomic grouping, the weaker the decrease phase.

Finally, some taxa should exist that do not decline at higher productivities. Where are they? As Tilman and Pacala point out (chap. 2), we do not yet know of any group whose diversity rises monotonically with productivity. North American trees may (Currie and Paquin 1987), but data suggesting they do are truncated by the southern political boundary of the United States. Would extending these data farther south reveal decreases in their diversities with even higher productivities? For the sake of this hypothesis, we hope not, because it is hard to imagine the taxon that would be responsible. Which taxon can compete successfully with trees at the highest productivities, and drive down their diversities?

On the other hand, the tropical forest data that we used to think exhibited the unimodal productivity relationship, may not. When Tilman (1982) suggested that the pattern exists among tropical trees (in two provinces, Malesia and the Neotropics), he used a measure of soil nutrient concentrations to stand in for productivity. But no one has yet shown how tropical forest productivity can be predicted from any surrogate variable. Although high precipitation leaches tropical soil and makes it very poor, tropical plants have evolved a root mat that buffers the loss of nutrients from the soil itself (Stark and Jordan 1978; Jordan 1983). This makes it easier for us to understand how ultra-poor, sandy white tropical soils can support immense plant diversity and abundance.

Also, controversy surrounds the correlation of tropical forest diversity to soil fertility in Neotropical and in African rainforests. In those forests, annual precipitation correlates well with plant diversity, whereas soil fertility does not (Hall and Swaine 1976; Gentry and Emmons 1987; Gentry 1988a, 1988b). The more precipitation, the more species. Now, if we only knew how productivity relates to precipitation, we would have the tropical forest pattern. But, we do not even know if productivity does relate to precipitation in the tropics.

Change in Competitive Structure

HYPOTHESIS: Interference competition prevails at intermediate productivities and adds considerably to the diversity maintainable at them.

Notice that this hypothesis also requires no other to predict the entire hump-shaped pattern.

REASONING: Brown (1971) noted that territoriality in chipmunks first increases, then declines as productivity increases. Territoriality commonly appears in asymmetrical competitive systems and can help to promote competitive coexistence in them (e.g., Pimm, Rosenzweig, and Mitchell 1985). Brown argues strongly that productivity should influence territoriality: poor situations are not worth defending; rich ones are too costly to defend because they support so many individuals that excluding them would take too much time. Perhaps species are less diverse in richer places because they cannot therein ameliorate the effects of competition by being territorial. In the terminology of Nicholson (1954), perhaps they are forced from contest competition into scramble competition.

We may be able to extend this reasoning to plants. Perhaps plants in poorer places cannot afford the costs of chemical defense (allelopathy). And perhaps plants in very rich places cannot preserve the areas around themselves for their own seedlings because if they did, they would encourage foraging by too many herbivores. We merely mean to suggest that such an argument is possible, not that we believe this one. We would welcome its improvement.

The plant argument uses herbivory to understand how interference competition could wane at high productivity. The animal argument uses the time cost of active territoriality. Nevertheless, we join them into one hypothesis because they make the same broad prediction: Interference competition should peak at intermediate levels of productivity.

EVALUATION: This hypothesis has three problems. They probably all stem from the fact that a rich place has plenty of levels of productivity on which to specialize.

First, interference (especially as exemplified by territoriality) is a well-documented phenomenon in some of the earth's most productive places (e.g., coral reefs and tropical rainforests). Any successful hypothesis concerning interference will have to predict that where productivity is high, interference should decline, but not very much.

Second, even in a region of high productivity, species may specialize on moderate habitats and so escape the pressure to forgo territoriality. All species in an ecosystem do not experience its productivity to the same extent. In a richer one, there tend to be species that specialize on wealth and others that can also use the sparser times and places that an environment offers. A good example is the bee community of the Santa Catalina Mountains (Schaffer et al. 1979). During each day, honeybees forage first, when nectar supplies are most plentiful. Then come bumblebees, and finally carpenter bees. Arizona hummingbirds give us another example (Pimm, Rosenzweig, and Mitchell 1985). One species lives exclusively in the rich riparian forest along canyon bottoms. The other two also prefer that habitat, but usually forage in poorer woods, arid slopes, or higher elevations. There are hundreds of other examples; they combine to form a common sort of community organization: shared preference (Rosenzweig 1987a, 1991).

Third, we certainly agree that species may stake out a range of habitat richnesses for their niche. But, although they may do so by interference, they need not (Brown 1986, 1989a, 1989b). Various sorts of morphological or physiological adaptations can take the place of interference if that gets to be an impractical strategy. Replacing interference as the means of subdividing a productivity axis does nothing to reduce the number of species that can subdivide it.

For example, some species of desert rodents forage during the poorer months of the year or in poorer patches of seeds. Others actually require the richer times and places that all prefer. But among heteromyid species, only the largest kangaroo rats use interspecific territoriality to protect their end of the habitat spectrum (the richest) (Frye 1983; Brown and Munger 1985). The niches of the others are determined by their positions in a trade-off continuum: the more efficiently a rodent deals with food (once

found), the less efficiently it travels in space and/or time to find it. (Hibernation [and torpor in general] provides the efficient way of traveling between times of abundance.) Efficient travelers tend to discover the rich patches first and so tend to monopolize them. But, usually because they are larger, efficient travelers are too inefficient at food use to exploit the poorer patches at all (Brown 1986, 1989b; Kotler and Brown 1988).

Someone needs to produce a quantitative model predicting the predominance of interference and its effects on diversity. Otherwise, we cannot believe that this hypothesis will help us to understand the diversity-productivity pattern.

Hypotheses That Reduce the Pattern to Another

Time

HYPOTHESIS: Richer patches have been around for a shorter time than poorer patches have. They have not reached equilibrium and are producing new species faster than they are losing them. The decrease phase is temporary.

REASONING: Speciation takes time. Considering the upheavals of the Pleistocene, many habitats and taxa may not have reached equilibrium. If the most productive habitats are also the newest or were the hardest hit, they should be the most depressed below equilibrium.

EVALUATION: We know of no evidence to support this hypothesis. It depends on richer patches being considerably younger than the time it takes evolution (and colonization) to fill them. It also depends on richer patches being considerably younger than poorer patches. Does anyone think this is generally true? How can the time hypothesis explain the pattern's existence and persistence for seventy-five million years during the Ordovician and Silurian? We mention this hypothesis because someone was bound to. But we also reject it.

Disturbance

HYPOTHESIS: Varying disturbance rates cause the productivity-diversity pattern. The higher the productivity, the more infrequent the disturbances. So, the productivity-diversity pattern is the disturbance-diversity pattern: As disturbance rate falls, diversity first rises, then falls (Grime 1973, 1979; Levin and Paine 1974; Connell 1978; Lubchenco 1978; Sousa 1979a; Paine and Levin 1981; Petraitis, Latham, and Niesenbaum 1989).

In a sense, this hypothesis also predicts the entire hump-shaped pattern. But the disturbance hypothesis itself requires two mechanisms. We shall discuss them when we evaluate this hypothesis.

REASONING: As rainfall increases, its coefficient of variation certainly does decline. Rainfall, necessary for actual evapotranspiration, helps set productivity (Rosenzweig 1968). Moreover, some ecologists believe that even in tropical biomes, the wettest, warmest habitats are the most stable. Why use two variables (disturbance and productivity) to do the work of one? Let's keep disturbance—

we have experimental proof of its role—and forget about productivity.

EVALUATION: Part of the reasoning is suspect. How well do productivity and disturbance correlate? After all, disturbance is the converse of stability, and few ecologists accept uncritical, undefined statements about stability any more. Few believe they know how to rank biomes, let alone habitats, as to their stability or freedom from disturbance. Take a walk in a Neotropical lowland rainforest. That popping and crashing you hear all around you is the sound of falling vegetation creating new open spaces. How is that open space so different from a patch of intertidal substrate newly opened by a violent wave?

Maybe it is far too soon to be sure that productivity and disturbance correlate well from middle to higher productivities. But there is no question that they do from lower to middle productivities. From the most extreme deserts to semiarid grasslands to mesic forests, productivity and disturbance surely correlate inversely. Isn't that enough to prove the point? In fact, it is not.

The disturbance-diversity pattern comes from a well-understood theory that makes several collateral predictions. As diversity increases from extreme deserts to richer places, do that theory's predictions come true? If it turns out that the richest places really do have the lowest frequencies of disturbance, do those collateral predictions also fit the data of the tropics?

The theory of the disturbance pattern has two component processes: First, species disappear from a patch by competition. Second, species disappear from a patch because something (a disturbance), like a figurative broom, sweeps them away. Against these tides of dynamic destruction, species move into the patch and replenish it.

Where disturbances happen often, the average patch accumulates only a few species before the broom returns. So, most such patches are depauperate.

At intermediate disturbance rates, more species accumulate before disaster strikes. But disturbance still comes often enough to interrupt the process of competitive exclusion. The average patch is quite diverse.

Where disturbances rarely happen, then competition can usually complete its work. Diversity is lost. The average patch suffering infrequent disturbance will be mature and depauperate.

What are the collateral predictions of this model? First, the species found in the least disturbed patches should be a subset of those in the most diverse patches. There will be no species restricted to the least disturbed patches. Second, although it will rarely be disturbed, once it is, the recuperation history of any one of the least disturbed patches should follow a course much like the pattern of the combined patches. The patch should begin by accumulating species, reach a diversity like that of a patch with an intermediate disturbance rate, and then decline in diversity owing to loss of species already present. In other words, patches with little disturbance should recapitulate in their own histories the entire diversity-disturbance pattern.

The diversity-productivity graphs themselves offer no information to help us decide whether they match the attributes of a disturbance pattern. But we know enough of the details of the mammal patterns to believe that they do not fit the disturbance hypothesis.

First, low diversity–high productivity points should contain a small, predictable subset of the species from peak diversity points. But the species lists from East Texas (high productivity) and from lowland rainforests in tropical Australia (high productivity) contain many species not found at all in peak-diversity places. This is glaringly so in Texas: only a few species of the rich Trans-Pecos rodent fauna also live in east Texas, and there they make up a small proportion of its impoverished rodent fauna.

Finally, at least for the sites of southern Arizona, we can say that there is no apparent historical pattern. Occasionally, we have seen the most species-rich patches devastated by a predator (such as a badger); but soon (in less than a year), the same species that were wiped out recolonize. And no one has ever seen a short grass patch (slightly high productivity) at an "intermediate stage of history" with lots of the species generally found in mixed desert scrub patches (of highest diversity and lower productivity). An army of students of mammals has sampled so many short grass patches for so many decades that by now someone should have reported such a high-diversity assemblage in a short grass patch. Reported one, that is, if they exist. No one has.

We now confront a most anomalous feature of the disturbance hypothesis. According to the disturbance hypothesis, productivity and disturbance are inversely correlated. But that is only if you are a terrestrial ecologist. Marine ecologists again and again claim the opposite (e.g., Sanders 1968, and many, many others). They assure us that the most productive, shallowest waters are the most disturbed. The marine paleoecologist sees the same pattern in the fossil data. Mikulic and Watkins (1981), for instance, make it clear that the shallow, productive community is the one that is most storm-influenced and unstable. Thus, if we believe the disturbance hypothesis, we must admit that the good correlation of disturbance rate with productivity is positive in the ocean and negative on land. Yes, this could be true. But, combined with the real failures of this hypothesis, the anomaly does not inspire confidence.

Notice the time scale at which the disturbance hypothesis operates. Species come from a preexisting pool. They disperse to empty patches; they do not evolve to fill them. Despite this, paleobiologists and marine biologists have appealed to the disturbance relationship for explanations of the patterns they find (e.g., Watkins 1979; Rex 1981). High-productivity, highly disturbed shallower water should generate high extinction rates, they theorize. But no one knows whether the rate of brief, local disturbances, like wave scouring, indicates anything about the sorts of disturbances that drive species extinct.

Even if the disturbance hypothesis works at an evolutionary time scale in marine environments, it cannot work at that scale in terrestrial regions. There it would be predicting the decrease phase by saying that the higher the rate of extinction-causing disturbances, the more species there are. On an evolutionary time scale, however, the opposite holds. The higher the rate of extinction-causing disturbances, the fewer the species. So, high-productivity, low-disturbance terrestrial locales ought to have the most species. They do not.

Reduction in the Covariance of Population Densities

HYPOTHESIS: High temporal covariance of population sizes among species leads to higher diversity. As productivity increases, the covariance diminishes. So the decrease phase of the productivity pattern is due to decreasing covariance, not increasing productivity.

REASONING: Population sizes fluctuate. The covariance statistic of this hypothesis summarizes the tendency for different species to simultaneously experience above-average populations and, at other times, to simultaneously experience below-average populations.

Some density-dependent optimal foraging theory predicts that individuals of competing species should restrict themselves to their special habitat(s) only if they and their competitors (of other species) are similarly common relative to their respective averages (Rosenzweig, 1979, 1987b). When species do restrict themselves in that way, they compete minimally and have the best chance to coexist. On the other hand, when one species is common and another is very rare, then the common one must use the rare one's special habitat, and they compete intensely (Rosenzweig 1981; Pimm and Rosenzweig 1981; Brown and Rosenzweig 1986). In sum, according to optimal habitat selection theory, high temporal covariance leads to dissimilar behavior, reduced competition, and lower extinction rates; low covariance to more similar behavior, increased competition, and higher extinction rates.

Covariance and productivity probably do correlate negatively (Rosenzweig 1979). In a truly harsh, unproductive place, some master variable (like water availability in deserts) should govern the fate of all (or most) species. A good year for one is likely to be good for all. Once productivities increase, other, more diverse variables should exert more control, thus reducing covariances.

EVALUATION: This mechanism seems too restricted. It applies only to unstable environments whose species have true habitat specialties and whose individuals cannot sense a patch's quality except by taking resources from it (Brown and Rosenzweig 1986). Desert rodents, for example, do not conform to these requirements. They can sense (probably by olfaction) seed quantities in their immediate vicinity (Brown 1989b). And they do not have true habitat specializations: some species merely get to the best places first (Brown 1989b) or are capable of defending the best places (Frye 1983); but all prefer areas with the highest seed abundances. Such a relationship among the niches in a guild is called "shared preference." Field ecologists discover shared preference relationships more than any other kind of community organization (Rosenzweig 1991), and the hypothesis of covariance reduction applies to none of them.

A second problem: this hypothesis is 90% theory. No one has good evidence for the connection between productivity and covariance. Several field experiments do support the isoleg models that make the optimal foraging predictions (Pimm, Rosenzweig, and Mitchell 1985; Rosenzweig 1986; Abramsky et al. 1990; Abramsky, Rosenzweig, and Pinshow 1991). But only theory says that higher covariance leads to lower extinction rates.

We include this hypothesis because we want to emphasize covariance relationships. This particular hypothesis may need so much revision that it becomes unrecognizable before it becomes very useful. But we will probably need to test a better hypothesis based on covariance before the problem gets settled. The models of Chesson may supply the improvements (Chesson and Huntly 1988).

Area

HYPOTHESIS: High-productivity habitat is scarce compared with intermediate habitat. Smaller areas usually harbor fewer species. The productivity pattern is just the species-area curve on different coordinate axes.

If the productivity of very large patches is distributed with a central tendency (say, normally or log-normally), then this hypothesis also predicts the entire hump-shaped pattern. Both rich and poor patches will be relatively scarce and relatively depauperate.

REASONING: Like all variables, productivity should have a central tendency. Deviations in each direction should be less and less probable as they depart from the mean. The rest is just the species-area curve in action.

EVALUATION: Much suggests that the area hypothesis does not account for the productivity pattern. First, if area is the answer, all taxa should peak over the same productivity (since that represents the most common state). But they don't. Compared with plants, desert rodents in Israel peak at half the productivity. Compared with carnivores, rodents in Texas peak at about one-tenth the productivity (Owen 1988).

Second, the area hypothesis cannot apply to United States rodents. They peak in the semidesert and fall to much lower diversities in grasslands. Who will defend the idea that grasslands are relatively scarce in North America compared with semideserts?

Last, Abramsky and Rosenzweig (1984) discovered that the productivity-diversity pattern remains in their data even after they take out the unequal-area effects. Going a step further, Rahbek shows that the productivity pattern doesn't even appear in tropical bird data until he actually removes the effect of area.

CONCLUDING OBSERVATIONS

We ecologists retain an abiding fascination with the question of what controls the diversity of species. We've made considerable progress in understanding some aspects of this control. Preston (1962), MacArthur and Wilson (1963, 1967), and Simberloff and Wilson (1969b) taught us the fundamentals of island diversity patterns. Three overviews of the species-area curve (Connor and McCoy 1979; Coleman et al. 1982; McGuinness 1984a) emphasize how much we now know about that. The puzzle of the latitudinal diversity gradient was solved in the mid-1970s (Terborgh 1973; Rosenzweig 1975, 1977c) (more about this below). And the relationship between disturbance and diversity (e.g., Grime 1973; Levin and Paine 1974; Lubchenco 1978)—at least for some associations—is also well modeled and documented.

Meanwhile, the connection between productivity and diversity—long thought to be intimate—remains murky. In this chapter, we hope we have convinced you of the pattern produced by that relationship within large regions: Diversity rises over low productivities and falls over high productivities. We also hope you have decided that no one knows why.

Not a single hypothesis successfully explains the pattern. The rising phase of the pattern is most likely caused by higher productivities being able to support larger population sizes. Larger population sizes should experience lower extinction rates. However, all of the nine hypotheses to explain the decrease phase have noteworthy weaknesses. Three of them—the time, area, and predator-victim ratio hypotheses—are so weak we need not discuss them further.

Two others—the dynamical stability and disturbance hypotheses—address scales of time or space too small to allow for evolution. Given enough time, evolution tends to restore dynamical stability. And in evolutionary time, the relationship between diversity and disturbance rate should be monotonically negative, not unimodal.

We also doubt the hypothesis that interference competition peaks over intermediate productivities, thereby causing the pattern. This hypothesis seems far too poorly worked out and has far too little empirical support.

The covariance hypothesis is more promising. However, at present, too many of its links are purely theoretical. And even the theories it relies on need fuller treatment.

The most attractive hypotheses are environmental heterogeneity and intertaxon competition. The heterogeneity hypothesis states that heterogeneity peaks at intermediate productivities. A lot of evidence agrees with that statement. But heterogeneity appears to be a coevolved response to species diversity. If so, the heterogeneity hypothesis is a tautology. To decide, we need a better theory of the evolution of habitat heterogeneity.

The other strong hypothesis, intertaxon competition, fits what few facts we have. It predicts great variety from taxon to taxon in the place of the peak diversity. Certainly, we see this. But it lacks a mechanistic model and is therefore poor in predictions.

The intertaxon hypothesis needs fuller explanation. Why should taxa have optima along a productivity gradient? Why can't they at least defend the share of the productivity they have already acquired when that optimum is exceeded?

Experiments

Correlations are among the best ways ecologists have of searching for patterns. Once found, however, patterns must be investigated further, preferably by experiment. That is particularly true of productivity patterns because productivity itself correlates with so many other important ecological variables. Proper control of ecological experiments does require a smaller than regional scale. But we can use the results if we are cautious about designing the experiments.

Ecologists have done many experiments on the relationship between productivity and diversity (Swingle 1946; Yount 1956; Patrick 1963; Kirchner 1977; Silvertown 1980; Tilman 1987; Schindler 1990; Goldberg and Miller 1990, and many others reviewed in the latter and in Huston 1979). All these experiments involved increasing productivity. Occasionally diversity failed to respond to the change, but far more often it decreased dramatically. Only one experiment (Abramsky 1978) reported an increase. Does that confirm the existence of the decrease phase? Does it mean we should distrust the increase phase?

We think the set of experimental results now available casts little or no light on the regional pattern. None of these experiments take evolution into account, although the evolution of species to take advantage of higher productivity must constitute an important part of the response of a region to higher productivity. Some experiments fail to allow even enough time for succession, although we know how important that can be. Whittaker (1975), using the data of Woodwell and Holt, shows that diversity and productivity both increase during the first few years of succession.

Experiments to test the increase phase should compensate for a lack of evolutionary time. For example, they might be sited near enough to potential sources of "new" species so that immigration could supply them. And at least some experiments should take place at sites where the regional pattern predicts a diversity increase. Before we develop enrichment experiments that compensate for lack of time, their relevance will remain restricted. We should use them solely to understand and predict the short-term consequences of local enrichments.

Productivity versus Biomass

Measurements of standing crop (i.e., biomass) are easier to find than measurements of energy flow (i.e., productivity). The two ought to correlate well, although not perfectly. Consequently, some ecologists use either one to study the relationship of productivity to diversity. We hope this does not turn out one day to have created confusion.

The two variables do differ. One of them may be causal. If we are to discover which one, keeping them sorted out seems a minimal strategy.

We found one data set (Haedrich, Rowe, and Polloni 1980) that offers measures of both biomass (grams caught per hour of sample) and productivity (ocean depth). The authors separate their data into fishes, echinoderms, and decapods. Both fishes and echinoderms clearly show the unimodal pattern for depth. Decapods probably also show it, but their peak occurs in relatively shallow water, so only the shallowest samples (40–264 m) attest to the decrease phase. None of the three taxa's diversities fit biomass as well as they do depth. Low biomasses are associated with almost any diversity, although higher biomasses always seem to be associated with at least moderately high diversities. We hope to see more such comparisons in future reports.

Productivity versus Energy

Wright (1983), realizing that a barren hectare has no species, extended the concept of species-area studies. He created a new variable by multiplying area times productivity. The result is gratifying. Adjusting area for productivity improves the fit of data to the independent variable. More productivity always seems to increase an area's capacity for species. But doesn't that monotonic relationship contradict the unimodal relationship we've championed in this chapter? No, it does not.

Please read Wright, Currie and Maurer's contribution to this volume carefully (chap. 6). You will see that their chapter and ours agree on the pattern at the regional scale: It is unimodal. The monotonic pattern appears at the global scale.

The Global Scale

So far we have limited ourselves to considering regions. But we do not want to avoid the big picture forever. What is the productivity-diversity relationship among biogeographical provinces? Is it monotonically positive? We are less than sure.

By combining the variables of area and productivity, Wright (1983) loses the ability to determine the influence of each variable separately. Now, we are not reductionists, but if a good mechanism exists to tell us that one of the two variables has a separate and powerful influence on diversity, then we prefer to treat that variable separately. Area does exert such an influence. Once we have taken it into account, we can go on and combine it with other variables in search of other patterns. But first we must reckon with the known.

The known mechanism is allopatric speciation. The larger a geographical range, the more likely that a geographical barrier will split it into isolates (Rosenzweig 1975, 1977c). So, other things being equal, large ranges produce more new species per unit time than small ranges do. Quite probably, larger ranges also reduce extinction rates by harboring more individuals and by reducing the risk that a perturbation will affect the entire range. But those influences are merely hypothetical, whereas the influence of area on allopatric speciation rate is inherent in the mechanism itself.

Terborgh (1973) observed that there is a marked gradient in area from tropical to polar climates. Tropical regions cover far more territory (and ocean, we might add). Terborgh's gradient coupled to allopatric speciation accounts for the latitudinal gradient in diversity (Rosenzweig 1992). At any diversity, tropical regions should experience higher speciation rates (and probably lower extinction rates too). Consequently, steady-state tropical diversities exceed those of any other latitude. Only if tropical biotas were significantly younger than biotas of other latitudes could they be less diverse.

Now you can appreciate why we remain unsure of the global relationship of productivity and diversity. Diversity does increase monotonically with Wright's variable. But we believe that stems from its area component rather than from its energy flow component. On a global scale, that

area component should be so important as to overwhelm and mask the influence of productivity.

A caveat: Although area's influence has to be teased apart from that of productivity, multiple regression cannot do it beyond the scale of the region. Multiple regression effectively reduces each sample to a common area and then looks for the influence of energy flow on the residuals. That takes care of the regional species-area curve, but not Terborgh's latitudinal gradient of area (which works in evolutionary time at the level of speciation and extinction rates). Convince yourself of the problem by imagining two bird lists, one from a square kilometer of Britain, the other from a square kilometer of Peru. Each receives immigrants from its region. Since the Neotropical region contains so many more species than the Eurasian region does, we expect the British area to have far fewer species. It is as if each square kilometer carries the evolutionary mark of its participation in the larger region to which it belongs. In fact, that is precisely the case. Our solution to the problem? Stay inside regions to find out how productivity affects diversity. If we are correct, then the tropics would be even richer if we could make them somewhat less productive for several million years.

Prospects

Maybe there will never be a hypothesis that fully explains all known instances of the decline phase. Maybe we are looking for something imaginary. Suppose the increase phase comes from the relationship of productivity to population size (as most believe it does). Suppose that once a system passes a critical point, further increases in population size bring insignificant returns in reduced extinction rates. Then, beyond that point, the diversity of any taxon may be little more than a balloon waiting for a pin. Any number of environmental changes associated with increased productivity would deflate the diversity. For now, however, we will pursue our own investigations as if a single successful decline phase hypothesis exists.

The unimodal curve that describes the effect of productivity on diversity may fit vascular plant diversities. It does fit the diversities of various taxa of marine invertebrates. It fits the diversities of mammals in tropical and subtropical latitudes, and the diversities of birds along a tropical elevation gradient. We have found it on five continents so far. Once you adjust for the effects of area and latitude, the hump-shaped pattern may pop out of your favorite taxon-region combination too. If ecologists are to do a good job conserving as many of the world's species as possible, we have to understand the mechanisms that cause this and other common diversity patterns.

SUMMARY

The number of species within a region usually varies unimodally with the rate of ecosystem energy flow. This hump-shaped pattern shows up in many biogeographical provinces. Plant and animal taxa, including vertebrates and invertebrates, follow it. We find it in marine and in terrestrial biomes.

Most ecologists will agree that the increase in diversity

that occurs over low productivities comes about because the total abundance of all species together increases over that range of productivities. We describe and evaluate nine hypotheses to explain the decrease phase of the pattern, i.e., why diversity declines as productivity grows past a certain point. We reject five of these: dynamical instability; change in predator-victim ratio; extreme youth of the most productive places; disturbance; and area. These hypotheses often explain other patterns of diversity, but not this one. We have little confidence in a sixth hypothesis: decrease in interference competition at the highest productivities.

The remaining three hypotheses all have problems. Two are still poorly modeled, and therefore have inadequate empirical support. One, the hypothesis that a competing taxon causes the decrease, predicts that peak diversities should vary among taxa, and they do. But it also predicts some taxa should not show the peak, but instead should increase monotonically with productivity. So far,

we have found no such taxa. However, that may well be because we cannot adequately examine the pattern in the tropics since no one yet knows what environmental variables control productivity among tropical ecosystems. The second hypothesis predicts that competition will be more intense when productivity is very high because more productive sites generate lower covariances between the competing species' population densities. But no one has tested this hypothesis.

The third surviving hypothesis looks deceptively like the strongest: Higher productivities reduce environmental heterogeneity. This one has been carefully modeled and has much empirical support. But it is probably a tautology. Heterogeneity, being an evolved response of life, should evolve to be higher where there are more species.

We discuss the relationship of the regional pattern to global patterns such as the latitudinal gradient. We believe the effect of productivity on diversity is best studied at the regional level.

6

Energy Supply and Patterns of Species Richness on Local and Regional Scales

David H. Wright, David J. Currie, and Brian A. Maurer

It has long been recognized that species richness shows enormous geographical variation at several different scales. The most obvious variation in number of species is the variation from the tropics to the poles (Wallace 1878; Simpson 1964a; Adams and Woodward 1989). Pronounced gradients also occur on very local scales, both within and among habitat types (Moore and Keddy 1989; Nilsson et al. 1989). These observations beg the obvious question: Why are there many species in some places and few in others? Recent reviews suggest that many processes can potentially affect species richness (Ricklefs 1987; Begon, Harper, and Townsend 1990): physical factors such as climate and energy supply, historical factors such as speciation rates and dispersal, and biotic interactions such as predation and competition.

Among the many different explanations proposed for variation in the diversity of species, the hypothesis that the supply of usable energy in the environment plays a role has recently received increased attention. In the 1920s, Lotka (1956) treated the development of ecological communities as "essentially the evolution of a system of energy transformers." Quantitative studies of ecological energy flow began with Lindeman's (1942) application of thermodynamic bookkeeping to ecosystems. Subsequently, some early authors presented verbal arguments or models predicting a positive relationship between "productivity" and diversity in certain situations (Hutchinson 1959; Connell and Orias 1964; MacArthur 1965; MacArthur 1972, 183; H. T. Odum 1970). By 1971, however, E. P. Odum summarized the prevailing view when he wrote: "While productivity or total energy flow certainly affects species diversity, the two quantities are not related in any simple linear manner" (Odum 1971, 150; see also Margalef 1968; Krebs 1972, 521). Commonly cited counterexamples with high productivity but low diversity (for example, salt marshes, hot springs) and low productivity but high diversity (deep-sea benthos, plant communities on some serpentine soils), as well as the results of enrichment experiments, served to reinforce this view. Nevertheless, in the past decade a number of papers have appeared advocating a reappraisal of productivity-diversity relationships (e.g., Huston 1979; DeAngelis 1980; J. H. Brown 1981; Tilman 1982; Wright 1983, 1987, 1990; Abramsky and Rosenzweig 1984; Turner, Gatehouse, and Corey 1987; Turner, Lennon, and Lawrenson 1988; Currie and Paquin 1987; Abrams 1988;

Currie 1991; and see Rosenzweig and Abramsky, chap. 5; but see Latham and Ricklefs 1993). Arguments for such a reappraisal have been made on both empirical and theoretical grounds, and reflect a variety of opinions, some authors proposing a positive species-energy relationship, others a negative one, and some both.

It is undoubtedly true that many factors influence species richness. While a compilation of such factors is of some interest, merely listing them generates few testable predictions. A more powerful approach would specify three things: the factors that account for the greatest amount of the natural variability in species richness, the mechanisms by which these factors control richness, and the perturbations to which richness responds. We treat portions of each of these topics below. In particular, we address patterns and possible mechanisms linking energy supply and species richness. We suggest that some of the disparity of opinion about the effects of the energy flow on species richness may be due to real differences in the patterns that exist at different scales.

A SURVEY OF THE LITERATURE

If much, or most, of the variability in species richness is related to a small number of characteristics of the environment, then one would expect to observe strong correlations between those factors and geographical variation in species richness. This is essentially an epidemiological problem: If one surveys extant studies, what characteristics have been observed to be most strongly correlated with patterns of species richness?

To answer this question, we compiled from the published literature studies that investigated the spatial variability of species richness and its relationship to environmental characteristics. We included studies that reported a statistical correlation (or presented data sufficient to calculate one) between richness and any aspect of organisms' environments, except as follows. We excluded studies whose principal focus was the species-area relationship, or the relationship between latitude and richness, since correlations with these variables are almost certainly indirect. This restriction eliminated most studies that examined the effect of variables traditionally associated with island biogeography, such as isolation. We excluded all studies of experimental manipulations, since those studies do not address natural variation in species richness (we return to the topic of enrichment experiments in a later section). We also excluded the large body of studies that described patterns without any statistical test of their significance.

This literature search yielded 53 studies reporting 82 statistical relationships between species richness and environmental factors (appendix 6.1). For each of these relationships, we tabulated the strongest observed correlation (excluding area or latitude). Additional, weaker correlations were not considered, even if statistically significant.

By far, the variables most frequently observed to be significantly correlated with richness were climatic: mean annual temperature, precipitation, and functions thereof, such as potential and actual evapotranspiration (table 6.1). Correlations with measures of primary productivity, food availability, or limiting-nutrient availability were also commonly observed. All of these factors have been interpreted as representing indices of the availability of energy to organisms; we will refer to them collectively as energy-related factors. The use of climatic or other variables as surrogates for a direct measure of the energy available to a particular set of species has previously been discussed by Wright (1983, 501).

In several analyses below we compare the results of the studies that reported greatest correlations with an energy-related variable and those that reported greatest correlations with some other variable. Four correlations between species richness and range in temperature ("seasonality" in table 6.1), calculated from J. W. MacArthur's study (1975), stand out in these analyses. Seasonality is strongly intercorrelated with many climatic variables, and so it is unclear whether MacArthur's results actually reflect an effect of temporal environmental variance on species richness, or whether perhaps they reflect covariation with energy. Therefore, we exclude MacArthur's correlations from our comparisons of energy-related versus other factors. Our conclusions remain unchanged regardless of their exclusion or inclusion, either as energy-related or non-energy-related variables.

Which energy-related variable best predicted richness varied according to the system and organisms studied, but in many cases, the best correlate was also an index of primary productivity. In studies having global or continental scope, for example, actual evapotranspiration or a similar measure of the simultaneous availability of water and radiant energy was closely related both to richness (Wright 1987; Owen 1988; Adams and Woodward 1989; Currie 1991) and to primary productivity (Rosenzweig 1968; Lieth 1975). In studies restricted to dry areas, precipitation was often a strong correlate of richness and primary productivity (Brown 1973; Davidson 1977; Abramsky and Rosenzweig 1984), and in wetlands, biomass accumulation—interpreted as net annual primary productivity—was related to richness (Al-Mufti et al. 1977; Moore and Keddy 1989). In freshwater studies, richness and primary productivity were both related to phosphorus supply (Vollenweider 1975; Eloranta 1986). On the other hand, in the case of animals, richness was often better correlated with a measure of simple heat (e.g., temperature, potential evapotranspiration, or solar radiation: Pianka 1967; Turner, Lennon, and Lawrenson 1988; Currie 1991). It is not clear why this is so; one of us has argued that heat may be a better measure of the energy physiologically available to animals than is primary productivity (Currie 1991).

The energy-richness relationship is clearly scale-dependent. On the global scale, richness increases monotonically with energy (Wright 1983, 1990; Adams and Woodward 1989; Currie 1991), whereas on smaller scales, richness is sometimes a peaked function of energy (Al-Mufti et al. 1977; Tilman 1982; Moore and Keddy 1989; Abramsky and Rosenzweig 1984; Owen 1988; see also Rosenzweig and Abramsky, chap. 5; Tilman and Pacala, chap. 2). This may result from different processes controlling richness at different spatial scales. The energy-richness relationship is also known to be dependent on the taxonomic scale of the investigation: correlations of richness with energy are weaker, and their signs more

Table 6.1 Principal Determinants of Species Richness

Factor cited	Correlations with species richness	
	Significant	Not Significant[a]
Energy-related factors		
Mean annual heat and/or humidity variables	29	0
Productivity (sometimes est. as standing crop)	9	0
Nutrient, food availability	3	0
Seasonality	4	0
Other factors		
Habitat complexity, subdivision, microrelief	9	3
Disturbance	5	5(2)
Environmental chemistry (pH, cations)	3	1(1)
Isolation, peninsula effects	2	1(1)
Diversity of food, prey	6	1
Regional richness	2	0
Time, historical factors	1	2(2)
Competition, predation	0	1
Other	3	1

Note: Factors investigated as the principal determinant of species richness in 53 studies investigating 82 correlations with species richness.
[a] Numbers in parentheses indicate the number of correlations that were omitted from further analysis because no correlation coefficient was reported.

variable, at the levels of genus and family than at the levels of order and class (Currie 1991; see also Brown and Gibson 1983, 493 on latitudinal gradients). However, we detected no significant effect of taxonomic scale on the strength of energy-richness correlations in our survey.

Although marine systems may appear to provide exceptions to the monotonic energy-richness patterns observed among terrestrial taxa at the global scale, the evidence is not conclusive. Marine zooplankton seem exceptional in that their richness is maximal at high latitudes. However, marine primary productivity is also maximal in nutrient-rich high-latitude waters (Food and Agriculture Organization of the United Nations 1972). The richness of marine benthic organisms reportedly peaks in waters of intermediate depth (Rex 1981), while energy supply declines approximately exponentially with depth (Pace et al. 1987). This may reflect the same sort of peak of richness at intermediate productivity that is observed on small to moderate scales in terrestrial systems, although it is curious that richness declines monotonically with depth in fresh water (S. Chiviliov and D. J. Currie, unpublished data).

Considering again our literature review, studies investigating energy-related factors typically found them to account for most of the observed variation in species richness (median 70%, $n = 41$, fig. 6.1A). In contrast, studies of other factors typically found them to account for much less (median 23%, $n = 37$, fig. 6.1B; Mann-Whitney $U = 225$, $n = 78$, $p < .00005$).

The strength of the correlations between environmental characteristics and richness depended on several other factors as well. Most strikingly, the proportion of the variance explained increased with quadrat size (Spearman rank correlation $r_S = .51$, $n = 81$ [one study did not report quadrat size], $P < .00005$). Furthermore, this was much more pronounced among the studies of energy-related factors (test for homogeneity of slopes: $P = .035$): the studies of other factors explained richness patterns equally well—or equally poorly—in large and small quadrats (fig. 6.2). Quadrat size was well correlated with the geographical scale of the investigation. These results indicate that measures of energy availability were very good predictors of the large-scale variation in richness, but were much less reliable on the local scale. One reason for this is that the large-scale studies included a greater range of energy values than did the small-scale studies, thus offering greater potential for observing an effect on species richness. Other factors usually accounted for small amounts of variance, regardless of the spatial scale. (If MacArthur's [1975] correlations with seasonality are included with "other factors," this changes somewhat: the fraction of the variance explained by other factors becomes significantly related to quadrat size ($r_S = .36$, $n = 41$, $P = .029$), but the relationship is still weaker than that seen for energy-related factors.)

Surprisingly, studies of animals explained nearly twice as much of the variation in richness (median 69%, $n = 51$) as did those of plants (median 30%, $n = 31$; Mann-Whitney $U = 420$, $n = 82$, $P = .0004$). No other factor was significantly related to the percentage of variance explained.

A. ENERGY-RELATED FACTORS

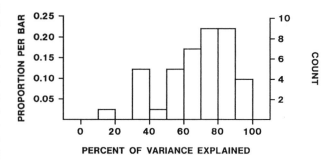

B. OTHER FACTORS

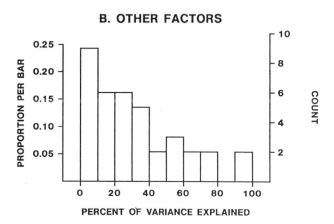

Figure 6.1 (A) Histogram of the percentage of spatial variation in species richness that was statistically explained in studies that investigated energy-related factors (as defined in the text) ($n = 41$). (B) Histogram of the percentage of spatial variation in species richness that was statistically explained in studies that investigated factors other than energy (e.g., habitat complexity, disturbance, history, etc.) ($n = 37$). The median percentage of variance explained by energy-related factors was significantly greater ($P < .00005$.)

There was a marked paucity of concrete evidence in support of many of the commonly cited explanations for variation in species richness. Several widely cited "general hypotheses of species diversity" invoke the influence of competition, predation, or disturbance (Huston 1979; Menge and Sutherland 1987; Menge and Olson 1990). Although there are unquestionable demonstrations that experimental perturbations of these processes affect diversity, we encountered strikingly little evidence that, for example, habitats with greater levels of predation contain more or fewer species (see table 6.1). There was also little evidence (beyond anecdote) that historical factors can statistically account for a significant amount of variation in species richness (see table 6.1; but see Adams and Woodward 1989; Schluter and Ricklefs, chap. 21). This paucity of evidence is undoubtedly due in part to the difficulty of quantifying biotic interactions and historical influences. However, since energy-related factors alone accounted for most of the variation in richness, biotic and historical factors must either be strongly correlated with energy levels, or they must account for little of the variation in species richness.

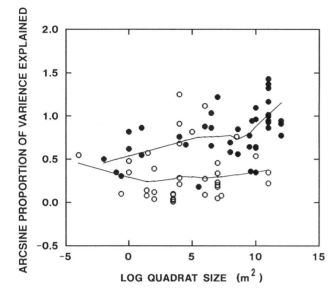

Figure 6.2   The proportion of the variance in species richness explained in 77 correlations of richness with some characteristic of the environment, expressed as a function of the common logarithm of the area of the quadrats used. The proportions were arcsine-transformed in order to normalize their distribution. For reference, the arcsine-transformed values 0.25, 0.5, 0.75, 1.0, and 1.5 correspond to 25, 48, 68, 84, and 99.7% of the variance, respectively. Solid symbols represent studies of energy-related factors; open symbols represent other studies. The upper line shows locally weighted sums of squares fit to the energy-related data; the lower line, the fit to non-energy-related factors. Energy-related factors explain significantly more variance in larger quadrats (Spearman $r_s = 0.58$, $n = 40$, $P < .00005$), but other factors do not (Spearman $r_s = .15$, $n = 37$, $P = .40$).

From this review of the literature, one may conclude that the simplest, most powerful (i.e., predictive) explanation of the spatial variation in species richness is that it depends upon levels of energy in the environment. Several important questions remain: Is there a causal link between energy and richness? If so, how does it operate? Why do energy and richness covary in different manners at different spatial scales?

STATISTICAL MECHANICS OF SPECIES RICHNESS AND AVAILABLE ENERGY

If there is a causal link between energy flow and species richness, its mechanism ultimately operates at the level of individual organisms foraging for sufficient energy to survive and reproduce. This process occurs at the local scale, but variation among species and in the environment can cause large-scale patterns to differ qualitatively as well as quantitatively from local ones.

Here, we develop a model based on these premises. Like all models, it represents a simplification of nature and is thus unrealistic in some ways, but it captures the following important ecological features:

1. Local habitats have many resources, and those resources embody varying amounts of energy production.

2. Local habitats differ in their production of energy associated with the various resources.

3. Species differ in their capabilities of using resources.

4. Species differ in their metabolic requirements—for example, due to differences in body size.

5. The success (abundance) of a species in a habitat is in part stochastic, and in part determined by the production of appropriate resources in the environment, competition with other species, and its own metabolic requirements.

6. There is a regional pool of species providing potential colonist to local habitats that is often much richer in species than any local habitat.

Our model is based on that of Maurer (1990). Species in the model are assumed to be of similar trophic position, not interacting as predators and prey, while the environment is subdivided into habitat patches within a landscape. We begin with a description of how individuals within species obtain energy, and then derive a probability distribution of population size within a patch of habitat. By relating this probability distribution to statistical patterns of incidence (frequency of occurrence or "distribution": Hanski, Kouki, and Halkka, chap. 10) of species within habitat patches and across a landscape of patches, we obtain an average relationship between productivity in the environment and both local and regional species richness. Finally, we show that these species-energy relationships are general features of the interaction of energy, abundance, and incidence, and are not dependent on the details of our underlying model of energy acquisition.

In the empirical survey presented above, we reported little evidence that competition affects species richness, yet our model includes competition. This is not necessarily contradictory, because competition can affect local species' abundances without explaining natural variation in species richness. It may be, for example, that competition shapes communities in roughly similar ways everywhere, but that the sizes of communities are determined by energy flow.

From Energy Acquisition to Abundance

The success with which individual organisms obtain resources is an important determinant of how many individuals occur locally. Suppose that E_j, the total energy production of all resources in patch j, occurs as a relatively large number of discrete packets, such as seeds or small insects. (Our argument would not be substantially different if we treated E_j as continuous.) For simplicity and without loss of generality, we assume that all packets have the same energy content: 1 energy "bit."

Let θ_{ij} be the probability that an individual of species i obtains a particular bit of energy production in patch j. Then the total amount of resource energy obtained by species i in patch j, call it U_{ij}, is the result of E_j trials, each with probability θ_{ij} of success. In other words, U_{ij} has a binomial probability distribution, with parameters $n = E_j$ and $p = \theta_{ij}$ (Maurer 1990).

Denote by r_i the number of energy bits required to support one individual of species i. The number of individuals of species i found on patch j, N_{ij}, is then

(6.1) $$N_{ij} = U_{ij}/r_i,$$

or, more precisely, the integer part of this quantity.

Clearly, the probability of energy acquisition θ_{ij} is a central feature of this model. What determines θ_{ij}? In nature, the chance that an individual of a species obtains a bit of resource production depends strongly on the appropriateness of the species' adaptations and constraints in the context of the local environment. Thus θ_{ij} depends on the match between the resource production characteristics of the patch and the resource utilization capabilities of the species. The presence of competing species also affects the probability of resource acquisition, so θ_{ij} is a conditional probability: it measures the probability that species i will obtain a bit of resource production in patch j given that other species have the opportunity to use the same bit. The probability of energy acquisition varies among species within patches, and varies between patches within species if the environment is heterogeneous.

From Abundance to Incidence and Local Species Richness

The probability of species i being found in patch j, p_{ij}, is the probability that i's energy utilization in the patch meets or exceeds the minimum requirement, so that at least one individual of the species is present:

$$(6.2) \quad \begin{aligned} p_{ij} &= \Pr\{N_{ij} \geq 1\} \\ &= \Pr\{U_{ij} \geq r_i\}, \end{aligned}$$

which, from the binomial distribution of U_{ij}, gives

$$(6.3) \quad p_{ij} = \sum_{u=r_i}^{E_j} \left(\frac{E_j!}{u!\,(E_j - u)!} \right) \theta_{ij}^u \, (1 - \theta_{ij})^{E_j - u}.$$

The expected number of species in patch j (local richness: $\hat{S}_j$) can be calculated as the sum of these local incidence probabilities over all species:

$$(6.4) \quad \hat{S}_j = \sum_{i=1}^{m} p_{ij}$$

where m is the number of species in the species pool (Wright 1985). Wright specified that species should occur independently of one another in order to ensure agreement with the particular form of the equation for the p_{ij}'s that he used; however, independence is not required in general for equation 6.4 to hold.

Equations 6.3 and 6.4 establish a relationship between the quantity of energy produced in the local environment (E_j) and the average number of species in a given patch of habitat ($\hat{S}_j$). For this model, the relationship is a positive one (fig. 6.3). The species' energy acquisition probabilities (θ_{ij}'s) and energetic requirements (r_i's) also enter into the expression in equation 6.3.

Note that local richness depends only on the probabilities of the species being present (equation 6.4), and not on the details of how energy is apportioned among species. Thus the model is modular in the sense that any submodel relating a probability distribution of population size to resources could be substituted for the binomial expression used above. For example, if a submodel were used in which the abundances of some species decreased with increasing productivity (perhaps modified from Tilman 1982), then very different results could be obtained, including hump-shaped or decreasing curves of local richness versus energy.

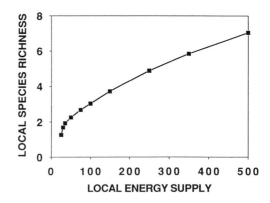

Figure 6.3 Expected local species richness (S_j) as a function of local energy supply (E_j; arbitrary units), for the model given by equations 6.3 and 6.4. The metabolic requirement and energy acquisition parameters (r_i and θ_{ij}, respectively) were chosen to simulate the community of eight seed-eating desert rodent species described by Brown and Zeng (1989). For further details, see appendix 6.2.

From Patterns of Incidence to Regional Species Richness

Regional species richness is the cumulative number of species present in all patches in the landscape. We consider a species to be present in the landscape if it is present in at least one patch. Therefore $p_{i\cdot}$, the probability that species i is present in the landscape, is

$$(6.5) \quad p_{i\cdot} = \Pr\{N_{ij} \geq 1 \quad \text{for some } j\}$$

Assuming patches are independent,

$$(6.6) \quad \begin{aligned} p_{i\cdot} &= 1 - \prod_{j=1}^{k} \Pr\{N_{ij} = 0\} = 1 - \prod_{j=1}^{k} (1 - p_{ij}) \\ &= 1 - \prod_{j=1}^{k} \Pr\{U_{ij} < r_i\}, \end{aligned}$$

where k is the number of patches. Once again, equations 6.5 and 6.6 express the general relation regardless of the details of how the probability distribution of N_{ij} is determined. Making use of the binomial distribution of U_{ij} presented above, we obtain the specific relationship:

$$(6.7) \quad p_{i\cdot} = 1 - \prod_{j=1}^{k} \left(\sum_{u=0}^{r_i-1} \left(\frac{E_j!}{u!\,(E_j - u)!} \right) \theta_{ij}^u \, (1 - \theta_{ij})^{E_j - u} \right).$$

The expected value of regional species richness, $\hat{S}_\cdot$, is then

$$(6.8) \quad \hat{S}_\cdot = \sum_{i=1}^{m} p_{i\cdot}$$

In order to calculate the relationship between regional richness and energy supply given by equations 6.7 and 6.8, we need to know how θ_{ij} varies both between species and between habitat patches. How resource acquisition varies among species depends on differences in their resource use capabilities. A pool of species in which some species exhibit strong dominance in energy use will tend to have a few abundant, ubiquitous species and lower landscape-level richness than a pool among which resources are divided more evenly.

How energy acquisition (θ_{ij}) varies across patches is of particular interest, since it reflects the heterogeneity of the environment. In a perfectly homogeneous environment, resource production characteristics are identical everywhere, so each species has only one value of θ_i in all patches. This means that species that are rare in one particular patch will be rare everywhere. In a heterogeneous environment, species that are rare in one patch may be abundant elsewhere, and vice versa. The net result is that heterogeneous environments support many more species at the regional level (fig. 6.4). This positive effect of environmental heterogeneity on regional species richness can soften or even reverse negative local effects of energy on

richness (fig. 6.5). In other words, there is an interaction between energy supply and heterogeneity, since the effect of energy on species richness is modified by environmental heterogeneity.

Generalization

Although we have made use of a specific model of energy acquisition which resulted in a binomial probability distribution of local population size, our results are generalizable. Local and regional richnesses are both functions of the incidence probabilities of species in patches, i.e., of $\Pr\{N_{ij} \geq 1\}$. This is simply probability theory; the only assumption implicit in equations 6.2, 6.4, 6.5, and 6.8 is that population size is at least partially stochastic.

For any reasonable probability distribution of population size, the likelihood that the species will be present increases as mean abundance increases (Wright 1991). Thus, to the extent that gross energy supply has a positive effect on species abundances, the effect of energy on local species richness will be positive, regardless of the form of the underlying model. The positive interactive effect of landscape heterogeneity and energy on regional richness is also independent of such details (see fig. 6.5).

ENRICHMENT EXPERIMENTS AND COUNTEREXAMPLES

In its more general aspects, the statistical-mechanical model of species richness that we have presented provides a framework for explaining some of the empirical complexities of the relationship between energy supply and species richness, such as differences in the patterns seen at local and at regional scales. Ultimately, however, any theory relating species richness to energy supply will have to address two additional areas that historically have pre-

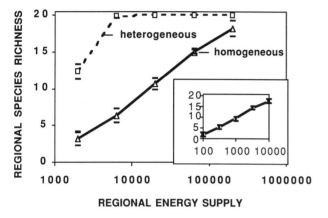

Figure 6.4 Expected regional species richness ($\hat{S}_\bullet$) as a function of regional energy supply (arbitrary units) in a homogeneous and a heterogeneous landscape (from equations 6.7 and 6.8). The inset shows the response of local richness to increasing local energy supply (equations 6.3 and 6.4). In the homogeneous case, regional richness is nearly a replica of the local pattern, but in a heterogeneous landscape, regional richness is greatly enhanced. Consult appendix 6.2 for further details about the figure.

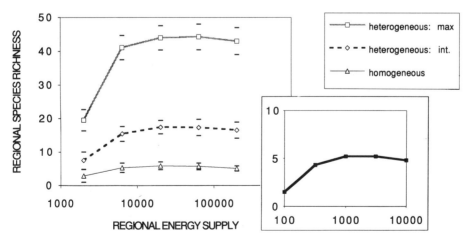

Figure 6.5 Expected regional species richness ($\hat{S}_\bullet$) versus regional energy supply (arbitrary units) when local richness is a peaked function of energy (inset). This example was constructed by assuming that the evenness of local energy use among species was lower under conditions of greater energy supply, as has been observed in experimental situations (Tilman 1982). Regional richness is shown for a homogeneous landscape and for landscapes of intermediate and maximal heterogeneity. In the homogeneous case, regional

richness is similar in form to the hump-shaped local richness pattern. In the maximally heterogeneous landscape, however, regional richness increases strongly with increasing energy supply. Moderately heterogeneous landscapes fall between these two extremes. This shows that regional species-energy relationships can differ qualitatively from those observed at the local scale, tending to be more positive and monotonic. Further details are given in Appendix 6.2.

sented an important challenge to productivity-diversity hypotheses: enrichment experiments and counterexamples.

Enrichment Experiments

Ecologists have not been idle about seeking manipulative tests of productivity-diversity relationships. Many and rigorous experiments have been performed, for the most part observing the effects of the addition of a limiting nutrient (commonly N or P) on the local diversity of terrestrial plants or aquatic organisms (much of this literature is reviewed in Tilman 1982, 1987; see also Carson and Pickett 1990; Goldberg and Miller 1990; Schindler 1990). The results of these experiments are quite consistent: productivity (of the vegetation or of phytoplankton) increases, but species richness declines. This has been termed "the paradox of enrichment" by Rosenzweig (1971), who described a theoretical basis for it in predator-prey systems. Theoretical studies of similar mechanisms in competitive systems have also been presented (Riebesell 1974; Huston 1979; Tilman 1982). Empirically, the reduction of diversity in enrichment experiments is associated with large increases in the density or biomass of a small number of species; other species are driven to extinction, apparently due to the depletion of other resources (including light) or to interference competition.

In terms of niche theory, these results can be interpreted as follows: Additions of a single limiting nutrient greatly increase the availability of energy, but only in a very limited portion of the resource spectrum. Species that happen to be particularly good at using that portion of the spectrum are drastically favored at the expense of the remaining species.

Several questions about the relevance of the results of enrichment experiments remain to be answered. First and least, there are methodological concerns about the measurement of species richness. Some studies report an index of diversity other than richness, such as the Shannon index, H', which is strongly affected by the evenness of abundances. Even counts of species richness may be biased if the experimental units are small and enrichment changes the *size* of individuals, as it often does in studies of plants. A second and more important question is whether enrichment experiments accurately mimic natural variation in energy availability. Most of the experiments are short-term and small in scale, whereas natural variation in energy supply exists on very large scales and has existed throughout evolutionary time. Furthermore, as noted above, experimental manipulations change not only the quantity of resources but also the relative availability of different resources. A third concern has to do with the applicability of the results to terrestrial animals.

Despite these reservations, the paradox of enrichment is real, and enrichment experiments can provide ecologists with useful information about regularities in the way that assemblages of species divide available resources. Clearly, the consistent experimental results indicate that increases in gross energy production do not always increase all species' populations. More generally, some populations increase, others decline, and the change in rich-

ness depends on the nature of the change in energy supply and the characteristics of the species in the species pool.

Counterexamples

Salt marshes and deep sea benthos are two classic examples of communities that have puzzled ecologists by defying intuition about how species richness and energy flow should be related. J. H. Brown (1981), building on earlier work by Terborgh (1973), proposed an explanation for such exceptional cases (see also Brown and Gibson 1983, 518–520). He suggested that some productive habitats, notably salt marshes, contain few species because they have unusual physical conditions compared with their surroundings (e.g., periodic flooding with seawater), which isolates them from potential colonists, and because they are small in area or ephemeral, which elevates their extinction rates. Conversely, habitats that are very large and persistent may support many species even though they are unproductive—for example, the ocean floor. Brown's model generalizes the controlling factors of island biogeography theory—area and distance—by replacing them with energy and isolation, where isolation can mean not only physical separation but also dissimilarity in environmental characteristics.

This hypothesis explicitly requires that energy be measured as an extensive variable: that is, as total energy production in the area in question, not as energy produced per m^2 (see also Wright 1983). In fact, in our model above we treat energy as an extensive variable throughout (energy production per patch, total energy production in the landscape). Most of the studies reviewed in the literature survey used an intensive measure of energy, although intensive and extensive energy measures only differ by a constant proportion if quadrats of equal area are sampled.

CONCLUSION

The evidence from our review of the literature demonstrates that effects of energy on species richness are a common, albeit complex, phenomenon, deserving of deeper and more systematic investigation by ecologists. We suggest, as was elaborated in our model, that a mechanism for species-energy relationships be sought in the way that usable energy is divided among species, and in how the resulting patterns of local abundance affect incidence and richness in the landscape.

Of particular interest is the very general model result that regional species-energy patterns have a stronger tendency to be positive than do local patterns, given a heterogeneous landscape. It is possible for the regional species-energy relationship in a heterogeneous landscape to be positive even when local relationships are hump-shaped or negative, a finding that agrees well with the empirically observed scale-dependence of the relation between species richness and measures of energy. This potential for diverse and conflicting yet mechanistic patterns at different scales may explain why it has been so difficult for ecologists to reach a consensus about the importance of energy in controlling species richness. Enrichment experiments and counterexamples have also complicated the picture, but are not incompatible with our results.

Thus, although there is considerable room for improvement in the predictive power of energetic as well as other explanations for the spatial patterns of species diversity on the face of the earth, we feel that it is time to acknowledge a significant place for energy in the pantheon of factors known to affect species richness, and to proceed to finding out more about how it works.

Appendix 6.1. Studies Included in the Literature Survey

Reference	Organisms studied	Environmental variable (s)
Abramsky and Rosenzweig 1984	Rodents	Precipitation
Adams and Woodward 1989	Trees	NPP[a] (estimated)
Arnold 1972	Snakes	Prey diversity
Askins, Philbrick, and Sugeno 1987	Birds	Range of temperature
Bailey 1988	Mollusks, macrophytes	Exposure
Birks 1980	Insects	Time
Bock and Lepthien 1975	Birds	Temperature
Bond 1983	Birds	Plant biomass
Brown 1987	Butterflies	Peninsular effect
Bruns, Hale, and Minshall 1987	Invertebrates	Food availability
Connell 1978	Corals	Disturbance
Cornell 1985a	Cynipid wasps	Host range
Cornell 1985b	Cynipid wasps	Regional richness
Currie 1991	Vertebrates	PET[b]
Currie and Paquin 1987	Trees	AET[c]
Davis and Wilce 1987	Attached algae	Disturbance
Day et al. 1988	Plants	Standing crop
Eloranta 1986	Phytoplankton	N and P availability
Fernandes and Price 1988	Galling wasps	Elevation
Fischer 1960	Mollusks	Temperature
Hawkins and Lawton 1987	Insect parasitoids	Host diet, host range × sample effort
Heggberget 1987	Birds	Temperature
Jackson and Harvey 1989	Fishes	pH
Johnson 1986	Mammals, ants, reptiles	Isolation, microrelief
Johnson, Mason, and Raven 1968	Plants	Relief
Lynch and Whigham 1984	Birds	Vegetation
MacArthur 1975	Birds	Range of temperature
McAllister et al. 1986	Fishes	Climate (PCA scores)
Milne and Forman 1986	Woody plants	Peninsular effect
Moore and Keddy 1989	Marsh plants	Standing crop
Nilsson 1987	Riparian plants	Current velocity
Nilsson et al. 1989	Plants	Substrate heterogeneity
Nilsson and Keddy 1988	Bryophytes, vascular plants	Total abundance
Nilsson and Nilsson 1978	Water birds	Lake pH
Owen 1988	Mammals	NPP[a] (estimated)
Pianka 1967	Lizards	Temperature
Pianka and Huey 1971	Birds	Foliage height diversity
Quinn and Harrison 1988	Various	Habitat subdivision
Rabinovich and Rapoport 1975	Passerines	Precipitation, temperature, elevation
Recher 1969	Birds	Foliage height diversity
Richerson and Lum 1980	Plants	Climatic variables, relief
Rydin and Borgegård 1991	Plants	Isolation, vegetation height, shore length, number of habitats
Savage and Gazey 1987	Gastropods	Cations
Schall and Pianka 1978	Vertebrates	Climate
Scheibe 1987	Lizards	Temperature, precipitation, competition
Siegfried and Crowe 1983	Birds	Vegetation structural diversity, plant species richness
Silvertown 1985	Trees	Time
Smith and Duke 1987	Mangroves	Climate
Strong, McCoy, and Rey 1977	Arthropods	Time
Turner, Gatehouse, and Corey 1987	Butterflies and moths	Temperature, hours of sunshine
Turner, Lennon, and Lawrenson 1988	Birds	Temperature
Wilson and Keddy 1988	Wetlands plants	Plant biomass
Wright 1983	Plants, birds	AET[b], 1° productivity (est.)

Note: A more complete list of the data and references is available, for a nominal fee, from the Repository of Unpublished Data, Canadian Institute for Scientific and Technical Information, Montreal Road M-55, Ottawa, Ontario K1A 0S2, Canada.
[a]Net primary productivity   [b]Potential evapotranspiration   [c]Actual evapotranspiration

Appendix 6.2.   Additional Details on the Construction of Figures 6.3 through 6.5

Figure 6.3. The minimum energy requirements r_i for the eight seed-eating species in Brown and Zeng (1989) were calculated using Nagy's (1987) allometric equation for rodent metabolic rate as a function of body mass. Energy acquisition probabilities θ_{ij} were calculated as $X_i/\Sigma X_i$, where $X_i = r_i \cdot$ (average population density); an estimator of population energy use by species i.

Figure 6.4. A hypothetical pool of $m = 20$ species occurring in a landscape of $k = 20$ patches is shown. Local energy supply was not varied among patches; in other words, for a particular level of energy supply, E_j was constant. Therefore the regional energy supply $E = \Sigma E_j$ was equal to $k \cdot E_j$. The r_i values were chosen to represent an idealized lognormal distribution of metabolic rates (geometric mean = 20, logarithmic variance (base e) = 0.47, range 8 to 50). A lognormal distribution of metabolic rates follows if body size is lognormally distributed and the allometric relationship of metabolic rate to body size is a power function. The θ_{ij} were taken to be the first 20 values of a geometric series with parameter 0.25. This meant that the dominant species got 25% of the total energy, the next got 25% of what was left, or 18.75%, and so on. Means and standard deviations of species richness for 10 randomizations of θ_{ij} and r_i are shown. In the homogeneous case, the θ_{ij} were not varied across patches, and each replication represented a different random matching of the θ and r values. In the heterogeneous case, the θ_{ij} were randomly matched with the r_i values in every patch. This represented a maximal degree of environmental heterogeneity in the sense that there was no interpatch correlation in a species' success in obtaining energy.

Figure 6.5. As in figure 6.4, there were $k = 20$ patches, and local energy supply E_j was not varied between patches. An idealized lognormal distribution of species' energy requirements (r_i values) with the same mean and variance was used. Unlike figure 6.4, a pool of $m = 50$ species was used, and the geometric distribution of the energy acquisition probabilities θ_{ij} was varied with energy. In order to generate a hump-shaped curve of local richness versus energy supply, the parameter of the geometric series used to determine the θ_{ij}'s was taken to be 0.25, 0.375, 0.5, 0.625, and 0.75 for $E_j = 100, 316, 1000, 3162$, and 10,000, respectively. At the highest energy supply, therefore, the dominant species obtained on average 75% of the total energy production, and the fraction remaining after each subsequent species took its 75% declined very rapidly. As in figure 6.4, in the homogeneous case each species had the same θ_{ij} in every patch, and in the maximally heterogeneous case θ_{ij} values were assigned to species at random in each patch. In the intermediate case the θ_{ij} values were reassigned in each patch in such a way that there was some correlation between patches in the θ_{ij} values for a particular species: the average value of the pairwise interpatch rank correlation in species' θ_{ij} values ranged from about 0.45 at low energy values (where there was less variation in θ_{ij} values) to about 0.31 at high energy values (where the θ_{ij} values were very disparate). Thus in the intermediate heterogeneity case, a species that was good at acquiring energy in one patch was likely (but not certain) to be good at acquiring energy in other patches in the landscape.

PART TWO

Coexistence at the Mesoscale

A major goal of community theory is to discover how interactions between species affect their coexistence and thereby limit the diversity of local ecological communities. For decades this theory has explored the issue of coexistence at the local level. Accordingly, the criterion for a species to exist at equilibrium is that it achieve a local balance between births and deaths. A species that does not satisfy this criterion would be excluded locally, and could not reinvade the community even though it might occur in another community nearby. Coexistence is assumed to be "pointwise" in that it is reached independently at each point in space regardless of external circumstances.

Local equilibrium theory is essential to understanding how species interactions tend to restrict community membership and yield a stable assemblage of species lower in diversity than that of the regional pool. However, it is equally clear that local equilibrium rarely can be achieved. For example, if coexistence were truly pointwise, then isolating the local community from the region within which it is embedded would have no effect on the number of species present. Yet, patterns of diversity on landbridge islands and in continental habitat fragments demonstrate that local diversity cannot be sustained in isolation and, therefore, that inputs from the region strongly influence the diversity of communities.

Chapters in this section explore the theoretical consequences of regional processes for local diversity, as well as the mechanisms by which these processes exert their influence. The models are extensions of local equilibrium theory, but address the broader geographical context of species interactions. As such, they represent a large step toward a more comprehensive theory of species diversity. As Holt explains in chapter 7, the circumstance addressed here is that of the "metacommunity," the cluster of local communities connected by dispersal. Because these models have smaller dimensions than that of the whole biogeographical province, "mesoscale" is a satisfactory term.

Holt classifies into five categories the mechanisms underlying the influence of regional diversity on local diversity. The first of these includes source pool effects, whereby diversity may be high in a locality merely because the regional pool of species is large, yet local populations of species are not sustained by continual immigration of individuals from outside. While unrealistic, this category might be regarded as a "null hypothesis" to be tested when studying mechanisms responsible for observed correlations between local and regional diversity. The second category of mechanism, termed autecological requirements, governs the presence of species that cannot satisfy their resource requirements within a single locality or habitat type. Mechanisms in the third category (source-sink effects) swell local diversity by adding species unable to maintain viable populations without immigration from surrounding productive habitats. A fourth category includes metapopulation dynamics, which emphasizes colonization from other similar localities of habitat patches in which a species has disappeared owing to stochastic fluctuation in local population size. Theoretical study of the last two mechanisms shows that the species richness within local communities may indeed be enhanced by dispersal between localities, even when species at a given site would not otherwise be able to coexist. Finally, Holt outlines the negative effects of habitat selection on local species richness, illuminating the effects of particular patterns of dispersal on community diversity.

McLaughlin and Roughgarden expand on the metapopulation theme. They provide a detailed comparison of theoretical models addressing coexistence within either a patchwork or a continuum of environments connected by dispersal. These analytical models address the diversity issue by looking at interactions and coexistence between a pair of species, usually two competitors or a predator and its prey. These models provide consistent results: moderate dispersal between patches can maintain local and regional diversity when extinctions would otherwise occur in isolated patches. This effect is enhanced when environments differ between patches, because such differences weaken the intensity of the interaction between the species.

Caswell and Cohen construct a probabilistic model for coexistence of a large number species in a metapopulation framework. Identical patches accumulate species following local disturbance events that eradicate species present earlier. They contrast two cases, one in which species do not interact, and a second in which one species eventually outcompetes and drives extinct all others in the patch. They focus on the consequences of this scenario for the species-area curve, and for the correlation between local diversity and the diversity of the regional pool. The results show that species-area curves are generally insensitive to the presence or absence of competition, and that local diversity varies in proportion to regional diversity in either case. Disturbance is an essential ingredient: without it, patches ultimately accumulate all species (with competition absent) or only one species (with competition present).

Mesoscale models provide several mechanisms to explain observed relationships between local and regional diversity. Distinguishing among them presents an important challenge. Part of the solution to this problem will come from a better understanding of the link between the dynamics of species in local communities, their patterns of resource exploitation, and their regional distributions. Prior efforts in this direction have uncovered an interesting pattern, which is that widely distributed species tend to be more abundant locally than species with more restricted ranges. Hanski, Kouki, and Halkka demonstrate this relationship for a variety of taxa and test three possible alternative explanations: (1) sampling error: locally rare species are less likely to be detected, biasing estimates of their distribution; (2) ecological specialization: some species feed on more abundant, widespread resources than other species do, possibly because the former are less specialized; (3) metapopulation dynamics: the range and abundance of a species reinforce each other through an elevated number of dispersers between habitat patches. By formulating predictions from explicit models, Hanski and his co-workers show that empirical abundance-range relationships are necessarily influenced by (1), but support is also found for (3).

Haydon, Radtkey, and Pianka add the time dimension and use computer simulation to explore the effects of both regional and historical processes on species diversity. Their system is a hypothetical archipelago formed through the stepwise fragmentation of a single large island. Between successive fragmentation events extant species disperse, go extinct, speciate, or evolve according to a set of specified probabilities. With this simulation tool they begin to address some fundamental issues arising from empirical studies of archipelagoes: the circumstances under which colonization and extinction events override the effects of geological history, the role of chance, and the potential for adaptive radiation. Their results are complex and yield few firm predictions about nature. Perhaps this indeterminacy and dependence on unique events and geographical situations is the major lesson to be learned.

7

Ecology at the Mesoscale: The Influence of Regional Processes on Local Communities

Robert D. Holt

A local community—to a first approximation, and viewed over a sufficiently long time span—is an ephemeral ensemble of species that originated somewhere else (Davis 1986). The species composition of local communities should thus reflect historical processes, such as speciation, vicariance, and dispersal, operating at very large spatial and temporal scales. Much of this volume is concerned with how community structure expresses the imprint of these biogeographical and evolutionary processes. I will instead examine a more strictly ecological problem: How do spatial processes acting over time scales shorter than that needed for speciation (i.e., $< 1 - 100$ generations) influence the structure of local communities? Because different species experience the spatial dimension of the environment in radically different ways (Wiens 1989b; Robinson et al. 1992), a community will reflect the compound action of many distinct spatial processes. I present a classificatory scheme for regional effects on local communities, in effect using local communities as a lens to examine regional processes. The issues I discuss below lie in a gray zone between the local mechanisms that are the traditional concern of community ecologists and the large-scale processes that are the province of biogeographers and systematists—hence, this chapter is an exercise

in ecology at the mesoscale (Roughgarden, Gaines, and Possingham 1988).

It is useful to begin by returning to a familiar theme from island biogeography: the weaker relationship between sample area and species richness within continents, as compared with the species-area relation among islands. This observation (Preston 1962) was a key stimulus in the development of the equilibrium theory of island biogeography (MacArthur and Wilson 1967), which took as its basic variable the total number of species on entire islands, and as its core processes long-distance colonization, extinction, and in situ speciation.

If one is concerned with the dynamics of local communities, one needs to know the number of species present within a defined area, rather than the total species list for an entire island or continent. Because the drawing of community boundaries is often a bit arbitrary (Underwood 1986), it is even more useful to ascertain how species richness scales with sample area for each island or continent being compared (Hart and Horwitz 1991; Holt 1992). Unfortunately, few island studies have constructed within-island, species–sample area curves across a range of island sizes such as that shown schematically in figure 7.1 (although investigators are converging on this topic

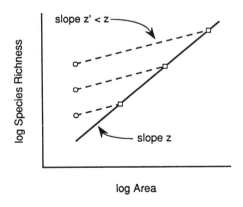

= total fauna of island or continent
= local communities of equal area
— = species-total area relation (among islands)
- - = species-sample area relation (within island)

Figure 7.1 Two kinds of species-area curves. The total species pool $S(A)$ of an island/continent is assumed to depend upon its total area A as described by the power law, which can be transformed to a straight line on a log-log plot: $\log(S) = \log(c) + z\log(A)$. The number of species $S(a,A)$ found within a sample area of size a on an island/continent of size A is also described by a power law: $\log(s) = \log(c')$ $+ z'\log(a)$. The species-sample area curve, for simplicity, is assumed to have the same slope z' across all islands. Given that $z' < z$, local communities increase in species richness with increasing A. For the purpose of illustration, the values of z and z' shown are at the upper limit of values reported in the literature (Connor and McCoy 1979).

with their increasing focus on the relation between local and regional species richness in continental samples: see, e.g., Compton, Lawton, and Rashbrook 1989; Lewinsohn 1991; Cornell and Lawton 1992; Lawton, Lewinsohn, and Compton, chap. 16).

For illustrative purposes (as in fig. 7.1), assume that the species-area relation within as well as among islands fits a power law (Sugihara 1981), extrapolated down to the spatial scale defining local communities. If $S(A)$ is the total number of species on an island/continent of size A, then $S(A) = cA^z$. If $s(a,A)$ is the number of species found in a sample of size a on an island/continent of size A, then $s(a,A) = c'(A)a^{z'(A)}$. In general, the coefficients c' and z' could themselves vary with total area (e.g., larger islands might harbor more habitat specialists). If we assume that these coefficients are independent of island size, Preston's observation implies $z > z'$.

If small islands were passive samples of larger islands or continents (Haila 1983), one would expect that $z = z'$; the two kinds of species-area curves would coincide, and the average number of species in a local community would be independent of regional species richness. If, by contrast, local communities were saturated (sensu Terborgh and Faaborg 1980; Cornell and Lawton 1992), the power law would break down for sample areas corresponding to the spatial scale of direct interactions such as exploitative competition; $s(a,A)$ would converge on a common value s' with decreasing a, independent of A. In the remainder of this chapter I assume, in accord with a growing body of evidence (Ricklefs 1987, 1989b; Cornell and Lawton 1992; Lawton, Lewinsohn, and Compton, chap. 16), that neither local saturation nor passive sampling adequately describe the relationship of within- and between-island (or region) species-area relations.

On each island, we can pick a focal community, defined as those organisms within a prescribed sample of area a, and then ask how local species richness varies with total island size. Noting that $s(A,A) = S(A)$ and manipulating the two species-area relationships leads to $\partial\ln(s)/\partial\ln(A)|_a = z - z' > 0$. The difference in z-values quantifies how increasing the size of the total species pool (correlated with island area) is reflected in an enhancement of local species richness.

A CLASSIFICATION OF REGIONAL EFFECTS ON LOCAL SPECIES RICHNESS

Different species in the same community differ greatly in the spatial scale required for the successful completion of their life cycles and in their ecological and evolutionary responses to spatial heterogeneity (Wiens 1989b). This implies that a number of distinct spatial mechanisms may jointly influence the species composition of a local community embedded in a larger landscape. These include: (1) source pool effects, (2) the spatial dimension of species-specific, autecological requirements, (3) source-sink population structures in heterogeneous environments, (4) habitat selection in heterogeneous environments, and (5) metapopulation dynamics. An important and as yet poorly documented aspect of community ecology is to de-

fine the role of such spatial mechanisms in determining local community structure. Rather than attempting a comprehensive review of the burgeoning literature pertinent to these topics, I will emphasize issues deserving more attention by ecologists.

Source Pool Effects

Brown and Gibson (1983, 444) propose the following thought experiment to study the short-term consequences of regional dynamics for a local community: construct a dispersal-proof fence (or the invisible "force field" of science fiction yarns) around that community, and wait. The null model, of course, is that nothing happens: a local community with a high species richness (e.g., on a large continent, relative to a small island, as in fig. 7.1) sustains its high species richness when cut off from the surrounding landscape matrix.

This null model seems rather implausible as a general rule. Most naturalists would expect some species to disappear rapidly from small patches, and a few to vanish from nearly all isolated patches. In the case of land-bridge islands, nature has constructed the fences in the Brown-Gibson thought experiment for us, and as habitat fragmentation accelerates globally due to anthropogenic habitat destruction, an inadvertent, massive fencing experiment is in effect under way. The data to date suggest that contemporaneous regional processes have substantial effects on local communities. For instance, on landbridge islands in the Gulf of California, the estimated extinction rates of mammals and reptiles are inversely correlated with island size (Case and Cody 1987; Richman, Case, and Schwaner, 1988), and many species seem to disappear rapidly from small habitat fragments (Terborgh 1990; Soulé, Alberts, and Bolger 1992).

Although I am mainly concerned with elucidating the spatial mechanisms underlying such effects, one should always consider the possibility that for some community members, spatial dynamics (coupling the community to the external landscape) is unimportant in explaining persistence, average abundance, and so forth. "Source pool effects" encompass all spatially explicit explanations for the presence of species in a local community *not* dependent on dispersal subsequent to the initial colonization event that "seeded" the local community with those species. As noted above, over sufficiently long time-scales, most current community members will have colonized from elsewhere, and so spatial processes must always be invoked in community assembly (Drake 1990b). Compared with a small isolated region or island, a large contiguous region or continent can generate more species via speciation, accumulate more species from other regions, and provide more avenues for long-distance dispersal. A rate of dispersal that is insignificant for local population dynamics may suffice to supply a local community amply with colonizing propagules. Enhanced local species richness on large islands or continents, as compared with small islands, might reflect in part the effect of total area on total species richness and the biogeographical importance of ecologically trivial dispersal rates. Although source pool effects are unlikely to be the sole explanation for local enhancement, it is entirely possible that for a core subset of the community, dispersal is of historical but not contemporaneous importance in accounting for species' presence and persistence (Williamson 1981).

Spatial Implications of Autecological Requirements

Together with species for which dispersal is dynamically irrelevant, most local communities also contain a few (and sometimes many) species that could not complete a single generation confined to the spatial bounds of those communities. For instance, top vertebrate predators often have enormous home ranges, and it is not surprising that they seem particularly prone to extinction on landbridge islands (Diamond 1984; Belovsky 1987). Yet terrestrial community studies often do not span even a single home range of top predators! Other examples include species with specialized resource requirements. A species may exploit seasonality by shuttling among distinct habitats or by utilizing different habitats at different stages in its life cycle. Such species cannot persist in an isolated local community for even a single generation.

The existence of species with large spatial requirements has multiple consequences for community structure and dynamics. Species that by virtue of their autecological requirements cannot persist in an isolated community will often be a biased subset of the nonisolated community (e.g., species with larger body sizes). A bias in short-term species losses following isolation may have predictable consequences for the residual community. For instance, the disappearance of a top predator that limited the abundance of its prey can produce ramifying shifts in abundance in the remaining community, including local extinctions or outbreaks (Soulé, Alberts, and Bolger 1992; Pimm 1991). Moreover, in a local community embedded in a large region, species with large spatial requirements can couple the dynamics of otherwise spatially separated communities. Thus, if the intensity of predation by mobile avian predators on a local rodent assemblage reflects local predator abundance, understanding the role of predation in structuring the rodent assemblage may require an analysis of broad, regional patterns of prey availability and productivity.

There can be predictable "bottom-up" as well as "top-down" effects: the disappearance of one species may make inevitable the extinction of other, dependent species. Consider, for instance, the phenomenon of sequential dependencies in food webs—schematic descriptions of the feeding relations among all organisms in a well-defined habitat. An enormous amount of work has been devoted to food webs (e.g., Cohen, Briand, and Newman 1990; Pimm 1991), yet surprisingly little attention has been given to the influence of spatial dynamics on local food web structure (Pimm, Lawton, and Cohen 1991; Drake 1990b; Yodzis, chap. 3).

A food web at the very least embodies one-way dependencies among organisms: species at high trophic levels depend on lower species for their continued existence (whether or not there exist strong reciprocal interactions). Extinctions at low trophic levels can drag down species

at higher levels; colonization at high levels must follow successful colonizations at lower levels.

Consider a food chain of "stacked specialists" in which species i occupies level i in the chain. Food chain length should be positively correlated with island area for two distinct reasons. First, trophic rank may predict population attributes that directly influence local persistence. For instance, if high trophic rank is correlated with small population size or large minimum home ranges (as in the "productive space" hypothesis of Schoener [1989]), high-ranking species may be unlikely to persist on small isolates. A second, subtler reason reflects the sequential dependencies per se of food webs: area (and other spatial) effects should be compounded up a food chain.

Diamond (1975) invented *incidence functions* to describe species' distributions on islands. An incidence function $p(i)$ for species i describes how the fraction of islands occupied varies with island area, species richness, or other island attributes. The concept of an incidence function can be broadened to express interdependencies of species in food webs. For a chain of stacked trophic specialists, let $p(1)$ be the incidence function for the basal species. For $i > 1$, the incidence of species i is constrained by the incidence of all lower-ranked species; if any of these are absent, so will be species i. Define the *conditional incidence function*, denoted $p(i|i-1)$, as the probability that species i is present, given that its requisite foodstuff, species $i-1$, is present. The incidence function for species i is a multiple of conditional incidence functions, one for each intermediate link:

$$p(i) = p(i|i-1)p(i-1) = p(1) \prod_{j=2}^{i} p(j|j-1)$$

In general, incidence functions will depend on both trophic rank and autecological factors that influence population persistence. One community-level attribute we might like to predict is the expected food chain length on an island, $E_n(L)$, given that a chain of n species exists in the source pool. The fraction of islands with just i species is $p(i)[1 - p(i+1|i)]$, so

$$E_n(L) = \sum_{i=0}^{n} ip(i)[1 - p(i+1|i)] = \sum_{i=1}^{n} p(i).$$

As a "null model" of food chain assembly, let each species have the same conditional incidence function, $p(i+1|i) = p$. Substituting into the above expression leads to

$$E_n(L) = p\frac{(1-p^n)}{(1-p)},$$

which in the limit of large n (a long food chain) converges on $p/(1-p)$. Gilpin and Diamond (1981) found that the simple form $p = A/(A+q)$ (where A is island area, and q is a fitted constant) described incidence functions for New Guinea birds. Using this for $p(i+1|i)$ leads to $E_n(L) \sim A/q$ at large n. Larger areas should thus sustain longer food chains of stacked specialists. Analogous results emerge from patch dynamic models that explicitly track colonization and extinctions (Holt, unpublished results). The static incidence and dynamic colonization-extinction models both predict that the slope of the species-area relationship should increase with trophic rank.

Little direct evidence is available to test this prediction. Glasser (1982) reanalyzed the classic data of Simberloff and Wilson (1969a) on arthropod colonization on mangrove islets and found some evidence for a temporal succession in trophic structure, with herbivores tending to colonize before their natural enemies. Although he does not remark on the fact, Glasser's figure 7 suggests that natural enemies are disproportionately underrepresented on small islands. Briand and Cohen (1987) compiled data on mean food chain lengths for a large set of food web data. In figure 7.2 the aquatic subset of this compilation is divided into three classes in accord with the size (volume) of the habitat providing the data (pond/stream < lake/river < bay/ocean). There appears to be a trend toward longer food chains in habitats of greater volume. Of course, many factors other than habitat volume that could influence food web structure vary along the pond-ocean axis (e.g., geological age, environmental variability), but the data at face value are consistent with the hypothesis that food chain length reflects the size of the region in which a local community is contained. A reanalysis by Schoener (1989) of this same data set buttresses this conclusion. However, a firmer affirmation of the potential influence of spatial dynamics on food chain length (and more generally on food web structure) must await better food web data (Cohen et al., 1993b).

The above model illustrates how one can start with a given species that, by virtue of its autecological requirements, persists in a local community only because that community is connected to a regional ensemble of communities, and then map out the community consequences entailed by the loss of that species without such spatial coupling. If a species' requirements extend beyond the spatial confines of the local community, then the scale used to define the local community does not adequately characterize even a single population of that species. I now turn to other modes of spatial coupling, in which it is sensible to consider that the local community does contain a species' population.

"Source-Sink" Population Structures in Heterogeneous Environments

Extinction in closed populations may occur in two ways. First, if in a given habitat the death rate always exceeds the birth rate, the population deterministically goes extinct; this is assumed in the source-sink models discussed in this section. Second, with temporal fluctuations in birth and death rates, extinction may result from runs of bad luck (due to demographic or environmental stochasticity), even though the expected birth rate exceeds the expected death rate; this is assumed in many metapopulation models.

A population can persist in a focal community despite a negative expected growth rate if there is regular immigration from other communities. This is the "mass effect" of Shmida and Wilson (1985): a "flow of individuals from areas of high success (core areas) to unfavorable areas" such that "some individuals of a species will become established in sites in which they cannot maintain viable populations." A mass effect is an important limiting case of the "source-sink" population structure considered by Holt (1985) and Pulliam (1988). The characteristic signature of a source-sink population structure is that local population growth rates (birth rate − death rate) are not spatially uniform, but instead are positive in the source and negative in the sink; the demographic equation is balanced in each habitat and in the whole population by dispersal, with net emigration from the source, and net immigration into the sink. In general, the existence of a source-sink population need not imply a mass effect. If a species persists locally because of a mass effect, then (by definition) its population inexorably disappears following isolation. But given negative density dependence in growth rates, the lower density produced by reduced immigration may lead to a compensatory increase of in situ growth rates, so that the sink population equilibrates at a lower density rather than going extinct. In the following remarks, I concentrate on the important special case in which immigration in fact maintains the sink population.

Single-Species Source-Sink Systems. Two general mechanisms that can generate source-sink structures in heterogeneous environments are interference competition and passive (viz., density-independent) dispersal. Pulliam (1988) has argued that intraspecific social interactions can force subordinate individuals to reside in suboptimal habitats, as in the "despotic distribution" of Fretwell 1972), that otherwise would not sustain a permanent population. A model illustrating this source-sink effect assumes that the source is saturated at a population size K; that these individuals continue to reproduce at a per capita rate r_{source}; and that new recruits are forced into a suboptimal sink habitat, with no mortality during dispersal. The rate of emigration from the source and immigration into the sink is $I = Kr_{source}$. In the sink, in the absence of immigration, the population declines at a per capita rate

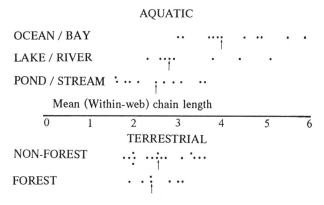

Figure 7.2 Mean food chain length for communities in different habitat types. The data is taken from the compilation of Briand and Cohen (1987). Means rather than maxima are used to give a more conservative assay of area effects. I have grouped the data from aquatic communities in accord with my estimate of the volume of the body of water from which the data is drawn. For completeness, the terrestrial data subset of Briand and Cohen is also shown; it is difficult to determine the area pertinent to these studies.

$r_{sink} < 0$, with no direct density dependence. The dynamics of the sink population, including both immigration and local growth, is described by $dN_{sink}/dt = Kr_{source} + r_{sink}N_{sink}$. The stable equilibrial density of the sink is $N^*_{sink^*} = Kr_{source}/|r_{sink}|$.

This simple model illustrates several general features also found in more complex source-sink systems. First, there are two distinct kinds of density dependence operating: (1) the direct density dependence in the source that determines carrying capacity (K) there, and (2) induced density dependence in the sink. The per capita growth rate in the sink is $I/N + r_{sink}$; the strength of density dependence (defined as the absolute value of d/dN [per capita growth rate]) is I/N^2, which is large at low N. A constant immigration rate in effect induces stabilizing density dependence in local population dynamics, particularly at low densities. Second, the size of the sink population maintained by immigration is directly proportional to source productivity, and to the characteristic time scale of the sink population's exponential decay toward extinction without immigration. A productive source can maintain a substantial sink population if there is a gentle rate of population decline in the sink.

In this model, emigration does not affect the size of the source population. Often, however, emigration lowers recruitment and can potentially depress local population size. Moreover, for many organisms, dispersal is governed by physical transport processes rather than by density-dependent interactions. This generates source-sink population structures if there is spatial heterogeneity in carrying capacity and if dispersal influences local population size (Holt 1985). Passive dispersal generates a net flux of individuals from high- to low-density areas. Emigration, which can lower local density, should characterize high-K habitats, whereas immigration, which raises local density, should predominate in low-K habitats, particularly in sink habitats where species persistence requires immigration.

Consider the following source-sink model with passive dispersal (Holt 1985). Assume that a population grows logistically in a source with intrinsic growth rate $r_{source} > 0$ and carrying capacity K_{source}; that it disperses at rate e between the source and a sink; and that in the sink it experiences a growth rate $r_{sink} > 0$ (and no direct density dependence). These assumptions imply the following equilibrial abundances:

$$N_{sink} = N_{source} \left(\frac{e}{e - r_{sink}}\right),$$
$$N_{source} = K_{source} \left[1 + \frac{r_{sink}}{r_{source}} \left(\frac{e}{e - r_{sink}}\right)\right]$$

With passive dispersal, a large sink population is maintained if the source population has high r and K, particularly if $|r_{sink}|$ is small. Without immigration, the sink population disappears, declining by a factor e^{-1} during a time period $1/|r_{sink}|$. Dispersal depresses N_{source}; low dispersal rates increase N_{sink}, but at high dispersal densities in the two habitats converge, and total as well as local densities

may be depressed. The entire population risks extinction at high dispersal rates if $r_{sink} + r_{source} < 0$.

As with interference competition, passive dispersal may permit a species to occupy habitats it otherwise would not. But dispersal is not a universal enhancer of local species richness, for the simple reason that when emigration depresses local recruitment, passive dispersal may endanger the persistence of the source population. Source populations of small areal extent are particularly at risk because the magnitude of loss due to passive dispersal into unfavorable habitat, relative to the capacity of the source population to replace those losses, scales as the perimeter:area ratio of the source habitat. This deterministic cause of extinction is the basic extinction process assumed in models to predict the minimum critical patch size permitting population persistence (e.g., for planktonic organisms [Okubo 1980] and territorial birds [Lande 1987]).

The best examples to date of source-sink population structures come from plant ecology (e.g., Kadmon and Shmida 1990). A convincing example of a population maintained by flows from a source into a sink has been provided by Keddy (1981), who studied a summer annual, *Cakile edentula*, along a gradient across sand dunes in Nova Scotia. Population density was greatest in the middle of the gradient, but analyses of fecundity and mortality revealed that only at the end of the gradient nearest to the sea were in situ birth rates sufficient to replace deaths; directional seed dispersal (due to both wind and waves) sustained a large population in parts of the gradient where, in the absence of dispersal, local extinction would be predicted. Moreover, emigration seemed to depress density in the source.

Source-Sink Effects and Interspecific Competition. A local community might be a sink for a species because of interspecific competition; immigration from a source can sustain a sink population in the face of competitive exclusion. The resident's competitive edge could reflect either its intrinsic individual superiority in that habitat or local abundance (including priority effects). Abiotic and biotic differences between habitats provide axes for niche differentiation; each of a set of species could be the superior competitor in a particular habitat, which could then be a source sustaining sink populations elsewhere. Theoretical analyses of competition in patchy environments demonstrate that local habitat specialization and priority effects can promote both local and regional diversity (Levin 1974; Yodzis 1978).

This mechanism for enhancing local species richness may break down with large spatial differences in productivity and/or high dispersal rates (Levin 1974). In the source-sink models sketched above, the number of individuals maintained in the sink at equilibrium is directly proportional to the source carrying capacity. Now consider a second competing species, specialized for the sink habitat of species 1, with a carrying capacity there of K_2. Using the usual Lotka-Volterra competition model, this species cannot increase when it is rare and species 1 is at equilibrium if $K_2 < \alpha_{21}N_{sink} = q\alpha_{21}K_{source}$, where α_{21} is the competition coefficient and q is a complicated function of the rate of dispersal and intrinsic growth rates of species 1. Species 2 may be excluded, despite its inherent local superiority, if species 1 has a sufficiently high carrying capacity in its own source habitat to sustain a high abundance in the sink habitat. Moreover, alternative stable states (with and without the resident competitor) may occur (Christensen and Fenchel 1977). The mass effect most effectively enhances local species richness if the habitat heterogeneity that permits each species to be superior in a particular habitat occurs without substantial spatial variance in productivity, and if dispersal rates are low.

Source-Sink Effects and Predator-Prey Stability. Predator-prey systems tend to be dynamically unstable when predators limit prey well below carrying capacity. Elsewhere (Holt 1985), I have analyzed a general model in which a food-limited predator occurs in two habitats between which it passively disperses. The prey population in the source habitat is dynamically responsive to predation, but the prey population in the sink is not; in the sink habitat, the predator is "donor-controlled." In order for the latter habitat to be a predator sink, the resident prey must be sufficiently low in availability or poor in quality that the predator has a negative growth rate. The predator can nonetheless persist in the sink because of coupling to the source, and back-migration to the source can stabilize an otherwise unstable predator-prey interaction.

A similar stabilizing effect of predator dispersal occurs if the prey populations in both habitats are dynamically responsive to predation. Elsewhere (Holt 1984) I have examined a two-habitat model in which the predator-prey interaction in each patch is described by the classic, neutrally stable Lotka-Volterra model. If the two habitats are equivalent (i.e., uniform parameter values), predator dispersal has no effect on stability. However, if the two habitats vary in any way—say in the predator's density-independent mortality or the prey's intrinsic growth rate—predator dispersal is *always* stabilizing. My interpretation of this result is that passive dispersal in a heterogeneous environment generates a source-sink population structure, and that back-migration from the sink dampens predator-prey cycles in the source (Holt 1984; see also St. Amant, cited in Murdoch and Oaten 1975; McLaughlin and Roughgarden, chap. 8). Comparable effects occur in more realistic models that allow limit cycle behavior in the absence of dispersal (Holt, unpublished results).

Prey sinks can also be stabilizing. To illustrate this, assume that a predator is restricted to habitat 1, where the dynamics are described by the classic Lotka-Volterra model, and that prey passively disperse between habitat 1 and a refuge, habitat 2. The habitat-specific growth rates for the prey in the two habitats are r_1 and r_2 respectively; the per capita dispersal rates are e in the predator habitat and e' in the refuge. There are three possible outcomes for this predator-prey interaction (fig. 7.3): (1) If prey growth rates are too low (the region bounded by the hyperbolic line through the origin) the prey alone cannot persist, and so neither can the predator. (2) If prey growth in the refuge is too great (i.e., $r_2 > e'$), the predator cannot regulate the prey population at all. (3) Finally, the system may per-

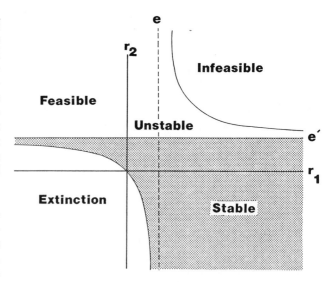

Figure 7.3 Conditions for stability in a predator-prey source-sink model. There are two habitats. The predator (density P) is restricted to habitat 1; prey occupy both this and a refuge (habitat 2). The prey have respective densities of R_1 and R_2, in the two habitats, and diffuse between them at constant per capita rates. The interaction in habitat 1 is described by the classic Lotka-Volterra model, and the prey grows or declines exponentially in habitat 2: $dP/dt = P(aR_1 - C)$; $dR_1/dt = R_1(r_1 - aP) - eR_1 + e'R_2$; $dR_2/dt = R_2r_2 - e'R_2 + eR_1$. The parameters are: a, attack rate; C, predator mortality; r_i, prey intrinsic rate of growth in habitat i; e, rate of dispersal from habitat 1 to 2; e', rate of dispersal from habitat 2 to 1. The three possible outcomes are: (1) extinction of both prey and predator; (2) unstable growth of prey, unregulated by the predator; (3) a locally stable equilibrium (the hatched region), determined by evaluating the eigenvalues of the 3×3 Jacobian matrix of the model at equilibrium. (A technical detail indicated in the figure is that for case (2), an unstable point equilibrium may [denoted "feasible"] or may not [denoted "infeasible"] exist.)

sist at a stable equilibrium. This is likely if the habitat with the predator is intrinsically a source habitat for the prey (i.e., $r_1 > 0$), and the refuge is a sink (i.e., $r_2 < 0$). By contrast, if the refuge is intrinsically a prey source, and the habitat with the predator intrinsically a prey sink, a much more delicate balancing of parameters is required for stability. Similarly, in host-parasitoid systems (which tend to be violently unstable), the stabilizing potential of refugees from parasitism is greatly enhanced if hosts in refuges have low intrinsic growth rates, so that refuge populations are intrinsically sink populations (Holt and Hassell, 1993).

There is thus a broad tendency for predator and prey dispersal that couples sources and sinks in heterogeneous environments to stabilize otherwise unstable predator-prey dynamics. I should stress that the mechanism involved here is quite different from that envisaged in metapopulation models for predator-prey persistence in patchy environments (e.g., Caswell 1978), in which global persistence depends upon a balance between local colonizations and extinctions.

If there are a limited number of refuges available, if prey compete for them, and if excess individuals are forced into the habitat containing the predator, then the prey population has a source-sink structure with strong density dependence in the source, and the predator-prey interaction tends to be stable (Holt 1987b, Sih 1987). Likewise, if a predator population has a source-sink population structure because of intraspecific interference competition in the source (along the lines of Pulliam's (1988) model), this can be strongly stabilizing. In both cases, the predator-prey interaction in the sink habitat is stabilized because of induced density dependence, and the system as a whole is stabilized due to direct density dependence in the source.

As a cautionary note, it should be pointed out that in special circumstances, dispersal in source-sink situations is destabilizing. Consider a predator-prey interaction in a source habitat that without dispersal would stabilize at low prey densities because the predator has a type III functional response (Murdoch and Oaten 1975). Predator emigration tends to increase prey density in the source; density dependence in prey mortality can thereby be weakened or even reversed in sign, reducing the stabilizing influence of the predator's functional response. This indirect destabilizing effect of dispersal can outweigh the stabilizing effect of back-migration from the sink (Holt 1985). In general, dispersal can be destabilizing if either population exhibits local, positive density dependence in growth rates ("diffusive instability," Okubo 1980).

Source-Sink Effects on Prey Communities. Considering only specialist predator-prey pairs, source-sink structures arising from passive dispersal in heterogeneous environments should often enhance local species richness by stabilizing strong interactions. But if predators are generalists, passive dispersal by either predator or prey may in some situations reduce local species richness. Different prey species that do not compete for resources can nonetheless indirectly compete via a numerical response by the predator—an interaction I call "apparent competition" (Holt 1984). In apparent competition, the winning prey species is usually the one that can withstand the highest predator density—and that is usually the prey species with the highest value for r/a (intrinsic growth rate/per capita rate of mortality due to predation) (Holt 1984; Holt, Grover, and Tilman, in press). At low rates of dispersal, habitat heterogeneity permits a multiplicity of prey species to coexist regionally if each is superior at withstanding predation in its own habitat (Holt 1984).

But if predators passively disperse among habitats (or if some predators are forced out of high-quality habitats by intraspecific interference), prey in low-productivity habitats can suffer an increase in predation and even be driven extinct. In contrast, prey in high-productivity habitats may enjoy a relaxation in predation if predators emigrate. At sufficiently high rates of predator dispersal, the single prey species with the highest regional value for r/a tends to displace other prey (Holt 1984; for an example see Settle and Wilson 1990). A mass effect at one trophic level thus tends to reduce local species richness at the trophic level below it. (Oksanen [1990] has recently extended the models of Holt [1985] to three trophic levels and reached broadly similar conclusions).

If each prey species in a region has an exclusive refuge from predation, or if prey productivities and attack rates are homogeneous across space, low rates of prey dispersal tend to increase local prey species richness via a mass effect (Smith 1972; Holt 1987b). But if prey do not have exclusive refuges, prey dispersal indirectly increases predator densities in unproductive habitats, where the prey may be overexploited (even to the point of local extirpation) without the predator endangering its own persistence.

A good example of the indirect effect of the dispersal of one prey species on the limitation of another by predation has been described by Flaherty (1969). In the vineyards of the San Joaquin Valley, the abundance of the Willamette mite, *Eotetranychus willamettei,* was more effectively limited by a predatory mite, *Metaseiulus occidentalis,* on grapevines interspersed with Johnson grass than on grass-free vines. The Johnson grass supported a second prey species, the two-spotted mite, *Tetranychus urticae,* (but not the predatory mite). The two-spotted mite dispersed into the vines in response to a deterioration in grass quality (often associated with overutilization by the mites themselves). The influx of this alternative prey species sustained the predatory mite on the vines at a higher level when the Willamette mite was low in numbers; this in turn permitted the predatory mite to depress the Willamette mite to lower levels than otherwise possible. This example demonstrates how indirect interactions between prey species due to shared predation can be influenced by refuges (e.g., Johnson grass for the two-spotted mite) and spatial flows of prey individuals.

Optimal Habitat Selection in Heterogeneous Environments

Habitat selection has important implications for both population persistence and community structure. Natural selection favors organisms that select habitats so as to maximize their relative fitness (Fretwell 1972). In a spatially variable but temporally constant environment, if individuals move freely among habitats, choose where to settle without interference from conspecifics, and are sensitive to density-dependent effects on fitness (as in the ideal free distribution model of Fretwell 1972), habitat selection tends to equilibrate fitnesses across space (Fretwell 1972; for an example, see Valladares and Lawton 1991). If the total population is in demographic equilibrium, then total births must match total deaths; given an ideal free distribution as well, local births must also match local deaths. This implies that each local population settles to its local carrying capacity. More broadly, each local community should be at an equilibrium structured solely by local processes (Holt 1984, 1987a). Hence, in the absence of temporal variability, optimal habitat selection seems to dilute the effect of regional processes on local communities.

Habitat Selection and Population Persistence. The above conclusion ignores the interplay of temporal and spatial variability. Diamond (1975) has argued that organisms that exhibit habitat selection can track local "hot spots" in resource availability and productivity, buffering populations against extinction. Following a disturbance that greatly reduces population size, intraspecific density dependence should be weak. In this case, habitat selection behavior maximizing individual fitness also maximizes the expected rate of population growth. When densities are higher and local density dependence occurs, optimal habitat selection does not necessarily maximize overall population growth rates (Holt 1987a). Because populations are most vulnerable to extinction at low densities, habitat selection can promote population persistence in variable environments by increasing population growth rates at low densities.

The Effects of Habitat Selection on Predator-Prey Stability. Habitat selection can be an important factor stabilizing predator-prey dynamics. There has been considerable interest for many years in the stabilizing influence of predator aggregation and prey refuges in predator-prey systems. For instance, Comins and Hassell (1979) analyzed a discrete-generation host-parasitoid model in which parasitoids sought out prey patches of high profitability, and showed that this behavior could be strongly stabilizing if the host has a moderate growth rate and exhibits sufficient spatial variance in local density. At the community level, if different prey species occupy different habitats, predator aggregation leads to prey "switching," so that relatively abundant prey are disproportionately represented in the predator's diet; theoretical models suggest that switching can stabilize otherwise unstable prey dynamics (Murdoch and Oaten 1975).

Spatial heterogeneity, alas, is not a universal stabilizer (Hochberg and Lawton 1990). Murdoch and his associates (Murdoch and Stewart-Oaten 1989; Murdoch et al. 1992; but see Godfray and Pacala 1992 for a contrary interpretation) have argued that in some circumstances, predator aggregation in continuous-time models of Lotka-Volterra form, contrary to the conventional wisdom, may be *de*stabilizing. These recent results suggest that habitat selection by predators may not always be stabilizing. Having said this, I think it is nonetheless fair to conclude that in the majority of circumstances, dispersal and habitat selection in heterogeneous environments will prove to have a stabilizing effect on predator-prey interactions.

Fennoscandian ecologists have documented a striking geographical pattern in the cyclic fluctuations of microtine rodents in northern Europe. These multiannual cycles decrease in regularity, amplitude, and interspecific synchrony along a geographical gradient from north to south (Hansson and Henttonen 1988). Hanski, Hansson, and Henttonen (1991) argue that the southern microtine populations are more stable because of aggregation by mobile predators in heterogeneous landscapes. Along the gradient, the landscape shifts from mostly boreal forest in the north to a mosaic of several distinct habitats in the south. In the north, the predators are mainly specialists that are ineffective long-distance dispersers (e.g., least weasels); these predators appear to drive microtine cycles. Further south, the predator community comprises nomadic, specialist bird predators (which tend to concentrate in regions with higher than average prey density) and general-

ist mammalian predators (whose numbers are supported by a number of prey populations distributed across a number of distinct habitats). Korpimaki and Norrdahl (1991a, 1991b) have shown that nomadic avian predators do have pronounced aggregative responses to microtine populations, and that this leads to sufficiently strong density-dependent mortality to dampen population fluctuations. If the suggestion of Hanski, Hansson, and Henttonen (1991) is borne out by further work, it would provide a dramatic example of the effect of a regional process—the maintenance of a pool of mobile predator species expressing habitat selection in a mosaic landscape—on local population dynamics.

If both prey and predator are mobile habitat selectors, the spatial manifestation of their interaction could become quite complicated, and in general must be analyzed as a dynamical game. Because predators should concentrate on patches of high relative prey density, and prey should flee patches of high relative predator density, it is clear that a potential for sustained oscillations exists unless there are other stabilizing forces acting. Schwinning and Rosenzweig (1990) have studied a simulation model for the within-generation spatial dynamics of a top predator feeding on two prey species, one of which also consumes the other, when all three species can move between two habitats (one being a relative refuge). They found that in some circumstances it was impossible for the system to settle into a stable distribution where each species' fitness was equilibrated across space; instead the system displayed sustained oscillations as predators chased prey between the two habitats. Stability was achieved by providing an absolute refuge or by allowing individuals to make "mistakes" in dispersal; these manipulations in effect introduced a modicum of source-sink stabilization into the system.

The Effect of Habitat Selection on Species Coexistence. Habitat selection promotes the regional coexistence of competing species by allowing them to sort out along stable environmental gradients, but it also reduces the number of species found within particular local habitats. Michael Rosenzweig and his associates (e.g., Rosenzweig 1987a; Abramsky et al. 1990; J. S. Brown 1990) have developed a systematic research program aimed at determining the effect of habitat selection on species coexistence. An interesting implication of this work is that if there are distinct habitat types, if potential competitors have distinct habitat preferences, and if there is no cost of habitat selection, then habitat selection at equilibrium can lead to complete species segregation. Two species may coexist regionally, but because each avoids the habitat containing the other, habitat selection reduces local, within-habitat species richness. Moreover, small perturbations in the density of one species do not affect the abundance of the other, so (at least by this measure) there appears to be no competition at equilibrium.

Optimal habitat selection by predators has two distinct effects on prey communities. First, if different prey species occupy different habitats, and if the system is demographically stable, then at equilibrium, optimal habitat selection by predators decouples the predation pressure experienced by different prey (Holt 1984); predator abundance in a given habitat matches the productivity of the prey resident there and is independent of prey productivity in other habitats. If the predator population is regulated by prey availability, and if different prey species are superior at withstanding predation in different habitats, an ensemble of prey species can sort out among habitats and coexist regionally. This result parallels the effect of habitat selection on direct competitors. Second, different prey species in the same habitat may experience apparent competition if predators show an aggregative numerical response to local increases in total prey abundance (Holt and Kotler 1987).

In short, optimal habitat selection by mobile predators promotes the regional coexistence of prey species, but tends to reduce the number of prey species coexisting within particular habitats. These effects are enhanced if the prey directly compete (Comins and Hassell 1987). An elegant, experimentally based example of predator-mediated habitat segregation, in which segregation between gastropods and bivalves in a subtidal community is driven by the aggregative responses of predators (lobsters, octopi, and whelks), has been provided by Schmitt (1987).

Metapopulation Dynamics

To recapitulate, habitat selection enhances population persistence because individuals, by virtue of their own behavior, avoid environments in which they have relatively low fitness. In source-sink population structures, a species persists by a mass effect at one place because elsewhere there exist sites that sustain a persistent population providing a source of immigrants. The final possibility I consider is a species that persists regionally in a metapopulation although it occupies no site permanently. Because other chapters in this volume deal with metapopulations (e.g., Caswell and Cohen, chap. 9) and a number of excellent reviews and books on this subject have recently appeared (e.g., Shorrocks and Swingland 1990; Turner 1989; Taylor 1990; Hassell, Comins, and May 1991; Gilpin and Hanski 1991; Schoener 1991), I give this topic less attention than it deserves.

A "metapopulation" in its most general sense is defined as a system of local populations linked by dispersal (Gilpin and Hanski 1991). Usually the term is used to describe systems in which populations go extinct and are recolonized. In contrast to source-sink systems, in which extinction in the sink occurs deterministically if immigration is prevented, in the empirical systems that motivated much of the work on metapopulation dynamics (e.g., landscapes with shifting mosaics of patches at different stages of succession) there is a strong stochastic component to local extinction and/or colonization.

In recent years there has been a great deal of interest in characterizing species persistence in metapopulations (e.g., Fahrig and Merriam 1985). The essential idea is that transient differences among sites may arise from localized environmental fluctuations or by chance ("phase differences," sensu Levin 1976b), and that species may exploit

these differences by dispersal, forestalling extinction over the entire metapopulation even though every local population potentially goes extinct. Expanding on a scheme proposed by Taylor (1990), we can distinguish several roles dispersal plays in promoting the local and regional persistence of a metapopulation. First, following the extinction of a local population in a given community, dispersal permits recolonization so long as there are other communities with that species elsewhere in the regional ensemble of communities. This is most likely if there are local communities in the region where that species permanently resides (as in the source-sink scenarios sketched above; see Harrison 1991), but it can also occur if there are simply a large number of replicate patches experiencing uncorrelated extinctions (a kind of "spatial storage effect"; Holt 1992). Second, dispersal may mask or prevent local extinction. This has been called the "rescue effect" by Brown and Kodric-Brown (1977) in the context of classic island biogeography, and an "internal rescue effect" when applied to local sites in a metapopulation (Hanski 1982; Gotelli 1992; Holt 1992).

Dispersal may influence the frequency distribution of local abundances over time (Vance 1980) and thereby alter the probability of local extinction. To examine this effect, K. Parker and I have carried out numerical studies of populations distributed over an archipelago of patches, in each of which there is logistic-like density dependence and random variation in density-independent growth rates. We shall report this work elsewhere (Parker and Holt, unpublished results) and here simply summarize some pertinent findings (fig. 7.4). In our model, following local population growth, a fraction of each population either enters a dispersal pool, which is redistributed among all patches, or disperses to neighboring patches in a cellular lattice. Even low dispersal rates can substantially reduce the overall magnitude of fluctuations in abundance and the frequency of excursions to critically low population levels (fig. 7.4).

Intuitively, migration in a metapopulation tends to moderate local population fluctuations for two distinct reasons. When a local population at high density has an unusually high growth rate (compared with the average over the metapopulation), realized population growth is reduced because more individuals leave the patch than enter it. Conversely (and more importantly), when a local population is perturbed to low densities, more individuals immigrate than leave, thus increasing the rate at which the population rebounds. In the pool model, the stabilizing effect of immigration at low densities becomes more pronounced with an increase in the number of patches (though it exists even for coupled pairs), and for a given single patch coupling to a large metapopulation, is similar to that in a single-patch model with a constant rate of immigration. The reason is that with many patches, the effect of any single patch on the dispersal pool becomes negligible; because one is averaging over numerous patches, the rate of immigration into a single patch for all practical purposes becomes a constant decoupled from local dynamics.

A number of the authors who have investigated meta-

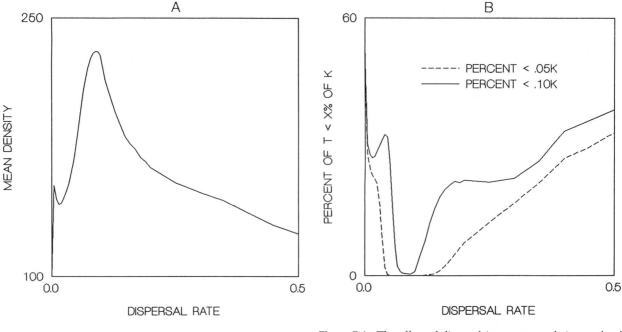

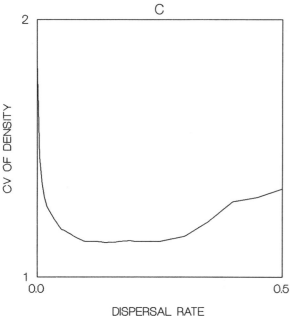

Figure 7.4 The effect of dispersal in a metapopulation on local abundance and variability. Modified from Parker and Holt, unpublished results. In a metapopulation of n patches, local density in patch i before dispersal is governed by $N_i'(t + 1) = N_i(t)\exp[r_i(1 - N_i/K_i) + u_i]$, where $\exp[r_i]$, K_i, and u_i are respectively the finite rate of increase at low densities, the local carrying capacity, and a normally distributed random variable with a mean of zero and a standard deviation m; the last term describes temporal fluctuations in density-independent growth factors. In a given generation, the realized intrinsic growth rate is $r_i + u_i$, and the realized carrying capacity is $(1 + u_i/r_i)K_i$. If the realized intrinsic growth rate is sufficiently large, the local population can overshoot K. In an isolated patch, the long-term average population density equals K_i. However, population fluctuations may drive the population to dangerously low densities. The figure depicts the outcome of allowing a fraction e of individuals following local growth to enter a dispersal "pool," which is then equitably distributed among all patches; local populations are censused after dispersal. In the example shown there are 30 patches, all with the same $r_i = 3.8$, $K_i = 100$, and $m = .5$ (uncorrelated among patches), and dispersal rates vary from 0 to 0.5. (*A*) Effect of dispersal on mean local abundance in a randomly chosen "focal" population. (*B*) Effect of dispersal on the percentage of generations in which a local population occurs at less than 5% or 10% of its carrying capacity. (*C*) Effect of dispersal on the coefficient of variation in local abundance in a "focal" population. At very low dispersal rates, the patterns are complex. The qualitative finding is that even low rates of dispersal can substantially moderate population variability and the frequency of excursions to low levels; higher rates of dispersal weaken these effects by synchronizing patches.

population dynamics (cited above) have explored its implications for interspecific interactions. Analyses of single predator–single prey interactions (e.g., Caswell 1978; Crowley 1981; Sabelis and Diekmann 1988; Reeve 1988) suggest that colonization-extinction dynamics can permit the regional persistence of strong predator-prey interactions that are locally unstable. Competing species may coexist regionally, both because colonization-extinction dynamics open up additional axes for niche diversification (Pickett 1976), and because patchiness tends to augment intraspecific density dependence, making competitive co-

existence easier (e.g., Hanski 1983; Nee and May 1992). Caswell (1978), Hanski (1981), and Hastings (1978) have analyzed patch dynamic systems in which a predator attacks two competing prey species, and argue that all three interacting species could coexist for a long time regionally, even if local extinction is inevitable.

The overall impression one might draw from these studies is that metapopulation dynamics could be a significant factor enhancing local diversity on large islands or continents. Although I suspect this is true, it is worth emphasizing that metapopulation dynamics also opens up additional mechanisms for species exclusion. For instance, one species may outcompete another because it has a dispersal strategy that gives it a head start in seizing newly available patches, rather than because of any advantage in head-to-head confrontations. In like manner, two prey species that could never directly interact because they occupy distinct patch types may nonetheless be locked in long-distance apparent competition if they support a regional pool of predators that can invade either patch type.

This raises a more general point. Because local species richness in practice tends to increase with island area, or more generally, the size of the regional species pool (see fig. 7.1), it is a natural temptation to concentrate on the enriching effect regional processes have on local communities. A consideration of the mechanistic bases for regional effects leads to a more complex view of the world. For instance, source-sink dynamics may permit a species to occur in a wider range of communities, but may also make that species more vulnerable to regional extinction, or alternatively, more able to exclude competitors from local habitats where it is an inferior competitor. Indeed, any mechanism involving dispersal which sometimes increases local species richness can, in other circumstances, have just the opposite effect. The search for regional mechanisms of persistence and coexistence should be balanced with a search for regional mechanisms of extinction and exclusion.

CODA

Let us return to our thought experiment, in which a local community has been freshly isolated from its surrounding landscape, and now try to characterize the net community response. The simplest description of a local community is a list of its members, so a useful profile of the community's response to isolation is given by the distribution among community members of expected time to extinction and of the variance in time to extinction. The regional mechanisms sketched above describe a range of first-order responses to the breakup of spatial coupling. For some species—those that are in the community because of source pool effects but are otherwise dynamically decoupled from the external landscape—there may be very long expected times to extinction. For others—which due to their autecological requirements straddle this and other local communities each generation—extinction will be swift and inevitable. For yet others—present in the nonisolated community as sink populations maintained by immigration—there will be a predictable time to extinction with low variance. Habitat selectors in a temporally varying environment will be present only when environmental conditions are appropriate, and so will go extinct on time scales driven by temporal variability. Species present because of metapopulation dynamics coupling the local community to many like communities may go extinct, but with high variance in time to extinction. On top of these first-order extinctions, second-order extinctions may occur due to shifts in the patterns and strength of interspecific interactions. The overall profile

of times to extinction in the newly isolated community is an assay of the importance of regional processes in determining the structure of the original nonisolated community.

A deep understanding of the local consequences of regional processes will require a melding of experimental, theoretical, and comparative techniques. I believe that an important item on the agenda for community ecology will be to grapple with the messy reality that local communities contain species that experience the world at vastly different spatial scales. The structure of a community will surely reflect the interplay of disparate regional processes. For instance, Schoener and Spiller (1987) suggest that in spider communities on small Bahamian islands, some species are highly persistent without immigration, whereas others persist only because of immigration. One could easily imagine that each group of species has a substantial impact upon the other; the abundance of the persistent species, whose presence is explained via source pool effects, might be strongly influenced (via competition or predation) by the collective flux of nonpersistent species.

Regional processes have an important methodological implication for community ecology: they make the detection of interspecific interactions in local communities by manipulative experiments more difficult, and indeed, cast doubt on the utility of detailed analyses of population dynamics in single local patches. Cooper, Walde, and Peckarsky (1990) reviewed the literature on predation effects in freshwater habitats and concluded that "the magnitude of prey exchange (=immigration/emigration) among substrate patches has an overwhelming influence on the perceived effects of predators on prey populations." I suspect that this conclusion applies quite generally to any interaction if the dynamics of any of the interactants is not circumscribed by the bounds of one's study. Moreover, regional processes can modify the qualitative character of interspecific interactions. For instance, Danielson (1991) has shown that two species may compete in each of an array of habitats, yet the overall interaction may be mutualistic when the interactions are averaged over space. Given the recent interest in indirect interactions in communities, it is also important to recognize that indirect interactions are often propagated by dispersal through space (e.g., prey species segregated into different habitats may experience strong apparent competition due to mobile predators; Holt 1984).

And finally, to the extent that regional ecological processes enhance local species richness, we might expect some species to show rather coarse, imprecise adaptation to many of the local environments they occupy (Futuyma 1986). If dispersal is important in facilitating the persistence of a species over ecological time in a local community, then that same dispersal iterated over evolutionary time scales could lead to a kind of adaptive averaging over space and communities, which in turn implies a degree of seeming maladaption in some local communities (relative to highly persistent resident species). This is particularly likely in species with persistent source and sink population; adaptive evolution is biased toward further adaptation to the source environment, and relatively impotent at improving adaptation to the sink (Holt and Gaines 1992). These observations suggest that evolutionary ecologists should begin to place adaptive analyses of traits into the context of the classic gene flow–selection problem of evolutionary genetics (Antonovics 1968; Slatkin 1987; Pease, Lande, and Bull 1989). Given that we need to develop an understanding of ecology at the mesoscale to understand population dynamics and community structure in local communities, our quest for an evolutionary understanding of phenotypic evolution and adaptation will ultimately need to be cast at the mesoscale, too.

8

Species Interactions in Space

John F. McLaughlin and Jonathan Roughgarden

The earth varies dramatically in space, and ecologists have long agreed that spatial heterogeneity is an important factor in the maintenance of community diversity. Yet most theories of species interactions either assume a constant environment or ignore space entirely. Many ecological models that do include spatial heterogeneity in some form, however, exhibit fascinating behavior and make interesting predictions about diversity. This chapter reviews some of these models and identifies common themes in their results that reflect the role of heterogeneity in maintaining diversity and shaping species distributions.

Theoretical descriptions of space often seem unrealistically simple. Many treat space as a straight line, a flat plane, or a collection of identical patches. Most assume that spatial and temporal variation in environmental conditions are distinct and independent. Many assume that dispersing individuals either move randomly or reach all parts of a patchy environment. The value of this work is appreciated when it is viewed as an initial step toward an understanding of the full scope of environmental variation.

Including space in models of species interactions complicates their analysis. Adding spatial heterogeneity in growth rates, interaction terms, and dispersal rates renders the study of interactions even more difficult. Consequently, researchers have begun by analyzing the effects of space and heterogeneity on models of very simple interactions. Then the behavior of more realistic models can be studied. Most spatial theory currently is limited to simple models; accordingly, most models discussed in this chapter describe spatial interactions among just two species. These models divide larger communities into pairs of interacting species, and they address issues of diversity by studying the properties of dispersal and heterogeneity that lead to pairwise coexistence.

Spatial models of species interactions treat space either as discrete patches or as a continuum (Levin 1978). Interactions in patchy environments are assumed to take place within discrete habitat fragments whose dynamics are coupled by interpatch dispersal. Patches may be identical or heterogeneous. Continuous environments permit movement and interaction throughout space, over which conditions may be constant or heterogeneous. Local and regional scales have different meanings in patchy and continuous environments. "Local" refers to events that occur

within a given patch or, in continuous models, within a small neighborhood of a particular location. "Regional" processes encompass an entire collection of patches or the full extent of continuous space.

The remainder of this chapter has six sections. The first four alternately discuss competitive and predator-prey interactions in patchy and in continuous environments. The fifth section analyzes the spatial behavior of interactions between species that respond to different scales of heterogeneity. Finally, the sixth section summaries the chapter's discussion of space, heterogeneity, and diversity.

COMPETITION IN PATCHY ENVIRONMENTS

Many environments consist of discrete habitat patches, and there is a large theoretical literature devoted to the effects of patchiness on the diversity and dynamics of competing species (reviewed in Levin 1976a, 1978; Yodzis 1978). The basic spatial unit in this theory is the patch, within which reproduction, population regulation, interactions, and the movements of most individuals take place. Environmental conditions may vary from patch to patch, but in order to simplify the analysis, most models assume that all patches are identical. Regional environments consist of many patches connected by dispersal, which may be from a given patch to adjacent patches only, or to all other patches in the system.

Competition theory has identified two ways that environmental patchiness can promote species coexistence. First, a system of patches may contain permanent spatial variation that favors different species in different patches. Species diversity in this case depends on the amount of heterogeneity in the environment. Second, a species residing in a patch may exclude potential invaders, so that the occupancy of a given patch depends on its colonization history (Yodzis 1978). By this mechanism, diversity can be maintained in an environment of identical patches if each species colonizes empty patches sufficiently early and if dispersal between patches is low enough to preserve the advantage of residents.

Most theoretical descriptions of competition in patchy environments have taken one of three different approaches to the treatment of population dynamics at local and regional scales. The three kinds of models focus on events at local, regional, and metapopulation levels. Local models describe in detail species abundances in a small number of patches, whose dynamics are coupled through interpatch dispersal. Regional models ignore intrapatch dynamics, and instead focus on the number of patches oc-

This chapter is Publication No. 3732 of the Environmental Sciences Division, ORNL.

cupied by each species. Regional models track changes in patch occupancy due to immigration and extinction events and determine when both competitors can persist in the system even though they cannot coexist within a given patch. Metapopulation models combine features of local and regional models by describing both intrapatch dynamics and interpatch dispersal in systems containing many patches. Results from local and regional models concur that low to intermediate rates of dispersal promote coexistence. Metapopulation models, being more intricate, lead to more complex conditions for regional coexistence.

Local Coexistence

Early work with models of competition in an environment consisting of a small number of patches relates local diversity to interpatch dispersal rates (Karlin and McGregor 1972; Levin 1974). These models assume that each patch favors a different species, so that in a two patch–two species system, each species occupies one patch and is excluded from the other by its competitor. When the patches are weakly coupled by low rates of interpatch dispersal, species coexist locally. In this case, each species persists where it is competitively inferior due to migration from its favored patch. Rapid dispersal destabilizes the community by enabling one species to exclude the other from both patches. In these models, rapid dispersal collapses a subdivided environment into effectively one low-diversity patch, but low to moderate dispersal rates preserve patch distinctiveness and maintain high local diversity.

Regional Coexistence in Multipatch Models

Regional models ignore the dynamics of local interactions to concentrate on patterns of extinction and colonization. They identify four patch categories: "empty" areas where neither species exists, patches occupied by one or the other species only, and patches occupied by both competitors. Most regional models implicitly assume that local competitive exclusion occurs rapidly compared with the rate of migration between patches within the region. They generally assume that coexistence within a particular patch is impossible and determine whether both species can persist by colonizing new patches before going extinct in their current locations.

Levins and Culver (1971) and Slatkin (1974) analyzed models of two species competing in an environment consisting of many identical patches. They showed that, with some constraints on rates of local competitive exclusion, regional coexistence is possible despite inevitable local extinction. They also found migration to be a strong promoter of coexistence, particularly among rare species. Horn and MacArthur (1972) modified the environment of the above models to include two kinds of patches. Although the ensuing constraints on rates of migration and local extinction were more complex, they obtained similar results.

Armstrong (1976) studied a variation of the model of Levins and Culver (1971) in which one species is locally dominant, the other species is a better disperser, and local coexistence is precluded. He determined the rates of migration and local extinction that lead to regional coexis-

tence, and he calculated the sensitivity of coexistence to changes in these parameters. Armstrong's model predicts coexistence unless the environment changes by a magnitude that exceeds the dispersal ability of the locally subdominant species relative to that of its dominant competitor.

Hanski (1983) generalized the regional models of Levins and Culver (1971) and Slatkin (1974) by relaxing the assumption that local and regional dynamics occur on different time scales. He found that large differences between local (fast) and regional (slow) scales favor coexistence. When the two scales are similar, regional coexistence or competitive exclusion depends on initial species abundances and distributions (Hanski 1983).

Caswell and Cohen (chap. 9) further discuss local and regional diversity, and they show how competition shapes community composition differently than does disturbance.

Competition between Metapopulations

Several researches have studied local and regional dynamics simultaneously. Like regional models, metapopulation models analyze a risk-spreading hypothesis, which suggests that species evade extinction due to competition or demographic disasters by dispersing among a system of habitat patches. These models approach the difficulties inherent in considering population dynamics on both local and regional scales by restricting growth and dispersal to different life stages. Individuals of one stage remain within the patch where they hatch or settle, and those of the other stage disperse among patches. Consequently, local dynamics are confined to the sedentary stage and regional dynamics unfold in the dispersing stage.

Ives and May (1985) studied a model of two species in an environment of ephemeral patches, using flies that compete for fruit, carrion, or dung as a biological reference. They assumed that larvae stay where they hatch and that adults disperse to other patches to lay eggs. They concluded that intraspecific aggregation of eggs leads to coexistence that is largely independent of the mechanistic details of competition: when one species tends to lay most of its eggs in a few patches, its competitor can persist by using other patches. Intraspecific aggregation also enhances the coexistence of two or many species by a similar mechanism in the models of Shorrocks, Atkinson, and Charlesworth (1979). Ives and May also found that habitat fragmentation promotes coexistence: both species persist more easily when a given amount of resource is divided among many small patches than when it is contained in fewer, larger ones. Although patchiness enhances diversity in this case, it also reduces population abundances (Ives and May 1985).

Chesson (1985) analyzed competition between metapopulations in stochastic environments. He distinguished three kinds of environmental variation: *spatial variation*, in which conditions differ permanently among patches; *temporal variation*, in which all patches change with time in the same way; and *pure spatio-temporal variation*, or spatial variation among patches that is statistically independent of temporal fluctuations. Chesson added the same three kinds of variation in birth and death rates to a model of lottery competition within many patches. All three kinds of variation lead to coexistence when competitive exclusion would have resulted otherwise, but their effects depend on the life histories of the competing species. Each kind of variation enhances coexistence similarly in species with overlapping generations. For short-lived organisms, however, temporal variation promotes coexistence less than does spatial or spatio-temporal variation.

The metapopulation model of Iwasa and Roughgarden (1986) resembles that of Chesson (1985), but it contains many species competing in an environment that varies only in space. Like Chesson's model, theirs is motivated by marine systems with sedentary adults and widely dispersing larvae. They found that many species can coexist regionally if there are at least as many different kinds of habitat patches as there are species. A second result of this model is that for any pair of coexisting species in the multispecies community, each member of the pair must have at least one larval source (a patch where the species has a greater net larval productivity than its competitor does). Species interactions and diversity in regions that contain source and sink habitats are discussed further by Holt (chap. 7).

PREDATION IN PATCHY ENVIRONMENTS

The tendency of predator-prey and parasitoid-host interactions to cause local extinctions and large fluctuations in species abundances has motivated a broad search for mechanisms that account for their persistence in nature. One well-studied hypothesis is that predators and prey or parasitoids and hosts can persist regionally by dispersing among discrete local populations, even if their interaction within a given patch is unstable (see reviews by Holt 1984, Appendix 2; Levin 1976a; McMurtrie 1978; and Taylor 1988, 1990). Models that address this hypothesis, like those of competition in patchy environments, can be classified according to their representation of population dynamics at local and regional scales. Models with explicit within-patch dynamics determine the potential of dispersal to dampen the oscillations in population densities that occur within isolated patches. Regional models assume that patches with the same complement of species undergo similar dynamics, and concentrate on the numbers of patches in each state of occupancy. Most regional models also assume that local extinction inevitably follows predator or parasitoid immigration to patches containing prey or hosts. Both kinds of models predict similar criteria for regional persistence, differing primarily in the rates of dispersal that cause stability.

Models with Internal Patch Dynamics

Chewning's (1975) model of predator-prey interactions in an environment consisting of many patches includes within-patch dynamics of arbitrary complexity. Regional stability in Chewning's model requires either local stability or spatial heterogeneity. In an environment of identical patches, dispersal eliminates asynchrony among population oscillations in different patches, so the system becomes no more stable than an individual patch. Dispersal among spatially heterogeneous patches can have a stabilizing influence, as Chewning illustrates in an example of two patches with internal Lotka-Volterra dynamics.

Holt (1984) also showed that dispersal among heterogeneous patches can stabilize predator-prey interactions by coupling the dynamics of populations in source and in sink habitats. Holt analyzed a two-patch Lotka-Volterra model with randomly dispersing predators and sedentary prey, and found that predator-prey oscillations in both patches were dampened by interpatch variation in any parameter. This stabilization (which is discussed more generally in Holt 1985) is due to net dispersal of predators from the more productive patch to the less productive habitat: from the source to the sink. Holt (1984) also showed that low to moderate predator dispersal can promote the coexistence of many prey species in a system of heterogeneous patches. If predator dispersal is slow enough to prevent homogenization of predator densities among patches, predation can act as multiple and partially independent limiting factors to multiple prey species.

Nachmann (1987) correlated stability with system size and dispersal rates in a stochastic simulation of host plants with herbivorous and predatory mites. Limited ratios of long- to short-range dispersal and larger numbers of plants enhanced regional stability in the system of locally unstable interactions. Greater mite mobility tended to synchronize local predator-prey oscillations, leading to extinctions.

Crowley (1981) obtained similar results with a model of identical patches whose internal dynamics contain a stable limit cycle. Three factors stabilize the model by preventing interpatch synchronization: low migration rates, sufficient temporal variation in the environment, and a large number of patches. Crowley also concluded from numerical results that if a system of contiguous "patches" is large enough, it divides into two more subsystems. Dispersal between subsystems is sufficiently low that they remain asynchronous and hence preserve overall system stability.

Parasitoid-host interactions also have motivated a number of models that study the effects of dispersal among patches on system stability (reviewed in Reeve 1990). Because many parasitoids and hosts have discrete generations, most parasitoid-host models employ difference equations rather than the differential equations typical of predator-prey models. Despite this and other technical distinctions between the two classes of models, both tend to concur that low to moderate dispersal and maintenance of asynchrony are necessary for the regional persistence of a collection of locally unstable interactions. Hassell and May (1988) analyzed a model in which parasitoid-host interactions in each patch exhibit unstable Nicholson-Bailey dynamics of increasing oscillations. Regional stability in the model depends on the dispersal patterns of parasitoids and hosts. When neither species disperses, or when both species disperse to all patches, the model's regional behavior mimics that of an individual patch. When only one species disperses, regional stability depends on the establishment of host refuges. If parasitoids are the dispersing species, stabilizing refuges result if they fail to reach all patches during each generation. When only hosts disperse, both species persist in a system of high- and low-density patches if the host growth rate is below a threshold value. In this case, the parasitoids survive in patches with high host density and perish in low-density patches. Thus, locally unstable interactions in the model of Hassell and May persist regionally when one species does not disperse and host refuges desynchronize patch dynamics.

Reeve (1988) compared the effects of migration, environmental variability, and density-dependent coupling of parasitoid and host populations, and found that regional persistence requires all three factors. A combination of low migration rates and spatial and temporal variability maintains the asynchrony among different patches that is necessary for persistence. Reeve's results concur with those discussed above that high migration rates reduce regional persistence.

Dynamics of Patch Occupancy

Models of patch occupancy, or regional predator-prey models, assume that prey inevitably become extinct in patches invaded by predators. Analysis of these models focuses on determining the rates of predator and prey dispersal that lead to regional persistence of both species. Most models of this kind also assume that patches differ only in the species they contain at a particular time; conditions within each patch are identical and the "dynamics" of all patches in a given state of occupancy are the same.

Most regional models conclude that both predators and prey persist if their dispersal rates are sufficiently great. Hastings' (1977) patch occupancy model behaves in three different ways, depending on the dispersal rates of predators and prey. Both species persist in a stable fraction of patches when each disperses rapidly. Similarly, Hilborn (1975) concluded from a numerical model based on Huffaker's (1958) laboratory system that high dispersal ability in both species leads to persistence. Hilborn also found that persistence is enhanced by a large number of patches, with persistence in smaller systems requiring high prey dispersal ability and moderate predator dispersal. High prey mobility and moderate predator mobility cause persistence in Zeigler's (1977) model of patches with pseudo-internal dynamics, in which prey do not disperse from a patch until their abundance in that patch reaches its carrying capacity. Sabelis and Diekmann (1988) simplified Hastings' (1977) model and differentiated sources of dispersing prey among patches with prey only and patches containing both species. They concluded that prey dispersal occurring between predator invasion and local prey extinction is stabilizing: it leads to predators and prey occupying constant numbers of patches when otherwise regional extinction or neutrally stable oscillations would result.

Persistence in Vandermeer's (1973) model requires that predators disperse slowly relative to prey. This divergent conclusion is probably related to Vandermeer's assumption that predators survive in patches without prey (perhaps by consuming alternate prey), an assumption that leads to other anomalous results (Hastings 1977).

Diversity, Asynchrony, and Dispersal Rates

Models that focus on the regional dynamics of patch occupancy share with those containing intrapatch dynamics a common stabilizing mechanism: dispersal among asynchronous patches. The two kinds of models differ, however, in the rates of dispersal that lead to coexistence, with species persisting at greater dispersal rates in regional models than in those with intrapatch dynamics. This may be an artifact on the different forms of asynchrony in the two kinds of models: asynchrony in regional models refers to different states of patch occupancy, but asynchrony in models with explicit local dynamics is due to phase differences among local predator-prey cycles. Persistence in regional models depends on predators and prey dispersing rapidly enough to balance local extinctions. Prey must find unoccupied patches and predators must find new prey patches before each becomes extinct in patches where they co-occur. In models with explicit local dynamics, however, dispersal must not be great enough to synchronize local cycles. Hence, asynchrony maintains diversity in both kinds of models, but its form and the dispersal rates that produce it differ between regional and intrapatch dynamic models.

COMPETITION IN CONTINUOUS ENVIRONMENTS

Models of competition between species in a patchy environment, as discussed above, show that dispersal among discrete habitat fragments can maintain regional diversity. Models of competition in continuous environments also study the relationship between dispersal and diversity, but they obtain coexistence by a different mechanism. Coexistence in these models is usually due to species using space in different ways. In many models, competitors divide the habitat along sharp boundaries, leading to a contrast in local versus regional diversities analogous to that of patchy models. Most continuous-environment models that predict coexistence without spatial segregation of competitors contain some other means whereby species utilize space nonuniformly. Both patchy and continuous environment models predict coexistence that is impossible in nonspatial models, but models with continuous environments possess the additional ability to describe spatial patterns in species densities.

The theory of spatial patterns in the distributions of interacting species builds on work with single populations. Growth and dispersal of single populations in continuous habitats were studied by Skellam (1951), whose models resemble those used by Fisher (1937) to describe clines in gene frequencies and the spread of introduced mutations. Skellam found a threshold for the size of favorable habitat, below which a population of dispersing individuals cannot persist in an otherwise hostile environment. Skellam's minimum habitat size requires strong boundary conditions, however: it rests on an assumption that the population growth rate approaches $-\infty$ outside of the favorable region. Minimum habitat size is also obtained in the models of competitive interactions discussed below, but it occurs with more realistic boundary conditions because competition acts as a limiting force that is absent in models of single populations.

Homogeneous Environments

Models with homogeneous environments follow the dynamics of species competing in the environmental analogue of a tabletop. Coexistence in these models requires some mechanism that induces heterogeneity in the interaction itself. Two such mechanisms are asymmetrical boundary conditions and the influence of a rapidly moving predator.

Gopalsamy (1977) found that coexistence among species in a homogeneous environment depends on the locations and conditions of habitat boundaries. Gopalsamy's model (equation 8.1), treats dispersal as simple diffusion: it implicitly assumes that individuals move randomly and independently of one another. (See Levin 1978; McMurtrie 1978; Okubo 1980 for details about describing dispersal as diffusion.)

$$(8.1) \qquad \frac{\partial n_i}{\partial t} = r_i n_i - a_i n_i n_j + D \frac{\partial^2 n_i}{\partial x^2}$$

In equation 8.1, $n_i(x,t)$ and $n_j(x,t)$ are the densities of species i and j ($i,j = 1,2; i \neq j$) at location x and time t, r_i and a_i are growth rates and competition coefficients, and D is the diffusion rate for both species. Because the model contains no intraspecific limitation on growth, coexistence cannot occur without some form of environmental variation. Gopalsamy introduced heterogeneity by specifying unequal population reservoirs at one end of a linear habitat and by setting population densities to zero at the other boundary. Coexistence in this model is essentially impossible if the environment is infinitely long, but in a finite habitat, species coexist in distributions that form standing waves.

Mimura, Kan-on, and Nishiura (1988) also assumed that species diffuse in a finite and homogeneous environment, and they obtained coexistence and spatial segregation between species in the presence of a predator. Their model contains three species, a predator and two competing prey, that interact locally according to simple Lotka-Volterra competitive and predator-prey mechanisms. Mimura et al. chose parameter values that precluded the coexistence of all three species without diffusion and that led to the extinction of one prey species in the absence of the predator. They found predator-mediated coexistence among prey species when the prey move slowly relative to the predators. In this case, coexisting prey species divide the habitat sharply along a boundary that oscillates in space.

Heterogeneous Environments

Many forms of environmental heterogeneity have been added to models of competition, and they have produced correspondingly diverse results. Although these results share a familiar theme, that species can coexist if the habitat reduces the intensity of competition, models with heterogeneous environments accomplish this reduction in novel ways.

Pacala and Roughgarden (1982) developed a model in which species diffuse in a linear environment with carrying capacities that vary in space and compete according to Lotka-Volterra dynamics. They analyzed the case of a "step-function" environment, consisting of two intervals with different carrying capacities. The basic result of this model resembles Skellam's (1951) minimum habitat size for single populations: a species can persist if it is favored in an interval that is sufficiently large relative to its dispersal rate. Pacala (1987) reproduced this result in a model of plant competition in which each plant competes for resources only within a small neighborhood of itself.

Namba (1989) introduced spatially varying growth rates to a model in which population diffusion is density dependent. Space is the limiting resource in Namba's model (equation 8.2), and competition is expressed through a species' effect on its competitor's movement.

$$(8.2) \qquad \frac{\partial n_i}{\partial t} = r_i(x)n_i + D_i \frac{\partial^2}{\partial x^2} [(n_i + a_i n_j)n_i]$$

In equation 8.2, $n_i(x,t)$ and $n_j(x,t)$ are species densities ($i, j = 1,2; i \neq j$), $r_i(x)$ are spatially varying growth rates, a_i are competitive influences on movement, and D_i are species-specific diffusion rates. Namba found that species coexist if the ratio of their growth rates is neither very large nor very small, and if interspecific competition is weaker than intraspecific competition. The second condition of coexistence is similar to the familiar Lotka-Volterra result, although Namba's model derives it from a different mechanism. Because competition occurs via density-dependent diffusion in this model, species can coexist even though they have similar distributions.

Czárán (1989) simulated competition among two annual plant species in which individuals disperse seeds only to adjacent cells on an environmental grid. With a homogeneous environment, Czárán obtained the usual result that one species excludes the other throughout the grid unless interspecific competition is weaker than intraspecific competition. Regional coexistence ensued, however, in the presence of spatially variable competitive abilities with different optimal locations for each species. Each species dominated the portion of the grid where it was competitively superior, so that areas of local coexistence shrank over time and a distinct species boundary formed.

Shigesada, Kawasaki, and Teramoto (1979) also obtained coexistence through local habitat segregation. Their model, which leads to regional competitive exclusion in the absence of heterogeneity, represents spatial variation as a nonlinear bias in the direction of movement. Dispersal occurs through biased and density-dependent diffusion: individuals tend to move toward favorable areas of the environment and away from one another. Coexistence in this model is both local and regional: the distributions of competing species are centered in different locations, but each species occurs throughout space in at least low density. Even though they favor the same areas, competitors reach different spatial distributions, whose centers and shapes are determined by the species' competitive strengths and initial abundances.

Shigesada (1984) generalized the model of Shigesada, Kawasaki, and Teramoto (1979) to include spatial heterogeneity in growth rates. She assumed that dispersal is rapid relative to growth and then determined the outcome of competition when species favor different places. If dispersal is sufficiently faster than growth, competitors coexist via habitat segregation; otherwise, one species excludes the other throughout the environment.

The work discussed above shows that spatially distinct species distributions can develop in heterogeneous environments, and Roughgarden (1974) further analyzed the relationship between patterns in species' distributions and patterns in their environments. Roughgarden's model considers patterns in species distributions as reflections of environmental heterogeneity, and it treats competition and dispersal as filters that shape those reflections. Competition without dispersal amplifies fine-grained environmental variation, and dispersal alone reflects coarse-grained heterogeneity while smoothing fine-grained patterns. When dispersal is combined with competition, the effect of dispersal dominates. These results suggest that regional diversity among competing species will be greatest in an environment containing much fine-grained structure and when competitors disperse slowly from different favored microhabitats. Such conditions also lead to low local diversity and strong habitat specialization. Both local and regional diversity can be maintained by moderate dispersal, however, the effect of which is similar to that of weak coupling on patchy systems, discussed above.

Further analysis of the relationship between environmental patterns and species distributions was performed by Roughgarden (1978), who studied the effects of competition on patchiness in multispecies communities. In this model, patchy distributions are caused by stochastic fluctuations in spatially heterogeneous carrying capacities, but the length and distinctiveness of patches are determined by dispersal rates, population growth, and the strength of competition. Roughgarden concluded that species distributions become less patchy with weak competition than in the absence of competition. Strong competition enhances patchiness, which is further increased by high growth rates and intermediate dispersal distances. These conditions also lead to segregation of species in different patches, particularly when environmental fluctuations that improve conditions for one species are detrimental to its competitor.

PREDATION IN CONTINUOUS ENVIRONMENTS

The theory of predator-prey interactions in spatially continuous environments, like its counterpart in patchy environments, studies the effects of dispersal on the maintenance of diversity. Unlike the theory of patchy systems, however, it also describes spatial and temporal patterns in species distributions and relates these patterns to diversity. Like models of competition, predator-prey models portray continuous environments in two ways: constant or spatially variable.

Homogeneous Environments

As in models of competition in continuous environments, adding space without heterogeneity to predator-prey models usually fails to stabilize population oscillations or to prevent the extinction of species involved in unstable interactions (McMurtrie 1978). This has been demonstrated repeatedly with the Lotka-Volterra model, which is a common starting point in studies of predator-prey interactions in continuous environments. The classic nonspatial Lotka-Volterra model features predator and prey oscillations that persist at amplitudes determined by their initial displacements from equilibrium. When a finite and homogeneous environment is added to the model, simple diffusion (equations 8.3–8.4) in predators and/or prey flattens initial patterns in their distributions, and the populations undergo spatially homogeneous oscillations with the same neutral stability of the nonspatial model (Chow and Tam 1976; Hadeler, an der Heiden, and Rothe 1974; Murray 1975; Rothe 1976, 1984).

$$(8.3) \qquad \frac{\partial n}{\partial t} = rn - anp + D_n \frac{\partial^2 n}{\partial x^2}$$

$$(8.4) \qquad \frac{\partial p}{\partial t} = bnp - cp + D_p \frac{\partial^2 p}{\partial x^2}$$

Similar to equation 8.1, $n(x,t)$ and $p(x,t)$ are prey and predator densities, r is the prey growth rate, a is the prey capture rate, b is the conversion rate of prey into predators, c is the predator starvation rate, and D_n, D_p are prey and predator diffusion rates.

Adding diffusion in homogeneous environments to models with logistic prey growth (Murray 1975; Rothe 1976, 1984) and/or many species (Murray 1975) also leads to uniform distributions with the dynamics of the original nonspatial models. Logistic prey growth, which is locally stabilizing, even accelerates regional homogenization of predator and prey densities (Murray 1975).

Predator and prey distributions do not become uniform in very large domains, however. Lotka-Volterra models with simple diffusion in constant and infinitely long environments generate heterogeneous distributions (Hadeler, an der Heiden, and Rothe 1974) or traveling waves of pursuit and evasion (Chow and Tam 1976; Dunbar 1983). Li (1989) showed that, for minimal restrictions on the mechanism of local interactions, diffusing predators and prey coexist if the environment is sufficiently large.

Heterogeneous distributions also result from a combination of autocatalytic prey growth and differential species mobilities (Levin and Segel 1976; Segel and Levin 1976). In this model, stable patterns in predator and prey densities occur in a homogeneous environment when predators diffuse sufficiently faster than prey.

Heterogeneous Environments

Although few have been studied, models of predator-prey interactions in heterogeneous environments support the hypothesis that spatial variability enhances diversity and stability. Rothe (1983) showed that predators and prey coexist in an infinitely long and spatially variable environment when only one species diffuses. Rothe did not restrict prey growth and predator death terms to specific functional forms, however, and this generality precluded consideration of patterns in species distributions.

We have adopted the alternative approach of specifying a Lotka-Volterra mechanism and analyzing patterns in predator and prey densities (McLaughlin and Roughgarden 1991a). We added spatial heterogeneity in the prey growth rate to equations 8.3–8.4, and found that any environmental variation more complex than a simple cline alters the model's behavior in both space and time. When predators move and prey are sedentary, the predator distribution mirrors heterogeneity in prey growth rate, and the prey distribution reflects this heterogeneity in proportion to the predator diffusion rate. Consequently, locations that support rapid prey growth become refuges where, because predators exhibit net dispersal away from areas of high predator density, prey suffer lower per capita mortality rates. When both species diffuse rapidly, patterns in their distributions are smoothed, but total abundances are maximized.

Movement in a heterogeneous environment stabilizes equations 8.3–8.4 in proportion to the amount of heterogeneity and the ratio of predator and prey diffusion rates (McLaughlin and Roughgarden 1991a), consistent with Holt's (1984, 1985) models of predator dispersal between source and sink patches. Oscillations that occur in the absence of movement or spatial variation decay exponentially to the stable patterns described above. When prey are sedentary, the rate of this decay is proportional to the predator diffusion rate and the amount of nonlinear variation in the environment. Population oscillations persist longer as prey mobility increases. The basic conclusion of this model is that environmental heterogeneity is manifested in species distributions and stabilizes their interaction, and that it does so most strongly when predators move rapidly relative to prey. The effects of differential mobilities will be discussed further below in connection with models of species that move on different scales.

Comins and Blatt (1974) obtained qualitatively similar results with a Lotka-Volterra model containing biased diffusion. They represented spatial heterogeneity as a tendency in both species to move toward one part of the environment. This stabilizes the model, but in contrast to the results of McLaughlin and Roughgarden (1991a), the decay in population oscillations is enhanced by slow diffusion, with greatest stability in the case of sedentary *predators*. A prey refuge develops in the model of Comins and Blatt, but its location in unfavorable habitat is reversed relative to the refuge location in McLaughlin and Roughgarden (1991a): since both species tend to move toward favorable areas, prey that remain in less favorable places suffer reduced predation mortality. The difference in refuge locations between the two models is an artifact of the difference in their depictions of environmental heterogeneity.

INTERACTIONS ACROSS SPATIAL SCALES

The results of many of the models discussed in this chapter concur that mobility differences between species can create patterns in their distributions and maintain diversity. This section extends the study of such differences by considering movement on different scales. There is a consensus that ecological communities are affected by patterns and processes on many scales and that recognition of these scales is important (Addicott et al. 1987; O'Neill et al. 1986; Powell 1989; Wiens 1986, 1989b), but theoretical understanding of scale issues is limited. The following discussion addresses this limitation by developing an approach to the study of communities that contain two or more scales of heterogeneity and dispersal.

We begin with a general model of two species that move on different scales and, hence, respond to different scales of environmental heterogeneity. We then select one scale as a frame of reference and show how scale differences affect the spatial and temporal dynamics of the species interaction. We illustrate with a specific predator-prey example the sharp spatial patterns that can result from interactions across scales. We conclude by extending the model to communities with multiple scales and many species, and by discussing how patterns in the environment can translate across ever larger scales.

Scales and Local versus Regional Interactions

Imagine that individuals of one species (n) move about their linear environment slowly or on a relatively fine scale, appropriately measured in units of length x. Members of a second species (p) move rapidly or on a coarse scale, measured in z. Consequently, the slowly moving species is more sensitive to fine-grained variation in the environment than is the rapidly moving species. This scenario accords with Addicott et al. (1987), who suggest that the movement of individuals often determines the appropriate scale on which to study a population. If each moves at random, then their interactions in space may be described by the following equations:

$$(8.5) \qquad \frac{\partial n}{\partial t} = nf(r_n, n, p) + D_n \frac{\partial^2 n}{\partial x^2}$$

$$(8.6) \qquad \frac{\partial p}{\partial t} = pg(r_p, n, p) + D_p \frac{\partial^2 p}{\partial z^2}$$

Local interactions are described by functions f and g, which are not restricted to any particular form, so equations 8.5–8.6 can represent many kinds of interactions or nonlinear dependencies on species densities $n(x,t)$ and $p(z,t)$. The model portrays environmental heterogeneity as variation in species intrinsic growth rates $r_n(x)$ and $r_p(z)$.

We will analyze the model's dynamics on one scale at a time, using an assumption that the scales are related by movement rates in the following manner:

$$(8.7) \qquad D_n x = D_p z$$

The essence of equation 8.7 is that organisms that take smaller "steps" (or other modes of dispersal) require, on average, more steps to get from one place to another.

The model's fine-scale behavior can be studied by using equation 8.7 to convert the dynamics of the rapidly moving species, equation 8.6, to the smaller scale of equation 8.5. If the scales are indeed distinct, i.e., if the mobility ratio D_p/D_n is sufficiently large, then fine-scale spatial changes in the density of species p are determined primarily by rapid dispersal instead of by spatial variation in growth. Shigesada and Roughgarden (1982) and Shigesada (1984) obtained similar approximations in competition models with rapid dispersal. Consequently, the density of the rapidly moving species appears constant at the fine scale (McLaughlin and Roughgarden 1991b). Thus, equations 8.5–8.6 simplify to the following when the fine scale is the frame of reference:

(8.8)
$$\frac{\partial n}{\partial t} = nf(r_n, n, P) + D_n \frac{\partial^2 n}{\partial x^2}$$

(8.9)
$$\frac{dP}{dt} = P \int g(r_p, n, P) dx$$

$P(t)$ is the sum of $p(x,t)$ over the length of the habitat. The structure of equations 8.8–8.9 differs qualitatively from that of other spatial models. The dynamics of interacting species in most models, including equations 8.1–8.4, are coupled locally in space. The dynamics of equations 8.8–8.9, however, are coupled on a regional level. When individuals of one species move widely enough so that their growth is not sensitive to fine-scale heterogeneity in either the environment or the distribution of the other species, we must shift to a coarser scale to observe spatial variation in the behavior of the interaction. We now turn to the patterns in species distributions that result from such an interaction.

Magnification of Environmental Patterns

We can analyze spatial patterns in an interspecific interaction by studying the model's equilibrium behavior, although "equilibrium" in this model is actually a spectrum of sites whose populations are in local disequilibrium. The equilibrium distribution of the relatively sedentary species, determined by setting the left sides of equations 8.8 and 8.9 to zero, is a balance between source and sink hab-

itats. The species grows to high density in favorable habitats and declines to low density or extinction in poor habitats. The balance between sources and sinks, and hence the sharpness of the patterns in the slow species' distribution, is determined by its diffusion rate.

We illustrate this behavior with a modified Lotka-Volterra model in which predators move on a large scale relative to their prey (McLaughlin and Roughgarden 1991b). Figure 8.1 shows the patterns in the prey distribution that develop when the prey population grows twice as rapidly in the center of the environment as it does at the edges. Slowly moving prey become narrowly confined to the most productive habitat(s). Analogous patterns are observed for clines, periodicity, and other kinds of heterogeneity in prey growth. Thus, mortality from a widely moving predator causes the prey distribution to greatly magnify the underlying spatial patterns in the environment. As in the models above that predict separation of competing species along sharp boundaries, prey become patchily distributed even though spatial variation in the environment is gradual.

Analysis of the model's coarse-scale behavior proceeds in a manner similar to that of the fine scale. Spatial patterns in the relatively mobile species, like those in the more sedentary species, reflect underlying heterogeneity in the environment. The distribution of the mobile species is also shaped by that of the sedentary species, however. Patterns in both the environment and the distribution of the slow species are manifested in the distribution of the rapidly moving species according to the form of g in equation 8.6. If g has a Lotka-Volterra predator-prey form (as

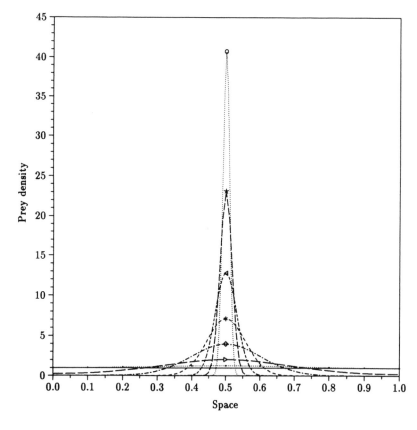

Figure 8.1 Effect of prey diffusion on the prey equilibrium distribution. Numerical solution of the Lotka-Volterra diffusion model with humped prey growth rate, $r(x) = (1/2)[3 - \cos(2\pi x)]$, and predator dispersal on a large scale relative to prey. The height of the prey peak oscillates in time with constant amplitude; this figure plots prey distributions at the midpoint of the oscillation. $\circ, D = 10^{-7}$; $\bigstar, D = 10^{-6}$; $\triangleleft, D = 10^{-5}$; $*, D = 10^{-4}$; $\diamond, D = 10^{-3}$; $\triangleright, D = 10^{-2}$; $\cdot, D = 10^{-1}$; $-$, prey initial distribution, $n_0(x) = 1$. (From McLaughlin and Roughgarden 1991b.)

in McLaughlin and Roughgarden 1991b), then the predator distribution (coarse scale) simply reflects patterns in the prey distribution (fine scale), although less sharply than the prey reflect patterns in their own growth rate. Thus, patterns can translate across scales, as we now consider in a model containing multiple scales.

Communities with Many Scales

Most communities contain numerous species and many scales of heterogeneity. Such complexity may be studied by expanding equations 8.5–8.6 to many species and multiple scales (equation 8.10) that are related according to species mobilities (equation 8.11), as follows:

$$
\begin{aligned}
\frac{\partial u_1}{\partial t} &= u_1 f_1(r_1, u_1, \ldots, u_n) + D_1 \frac{\partial^2 u_1}{\partial x_1^2} \\
\frac{\partial u_2}{\partial t} &= u_2 f_2(r_2, u_1, \ldots, u_n) + D_2 \frac{\partial^2 u_2}{\partial x_2^2}
\end{aligned}
$$

(8.10)

$$
\frac{\partial u_n}{\partial t} = u_n f_n(r_n, u_1, \ldots, u_n) + D_n \frac{\partial^2 u_n}{\partial x_n^2},
$$

where $u_i(x_i, t)$ is the density of species i, and x_i measures distance on the ith scale.

(8.11) $\qquad D_1 x_1 = D_2 x_2 = \cdots = D_n x_n$

As in the model with two scales, community dynamics on a given scale in equations 8.10–8.11 are isolated by choosing that scale as a frame of reference. This approach divides community complexity into pieces and studies the dynamics of each piece semi-independently, just as hour, minute, and second hands on a clock record the passage of time on different scales.

Preliminary analysis of equations 8.10–8.11 with simple Lotka-Volterra predator-prey interactions (McLaughlin, unpublished) suggests that scale differences shape species distributions as in the two-scale model. Sharp patterns form in the distribution of each species on its scale if each scale is sufficiently discrete from those above and below it. Patterns are less distinct when scale differences are small, just as the prey distribution in figure 8.1 flattens when prey diffuse rapidly ($D > 10^{-2}$). More complex patterns are anticipated when two or more species respond to each scale of environmental heterogeneity.

We caution against inappropriate application of these models to field systems. Many organisms that move on large scales are selective in their utilization of space and hence violate the assumption of unbiased random movement. More complex dispersal patterns can significantly alter the dynamics of species interactions, as shown by Comins and Blatt (1974), Namba (1989), and Shigesada, Kawasaki, and Teramoto (1979).

When critical assumptions are met, however, these models can guide field study of interactions among species that respond to different scales of heterogeneity. Manipulative experiments with local interactions should be complemented with work that considers spatial patterns in the interaction(s) on the scale of the most mobile species. Holt (chap. 7) discusses this issue further.

CONCLUSIONS

Just as many forms of heterogeneity affect ecological communities in myriad ways, theoretical research has taken diverse approaches toward understanding the role of spatial variation in maintaining community diversity. Nevertheless, several themes emerge from this chapter's limited review of the theory of species interactions in spatially explicit environments.

First, many spatial models develop a concept of equilibrium different from that of most ecological theory. "Equilibrium" in these models is regional instead of local: species persist and stable patterns develop regionally, even though local interactions are unstable or in disequilibrium.

Second, moderate rates of dispersal in heterogeneous environments can maintain local and/or regional species diversity when extinctions would otherwise occur in isolated or homogeneous habitats. Environmental heterogeneity, in the form of patchiness or patterns in continuous habitat, induces asynchrony among interactions at different locations. Moderate rates of dispersal enable species to exploit asynchrony, so that each species always experiences favorable conditions somewhere in the system. Rapid rates of dispersal often synchronize population dynamics throughout the environment, leading to extinctions if local species interactions are inherently unstable.

Third, species that interact in continuous environments may acquire distributions with interesting spatial patterns. Moderate dispersal can lead to the formation of refuges whose locations are transient, as in traveling waves of pursuit and evasion, or fixed, as in spatially distinct distributions of competitors or predators and prey. When dispersal is not too rapid, species distributions may reflect or magnify underlying heterogeneity in the environment.

Fourth, the spatial structure of interactions is significantly altered when species respond to different scales of environmental heterogeneity. The population dynamics of a pair of species become coupled at the coarser scale of the more widely dispersing species. The distribution of the more sedentary species magnifies fine-scale heterogeneity in the environment to an extreme degree, and the species may become patchily distributed over smoothly varying habitat. This mechanism for the origin of patchy distributions is an alternative to traditional views that emphasize disturbances (Paine and Levin 1981; but see Levin and Segel 1976) or discontinuities in the physical environment (Shmida and Wilson 1985). In complex communities that contain a well-defined hierarchy of scales, fine-scale patterns in species distributions and the environment are manifested at ever coarser scales. In communities that lack discrete scales, pattern translation across scales tends to be less clear.

The general conclusion of this chapter is that spatial heterogeneity in the environment can enhance diversity by reducing the intensity of species interactions. Although coexistence of species in nonspatial models requires some mechanism that stabilizes their interaction, species can coexist in heterogeneous environments regardless of the nature of local interactions. Heterogeneity, in the form of patchiness or patterns in continuous habitat, maintains diversity by creating refuges of reduced competition or predation. These refuges, which may be ephemeral or persistent, are exploited and preserved by moderate rates of dispersal. If dispersal is too low, then prey or inferior competitors fail to reach the refuges. If dispersal is too rapid, then they are overrun or colonized too quickly by predators or superior competitors, and they cease to function as refuges.

Regional species diversity is strongly influenced by an interaction between environmental heterogeneity and dispersal. Where dispersal is very slow, diversity is low and the occupancy of a given site is determined by both its colonization history and the dominance relationships among colonists. Where the dispersal rate is moderate, local and/or regional diversity are maximized and species distributions can augment environmental heterogeneity. Where dispersal is rapid, local and regional diversity are again low and the regional persistence of species is determined by local dominance relationships.

Spatial heterogeneity can maintain species diversity, but it also may determine the capacity of an environment to support additional species. Environmental heterogeneity is reflected in species distributions (Czárán 1989; McLaughlin and Roughgarden 1991a, 1991b; Roughgarden 1978; Shigesada, Kawasaki, and Teramoto 1979), and it can determine an upper limit on species diversity (Holt 1984; Iwasa and Roughgarden 1986). Richly variable environments can support a greater number of species than more homogeneous ones, but this number is bounded by underlying habitat diversity. Habitat diversity itself, however, may be enhanced by the species that utilize it (e.g., Boettcher and Kalisz 1990), but discussion of how species alter their environments is beyond the scope of this chapter.

9

Local and Regional Regulation of Species-Area Relations: A Patch-Occupancy Model

Hal Caswell and Joel E. Cohen

Local species diversity (i.e., diversity in a small homogeneous patch of habitat) may be determined by local processes such as competition, predation, mutualism, and disturbance, or by regional processes such as speciation and biogeographical dispersal (Ricklefs 1987). Most recent ecological theory has focused on the local processes that determine membership in a community, rather than the regional processes that determine the species pool from which members may be drawn. One approach to distinguishing local from regional control of diversity is to examine a number of communities and plot local diversity as a function of the regional species pool (Terborgh and Faaborg 1980; Cornell 1985a, 1985b; Ricklefs 1987). In saturated communities local diversity should be independent of regional diversity. A direct relation between local and regional diversity, however, is interpreted as evidence that "in these cases, local communities are not saturated, diversity is not prescribed by local conditions, and the number of species found within small areas is sensitive to such regional processes as geographic dispersal . . ." (Ricklefs 1987, 168). The results of such tests have been mixed. Local diversity of cynipine wasps on oak trees (Cornell 1985a, 1985b) and songbirds on Caribbean islands (Ricklefs 1987) is more or less linearly correlated with regional diversity, thus providing no evidence of saturation of local communities. In another study of island birds in the Caribbean, however, Terborgh and Faaborg (1980) found that local diversity was independent of regional diversity above some minimum number of species.

In this chapter we construct a model that includes both local processes and a regional species pool, and which makes predictions about local diversity. We find that local-regional diversity regressions must be interpreted with care, because the interaction of disturbance, colonization, and competition in locally saturated communities can produce patterns that completely obscure the effects of saturation.

Any model intended to study local and regional regulation of species diversity must contain at least two spatial scales. Most of the classic theories of diversity, however, are based on species-interaction models (the Lotka-Volterra equations and their relatives) which include only a single scale. They focus on local processes to the exclusion of regional effects.

This chapter is Woods Hole Oceanographic Institution Contribution 7641.

Patch-occupancy models, which include two spatial scales, are a next step in complexity and realism. They picture the world as a set of patches in which local species interactions take place. The state of a patch is defined by species presence or absence. If the collection of patches is well mixed, so that each patch interacts equally with all others, the system can be described by a set of differential or difference equations in the *proportions* of patches in each state. This assumption of mixing limits the scales to two: the local scale within a single patch and the regional scale of the entire set of patches. Patch-occupancy models were introduced by Cohen (1970) and Levins (1970), and have since been applied to a variety of ecological interactions (Slatkin 1974; Hastings 1977, 1978; Caswell 1978; Crowley 1979; Hanski 1983; Caswell and Cohen 1991a, 1991b).

Caswell and Cohen (1991b) used patch-occupancy models to examine coexistence and diversity in competitive and predator-prey interactions. In those models, coexistence is determined by the interplay of the rate of approach to local equilibrium (competitive exclusion) and the rates of disturbance and dispersal. Disturbance can maintain nonequilibrium coexistence and increase both alpha and beta diversity (Caswell and Cohen 1991b).

It is difficult to analyze patch-occupancy models with large numbers of species because the number of possible patch states increases exponentially with the number of species. Thus patch-occupancy models have been used primarily to describe two- and three-species interactions. Conclusions about species diversity rest on the usual ecological inference that diversity is determined by the outcome of such interactions. In this chapter, we develop a patch-occupancy model for large numbers of species, and examine the relation between local diversity and the regional species pool.

THE MODEL FRAMEWORK

We approximate the world by an effectively infinite set of effectively identical patches. This boring landscape is home to S_{tot} species. Each patch is independently subject to disturbance, with probability p, at each time step. A disturbance eliminates all species from a patch. The probability that species i will colonize a suitable patch is a species-specific constant c_i. By assuming that the c_i are constant, we ignore both neighborhood effects (which would make colonization probability depend on the proximity of occupied patches; see Caswell and Etter 1993) and frequency effects (which would make colonization probability depend on the abundance of the species). Which patches are available for colonization, and what happens to the species once they have arrived, depend on the hypotheses made about local interaction.

We want to describe the patterns of diversity resulting from this simple model. To do so, we focus not on a single index but on the species-area curve. This curve gives not only the expected species richness in a single patch (the most local kind of diversity admitted in the model) but also the expected richness in two, three, . . . randomly selected patches. The slope of the species-area curve is one measure of beta diversity. Species-area curves are frequently reported, and there is a large literature on the functions suitable to describe them (McGuinness 1984a).

We will present two models, which differ in their assumptions about local processes. The first contains no competition of any kind. Communities in this model are never saturated, and there is no local limitation on either coexistence or diversity. The number of species in a patch depends only on the history of colonization since the most recent disturbance. At the other extreme, we model strong local saturation by supposing that there is one species that eventually excludes all the others from any patch that it colonizes.

Based on our experience with patch-occupancy models for several species (Caswell and Cohen 1991a, 1991b), we expect the time scales of disturbance and competition to play important roles in these models. If disturbance is frequent and competitive exclusion is slow, we expect the model with strong local saturation to behave like the noncompetitive model, because only rarely will a patch be undisturbed long enough for exclusion to occur. On the other hand, if disturbance is rare or exclusion is rapid, we expect local diversity to reflect competition, and local communities to be saturated with species.

NONCOMPETITIVE COMMUNITIES

A Single Patch

We begin with a single patch, without competition. Let c_i be the probability of species i arriving, per unit time, independently of all other species. Let p denote the probability of disturbance and t the time since the last disturbance ($t = 0$ means the patch is currently disturbed). We assume $p > 0$. Then the probability of finding species i in the patch is

$$(9.1) \qquad P[\text{sp. } i \text{ present now}|t] = 1 - (1 - c_i)^t$$

Let S denote the number of species in the patch. Since species colonize independently,

$$(9.2) \qquad E[S|t] = \sum_{i=1}^{S_{tot}} [1 - (1 - c_i)^t]$$

and thus

$$(9.3) \qquad \begin{aligned} E[S] &= \sum_{t=0}^{\infty} E[S|t] \, P(t) \\ &= \sum_{t=0}^{\infty} E[S|t] p(1 - p)^t \end{aligned}$$

To proceed, we ignore variation in the colonization rate, setting $c_i = c$ for all i. Then, combining equations 9.2 and 9.3, we get

$$(9.4) \qquad E[S] = S_{tot} \left[1 - \frac{p}{p + c - pc} \right]$$

Thus, in this simple single-patch model, local diversity is directly proportional to the size of the regional species pool.

Species-Area Relationships

We turn now from single patches to collections of patches. We want to derive a species-area relation. Our approach is to calculate the expected species richness of a set of k patches, conditional on its disturbance history, and then take an expectation of this quantity over the probability distribution of disturbance histories.

Consider a random sample of k patches. We want to calculate the "age" distribution of this set of patches, where age is measured since the most recent disturbance. Let $D(t_1, t_2, \ldots, t_k)$ denote the event that patch j was last disturbed at time $-t_j$, for $j = 1, \ldots, k$. Because patches are disturbed independently with a common disturbance probability p, we know that

$$(9.5) \qquad P[D(t_1, \ldots, t_k)] = \prod_{j=1}^{k} p(1 - p)^{t_j}.$$

Consider species i. The conditional probability that species i is present in at least one of the k patches is

$$P[\text{sp. } i \text{ present in some patch}|D(t_1, \ldots, t_k)] = $$
$$1 - P[\text{sp. } i \text{ absent from every patch}|D(t_1, \ldots, t_k)] = $$
$$1 - \prod_{j=1}^{k} (1 - c_i)^{t_j} = 1 - (1 - c_i)^T$$

where $T = t_1 + \cdots + t_k$.

From this we can calculate the conditional expectation of the species richness in the collection of k patches

$$(9.6) \qquad E[S(k)|D(t_1 \ldots, t_k)] = \sum_{i=1}^{S_{tot}} [1 - (1 - c_i)^T]$$

so that the species-area relationship is given by

$$(9.7) \qquad \begin{aligned} E[S(k)] &= \sum_{t_1=0}^{\infty} \cdots \sum_{t_k=0}^{\infty} E[S(k)|D_i(t_1, \ldots, t_k)] \\ &\quad \times \prod_{j=1}^{k} p(1 - p)^{t_j} \\ &= \sum_{t_1=0}^{\infty} \cdots \sum_{t_k=0}^{\infty} \sum_{i=1}^{S_{tot}} \left[1 - (1 - c_i)^T \right] p^k (1 - p)^T \end{aligned}$$

This expression can be simplified, because we can collect all those terms in which $t_1 + \cdots + t_k = T$, with $T \geq t_j \geq 0, j = 1, \ldots, k$, and write

$$(9.8) \qquad E[S(k)] = \sum_{T=0}^{\infty} H(T,k) \sum_{i=1}^{S_{tot}} \left[1 - (1 - c_i)^T \right] \cdot \\ \times p^k (1 - p)^T$$

where

$$(9.9) \qquad H(T,k) = \underbrace{\sum_{t_1=0}^{\infty} \cdots \sum_{t_k=0}^{\infty}}_{t_1 + \ldots + t_k = T} 1$$

(9.10)
$$= \binom{T + k - 1}{k - 1}$$

Thus

(9.11) $E[S(k)] = \sum_{T=0}^{\infty} p^k (1 - p)^T \sum_{i=1}^{S_{tot}} [1 - (1 - c_i)^T]$
$$\times \binom{T + k - 1}{k - 1}$$

If, as before, we simplify this by assuming that $c_i = c$ for all i, we obtain

(9.12) $E[S(k)] = p^k S_{tot} \sum_{T=0}^{\infty} (1 - p)^T [1 - (1 - c)^T]$
$$\times \binom{T + k - 1}{k - 1}$$

Now for some combinatorial sleight-of-hand. We note that

(9.13)
$$\binom{T + k - 1}{k - 1} = \binom{k + T - 1}{T}$$

and that these are the *figurate numbers* (Riordan 1958, 25). For any $r < 1$, these numbers satisfy

(9.14)
$$\sum_{T=0}^{\infty} \binom{k + T - 1}{T} r^T = \frac{1}{(1 - r)^k}$$

(Riordan 1958, 10). Substituting this into equation 9.12 and simplifying, we obtain a final simplified expression for the species-area curve as a function of colonization, disturbance, and the species pool:

(9.15)
$$E[S(k)] = S_{tot} \left[1 - \frac{p^k}{(p + c - pc)^k} \right]$$

Results for Noncompetitive Communities

Species-Area Relationships. The species-area curves produced by equation 9.15 look surprisingly realistic (fig. 9.1). They are nearly linear on a log-log scale for small k, eventually reaching an asymptote at S_{tot} as $k \to \infty$. Log-log species-area curves, of course, are a familiar sight in ecology, and have been interpreted in terms of competitive equilibrium, island biogeography theory, and the canonical lognormal distribution of species abundances (May 1975a; McGuinness 1984a). In particular, the canonical lognormal predicts a slope of 0.25, while reasonable but noncanonical lognormal distributions yield slopes between about 0.15 and 0.4 (May 1975a).

In our model, the slope and intercept of the species-area curve are largely determined by the ratio of the disturbance rate p and the colonization rate c. Figure 9.2A shows the intercept (i.e., $E[S(1)]$) as a function of p/c. There is a sharp threshold in the neighborhood of $p/c = 1$; when $p/c \gg 1$, an individual patch tends to contain only a fraction of the species pool ($E[S(1)] \ll S_{tot}$). When $p/c \ll 1$, each patch is expected to contain most of the available species and $E[S(1)] \approx S_{tot}$. This threshold can be shown analytically by rewriting equation 9.15 as

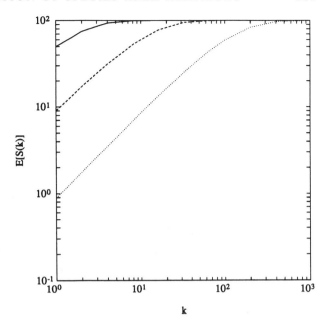

Figure 9.1 Species-area curves for the noncompetitive model. $c = .001$ and $S_{tot} = 100$ throughout. Solid line, $p = .001$; dashed line, $p = .01$; dotted line, $p = .1$.

(9.16)
$$E[S(1)] = S_{tot} \left(\frac{1 - p}{p/c + 1 - p} \right)$$

from which it follows that

$$p/c \gg 1 - p \Rightarrow E[S(1)] \ll S_{tot}$$
$$p/c \ll 1 - p \Rightarrow E[S(1)] \approx S_{tot}$$

Let z denote the slope of the log-log species-area curve, evaluated at $k = 1$. By differentiating equation 9.15, we obtain

(9.17) $z \equiv \dfrac{d \log E[S(1)]}{d \log k}$
$$= -\log \left(\frac{p}{p + c - pc} \right) \frac{p}{c(1 - p)}$$

Figure 9.2B shows that this is also determined primarily by the ratio p/c. When $p/c \ll 1$, the slope approaches 0, and species number accumulates only slowly with increasing area. The "canonical" value of $z \approx 0.25$ corresponds to $p/c \approx 0.1$; i.e., to a time scale for disturbance an order of magnitude larger than that for colonization.

Local versus Regional Regulation of Diversity. Consider now the slope b of the regression of local diversity on regional diversity. This regression has been used as a test for the importance of local processes (Ricklefs 1987). If we measure local diversity by $E[S(1)]$ and regional diversity by the species pool S_{tot}, this slope is

(9.18)
$$b = \frac{\partial E[S(1)]}{\partial S_{tot}}$$

Because this model contains no within-patch limitations on coexistence, it comes as no surprise that equation 9.16 shows that local diversity is directly and linearly propor-

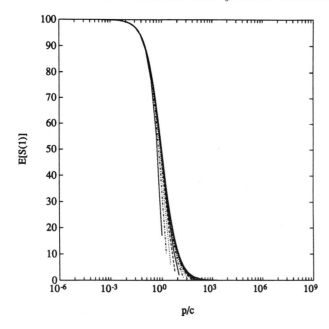

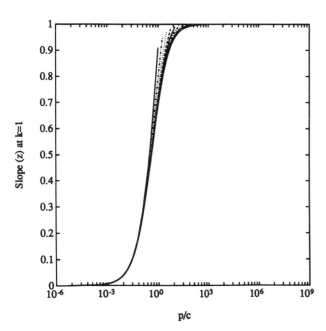

Figure 9.2 *Top:* The intercept, defined operationally as $E[S(1)]$, of the species-area curve for the noncompetitive model as a function of the ratio p/c, for $p,c \in [10^{-6},1]$. *Bottom:* The slope of the log-log species-area curve for the noncompetitive model, evaluated at $k = 1$. The slope is shown as a function of the ratio p/c, for $p,c \in [10^{-6},1]$. $S_{tot} = 100$ in both graphs.

tional to the regional species pool S_{tot}. Thus data showing that local diversity is proportional to regional diversity (Cornell 1985a, 1985b, Ricklefs 1987) are indeed compatible with an unsaturated community.

Because the slope b is given by the right-hand side of equation 9.16 when $S_{tot} = 1$, figure 9.2A can be interpreted as a plot of b, by rescaling the ordinate from zero to one. The empirical regressions interpreted in the literature

as showing lack of saturation have slopes[1] ranging from $b = 0.14$ to $b = 0.49$. In our model, these slopes correspond to $p/c \approx 0.5$.

COMPETITIVELY SATURATED COMMUNITIES

We turn now to a model with strong local competitive saturation, with the hope of distinguishing it from the noncompetitive model. Suppose that local niche space becomes saturated as one species (call it species 1) excludes all other species. In addition to the terms defined in the previous section, we define v as the rate of competitive exclusion, by species 1, of all other species present in a patch. We begin with the basic relationship

$$(9.19) \quad E[S(k)] = \sum_{t_1=0}^{\infty} \cdots \sum_{t_k=0}^{\infty} E[S(k)|D(t_1, \ldots, t_k)]$$
$$\times P[D(t_1, \ldots, t_k)]$$

The conditional expectation $E[S(k)|D]$, where D abbreviates $D(t_1, \ldots, t_k)$, is

$$(9.20) \quad E[S(k)|D] = \sum_{i=1}^{S_{tot}} P[\text{sp. } i \text{ present in} \geq 1 \text{ patch}|D]$$

$$(9.21) \qquad\qquad = \sum_{i=1}^{S_{tot}} (1 - P[\text{sp. } i \text{ absent from all}$$
$$k \text{ patches}|D])$$

The conditional probability that species 1 is absent from all k patches is

$$(9.22) \qquad P[\text{sp. } 1 \text{ absent}|D] = (1 - c_1)^T$$

where $T = \sum_i t_i$ as before. For the other species,

$$(9.23) \quad P[\text{sp. } i, i \geq 2, \text{ absent}|D] =$$
$$\prod_{j=1}^{k} P[\text{sp. } i \text{ absent from patch } j|D]$$

The probabilities $P[\text{sp. } i \text{ absent from patch } j|D]$ for $i \geq 2$ depend on whether species i has colonized, whether species 1 has also colonized, and if so, whether competitive exclusion has occurred.

Consider patch j, $j = 1, \ldots, k$. Three mutually exclusive and collectively exhaustive events may be identified: (1) species 1 has not colonized, (2) species 1 colonized at some time $-\tau$ and competitive exclusion has occurred, and (3) species 1 colonized at some time $-\tau$, but competitive exclusion has not yet occurred. The probabilities of these three events are, respectively,

$$P[\text{event 1}] = (1 - c_1)^{t_j}$$
$$(9.24) \quad P[\text{event 2}] = \sum_{\tau=0}^{t_j-1} c_1(1 - c_1)^{t_j-\tau-1}[1 - (1 - v)^\tau]$$
$$P[\text{event 3}] = \sum_{\tau=0}^{t_j-1} c_1(1 - c_1)^{t_j-\tau-1}(1 - v)^\tau$$

The conditional probabilities of absence of species i, $i \geq 2$, are

---

1. Cynipid gall wasps (rare), $b = 0.14$; cynipid gall wasps (common), $b = 0.35$; cynipid gall wasps (total), $b = 0.49$ (Cornell 1985b); Caribbean island birds, $b \approx 0.22$ (Ricklefs 1987)

$$P[\text{sp. } i \text{ absent}|\text{event 1}, D] = (1 - c_i)^{t_j}$$

(9.25) $P[\text{sp. } i \text{ absent}|\text{event 2}, D] = 1$

$$P[\text{sp. } i \text{ absent}|\text{event 3}, D] = (1 - c_i)^{t_j}$$

To make the notation easier, define

(9.26) $A(i,j) = P[\text{sp. } i \text{ absent from patch } j|D]$

Then we combine equations 9.24 and 9.25 to obtain, for $i \geq 2$,

(9.27) $A(i,j) = (1 - c_1)^{t_j}(1 - c_i)^{t_j} + \sum_{\tau=0}^{t_j-1} c_1(1 - c_1)^{t_j-\tau-1}$

$$\times [(1 - v)^\tau(1 - c_i)^{t_j} + 1 - (1 - v)^\tau]$$

which, after some tedious algebra, simplifies to

(9.28) $A(i,j) = 1 + [(1 - c_i)^{t_j} - 1]$

$$\times \left[\frac{c_1(1 - v)^{t_j} - v(1 - c_1)^{t_j}}{c_1 - v} \right] \quad i \geq 2$$

In the case where $c_1 = v$, the result is

$$A(i,j) = 1 + (1 - c_1)^{t_j}$$

(9.29) $\times \left\{ (1 - c_i)^{t_j} + \frac{t_j c_1}{1 - c_1} [(1 - c_i)^{t_j} - 1] - 1 \right\}$

$$i \geq 2$$

The probability of absence for species 1 is given by equation 9.22:

(9.30) $A(1,j) = (1 - c_1)^T$

We finally get our species-area relationship

(9.31) $E[S(k)] = \sum_{t_1=0}^{\infty} \cdots \sum_{t_k=0}^{\infty} [S_{\text{tot}} - \sum_{i=1}^{S_{\text{tot}}} \prod_{j=1}^{k} A(i,j)]$

$$\times P[D(t_1, \ldots, t_k)]],$$

where $P[D(t_1, \ldots, t_k)] = p^k(1 - p)^T$.

Results for Competitively Saturated Communities

Equation 9.31 reveals little about the form or behavior of the species-area relationship. To study it numerically, we used a Monte Carlo approach. For each value of k, disturbance histories were sampled by drawing k independent random variables from a geometric distribution with parameter p. The conditional expectation of $S(k)$, given this disturbance history, was calculated and the unconditional expectation obtained by averaging over a large number of disturbance histories.

Species-Area Relationships. The species-area relationship in this model is similar to that in the noncompetitive model (fig. 9.3). The curves are nearly linear on a log-log plot for small sample sizes, and asymptotic at S_{tot} as sample size increases. The parameters c, v, and p interact to determine the location of the curve. As might be expected, increasing v reduces diversity, regardless of the values of the other parameters. When disturbance is low ($p = 0.01$) increases in c increase diversity when $v = 0.011$ and decrease diversity when $v = 0.11$. This reflects the fact that c is the rate of colonization for the superior competitor as well as for all other species. When competitive exclusion is slow, the increase in colonization by competitively inferior species makes up for the increase in colonization by the competitive dominant. But when exclusion is fast, the primary effect of increased c is more rapid colonization by the superior competitor, with a consequent reduction in diversity. At a higher disturbance probability ($p = 0.1$), diversity increases with increasing c, regardless of the value of v.

Disturbance has a marked effect on diversity for any value of k (fig. 9.4). Diversity is maximized at an intermediate frequency of disturbance, and increases with the colonization rate c and decreases with the competitive exclusion rate v. As v increases, the frequency of disturbance

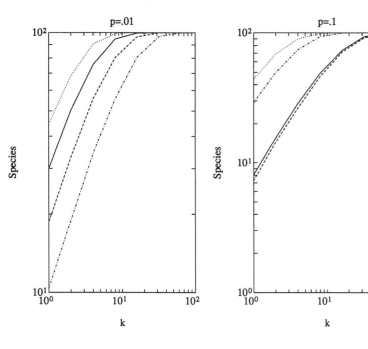

Figure 9.3 Species-area curves for the model with local competitive saturation. Parameter values (c,v) are: solid line = (.01, .011); dashed line = (.01, .11); dotted line = (.1, .011); dash-dot line = (.1, .11). $S_{\text{tot}} = 100$ throughout.

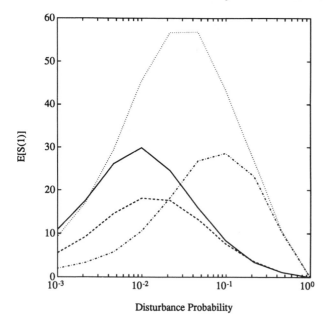

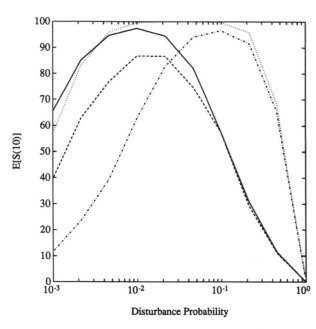

Figure 9.4 Effects of disturbance probability on species richness ($E[S(k)]$) for $k = 1$ (*upper*) and $k = 10$ (*lower*). Parameter values (p,c) are: solid line = (.01, .011); dashed line = (.01, .11); dotted line = (.1, .011); dash-dot line = (.1, .11). $S_{tot} = 100$ throughout.

needed to maximize diversity increases. These results are similar to those of the dynamic patch-occupancy models in Caswell and Cohen (1991a, 1991b).

Local versus Regional Regulation of Diversity. In the absence of disturbance, competitive saturation in this model limits local diversity to $E[S(1)] = 1$, independent of S_{tot}. Thus the slope b of the regression of local diversity on the regional species pool should equal zero. Disturbance perturbs this situation, perhaps sufficiently to obscure the

action of competition. We can investigate this by examining how b varies with p, c, and v.

Equation 9.31 can be simplified to give an analytical expression for $E[S(1)]$. Let $\bar{p} = 1 - p$, $\bar{c} = 1 - c$, and $\bar{v} = 1 - v$. Then some tedious algebra yields

$$(9.32)\quad E[S(1)] = 1 - \frac{p}{1 - \bar{p}\bar{c}} + \frac{p(S_{tot} - 1)}{c - v}$$
$$\times \left(\frac{c}{1 - \bar{p}\bar{v}} - \frac{c}{1 - \bar{p}\bar{c}\bar{v}} - \frac{v}{1 - \bar{p}\bar{c}} + \frac{v}{1 - \bar{p}\bar{c}^2} \right)$$

Figure 9.5 plots local diversity ($E[S(1)]$) as a function of both disturbance probability p and the regional species pool S_{tot}. Four combinations of the colonization probability c and the exclusion probability v are shown; in every case local diversity depends on the regional species pool (i.e., $b > 0$) except at the lowest disturbance rates. At some intermediate disturbance rate, there is a relation between local and regional diversity every bit as strong (i.e., with a value of b as large) as that produced by the noncompetitive model (equation 9.16).

The slope b of the relation between $E[S(1)]$ and S_{tot} can be obtained directly from equation 9.32; this slope is shown as a function of disturbance probability in figure 9.6. At sufficiently small values of p, $b \to 0$ as local diversity is limited by competition. As p approaches 1, $b \to 0$ as all species are eliminated by disturbance. At some intermediate frequency of disturbance, b is maximized.

When b is maximized, the community appears as unsaturated as it can be, given its values of c and v. How much disturbance is required to produce this result? The disturbance frequency p_{max} that maximizes b depends on c and v. To study this dependence we define a new rate, ρ, for the combined processes of colonization *and* competitive exclusion. A patch reaches equilibrium after species 1 has colonized and excluded all other species. The expected time required for these two processes is $1/c + 1/v$. The rate for the combined process is the inverse of this time scale:

$$(9.33)\quad \rho = \left(\frac{1}{c} + \frac{1}{v} \right)^{-1}$$

We used a Monte Carlo procedure to evaluate the relationship between p_{max} and ρ. We generated 500 random combinations of c and v, each log-uniformly distributed over the interval $[10^{-6},1]$, and calculated the resulting p_{max}.

To a good approximation, $p_{max} \approx 2\rho$ (i.e., the median value of p_{max}/ρ was 1.867, and in 78% of the cases, $1.2 < p_{max}/\rho < 10$). Thus the community appears most *un*saturated when the rate of disturbance is of the same order of magnitude as the rate of transition from an empty patch to competitive equilibrium. This is quite a low disturbance rate (measured, as it must be, relative to the other time scales in the system). For example, if $p_{max} = 2\rho$, in fully one-third of all cases, colonization and exclusion will proceed to equilibrium before the first disturbance occurs.

The slopes at the critical disturbance probability p_{max} in figure 9.6 all fall within the range of values accepted as evidence for nonsaturation by Cornell (1985b) and

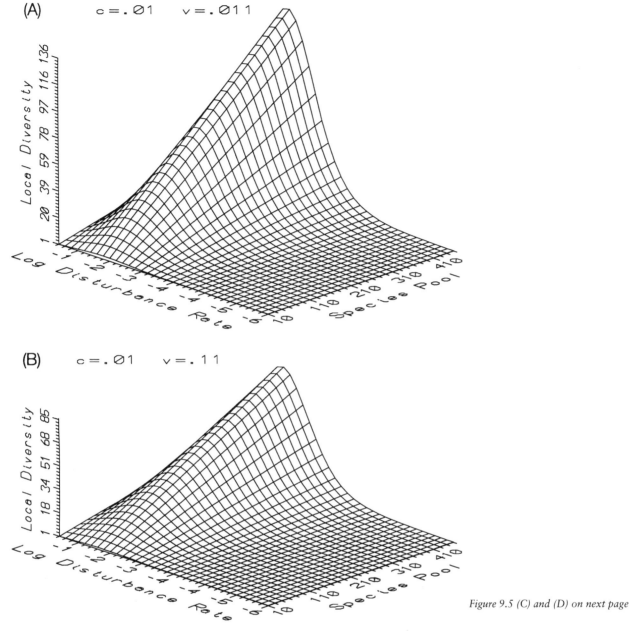

Figure 9.5 Plots of local species diversity ($E[S(1)]$) as a function of the disturbance probability p and the regional species pool S_{tot} for four combinations of the colonization rate c and the competitive exclusion rate v. (A) $(c,v) = (.01, .011)$; (B) $(c,v) = (.01, .11)$; (C) $(c,v) = (.1, .011)$; (D) $(c,v) = (.1, .11)$; $\log = \log_{10}$.

Ricklefs (1987). In fact, the slope at p_{max} in our 500 Monte Carlo samples is bounded below by 0.17, implying that, at p_{max}, practically any combination of c and v will produce a slope indicative of a noncompetitive community.

An alternative is to ask how large p must be, relative to ρ, to yield a slope large enough that it would be accepted as evidence of regional rather than local control (say, $b = 0.1$). In our Monte Carlo calculations, the ratio p/ρ producing a slope of $b = 0.1$ ranged from 0.1 to 0.4. The corresponding probabilities of reaching equilibrium before the first disturbance are 0.91 and 0.71.

In summary, even though this model includes a strong form of competitive saturation, even low rates of disturbance can completely obscure the role played by competitive saturation.

Our conclusions here are based on expected species richness. Patterns in the variance of species richness, or in the similarity among patches, may help to distinguish the effects of competition and disturbance. We have made some progress in this direction and will report the results elsewhere.

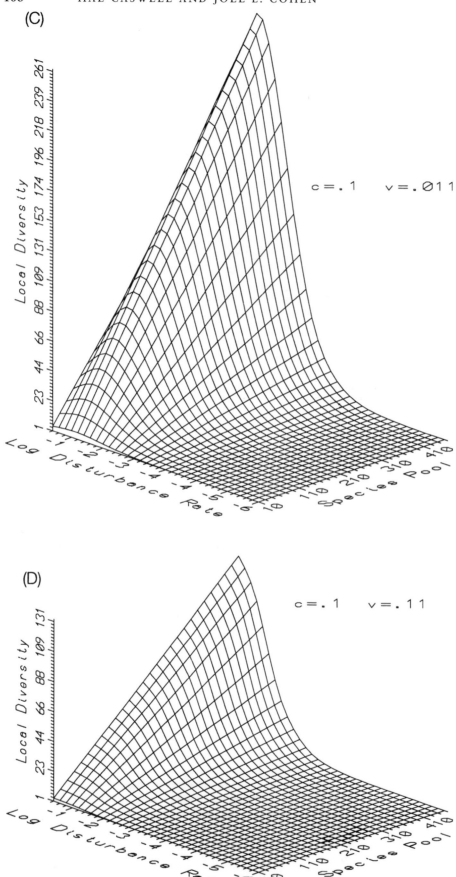

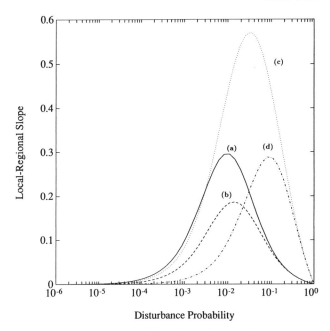

Figure 9.6 The slope of the line relating local diversity ($E[S(1)]$) to the regional species pool S_{tot}, as a function of the disturbance probability p. For (a), $(c,v) = (.01, .011)$; (b), $(c,v) = (.01, .11)$; (c), $(c,v) = (.1, .011)$; (d), $(c,v) = (.1, .11)$.

Conclusions

1. A simple path-occupancy model produces quite realistic-looking log-log species-area curves at small sample sizes, eventually becoming asymptotic to the regional species pool as the sample becomes large enough to include all the species.

2. In communities without competition, the slope of the species-area curve is an increasing function of the ratio of the disturbance rate to the colonization rate. The intercept, which measures diversity in a single patch, is a decreasing function of this same ratio.

3. In the absence of competition, local diversity is directly proportional to that of the regional species pool.

4. When competitive saturation is added to the model, the relation between local and regional diversity depends on the disturbance rate. Local diversity is independent of the regional species pool, provided that the disturbance rate is sufficiently low, but the effects of even strong competitive saturation can be obscured by extremely low rates of disturbance. Rates of disturbance one-half to one-tenth of the rate of approach to local equilibrium produce communities where local diversity is directly proportional to that of the regional species pool, with slopes comparable to those found in empirical studies interpreted as evidence of unsaturation.

5. Empirical studies relating local and regional diversity must be interpreted with caution, because the absence of competition has effects that are indistinguishable from those of strong competitive saturation in the presence of modest levels of disturbance. Future empirical studies of local and regional diversity should include quantitative measurements of the rates of disturbance, competition, and other dynamic processes.

10

Three Explanations of the Positive Relationship between Distribution and Abundance of Species

Ilkka Hanski, Jari Kouki, and Antti Halkka

Ecology has been defined repeatedly as the study of the distribution and abundance of species (Andrewartha and Birch 1954; Krebs 1972), but ironically, little notice was given to the relationship between distribution and abundance until the 1980s. McNaughton and Wolf (1970) and Hanski (1982) were perhaps the first to emphasize and document that, almost without exception, species with more extensive distribution tend to be more abundant locally than species with more restricted distribution. The purpose of this chapter is to discuss three explanations of the positive relationship between distribution and abundance: *sampling, metapopulation dynamics,* and *ecological specialization.* We stress at the outset that we are not concerned here with why some species are more widely distributed than others, or why some species tend to be more abundant than others; we are asking specifically why the more widely distributed species should be more abundant than species with a more restricted distribution.

The relationship between distribution and abundance has important implications for the study of community diversity. Species turnover among local sites and along geographical gradients, or beta and gamma diversities, critically depend on how widely different species are distributed. If distribution and average abundance are correlated, we can expect different patterns of beta and gamma diversities in abundant and rare species. The species-area relationship and the patterns of species' joint occurrences at sets of sites are two other aspects of community diversity that are closely linked with the distribution-abundance relationship. We shall return to these issues in our discussion.

We start by defining distribution and abundance. The three explanations of the positive distribution-abundance relationship are next described and discussed, with the emphasis on how one can distinguish between the three alternatives in practice. Next we analyze a range of examples, including plants and animals, which cover spatial scales from the distribution of species within a small locality to the geographical ranges of species. Finally, we draw three main conclusions. First, the positive distribution-abundance relationship has no single, universal explanation. Second, although the sampling model generally fits the data well, it may not reveal the key processes that generate the distribution-abundance relationship. Third, we

find less support for the hypothesis of ecological specialization than for metapopulation dynamics as an explanation of the positive relationship between distribution and abundance of species.

DEFINITIONS

A source of confusion is that the term *distribution* has been used with at least two different meanings in the literature. The first meaning is a species' *geographical range,* the smallest area with a reasonably smooth outline that includes all observations of the presence of the species. We recognize that, given a set of observations with geographical coordinates, a range boundary may be drawn in different ways (Gaston 1990). We assume that some simple and biologically meaningful procedure can be agreed upon.

The second kind of distribution is the *fraction of area used by the species* within its geographical range or within some smaller naturally or arbitrarily defined region. As not all of this area is likely to be suitable for the species, it is more informative to calculate the fraction of the habitable rather than the total area that is occupied by the species. Often the habitable area comes in patches, in which case one may use the fraction of occupied patches as a measure of distribution (p in Levins's 1969 metapopulation model). When calculated for roughly equally sized patches, this fraction has been called the *incidence* of the species by Diamond (1975).

If different terms are needed for the two kinds of distribution, we suggest calling the first kind *geographical distribution* and the second kind *regional distribution.* Many examples of the distribution-abundance relationship (e.g., Gaston and Lawton 1990a) represent spatial scales smaller than the regional scale, "distribution" now being the fraction of samples taken from one local population. This kind of distribution, which is often called the *frequency* of the species (Hanski 1982), could be referred to as *local distribution.* The actual spatial scales naturally vary among different kinds of organisms.

Abundance is a more straightforward concept than distribution. By *local abundance* we mean the size (or density) of a local population. In practice, "local" often refers to an arbitrarily defined study area. While examining the relationship between distribution and abundance, one is expected to use the average or the median size of many local populations as a measure of abundance, and to include in that average only those sites at which a local population was recorded (the latter to avoid an obvious artifactual correlation between distribution and abundance; Wright 1991). In some analyses, however, the average is calculated across all sites, including the ones in which the species was absent. To avoid confusion, we have used the symbols x_{all} and $x_{present}$ for the average abundance at all sites and at occupied sites, respectively. In studies dealing with geographical distributions, abundance data are often available for only one locality. Gaston and Lawton (1990a) demonstrate how the distribution-abundance relationship is affected by how representative this "reference locality" is of the study area as a whole.

One further methodological point is worth stressing. As both distribution and abundance may change over time, they should both be estimated within a sufficiently short period of time. This is a potentially serious problem with many atlas studies, on which estimates of distribution are often based, because such studies typically accumulate observations over long periods of time.

Assuming now that problems of measuring distribution and abundance have been solved, we turn to the main issue of this chapter. Empirically, ecologists have observed a positive relationship between the distribution and abundance of species, regardless of which definition of distribution is used, regardless of the spatial scale studied, and regardless of whether data for one or several species are analyzed (Hanski 1982; Bock and Ricklefs 1983; Brown 1984; Gaston and Lawton 1990a; Kouki and Häyrinen 1991). In the next section we examine three possible mechanisms that may yield the positive distribution-abundance relationship. In anticipation of our conclusions, we submit at this point that it may be very difficult in practice to discriminate among the three explanations. But one may make some progress by asking what other predictions apart from the positive distribution-abundance relationship the different hypotheses make, and by using such extra information while studying specific examples. In particular, we focus on what the three hypotheses say about deviations from the generally positive relationship between distribution and abundance. Two of the three, the sampling model and ecological specialization hypotheses, are more widely recognized (Brown 1984; Gaston and Lawton 1990a) than the third, the metapopulation dynamics hypothesis (Hanski 1991). We shall therefore spend more space in describing the latter one, although we do not intend to imply that metapopulation dynamics is a more likely explanation of the distribution-abundance relationship than is sampling or ecological specialization.

THE THREE EXPLANATIONS

The Sampling Model

The positive relationship between distribution and abundance may be a sampling artifact. Because locally rare species are more difficult to detect than are locally abundant species (McArdle 1990), the number of sites at which a species is found with some fixed scheme of sampling is a monotonically increasing function of the average abundance of the species. Wright (1991) discusses how observed distribution-abundance patterns can be tested against the simplest sampling model, the Poisson model. Unfortunately, this model is of little practical value, as nearly all species show an aggregated spatial distribution at all spatial scales (Taylor, Woiwod, and Perry 1978).

Assuming that the distribution of local abundances is negative binomial, the fraction of empty sites p_0 is given by

$$(10.1) \qquad p_0 = \left(1 + \frac{x}{k}\right)^{-k},$$

where x is the average abundance calculated across all sites (x_{all}) and k is a parameter describing the degree of aggregation. k can be expressed in terms of x and CV, the coefficient of variation of abundances (standard deviation divided by the mean, calculated across all sites),

$$k = 1/(CV^2 - 1/x).$$

Substituting this expression for k in equation 10.1, and taking logarithms, yields the equivalent equation

$$(10.2) \qquad -\ln p_0 = \frac{1}{CV^2 - x^{-1}}(\ln x + 2\ln CV).$$

Because the dependence of $1/CV^2$ on CV is stronger than the dependence of $\ln CV$, we may approximate this equation for large values of x by

$$(10.3) \qquad \ln(-\ln p_0) \approx \ln(\ln x) - 2\ln CV.$$

This equation indicates that the fraction of empty sites increases, and hence the regional distribution (p) decreases, with decreasing average abundance (x_{all}) and with increasing spatial variation (CV_{all}). We use equation 10.3 to test whether CV has a significant positive effect on p_0 (negative effect on p) when the effect of x is controlled for.

Ecological Specialization

Brown (1984) suggested that interspecific differences in ecological specialization may explain the positive relationship between distribution and abundance. There appear to exist two main versions of this hypothesis (McNaughton and Wolf 1970): either "species able to exploit a wide range of resources become both widespread and locally abundant" (Gaston and Lawton 1990a); or there are no interspecific differences in the degree of ecological specialization, but the resources used by some species are both more widespread and locally more abundant (or productive) than are those used by some other species, yielding the positive distribution-abundance relationship. The latter version of the hypothesis effectively moves the problem one trophic level downward, to a question of why some resources are both more widespread and locally more abundant than others. The first version of Brown's (1984) hypothesis is straightforward to test: it predicts that ecological generalists are *both* widely distributed and locally abundant, while specialists have *both* a restricted distribution and low average abundance (fig. 10.1). While testing this prediction one should be careful not to use a measure of specialization that is affected by distribution or by abundance.

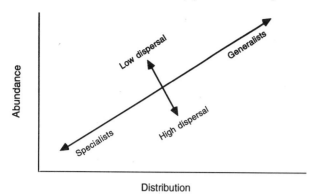

Figure 10.1 A schematic representation of the main predictions of the hypotheses of ecological specialization and metapopulation dynamics.

Metapopulation Dynamics

Andrewartha and Birch (1954) emphasized how "a natural population occupying any considerable area will be made up of a number of . . . local populations." Following Levins (1969), an assemblage of such local populations connected by dispersal is called a *metapopulation*. The question about the relationship between distribution and abundance arises naturally in the context of metapopulation dynamics, but here distribution is used strictly in the sense of regional distribution, the fraction of habitat patches occupied, p. Throughout this section, the average size of local populations refers to the average at occupied sites, $x_{present}$.

The pioneering model of single-species metapopulation dynamics by Levins (1969, 1970) assumed that the environment consists of infinitely many habitat patches of the same size and quality, and that the size of local populations occupying these patches is either 0 (extinct) or K (local carrying capacity). In other words, the simplest metapopulation models (patch models) ignore local dynamics, apart from the extinction and colonization events, and these models cannot therefore say anything about the relationship between distribution p and abundance x. To examine this question, we need to turn to structured metapopulation models, in which variation in local population size is taken into account in one way or another. Such models range from a three-stage extension of the Levins model (Hanski 1985) to fully structured metapopulation models, which model the complete distribution of local population sizes (Hastings and Wolin 1989; Gyllenberg and Hanski 1992).

We shall describe here a model due to Hanski (1991) and shown to be a limiting case of a fully structured model by Gyllenberg and Hanski (1992). The model assumes that the time scale of local dynamics is much faster than the time scale of metapopulation dynamics. All local populations are therefore assumed to be of equal size x, but unlike in the Levins model, x is affected by dispersal as well as by local processes.

Let us denote by D the number of dispersers per habitat patch. The following three ordinary differential equations give the rates of change in p, x, and D:

(10.4A) $dp/dt = \alpha\beta D(1 - p) - ep$
(10.4B) $dx/dt = -mx + \alpha D + rx(1 - x)$
(10.4C) $dD/dt = mpx - vD - \alpha D.$

The colonization rate of empty patches is assumed to be proportional to the number of dispersers and the fraction of patches that are empty. Most individuals arriving at empty patches are assumed to perish without giving rise to a new population, hence the rate of establishment of new populations (determined by β) is assumed to be small in comparison with immigration rate (α). For simplicity, we assume that all populations have the same extinction probability (parameter e). The number of dispersers, or individuals moving between habitat patches, increases due to emigration (described by m) from the existing local populations and decreases due to mortality (parameter v) and immigration to occupied and empty patches. Average local population size decreases due to emigration, which is assumed to be density independent, and it increases due to immigration and local population growth, the latter modeled by the logistic equation with the carrying capacity set to unity and the intrinsic growth rate given by r. Several empirical studies have reported that immigration may significantly increase the growth rate of small populations (Smith 1974; Holliday 1977; Gottfried 1979; Rey 1981; Rey and Strong 1983; Connor, Faeth, and Simberloff 1983; Fahrig and Merriam 1985).

At equilibrium, the following two relationships exist between x and p:

(10.5) $x = (1 - a_1) + a_1 a_2 p$
(10.6) $x = (a_3/a_2)/(1 - p)$

where $a_1 = m/r$, $a_2 = \alpha(\alpha + v)$, and $a_3 = e/(\beta m)$. The intersection points of equations 10.5 and 10.6 give the model equilibria. We note in passing that the model has one stable equilibrium if a_3 is less than $(1 - a_1)a_2$; there is no positive equilibrium if a_3 is substantially greater than $(1 - a_1)a_2$; while in the intermediate cases there are two alternative equilibria (for details see Hanski 1991; Gyllenberg and Hanski 1992). It is especially noteworthy that alternative equilibria are not possible if immigration has no effect on the dynamics of the existing local populations.

Let us now restrict our attention to those parameter combinations that yield a positive stable equilibrium point (regardless of whether the trivial equilibrium is also stable or not). Assume that there is an assemblage of species with variation in the values of a_1, a_2, and a_3. What does the model predict about the relationship between x and p in such an assemblage?

If there is variation in only one of the three parameters a_1, a_2, and a_3, there is always a positive relationship between x and p. This is obvious in the case of a_1 and a_3, because they occur in only one of the two equations 10.5 and 10.6, hence one isocline remains unchanged, and all the equilibria are located along this isocline, which gives a positive relationship between x and p. If there is variation in a_2 only, both isoclines are affected, but a scatter of equilibria for different values of a_2 still shows a positive relationship between x and p. Our first conclusion is that if the species in our hypothetical assemblage show variation in only one biological parameter (actually, in one of the three combinations of parameters), we would expect a positive relationship between distribution and abundance due to metapopulation dynamics.

In many natural communities, species differ from one another in several respects, and one may ask whether the model would also predict the positive distribution-abundance relationship in these cases. Figure 10.2A shows the answer for a range of parameter values. There is a significant positive relationship between x and p, though there is much scatter around this relationship. The results in table 10.1 and in figure 10.2B indicate that much of this variation can be explained by a_1 and a_2 (the effect of a_3 was much smaller). Thus, species that have

Table 10.1 Multiple Regression Analysis of the Residual from the Regression of x on p in figure 10.2a against the parameter combinations a_1 and a_2

Variable	Estimate	SE	Student's t	P
Constant	0.012	0.009	1.32	.1847
a_1	−0.500	0.016	−31.87	<.0001
a_2	0.322	0.012	26.65	<.0001

Note: See equations 10.5 and 10.6.
Model: $R^2 = .83$, $F = 832.9$, $df = 337$, $P < .001$.

large local populations (large x) but restricted distribution (small p) are predicted to have a high local growth rate (r) in relation to the emigration rate (m), and low mortality during dispersal (small v). The opposite attributes characterize species with low average abundance but wide distribution. Our second conclusion is thus that we may also expect a positive relationship between distribution and abundance in species assemblages in which the species show several kinds of differences, though now we expect more scatter around the relationship than when the species differ in only one respect. This scatter may be used to test the model, because it makes specific predictions about which kinds of species should deviate and in which direction from the general distribution-abundance relationship. In practice, the most useful prediction is that species with high dispersal rates should have a negative deviation (see fig. 10.1). Finally, we emphasize that although the metapopulation model predicts a generally positive relationship between distribution and abundance of species, it is possible to construct examples in which there is no relationship or a negative relationship (Gyllenberg and Hanski 1992). We shall return to this point in the concluding section.

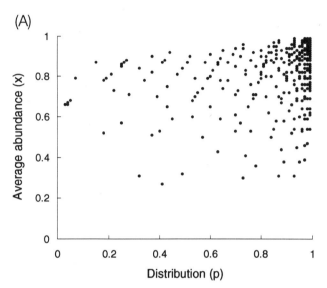

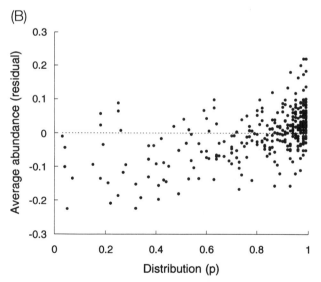

Figure 10.2 (A) The relationship between average abundance x and distribution p in the model of equations 10.4A–C. The stable equilibrium point (if it existed) was solved for several hundred combinations of parameter values ($a_1 = 0.1 \ldots 0.8$, $a_2 = 0.1 \ldots 1.0$, and $a_3 = 0.002 \ldots 1.0$). Regression of x on p is significant ($F = 21.80$, $df = 338$, $P < .001$, $R^2 = 06$). (B) As (A) but first removing the effect of parameters a_1 and a_2 on average abundance x. Regression of the residual on p is significant ($F = 130.2$, $df = 338$, $P < .001$, $R^2 = .28$).

EMPIRICAL RESULTS

Our aim in this section is to test the predictions of the three hypotheses just described. We shall first examine in detail one example, the distribution and abundance of butterflies in Britain, and we shall then summarize the corresponding results for a range of other taxa.

British Butterflies

Pollard, Hall, and Bibby (1986) present extensive transect count data on the abundance and distribution of British butterflies. We have used data collected in 1984 from 42 sites located throughout Britain (England, Scotland, and Wales; excluding sites with clearly atypical habitat, e.g., seashores and mountaintops). These data include 49 of the 58 butterfly species known from Britain. The data have not been properly standardized for between-site comparisons (Pollard, Hall, and Bibby 1986), but this should add to unexplained variance in our analysis and should not yield spurious relationships. The following additional information was taken from Thomas (1986): size of the species (average wingspan in mm); dispersal tendency (table 33.2 in Thomas 1984); habitat and foodplant specialization (our classification based on the descriptions in Thomas 1986); and family. As the degrees of habitat and foodplant specialization were correlated, we calculated a composite index of specialization as the average of these two variables.

There is a highly significant positive relationship between distribution and abundance ($x_{present}$) in British butterflies (fig. 10.3). The sampling model predicted that average abundance would have a positive, but the coefficient of variation a negative, effect on distribution (equation 10.3), and these effects were observed (table 10.2). The hypothesis of ecological specialization predicted that generalists would have a wider distribution and greater average abundance than specialists. In British butterflies, there is support for the first, but not for the second, prediction (table 10.3). The same result emerged from a study by Thomas and Mallorie (1985) on the butterflies of the Atlas Mountains: the species that occupied more habitat types in Morocco (generalists) tended to have larger geographical distributions in Europe ($R^2 = .18$, $P < .001$), even if there was no correlation between abundance (in Morocco) and geographical distribution (perhaps for the reason explained in Gaston and Lawton 1990a).

The metapopulation model predicted that, for a given average abundance, more dispersive species should have a wider distribution than less dispersive species. This prediction is supported by the data for British butterflies (ta-

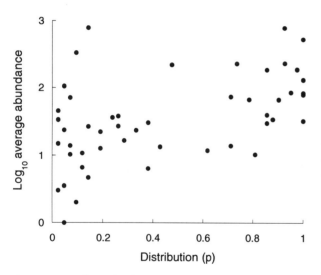

Figure 10.3 Relationship between distribution and abundance in British butterflies. Distribution was measured by the fraction of 42 localities distributed throughout the Britain in which the species was found in 1984. Abundance is the logarithm of the average size of local populations ($x_{present}$, annual counts; data from Pollard, Hall, and Bibby 1986).

Table 10.2 Test of the Sampling Model for British Butterflies Using Regression Analysis.

Variable	Estimate	SE	P
Intercept	0.663	0.225	.0059
ln (ln (x))	0.378	0.102	.0008
$-2 \times \ln (CV)$	0.945	0.093	.0001

Note: Coefficient of variation *(CV)* and average abundance (x_{all}) were calculated across all sites. The analysis tests whether the parameters of the sampling model differ from zero. The dependent variable is $\ln [-\ln (p_0)]$. See equation 10.3.
Model: $R^2 = 0.90$, $F = 132.16$, $df_{model}/df_{total} = 2/33$, $P < .0001$.

Table 10.3. Effect of Ecological Specialization on the Distribution and Average Abundance of British Butterflies.

Category	Average $\log_e$-abundance	Distribution	n
Generalists	3.78 ± 1.38	0.89 ± 0.45	23
Intermediate	3.42 ± 1.68	0.70 ± 0.51	20
Specialists	3.09 ± 1.42	0.35 ± 0.17	6
ANOVA models:			
R^2	0.03	0.13	
F	0.62	3.56	
P	0.54	0.04	
df_{model}/df_{total}	2/48	2/48	

Note: The classifying variable in the analysis of variance is a compound index of ecological specialization, which takes into account habitat and food plant selection. Distribution was arcsin-sqrt-transformed and abundance was $\log_e$-transformed. The table gives means ± standard deviations.

Table 10.4 Effect of Species Dispersal Abilities on the Distribution and Abundance of British Butterflies

Variable	Estimate	S.E.	P
Intercept	−0.167	0.168	.326
Abundance	0.190	0.037	<.001
Dispersal ability	0.256	0.071	.001

Note: Multiple regression analysis with two independent variables, average abundance ($x_{present}$, $\log_e$-transformed) and dispersal ability ($\log_e$-transformed). The independent variable is distribution (arcsin-sqrt-transformed). Using partial sums of squares, the dispersal effect tests whether residuals from the distribution-abundance relationship can be explained by dispersal ability. Model: $R^2 = .42$, $F = 16.54$, $df_{model}/df_{total} = 2/48$, $P < .0001$.

ble 10.4). A complicating factor is that most of the dispersive species belong to two families, Pieridae and Nymphalidae. However, a stepwise multiple regression picked up species' dispersal behavior as the significant predictor, suggesting that taxonomy may not be important independently.

To summarize, the sampling model (equation 10.3) fits the data well, but it leaves open the crucial question of why some species have more aggregated distributions than others. Because the level of aggregation may depend on dispersal rate, the good fit of the sampling model may partly reflect the effect of dispersal on distribution, as predicted by the metapopulation model. We found little support for the hypothesis of ecological specialization.

Results for Other Taxa

Empirical analyses and results for ten taxa are summarized in tables 10.5 and 10.6. Depending on the nature of the variables, we used either analysis of variance, analysis of covariance, or regression analysis (table 10.5). The analyses were done using the same procedures as for the British butterflies. The taxa studied included dung beetles from Africa (Hanski and Cambefort 1991), terrestrial gastropods from northern Finland (Fosshagen, Palmgren, and Valovirta 1972), freshwater aquatic plants from southern Finland (Maristo 1941), coastal meadow plants (T. Ryttäri, unpublished data) and rich fen plants from southern Finland (H. Heikkilä, unpublished data), carabids from southern Norway (Andersen et al. 1990), parasites of the lesser scaup duck (Bush and Holmes 1986a), and dytiscid beetles from central Finland (Lindberg 1948; data compiled by E. Ranta).

The relationship between distribution and abundance

Table 10.5   Ten Data Sets Used to Test Three Hypotheses about the Distribution-Abundance Relationship (DA)

Group	Number of species	Number of localities	R^2 of DA relationship	Ecological specialization Class[a]	Ecological specialization Analysis[b]	Dispersal Class[a]	Dispersal Analysis[b]
Butterflies	49	42	0.25	3	ANOVA	cont.	REG
Dung beetles, local	35	20	0.73	cont.	REG	—	—
Dung beetles, regional	119	6	0.33	cont.	REG	—	—
Terrestrial gastropods	12	32	0.65	cont.	REG	3	ANCOVA
Freshwater plants	56	20	0.04	cont.	REG	—	—
Coastal meadow plants	51	22	0.20	cont.	REG	—	—
Carabids	25	9	0.38	cont.	REG	3	ANCOVA
Parasites of lesser scaup	15	45	0.58	2	ANOVA	—	—
Rich fen plants	18	35	0.00	cont.	REG	2	ANCOVA
Dytiscids	20	18	0.63	—	—	2	ANCOVA

Note: Data sources (DA, distribution-abundance data; S, specialization data; D, dispersal ability data): butterflies, DA: Pollard, Hall, and Bibby 1986, S, D: Thomas 1986; dung beetles, local and regional, DA, S: Hanski and Cambefort 1991; terrestrial gastropods, DA: Fosshagen, Palmgren, and Valovirta 1972, S, D: Baur and Bengtsson 1987; freshwater plants, DA, S: Maristo 1941; coastal meadow plants, DA: T. Ryttäri, unpublished data, S: Hämet-Ahti et al. 1986; carabids, DA: Andersen et al. 1990, S, D: Den Boer 1990; Lindroth 1949, 1985, 1986; parasites of lesser scaup, DA, S, D: Bush and Holmes 1986a; rich fen plants, DA: H. Heikkilä, unpublished data, S: Eurola, Hicks, and Kaakinen 1984, D: Jalas 1958, 1965, 1980; dytiscids, DA: Lindroth 1949 (abundance data compiled by E. Ranta), D: Jackson 1956; Ericsson 1972.
[a]Either number of classes or an indication that the independent variable is continuous.
[b]Type of analysis: ANOVA, analysis of variance; ANCOVA, analysis of covariance; REG, regression analysis.

Table 10.6.   Results of Tests of Three Hypotheses about Distribution-Abundance Relationships

Group	Scale[a] L	Scale[a] M	Scale[a] G	Sampling available	Sampling Support[b] x	Sampling Support[b] CV	Specialization available	Specialization Support[b] x	Specialization Support[b] p	Dispersal Available	Dispersal Support[b]
Butterflies		x	x	Yes	Yes	Yes	Yes	No	Yes	Yes	Yes
Dung beetles, local	x			Yes	Yes	Yes	Yes	No	No	No	—
Dung beetles, regional			x	(Yes)	Yes	Yes	Yes	No	No	No	—
Terrestrial gastropods	x	x		Yes	No	Yes	Yes	No	No	Yes	No
Water plants		x	x	Yes	No	Yes	Yes	No	Yes	No	—
Coastal meadow plants	x	x		Yes	Yes	Yes	Yes	Yes	No	No	—
Carabids		x		Yes	No	Yes	Yes	No	No	Yes	No[c]
Parasites of lesser scaup		x		No	—	—	Yes	No[d]	No[d]	No	—
Rich fen plants		x	x	Yes	No	No	Yes	No	No	Yes	No
Dytiscids	x	x		Yes	Yes	Yes	No	—	—	Yes	No

[a]Indicates whether the study is primary concerned with local (L), metapopulation (regional, M) or geographical (G) scale.
[b]Yes, significant ($P < .05$) effect; No, no significant effect; —, analysis could not be done.
[c]Worst dispersers have largest negative deviation from the DA relationship.
[d]Significant but to other direction than predicted.

was positive and significant in all taxa except freshwater aquatic plants and lush fen plants (table 10.5). The R^2 of the distribution-abundance relationship varied between zero and .73. Weak distribution-abundance correlations in the plant data may partly result from the use of cover as a measure of abundance. As Harper (1981) has observed, "a (plant) species may appear abundant in a community because its individuals are of large biomass or because, although they have small biomass, they are present in very large numbers." Harper uses as an example the bracken *Pteridium aquilinium,* which is one of the five most abundant plant species on earth but which in most places is abundant by virtue of the enormous vegetative extent and biomass of a few genetic individuals. It is not clear how abundance of plants should be best measured in this context.

The sampling model was supported for nearly all taxa (table 10.6), and often the residual variation was negligible. The order of the taxa according to the R^2 of the sampling model was as follows: dytiscids ($R^2 = .99$), aquatic plants (.96), gastropods (.96), coastal plants (.91), butter-flies (.90), dung beetles at the local scale (.83), carabids (.64), lush fen plants (.45), and dung beetles at the regional scale (.39). For the parasites, the sampling model could not be tested because appropriately calculated averages and CV's were not available. We conclude that the sampling effect cannot be generally ruled out as a possible cause of the positive distribution-abundance relationship.

To test the hypothesis of ecological specialization, we classified the species into different categories of specialization whenever suitable data were available. Seven taxa could be analyzed, but none of these supported the hypothesis (table 10.6). In some cases (gastropods and coastal plants) there was a significant relationship between specialization and either distribution or abundance, but not with both of them as predicted by the hypothesis. The helminth parasites of the lesser scaup duck are a notable exception: there was a significant relationship between specialization and distribution, and between specialization and abundance, but in the opposite directions than predicted, generalists having smaller "distributions" (numbers of host individuals occupied) and lower average abundances than specialists. Such a situation seems to be typical for other parasites as well (e.g., Goater and Bush 1988; Haukisalmi 1986).

Measures of dispersal ability were available for only four taxa other than British butterflies, and none of them supported the metapopulation hypothesis (table 10.6). In carabids, dispersal ability (as measured by the presence of long wings) was actually negatively correlated with distribution. Note, however, that our data on dispersal ability is biased toward the local scale (table 10.6), and hence these results are not sufficient to reject the metapopulation hypothesis as an explanation of the distribution-abundance relationship at the regional scale. For instance, even short-winged carabids may be good dispersers at short distances (E. Halme, personal communication), and it is in fact questionable whether flight is the best means of dispersing short distance (Den Boer 1990). The gastropod dispersal data were mostly derived from the order of colonization of two Finnish archipelagoes, and they may reflect species' densities on the mainland more than they do the species' dispersal tendencies. Finding suitable correlates of dispersal ability or rate is also difficult in plants. For example, Forcella (1985) could find no such correlates in seed characteristics of weeds, for which good data on dispersal rate were available. In the lush fen plants (table 10.5), the best dispersers are the ones with good vegetative growth, but as abundance was measured by cover, the best dispersers probably received disproportionately high abundance values.

The metapopulation model is supported by Söderström's (1989) recent study of epixylic bryophyte species in stands of late-successional spruce forests in northern Sweden, even though his data were not suitable for our analyses. Söderström (1989) found a positive relationship between distribution and abundance, the latter being estimated as percentage of cover on available logs in occupied forest stands. There were three deviating species with large x but small p, aptly called the "urban species." Among all the species, only the three urban species regularly produced gemmae (asexual reproduction) but were not observed to reproduce sexually. Most likely, these species have high local growth rates but low dispersal rates, a combination of traits expected to yield high x but low p.

Discussion

The sampling model is necessarily correct, in the sense that it must make a contribution to the positive relationship between distribution and abundance, whenever distribution is underestimated. The sampling model is least likely to make a significant contribution at the geographical scale because in many taxa, species' geographical distributions are relatively well known. In contrast, the effect of sampling can hardly ever be dismissed at smaller spatial scales, when dealing with species' local or regional distributions, because many small populations will remain unrecorded. Our results also unequivocally support the effect of aggregation on distribution: for a certain level of average abundance, species occur at more sites if they are less aggregated than if they are more aggregated. The sampling model fits so well in most cases (see table 10.6)

that there remains little variance to be explained. In this sense, the positive distribution-abundance relationship may be considered an "artifact" of sampling.

Nonetheless, this is not necessarily the most helpful perspective on the distribution-abundance relationship, especially at the regional scale. One may ask why some species are more aggregated than others. The metapopulation model incorporates one factor that is often likely to be important, species' dispersal rates. A high dispersal rate both decreases the variance in the numbers of individuals among occupied sites and increases the number of occupied sites. British butterflies, for which relatively good data are available, are a case in point. The sampling model fits these data well, but so does a model that explains distribution with average abundance and species' dispersal rates. The latter model is biologically more informative. We suspect that the reason why dispersal rate had no effect in the other taxa in table 10.6 was the poor quality of the data on dispersal that was available for our analyses.

Some other results and observations support the metapopulation hypothesis. Figure 10.4 shows the final distribution of alien weeds in the northwestern United States as a function of their rate of spread. It is clear that the species with higher dispersal rates finally occupied more localities than did species with lower dispersal rates. Unfortunately, no suitable data on local abundances were available in this case.

It is often found that local density decreases with increasing isolation of habitat patches, presumably because of the low rate of immigration to isolated patches. Huffaker's (1958) well-known experiment with mites feeding on oranges and the study by Fahrig and Merriam (1985) on *Peromyscus leucopus* inhabiting more or less isolated woodlots are examples from the laboratory and the field, respectively, demonstrating how isolation influences average population size in metapopulations. These observations are predicted by metapopulation dynamics, but not

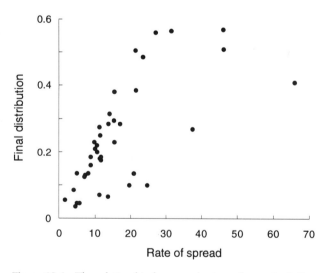

Figure 10.4 The relationship between the rate of spread of alien weeds in the northwestern United States and their final distribution, measured as the fraction of counties occupied. (Data from Forcella 1985.)

by the static sampling model nor by ecological specialization.

The metapopulation model does not predict that there is always a positive relationship between distribution and abundance. We cite two examples in which there is no positive relationship between distribution and abundance, but which both appear to be compatible with the metapopulation hypothesis. First, Gaston and Lawton (1990b) found no relationship between distribution and abundance among the freshwater fishes of Britain. The metapopulation hypothesis predicts no relationship between distribution and abundance when migration plays no role in local dynamics, which may well be the case with freshwater fishes. Second, Arita, Robinson, and Redford (1990) reported a negative relationship between distribution and abundance in Neotropical forest mammals. Their result is essentially due to large mammals having low densities but wide distributions, while smaller species tend to have more restricted distributions but are often locally abundant. This is consistent with the metapopulation hypothesis, because large species have larger m/r (high dispersal rate in relation to the intrinsic growth rate) but probably smaller e/β than do small species (large species are less affected by environmental stochasticity; see the discussion of metapopulation dynamics above; Gyllenberg and Hanski 1992). As Brown (1984) has observed, the positive distribution-abundance relationship may not hold for species assemblages in which the species show many kinds of differences in their biology.

The metapopulation hypothesis is based on a dynamic, single-species model of population dynamics, and it predicts that temporal changes in the regional distribution and average abundance of any species are correlated. The sampling model makes the same prediction, but it is difficult to see how Brown's (1984) hypothesis about ecological specialization would explain such intraspecific correlations, unless one assumes that both distribution and abundance passively reflect temporal or spatial variation in resource availability (which is possible, but difficult to test). Figure 10.5 shows one example from *Gyrinus* beetles, in which both regional distribution and average abundance decrease toward the species' geographical range limit.

To conclude our discussion of the metapopulation hypothesis, we found several lines of evidence supporting it at the regional scale. The model described above is not appropriate for the local or geographical scales, but other related models including species' movement behavior could be constructed for these spatial scales.

We found no examples in which generalists had both wider distribution and greater average abundance than specialists. Since the hypothesis of ecological specialization predicts that the degree of specialization is negatively correlated with *both* distribution *and* abundance, refutation of either of these correlations should suffice to refute the hypothesis of ecological specialization as an explanation of the positive distribution-abundance relationship. Thus our results give little support to the hypothesis of ecological specialization (Brown 1984). Gaston and Lawton (1990a) reported that only half of the studies comparing niche width and local abundance found a positive re-

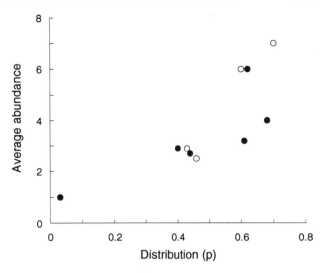

Figure 10.5   The relationship between regional distribution (fraction of temporary pools occupied) and average abundance in two different species of *Gyrinus* beetles (open and filled circles) at distances from 50 to 600 km from their geographical distribution limit. When the distribution limit is approached, both regional distribution and average abundance decrease. (Data from Svensson 1992.)

lationship between these two variables. Generally, we would not expect that generalists are locally more abundant than specialists, which would imply that ecological specialization has no advantages. On the other hand, we would be less surprised by a generally positive relationship between niche width and the size of geographical distribution (e.g., Kouki and Häyrinen 1991). Unfortunately, a problem in testing Brown's (1984) hypothesis is how to independently measure niche width, abundance, and distribution. If abundance is measured across several habitat types, and niche width measures habitat selection, there may be positive correlations between niche width, abundance, and distribution (e.g., Gallé 1986), but the reasoning may be circular.

Finally, we shall turn to wider issues in the study of community diversity. An important consequence of the positive distribution-abundance relationship is that the likelihood of pairs or larger sets of potentially interacting species occurring together at a randomly selected site is positively correlated with their average abundance. Because the potential for evolutionary and coevolutionary changes in interacting species increases with the frequency of their co-occurrence at the same sites as well as with increasing abundance (Thompson 1982; Thompson and Moody 1985; Brown and Kurzius 1987), the effect of distribution and the effect of abundance reinforce each other. This sort of amplification means that, over evolutionary time scales and over large spatial scales, a subset of species may interact strongly among themselves and coevolve in response to interspecific interactions, while the remaining species face continuously fluctuating and unpredictable interactions with one another. The latter species may evolve in response to the presence of the widely distributed and abundant species, but interactions among the rare species may be inconsequential. (For related discussions on community diversity see Caswell 1976; Hanski 1982.)

The almost universal positive relationship between distribution and abundance reminds us of another generally valid "law" in community ecology, the species-area relationship. There is a striking similarity between the two generalizations: the three explanations of the distribution-abundance relationship that we have discussed in this chapter match the three explanations examined by Connor and McCoy (1979) for the species-area relationship, namely, passive sampling (our sampling hypothesis), area per se hypothesis (metapopulation dynamics) and habitat-diversity hypothesis (ecological specialization). Connor and McCoy (1979) concluded their investigation about the mechanisms of the species-area relationship by noting that "there may be at least a grain of truth in each of these mechanisms. Each of these three, and possibly others, may play a role in producing the observed positive correlation between species number and area." We have reached very similar conclusions about the distribution-abundance relationship. The two "laws" of species-area and distribution-abundance relationships are not logically connected, but they may reflect something about the distribution and abundance of species that we do not yet perceive clearly.

11

Experimental Biogeography: Interactions between Stochastic, Historical, and Ecological Processes in a Model Archipelago

Daniel Haydon, Ray R. Radtkey, and Eric R. Pianka

If it had not been so hot on 20 July 1944 in Rastenburg, Germany, the geopolitical face of postwar Europe might have been completely different. Because of the heat, Adolf Hitler moved his daily conference from the underground bunker in which he normally received briefings to a wooden hut. When the bomb hidden in Colonel Stauffenberg's briefcase exploded, the force dissipated much more readily than it would have in the underground bunker; Hitler escaped. If the assassination attempt had been successful, World War II probably would have ended earlier and with a profoundly different peace settlement. Hitler was unsure whether to attribute his survival to chance, calling it "incredible luck," or to determinism, saying, "More proof that Fate has selected me for my mission" (Toland 1976, 799–800).

The Second World War will never be refought, and we will never know what might have happened had 20 July been a cold day. Similarly, biogeographical history cannot be replayed, leaving biogeographers with a dilemma: are today's biogeographical patterns the product of essentially deterministic processes, or are they contingent on a series of stochastic events? For example, how many species of *Anolis* would evolve if the last ten million years were played over again? What if the relative timings of eustatic and geological events in the Caribbean had been different? How would their distributions differ had *Anolis* populations been more extinction-prone or less constrained by interspecific competition? Colwell and Winkler (1984) attempted to address these problems: "If we could seed a series of virgin, replicate earths with primordial life and then set the level of interspecific competition differently in each, could we tell them apart three billion years later by looking at biogeographical patterns?" (p. 344). The processes of replication and repeatability, usually fundamental to the testing of scientific hypotheses, are not available to biogeographers. They must instead rely on comparative studies in order to arrive at general theories of diversity.

Most modern-day communities have developed side by side with changes in their physical environments, interacting with them in a necessarily stochastic manner. After millions of years, over which regional biotas are exposed to a unique sequence of physical changes, over which his-torical events build on one another, and during which reaction to the changing physical environment is a probabilistic process, one is left with a sample size of one community. In order to increase the sample size, it is necessary to incorporate communities that differ ecologically, historically, or both.

It is therefore pertinent to ask: To what extent are the relevant properties of biotas susceptible to these structuring influences? How resistant are various biotas to the vagaries of stochastic interactions? To what extent are biotas shaped by particular regional geographical history? Are there methods for ascertaining the influence of such processes? Every instance of faunal buildup generates only one result; therefore the nature of the distribution of all possible outcomes must be established before the result can be usefully compared with others. The species diversification process may feed back on itself positively (generating a large variance in the final species count), negatively (generating a smaller variance), or the process may be "memoryless" (of intermediate variance). By understanding the answers to these questions in an artificial setting, conclusions from comparative studies may be interpreted more meaningfully.

Given the differences in the life history strategies and dispersal abilities of different taxonomic groups, it seems likely that ecology, chance, and history will play different roles in shaping their respective biogeographies. Elucidation of the factors determining biogeographical distributions would be aided by knowledge of the influence chance, history, and ecology have on different types of organisms. Field biologists should be aware a priori of the possible biases in the influence of these three structuring forces, and take them into account when attempting to establish generalizations from comparative studies. Information of this nature is a prerequisite for the study of community patterns such as the species-area relationship, patterns of endemism, and niche packing.

Motivated by the challenges outlined by Ricklefs (1987) and inspired to "play God" by the work of Colwell and Winkler (1984), we propose to experiment with biogeography, exploiting the computer to model faunal buildup over ecological and evolutionary time scales in an omniscient fashion. We create and examined a model system in which vicariant events, phyletics, and ecology are all perfectly known; we exploit this hypothetical "archipelago" to explore interactions between ecological and historical processes and to determine their influence on local and regional diversity. In our model, geological history is an independent variable overseeing the interaction of ecologically prescribed taxa with chance events. Mechanisms are developed that allow the evolution of faunal buildup to be replayed repeatedly over a fixed geological history. This enterprise allows us to overcome the problems of unit sample size and to investigate the variability inherent in historical and ecological processes. In so doing, we are forced to adopt a broader perspective than commonly required in empirical studies and to formulate a synthesis from a wide spectrum of disparate processes operating on different scales. This process has generated some observations and questions for which we have no satisfactory explanations. Since our approach remains embryonic, we have been tempted into conjectures that, if not entirely justified, we hope will at least be provocative.

DESCRIPTION AND JUSTIFICATION OF OUR APPROACH

The Model

The functions and parameters used for our simulations are laid out in table 11.1. The model simulates the fragmentation of one large island of area (A_S) over an arbitrary number of "vicariant" time units (V_T). The probability of island i splitting in any one "vicariant" time unit is a decreasing function, $V(a_i)$, of island area a_i. Total area is conserved throughout the simulation, with the size of a "daughter" island determined as a randomly selected fraction of the "mother" island. Upon splitting, the two new islands are allotted randomly selected speeds and directions, which are maintained until the occurrence of a subsequent split. It is thus possible to model the develop-

ment of a random archipelago with a completely known geological history.

The simulation models the evolution of a radiation within an intermediate-ordered taxon. The original island is seeded with one species representative of the taxon being modeled. Properties of taxa that are considered to remain constant within a radiation are vagility, propensity for interspecific competition, and proneness to extinction.

Vagility is modeled using a function, $D(d_{i,j})$, that yields the probability of dispersal between islands i and j, separated by a distance $d_{i,j}$. Vagility is increased by decreasing the taxon-specific constant k.

Each species has coordinates describing its *mean* position in a two-dimensional niche space that correspond to the species' use of two hypothetical resource distributions. The original seed species is assigned to the point (0,0). Associated with the mean niche position is a function, $C(i, x, y)$, describing in what region of niche space species i (with niche coordinates x_i, y_i) would *potentially compete* for resources x, y in the vicinity of its mean niche position. Increasing q effectively broadens the niche and increases the proportion of resources sequestered by a species on any one island. This function permits an estimate of the level of interspecific competition experienced by a population of a species on a particular island. The function $N(a_i)$ relates the available niche space on island i to its area a_i; niche space available on smaller islands are nested subsets of larger ones. The mean niche position of a species randomly "walks" within the available niche space at a random pace, unless it wanders into an area of niche space for which interspecific competition for resources is above a certain threshold (T_C). In this case the niche position moves in the direction that minimizes competition at a pace proportional to the level of competition encountered.

An extinction function, $E(a_i, s_{ij})$, defines the probability that a population of species j on island i will go extinct, where s_{ij} is the level of interspecific competition species j is experiencing on island i. The exponent c allows us to alter the "proneness" to extinction of the taxa involved in any particular run.

Every vicariant time unit is subdivided into an arbitrary number of "ecological" time units (E_T) in which each population of extant species on every island has a probability of: (1) dispersing from the islands on which they are found to all other islands; (2) becoming extinct; and (3) niche position evolution. When a population of a particular species becomes sufficiently isolated (i.e., the probability of immigration from other conspecific populations falls below a specified minimum T_S), the population is considered to undergo speciation. A species can only successfully disperse to an island if its required resources (as defined by its niche coordinates) are present on that island (as defined by $N(a_i)$) and if the level of competition for those resources is below a specified threshold T_{ex}.

Unless two islands have identical species compositions, conspecific populations of a species on different islands will be exposed to different *local* interspecific conditions. The change in the species niche position is therefore the average of the forces acting on the niche position on each island population. When an island breaks up, it obviously

Table 11.1 Functions and Parameters Used in the Simulations

$V_T = 100$
$A_S = 100$
$V(a_i) = 0.2 a_i / A_S$ if $a_i > 5$; 0 if $a_i < 5$
$D(d_{ij}) = e^{-kd_{ij}}$
$k = 10, 1, 0.75, 0.5, 0.1$
Extent to which species i with niche coordinates (x_i, y_i) would potentially compete for resources (x, y):

$C(i, x, y) = e^{-2q\sqrt{(x_i - x)^2 + (y_i - y)^2}}$
$q = 1, 0.75, 0.5$
$N(a_i) = 20\, a_i / (50 + a_i)$ Niche radius $= \sqrt{N(a_i)/3.14}$

Probability of species i with niche coordinates (x_i, y_i) becoming extinct on island j with n species present in one 'ecological' time unit:

$E(a_i, s_{ij}) = e^{-ca_i}(1 + s_{ij})$
$c = 5, 0.9, 0.5$
where $s_{ij} = \sum_{r=1, r <> i}^{n} e^{-2q\sqrt{(x_r - x_i)^2 + (y_r - y_i)^2}}$
$T_c = 0.367$
$E_T = 20$
$T_s = 0.1$
$T_{ex} = 0.54$

becomes smaller, and the available niche space, defined by $N(a_i)$, declines. The biota on the two new islands are replicates of the original biota on the island before vicariance, except for those species whose required resources fell within the lost niche space (and therefore go locally extinct). Probabilistic events are deemed to have occurred if a random number (0–1) is selected less than the probability of the event occurring. Uniformly distributed random numbers are generated from a function utilizing a predefined seed.

Assumptions

No attempt was made to estimate parameter values from empirical data; the model was developed entirely abstractly. The time scale is assumed large enough to allow at most two thousand sequential speciation events in any one lineage. In any simulation attempting to integrate an ecological with a geological time scale, it is necessary to view ecological time in a very coarse grain. It should be remembered that the time units associated with dispersal, extinction, and evolutionary rates correspond to large lengths of time. Parameter values were chosen that compromised parameter space sampling with computer time and storage limitations. For example, combinations of parameter values that allowed the size of phylogenies to exceed five hundred were avoided. It was not at all clear when the simulations should be halted; In time, the original contiguous land mass would break up into a fine dust and support no diversity at all. An arbitrary decision was made to stop the program shortly after the appearance of the fourteenth island. For the parameter values sampled here, it was not uncommon to see the total number of extant species peak and begin to drop off toward the end of the one hundred units of "geological" time (see fig. 11.3).

When a population of a particular species undergoes speciation, its evolutionary trajectory through niche space becomes decoupled from that of populations of its former conspecifics, and it will drift randomly (or shift in accordance with local interspecific competition should it be sufficiently intense) through the two-dimensional niche space. There is, however, no arbitrary shift of its niche coordinates purely as a result of speciation per se, although it is very likely that subsequent divergence will occur. This assumption is in contrast to other published work of this type (e.g., Endler 1982; Colwell and Winkler 1984; Raup and Gould 1974). Our model considers many fewer characters (be they morphological, behavioral, or whatever), and we assume that the cause of reproductive isolation lies in some other "dimension" of the population.

That species may under some conditions of low competition wander randomly around niche space is not to assume that populations are not exposed to directional selection processes. The simulation implicitly assumes the existence of *local* selection forces other than interspecific competition, which, over the scale of geological time, will not be unidirectional. Thus conditions are homogeneous on each island but implicitly heterogeneous over the archipelago.

For convenience, this simulation is presented in the context of an island archipelago developing over a very long period of time. It could easily apply to any sort of large-scale, long-term fragmentation of habitat. An important additional assumption is that the gross biology of the taxa under consideration does not evolve; it is assumed that the probability of dispersal over specified distances remains constant throughout the simulation, as does the probability of the population going extinct under specified interspecific conditions on islands of a particular size. Although these assumptions are not completely true, we believe they are not unreasonable when considering such disparate taxonomic groups as snails, lizards, finches, gulls, etc. Our model can be thought to operate over any taxonomic scale within which these constraints are met.

The ecological functions ($D(d_{ij})$, $C(i, x, y)$, $E(a_i, s_{ij})$) describe only the numerical properties of the functions without implying any causal process. Therefore, low extinction rates could be due to inherent properties of the population dynamics of the species (e.g., good bottlenecking ability: Roughgarden 1986), the nature of the environment, or maybe some interspecific interaction other than competition. Dispersal probabilities could derive from the spatial scale of the vicariant process or the vagility of the taxa. The extent of competition could be the result of differences in niche width, "keystone" predation (Paine 1966), or the absence of resource limitation.

It should be apparent from the description of the model that the extent to which competition occurs on the islands is not directly controlled. The parameter q determines the intensity of competition between two sympatric species separated by a given distance in niche space (but makes no assumptions as to the nature of the competition or how the species arrived at their respective positions). This parameter also plays a critical role in determining whether or not a species can successfully colonize an island (through T_{ex}), the probability of extinction (through $E(a_i, s_{ij})$) and the evolutionary trajectory through niche space. The parameter q is instrumental in the generation of diversity over the entire developing archipelago, but only indirectly influences the overall frequency of competitive interactions or the importance of competition in determining the nature of the resultant communities.

The function $N(a_i)$ relating island area to available niche space on island i has no empirical basis. It only assumes that new niche area is added progressively more slowly as island area increases. Niche space is assumed to be circular, and the carrying capacity is assumed constant at all points under the resource distribution. It follows from this assumption that the risk of extinction is not *directly* dependent on a species' position in niche space.

The map of the archipelago over one hundred vicariant time units is illustrated in figure 11.1. The area cladogram corresponding to the development of the archipelago is shown in figure 11.2 together with the island sizes.

Combining the function $N(a_i)$ with $C(i, x, y)$ (the function describing the extent to which species compete in niche space) and T_{ex} allows calculation of a theoretical ceiling on diversity. Assuming species are distributed evenly over the niche space available on each island at a sufficient density that the level of competition for all re-

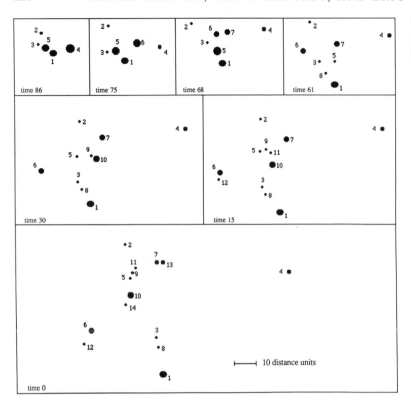

Figure 11.1 Map of the archipelago through geological time (100 time units ago to present). All diagrams are to the same scale. Size of island marker is approximately proportional to island area.

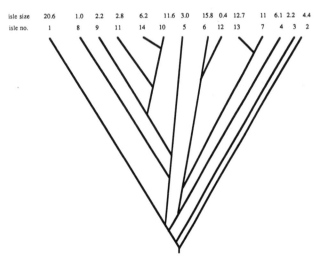

Figure 11.2 Area cladogram of the archipelago, with island sizes. Bifurcations are marked on a vertical scale of 100 time units.

sources just exceeds T_{ex}, then the island is invasion-proof to further immigrants, and an upper limit on the number of species supported by the island can be calculated. If this calculation is carried out for all the islands in the archipelago at different times in its development, and if endemism is assumed to be 100%, then one arrives at the total possible diversity, as illustrated in figure 11.3. Overlaid on the graph are typical curves for the buildup of extinct and extant species and the timing of the thirteen vicariant events.

Procedure

Different seeds were used for different random processes (i.e., vicariant and ecological processes). By holding the

seed responsible for the generation of the random archipelago constant, the vicariant history was made identical for all simulations. Each simulation yielded a complete record of the biogeographical history of the archipelago with respect to predefined levels of extinction, dispersal, and competition. One simulation was performed for each combination of extinction ($c = 0.5, 0.9, 5$), dispersal ($k = 0.1, 0.5, 0.75, 1, 10$) and competition ($q = 1, 0.5$) parameters. In order to investigate the influence of chance on these simulations, ten additional runs were performed for *each* combination of extinction and dispersal parameters with an intermediate level of competition: $q = 0.75$, $c = 5, 0.9, 0.5$; $k = 0.1, 0.5, 0.75, 1, 10$, changing only the seed responsible for nonvicariant events. The program was written in Pascal and run on a Sun 4 computer. Each run lasted from 1 to 30 hours, depending on the parameter settings; a total of 2000 hours of computer time was expended.

Sharpening the Concepts

For the purposes of clarity and conciseness, we will restrict the meaning of certain terms to particular aspects of the model. By *history* we mean the geological vicariant history of the islands (age, ancestry, and physical positions through time). The *effects of history* are defined as the detectable imprint of history on specified aspects of the biota of the archipelago. *Chance* is the phenomenon whereby a particular event with a specified probability occurs or does not. The *effects of chance* are those that result from changing the sequence of random numbers that the computer uses to determine the occurrence of chance events (*not* the probability of the events themselves). *Ecology* is restricted to the vagility, extinction, and competi-

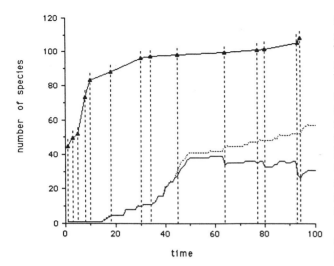

Figure 11.3 The theoretical upper limit to species diversity changes with the development of the archipelago. The curve shown (triangles) is for a taxon prone to high levels of interspecific competition (the curves for low and medium competition are exact multiples of this). The lower two curves illustrate gamma diversity (solid line) and the total number of species evolved (dotted line) for a highly competitive, highly vagile taxon that has a low extinction probability. The thirteen vicariant events (dashed lines) are superimposed on the graph.

tion functions of the species (outlined in table 11.1). The *effects of ecology* are those that result from changing the parameters in the aforementioned functions. In this model, history is fixed, ecology and chance are variables. The *effects* of history, chance, and ecology on one another and on the emergent biota are a complex interplay of the three phenomena.

The model clearly recognizes a "geological" time scale and an "ecological" time scale through the ratio of V_T: E_T. By manipulating these two constants, emphasis can be placed on geological or ecological time as required. For computational convenience, it is useful to make E_T moderately small and to increase dispersal and extinction probabilities and evolutionary rates accordingly (for example, by decreasing E_T from 1000 to 1 and increasing the probability of extinction from 0.00001 per year to 0.01 per 1000 years). This procedure does not affect the mean number of times such probabilities are realized, but it does affect the variance around the mean (a major focus of this study). Difficulties of integrating ecological and geological time are severe (Ricklefs 1989b). The manner in which the "paradox" between these time scales is resolved is likely to have a significant impact on the results obtained from such models. The problem, however, is not easily addressed in our model, because changing E_T would certainly change the random number sequence; any subtle trends would then be difficult to distinguish from those of chance.

RESULTS

The Effects of Ecology

The total number of species extant over the entire archipelago (gamma diversity) for different parameter settings

is illustrated in figure 11.4. The gamma diversity behaved much as expected, being highest at intermediate vagilities. There were, however, some occasions when an increased rate of extinction resulted in higher gamma diversity. When levels of endemism are sufficiently low, a high *local* extinction rate *could* vacate niche space on an island, enabling additional dispersal-speciation events that would result in the generation of a higher number of species over the whole archipelago. When endemism is at a very high level (mostly very low vagility taxa), local extinction is often accompanied by global extinction, and the above explanation is insufficient. These cases, however, show rather high coefficients of variation (see fig. 11.11), and chance alone could account for them.

Faunal diversification occurs either through vicariance and subsequent speciation of island faunas or through interisland dispersal and subsequent speciation. Dispersal probabilities decrease exponentially with distance between islands, but speciation only occurs after the probability of dispersal drops below T_S. There is, therefore, a limited range of interisland distances, falling within a relatively narrow "window," in which dispersal is at all likely and speciation possible. Island pairs whose distance from each other lies within these windows are likely to experience a local proliferation of species resulting from reciprocal dispersal and speciation (we call this the "ping-pong" effect).

For taxa of very low vagility, these windows will probably occur only between islands that have recently split; Hence, there will be relatively few such windows. For taxa of very high vagility, these windows will occur only between the most distant islands—of which there are also relatively few. But for taxa of intermediate vagility, there may be many such island pairs over geological time, allowing for much greater diversification. It follows that pairs of islands drifting apart may move through a window corresponding to a lower-vagility taxon and then, at some later time, move through the window corresponding to a higher-vagility taxon. The archipelago thus nurtures the proliferation of taxa with different vagilities at different stages of its geological development. Alternatively, archipelagoes with very different histories could generate equivalent sequences of such windows for very disparate types of taxa. Similarities in ecological and biogeographical patterns may emerge as a result of corresponding ratios of vagility to vicariant spatial scale, and not necessarily from any taxonomic proximity of the study organisms. Diversity is not a function of vagility alone, but is influenced by the creation of these speciation windows, and this is the probable explanation for the approximately equal gamma diversity of the taxa with low, medium, and high vagilities.

Total diversity (extant + extinct species) behaved similarly to gamma diversity with respect to vagility, but increased with extinction proneness (fig. 11.5). High extinction rates generated increased species turnover rates, vacating niche space for additional dispersal-speciation events. From figures 11.4 and 11.5, we see that, although taxa prone to high extinction rates are more proliferous, taxa prone to low extinction rates have higher extant diversity. (Note that, in the interests of clarity of graphic

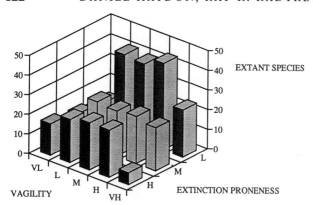

Figure 11.4 Mean gamma diversities for taxa of medium competition proneness. For taxa of very low (VL), low (L), medium (M), high (H), and very high (VH) vagility and high (H), medium (M), and low (L) extinction proneness.

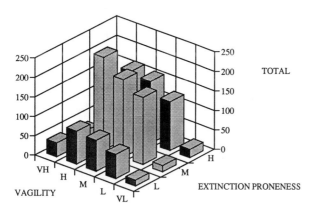

Figure 11.5 Mean total diversities (extinct + extant species) for taxa of medium competition proneness. (Abbreviations as in figure 11.4.)

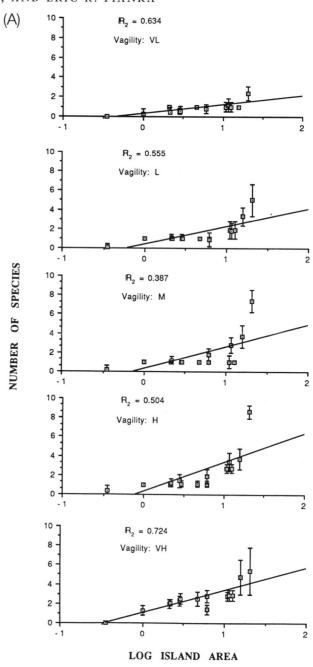

Figure 11.6 Species diversity (alpha diversity) plotted against the log of island area for taxa prone to medium levels of interspecific competition and to (A) high, (B) medium, and (C) low levels of extinction. Each point is the mean of ten runs in which all parameters were kept constant except for the seed used to determine the sequence of random numbers. One standard deviation is shown on either side of the point. Scanning down each column, the vagility increases from very low (VL) through low (L), medium (M), high (H), to very high (VH).

presentation, the orientation of the horizontal axes are frequently reversed in the figures.)

Alpha diversity was defined as the number of species on each island. Because larger islands have a greater total niche area, we expected larger islands to support more species. Given the arbitrary assignment of constants to the function relating niche space to island area, $N(a_i)$, no particular mathematical relationship between local alpha diversity and area was expected, however. Figure 11.6 shows the observed species-area relationship plotted semilogarithmically. Each point is the mean of ten runs; one standard deviation around each point is shown. The slope of the least squares fit steepens with increasing vagility at all three extinction settings, which is indicative of the progressively more influential role of the larger islands in "housing" the diversity of the archipelago. This suggests that taxa with higher vagilities may exploit the additional niche space on larger islands more efficiently than do lower-vagility taxa. The goodness of fit is worst at intermediate vagilities (maximum gamma diversities). The percentage of endemic species varied from 33% for very high vagility, high-extinction, low-competition taxa to 100% for almost all very low vagility taxa. Beta diversity was defined as gamma diversity divided by mean alpha diversity over the fourteen islands (Whittaker 1972). Thus, beta diversity behaves very similarly to percentage of endemic species (fig. 11.7). As expected, beta diversity decreased with increasing vagility. For reasons similar to those discussed above, we expected a positive correlation between extinction rates and beta diversity, especially

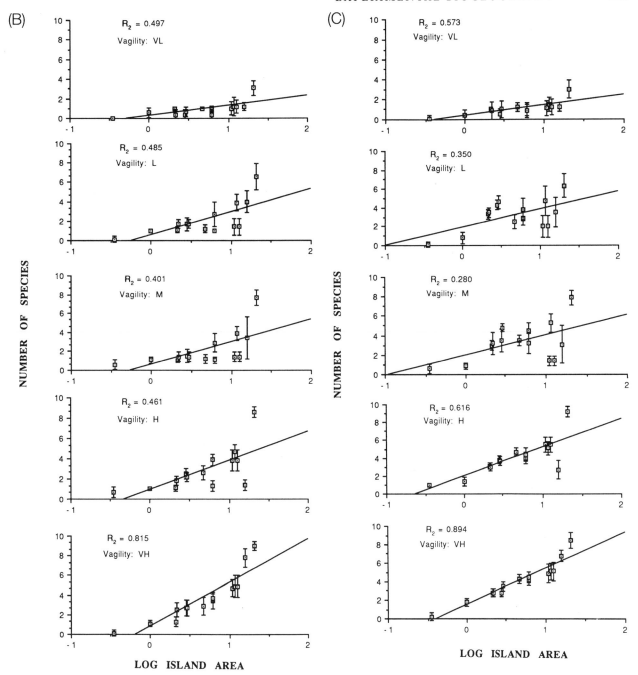

when endemism levels are low; however, proneness to extinction had virtually no effect on beta diversity.

Attempts to control the severity of competition in the simulations were largely unsuccessful due to compensatory adjustments in diversity. The process of diversification merely acted to fill up the niche space until a high level of competition inhibited it from doing so further; this required a higher species density when species' competitive capabilities were low than when they were high. It would seem that attempts to create low levels of competition in communities created this way are bound to fail if competitive constraints are imposed on a "per species"

basis. The competition parameter could only be used as an indirect control of species diversity.

The Effects of History

Chance, history, and ecology all combine to mold the size and topology of a taxon's phylogeny. On its own, the phylogeny provides no historical information. In combination with data on species distributions or elements of geological history, however, phylogenies are powerful and essential tools for analyzing the effects of history on the evolutionary development of any particular biota. An index quantifying the influence of history should obviously

124 DANIEL HAYDON, RAY R. RADTKEY, AND ERIC R. PIANKA

be related to the amount of historical information present in the distributions of species over the archipelago and be proportional to the accuracy with which the geological history can be reconstructed from biogeographical information. Our data were analyzed for the presence of historical impact in two different ways. The first assumes complete knowledge of the geological and biological history, the second, only the phylogeny and distributions of the species on the islands.

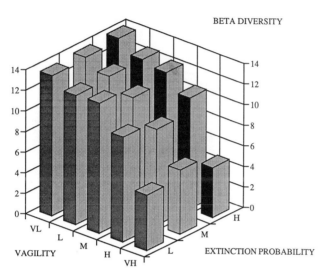

Figure 11.7 Mean beta diversities over ten runs with different seeds for taxa of medium competition proneness. (Abbreviations as in figure 11.4.)

If the island cladogram and the distribution of species on the islands are fully known, then the importance of history in shaping the phylogeny will be reflected in the substantial intersection between sets of species formed by the union of species assemblages on islands belonging to the same "geological" clade, and sets of species belonging to a phylogenetic clade. Figure 11.8 and Table 11.2 demonstrate the development of an index that measures the degree of concordance between the island area cladogram and a given phylogeny. Figure 11.9 shows the application of this index to the phylogenies resulting from replication at medium levels of competition. Each bar is the mean of ten simulations and represents the concordance above and beyond that expected to occur by chance alone. Observed concordances vary between 20% and 96% above expected values (i.e., those expected to occur if species were distributed randomly over the islands). In general, historical constraints become more important as vagility decreases and extinction rates increase. High extinction rates (resulting in a modern extant fauna) have the effect of continually updating the relationship between geological past and biogeographic distributions, eliminating the influence of "deep history" but deepening the imprint of more contemporary history.

In practice, an accurate vicariant history is rarely available; when this is the case, an alternative indication of historical constraint would be a higher-than-expected relatedness between sympatric species. Accordingly, the mean relatedness (as measured by the number of nodes between two species on the phylogeny) was calculated between all extant sympatric species pairs on each island, and then

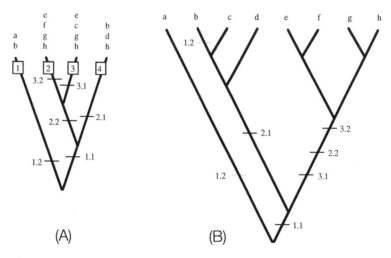

(A)   (B)

Figure 11.8 Calculation of a concordance index (H) between area cladograms coupled with species distributions and the species' phylogenies for a hypothetical data set. (A) Area cladogram for four islands (1–4). Species present on each island are indicated at the termini (a-h). Six area clades are denoted (1.1–3.2). (B) Species represented in these six area clades are "mapped" onto the known species phylogeny. The objective is to match sets of species represented in area clades with sets of species constituting clades within the known species phylogeny, and to establish to what extent the similarity of each pair of sets exceeds that expected from a random distribution of species on islands. For each of the six sets formed by the union of species assemblages of islands belonging to area clades (1.1–3.2), the species phylogeny is scanned for the species clade containing the most similar set of species. These

"most similar" clades are indicated on the known species phylogeny (by a dotted line when there is more than one "most similar" clade.) The similarity of the sets to each other is quantified using Sorenson's index (Pielou 1969) (see table 11.2). Because the size of the species pool is known, one can also calculate analytically the expected value of Sorenson's index for two sets of known sizes if the sets were assembled randomly from the species pool. These observed and expected values of concordance are averaged over all clades, and the ratio between them is taken as an estimate (H) of historical constraint on the species' biogeographical distributions. In this example, the concordance between the island cladogram and species phylogeny is 30.5% over what would be expected by chance alone, indicating a substantial degree of historical constraint.

Table 11.2. Calculation of Observed and Expected Sorenson's Index

Clade	Spp. common to both clades	Spp. unique to area clade	Spp. unique to phylogenetic clade	Observed Sorensen's index	Expected Sorensen's index
1.1	7 (b,c,d,e,f,g,h)	0	0	1.000	0.877
1.2	1 (a), or (b)	1 (b), or (a)	0	0.667	0.667
2.1	2 (b,d)	1 (h)	1 (c)	0.667	0.523
2.2	4 (e,f,g,h)	1 (c)	0	0.889	0.603
3.1	3 (e,g,h)	1 (c)	1 (f)	0.750	0.571
3.2	4 (e,f,g,h)	0	0	1.000	0.571
			Means	0.829	0.635

Note: Calculations for hypothetical data set presented in figure 11.8.

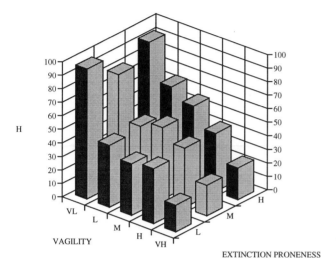

Figure 11.9 Mean concordance indexes (*H*, see figure 11.8) over ten runs with different seeds for taxa of medium competition proneness. (Abbreviations as in figure 11.4.)

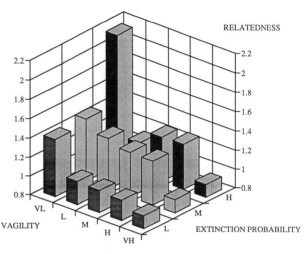

Figure 11.10 Mean ratios of expected to observed relatedness of sympatric species (R) over ten runs with different seeds for taxa of medium competition proneness. (Abbreviations as in figure 11.4.)

this quantity was averaged over the entire archipelago. The ratio of this figure to the mean relatedness of all extant species pairs inhabiting the archipelago was taken as an index of historical constraint. When the index is greater than 1, it implies that sympatric species are more closely related to each other than expected; when it is less than 1, sympatric species are less related than expected. It is evident from figure 11.10 that this measure is not as sensitive to the ecological context as is the concordance index. The broad trends shown by the two measures are, however, entirely consistent, despite the use of less information in the relatedness index. It appears that by taking into account the geological ancestry of the islands, the concordance index is made more sensitive to subtle associations in species distributions arising from long past geological associations.

Various idiosyncrasies in the biogeography of this archipelago, attributable to history, can be observed by matching the species-area relationships (see figure 11.6) to the map of the archipelago over time (see figure 11.1). For example, the largest island (island 1) consistently floats above the predicted alpha diversity; for low extinction rates, the second largest island (island 6) is often found substantially below the predicted diversity. The two most recent islands (13 and 14) often fall below the predicted diversity. These idiosyncrasies reflect the predictable effects of island isolation, size, and age on biogeography, as developed by MacArthur and Wilson (1967).

Effects of Chance

The effects of chance on the output of the simulations were assessed in terms of the standard deviations (SDs) and coefficients of variation (CVs) surrounding the means of the variables of interest. Figure 11.11 illustrates how the CVs around the mean gamma diversities vary with the ecological parameters. Chance has the greatest influence (CVs around 20%–35%) on those taxa with very low and very high vagilities; the effect is enhanced at the extremes of extinction proneness. These parameter settings correspond to those communities with a natural tendency to low diversities.

In contrast, chance had a relatively low impact on beta diversity. The CVs surrounding the means illustrated in figure 11.7 never exceeded 7.9%, except in the case of the high-extinction, very high vagility taxa, whose CV was 33.7%. The impact of chance on alpha diversity and the species-area relationship can be seen from inspection of the error bars (representing one standard deviation each way) on the graphs presented in figure 11.6. There are no very clear or simple patterns: some islands seem to be particularly vulnerable to the effects of chance (for example, the second largest island), while others are very predictable (many of the intermediate-sized islands). The error bars are broadest at intermediate vagilities (corresponding to the highest gamma diversities).

Chance not only affects the observed ecological pat-

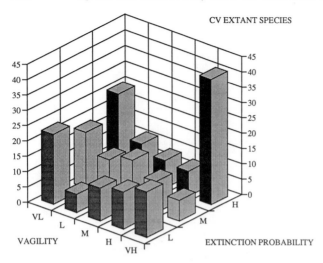

Figure 11.11 Coefficients of variance (SD/mean × 100) of the gamma diversities over ten runs with different seeds for taxa of medium competition proneness. (Abbreviations as in figure 11.4.) The means corresponding to these coefficients are illustrated in figure 11.4.

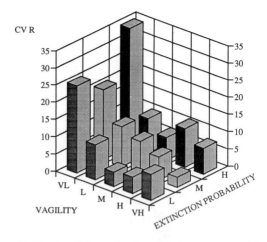

Figure 11.13 Coefficients of variance (SD/mean × 100) of the ratio of expected to observed relatedness of sympatric species (R) over ten runs with different seeds for taxa of medium competition proneness. (Abbreviations as in figure 11.4.) The means corresponding to these coefficients are illustrated in figure 11.10.

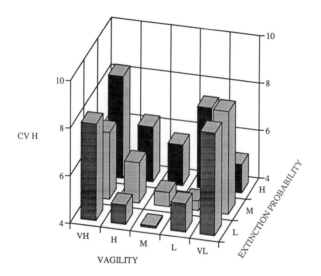

Figure 11.12 Coefficients of variance (SD/mean × 100) of the concordance indexes (H) over ten runs with different seeds for taxa of medium competition proneness. (Abbreviations as for figure 11.4.) The means corresponding to these coefficients are illustrated in figure 11.9.

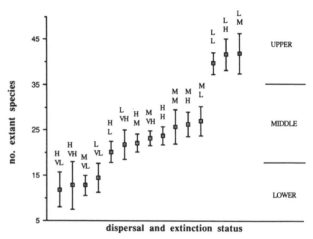

Figure 11.14 Mean gamma diversities of taxa of medium competition proneness with one standard deviation plotted on each side of the mean for the 15 different combinations of vagility and extinction proneness (each replicated ten times). Abbreviations as in figure 11.4: the top abbreviation corresponds to the level of extinction proneness; the lower, to the level of vagility.

terns but could also affect the imprint of history. Figure 11.12 illustrates the coefficients of variation around the mean values of the concordance index presented in figure 11.9. The percentages are all impressively low (2%–8%), yet there is a well-defined "valley" carved into the surface at intermediate parameters. In contrast, the CVs around the mean values of the expected: observed relatedness ratios (figure 11.3) decline almost monotonically from quite high levels (20%–30%) for the low vagilities down to the very high vagility taxa (3%–5%).

Figure 11.14 shows that for the fifteen different combinations of vagility and extinction proneness, there are only three readily distinguishable levels of gamma diversity. The three taxa among the group with the highest

gamma diversity are all of low extinction proneness with low, medium, and high vagilities. The two groups with lower gamma diversities are composed of an apparently miscellaneous combination of vagility and extinction proneness.

Evolution

Figure 11.15 illustrates the temporal development of phylogenies embedded in the two-dimensional niche space for three values of vagility at intermediate extinction rates. The phylogenies illustrate two major features. First, there is a high degree of variability in lineage stasis with time. The very low vagility taxa demonstrate very erratic and convoluted evolutionary trajectories through time. A high percentage of endemism coupled with a low density of species packing results in a measure of "freedom" with

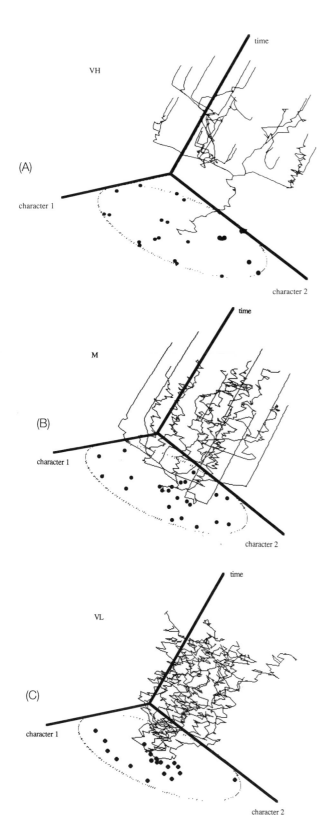

which to respond to the local conditions. At intermediate vagilities, the systems become more diverse, and "wandering" of a lineage is confined by the presence of interspecific competition. At very high vagilities, gamma diversity drops off, but the trajectories are stabilized by the fact that endemism is now at a very low percentage, and gene flow from other islands (on which conditions may be quite different) inhibits local response to changing conditions.

The second feature is the general level of "bushiness," which results from the geography becoming conducive to speciation at different times for taxa with different vagilities. For low-vagility taxa, speciation starts and continues with each vicariant event. At high vagilities, such "windows" for high speciation rates (as referred to earlier) do not open until much later in the development of the archipelago.

These ideas are supported by inspection of the correlation coefficient between phylogenetic relatedness (as measured by the number of nodes between any particular species pair) and morphological separation (the euclidean distance between a pair of species in niche space). These correlation coefficients (illustrated in figure 11.16) are especially high for very high vagility organisms and fall off very steeply to surprisingly low levels as vagility decreases. This relationship arises as a result of the negative correlation between vagility and endemism. Species of highly vagile taxa are often found on more than one island and are therefore less likely to be subjected to unidirectional selection. Rapid character divergence away from sister species is unlikely to occur under these circumstances, resulting in a higher correlation between phylogenetic relatedness and morphological separation. When most species are endemic, there is no such inhibition on the evolutionary trajectories, and the correlation is substantially weakened.

Figure 11.15 The temporal development of phylogenies (extant species only) embedded in the two-dimensional niche space. The "shadows" of the tips of the phylogenies are shown on "floor" of the graph. The dotted circle delineates the niche space available on the largest island (island 1). VH, very high vagility; M, medium vagility; VL, very low vagility. The phylogenies correspond to taxa prone to medium levels of competition and extinction.

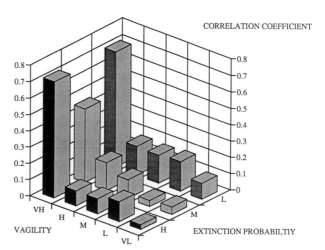

Figure 11.16 Mean correlation coefficients between phylogenetic "relatedness" as measured by the number of nodes between any particular species pair, and morphological separation, as measured by the euclidean distance between a particular pair of species in niche space for taxa prone to medium levels of competition (each replicated ten times).

DISCUSSION

The well-known equilibrium model of MacArthur and Wilson (1967) addresses the effects of static physical factors and current ecological conditions on species diversity and distributions. Species diversity patterns are a result of the effect that physical factors, such as island area and isolation, have on the probability of colonization or extinction of species derived from a mainland pool. Because local conditions are island-specific, colonization and extinction probabilities are unique to each island. Using the intuitively simple idea that ecological conditions cause colonization rates to decrease and extinction rates to increase with species number, an equilibrium species number is predicted for each island.

We have employed much of the same simple logic as MacArthur and Wilson did in designing the dynamics of the events that take place over our ecological time scale. The functions used to determine whether species successfully colonize mimic the assumptions of the equilibrium model. The probability of successful island colonization decreases directly with distance and indirectly with island area. The decrease with area and number of resident species (alpha diversity) is implicit because, as area decreases and alpha diversity increases, the probability that an island contains invadable niche space also decreases. In addition, the probability of extinction increases directly with a decrease in island area.

There are, on the other hand, some important differences between the fundamentals of our model and the equilibrium model. First, because the ecological events are embedded within a broad geological time scale, components of the MacArthur-Wilson model that are held constant become variables in our simulations. This nested structure allows us to see how drastically the dynamical nature of the equilibrium model overwrites the effects of history. Second, by assuming that a species requires a fixed proportion of two-dimensional niche space, we are specifically defining why colonization probabilities decrease with decreasing island area and increasing alpha diversity. Third, we have no preexisting species pool from which colonizing species are randomly drawn. Instead, the species pool expands over time in a process analogous to adaptive radiation. Access to components of the species pool may be direct or through a complex sequence of stepping-stone islands. Finally, the equilibrium species number on which the MacArthur-Wilson model should eventually converge is itself a dependent variable in our system, determined by a complex interaction of various independent variables.

The role of chance in structuring the communities developed in these simulations has been assessed using the variance and coefficient of variation generated around the means of various community properties observed in repeated random simulations under a fixed set of conditions. Of particular interest were the conditions under which chance might be expected to be more or less important. Purely statistical processes may explain the observed trends in the variance and coefficient of variation; these may be suggested by the construction of extremely simple probability models that emulate certain processes occurring within the simulations.

For example, for any one geological time unit, dispersal between two islands can be viewed as a series of n Bernoulli trials, each with a probability p of success (where p is clearly directly proportional to vagility). If it is assumed that p is sufficiently low so that dispersal results in speciation, the variance ($np\{1 - p\}$) around the mean number of species increases with time (n) and the probability of dispersal (vagility), while the CV decreases with both these variables. If the mean number of successes (np) is held constant but the ratio $n{:}p$ is increased (i.e., decreasing the "grain" of ecological time), the variance converges on the mean and the CV increases. The use of this type of model would therefore suggest that any attempt to integrate such contrasting time scales will result in underestimating the role of chance in structuring these biogeographical distributions.

Alternatively, if an island is drifting away from a major species pool, then the probability of a successful dispersal-speciation event, P{d-s}, could be viewed as a decreasing function over geological time during which the distance (s) to the mainland increases:

$$(11.1) \qquad P(d\text{-}s) = \int_{a}^{b} e^{-ks}ds$$

where k is inversely proportional to vagility, a is the distance at which the island populations are sufficiently isolated for speciation to occur, and b is the present distance. Once again the variance around the mean number of species increases with the probability of dispersal (vagility) while the CV decreases. These simple models conspicuously fail to predict the increased role of chance in determining the distributions of taxa with very high vagility. The generation and analysis of more sophisticated models of this nature would seem to be an area worthy of further investigation.

The model presented here permits the development of phylogenies in a highly nonrandom manner. It is of interest to compare the results with those that would have been obtained had the phylogenies developed in a purely random way (as exemplified by Raup et al. 1973; Raup and Gould 1974; Gould et al. 1977; Colwell and Winkler 1984). Using the number of extant and extinct species in our phylogenies, it is possible to calculate the probabilities of speciation (ρ) and extinction (ε) that would generate completely random phylogenies with the same number of extant and extinct species in the allotted time (τ). Kendall (1948) gives formulae for the mean and the variance of such random "birth and death" processes:

$$(11.2) \quad \text{Mean no. extant species} = (1 + \rho - \varepsilon)^{\tau}$$

$$(11.3) \quad \text{var} = \frac{\rho + \varepsilon}{\rho - \varepsilon} \varepsilon^{(\rho - \varepsilon)^{\tau}} (\varepsilon^{(\rho - \varepsilon)^{\tau}} - 1)$$

The means are by definition the same, but the coefficients of variation of the random process (which we have found to be typically 100%–120%) are consistently five to twenty times greater than those of the process that generates the phylogenies in our model. This suggests that the processes producing faunal buildup in our model are subject to some negative feedback, resulting in a degree of convergence above that expected from pure chance alone. Since the discrepancy between the observed and expected variances is greatest at intermediate values of extinction proneness and vagility, this is where the feedback operates maximally (i.e., at parameter settings that correspond to high gamma diversity).

The source of the negative feedback is undoubtedly competition. For low-diversity taxa, or for those high-diversity taxa for which interspecific competition is of little importance, the system has the potential to reflect, indeed amplify, the stochastic component of the buildup process. At higher diversities (or where interspecific competition is a regulating factor), added constraints on the stochastic process restrict the variance, preventing the reflection of chance events to the same degree. When competition is not likely to be a regulating factor, or when diversity is low (particularly if as a result of low vagility), there are straightforward statistical reasons for expecting chance to play a greater role in the process of faunal buildup. Under these circumstances of low feedback, the imprint of history should be most variable, and this view is at least partially supported by inspection of figures 11.12 and 11.13. This hypothesis may explain the discrepancy between the observed results and those expected from the simplified probability models.

The temporal buildup of diversity under a random scheme is expected to result in an exponential increase in the number of extant (and extant + extinct) species. The temporal buildup of diversity in our simulations was hardly ever exponential (a typical example is illustrated in figure 11.3) and this difference indicates that the nonrandom processes contributing to the faunal buildup in our model *are* distinguishable from random ones.

Recently, methods have become available to predict the approximate structure of phylogenies grown under the random scheme referred to above. If a clade is "picked" from a phylogeny containing n species and the basal node removed, one is left with two clades containing r and $n - r$ ($= s$) species. This process is known as partitioning and generates a partition of type $r - s$. Slowinski and Guyer (1989) have shown that for clades in which the terminal taxa are labeled, there are the same number of possible trees for each type of partition (except for the cases when $r = s$, in which case there are only half as many). This property holds for randomly grown trees (personal observation). So, for a large number of random clades of identical size, one would expect to see a uniform distribution of clades over all possible partition types (bar the case when $r = s$, when one expects half as many as in every other type class).

It is of interest to compare the distribution of partitions of clades from our model with the expected distribution had the phylogenies developed randomly. In order to assemble a sample size that would allow the statistical comparison of the observed and expected distributions, it was necessary to pool all the clades of a particular size generated from all the parameter combinations (a total of 180 phylogenies). The comparison revealed that clades consis-

tently (and highly significantly) partition in a very unbalanced way (i.e., that partitioning of a clade resulted in a small and a large subclade many more times than one would expect). This result holds for clades of size 4 (the minimum in which such partitioning is meaningful) to 16 (the largest clades for which an adequate sample size was available). Unfortunately it was not possible to detect for what type of taxa this nonrandomness would be greatest because of the large number of clades required to carry out a statistically meaningful analysis. Whether this is a legacy of the region's geological history or a by-product of ecological interaction remains unclear. Guyer and Slowinski (1991) have recently documented empirical evidence confirming the prevalence of unbalanced clade structures in real phylogenies.

Inspection of the island cladogram (fig. 11.2) and the map of the developing archipelago (fig. 11.1) suggests that species found on some islands will have a greater potential to undergo adaptive radiation than those on others. This may be due to the differential proximity of neighboring islands or the frequency of vicariant events effecting a former land mass, both in combination with the biology of the taxon. Another possibility is that development of one phylogenetic clade competitively suppresses the proliferation of another. It will be a long time before enough real phylogenies are established to test this hypothesis empirically.

Using the model presented in equation 11.1, it is possible to illustrate the window effect discussed above. In our simulations, speciation was assumed to occur when the probability of conspecific immigration was reduced to below 0.1 per ecological time unit. This probability corresponds to a distance of $-\ln (0.1)/k$ and represents the near side of the window. If b is now assumed to be large, it is possible to estimate the distance (intermediate between $-\ln (0.1)/k$ and b) at which 80% of the mean number of dispersal-speciation events would have occurred. This figure will be the far side of the window. For the same values of k used in this simulation, the windows for taxa of very high, high, medium, low, and very low vagility become from 23.02 to 44.05, 4.6 to 8.81, 3.1 to 5.9, 2.3 to 4.4, and 0.23 to 0.44 distance units, respectively. A glance at the map in figure 11.1 confirms that the broadest and narrowest windows will only be encountered rarely, explaining the lower gamma diversity observed in these vagility classes. That the majority of the diversification should be compressed into these relatively narrow windows is an inevitable consequence of using an exponentially declining dispersal function.

Erwin (1981, 171) states that "an understanding of a group's vagility must be a forerunner to any biogeographical study." Unfortunately, the dispersal abilities of many organisms (especially over geological time scales) are still shrouded in considerable mystery (but see Diamond, 1987 and references therein). Erwin argues that "founderism" (sensu Mayr 1942) will be rare in nature because ". . . weather patterns, ocean currents, or other means of dispersal are not 'freakish' when viewed on the geological time scale or especially on the evolutionary-rate scale." Such a view is open to two interpretations. The first is that dispersal is not a stochastic process—if it *could* happen it always will (i.e., dispersal with probability 1.0 over a finite range of distances and 0.0 otherwise); this view would virtually eliminate the role of chance in the development of biogeographical patterns. Alternatively, it could suggest the use of dispersal functions that generate uniformly low probabilities of dispersal over a range of distances. This interpretation permits the influence of chance but results in the contemporary past increasingly masking the influence of deep history. Our particular dispersal function and its relationship to the speciation mechanism is the product of an especially blind grope. Clearly, we must understand the nature of "epic" dispersal before we can seek consensus in the ongoing debate over the role of history and ecology in biogeography.

We are not aware of any previous attempts to model biogeography in this way. It is not clear to us how much can be learned from simulations of this type; however, several lines of investigation appear worthy of further pursuit. First, there is some heuristic value to abstract simulations of this type in that they demand the encompassing of an array of processes and patterns about which fairly little is known, thus highlighting areas for future empirical investigation. They also help by suggesting answers to questions for which we will never be able to get definitive empirical answers; for example: Were the entire process of faunal buildup over an archipelago to occur again, what is the probability of obtaining a similar result? How reliable are the distributions of different sorts of organisms in reconstructing geological history? Even though we have only considered these biogeographical patterns one taxon at a time, in reality faunas and floras develop concomitantly (although various temporal displacements may occur). Trends for hypothetical taxa reported here should be compared to real biogeographical patterns (e.g., snails, lizards, and birds in the Caribbean). Finally, highly "tactical" versions of such models that simulate the biogeography of particular real areas, defining in the program a hypothesized geological history and species source pool's could easily be produced (snails, fruit flies, and birds in Hawaii or the Galapagos are obvious candidates). The output from such models could illuminate possible inconsistencies in proposed biogeographical "stories." The failure of such tactical models to explain observed species distributions could raise questions even more interesting than those that result from successful models.

CONCLUSIONS

In its entirety our model is complex; it is, however, composed of a set of simple and intuitively appealing concepts. The output of diversity measures, species-area relationships, niche packing structures, etc., are sufficiently consistent with reality to generate a measure of confidence in some of the less obvious results. The fact that different organisms are differentially impressionable to the array of forces molding ecological and biogeographical patterns is an important dimension of the analysis and interpretation of ecological data.

Analysis of this model suggests that the nature of the study organism has a profound impact on the visibility of the historical effect, differentially affecting the rates at which the impressions of deep and contemporary history are obscured. Empirical quantification of this phenomenon remains a substantial challenge. Chance is expected to play a greater role in determining the diversity of inherently low-diversity, less competitive, and less vagile taxa. The detection of nonrandom processes through the analysis of patterns is often a formidable task. Techniques remain relatively unsophisticated, and difficulties are compounded because of low sample sizes. Our results suggest that the processes of faunal buildup and phylogenetic development may not be accurately approximated by a random algorithm.

Gould has remarked: "And so ultimately, the question of questions boils down to the placement of the boundary between predictability under invariant law and the multifarious possibilities of historical contingency" (Gould 1989, 290). We suggest that in the context of biogeography, the placement of this boundary depends on the general characteristics of the taxon under consideration. Further studies like this one might offer promising initial leads in establishing the details of this dependency.

SUMMARY

1. Ratios of vagility to vicariant spatial scale dictate ecological and biogeographical patterns.

2. The bulk of dispersal-speciation events occur within fairly well defined windows of interisland distance; within these windows, reciprocal dispersal and speciation events generate diversity rapidly.

3. Archipelagoes may provide similar sequences of such windows for quite disparate taxa at different times, thus nurturing the proliferation of taxa with different vagilities at different stages in the geological development of the archipelago. The resulting biogeographical patterns may be remarkably similar.

4. Extinction-prone taxa proliferate species, many of which become extinct; non-extinction-prone taxa generate a higher extant diversity.

5. Proneness to extinction has a limited influence on beta diversity.

6. High extinction rates continually update the relationship between the geological past and geographical distributions, eliminating the influence of "deep history" but deepening the imprint of contemporary history.

7. Chance has its greatest influence on taxa with very low and very high vagilities (i.e., low-diversity taxa); its effect is enhanced at extremes of extinction proneness.

8. The imprint of history on the biogeography of taxa is very predictable.

9. While alpha and gamma diversity may be quite unpredictable under some circumstances, their ratio (beta diversity) is generally very predictable.

10. Several measures of phylogenetic clade structure and development indicate that patterns within real clades are not likely to be adequately characterized by random structures or processes.

Regional Perspectives

The chapters in Part Three emphasize geographical comparison as a means of evaluating the influence of local and regional processes on community diversity. Most of the chapters compare local diversity within particular ecological settings between regions having independent or partially independent histories; most of the comparisons involve different continents. A consistent pattern across regions in the relationship of local diversity to environmental conditions would support the hypothesis that local processes principally determine local diversity. Conversely, differences in diversity between regions despite environmental similarity (diversity anomalies) would suggest a role for geographical and historical factors.

The general picture that emerges is that the species richness of both large regions and small areas of homogeneous habitat bears the unique imprint of the region. Whether this imprint reflects historical events or unique geographical configurations cannot always be determined from these analyses. The results also reveal strong habitat effects on diversity, highlighting the influence of local ecology. These studies demonstrate that broad comparisons provide information about the types of processes generating and maintaining species richness and about the unique history of species assemblages.

Blondel and Vigne lead off this section of the book by providing a detailed account of the unique history of the Mediterranean region and its biota. They relate species richness to the particular biogeographical setting of the region, the effects of tectonic events (for example, the formation of the Strait of Gibraltar about five million years ago), climate change during the Pleistocene glacial cycles, the long history of occupation of the region by humans, and fire and other ecological disturbances. In addition, we see that each group of organisms responds differently to a particular physical setting, depending on its sensitivity to barriers to dispersal and the extent of habitat it requires for proliferation and diversification of taxa.

Cody compares the distribution of birds over a habitat gradient and over distance within habitats in northeastern and southwestern Australia. The structurally similar mesic habitats of these two subregions are separated by the vast arid interior of the continent. Cody found that more species of birds inhabit the northeastern region, and he examines how different components of diversity contribute to the greater richness there. Both alpha and beta diversities are higher in northeastern Australia, but gamma diversity also is higher in most habitats there. Gamma diversity measures turnover of species within habitats over distance and represents replacement by ecological counterparts and extension of species' distributions from different adjacent habitats in different parts of the region.

Morton compares diversity in arid habitats of Australia to that in similar habitats in other parts of the world, particularly North America. Differences abound: Australia has fewer rodents, but more insectivorous small mammals, granivorous birds, termites, and lizards. The number of species of all types of birds is similar; Australian ants exhibit a higher regional diversity but similar local diversity to those in North America. Even in these latter two taxa, the niche space occupied differs strikingly between the two regions. Morton relates these differences to contrasting environmental conditions (Australia has more variable rainfall and poorer soils) and historical contingency (Australian plants and animals have unique phylogenetic origins).

Following up on Cody's and Morton's Australian perspectives, Westoby makes more general comparisons between Australia and other regions by examining the relationship between species richness in small (local) and large (regional) areas within matched habitats. Vascular plants exhibit local determinism: local diversities are similar on different continents despite regional differences in diversity. In contrast, differences between regions in local diversity parallel regional differences in lizards and in coral reef fishes, indicating regional or historical effects. Westoby suggests that local processes more strongly control diversity in taxa that disperse poorly, such as plants, or are highly specialized or endemic; in such taxa, local communities contain a small proportion (ca. 5%) of the regional species pool, and species-area relationships are steep. In generalized, broadly dispersing taxa, local communities contain a much larger fraction (perhaps 40%) of the regional pool of species, and local diversity more strongly mirrors regional diversity. Thus, according to Westoby, local communities may approach local equilibria more closely in the absence of high migration from other habitats. This conclusion agrees with the theoretical arguments in several chapters of Part Two.

The chapter by Lawton, Lewinsohn, and Compton underscores the geographical and historical contexts of the local community in an intercontinental comparison of insect herbivores on bracken fern. The niches occupied by these herbivores differ between regions, and most localities appear to have vacant niches, suggesting that local communities are neither saturated nor are their diversities governed by local interactions. Lawton and his colleagues find a strong species-area effect on regional richness; in addition, local and regional richness covary. They conclude that variation in species richness of herbivores on bracken reflects colonization by species from the distinct insect faunas inhabiting the different regions.

Aho and Bush tackle the problem of community boundaries in their analysis of the parasites of freshwater fishes. They distinguish the parasites residing in an individual host (infracommunity) and in the entire host population (component community), but also recognize the geographical dimension of the host range, over which physical conditions and intermediate hosts may vary. As in other chapters, Aho and Bush find empirical evidence for both local (host body size, feeding guild) and regional/historical (host range) effects on local diversity. Their analysis also shows evidence of community saturation: local parasite diversity in a particular host population is unrelated to the total parasite diversity of the host throughout its range. This analysis has the same design as that of H. V. Cornell (chapter 22) on cynipine gall wasps that infest oaks, but it produced a strikingly different result. Aho and Bush's call for more study of the autecology of individual species should be extended to host-parasite systems more generally, including the defensive responses of the host and cross-resistance between parasites.

Pearson and Juliano find that the distribution of body and mandible sizes among tiger beetles differs between continents and reflects phylogenetic origins of taxa inde-

pendently of habitat. Similar effects have been noted in this volume by Morton, Lawton and co-workers, Cadle and Greene, and Ricklefs and Latham. These observations raise several issues concerning community saturation: Does the filling of ecological space map closely onto morphological space? Do other taxa fill unoccupied niches? Comparisons of local and regional diversity of tiger beetles among regions also provide support for both local and regional/historical influences on local beetle diversity.

Pearson and Juliano apply a device frequently used to portray diversity patterns: the tabulation of diversity within grid squares (in this case, 275 km on a side). Such grids fall between the local and regional scales of most of the studies in this volume. Pearson and Juliano's analysis demonstrates that local diversity parallels grid diversity, and that the latter can be related to climate (precipitation). Grid analysis provides a graphic portrayal of the spatial pattern of diversity within a region. However, grids pose difficulties as units of analysis: adjacent grid cells are pseudoreplicated samples from a region, and it is difficult to apply statistical tests to grid data. Furthermore, grid analyses do not incorporate habitat heterogeneity within grids, and grid diversity does not distinguish alpha and beta components of species richness.

McGowan and Walker discuss what may constitute the world's largest communities, the plankton of oceanic gyres. Mixing of water over huge spatial scales produces homogeneous assemblages distributed over significant proportions of the globe. These assemblages have many fewer species than, for example, insects in terrestrial habitats, which may indicate the strong influence of a local factor—habitat heterogeneity—on species richness. However, as McGowan and Walker point out, conventional explanations of diversity patterns do not apply well to the pelagic environment. How does one account for the maintenance of species that are locally (and globally) extremely rare? How are species added to the system when there are few possibilities for allopatric speciation? What is the relationship of relatively small, peripheral bodies of water to the central gyre? The high zooplankton diversity of the California Current, for example, seems to reflect the mixing of taxa from bordering water masses, brought about by high levels of "mesoscale" eddying.

Ricklefs and Latham explore a striking diversity anomaly in mangrove vegetation: the Indo–West Pacific region has 6–7 times as many species as the Atlantic–Caribbean–East Pacific region. Local diversity parallels regional diversity; part of the niche space is vacant in the species-poor communities of the Western Hemisphere. The fossil record suggests that the diversity anomaly did not arise by differential extinction; rather, it appears to have resulted from differences in the origination of new mangrove taxa after the two regions were cut off by the closure of the Tethys Sea during the Eocene. Ricklefs and Latham conclude that origination was facilitated in the Indo–West Pacific, where wet climates created a gradual transition between terrestrial and mangrove habitat. The mangrove system emphasizes the role played by barriers to physiological adaptation in restricting diversity in novel habitats. It also suggests a role for regional climate and habitat configuration in the rate of origination of new forms.

Schluter and Ricklefs formalize and quantify the comparative approach taken by Westoby and others. They develop an ANOVA technique for distinguishing local and regional effects by comparing diversity between matched habitats or ecological roles (niches) within habitats on different continents. The habitat effect includes the influence of local processes on local species richness, whereas the continent effect and the habitat-times-region interactions reflect the influences of regional and historical factors. Results show strong habitat effects on local diversity that are consistent across continents (convergence). Local diversity also differs between regions in most comparisons, by a factor of about 1.7 on average.

We note, however, that habitat effects themselves may include the influence of regional processes and historical events when certain habitats are consistently older or occupy greater areas than other habitats in all regions. For example, an ANOVA of terrestrial and mangrove tree species diversity between Southeast Asia and the Caribbean basin would reveal a strong habitat effect and a weaker regional effect. Nevertheless, the mangrove habitat is not saturated, and mangrove species apparently have accumulated over evolutionary time. The habitat effect in this case partly reflects the origination of higher angiosperm taxa in terrestrial habitats combined with the difficulty of crossing the adaptive barrier separating mangrove from terrestrial habitats.

The chapters in this part of the book emphasize the need for ecologists to consider history and biogeography, and to address simultaneously the ecological positions of species within communities and their phylogenetic, geographical, and adaptive histories. Although the information presently available convinces us of the importance of regional and historical effects for local community diversity, the available comparative data are still insufficient to permit a global comparison of diversity at local and regional scales. The analyses presented in these chapters also cry out for the phylogenetic and historical information needed to determine the history of community development. Finally, although community ecology will inevitably pass through a phase in which the histories of communities are reconstructed, the more difficult challenge will be to develop principles concerning the interaction of regional and local processes within the contexts of local environments and regional geography.

12

Space, Time, and Man as Determinants of Diversity of Birds and Mammals in the Mediterranean Region

Jacques Blondel and Jean-Denis Vigne

Two factors explain why the Mediterranean region (delimited in fig. 12.1) is well suited for analyzing patterns and processes of diversity. First, the landmasses encircling this "sea-among-the-lands" (*Medi-terranean*) are parts of three different continents (Europe, Asia, and Africa), which naturally leads to biogeographical diversification. Second, the unusual geotopographical and geobotanical diversity of habitats around or on the region's mountains (up to more than 4,000 m elevation), peninsulas, and islands makes Mediterranean biotas extremely diversified. Habitats range from arid steppes in North Africa and the Near East to moist fir-beech forests in the mountains of mainlands and some larger islands. This fact explains why the Mediterranean realm has been one of the most important biogeographical centers of the Palaearctic region for speciation in plants; nearly half of the 23,000 species of this region are Mediterranean endemics (Quézel 1985).

Such features also may have produced high vertebrate diversities. We will investigate local and regional diversities, using case studies from birds and nonvolant terrestrial mammals, and dealing in a time-space perspective that takes into account palaeobiology, biogeography, ecology, systematics, and the impact of human beings, because no other region in the world has been so dramatically modified.

HISTORICAL DETERMINANTS OF VERTEBRATE MEDITERRANEAN FAUNAS

One difficulty in explaining causal relationships between distributions of animals and history is that historical events are unique and lack a theoretical framework for testing alternative hypotheses (Illies 1974; Cracraft 1980). However, because biological systems always reflect a past history (Mayr 1983; Ricklefs 1989a) we will try to explain modern diversities in light of the multiple historical events of the last two million years.

The Origin and Development of Mammalian Faunas

Cheylan (1991) tentatively estimated the present number of nonvolant land mammal species in the Mediterranean region at 158 (with only 25% endemic species). The pres-

ent diversity of this biota can be ascribed to three causes: (1) the multiple biogeographical origin of the mammals, which entered the Mediterranean from Europe, Asia, and Africa; (2) Pleistocene climatic changes, which produced periodic faunal turnovers, and hence large beta diversities, over time because of large intercontinental faunal exchanges; and (3) severe and ancient (Neolithic) human pressures (modifications of habitat, animal husbandry, hunting, and introductions of alien species) that modified natural distributional patterns and produced a decrease in the previous, much richer, glacial fauna.

Because such weak dispersers as nonvolant mammals are sensitive to geographical barriers, the mammalian faunas of the three main subregions of the Mediterranean (i.e., Mediterranean Europe, the Middle East, and North Africa) are rather distinct. Exchanges between Europe and Africa have been precluded by two impassable barriers. The first is the Strait of Gibraltar (14 km wide), which has isolated Europe from Africa since the Messinian tectonic crisis (Jaeger et al. 1987) more than five million years ago. The second is the extension of the Sahara, which has blocked the eastern route (through eastern North Africa), except for short "pluvial periods" during the Pleistocene glaciations.

In Mediterranean Europe, only a few subtropical taxa (*Hystrix, Macaca*) from the early Pleistocene fauna (ca. 2 mya) found refuge and survived after the large extinctions of the first glaciations (Günz, Mindel) (Kurtén 1968). Other genera of boreal origin, such as *Sus, Cervus,* and *Ursus,* survived until the present and are basic components of the European fauna, including that of the Mediterranean. Few subsequent speciation events occurred except among shrews (Catzeflis, personal communication) and rodents such as Arvicolidae (Chaline 1974). Present diversities have been determined by several immigration waves that brought new faunal elements from temperate Asia during the main glacial episodes. However, the many extinctions in both Mediterranean Europe and North Africa (Cheylan 1991) at the end of the Würmian glaciation (14,000–12,000 B.P.), probably as a result of human impact, produced a sharp decrease in species richness.

In western Asia (Middle East), species richness of large mammals sharply increased during the Riss-Würm inter-

135

glacial (ca. 110,000–70,000 B.P.), probably because of climatic improvement (Tchernov 1984). The new immigrants mainly originated in Eurasia and, to a much lesser extent, in the Afrotropical region. However, during the Holocene (last 11 millenia), humans brought about the extinction of many of them (except rodents) (Tchernov 1984), especially large hoofed mammals.

In North Africa, Afrotropical elements always dominated the fauna (Jaeger 1975). The richness of large hoofed mammals and carnivores increased from 17–20 to 29 species during the late Riss–late Würm (110,000–14,000 B.P.), thanks to an immigration of Eurasiatic species through the eastern route. However, species richness subsequently decreased, especially that of hoofed mammals, of which only 5 species remain (*Sus scrofa, Gazella dorcas, G. cuvieri, Ammotragus lervia,* and *Cervus elaphus*). These extinctions resulted from the combined effects of desertification and human activities (Jaeger 1975).

The three main subregions of the Mediterranean basin experienced massive post-Pleistocene invasions. Some species invaded Mediterranean Europe from the Middle East (examples are *Vormela peregusna, Canis aureus, Microtus guentheri*) and North Africa (*Herpestes ichneumon, G. genetta*). Others, such as *Oryctolagus cuniculus* and *Cervus elaphus* were introduced by humans from Europe to North Africa or to the Middle East. Three examples of direct introductions by humans are worth citing.

The mouflon, *Ovis ammon orientalis,* was first domesticated in the Near East (Djezireh) about 8500 B.P. (uncalibrated dating). The resulting domestic sheep, *O. aries,* was rapidly spread by humans throughout the Mediterranean basin. In some places, especially on large islands, some individuals escaped human control and gave rise to feral populations. Much later (during the twentieth century) these new feral "species" were introduced in many places throughout the Mediterranean region.

Two commensal murid species, the mouse (*Mus mus-* culus) and the black rat (*Rattus rattus*), invaded human settlements from Asia to the Near East at the beginning of the Holocene. The domestic mouse rapidly invaded the whole western Palaearctic, where it differentiated into two semi-species (Auffray, Vanlerberghe, and Britton-Davidian 1990). The black rat colonized the whole Mediterranean basin during classical antiquity (Armitage, West, and Steedman 1984) using human boats, as it did later throughout the world (Atkinson 1985).

Finally, several species (*Myocastor coypus, Ondatra zibethicus, Sylvilagus floridanus*) have been introduced by humans from America to the Mediterranean region.

The History of Bird Faunas, or Why Are There So Few Species of Mediterranean Origin?

Bird diversity in the Mediterranean region is extremely high: 343 species presently breed in an area of three million square kilometers, as compared with the 419 species that breed in the ten million square kilometers of all Europe (Voous 1960). After the extinction in Eurasia of many tropical elements such as Cracidae, Psittacidae, Musophagidae, Coliidae, Trogonidae, and Bucerotidae at the end of the Pliocene (Brodkorb 1971; Bochenski 1985), the Pleistocene-Holocene development of avifaunas involved faunal elements from nine biogeographical units (fig. 12.1; see also Blondel 1988), among which two played the most important roles: Eurasia (150 species) and the eremian belts of the south and southeast margins of the region (85 species). In contrast to what happened with mammals (see above and Heim de Balsac 1936; Moreau 1966) the relationships of the Mediterranean avifauna with tropical Africa are rather weak (fig. 12.1). In contrast, the bird faunas of the southern Mediterranean margin have close affinities with those of Asiatic steppes because of an active radiation within the wide belt of semiarid habitats that encircles the western Palaearctic from the Atlantic Ocean to the Arabian Plate and the

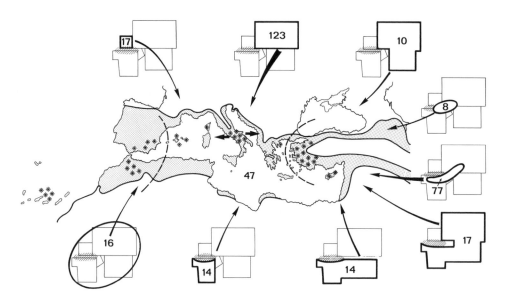

Figure 12.1 Biogeographical origin of the bird faunas of the Mediterranean region (delimited by the shaded area). Pictograms around the figure represent the Old World biogeographical origins of the species; the numbers therein are the numbers of species originating in each area. 47 species are of Mediterranean origin and 36 of them (asterisks) are regional endemics. (From Blondel 1988).

steppes of south-central Asia. A striking feature of the Mediterranean bird fauna is its very low level of endemism: no more than 47 species (14% of the fauna) are of Mediterranean origin.

Most Mediterranean birds belong to four broad ecological categories: steppe species (88 species), forest species (76), water birds (69), and shrubland species (41). Although we would expect the latter to be numerous and dominant in the many types of matorrals that extend over more than half the region, they are poorly represented (12%), whereas forest birds of boreal origin are widespread and dominant everywhere. Such a low endemism rate and low species richness in Mediterranean shrublands were unexpected.

Why did the high habitat diversity of the Mediterranean region, with its many types of endemic vegetation and geographical discontinuities, not provide more opportunities for speciation during the Plio-Pleistocene, such as occurred in other parts of the world, for instance, in North America (Mengel 1964), South America (Simpson and Haffer 1978; Vuilleumier and Simberloff 1980), tropical Africa (Hall and Moreau 1970; Snow 1978) and Australia (Keast 1961)? Examining the story of bird faunas across different scales of space and time gives some clues, which may be summarized in four explanations (see Blondel 1988 for more details):

First, at the height of the pleniglacial periods, virtually no arboreal vegetation could persist to the north of the Pyrénées, the Alps, and the Carpathians (Pons 1981; Huntley and Birks 1983). Accordingly, all forest biotas and their associated vertebrate faunas must have withdrawn somewhere to the south of the 45° parallel.

Second, contrary to what has been believed for a long time (e.g., Moreau 1954), the Mediterranean life zone did not shift as a whole to the south in what is now the Sahara during the glacial episodes, but remained in local refugia mostly within the limits of the Mediterranean region (fig. 12.2), which was larger than it is today because of a 100–150 m regression in sea level (Pons 1981). Paleobotanical and paleontological records have shown that the diversity of both geography and climate within the Mediterranean during both glacial and interglacial periods would have allowed for the coexistence on a regional scale of all the present European vegetation belts and their associated faunas (Blondel 1988). One example is that of the remains, found together in the same deposits of Würm II in southern France, of such species as *Falco naumanni* (a thermophilous Mediterranean species), *Lagopus mutus*, *Nyctea scandiaca* (birds of the tundra and the alpine belt), and *Monticola saxatilis* (an oromediterranean saxicolous turdid) (Mourer-Chauviré 1975). Similar composite faunal assemblages have been found for micromammals and amphibians in deposits of the Riss glaciation by Chaline (1972) and Rage (1972) respectively. Such puzzling and apparently aberrant assemblages suggest that local landscapes in the Mediterranean region during pleniglacial times might have been a kaleidoscope of such habitats as tundras, steppes, and both coniferous and broad-leaved forests. This kaleidoscope caused a striking telescoping of the European faunas within the Mediterranean basin. It is difficult to imagine how these landscapes looked, how-

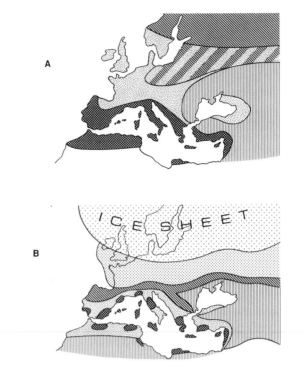

Tundra   Conifers   Deciduous   Steppe   Mediterranean

Figure 12.2 Extension of the major vegetation belts (*A*) at the present time and (*B*) during the last glaciation (Würm). (The location of the refugia of Mediterranean biotas during the Würm is no more than indicative and does not claim to represent their true location and extension.)

ever, because late glacial vegetational types have no convincing modern analogies (Huntley and Birks 1983). Local alpha diversities must have been very high, but no clues are available about their range of magnitudes.

Third, during interglacial episodes such as the present one, most lowland and mid-elevation landscapes were forested primarily by deciduous broad-leaved trees (e.g., *Quercus pubescens*), not sclerophyllous ones (Vernet 1973; Pons 1981, 1984). The relative extension of deciduous and sclerophyllous trees suddenly changed as a result of human activity in the late Holocene, which is reflected in pollen diagrams by a sharp increase in pollen of the sclerophyllous *Q. ilex* and of domesticated cereals and plants of secondary growth habitats (Pons 1981). Bird faunas of these forests must have been of mid-European character and were not typical of the fauna we imagine when we think of evergreen shrublike vegetation.

Last, different types of more or less isolated patches of matorral developed in the Mediterranean region as early as the beginning of the Pleistocene (Suc 1978) and persisted during the whole Quaternary in local areas where climatic, edaphic, and topographical conditions allowed for the existence of only a shrubby vegetation (Pons 1981, 1984; Reille 1984; Pons and Reille 1986). Contrary to that in other Mediterranean zones of the world (California, Chile, South Africa, Australia), the climactic vegetation of most of the Mediterranean region would be not shrubs, but different types of forests, if humans had not systematically destroyed Mediterranean forests. The present extension of matorrals is a modern and secondary feature caused by human deforestation, which was continuous after the early Neolithic.

Thus the Quaternary story of Mediterranean biotas is that of a pervasive importance of forests, without any clear geographical delimitation between them and mid-European forests. Forest habitats were extensive but did not provide the geographical isolation that would have been a prerequisite for allopatric speciation in birds. This explains both the very low number of endemic bird species in Mediterranean forests (three nuthatches, one tit mouse, and three pigeons) and the homogeneity of forest bird faunas everywhere in Europe, as well as the large number of "core species" (Hanski 1982), i.e., species that are widespread and that tend to be abundant where they occur.

In contrast, some species evolved in the patches of matorral that persisted during the whole Plio-Pleistocene (*Sylvia spp., Alectoris spp., Monticola spp., Phoenicurus moussieri*). The hypothesis of successive speciation events in this type of habitat is consistent with palaeobotanical findings (Florschutz, Menendez Amor, and Wijmstra 1971; Pons 1984; Reille 1984), which demonstrated that the spatial extension of these patches of matorral varied in time according to shifts of temperature and moisture. The best example of such a radiation is that of the genus *Sylvia*, which includes nineteen species, of which fourteen are Mediterranean endemics. However, all these warblers reached full species status, so that the absence of secondary contact belts with hybridization cannot tell us anything about the spatio-temporal processes of differentiation (see Blondel 1988).

As a consequence of this history, bird communities of different regions in Europe should be more similar in forest habitats than in open and shrubby habitats, with no "Mediterranean syndrome" in the bird communities of mature Mediterranean forests. This hypothesis has been tested with an analysis of the turnover of bird communities along successional habitat gradients located in different parts of Europe. Blondel and Farré (1988) used four matched habitat gradients of six habitats each, two in the Mediterranean region (Provence and Corsica) and two in central Europe (Burgundy and Poland). Using a correspondence analysis to display the trajectories of bird communities in a multivariate space, they showed that each local gradient starts from a different position in the multivariate space and then each of the four sets of bird communities converges toward a similar position in the last forested stage (fig. 12.3). In other words, there is less and less difference between the four sets of communities in matched habitats as one progresses from the shrubby habitats of the first stage to the last forested stage because habitats share more and more species. The communities of the four forests are composed of the same species that are the core species of forest habitats located everywhere in Europe.

To conclude this section, the Quaternary stories of mammals and birds share many features, but differ in several aspects. For both groups, the Mediterranean is a melting pot into which species dispersed from different

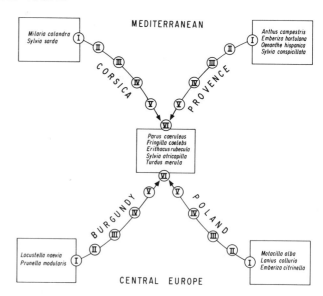

Figure 12.3 Convergence of the ecological trajectories of bird communities along successional habitat gradients (from low matorrals [I] to mature forests [VI]) located in different parts of Europe. Examples of typical species for the first stages of the successions and for the mature forest are given in the squares. There is no difference between the communities in mature European forests. (Data from Blondel and Farré 1988.)

biogeographical origins. Because mammals are more sensitive to geographical barriers than birds are, endemism rates are higher for the former (25%) than for the latter (14%), and each of the three main subregions in the Mediterranean has its own mammalian particularities, especially North Africa. In contrast, bird faunas are more homogeneous throughout the whole Mediterranean region. For both mammals and birds, most faunas (with the exception of North African mammals) are not fundamentally different from those of cold and temperate Europe, which mainly belong to Holarctic faunal types. Consequently, all of these faunas have been associated with the same geographical and climatic history, especially the climatic vicissitudes of the Quaternary, which produced both extinction and immigration processes resulting in large faunal turnovers. But the most important effects on the faunas have been the dramatic and recent effects of human activity, which has had and continues to have different consequences for mammals and for birds. For mammals, human activities decreased species diversities and caused many changes in the species composition and distributional patterns of the fauna. For birds, the main consequence has not been so much a decrease in overall species richness; rather, it has been a tremendous advantage for species from arid and shrubby habitats, which benefited from large-scale deforestation and habitat degradation at both the scale of their distributional ranges and that of their population sizes.

PATCHINESS, HABITAT STRUCTURE, DISTURBANCE, AND THE DYNAMICS OF BIRD COMMUNITIES AT LOCAL AND REGIONAL SCALES

From the historical background that determined the setting of modern faunas, we will turn to the investigation of their diversities at finer scales of space.

Understanding relationships between habitats and communities, and explaining them in terms of controlling processes, in environments as patchy as those of the Mediterranean region are undertakings plagued by difficulties of both spatial and temporal scaling. Because edge effects are so widespread at a fine scale, the concept of homogeneous habitats with well-defined communities, which was so popular in the 1960s and early 1970s because of the attractiveness of the equilibrium theory, does not make much sense in Mediterranean habitat mosaics. Therefore, community ecology must be approached at a landscape level, incorporating the permanent role of disturbances at various scales of time and space. Two examples, one at the community level and the second at the population level, will illustrate the importance of disturbances and the role of habitat patchiness in local diversities, respectively.

Fire and the Dynamics of Bird Communities

A good example of the effects of disturbance has been provided by Prodon (Prodon and Lebreton 1981; Prodon, Fons, and Athias-Binche 1987), who studied the influence of fires on the dynamics of patchy habitats. Fires, at the present time and at the scale of community dynamics, are important disturbance events in the Mediterranean. Prodon first surveyed bird-vegetation relationships in eleven habitats ranging over an undisturbed habitat gradient from grasslands to mature forests of the sclerophyllous holm oak *Quercus ilex*. The experimental design involved 186 spot sites evenly distributed among these habitats. At the scale of this landscape, 51 bird species were censused using a point count technique. They ranged from species of open vegetation (e.g., *Lullula arborea, Emberiza calandra, Carduelis cannabina*) to forest species (e.g., *Erithacus rubecula, Sylvia borin, Fringilla coelebs*), with species of matorrals (e.g., *Sylvia* warblers, *Luscinia megarhynchos*) between them. At each census spot, the dominant plant species were recorded, as were vegetation profiles using cover percentages within seven layers plus the cover percentage of bare and stony ground. These data were used to model the relationships between vegetation structure and bird communities in the eleven habitats, using correspondence analysis (CA) (reciprocal averaging) and multiple regression (Prodon and Lebreton 1981). The first axis of CA organized the data as a habitat gradient and indicated a strong turnover of species along the gradient. The coordinates of samples on the first axis of CA were correlated with the eight cover values of the vegetation profiles. This gradient model allowed prediction of the position on the gradient of any sample of birds. An "Index of Avifauna Gradient" (IAG) represented the mean position of the set of species of a sample bird community on the gradient. By calculating the prediction of IAG from the eight vegetation values, Prodon and Lebreton (1981) calculated an "Index Structure of the Gradient" (ISG) that allows prediction of the position of any habitat site on the gradient from its vegetation and rock profiles. IAG and ISG allow one to measure independently the extent of vegetation and community changes in the same system of units. This modeling, carried out in undisturbed habitats, was used as a control to measure changes in communities and habitats as a result of disturbance by fire.

Then, birds were sampled and vegetation profiles were recorded for several years after a fire. Bird censuses were used as supplementary (or extra) samples in the correspondence analysis by averaging relationships inherent in CA. Birds were positioned on the first axis of CA using the IAG function. Similarly, the values of ISG after fire were obtained by applying the multiple regression equation to the vegetation profiles.

Figure 12.4 illustrates the shift of the IAG value for the forest part of the control gradient (C in fig. 12.4) at the first breeding season after a fire, and then the recovery dynamics of the bird community. These dynamics exhibited three main features:

First, at the scale of the whole range of habitats within the mosaic, all 51 species found in the post-fire gradient were already present somewhere within the landscape. Accordingly, the whole gradient is a closed system within which processes of local extinction and recolonization operate.

Second, species of open vegetation, such as larks (*Lullula arborea*), partridges (*Alectoris rufa*), and linnets (*Carduelis cannabina*) colonize habitats immediately after fire and then are replaced by species of matorrals, such as warblers (*Sylvia* spp.) and the nightingale (*Luscinia megarhynchos*) as habitats recover. In turn, matorral species will be replaced, and so forth, until the recovery process is completed (return to a forest stage). This process of turnover starts from a sharp decrease of IAG values on the first axis of CA and then a progressive increase with time since the disturbance. Alpha diversities are rather high in the first stages of the recovery process, then decrease as the vegetation gets more uniform, and finally increase again as the arboreal vegetation gets higher and more complex.

Last, there is a surprising persistence after fire of some species of forest stages (core European species such as *Erithacus rubecula, Sylvia atricapilla, Regulus ignicapillus, Parus major, Fringilla coelebs*; (fig. 12.4). They contribute to reducing the downward shift of IAG just after fire, so that ISG drops more markedly on the CA axis than does IAG. Such a discrepancy is illustrated in figure 12.5, where variations of both IAG and ISG for a larger series of samples than in figure 12.4 are plotted over seven years after fire. The high values of IAG relative to those of ISG express the persistence of forest species that keeps the communities from being bird assemblages typical of matorrals. Such persistence of forest species in disturbed habitats may result from the fidelity of territorial males to earlier breeding sites (Wiens 1986) despite major habitat changes (Hilden 1965). Wiens and Rotenberry (1985) have similarly shown in their studies of breeding birds in North American shrub-steppes that densities of territorial individuals may remain unchanged in response to such changes in habitat structure as range fires or physical removal of shrubs.

This example highlights the importance of scale effects in community investigations (Wiens 1981, 1986, 1989a; Ricklefs 1987). Because environmental variability plays a prominent role in structuring communities, ecological processes operate on spatial and temporal scales far larger than those usually used in field studies. At the scale of a landscape and over a long time, the survival of all the spe-

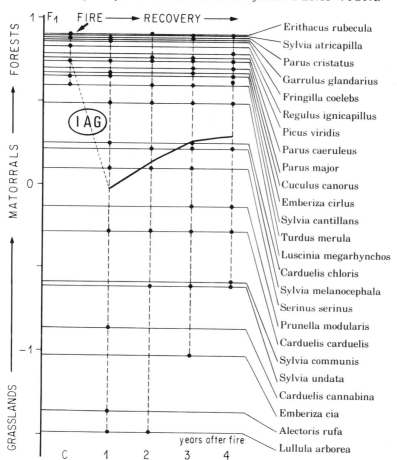

Figure 12.4 Post-fire succession of bird communities from a grassland to a mature holm oak forest (vertical axis). IAG = "Index of Avifauna Gradient." C = control gradient (unburnt forest). The right-hand column gives examples of some species indexed on the first axis of correspondence analysis. Black dots indicate their presence in the gradient according to the number of years after fire. (After Prodon, Fons, and Athias-Binche 1987.)

cies of a mosaic of habitats involves the existence of a disturbance regime that is unpredictable in time and space in the short term but predictable in the long term. This regime periodically moves the position of any given habitat patch up and down on the grasslands → matorrals → forest system. At a broader geographical scale, the combination of a disturbance regime with community dynamics, resilience, and inertia results in an equilibrium that may be fairly stable in the long term. Such a system is characterized by a given pool of species that is a legacy of history and a regime of disturbance that is specific to each region (Sousa 1984; Pickett and White 1985). To be sustainable in the long term, such a system requires areas large enough for the spontaneous occurrence of stochastic and chance events. Blondel (1986, 1987) coined the term "metaclimax" to define both the space scale required for maintaining a self-sustaining system and the disturbance regime that guarantees the existence of all the habitat patches required for dispersal patterns of all the species legated by history. In all habitat gradients so far studied in forest systems in Europe, be they located in Finland, in Poland, or in France, including the Mediterranean region, a regional metaclimax includes a similar number of species: about fifty (Blondel 1986). This similarity is a consequence of the common history of European faunas.

Blue Tits in Sclerophyllous and Deciduous Oaks

Disturbances such as those discussed above result in local extinction-recolonization processes of species within a

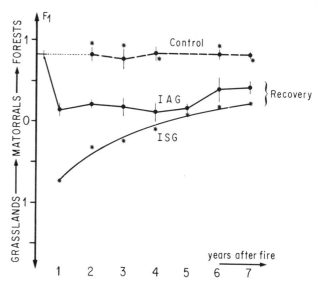

Figure 12.5 Post-fire recovery of bird communities (IAG) and of vegetation (ISG) in a holm oak forest, compared with a control (unburnt) forest. There are more changes in the vegetation structure than in bird communities after fire. The IAG trajectory differs slightly from that in figure 12.4 because several sites have been combined here. (After Prodon, Fons, and Athias-Binche 1987.)

landscape. These processes, however, depend on the demography of individual species. Community diversity at the scale of a landscape may result from differential exchanges of propagules among habitat patches; that is, some local populations work better than others, but the latter persist, provided that they are periodically replenished by the former. Such differences are determined by local variation in habitat quality, predation, or parasitism, resulting in a "source-sink" system (Pulliam 1988).

An example is provided by the population dynamics of the blue tit, *Parus caeruleus*, in a kaleidoscope of three Mediterranean habitats of different quality (Blondel et al. 1992) (fig. 12.6). One habitat is a rich (in terms of food resources) deciduous oakwood on the mainland, the second is a poor sclerophyllous oakwood on the mainland, and the third is a poor sclerophyllous oakwood on the island of Corsica. The population in the deciduous mainland habitat breeds first (laying date 11 April), then the sclerophyllous mainland one (19 April) and, much later, the sclerophyllous island population (9 May). The spring development of caterpillars, the most important food resource for the tits, starts on average three weeks earlier, with a much higher peak of abundance in deciduous than in sclerophyllous woodlands, whatever the location of the latter. If one assumes that the best time to start to breed is such that young will be raised when food availability is at its maximum (Perrins 1965; van Balen 1973), two out of the three populations are rightly timed with respect to the food supply: the deciduous mainland one and the scle-

rophyllous island population. In contrast, the sclerophyllous mainland population is mistimed; young are raised long before the caterpillar peak, and many pairs fail to produce fledglings. To explain this mistiming, one must look at the scale of the whole landscape, including both rich deciduous and poor sclerophyllous habitats. Unlike Corsica, where all the habitats within the dispersal range of the population are poor for tits because the vegetation is entirely sclerophyllous, the mainland is a patchwork of poor sclerophyllous and rich deciduous oakwoods. One explanation of the mistiming in poor mainland habitats is the gene flow hypothesis. Blue tits on the mainland are primarily adapted to optimal deciduous oakwoods, where they produce many fledglings of good quality. Those individuals that settle in poor oakwoods breed there as if they were in a rich habitat. They are mistimed because asymmetrical gene flow from deciduous oakwoods prevents them from becoming adapted to the sclerophyllous habitat. One the other hand, the absence of deciduous forests on Corsica, where tits are isolated from any mainland population, has allowed the birds to evolve life history traits fairly well adjusted to local sclerophyllous habitats. This hypothesis is supported by the demonstration of a strong genetic component for laying date (van Noordwijk, van Balen, and Scharloo 1980; Blondel, Perret, and Maistre 1990).

This case study demonstrates the need to extend studies of populations (and communities) to a larger scale than just one habitat, because the survival of populations in poor habitats, and hence some components of community diversity, strongly depends on neighboring habitats.

COMMUNITY DIVERSITY ON MEDITERRANEAN ISLANDS: A MATTER OF SCALE, HISTORY, AND MAN

A recent and stimulating trend in ecology has been the recognition that the organization of living systems is hierarchical and that processes in nature are sensitive to the scales of time and space on which they are considered (Wiens 1981, 1986; Ricklefs 1987). This is especially true in island biogeography, a field in which too many studies in the past have yielded ambiguous results because they focused on local communities without considering their spatial and temporal context. Spatial and historical determinants of vertebrate diversities will be discussed in this section, using as examples birds and mammals on the island of Corsica.

Spatial Scaling and Bird Diversities on Corsica

Mainland-island surveys and censuses conducted at the scales of the whole island of Corsica (8680 km²) compared with three mainland areas of similar size (fig. 12.7A), mosaics of six habitats along matched habitat gradients on Corsica and on the mainland (fig. 12.7B), and local habitat patches (fig 12.7B) have shown that species impoverishment strongly varies depending on the size of the area and the type of habitat considered (Blondel, Chessel, and Frochot 1988, Blondel 1990a).

First, at the scale of the whole island, species impoverishment is about 37%, since 109 species regularly breed on Corsica as compared with 170 to 173 in three areas of similar size and similar physiographic diversity on the

MAINLAND

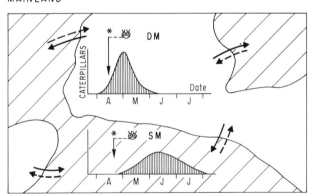

CORSICA

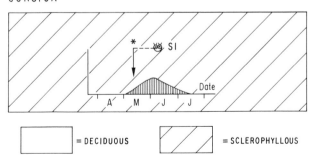

□ = DECIDUOUS   ▨ = SCLEROPHYLLOUS

Figure 12.6 Breeding of the blue tit (*Parus caeruleus*) in relation to food supply (caterpillar biomass) in a mosaic of deciduous and sclerophyllous Mediterranean habitats and in Corsica, where all the vegetation is sclerophyllous. DM, = deciduous mainland, SM, sclerophyllous mainland, SI, sclerophyllous island. Asterisks indicate laying date and the symbols to their right indicate hatching date. Pairs of arrows indicate assumed dispersal rates, whereby more individuals immigrate into sclerophyllous poor habitats from rich deciduous habitats than the opposite. (From Blondel et al. 1992.)

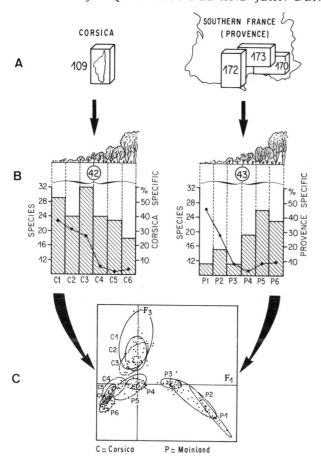

Figure 12.7 Bird diversities on Corsica compared with those on the mainland (southern France). (*A*) The whole island, compared with three areas of similar size in southern France (Provence). Figures are species richnesses. (*B*) Local habitat along matched habitat gradients on Corsica (left) and on the mainland (right). Figures encircled are the total numbers of species at the scale of the whole gradient. Bars indicate the number of species per habitat ranging from a very low matoral (C1, P1) to a mature forest (C6, P6). Solid lines are the percentages of region-specific species, i.e., species that occur in only one region. (*C*) Display of the trajectories of the bird communities of the two gradients on the bivariate space $F_1 \times F_3$ of a correspondence analysis. C1–C6, Corsica; P1–P6, mainland (Provence). Dispersion ellipses summarize the location of the communities (each dot is one census in each habitat type) in the bivariate space. (Data from Blondel, Chessel, and Frochot 1988; Blondel 1990a).

est habitat into shrubby habitats (Blondel, Chessel, and Frochot 1988). Hence, beta diversities are much lower in the insular gradient than in the mainland one. Moreover, the number of species that are "region-specific," i.e., which were found in only one gradient (17 on Corsica and 15 on the mainland), steadily decreases from the first habitat to the last, forested one (solid lines in fig. 12.7B). Thus, species richness decreases over the Corsican gradient, which does not fit the classic trend of species richness variation along successions, and increases over the mainland one.

When summarized and displayed on the bivariate space of a correspondence analysis where the two data sets have been combined (fig. 12.7C), two patterns clearly emerge (Blondel, Chessel, and Frochot 1988). The first is a striking convergence of the ecological trajectories of communities toward the last forested stages. This is a consequence of the history of Mediterranean biotas, as discussed in the first section of this chapter. The second pattern is a much looser structure of the habitat gradient on Corsica, i.e., dispersion ellipses that display each community on the multivariate space have much more overlap on the island than on the mainland because of the habitat niche expansion of forest species, explained above.

Such results show that the processes of species impoverishment on an island may be much more complicated than predicted by the classic assumptions of island biogeography, and that the level of resolution at which systems are studied, as well as their history, has a powerful influence on current patterns of diversity.

Habitat Utilization by Birds, Life History Traits, and Community Structure on Corsica

One may wonder whether changes in habitat utilization on Corsica, i.e., niche enlargement, are the same for all species. Species impoverishment is much more severe in forest communities than in shrubland communities (see fig. 12.7B). Since forest birds such as titmice presumably evolved in large tracts forest of habitat over Eurasia, one may hypothesize that their life history traits, especially their dispersal patterns and population densities, are adapted to large areas, and that their survival on islands would require changes in some of those life history traits. On the other hand, since shrubland species such as warblers presumably evolved in more or less isolated patches of matorral, their survival on islands should not require any further particular adaptation. If this is true, there should be a much more pronounced insular syndrome (Blondel 1985, 1986) in forest species such as tits (*Parus* spp.) than in shrubland species such as warblers (*Sylvia* spp.). This hypothesis has been tested using habitat occupancy, densities, and clutch size as clues (tables 12.1 and 12.2).

The range of habitats occupied by those titmouse species that occur in both the mainland and the island gradients is much wider than that occupied by the three species of warblers, since tits occur in three additional habitats on the island, whereas warblers do not. Moreover, the tit have higher densities on Corsica, demonstrating the classic process of density compensation on islands (or density inflation, see Crowell 1983; Blondel, Chessel, and Fro-

mainland. The missing species are not a random sample of the source mainland fauna (Blondel 1985); the species that are the most likely to be absent from Corsica are forest species.

Second, at the scale of the combined series of six habitats in matched gradients, there is hardly any impoverishment (43 species on the mainland, 42 on Corsica).

Finally, at the habitat level, some habitats, especially old forests, are heavily impoverished on Corsica, whereas other habitats, such as matorrals, have many more species than do their mainland counterparts. Thus, contrary to theoretical expectation, there are on average much higher alpha diversities on the island than on the mainland for a similar overall richness at the scale of the whole gradients. This is because most forest species tremendously enlarge their habitat niche and spill over from their preferred for-

Table 12.1. Distribution and Densities of Warblers (*Sylvia* spp.) and Titmice (*Parus* spp.) in Matched Habitat Gradients in Provence and on Corsica

Successional stage	Density[a]					
	I	II	III	IV	V	VI
Distribution						
Provence						
S. undata	1.05	4.44	1.53		0.24	
S. melanocephala		0.38	6.45	0.35	2.27	
S. cantillans		1.82	4.18	8.30	2.97	
S. atricapilla				6.11	6.95	5.55
Total	1.05	6.64	12.16	14.76	12.43	5.55
Corsica						
S. undata	0.38	1.40	1.46	0.07		
S. melanocephala	0.29	1.80	3.56	0.17	0.06	
S. cantillans		0.99	4.24	1.98	0.08	
S. atricapilla	0.25	0.29	1.22	7.15	6.84	6.66
Total	0.92	4.48	10.48	9.37	6.98	6.66
Provence						
Parus major				2.16	3.08	3.20
Parus caeruleus						11.55
Parus ater						0.18
Total				2.16	3.08	14.93
Corsica						
Parus major	1.60	1.70	2.55	3.59	2.62	4.67
Parus caeruleus		0.52	0.20	3.33	7.86	14.15
Parus ater				1.25	2.10	4.10
Total	1.60	2.22	2.75	8.17	12.58	22.92

Source: Blondel 1985.
[a]Pairs per 10 ha.

Table 12.2. Clutch Sizes of Warblers (*Sylvia* spp.) and Titmice (*Parus* spp.) in Provence and on Corsica

	Average Clutch Size[a]		
	Provence	Corsica	P
S. sarda		3.2 (33)	
S. undata	3.9 (42)		
S. melanocephala	4.1 (70)	4.0 (14)	NS
S. cantillans	4.3 (32)	4.2 (10)	NS
Parus major	8.3 (193)	6.6 (46)	< .001
Parus caeruleus	8.6 (213)	6.6 (470)	< .001
Parus ater	8.4 (195)	5.7 (103)	< .001

[a]Sample size is given in parentheses.

chot 1988). On the other hand, the warblers, except the forest-dwelling *Sylvia atricapilla,* are slightly less abundant on the island, which could result from some kind of interspecific competition in the species-rich shrubland insular habitats. Finally, the clutch size of tits on Corsica is much smaller, which is also a classic trend on islands (Crowell and Rothstein 1981; Cody 1966b; Blondel, Pradel, and Lebreton 1992). On the other hand, the similar sizes of the clutches of warblers on Corsica and on the mainland support the hypothesis that the evolutionary history of warblers in small patches of matorral has preadapted them to live in insular isolates.

As for community structure, the large differences between the two gradients result in an inversion of some basic community attributes between matorrals and old forests, depending on whether they are on the island or on the nearby mainland. Let us take habitats III and VI of the series as an illustration (fig. 12.8). The insular matoral (habitat III) has three times as many species as its main-

land counterpart, and total densities are nearly twice as high, but average population sizes (D/S) are on average lower. Moreover, the average weight of the species is much lower on Corsica than on the mainland, which is a classic trend for small land vertebrates on islands (see Foster 1964; van Valen 1973b; Case 1978; Williamson 1981 for a discussion on changes of body size of organisms on islands).

On the other hand, the Corsican forest (habitat VI) includes five fewer species than its mainland counterpart, but total population densities are slightly higher because average population sizes of the species present are much higher (density inflation). The total biomass is slightly lower on Corsica than in Provence because, here again, birds are on the average smaller on the island. Finally, species did not markedly enlarge their habitat niche in the species-rich matorral (see habitat occupancy index in fig. 12.8), whereas they did so greatly in the species-poor Corsican forest because the core forest species discussed in this chapter spill over from forests into shrublands (see Blondel, Chessel, and Frochot 1988 for more details).

Thus, two biologists working independently in two different habitats, for instance, a low matorral and a forest, would come to completely different conclusions. All the ingredients of an insular syndrome are found in the forest, while the opposite is true in shrubland habitats. These findings show that processes of habitat colonization and community organization are highly sensitive to the species-specific history of their components as well as to the scale at which they are studied.

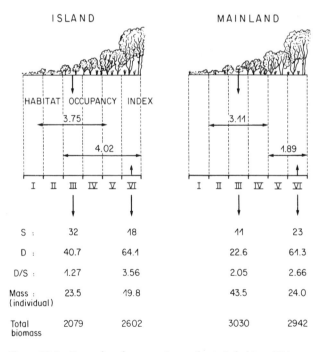

Figure 12.8 Examples of community attributes in habitats III (matorral) and VI (mature forest) in habitat gradients on Corsica and on the mainland. S, species richness; D, density (pairs per 10 ha). Habitat occupancy index is calculated on densities in each habitat from $e^{H'}$ where H' is the Shannon function (natural logarithms). Note the opposite trends in the "insular syndrome" between the matorral and the forest.

The Story of Mammals on Corsica

The story of terrestrial nonvolant mammals on Mediterranean islands illustrates the need for a careful analysis of history to interpret current diversities. The highly endemic and disharmonic[1] upper Pleistocene faunas of the Mediterranean islands have been completely replaced by the present faunas, which, compared with the mainland faunas, are still impoverished and disharmonic with a low endemic rate. A number of paleontological and archaeological studies starting from the Neolithic (7000 B.P.) yield a large series of bone samples that provide data on both faunal evolution and the relationships between humans and other mammals. Data from more than thirty-five archaeological sites spread over all cultural periods from the early Neolithic to the seventeenth century allow one to reconstruct extinctions, immigrations, and evolution. This story can be summarized in four main features (fig. 12.9).

The Pre-Holocene Story. The Corsico-Sardinian complex is a "continental" Tyrrhenian microplate that definitely disconnected from the European plate at the Oligocene-Miocene (30 mya). However, two important regressions of sea level allowed some species to invade the islands: the Messinian salinity crisis and a strong Middle Pleistocene (Günz or Mindel) glacio-eustatic recession (see review in Palombo 1985). Some taxa that evolved to endemic species and survived until the Pleistocene are associated with the first regression: insectivores (*Talpa, Episoriculus*), lagomorphs (*Prolagus*), rodents (*Tyrrhenoglis, Rhagapodemus*), artiodactyls (*Nesogoral, Sus nanus*), and a monkey (*Macaca majori*). Some species invaded the island during the second regression and evolved locally to endemic forms: one or two carnivores (*Cynotherium sardous* and probably *Enhydrictis galictoides*), a lagomorph (*Prolagus sardus*), a vole (*Tyrrhenicola henseli*), a mammoth (*Mammuthus lamarmorae*), and a deer (*Megaceros cazioti*). Some of these taxa became extinct during the climatic changes of the Upper Pleistocene so that the fauna of the late Pleistocene included a very small number of endemic taxa: *Episoriculus similis/corsicanus, Cynotherium sardous, Prolagus sardus, Rhagamys orthodon, Tyrrhenicola henseli,* and *Megaceros cazioti*. A shell-eating otter (*Cyrnaonyx majori*) and perhaps two other otter species might be added to the list. Humans probably did not establish permanent populations on the Corsico-Sardinian islands before the Mesolithic (9000 years B.P.) (Cherry 1990), although they could have done so (Held 1989). This Pleistocene fauna of Corsica is now extinct, and not a single nonvolant mammal of the present fauna has been found in fossil deposits older than the early Holocene.

The Holocene Human Colonization and Mammal Extinctions. In the pre-Neolithic period (ca. 10,000 B.P.), mammal assemblages were still very poor (fig. 12.9), with no more than four taxa (humans excluded): *Prolagus,*

Rhagamys, Tyrrhenicola, and *Episoriculus.* Hunting pressures rapidly (by 8000 B.P.) led the *Megaceros* deer and probably *Cynotherium* to extinction (Sondaar et al. 1986), an observation that supports the prehistoric overkill theory (Martin 1984). *Prolagus, Episoriculus,* and two rodents (*Rhagamys* and *Tyrrhenicola*) succeeded in resisting hunting and competition with new invaders during eight millennia and in coexisting with the complete modern fauna after the sixth century (Vigne and Marinval-Vigne 1983). They finally became extinct, however, because they could not resist the combined effects of predation by humans and by the new immigrants (dogs, foxes, weasels), competition with invaders, and vegetation changes due to pastoralism and agriculture. Their extinctions probably did not result from a single factor, because all the primary causes existed as early as the Neolithic, but from the combined effects of several of them and their generalization over the whole island.

Domestic Immigrants and the Process of Feralization. Humans introduced domestic species in Corsica (sheep, goats, pigs, cattle, and dogs) as early as 7000 B.P. These species, as well as those introduced later (horses, donkeys, cats), became strong competitors with the autochtonous endemic fauna. They became an important part of the present fauna because of feralization, as examplified by the mouflon (*Ovis ammon musimon*), which was absent from western Europe during the Pleistocene (Poplin 1979). Genetic data (see Geddes 1985 for a review) suggest that this species is closer to the present domestic sheep than to any wild sheep of the Near East. Thus, the present Corsico-Sardinian mouflon is nothing other than a relictual Neolithic domestic sheep that escaped human control before the late Neolithic (Poplin 1979; Vigne 1988c). A similar process of feralization probably occurred in the Corsican boar (Popescu, Quere, and Francheschi 1988) and the wild cat (*Felis silvestris libyca* var. *reyi*).

Origin and Evolution of Modern Wild Mammals. Corsica has been isolated from the mainland (80 km) throughout the Holocene, so spontaneous immigration by most mammals is most improbable. Such hypotheses as active swimming or travel on driftwood, as advocated for instance by Reyment (1983) for Mediterranean islands, are far from convincing for three reasons: (1) a colonization rate of fourteen species within less than 10,000 years is higher than expected from dispersion rates of land mammals; (2) the present wild species are much poorer overwater colonists than are the more or less amphibian mammals, such as elephants, hippos, otters, and deers, that characterized natural dispersion into isolated Mediterranean islands during the Pleistocene (Sondaar 1977); (3) mammals of Corsica are ecological generalists (Cheylan 1984) strongly linked with humans from both ecological and cultural points of view (Vigne 1988c). Some of them, such as the red deer, the hare, and the rabbit, are current game species, as was probably also the case in the Neolithic for the red fox (Vigne 1988a). Small mammals (*Crocidura, Suncus, Apodemus, Mus, Rattus, Eliomys,* and *Glis*) presumably were introduced as stowaways on ships. Among

---

1. Disharmony refers to changes in the relative proportions of different taxa or trophic levels on islands as compared with those on the mainland.

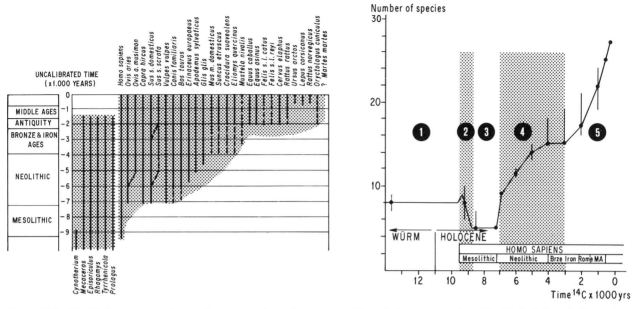

Figure 12.9 The Holocene story of nonvolant mammals on Corsica. *Left:* Stratigraphic repartition of the species. Vertical lines indicate the duration of presence of the species. *Right:* Variation in species richness between the Würm glaciation, late Pleistocene (1) and the present time (5). Vertical bars indicate the possible variation in species richness according to uncertainty on the stratigraphic distribution of some species. (After Vigne 1990.)

shrews, only *Crocidura* and *Suncus* immigrated; *Sorex* and *Neomys* are missing. Although the diversity of carnivores on Mediterranean islands is high (Cheylan 1984), such large species as the wolf, the wildcat, and the lynx never immigrated, probably because they are potential competitors with humans. Among the small mammals, the only species absent from Corsica are those that never had contact with humans (*Muscardinus,* squirrels, the beaver, the marmot, and voles). Human introduction of all the modern nonvolant mammals on Corsica resulted in a very disharmonic fauna and a complete turnover of species since the Pleistocene, with no surviving endemics. Such huge extinction-immigration processes resulted in a three- to fivefold increase of mammal diversities. The resulting modern fauna is much richer than predicted by the MacArthur-Wilson (1967) theory of island biogeography (Cheylan 1984), and, therefore should be considered as supersaturated, i.e., including more species than predicted by the dynamic equilibrium between extinction and colonization.

Mammals and Man on Other Mediterranean Islands

Can the story of Corsica be generalized to other Mediterranean islands? Although each island must be considered as a singular case with local idiosyncrasies, the answer is definitely yes. The replacement of an impoverished, endemic, and disharmonic mammalian fauna by a supersaturated, poorly endemic, and monotonous fauna during the Holocene occurred on all the larger Mediterranean islands, e.g., Cyprus (S. M. J. Davis 1989), Crete, Malta (Boessneck and Küver 1970; Storch 1970), the Balearic Islands (Alcover, Moya-Sola, and Pons-Moya 1981), and even smaller islands (Felten and Storch 1970; Cheylan 1984; Vigne 1988b).

However, there are local variations. For instance, together with Sardinia and other satellite islets, Corsica is part of the largest island microplate in the Mediterranean, which extends over more than 32,000 km², an area 7,000 km² larger than Sicily and nearly four times as large as Crete or Cyprus. It is also the nearest to the mainland (very high target-distance ratio: Held 1989), which explains some faunal specificities, especially the absence of the very particular "elephant-deer" assemblages (Sondaar 1977) that characterize other large islands such as Cyprus. The Corsico-Sardinian Pleistocene fauna was less endemic, with the presence of a carnivore (*Cynotherium*) limiting the insular evolution of deers and *Prolagus* (Sondaar et al. 1986). On Cyprus, and to a lesser extent on Crete, Mesolithic people also rapidly brought large mammals to extinction (Cyprus: Simmons 1988; Crete: Jarman and Jarman 1968). The situation is slightly different in the more remote Balearic Islands because human colonization was delayed to the Neolithic (Waldren 1982) and because a possible domestication of an endemic antelope (*Myotragus balearicus*) might have saved it from extinction until the Bronze-Iron Ages (Alcover, Moya-Sola, and Pons-Moya 1981).

The story of mammals on Mediterranean islands is that of a dramatic overall decrease of genetic diversity, since all the highly endemic mammal species but two became extinct as a consequence of human invasion. The only relictual Pleistocene endemic species in the whole Mediterranean are two shrews: *Crocidura zimmermani,* refuged in the mountains of Crete (Reumer and Payne 1986), and *C. sicula,* which probably evolved on Malta and Gozo (Vogel, Maddalena, and Schembri 1990). All the larger islands presently share four features: (1) a very low endemism rate, with a monotonous supersaturated fauna; (2) some feral mammals, such as mouflons on Sardinia and

Cyprus (Poplin 1979; Davis 1987), goats on Crete and several Aegean Islands (Uerpmann 1987), and cats on Sardinia and Mallorca (Ragni 1981 and Uerpmann 1971 respectively); (3) some game species, such as the fallow deer on Cyprus (Davis 1987), the fox in the Anatolian and Tyrrhenian areas (Vigne 1988a) and, more recently, deers, hares, and rabbits; and (4) a number of commensal small species such as *Erinaceus, Crocidura, Suncus, Glis, Eliomys, Apodemus, Mus,* and *Rattus* (Vigne 1988a).

This story of the role of humans in determining modern insular diversities is not specific to Mediterranean islands. The birds of the Hawaian archipelago (Olson and James 1982; James et al. 1987) and the vertebrates of New Caledonia (Balouet 1987) are only a few of many other examples of island faunas that have been strongly affected by human activities.

Concluding Remarks

Most ecologists agree today that there is no primacy for one or a few factors in determining distribution patterns of species and community diversities on a regional and on a local scale. Much emphasis has been placed in this chapter on the role of history, including that of humans. This does not mean, of course, that we assign a greater importance to history than to other factors. Simply, we argue that ignoring history may lead to misleading approaches, and hence to false conclusions. One example is the work of some biogeographers who tried to explain mammalian diversities on Mediterranean islands without considering the long and profound effect of the human story in the Mediterranean basin. Historical events are important because, in a sense, they establish the faunistic and evolutionary foundations on which present and future ecological and evolutionary processes operate. Spatio-temporal effects on diversities are the results of events ranging from those that occur over very long evolutionary scales of time to recent events that occurred one or a few years ago. Many convincing examples range from climatic Pleistocene events, which, in a large part, condition modern faunistic and floristic diversities on a regional scale, to the ecological and evolutionary consequences of very recent events for populations and communities. One of the best

examples of the effects of short time-scale events is that of the consequences of exceptionally dry and wet years, associated with El Niño events, for such phenomena as niche relationships, population dynamics, morphology, distribution, and local extinctions of different species of Darwin's finches on the Galapagos Islands (Grant and Grant 1989).

Most of the so-called "natural communities" where so many studies on population dynamics and community ecology have been carried out, especially in temperate regions, are far from natural because they bear to a varying extent the imprint of human activity. Yet, a large body of ecological theory has been constructed from studies of such communities. Long ago, David Lack (1965) warned against too much generalization from studies on evolutionary biology in artificial habitats because adaptations may be hard to recognize. Why is a blackbird foraging on a green meadow in Hyde Park black? Go into the primeval forest of Bialowieza (Poland) and you will immediately understand why this bird is black. Bird communities are widely believed to be more resistant to fragmentation in the temperate zone than in the tropics because species are claimed to occur in higher densities, to be more widely distributed, and to have better dispersal abilities (Wilcove, McLellan, and Dobson 1986). However, this view is probably fallacious, as shown by comparative work in woodlots of western Europe and in the primeval forest of Bialowieza (Tomialojc, Wesolowski, and Walankiewicz 1984; Blondel 1990b). Habitat fragmentation seems to have less severe effects on community ecology in the temperate zone only because damage to the area had already occurred long before the present stage of human ecological awareness. Integrating historical and spatial components of diversities with population biology and community ecology leads to more realistic ideas of the real world, which is no longer the orderly, predictable, and deterministic world of the sixties, but a world where factors such as competition, predation, parasitism, species-specific life histories, behavior, population structure and geographical variation of both phenotypes and genotypes share their roles in the shaping of biodiversity with disturbances, patchiness, historical events, and the impact of humanity.

13

Bird Diversity Components within and between Habitats in Australia

Martin L. Cody

Studies on bird species diversity have progressed considerably since the breakthrough made thirty years ago by Robert MacArthur in relating local species number to vegetation structure in northeastern North America (see, e.g., MacArthur and MacArthur 1961; MacArthur, MacArthur, and Preer 1962; MacArthur, Recher, and Cody 1966). In this chapter I will discuss various influences on the three main components of regional diversity:

1. Alpha diversity, defined as the number of species within a habitat type at a particular locality or site, and a variable function of the vegetation structure of the habitat as well as of other environmental influences

2. Beta diversity, the rate of species turnover between habitat types, a function mainly of difference between habitats, their areal extent, and their contiguity

3. Gamma diversity, the rate of species turnover within a habitat type between different sites, a product of overall speciation patterns and generally a function of site separation and of the intervening barriers to species dispersal.

This chapter presents data on bird diversity components, their magnitudes, and their variations over landscapes in southwestern and northeastern Australia. It is concerned specifically with the elucidation of the factors controlling diversity components in the two regions: vegetation structure, habitat area and contiguity, habitat patchiness and relative isolation, and the effects history and chance. I have collected, and reported on, similar data sets from Mediterranean-climate sites around the world (Cody 1975), and from habitat gradients in southern Africa (Cody 1983). These previous studies have shown that bird communities in matched habitats on different continents are quite similar in alpha diversity, while often dramatically different in beta diversity. Comparisons between replicated communities within a habitat type over a particular geographical region measure gamma diversity, and ecological counterparts often constitute a major contribution to regional diversity. Yet there is often a set of "core species" in the assemblages of a particular habitat type that is constant, and species turnover is contributed mainly by rarer species that invade from adjacent habitats. This phenomenon appears to be a function of habitat fragmentation and the character of the surrounding vegetation; I call such influences "spillover effects" (cf. the "mass effect" of Shmida and Wilson 1985; Auerbach and Shmida 1987). Diversity components can be clearly re-lated to parallel aspects of the rarity of species and to landscape patterns at various scales, and have pertinence to the conservation of biodiversity (Cody 1986).

For a variety of reasons, the Australian continent is particularly conducive to studies of diversity components. First, topographical relief is low everywhere except in the southeast, and barriers to bird dispersal are much more likely to have a biotic component (i.e., to be related to the characteristics of habitats, their resources, and their occupants) than to be purely physical. Studies of bird speciation in Australia have emphasized both the repeated invasions of species from forested New Guinea into the more arid southern continent during cooler, wetter periods of the Pleistocene, and autochthonous speciation in the continent's arid interior (Keast 1961, 1981b; Schodde 1982). Second, patterns of vegetation are well known and have been described in detail (see, e.g., Groves, 1981; Specht 1981b); vegetation varies in the usual fashion with climatic gradients across the continent and, more especially in Australia with its preponderance of nutrient-poor soils, with substrate type. Further, several exemplary studies have related the bird fauna to vegetation structure: the work of Kikkawa and associates (Kikkawa 1968, 1974, 1982; Brereton and Kikkawa 1963; Kikkawa and Pearse 1969; Kikkawa et al. 1981) in the northeastern forests, and of Recher and associates on alpha diversity (1969) and on the birds of eucalypt forests and woodlands (Keast et. al. 1985).

Of particular relevance to this chapter is the relative continuity of inland vegetation (mulga bushlands) and of the northeastern woodlands, in contrast to the discontinuity of the northeastern rainforests and of the forests and woodlands between eastern and southwestern Australia. Similar vegetation occurs in places with similar substrate and/or climate, in disjunct sites within the continent: sites with different areas, histories, and often differences in the surrounding vegetation.

FIELD SITES AND METHODOLOGY

In 1984–1985 I conducted bird censuses on fifty-five sites in southwestern Australia, over the range of vegetation types, forest to scrub desert, along three transects. The transects began in Mangimup at the southern tip of Western Australia, and ran 1000 km N, NE, and E, up the west coast, into the arid interior, and along the southern coast respectively. In 1989–1990, I collected similar census data on ninety-nine sites in northeastern Australia, in all common vegetation types along four 1500-km transects. Three of these ran west from the east coast of Queensland (Gladstone and Brisbane) and northern New South Wales (Grafton) into the interior (to Boulia, the Diamantina, and Thargomindah respectively). The fourth extended south from the tip of Cape York to Hughenden, thence west to Mount Isa.

Birds were censused on ±5-ha plots in forest and woodland habitats and on larger plots (up to 20 ha) in more open habitats (e.g., desert and grassland) with lower vegetation and bird density. The censuses aimed to measure the numbers and relative densities of breeding bird species. Data were accumulated during successive visits of 2–6 hours, each site being visited 2–12 times over 2–5 days, with repetition depending on the ease or otherwise of recording the birds and on the rate of accumulation of new information. Raptorial and nocturnal birds were excluded from the censuses. To compare census data between sites, I calculated species turnover as $[1 - C(T_1 + T_2)/2T_1T_2]$, where two censuses have species totals T_1 and T_2 respectively, and hold C species in common. If $T_1 > T_2$, then $0 \leq C \leq T_2$, and with $T_2 = aT_1$ and $C = bT_2$, this turnover index becomes $[1 - b(1 + a)/2]$ and increases more or less linearly with decreasing a, b, with a slight bias relative to other indices toward emphasizing turnover if a, b are simultaneously low (see, e.g., Pielou 1979 for a discussion of other turnover indices).

At each site a vegetation profile was computed by plotting mean vegetation density (measured as the inverse of the horizontal distance to a board half-covered by the vegetation) against height above ground, averaged along 20 randomly selected azimuths; the site composition by dominant plant species was also recorded. Habitats and vegetation types for the northeastern sites are shown in figure 13.1 and for the southwestern sites in figure 13.2. I measure structural differences between habitats as the euclidean distance between points in these figures. Note that there are both similarities and differences in habitats between the two regions. First, there is no rainforest or grassland in the southwest; second, otherwise equivalent habitats vary somewhat in structure: woodlands, heaths, and chenopode scrublands tend to be denser (higher profile area), and mulga to be taller, in Queensland (which has summer rainfall, in contrast to the winter rainfall region of the southwest).

The complete data base consists of about 3500 nonzero entries in a matrix of 309 bird species × 154 sites, with bird species characterized by their various ecological and biogeographical attributes and sites by various components of vegetation structure. During the course of this work, some 126 new breeding records were made (viz. Blakers, Davies, and Reilly 1984), and are published elsewhere (Cody 1991). Only a part of the data base is analyzed below, where I shall discuss in turn general trends in alpha diversity, differences in beta and gamma diversity in forests and woodlands between the southwest and the northeast, and diversity components in two contrasting habitat types, rainforest and mulga.

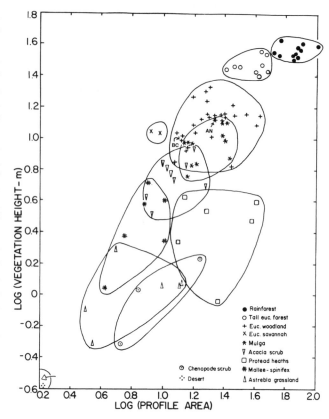

Figure 13.1 Vegetation structure of habitats censused in northeastern Australia, in terms of profile area and vegetation height ($n = 99$ sites). Each symbol on the graph represents a census site, with habitat types as indicated by the key. Mulga sites AN: Morven and BC: Dunkeld are indicated.

BIRD DIVERSITY IN GENERAL TERMS

Species-area curves give information on the number of species to be found within circumscribed regions, but this information is hard to interpret in terms of the different components of diversity. The form of such curves obviously depends on the number of habitat types, as well as their range and extent, and includes species accumulated with alpha diversity within habitats, beta diversity among habitats, and gamma diversity within habitats between different areas within the region. Overall species number ("regional diversity") will be some sort of composite figure reflecting contributions from these three components.

Table 13.1 gives regional species totals for Australia, along with comparative data from the continental United States. Species richness, relative to the land area under consideration, is comparable between the continents in overall bird species, but is higher for nonraptorial land birds in Australia. The Pacific states fall between Queensland and southwestern Australia in species richness, with higher values in California. The lower section of the table shows that just 30% of the combined northeastern and southwestern Australian birds are common to both regions, and that the censuses sample about 80% of each region's avifauna.

Note that Queensland supports over twice the number

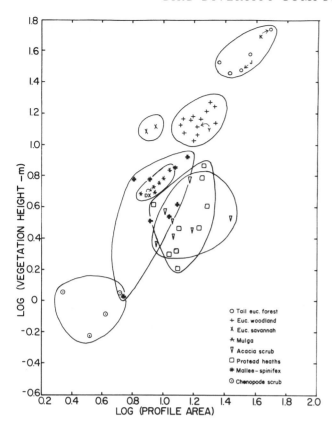

Figure 13.2 Vegetation structure of habitats censused in southwestern Australia ($n = 55$ sites). The mulga site DX: Hamelin Road is indicated by an arrow; karri, jarrah forest, and york gum woodland, are indicated by the letters K, J, Y.

Table 13.1. Regional and Census Bird Species totals: Northeastern and Southwestern Australia

	Area[a]	# bird spp.	Spp. richness[b]
Australia	7.61		
All birds[c]		726	0.95
Native, breeding spp.		668	0.88
Nonraptorial land birds[d]		531	0.70
Queensland	1.50[e]	330[f]	1.92
S and C Queensland	0.95[g]	210[f]	2.21
SW Australia	0.85[h]	154[f]	1.81
Continental U.S.	7.60		
All birds[c]		830	1.09
Native, breeding spp.		512	0.67
Nonraptorial land birds[d]		367	0.48
Pacific states[i]	1.12	226	2.02
California	0.41	175	4.27

Region (Australia):	NE∪SW[j]	NE	SW	NE∩SW[j]
Species in region:	369	330	154	115
Species censused:	311	268	121	78

[a]Land area in 10^6 km^2
[b]Calculated as (bird species number)/(land area in km^2 × 10^{-4})
[c]All species recorded in area; sources are lists and range maps in National Geographic (1987) and Pizzey (1980).
[d]Excludes all marine, freshwater, raptorial, and nocturnal birds
[e]Land area covered by Queensland censuses (excludes Gulf of Carpenteria)
[f]Diurnal, nonraptorial land birds only
[g]Land area covered by southern three NE transects
[h]Land area within 1000 km of Manjimup, W.A.
[i]Washington, Oregon, California, and Arizona
[j]∪ indicates union (in either region), ∩ indicates intersection (in both regions)

of bird species (330; 268 species censused) as does southwestern Australia (154; 121 species censused). But the northeast has about twice the land area (1.5 versus 0.85×10^6 km^2), has nearly twice as many census sites (99 versus 55), and has 5000 km of transect versus 3000 km in Western Australia. Does the difference in regional diversity represent a shift along the same species-area curve, or is overall diversity higher in the northeast? The latter is correct; the southern three northeastern transects cover 3150 km and an area 0.95×10^6 km^2, and still contain 210 species, over one-third more than the similar-sized area in the southwest. What accounts for the higher regional diversity in the northeast, and how are the extra species accommodated and distributed? And how is this higher regional diversity allocated among the different diversity components?

COMPARISONS IN ALPHA DIVERSITY

The first possible explanation is that higher regional diversity in Queensland is accommodated solely in higher alpha diversity counts there. To examine this possibility, I use (arbitrarily) a "Californian baseline," a curve fitted to Californian data (Cody 1975) to reflect alpha diversity (ordinate) as a function of vegetation structure (abscissa). The results of computing abscissa values for the fifty-five southwestern Australian sites are depicted in figure 13.3. There is considerable scatter ($R^2 = 66\%$), but the residuals are approximately normally distributed (shown in the histogram at the lower right of the figure). With the statistics computed in table 13.2, it can be seen that the southwestern sites average about one species lower in alpha diversity than structurally comparable Californian sites.

The same treatment of the Queensland data shows that residuals there average about seven species higher in alpha diversity, significantly higher (t-test, $p < .05$) than the southwest (or California; see table 13.2). This higher alpha diversity remains apparent even when the rainforest and tall-forest data (at the high end of the abscissa) are excluded from the comparison ($PC < 1.3$), and likewise when only vegetation below taller scrub and woodlands is considered ($PC < 1.0$).

In light of the lower bird alpha diversities found in the southwestern habitats, it is still remarkable that here vegetation structure constitutes a relatively poor predictor of alpha diversity (cf. R^2 values of about 85%–95% for sites in California, Chile, and South Africa, albeit with fewer sites and a somewhat reduced geographical range; Cody 1975). Various factors might contribute to this scatter. As one example, I show in figure 13.4 the effect of distance from the habitat epicenter on alpha diversity residuals for censuses in eucalypt-dominated woodlands. The abscissa of this plot is distance (km) of the census site from the geographical center of the southwestern woodlands, calculated from the vegetation maps of Beard (1981b). The residuals decrease with distance from woodland epicenter: $\text{RESID} = 6.08 - 0.025(\text{DIST})$, $r = .515$ ($n = 14$, $p = .06$). Further, part of the residual variation in alpha diversity is attributable to the small patch size of some woodlands, which now exist as relictual patches within

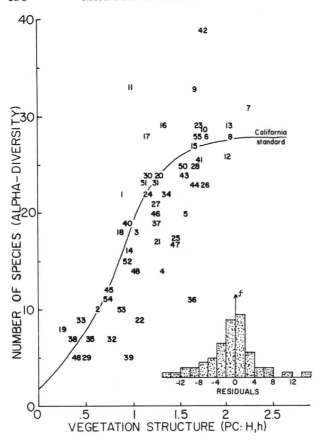

Figure 13.3 Numbers of bird species found in 55 southwestern Australian sites plotted against a measure of vegetation structure (a principal component of vegetation height H and half-height h at which half the profile area lies above and half below). The regression line is that derived from Californian data (Cody 1975). Residuals from the plot for the southwestern Australian sites are plotted at lower right.

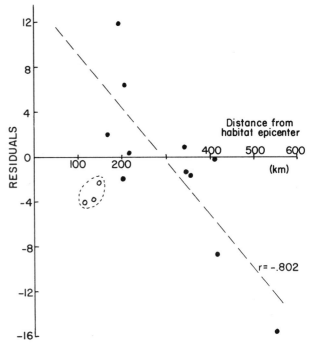

Figure 13.4 Residuals from fourteen eucalypt-dominated woodland sites from figure 13.3 are plotted against the distance of the census site from the geographical center of the southwestern woodlands, showing how alpha-diversity declines with distance from habitat epicenter. Three sites in wheat belt reserves are shown as open circles, illustrating a decline in alpha-diversity with isolation and patch size of the habitat.

Table 13.2. Alpha Diversity Residuals in Southwestern and Northeastern Australian Habitats

Abscissa	Full range	PC < 1.3	PC < 1.0
SW residuals	$n = 55$ -1.17 ± 5.50	$n = 31$ -1.10 ± 5.31	$n = 18$ -0.71 ± 5.74
NE residuals	$n = 99$ 7.35 ± 6.71	$n = 79$ 6.81 ± 6.48	$n = 50$ 7.35 ± 5.76

Note: Residuals are to the Californian baseline shown in figure 13.3.

the mostly cleared wheat belt of the southwest. Without the three sites in these wheat belt reserves (circled in fig. 13.4), RESID = 14.5 − 0.048(DIST) and $r = .802$ ($n = 11$, $p < .01$). That is, alpha diversity decreases by about five bird species per 100 km from the habitat epicenter.

Species Turnover among Eucalypt Forests and Woodlands: Beta or Gama Diversity?

In this section I use the census data from forest and woodland habitats dominated by *Eucalyptus* to examine species turnover among sites, and compare turnover rates between the northeastern and southwestern regions. Species turnover between sites is influenced potentially by two factors: difference in vegetation structure, which is the

beta component of diversity, and the incidence of ecological replacements or counterparts, or the gamma diversity component. The former is influenced by differences in resource availability and perhaps by a variety of other factors (both biotic and abiotic) that might mediate niche opportunities between sites; it is to these differences among sites that measures of vegetation structure are assumed to correlate. The latter is generated proximally by the occurrence of allopatric counterparts within, for example, genera or by checkerboard distributions of ecologically similar species, to which interspecific interactions over, for example, sites or territories are likely to contribute on a local scale. Gamma diversity is influenced in the midrange by distance and/or other topographical (e.g., habitat, biotic) barriers and is ultimately determined by the sorts of events that lead to or from speciation—that is, by historical factors rather than by the contemporary environmental conditions.

The data set includes twenty-one eucalypt-dominated forests and woodlands in southwestern Australia, and each site and census can be contrasted with each other site in turn. In figure 13.5, each of three sites is compared with the southwestern forests and woodlands: the tallest forest habitat (karri, *Eucalyptus diversicolor*) an intermediate forest in jarrah (*E. marginatus*), and the lowest woodlands, york gum (*E. loxophleba*). Species turnover increases with difference in vegetation structure, but increases faster between jarrah and other sites than between karri and other habitats and sites; turnover rate is lower (lower slope) between york gum and other sites, but is

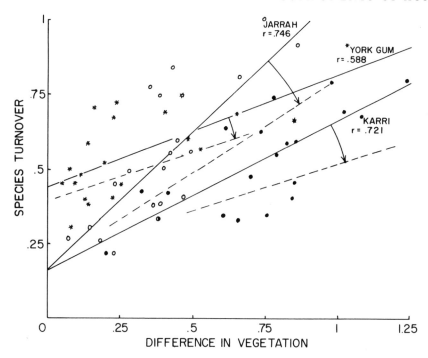

Figure 13.5 Bird species turnover in southwestern eucalypt-dominated forests and woodlands is shown as a function of difference in vegetation structure between sites. Comparisons are made between three sites (karri forest, jarrah forest, york gum woodland) and all other forest or woodland sites. Note the steeper slope for jarrah and the higher intercept for york gum. The solid lines are regression lines for all intersite comparisons; the dashed lines are those for comparisons made between sites that are < 500 km apart.

higher among york gum habitats (higher intercept on ordinate for small differences in vegetation structure). Note that the value of the intercept on the ordinate can be regarded either as a sampling error or as a measure, hopefully, of the unpredictability of censuses between like habitats. These findings on intersite differences are reinforced in other southwestern tall open forests of restricted distribution (tingle, *E. jacksonii*; tuart, *E. gongylocarpus*), which likewise have intermediate slopes and low intercepts, and in other lower, open-woodland sites (salmon gum, *E. salmonophloia*; morrell, *E. longicornis*), with low slopes and high intercepts.

A tempting interpretation of these results is that the low areal extent of the forests precludes the evolution of habitat-specifics that would otherwise contribute to beta diversity, whereas the widespread, high-area jarrah has more habitat-specifics, and therefore produces a higher beta component. The lower woodlands have been extensively cleared, and their current islandlike patches show the islandlike characteristics of low predictability of species identities in similar vegetation types and a lower slope of species versus vegetation turnover, indicating that site composition is as much affected by isolation and patch size as by vegetation structure.

But the effects of intersite distances have yet to be considered. If intersite comparisons are restricted to those sites less than 500 km apart, the species turnover rate decreases (dashed lines in fig. 13.5), considerably in jarrah and karri, but much less markedly in low woodlands. To determine whether species turnover is a function of vegetation difference per se or a consequence of site separation, I calculated regression statistics for the intersite comparisons using both difference in vegetation structure and distance apart of the sites in the comparisons. The residual species turnover from each plot was then used as the dependent variable for the other independent variable,

distance and vegetation difference respectively. When this analysis is completed, it is clear that the species turnover in the southwestern forests and woodlands is wholly determined by intersite distance, with no significant contribution by vegetation structure: $SPTURN = 0.354 + 0.00058DIST$ $(+ 0.0000VGTURN;$ $F_{2,207} = 88.79,$ $p < .001)$. Thus species turnover here measures gamma diversity, which increases on average by 0.1 (or 10%) per 172 km of site separation.

A similar analysis of twenty-nine eucalyptus-dominated habitats in south and central Queensland (in the three southernmost transects) gives quite different results. Both vegetation structure and intersite distance contribute to species turnover, which, in all pairwise site comparisons, increases by 0.1 per 417 km or per 0.56 of difference in vegetation structure: $SPTURN = 0.511 + 0.00024DIST + 0.177VGTURN$. For this regression, $F_{2,403} = 67.7,$ $p < .001$; both the DIST effect and VGTURN effect are significant ($F_{1,403} = 86.5, p < .001$, and $= 34.0, p < .001$ respectively). Species turnover contributes significantly to both beta and gamma diversity in the richer northeastern habitats.

In comparisons involving each of these Queensland sites in turn with every other site, 21 of 29 multiple regressions show significant effects of distance on species turnover, and 10 of 29 show significant effects of vegetation structure on turnover (with $n = 28$, $r > .37$ for $p < .05$). There is, however, a wide range of slopes (gamma diversity, coefficients of DIST) among the twenty-one sites with significant distance effects, from a low of 0.00009 to a high of 0.00054, and averaging $0.00031 \pm .000010$ SD (for an average species turnover of 0.1 per 323 km). When the residuals of these site-specific regressions of turnover versus distance are tested against VGTURN, one-half (14 of 29) show significant ($p < .05$) effects of vegetation structure. These fourteen sites again show a wide range of slopes (beta diversity, coefficients of VGTURN), from 0.124 to 0.436, and averaging 0.261 ± 0.098 SD.

The interpretation of intersite variations in beta and gamma diversity is a challenge not yet fully met, although some influences are decipherable. The lowest partial correlations of species turnover with vegetation structure come from sites relatively isolated from similar vegetation elsewhere, presumably because of spillover effects. Examples are Gwydir River (V), an island site of tall, riparian red river gum woodland well isolated from the coastal forests of comparable structure, and Alexander Bay (AR), a low *Eucalyptus-Banksia* woodland on the coast far from the similarly structured low inland woodlands and separated from them by much taller forests. Distance effects also are minimal in the most isolated sites, for which distant sites (with more similar vegetation structure) may provide the closest matches in bird species, but are also low at sites located in the most contiguous habitats, such as in woodlands on the inland slopes of the Central Highlands (e.g., AO: Mungindi; AP: Bollon; AU: Mundalee). It is apparently at intermediate levels of habitat patchiness, where habitats are contiguous but variable in structure over large areas, where beta and gamma diversity are highest.

In figure 13.6 the beta diversity relations among individual sites are seen more clearly through site ordination. Figure 13.6A shows the relations among sites in terms of vegetation structure (positions of the sites are shown in fig. 13.1), and figure 13.6B shows site ordination, based on the beta diversity among sites, via multidimensional scaling (MDS; see Kruskal 1964). If beta diversity were purely a function of intersite differences in vegetation structure, the two parts of the figure would be closely congruent, but they are not. In particular, the influences of positional (topographical) effects, and of the type and area of adjacent habitats, seem clearest. For example, the forest sites T, S, and V are structurally similar, but segregated by census differences into montane forest, Great Dividing Range, and interior riparian clusters. Similarly, the low woodland sites represented by CA, BJ, and BQ are all structurally similar, but show extensive differences in bird species. These correspond to obvious topographical and other site differences among the sites: CA is a far-interior (Jericho) sand-plain woodland in mostly *Acacia*-dominated bushland, BJ is a floodplain woodland in the interior foothills (south of Roma), and BQ is woodland high in the forested Bunya Mountains, very close to rainforest. I believe that the disentangling of the various influences on beta (and gamma) diversity is one of the more difficult tasks in ecological biogeography, but it seems an achievable goal, provided sufficiently large and comprehensive data sets can be assembled.

RAINFOREST: A NORTHEASTERN SPECIALITY

Rainforests are restricted to Australia's northeastern coasts and mountains. I conducted eleven censuses in this habitat from the northern tip of Cape York south through Queensland to Dorrigo in the tableland of northern New South Wales. The rainforest occurs in patches (and did so even before the extensive clearing—from two-thirds to

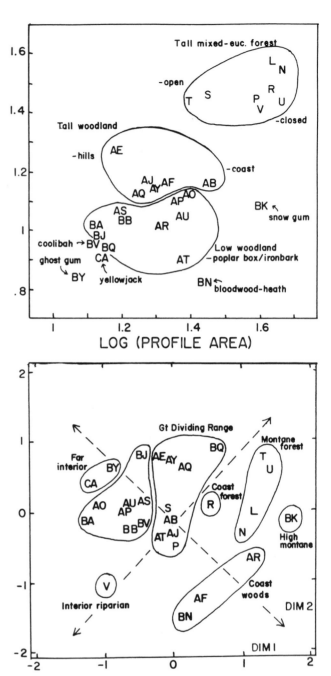

Figure 13.6 *Upper:* Relative positions of twenty-nine eucalyptus-dominated forests and woodlands in the plane of vegetation structure defined by vegetation height and profile area (from figure 13.1). *Lower:* Relations among the same sites according to their similarity in bird species (i.e., the complement of beta-diversity; the effects of intersite distance have been removed). Here the sites are ordinated in two dimensions using a multidimensional scaling procedure that optimizes the dispersion of sites at the plane that best reflects their similarities in bird species. Residual stress (= 0.202) measures the conflict in positioning that remains in the optimal two-dimensional ordination. Differences between arrangement of sites in upper and lower reflect positional and topographical influences.

three-fourths of the original area—that has taken place). Some of the patches are quite isolated from others, and are quite small in area; see the map in figure 13.7 for rainforest and census site distribution. Brereton and Kik-

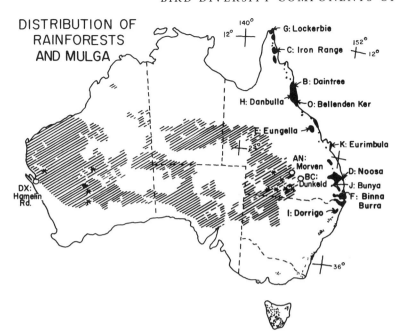

DISTRIBUTION OF
RAINFORESTS
AND MULGA

G: Lockerbie
C: Iron Range
B: Daintree
H: Danbulla
O: Bellenden Ker
E: Eungella
K: Eurimbula
AN: Morven
BC: Dunkeld
D: Noosa
J: Bunya
F: Binna Burra
I: Dorrigo
DX: Hamelin Rd.

Figure 13.7 Distribution of two habitat types in Australia, the northeastern rainforests and the interior and nearly continent-wide mulga bushlands. Habitat data are assembled from a number of sources, but particularly Webb and Tracey (1981) for rainforests, for which lettered census sites are indicated. Mulga sites in the east and west that are least contiguous with other habitat are indicated by circles; other mulga sites are indicated by asterisks.

kawa (1963) pointed out that the southern rainforests are relatively impoverished in bird species compared with (eucalyptus-dominated) forests and woodlands, and Kikkawa (1982) has shown that lowland rainforest birds and those of the (high-elevation) Tableland rainforests in the Atherton region of northeastern Queensland (sites B, H, O in fig. 13.7) differ considerably.

A total of 116 bird species was censused at the eleven rainforest sites, and about half of these (57 species) can be considered rainforest specifics. Alpha diversity at the sites varies from 31–51 species. One-third of this variation is accounted for (but with $n = 11$, $p > .05$) by the size of the relictual patches, but latitude contributes nothing to this variation (SPP $= 36.533 + 0.233$AREA $- 0.06$LAT; $R = 0.575$, $p = .06$).

Given that a total of 116 species was recorded, and that the average census produced 38.73 species, we can ask whether each census is just a "random grab" from the available species pool, assuming equal species "availability." A binomial model can answer this question: probability p of a species being selected in a random sample $= 38.73/116 = 0.334$; $q = 1 - p = 0.666$. The binomial model gives the probability that, in 11 successive random samples of the total 116 species, a given species will show up (as a success, k) a certain number of times. This can be compared with the observed number of times each species in fact shows up, and the expected numbers contrasted with the observed numbers by chi-square tests. Figure 13.8 shows the observed and expected cumulative distributions of X: proportion of sites occupied by Y: proportion of the species total. In fact the censuses do not represent random 39-species samples, in either the actual binomial (with a zero category) or an "adjusted" binomial (without the zero category).

A somewhat nicer analysis leaves in the zero category by estimating the number of rainforest species *not* seen in the eleven censuses. The species total accumulated in twenty random runs through the eleven sites produces a

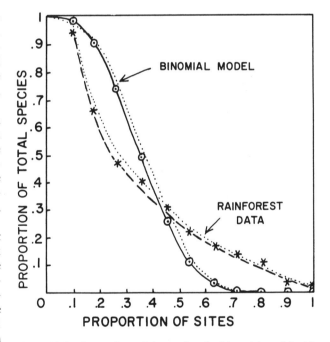

Figure 13.8 Comparison of the results of a binomial model with rainforest census data. The binomial model is based on the proportion $p = S_\alpha/S_T$, the ratio of mean alpha-diversity to the total number of species censused. Few of the total rainforest species occur in all sites (lower right extremity of the curves), but an increasing proportion of the total species is accumulated as species are included that occur in fewer and fewer sites (working up the curve to the upper left). The model assumes that censuses are random samples from the rainforest species pool, and predicts both significantly fewer ubiquitous (core) species and fewer casual (satellite) species than were actually observed.

total species accumulation curve, which can be modeled with an exponential assumption; the asymptote of this curve predicts an actual total of 122.5 species. Using this new (estimated) total of 122.5, the new binomial probability $p' = .3216$ is used in a second binomial test, with exactly the same results: the censuses are not random samples, and depart from random in that too many casual species are found (numbers in one or fewer sites: observed, 41.5; expected, 10.7), and that too many species occur regularly in many samples. Table 13.3 lists the commoner rainforest bird species, those 17 that occur in eight or more (>72%) of the sites; a single species, rather than 17, is expected in this category under the binomial model.

Species turnover between rainforest sites varies from a high of 80% (Dorrigo to Iron Range) to a low of 18% (Dorrigo to Bunya). Does this turnover constitute beta diversity related to vegetation structure? In fact it does not; there is no significant contribution to turnover by the differences in vegetation among sites, but distance between sites is a very important factor. Figure 13.9 shows a best-fit nonlinear model for all rainforest sites combined: SPTURN = $A*DIST^B$ with $A = 0.536$ and $B = 0.282$; 56% of variance in SPTURN is accounted for ($p < .001$); a less realistic linear model can do slightly better ($r^2 = 61\%$). Topography (in the form of elevational differences among sites) accounted for a further 15% of the total variance, and a further 3% was accounted for by the character of the intervening vegetation. All of the rainforest diversity is thus alpha or gamma diversity, and there is no beta component. Similarly, in multiple regressions at each site separately, distance is a statistically significant contributor to species turnover in 9 of 11 cases, and vegetation structure a significant contributor in no instance. Both sites with nonsignificant distance effects are high-elevation Tableland rainforests (H: Danbulla, E: Eungella) with faunistically different lowland rainforest in close proximity. In these regressions R averages 0.79 ($R^2 = 62\%$), and gamma diversity (coefficient of DIST) averages .000166 ± 0.000064, for a species turnover of 10% per 625 km.

Table 13.3. Core Rainforest Bird Species Occuring in More Than 72% of Rainforest Sites

Species		No. of Sites
Brush turkey	*Alectura lathami**	11
Rainbow lorikeet	*Trichoglossus haematodus*	11
Spangled drongo	*Dicrurus hottentottus**	10
Brown cuckoo dove	*Macropygia amboinensis**	10
Rufous shrike thrush	*Colluricincla megarhyncha**	10
Grey-breasted white-eye	*Zosterops lateralis*	9
Large-billed scrubwren	*Sericornis magnirostris**	9
Eastern whipbird	*Psophodes olivaceus*	9
Lewin's honeyeater	*Meliphaga lewinii**	9
Varied triller	*Lalage leucomela**	9
Black-faced monarch	*Monarchus melanopsis**	9
Spectacled monarch	*Monarchus trivirgatus**	9
White-browed scrubwren	*Sericornis frontalis*	9
Cicadabird	*Coracina tenuirostris*	8
Golden whistler	*Pachycephala pectoralis*	8
Wompoo pigeon	*Ptilinopus magnificus**	8
White-throated treecreeper	*Climacteris leucophaea*	8

*Species of chiefly rainforest

It is thought that many of the rainforest bird species gained access to northeastern Australia from New Guinea at times when the Torres Straits were narrower or even dry, and that speciation in Australian rainforests received considerable impetus via repeated invasions from the north. In line with this proposal, many species and super-species groups display partial or complete allopatry from north to south through the rainforest patches. These patterns now contribute to the high gamma diversity in this habitat. Some examples of such species groups, at various degrees of taxonomic divergence, listed from largely northern to central to southern distributions down the northeastern rainforests, follow:

Black to pied to grey butcherbirds (*Cracticus* spp.)
Tropical to yellow-throated to large-billed to white-browed scrubwrens (*Sericornis* spp.)
Yellow oriole to figbird to olive-backed oriole (*Oriolus* spp.)
Magnificent to paradise riflebird (*Ptiloris* spp.)
Northern to southern logrunner (*Orthonyx* spp.)
Graceful to yellow-spotted to Lewin's honeyeater (*Meliphaga* spp.)

Often the more northerly, first-named species are those most restricted to rainforests, whereas the more southerly, last-named species often range further into non-rainforest and woodland habitats.

MULGA: A CONTINUOUS MIDCONTINENT VEGETATION

Mulga is a bushland or low woodland dominated by a single species, *Acacia aneura*, and extends nearly continuously across central Australia, broken only by patches of *Triodia* or chenopode scrub (see fig. 13.7). It is structurally and floristically homogeneous at five census sites in Western Australia, tending into an open mixed-*Acacia* bushland near the west coast (a sixth site), and into denser acacia scrub in higher-rainfall areas toward the south. Mulga is taller at most census sites in the northeast. Near the eastern limits of its range in central Queensland, mulga occurs with a mixed overstory of eucalypt trees, an overstory that declines and eventually disappears to the west with declining rainfall. *Acacia aneura* is replaced by *A. excelsa* under taller and wetter woodlands, by structurally similar gidgee (*A. cambagei*) bushlands to the north and northwest of its Queensland stronghold, and by brigalow (*A. harpophylla*) on clay soils.

Eight Queensland census sites have more than two-thirds *A. aneura* cover (five with 90% or more), and five of six sites in the southwest are nearly pure mulga. A total of 82 bird species occurs in these fourteen sites, 8 of which are found in twelve or more sites. These are mulga core species, extremely predictable in this habitat type across the whole continent (table 13.4). Additional species are quite predictable, especially in the southwest, and still other species have clear counterparts in southwestern and northeastern sites. Besides species identities, species numbers (alpha diversity) are also relatively constant throughout the mulga across Australia, with $S_\alpha = 24.00 \pm 5.83$ in the northeast, and $S_\alpha = 23.93 \pm 4.53$ over all fourteen sites. This constancy is perhaps not surprising, given the

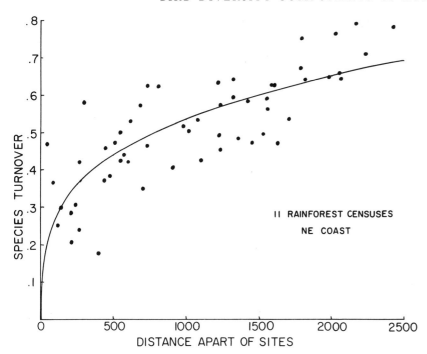

Figure 13.9 Bird species turnover among the rainforest sites is not related to any differences in vegetation structure (beta-diversity), but is strongly correlated with distance apart of the site (gamma-diversity).

Table 13.4. Core Bird Species of Mulga (*Acacia aneura*) Bushland

	Sites			Species rank $\sum D^a$
Species	SW	NE	Both	
Rufous whistler (*Pachycephala rufiventris*)	6/6	8/8	14/14	4:27.4%
Crested bellbird (*Oreoica gutteralis*)	6/6	8/8	14/14	6:35.5%
Splendid wren (*Malurus splendens*)	6/6	7/8	13/14	2:17.4%
Red-capped robin (*Petroica goodenovii*)	5/6	8/8	13/14	3:22.8%
Galah (*Cacatua roseicapilla*)	6/6	7/8	13/14	5:32.5%
Chestnut-rumped thornbill (*Acanthiza uropygialis*)	5/6	7/8	12/14	1:8.7%
Spiny-cheeked honeyeater (*Acanthogenys rufogularis*)	6/6	6/8	12/14	8:40.5%
Grey shrike thrush (*Colluricincla harmonica*)	6/6	6/8	12/14	7:38.4%
Butcherbirds (*Cracticus* spp.) (pied-2, grey-3) (pied-4, grey-4)	5/6	8/8	13/14	
Babblers (*Pomatostomus* spp.) (white-br.) (white-br.-3, Hall's-1, Grey-cr.-1, ch.cr-1)	5/6	6/8	11/14	
Little crow (*Corvus bennetti*)	6/6	—		
Austr. raven (*Corvus coronoides*)	—	8/8	14/14	
In the west:				
Southern whiteface (*Aphelocephala leucopsis*)	6/6	(2/8)		
Redthroat (*Sericornis brunneus*)	5/6	(0/8)		
Slate-backed thornbill (*Acanthiza robustirostris*)	5/6	(0/8)		
Crested pigeon (*Ocyphaps lophottes*)	5/6	(2/8)		
In the east:				
Willy wagtail (*Rhipidura leucophrys*)	(1/6)	6/8		

^{a}Species rank, followed by cumulative (summed) density in community.

homogeneity of the vegetation with its single dominant, *Acacia aneura*. Binomial analysis of accumulated species over sites reflects the high constancy of mulga species in the southwest (fig. 13.10A) the pattern is somewhat less constant but still significantly different from random in the northeast). More remarkably, the analysis also shows the same deviation from the binomial model for the combined mulga sites (fig. 13.10B).

Again predictably, given the narrow differences in vegetation structure among sites, there are no discernible effects of vegetation differences on species turnover among sites. There is, however, significant gamma diversity among sites, generated largely by the numbers of satellite species, the rarer and less predictable peripheral species. Distance apart of sites accounts for ±60% of this turnover in both the southwest and northeast, but gamma diversity is four times higher in Queensland than in the southwest. Figure 13.11 illustrates this difference, with each pair of sites in each region providing a comparison in both species turnover and intersite distance.

The explanation for the greater gamma diversity in Queensland mulga might lie in the differences in the continuity of the habitat type between northeast and southwest. In the southwest, the mulga forms a nearly unbroken vegetation inland of the woodlands, interrupted only occasionally by patches of chenopodes or other scrub types such as heath or casuarina (see fig. 13.7). The intersection of non-mulga vegetation by straight lines drawn between census sites in the southwest can be quantified from Beard's (1981b) vegetation maps; table 13.5 gives a breakdown of this interstitial vegetation. Measuring the extent to which the intervening vegetation constitutes a barrier to mulga bird distribution by the mean difference in vegetation structure in Figures 13.1 and 13.2, the average overall barrier among southwestern mulga sites is 0.125. This procedure must in fact overestimate intersite isolation, since one can walk in mulga among all sites except one (Hamelin Road, near Shark Bay). Residuals in figure 13.11 (lower line) and barrier (interstitial) vegetation are correlated ($r = .30$), accounting for a (statistically significant) extra 9% of the variance in species turnover.

In the northeast, the mulga is intersected by some powerful rivers (the Balonne and Warrego, running south to the Darling river system, and the Paroo and Bulloo, running southwest to Lake Eyre), with attendant habitat dissection by floodplain and riparian woodlands, grasslands, and other non-mulga habitat. The composition of habitats along straight lines between sites was read from vegetation maps (Queensland Department of Primary Industries), and is shown in table 13.5. Here the overall intersite habitat barrier computes as 0.152, 22% greater than in the southwest, but in the northeast some sites are quite isolated from other mulga by intervening vegetation of different structure. Although barriers are again positively correlated ($r = 0.37$) with residual species turnover, the contribution is not statistically significant, and intersite distance alone accounts for the variance.

Clearly my measure of species turnover, using only presence/absence information, exaggerates the differences among the mulga bird communities, in that it is enhanced by many casual or "satellite" species. Yet the core species account for the major part of the mulga bird biomass and, because the core species are predictable from site to site,

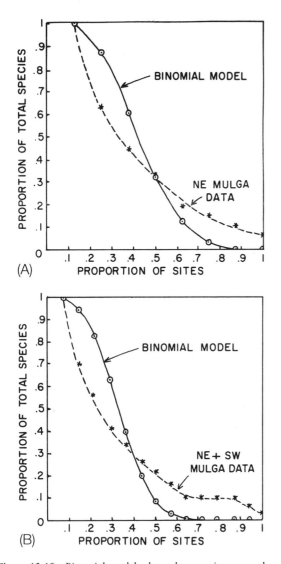

(A)

(B)

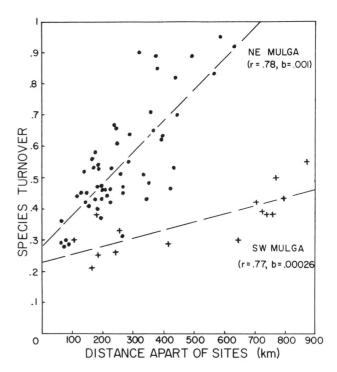

Figure 13.10 Binomial models show that certain core mulga species are more predictable elements of the censuses than is expected by chance, in both in the southwest (not shown) and in the northeast (A), and in cross-continent comparisons (B).

Figure 13.11 Species turnover with intersite separation is four times higher in mulga in the northeast than in the southwest, even though alpha-diversities are more or less constant. Factors that contribute to this difference are discussed in the text.

Table 13.5. Breakdown of Interstitial Vegetation between Mulga Census Sites in Southwestern and Northeastern Australia.

Southwest	Succ.scrub		Heath/mallee		Mulga		Woodland
Proportion[a]	.100 ± .248		.098 ± .136		.782 ± .151		.020 ± .028
Barrier[b]	.823		.378		0		.286

Northeast	Grassland	Succ.scrub	Gidgee	Mulga	Brigalow	Woodland
Proportion[a]	.098 ± .073	.005 ± .011	.077 ± .060	.551 ± .199	.006 ± .021	.263 ± .231
Barrier[b]	.885	.995	.269	0	.036	.151

[a]Indicates the proportion of each vegetation type along straight lines drawn between all pairs of mulga sites (mean and standard deviation).
[b]Difference in vegetation structure between mulga and each intervening habitat type.

a large proportion of the bird density is predictable from site to site. The eight most common core species (<10% of the total) account for more than 40% of the total bird density in the fourteen sites (see table 13.4). Satellite species are apparently drawn from neighboring habitats in which they are common, and are most prominent at sites farthest removed from contiguous mulga stands. I shall illustrate this next using the most isolated sites, east and west.

In the west, the Hamelin Road site (DX; mulga in vegetation structure [fig. 13.2] but dominated by other acacias such as *Acacia galeata, A. grasbyi, A. ramulosa,* and *A. xiphophylla*) is coastwards of contiguous mulga. The site supports 8 species (28% of the total 29 species at the site) that I recorded collectively just five other times in mulga (west plus east, fourteen sites). They occur here at modest densities, combining to 2.55 individuals/ha, comprising 22% of the total bird density at the site and having densities three times lower than those in typical mulga habitat. These satellite species include two species, red wattlebird (*Anthochaera carunculata*) and grey-breasted white-eye (*Zosterops lateralis*), that are more typical of nearby coastal heaths and acacia scrub (*Acacia rostellifera, A. xanthina*); two species, crimson chat (*Ephthianura tricolor*) and blue-and-white wren (*Malurus leucopterus*), that are abundant in the succulent (chenopode) scrub common near the west coast; three further species, masked woodswallow (*Artamus personatus*), white-winged triller (*Lalage sueurii*), and cinnamon quail thrush (*Cinclosoma cinnamomeum*), characteristic of lower, denser mixed heath/acacia bushland (dominated by *A. acuminata, A. ramulosa, A. rhodophloia, A. saligna,* plus various proteads); and the Port Lincoln parrot (*Barnardius zonarius*), found commonly in nearby mallee-woodlands.

The farthest eastern mulga site (BC: Dunkeld), while supporting 88% mulga cover, has a scattered eucalyptus overstory of *Eucalyptus populnea* and *E. melanophloia,* and is quite isolated from contiguous mulga further west by low eucalypt woodland. Six species (of a total twenty-two) at the site are typical woodland birds: brown-headed honeyeater (*Melithreptus brevirostris*), grey-breasted white-eye, grey fantail (*Rhipidura fuliginosa*), bar shouldered dove (*Geopelia humeralis*), noisy miner (*Manorina melanocephala*), and variegated wren (*Malurus lamberti*). The first three forage high in the vegetation, and could be present because of the eucalypt overstory rather than as "spillovers" from the woodland. The last three have beta replacements in lower scrub vegetation: crested pigeon (*Ocyphaps lophotes*), white-rumped miner (*Manorina flavigula*), and splendid wren (*Malurus splendens*), which but for the miner are typical mulga birds. Since most mulga sites at its eastern extreme in central Queensland have a few scattered eucalypts, it is difficult to separate the satellites that are woodland spillovers from those species present because of the sparse overstory. This is possible at one eastern site (AN: Morven) that is virtually pure mulga with no overstory trees, but is within a few km of extensive woodland. At the Morven site a number of typical woodland bird species are present: varied sitella (*Daphoenositta chrysoptera*), red-backed kingfisher (*Halcyon pyrrhopygia*), noisy friarbird (*Philemon corniculatus*), brown-headed honeyeater, red-winged parrot (*Aprosmictus erythropterus*), grey fantail, and striated pardalote (*Pardalotus striatus*). None of the first five species occured in six mulga censuses farther west, which do include one fantail and two pardalote records.

Besides the influence of species spillover from adjacent habitats, some of the species turnover among mulga sites is attributable to a checkerboard effect. By this I mean that species groups that include a number of close ecological counterparts are represented by several species over all mulga censuses, but by fewer than all species in the group at any one census site. Often species in the group are beta counterparts with overlapping habitat utilization despite their general differences in habitat preferences. Thus in mixed mulga-woodland habitats where the two interdigitate in central Queensland, one finds usually either variegated wrens (typical of the woodland) or splendid wrens (typical of mulga), and occasionally both species. Table 13.4 shows that 13 of 14 mulga sites have a butcherbird species: pied butcherbird (*Cracticus nigrogularis*) at 6 sites and grey butcherbird (*Cracticus torquatus*) at 7 sites. Of the two, the former is more typical in lower, drier, and more open habitats, and the latter predominant in taller, more mesic forest and woodland. But the two species have widely overlapping ranges and habitat preferences, differing in body size by a factor of two and coexisting commonly in tall woodland and open forest. Clearly both species find mulga acceptable, but they do not coexist in mulga at the same site. With incidences of 0.429 and 0.5 respectively, one expects to find by chance neither, one, and both species in 4, 7, and 3 sites respectively of the 14 censused; instead one finds 1, 13, and 0 sites in these categories (chi-square = 10.4, $p < .01$ that this occurs by chance).

In a similar fashion, 4 species of babblers (*Pomatostomus* spp.) are found in the northeastern mulga (see table 13.4), but no site has more than one species. Chestnut-rumped thornbill (*Acanthiza uropygialis*) is the most ubiquitous thornbill of the mulga sites (see table 13.4), but in fact 6 species of *Acanthiza* were censused in mulga. The average number of species per site is 2.5; most sites (13 of 14) had 2–4 species, and one site had a single species. Based on random (Poisson) occurrence, sites with 0, 1, 5, or 6 species should number 5.46, those with 2–4 species 8.54. These expectations are significantly different from the observations (chi-square = 5.97, $p < .01$).

Of Australia's butcherbirds, four species occur in the northeast and two of these occur in the southwest; four babbler species breed in Queensland, but only one of them reaches the southwest. While four species of thornbills have habitat preferences including mulga and range from southwest to northeast, there is but one additional woodland/forest species in the southwest, compared with five additional species in the northeast. There are many additional genera that are more species-rich in the northeast than in the southwest; presumably higher species numbers per genus in the northeast increase the incidence of checkerboard replacements, and open more possibilities for spillovers and mass effects. These appear to be some of the factors contributing to the higher gamma diversity of mulga birds in the northeast.

CONCLUSIONS

The higher regional species richness in northeastern Australia is due partly to an additional range of habitats (rainforest, grasslands) not found in the southwest, but even without these extra habitats the northeast is significantly richer in bird species. Historical factors, especially speciation in forest and eventually woodland birds, via faunal exchange with and repeated invasions from New Guinea, presumably contribute to the regional difference. Many genera, and many species within a genus, do not reach the southwest, although habitat comparable to that in which they live in the northeast occurs there.

The higher northeastern diversity is accomodated in various diversity components. Higher regional diversity in the northeast corresponds to an across-the-range increase in alpha diversity, especially in woodland and forest (which is isolated in the southwest), but specifically not in

mulga (which is nearly continuous between northeast and southwest, and in which autochthonous speciation is the mode; Schodde 1982). Besides being higher in alpha diversity, northeastern woodlands and forests are higher in beta diversity, being more finely subdivided by ecologically related species such as congeners. In contrast, species turnover related to differences in vegetation structure among woodland and forest habitats is a negligible component of diversity in eucalypt-dominated habitats in the southwest. Within habitats and between sites, gamma diversity seems to account for the major part of the differences between the two regions; even in mulga, which is rather constant across the continent in bird alpha diversity, gamma diversity is four times higher in the northeast. This corresponds to an increased dissection of the northeastern mulga by woodlands, a greater potential for spillovers from adjacent woodland and other vegetation types, and a higher incidence of checkerboard distributions within the mulga of species sets with more ecologically similar species.

14

Determinants of Diversity in Animal Communities of Arid Australia

Stephen R. Morton

Australia is a special land. Its long period of isolation following detachment from Gondwanaland, and its recent collision with Asia, have resulted in the development of many unique taxa (Keast 1981a). But how many of Australia's suites of species are unique ecologically, and to what extent is their uniqueness attributable to the continent's history?

One way to approach this question is to focus on the Australian arid zone, a vast area of dry country that covers 70% of the continent and which is a treasure-house of biological diversity. Several investigators, myself among them, have compared measures of diversity of various animal groups between arid Australia and deserts elsewhere. Here, I review these comparisons and explore their consequences for the general understanding of the relationships between physical environments and the diversity of biological communities within them.

Before proceeding, I must refine the question to be asked of intercontinental comparisons involving arid Australia. Ricklefs (1987) pointed out that we cannot always account for local diversity solely by local processes. He noted that the starting point for this realization was the fact that communities in similar environments on different continents did not always show similarity in local diversity. My inquiry will follow this path. I begin by asking whether assemblages of animals in arid Australia differ in local diversity from their equivalents in other deserts, particularly the well-studied North American deserts. If they do so, we must ask to what extent the disparity in diversity is attributable to the ecological characteristics of each system, or to other circumstances. This step inevitably takes us into consideration of other levels of diversity, because some communities of animals may have higher local diversity due to greater regional diversity (Cornell 1985b; Ricklefs 1987; Westoby 1988). If this is the case, then the investigation must attempt to determine whether greater regional diversity is best interpreted in terms of unique historical circumstances, or whether it is in turn a result of ecologically explicable regional processes.

Let us be clear about some important terms. Local species richness is the number of species living in a small, ecologically homogeneous area, and local diversity (alpha diversity) is a measure of the number of species weighted by their relative abundances, usually expressed as the Shannon-Wiener function (MacArthur 1965). Beta diversity is a measure of the rate at which new species appear in local communities as one moves from one habitat to another in the same region (MacArthur 1965); it may be expressed quantitatively as the inverse of the coefficient of similarity (e.g., Morton and Davidson 1988). Following Ricklefs (1987), I define local processes as those ecological forces at work in a community over a time scale within an order of magnitude of the generation time of the organism concerned, notably production of food, competition, predation, and disturbance; for arid Australia, Stafford Smith and Morton (1990) have provided a framework for consideration of these forces. I define temporal regional processes as the forces that produce species through evolution, or which allow or prevent exchange of species between regions. Spatial regional processes are the factors that may result in local diversity being significantly affected by regional diversity. It is difficult to arrive at an exclusive definition of historical processes in this context. The regional processes mentioned above have historical components, but each also has an ecological aspect, because species will only be produced and become distributed if ecological conditions are favorable to those particular types of organisms. Thus, I will use the term "history" only when referring to unique circumstances outside the realm of local and regional processes that prevent, attenuate, or allow the production or dispersal of groups of species.

In summary, this chapter asks the following questions:

1. Are communities of animals in arid Australia similar in richness and diversity to their equivalents in other desert regions?

2. If not, are the differences interpretable in terms of ecologically explicable regional processes, either temporal or spatial?

3. If not, to what extent do unique historical circumstances appear to have governed the richness and diversity of particular animals in deserts around the world?

The Historical and Physical Background

Aridity probably developed in Australia throughout the Pliocene, reaching conditions like those of the present day about a million years ago (Bowler 1982). Nevertheless, the continent has swung through arid and lacustral phases at least four times during the last 500,000 years, and these changes led to the creation of many of the dramatic landforms of inland Australia, such as the vast sandridge deserts and the massive lake beds (Bowler 1982). In particular, aridity intensified from about 30,000 years before present (B.P.) to a maximum at about 18,000 B.P. (Bowler and Wasson 1984), when it is estimated that rainfall was approximately 50% of modern precipitation (Wasson 1984; Dodson 1989). At about 13,000 B.P., the northern Australian monsoon began once more to penetrate intermittently into the heart of the continent, thereby ending the glacial-age climate (Wasson 1984; Singh and Luly 1989). The climate has not been stable since that time, but the wetter and drier phases have been more moderate.

As the present-day vegetation persists through periods of wide fluctuations in annual rainfall, it seems likely that the major vegetation formations of the inland (such as acacia shrublands and spinifex grasslands) would have been present to some extent during the glacial oscillations. However, it is probable that the vegetation of the semiarid zone was markedly affected as the perimeter of the truly dry country advanced and retreated with the climatic fluctuations. Human beings arrived in inland Australia before the height of the last arid phase, about 22,000 B.P. (Smith 1989).

Today, Australia is the world's driest continent in terms of the proportion of arid land area (fig. 14–1). Seventy per cent of Australia is arid; however, all of it receives more than 100 mm mean annual rainfall, so the aridity is not extreme. Rainfall is highly variable, even in comparison with other deserts. The cause of this heightened variability is the powerful El Niño–Southern Oscillation phenomenon (Nicholls and Wong 1990). Thus, any part of inland Australia can swing from prolonged drought to flooding rains when incursions of cyclonic or monsoonal depressions occasionally and unpredictably break dry spells with massive rainfalls (Bell 1979; Stafford Smith and Morton 1990), and rainfall can fail in every month at virtually every station in arid Australia (Williams and Calaby 1985). Thus, Australian organisms have to deal with both long dry periods and occasional spells of bountiful production. In addition, the Australian deserts are hot, and none of the area experiences the extreme cold temperatures of the more northerly North American deserts.

The second fundamental ecological feature of arid Australia is the infertility of its soils. Inland Australia is an ancient landscape in which the highly weathered soils contain less than half the mean levels of phosphorus and nitrogen observed in regions of comparable aridity on other continents (Lindsay 1985; Stafford Smith and Morton 1990). These differences are especially marked in comparison with the geologically youthful deserts of North America (West and Klemmedson 1978).

The ecological consequences of these physical features are manifold (Stafford Smith and Morton 1990). The lack of predictability of the timing and severity of drought is of major significance to plant life histories and production. The occasional massive rains trigger major biological events, such as establishment of long-lived perennial plants. The broad, infertile sweep of the landscape usually supports sclerophyllous and relatively indigestible plants, and only in the sparse fertile areas is plant growth regularly of high nutritional quality. Because of occasional heavy rain, and perhaps also because of the nutritionally limited rates of removal of plant biomass by consumers, fire is an important element in certain vegetation formations. Thus, there are strong grounds for suspecting that assemblages of animals in arid Australia might show characteristics different from those seen in other deserts (although I note in passing that local species richness in plants appears to be similar to that of other continents; Rice and Westoby 1983). As a corollary, matching of environments to test convergence between Australia and other arid regions is difficult.

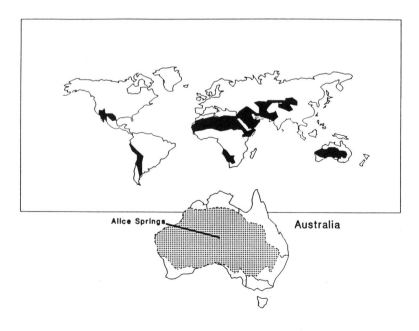

Figure 14.1 Deserts of the world (black), and map of Australia showing the high proportion of the continent (stippled) that is arid. (From Stafford Smith and Morton 1990.)

PATTERNS OF DIVERSITY AMONG ANIMALS IN ARID AUSTRALIA

Small Mammals

Similarity. The mammalian faunas of Australian and North American deserts are taxonomically distinct because the former has a marsupial component. However, differences between the mammals of the two faunas run much deeper than systematics. There are 109 species in North American deserts and 82 in Australian ones (Morton 1979; 9 further Australian species have been recognized in the arid zone since that paper appeared), despite the fact that the former region is only about a third the size of the latter. The difference between the two faunas is even more pronounced when the comparison is restricted to terrestrial small mammals (i.e., rodents, shrews, and dasyurid marsupials): North American deserts contain 71 such species and Australian ones only 39. The difference is attributable primarily to the paucity of granivorous rodents in Australia (Morton 1979).

Local species richness is also lower in Australia (tables 14.1 and 14.2). On average, only one species of granivorous (or principally granivorous) rodent occurs locally in arid Australia, but insectivorous dasyurids are prominent. At North American sites, there is usually no or only one species of insectivorous small mammal. Arid Austra-

lia contains the world's most substantial radiation of desert-dwelling insectivorous mammals (Morton 1979, 1982; see also Dickman 1989). In contrast, North American deserts possess a dramatically rich fauna of rodents, especially the granivorous heteromyids and omnivorous cricetids and sciurids, with the former dominating in richness and abundance (Brown and Kurzius 1987, 1989). It seems probable that this fauna of rodents is the richest of any in the world's deserts (Mares 1980). In summary, North American deserts contain both high local diversity and high regional diversity of small mammals, whereas Australia exhibits low local and regional diversity; however, the insectivorous trophic group runs against this trend in Australia.

One problem with the analysis presented above is that the Australian fauna has undergone a recent spate of extinctions. At least four species of rodents have become extinct in arid Australia, and another six species of rodents and three of dasyurids have declined dramatically in range since European settlement 200 years ago (Morton and Baynes 1985; Burbidge and McKenzie 1989; Morton 1990). However, Morton and Baynes (1985) concluded that although local richness was originally higher than it is today, it still could not have been as great as in much of North America; in any case, it appears unlikely that the extinct species were the supposedly missing granivores. It is also evident that the local richness of insectivorous mammals would have been even greater in the past than today, thereby enlarging the disparity between the two continents.

Ecologically Explicable Processes. The answer to the first question is, therefore, that communities of small mammals are dissimilar between arid Australia and North America. Now I ask whether the differences are due to local processes. Two such explanations may account for the disparity in granivory and omnivory. First, the possibility exists that the deserts of North America provide more regular food supplies for seed-eating mammals because the climate is more dependable and the soils more fertile. Relatively regular rainfall and lack of inhibition of growth by nutrients may result in the type of reliable

Table 14.1. Species Richness of Granivorous or Principally Granivorous Rodents in North American and Australian Deserts, and of Insectivorous Dasyurid Marsupials in Australia

Group	Number of study sites	Local species richness — Mean	Local species richness — Range	Total number of species
North American rodents[a]	202	3	1–9	29
Australian rodents[b]	84	1	0–2	3
Australian dasyurids[b]	84	2	0–5	10

[a]Data from Brown and Kurzius 1987.
[b]S. R. Morton and J. R. W. Reid, unpublished data.

Table 14.2. Species Richness in Arid Australia and in North American Deserts

	Local (mean) — Australia	Local (mean) — North America	Local (mean) — Australia cf. North America	Regional (total) — Australia	Regional (total) — North America	Regional (total) — Australia cf. North America
Small mammals						
Granivorous rodents	1	3	Low	12	35	Low
Insectivorous species	2	<1	High	30	7	High
Birds						
All species (breeding)	6	6	Similar	140	130	Similar
Granivores	No data	No data	High	26	4	High
Lizards	30	7	High	170	57	High
Ants						
All species	No data	No data	High	2000?	161	High
Harvesters	8	5	Similar	400?	52	High
Termites	10?	2?	High	200?	10	High

Note: Sources for the estimates of richness are given in the text.

production of seeds that is essential for the persistence of diverse communities of small mammals. The reverse seems true in Australia, where fluctuation in seed supply may be far greater even if mean abundance of seeds is similar. Thus, local diversity of granivorous small mammals may be limited. This potential explanation in turn has three implications.

First, if it were true, then one might predict that groups of granivores better able to withstand environmental unpredictability would be more diverse and more abundant in Australia than their equivalents in North America. If mean seed supplies in the soil are of similar magnitudes in the deserts of the two continents, the prediction suggests that consumers such as ants, which can harvest and store large quantities of food, and birds, which can readily fly from areas of declining production to satisfactory feeding locations great distances away, would show higher local diversity and abundance. In short, one could see taxonomic replacement within this trophic group. This prediction, by and large, has not been borne out. The available evidence indicates that mean seed populations in soils are of the same order of magnitude on the two continents (Westoby, Cousins, and Grice 1982; Morton 1985). However, feeding rates at experimental baits are lower overall in Australia than in North America, suggesting that the lack of granivorous rodents has not allowed directly comparable expansion in the biomass or diversity of ants or birds (Morton 1985; the case for birds remains less clear than that for ants). Consequently, the direction of change in communities of granivores is consistent with the prediction, but it does not seem likely that a simple replacement of one taxonomic group with another has taken place.

The second implication is that one would expect to see an increase in the proportion of small mammal species that are omnivorous rather than granivorous in Australia relative to North America. If production were more uncertain in Australia, then perhaps rodents would find it impossible to specialize upon only one food source, even if it were as dependable as seeds. Morton and Baynes (1985) suggested that most Australian species were omnivores and that few, if any, could be considered granivores equivalent to North American heteromyids; more recent work supports this conclusion (S. R. Morton and C. H. S. Watts, unpublished data). Thus, dietary patterns are consistent with the suggestion that environmental variability has had a substantial impact on the biology of the Australian rodents.

The preceding discussion supports suggestions by Stafford Smith and Morton (1990) that Australia's environmental uncertainty has a major impact on ecological patterns, but the third implication does not seem to do so unambiguously. We saw above that insectivorous small mammals are more diverse in Australia than in any other desert, a pattern inconsistent with the prediction. How could environmental unpredictability prevent one group of mammals diversifying while not depressing another? I am unable to answer this question. It is possible that some ecological feature of arid Australia—particularly the increased abundance and diversity of certain invertebrate animals (see below)—has allowed the insectivorous mammals to overcome this unpredictability, but current information is not adequate to allow more than speculation.

The discussion so far has focused on local processes. Now I turn to another possibility: the potential role of ecologically explicable regional patterns. The North American fauna has a greater regional diversity than the Australian, and it is important to determine whether this difference is sufficient to explain the disparities in local diversity. The greater regional diversity is attributable to a well-recognized pattern of speciation resulting from an oscillating series of Pleistocene boreal expansions and desert contractions throughout the mountainous southwestern United States (Findley 1969; Blair, Hulse, and Mares 1976; Riddle and Honeycutt 1990). The physiographic structure of this environment is such that many desert-dwelling species could be "budded off" during the desert contractions, thereby producing the large species pool that exists today (Glazier 1980; M. A. Bowers 1988; Brown and Zeng 1989). Thus, there is a ready explanation for the high regional diversity in North American deserts.

The situation in arid Australia is different. Inland Australia consists largely of plains and plateaus; the uplands are of only moderate height and are similar in climate to the surrounding deserts (Wasson 1982). It appears, therefore, that only limited speciation took place among mammals around montane retreats during glacial cycles in arid Australia. Instead, species of rodents seem to have been produced predominantly at the margins of the arid zone, where glacial climatic shifts pushed relatively steep rainfall gradients back and forth across the semiarid fringe (Baverstock et al. 1981; Baverstock 1982, 1984). Thus, there is at first sight a clear explanation for the relatively small species pool of rodents within the arid zone.

This clarity breaks down, however, when the dasyurid marsupials are examined. Precisely the same pattern of speciation at the semiarid fringe is observed (Baverstock 1982; Baverstock et al. 1982), and yet in this circumstance the group has not been prevented from achieving diversities within the arid zone higher than those seen among insectivorous small mammals anywhere else in the world's deserts. Consequently, there is no unambiguous explanation based on temporal regional processes for the disparities in diversity between arid Australia and North America.

In conclusion, there is a strong possibility that the lower local and regional diversity of small mammals in arid Australia is attributable to the uncertain production brought about by climatic unpredictability and edaphic infertility. However, there are important cases where this explanation fails, and so the likelihood must be considered that unique historical circumstances have played a part.

Historical Circumstances. Rodents have been in Australia for at least five million years (Lee, Baverstock, and Watts 1981). They have been present during the major arid phases experienced by the continent, and their evolutionary radiation has proceeded far enough to produce a genus of bipedal and saltatorial animals (*Notomys*) convergent in morphology with the "classic" desert-dwelling rodents of North America and the Old World. Having leapt all the hurdles along the path into desert life, why then did these rodents not diversify to the degree seen elsewhere?

One nonecological factor that might have attenuated the radiation of these rodents is the fact that they come from a stock in which the physiological capacity to enter torpor seems to be absent. No native Australian rodent is known to use this energy-conserving device (Lee, Baverstock, and Watts 1981), and as far as I am aware, no member of the Hydromyinae (the murid subfamily to which all Australian rodents except *Rattus* belong) possesses the capacity for torpor. In contrast, torpor is prominent among the heteromyid and cricetine murids of arid North America and the dipodids of the Old World deserts (Cade 1964); it even occurs in introduced *Mus domesticus* (Murinae) in Australia (Morton 1978). Further, it is almost certainly a feature of every species of dasyurid marsupial in arid Australia (e.g., Wallis 1982; Geiser 1986; Geiser and Baudinette 1987, 1988). Is it possible that the inability of the hydromyine murids to use torpor limited their diversification in the face of unpredictability of production in the Australian deserts? If so, this inability would represent a truly historical circumstance that restricted the production of species in arid Australia. It is also possible that other features, perhaps as-yet-unidentified peculiarities of population structure or genetic architecture, might have affected speciation to produce the disparities we see today.

Although a potential historical explanation exists for the lower diversity of rodents in arid Australia, none is apparent yet for the lack of insectivorous small mammals in North American deserts. I suggested earlier (Morton 1979) that the incapacity of shrews to use torpor had hindered their occupation of these deserts, but later work has shown that, in fact, several shrews do have this capacity (Genoud 1988). In any case, the suggestion does not explain why cricetine murid rodents, which *are* capable of becoming torpid, should not have become specialized insectivores in convergence with the Australian dasyurids. Thus, no concrete conclusion can be reached.

Other Deserts. Quantitative comparisons among other deserts are generally lacking, but my interpretation of the literature is as follows. South American deserts, both east and west of the Andes, are poorer in local species richness than North American deserts and contain few granivorous species; they are similar to those seen in Australia but possess fewer insectivorous species (Fulk 1975; Mares 1976; Meserve and Glanz 1978). In North Africa, the Middle East, and central Asia, rodent diversity approaches that of North American deserts, and there is a tendency for most species to be granivorous (Daly and Daly 1975; Abramsky and Rosenzweig 1984; Abramsky, Brand, and Rosenzweig 1985; Rogovin and Surov 1990); local richness may not be universally high, however (Mares 1980). In the semiarid Karoo of southern Africa, rates of seed removal by rodents are similar to those in Australia, and most species are omnivorous (Kerley 1991). In the nearby Kalahari, small mammals are apparently as diverse as in North America; food habits are also

diverse, but although insectivores are present they do not seem as prominent as in Australia (Nel 1975). Thus, a range of patterns in species richness is apparent. Unfortunately, information on the capacity or otherwise of small mammals in these deserts to use torpor is presently inconclusive (e.g., McNab 1982).

Mares (1980) noted that there was limited ecological similarity between the small mammals of arid Australia and North America, but proposed that there was considerable convergence in morphology of most of the world's desert-dwelling rodent faunas. He was unable to consider the question of diversity in detail. Given my conclusion that patterns of diversity are dissimilar between deserts, it is essential that I look briefly at the phenomenon of morphological convergence.

Although Mares (1980) showed that convergence in structure may be widespread among rodents from an array of trophic groups, the most dramatic and widely quoted example is that of the bipedal rodents. Such animals occur in all except two of the world's deserts (southern Africa and South America), and belong to three families (Heteromyidae, Muridae, and Dipodidae). Because of their unusual appearance, and because the heteromyids are abundant and diverse in North America, the bipedal rodents have come to be seen as the classic desert-dwelling small mammal. In particular, this archetypal desert-dweller is perceived to be granivorous (e.g., Reichman and Brown 1983). However, Mares himself (1980, 1983) has pointed out that the reality is more complex. There is no link between bipedality and seed eating; in fact, bipedal rodents cover the entire range from granivores, through herbivores eating both above-ground and below-ground plant parts, to insectivores (see Mares 1983). I suspect that these dietary disparities are only the tip of the iceberg when it comes to ecological divergence among the bipedal rodents. The North American *Dipodomys* are solitary, sedentary, and territorial animals, for example (Reichman 1983). In contrast, our sparse knowledge of the Australian *Notomys* indicates that they are highly social rodents with some singular reproductive characteristics, and that they frequently are not sedentary (Happold 1976; Breed 1979, 1981, 1986, 1990); in addition, they are omnivorous (Morton and Baynes 1985; S. R. Morton and C. H. S. Watts, unpublished data). Thus, morphological convergence has not necessarily sprung from or led to ecological convergence.

The issue of morphological versus ecological convergence among desert-dwelling small mammals touches upon determinants of diversity particularly in the Monte Desert of South America. Mares and Rosenzweig (1978) showed that the Monte Desert is inhabited by many fewer granivorous animals than are North American deserts, and contains especially few rodents. Several possible explanations for the difference were examined, one being that the argyrolagid marsupials had been granivores but had become extinct during the Pleistocene (Mares and Rosenzweig 1978). The argyrolagids were a group of bipedal animals with striking resemblance in body form to heteromyid and dipodid rodents. But were they granivores? Simpson (1970, 49) makes it clear that there is a great deal of room for speculation about their diets: "In the dentition, resemblance of the argyrolagids [to granivorous rodents] is hardly more specific than to rodents in general. . . . The numbers of incisors and of cheek teeth are different. . . . The convergence here is just that the animals in question all have gnawing-grinding dentitions." Given the wide range of diets exhibited by present-day bipedal rodents, it takes a leap of faith to conclude that the argyrolagids were specialized seed eaters. Thus, I find it difficult to accept completely the historical explanation advanced by Mares and Rosenzweig (1978) for divergence in richness between small mammals of the Monte and of the North American deserts. The reasons for this disparity remain obscure (Orians and Paine 1983; Brown and Ojeda 1987).

Birds

Similarity. The avifaunas of Australia and North America are divergent in origin because the former has a substantial Gondwanan element (Cracraft 1973; Keast 1981b; Schodde 1982). Nevertheless, regional species richness is similar; about 130 species (excluding water birds) may be found in North American deserts, and about 140 in Australia.

Only one methodical comparison exists between avian communities in arid Australia and North America. Wiens (1991) showed that sites on the two continents in shrub deserts dominated by *Atriplex* and (in North America) *Artemisia* contained indistinguishable numbers of species of birds (on average, about six species each). It remains to be seen whether this equivalence in species richness holds up in other vegetation formations. It will also be important to determine the relationship between short-term estimates of breeding species richness and multiyear measures, because year-to-year variation in breeding species in Australia due to climatic conditions can be dramatic. However, the available information suggests that both regional and local species richness are similar on the two continents (see table 14.2).

Wiens (1991) analyzed a matrix derived from fourteen ecological and life history characteristics of each species in his cross-continental comparison, and found little close matching of Australian with North American species. A further difference not immediately evident from Wiens's (1991) study is the relative importance of granivorous birds in arid Australia. There are twenty-six species of specialized seed eaters in Australia (Morton and Davies 1983), some of which can sometimes be very abundant and which cover the entire range from sedentary through nomadic (e.g., Wyndham 1983). There is no parallel to this assemblage in North American deserts. Thus, there are several important ways in which the avifaunas of the two continents differ.

Ecologically Explicable Processes. The answer to the first question—concerning similarity of species richness in local communities on the two continents—is that local species richness seems to be similar between the two continents. However, the similarity could well be due to the nearly identical size of the two regional faunas, and not to convergence. In addition, there are substantial differences between communities on the two continents, which leads

one to suspect that dissimilar environments have produced different ecological forces. I noted above that among mammals, the disparity between the continents was greatest between granivores; with birds again this seems to be true, although in this case Australia has the greater richness. If diversification of granivorous mammals was limited by the more variable climate of inland Australia, then perhaps birds were less affected by this constraint because their greater mobility allows them to escape the worst effects of drought (Morton and Davies 1983). However, this explanation is not completely convincing, first, because not all small mammals display reduced diversity in Australia (see above), and second, because there is no firm evidence that birds replace rodents in terms of the biomass of seeds that they consume (Morton 1985). Thus, some features of the Australian avifauna can be interpreted in terms of responses to a highly variable climate, as suggested by Stafford Smith and Morton (1990), but there are other grounds for concluding that further factors have affected the development of arid Australia's avian communities.

What of patterns of beta diversity and speciation? As far as I am aware, there are no data available allowing comparison of beta diversities between the two continents. It is notable that the relative lack of montane refuges prevented large-scale in situ speciation among arid Australian birds (Schodde 1982; Cracraft 1986) just as it did for mammals, but that in contrast to mammals, the birds do not show lower regional or local diversity in Australia than in North America. Thus, there is no necessary connection in Australia between speciation at the periphery of the arid zone during Pleistocene fluctuations and low present-day species richness within it, as I pointed out when discussing the mammalian pattern. In conclusion, it seems unlikely that temporal regional processes can tell us anything about the preponderance of granivorous birds in arid Australia.

Historical Circumstances. The parrots and pigeons are believed to be Gondwanan groups (Cracraft 1973; Schodde 1982). It is possible, therefore, that their long presence in Australia increased the range of source stocks for speciation and invasion of the arid zone during its formation. If so, the greater richness of granivorous birds in arid Australia may have a historical component, but it will be difficult to untangle the relative roles of history and ecological opportunity in this case.

Other Deserts. Pianka and Huey (1971) found the species richnesses of Australian and Kalahari birds to be similar at about nine sites each. They tentatively identified some differences between the trophic structures of the communities, but were unable to provide detailed explanations for them. I am not aware of data from other continents that could be used to compare patterns of species richness in arid regions.

Ants

Similarity. I move now to harvester ants because they are the third major taxon among a particular trophic group— the granivores—that has featured so prominently in comparisons between Australia and North America. The discussion will also consider ants in general.

All the major subfamilies of ants occur in both Australia and North America (Hölldobler and Wilson 1990). Nevertheless, Australian ants are taxonomically distinctive; they are dominated by certain formicine and dolichoderine genera, and constitute a fauna rich in endemic lineages (Brown and Taylor 1970; Greenslade 1979). Based upon estimates provided by G. C. Wheeler, Morton (1979) reported that 52 species of harvester ants (and 161 of all ants) occurred in North American deserts (these estimates may now be somewhat low). The alpha taxonomy of Australian ants is in a woeful state, largely because the sheer diversity of the fauna has daunted systematists. Despite the lack of specific data, we believe that there are several hundred species of harvesters in arid Australia (Morton and Davidson 1988), and that the total number of all ant species is likely to be in excess of 2,000 (Greenslade and Greenslade 1983; A. N. Andersen, personal communication). Clearly, there is a dramatic difference between these two faunas (see table 14.2).

The greater regional species richness of harvester ants in Australia does not correlate with a higher local richness. On both continents, local richness of common species varies from one to eight, although total richness tends to be slightly elevated by more rare species in Australia (Morton and Davidson 1988). Thus, there does appear to be similarity in local species richness. However, this conclusion is tempered by the fact that numerous other features of the communities are divergent. Local diversity does not seem to vary with measures of productivity in Australia, whereas it does in North America. The Australian ants are smaller, are more tightly distributed along the size gradient, forage over a much broader temperature range, and display temporal displacement of foraging. There are strong hints that, as with birds, there are important disparate ecological forces at work.

This cautionary note is supported by information on patterns of local diversity in non-harvester ants. No detailed comparisons of total ant faunas are available, but the maximum local richness in North American deserts seems to be about 33 species in desert scrub (Gaspar and Werner 1976) and most sites seem to contain fewer species (Whitford 1978; Bernstein 1979). Richness in chenopod shrub environments in Australia is about 37 species (Briese and Macauley 1977), but in *Acacia* shrublands in central Australia (roughly equivalent in structure to North American desert scrub) it can approach 60 species (Greenslade 1978). In *Eucalyptus* shrubland the number is about 100 species (Andersen 1983; Andersen and Yen 1985). Ryti and Case (1992) reported from 5 to 11 species at sites in North American deserts and from 9 to 63 in Australia. Ant faunas of patches of habitat several square kilometers in area support 200 to 300 species in Australia (Greenslade and Greenslade 1983), i.e., more species than in the entire North American deserts. It appears that the trophic groups that are particularly speciose in Australia are predators, scavengers, and collectors of honeydew and plant exudates (Greenslade and Greenslade 1983, 1984). It is notable that the Australian ants are more diverse despite the lack of fungus-growing ants, which oc-

cur in both North and South America (Hunt 1977). Thus, local ant communities are more diverse in Australia, and show different trophic composition.

Ecologically Explicable Processes. The answer to the first question is, therefore, that communities of harvester ants are similar in local species richness, but that communities of ants in general are not. Are these differences due to local or regional processes?

Morton and Davidson (1988) suggested that many of the disparities between communities of harvester ants on the two continents, especially the lack of correspondence between local richness and measures of productivity in Australia, reflect the greater degree of fluctuation in Australian rainfall as well as in soil differentiation and, therefore, in runon/runoff patterns (see also Briese 1982). The question remains as to the applicability of this suggestion to the greater local richness of all ant species in Australia. Stafford Smith and Morton (1990) argued that it *is* pertinent. They suggested that social insects are particularly prominent in arid Australia because they represent the extreme development of a persistent life history strategy in the face of unpredictable climatic variability. A social insect colony acts as a storage organ that can buffer the pulses of production, and its foragers can hunt more widely than individual insects for sparse and intermittent resources (E. O. Wilson 1987). Consequently, social insects might be expected to dominate environments where production is unreliable, or where it is persistent but of poor quality. Stafford Smith and Morton (1990) suggested that the ants and termites (see below) of arid Australia examplified these trends; thus, local ecological processes appear to favor ants.

The hypothesis outlined in the previous paragraph has not been rigorously examined, although important elements are testable. But what of beta diversity? Figure 14.2 shows the only quantitative comparison available (Morton and Davidson 1988). The cumulative number of species of harvester ants found at different sites levels off rapidly in North America. However, in Australia, about four previously unrecorded species are found at each new site, and there is no indication that the line is even approaching an asymptote. It is virtually certain that other trophic groups of Australian ants will be found to display similar trends.

How and why have all these species come into existence? First, the suggestion that ants are particularly suited to environments with unpredictable and sparse production may help explain their radiation in arid Australia. However, we are not dealing just with an unsorted mass of species: most ants display the same marked species turnover in relation to soil type observed with harvester ants (Greenslade and Greenslade 1984). Hence, species appear to have very specific soil requirements, and the high beta diversity reflects the constantly shifting edaphic environment as one moves across arid Australian landscapes (in comparison with the relative uniformity of North American desert soils; Stafford Smith and Morton 1990). In conclusion, there are potential ecological explanations, either at the local or regional level, for high local and beta diversity in Australia.

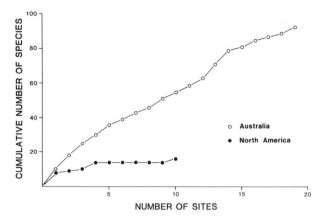

Figure 14.2 Cumulative numbers of harvester ant species found as one moves from wetter to drier sites in arid Australia and North America. (From Morton and Davidson 1988.)

The historical background of this exceptional radiation is obscure. Although species diversity is high, generic diversity is not, and important genera are not endemic to the arid zone; it seems that the ant fauna of the arid zone stems largely from groups centered on higher-rainfall areas (Greenslade 1979; Greenslade and Greenslade 1983, 1984). Because alpha taxonomy is poor, detailed patterns of speciation are a matter of conjecture. Greenslade and Halliday (1982) suggested that climatic change may have driven speciation if it resulted in the movement of zones of favorable conditions across a mosaic of soil types; they argued that speciation would have been most intensive at the semiarid margin.

Historical Circumstances. Although the Australian and North American faunas differ in taxonomic makeup, ants have very long histories on both continents (Hölldobler and Wilson 1990). I cannot discern any unique historical circumstances that might have caused ants to speciate more on one continent than the other.

Other Deserts. Quantitative comparisons are lacking, but no other desert is known to display similarity with Australia in ant diversity. The literature suggests that local diversity is similar to that of North American deserts in Israel (Ofer, Shulov, and Noy-Meir 1978) and the Namib Desert (Marsh 1986), but marginally greater in the Monte Desert of South America (Hunt 1977). It would be of great value to determine patterns of diversity in the Kalahari Desert, where infertility (Buckley, Wasson, and Gubb 1987) might be expected to lead to dominance by ants, but where the opportunities for speciation may have been fewer than in Australia.

Termites

Similarity. In the previous section, I argued that social insects may be prominent in arid Australia because of their ability to succeed under conditions of uncertain or poor-quality production. Although there are no rigorous studies of local termite diversity in arid Australia, it is worth examining the available information to see if it is consistent with this suggestion.

Both the Australian and North American faunas are dominated by kalotermitid, rhinotermitid, and termitid groups, sometimes of identical genera (Krishna and Weesner 1970). Ten species in two genera of termites occur in North American deserts (Snyder 1954), and up to 4 may occur in local communities (Haverty and Nutting 1975). The fauna of arid Australia is inadequately characterized, especially because of large numbers of undescribed species in *Nasutitermes* and *Tumulitermes* (Termitidae), but it contains in excess of 200 species (L. R. Miller, personal communication). In central Australia, local communities contain on average about 10 species, and some more than 14 (L. R. Miller, personal communication). The greater diversity appears to be due to the large number of species that harvest grass and litter or which subsist on subterranean plant material (Watson and Gay 1983). More studies are urgently needed, but it seems certain that both local and regional species richness are considerably higher in Australia, and that the continents are dissimilar (see table 14.2; see also Whitford, Ludwig, and Noble 1992).

Ecologically Explicable Processes. The sparse data presented here suggest that patterns of diversity among termites are consistent with those seen in ants. This matching of pattern may well reflect similarities in the ways that temporal regional processes have affected the two groups through evolutionary time. The paucity of termites in North America might reflect the fact that the comparatively fertile soils favor other herbivores and detritivores, or that the North American deserts lie too far north of the tropics for occupation by diverse communities of termites. Either way, and in line with the arguments of Stafford Smith and Morton (1990), the patterns appear to reflect ecologically explicable processes.

Little is known about speciation in Australian termites, but Watson (1982) suggested that at least one genus (*Drepanotermes*; Termitidae) diversified within the arid zone.

Historical Circumstances. I am unable to identify any unique historical circumstance that might have resulted in the differences evident between the two continents.

Other Deserts. Systematically collected data describing local diversity are nonexistent, but checklists provide some clues to regional patterns. The Arabian peninsula contains 33 species (Cowie 1989); this number could be bettered within an area of only 100 km² in central Australia (Watson, Barrett, and Lendon 1978). In contrast, the Kalahari has a rich fauna; although detailed species lists do not seem to be available, at least 14 genera occur, many of them speciose (Coaton 1963). Studies in nearby savanna indicate local richnesses similar to those seen in arid Australia (Ferrar 1982). Thus, there is some evidence that the two subtropical and infertile deserts experiencing only mild aridity—Australia and the Kalahari—may be similar.

Lizards

Similarity. The lizards of Australia and North America have been compared in detail in a monumental series of studies by Pianka, summarized in his 1986 monograph. Because this work has been extensively discussed in the literature, I present only a brief summary here.

The desert-dwelling lizard faunas of the two continents differ in their taxonomic composition, but in each there are five families represented. Despite the similarity in familial richness, however, local species richness differs dramatically: in Australia, Pianka (1986) recorded an average of 30 species, and in North America, 7. The greater local richness in Australia is matched by high regional diversity. The North American deserts contain 57 species of lizards, whereas arid Australia possesses about 170 species, with the number climbing every year as herpetologists locate new taxa (see table 14.2). Thus, in contrast to the rodents, and in agreement with the ants, the lizards of arid Australia show both high local and regional diversity relative to North America.

Ecologically Explicable Processes. What ecological processes might have brought about the dissimilarity between the lizard communities of Australia and North America? Pianka (1973, 1986) and other authors considered a variety of possible explanations. Morton and James (1988) argued that none of these postulates provided a complete explanation for the phenomenon, and attempted to fashion a more comprehensive synthesis (fig. 14.3).

The new approach was to ask why lizards were so diverse in spinfex grasslands, the least productive of all environments in arid Australia. Morton and James (1988) suggested that the infertility of the soil resulted in nutritionally poor production in spinifex, thereby limiting densities of herbivorous animals. However, termites are capable of using poor forage, and so are exceedingly common and diverse. Because termites are abundant, but individually small and often subterranean, they provide food more suited to lizards than to homeothermic birds and mammals. Thus, lizards are diverse in spinfex, especially where this formation grows on sandy soils suitable for burrowing. Morton and James (1988) postulated that, within different vegetation formations of the Australian arid zone, lizard diversity would decline with decreasing ter-

mite diversity as the fertility of the soil increased. In addition to specifying the role of infertility, this attempted synthesis brings together three features previously noted by Pianka: lizards are favored over homeotherms because of uncertain production brought about by highly variable rainfall; because many termites (and other invertebrates that prey upon them) forage nocturnally, abundant resources are available for exploitation by the prominent assemblage of nocturnal lizards; and arboreal lizards are widespread because occasional unusually heavy rains allow trees to become established and grow sparsely throughout arid Australia.

Pianka (1989a) concurred with some but not all of the synthesis outlined above. He argued that the differences in termite abundance and diversity between Australia and North America were consistent with the hypothesis, but that termites were in fact more diverse and abundant in southern Africa, where lizards are of intermediate richness, than in Australia. He also suggested that patterns of resource usage in different vegetation formations within Australia were not consistent with the predictions of Morton and James (1988). In addition, Pianka (1989a) advanced another ecological factor for consideration: frequent fires in spinifex grasslands cause dynamic habitat change which, he suggested, may allow greater lizard diversity. Despite the disagreements over detail, both Pianka (1989a) and Morton and James (1988) agree that there are powerful ecological differences between Australia and North America that could have caused, at least partially, the disparities in their lizard communities.

Why was speciation so dramatic in Australia but not North America? Several authors (Horton 1972; Pianka 1972; Cogger and Heatwole 1981) have suggested that in situ speciation took place as the Australian arid zone expanded and contracted throughout the Pleistocene. If lizards were particularly suited to arid conditions (as mammals were not), then certain taxa that had developed specific preferences for particular vegetation formations, or the soil types supporting those formations, could have undergone isolation and speciation as climatic changes swept back and forth across the landscape. Until the mo-

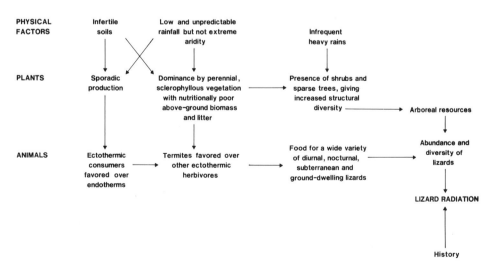

Figure 14.3 The web of forces postulated to account for the high diversity of lizards in spinifex grasslands of arid Australia. (After Morton and James 1988.)

lecular relationships of Australian lizards have been determined, it will be difficult to assess these suggestions.

Whatever the outcome of analyses of patterns in relationships, however, it is clear that lizards were able to speciate dramatically in arid Australia whereas homeothermic vertebrates in general were not. I would argue that, to a substantial extent, these differences are ecologically explicable. The uncertainty and poverty of production mentioned during the foregoing discussion would have pertained to a greater or lesser degree over the past 50,000 years, thereby providing much raw material for speciation among lizards. It is revealing that lizards appear to display a similar high beta diversity to that seen in ants in response to subtle changes in vegetation and soil type, suggesting that the capacity of both groups to prosper has brought with it high rates of speciation and habitat specificity.

Historical Circumstances. None of these arguments categorically dismisses historical accident as a contributor. Although Pianka (1986, 1989a) argued that it is facile to assert that Australian deserts have more species than North American deserts simply by virtue of history, he suggested that historical accident may have contributed to the disparities between the two faunas. To give an important example, I have interpreted the lack of nocturnal lizards in North America as a result of the relative paucity of termites or other invertebrate prey. However, *all* habitats in the New World exhibit low diversities of geckos, not just the North American deserts (Duellman and Pianka 1990). It is possible that some unique historical circumstance may have limited the spread or speciation of this group. Again, analysis of the relationships of lizards, through both traditional and molecular methods, will yield valuable results.

Other deserts. Local diversity of lizards is slightly higher in the Kalahari than that observed in North America, but lower than in Australia (Pianka 1986). Beta diversity is similar in the Kalahari and North America (Pianka 1973). Local diversity appears similar in North America, the Monte Desert of Argentina (Hulse, Sage, and Blair 1977), and the deserts of central Asia (Shenbrot, Rogovin, and Surov 1991). As with ants, therefore, the Australian deserts contain lizard communities that have no apparent parallels elsewhere in the world.

DISCUSSION

My approach to the question of determinants of diversity has been pragmatic. I have not attempted to test any particular proposition, but rather have surveyed as many animal groups as possible to see whether generalities emerge for the environment with which I am most familiar. The results are summarized in table 14.2 and figure 14.4.

Table 14.2 suggests a striking similarity between patterns in local and regional diversity for ecologically disparate groups. In all but one case, comparison between Australian and North American deserts indicates that the two aspects of diversity are correlated. Of course, the sample size is small, and one would like to see another twenty cases before being convinced that the results are not random. Nevertheless, the trend of the data is toward confirmation of two postulates: that regional species richness is frequently dissimilar between continents; and that local richness frequently reflects diversity at the regional level.

My investigations lead me to the conclusion that in most cases high diversity at all levels is primarily a result of the particular suitability of the body plan and life history of an organism to the opportunities provided by the environment under consideration (fig. 14.4). I agree with Westoby (1988), therefore, that many characteristics of biological communities can be understood in terms of their physical environments, especially climate and soil fertility.

I have argued throughout this chapter that the presence of a large regional species pool reflects the success of a taxonomic group in dealing with the hazards and opportunities of the environment under consideration. I would argue that, at the continental scale, climatic change and consequent habitat isolation frequently provide opportunities for speciation in "successful" groups (fig. 14.4), but that it is only secondarily that the continued persistence of diverging lineages within such groups results in sympatry and syntopy of relatively large number of species. The reason for the persistence of these lineages through time—ecological opportunity—is, by and large, the same as the reason why many species are able to find ways of making a living in a local community. If this view is correct, high regional diversity and high local diversity frequently spring from the same conditions.

Despite the above conclusion, one important issue is still not clear to me. Local diversity may be maintained at a high level in two ways: by swamping of communities by species from the large regional pool; or, if the environment is particularly suitable for the type of animal, as my argument implies, by more stringent partitioning of resources compared with that of animals living in less suitable environments. I cannot distinguish between these two generalities because most comparisons suggest that the physical environments of the Australian arid zone and the North American deserts are not, in fact, well matched. This lack of matching means that it is difficult to be convinced that convergence has taken place when one *does* see similarity in local diversity. It also means that it is impossible to determine the relative degrees to which fine niche partitioning in diverse Australian communities is due to suitable local conditions or to high regional production of species. Some resolution of these problems might come about through comparisons between arid and mesic habitats within Australia and between the Kalahari and Australia. In the first case, cladistic analysis of the Australian lizard, termite, and ant faunas would reveal a great deal about the patterns of species production and the origins of diverse arid-zone taxa. In the second, the Kalahari is especially tantalizing because it seems to be more like arid Australia than anywhere else on earth, yet it apparently displays some dissimilarities. Detailed comparison between Australia and the Kalahari may well reveal that history has played a greater role than I have allowed.

Several important riders are necessary. First, none of this implies that the communities under consideration are

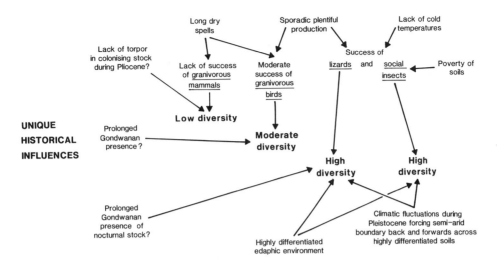

Figure 14.4 A summary of the web of forces postulated to have produced differing levels of diversity among certain groups of mammals, birds, lizards, ants, and termites in arid Australia. This flowchart is constructed in relation to the North American deserts, where more reliable rainfall and richer soils in a different biogeographic and historic context have produced a different set of communities. Not all connections discussed in the text are summarized here.

in equilibrium; maps of species density of Australian lizards, for example, show that some taxa exhibit centers of diversity that are apparently independent of present ecological conditions (Pianka 1972; Cogger and Heatwole 1981). Second, not all examples followed the trend (see table 14.2). At present, I cannot determine why harvester ants converge in local richness whereas other trophic groups of ants do not. Third, my survey does not dismiss history: in several examples, it is difficult to discern any other explanation for disparities. Nevertheless, I suggest that the evidence places unique historical circumstance in a subordinate position to ecological forces.

Finally, caution is always necessary during discussion of evolutionary phenomena. In suggesting close correspondence between patterns of diversity and environmental characteristics, I am aware of the danger of circularity and the Panglossian illusion (Gould and Lewontin 1979), as well as the possibility of random rather than deterministic effects (Slowinski and Guyer 1989). Indeed, there are good Australian examples of apparent phylogenetic effects on biological traits that had previously been interpreted as adaptations: Yom-Tov (1989) showed that the low reproductive rates of many Australian birds and mammals may be a consequence of the origins of those groups rather than a response to the erratic Australian climate. Thus, my conclusions must be considered tentative until a wider range of data becomes available.

SUMMARY

In this chapter I examine patterns of local and regional diversity in small mammals, birds, ants, termites, and reptiles. In all but two cases, Australian communities are not similar to their North American equivalents. Communities of Australian rodents are lower in both local and regional richness compared with their North American counterparts, but Australian insectivorous small mammals, granivorous birds, ants, termites, and lizards are higher at both levels of diversity. Birds in general may be similar between continents in local and regional diversity. Thus, the evidence suggests a strong connection between regional and local diversity; the only exception is harvester ants, which show greater regional diversity but similar local diversity. I suggest that the regional-local connection reflects the degree of matching between the environment and the life histories and body plans of the various taxa; lineages that are closely matched in this way radiate more, and more species are able to develop ways of making a living and persisting in local communities. The role of unique historical circumstance in affecting patterns of diversity remains obscure but potentially important.

15

Biodiversity in Australia Compared with Other Continents

Mark Westoby

This chapter considers three topics. First, the purpose of comparing diversity between continents is summarized, and some important issues that have arisen in comparing Australia with other continents are outlined. Second, data are selected and arranged from a particular perspective. The argument is as follows: Comparisons of local diversity alone are difficult to interpret, because it is hard to exclude the possibility that the physical environments being compared are different in some significant way. On the other hand, relations between regional and local diversity allow alternative interpretations to be distinguished. The relationships that are logically possible are outlined, then the available data are considered in relation to those possibilities. Thus the chapter considers only cases in which both local and regional diversity can be compared between Australia and elsewhere, and does not attempt a review of all comparisons that have been undertaken. Third, the chapter discusses the characterization and conservation of Australia's biodiversity, and the possible impact upon it of global change.

COMPARING DIVERSITY

Because traits of plants and animals are shaped by natural selection, similar solutions to common environmental challenges have often arisen in different lineages. Convergent adaptations in similar environments are well known: for example, leaf size and shape (Wolfe 1979), plant growth form (Schimper 1903; Box 1981), and diets and dentition of herbivorous mammals (Tyndale-Biscoe 1973). Ecologists have always hoped that convergence in relation to environment would prove to apply not only to traits of individual species, but to traits of assemblages, such as diversity. If diversity and other assemblage traits were strongly determined by particularities of evolutionary history, ecologists would have to relinquish their claim to understand communities and hand the discipline over to paleontologists, or to philistines who relish everything being a special case.

Comparisons between biogeographical regions offer the only available means to determine whether evolutionary history has an important influence on diversity. Australia has figured in such comparisons because it overlaps

in latitude with North America and Europe, and because much of its flora and fauna has had a separate phylogeny for a considerable period of time (Dodson and Westoby 1985; Westoby 1988).

MATCHING ENVIRONMENTS BETWEEN AUSTRALIA AND ELSEWHERE

Most comparisons between Australia and elsewhere have not taken methodical steps to match the physical environments. For example, Morton (chap. 14) reviews studies comparing various guilds in the arid zones of Australia and North America. None of these studies set out to match sites systematically in relation to latitude, rainfall seasonality, position within the valley system, or other variables. Some researchers, such as Milewski and Cowling (1985) have made very determined efforts to choose matched sites. But even in their study some measured aspects of the physical environment were different between Australia and South Africa.

Two widespread features of the Australian physical environment have often been invoked as prospective explanations for the uniqueness of Australian ecosystems. Compared with North America and Europe, Australia has had little mountain building and glaciation during recent geological history. In consequence, a much larger proportion of Australian soils are very old and infertile (Williams and Raupach 1983). Beadle (1954, 1966) argued that low phosphate levels are the essential cause of the sclerophylly (long-lived, hardened leaves) that characterizes much Australian vegetation. The sclerophylly in turn has been invoked as a potential cause of a wide range of effects. Australian forest insects show less seasonality than insects elsewhere, which might be a cause of lower clutch sizes in insectivorous birds (Woinarski and Cullen 1985). The hummock grasses or spinifexes in the Australian arid zones are thought to be sclerophyllous in response to low soil nutrients (Beard 1981a); this may lead to termites substituting for herbivores in the arid hummock grasslands, which in turn may lead to diversification of lizards (Morton and James 1988). The world's highest incidence of myrmecochory (plants with seeds adapted for dispersal by ants) is found in sclerophyllous vegetation on the very infertile soils in Australia and the Cape Province of South Africa (Willson, Rice, and Westoby 1990). At least within Australia, this phenomenon is largely associated with an

This chapter is Contribution No. 143 from the Research Unit for Biodiversity and Bioresources, Macquarie University.

abundance of low shrub species on infertile soils (Westoby, Rice, and Howell 1990).

Sclerophylly can also be an adaptation to drought, and it must be admitted that no work has attempted to partition what proportion of Australian sclerophylly should be attributed to dry climates and what proportion to low soil nutrients (I am indebted to Mike Austin for emphasizing this point to me). However, there are many locations in Australia where one can move from one lithology to another and from sclerophyllous to mesophyllic vegetation within a few tens of meters, so macroclimate can hardly be responsible. Further, the argument that plants should replace their leaves less often in an environment where mineral nutrients are harder to procure is simple and convincing, and has been successfully used to explain leaf longevity in a number of environments outside Australia that are not drought-prone (e.g., Loveless 1961; Chapin 1980). For this reason Australian sclerophylly continues to be regarded as mainly a response to infertile soils (Specht 1970; Beard 1983).

A second feature of the Australian environment compared with other continents is a greater probability of protracted droughts (at a given mean rainfall) and a balancing greater probability of very heavy rains. This feature has been mentioned as a possible explanation for the adaptive superiority of marsupial reproduction (Low 1978), for the virtual absence of stem-succulents (Stafford Smith and Morton 1990), and for the extension of tree growth forms into areas of relatively low mean rainfall (Milewski 1981b). This high year-to-year variability is now known to be the result of strong El Niño–Southern Oscillation influence (Nichols 1988). ENSO-driven droughts and wets typically last for twelve months or a little longer, and there is a tendency for strong droughts to be followed by strong wets. Droughts extend over much of the continent (Nicholls 1991).

THE LOGIC OF COMPARISONS BETWEEN CONTINENTS

When local assemblages in similar physical environments are found to have different diversities in Australia than elsewhere, it is difficult to decide whether the difference should be attributed to evolutionary history. Comparisons between continents are not fully controlled experiments. If they were, evolutionary histories would have been assigned to continents at random, independent of the effects of the physical environment, and repeated on replicate continents. In such a properly designed experiment, it would be possible to measure the probability that the diversity difference resulted from some unconsidered difference between the physical environments. If the probability were less than the conventional 5%, the diversity difference would be called "significant." But evolutionary histories have not been assigned independently of environmental differences between continents, and have not been replicated—repeat samples within continents are pseudoreplicates for this purpose. A formal statistical approach is not capable of eliminating the possibility that diversity differences between two continents are due to environmental differences.

Even if some variance in diversity is statistically attributed to a "regional effect" rather than to differences between environments within regions, we need to be careful not to leap to the conclusion that an effect of region is equivalent to an effect of evolutionary history. The hypothesis that diversity is shaped by evolutionary history ought not to be regarded as equivalent to chance effects, a category of explanation that can be attached to any unexplained variation. There ought to be specific propositions as to what sorts of evolutionary history differences occurred, and hypotheses as to the direction of difference in local diversity which should have resulted.

Useful steps for improving the interpretation of comparisons between Australia and elsewhere can start from either of these possible interpretations. If a difference *between* continents is due to the physical environment, a similar difference in the physical environment *within* a continent ought to have the same effect on diversity. Therefore, comparisons that include within-continent variation in relation to environment are helpful (Schluter and Ricklefs, chap. 21). As for prospective interpretations involving evolutionary history, in this chapter I will consider patterns whereby differences in local diversity are associated with regional diversity in a particular way. Such patterns might plausibly result from evolutionary history, as will be spelled out in more detail in the next section.

For these reasons the present chapter deals with groups for which regional as well as local diversity data are available, and for which some information is available as to variation in response to physical environment within continents.

COMPARING LOCAL DIVERSITY BETWEEN BIOGEOGRAPHICAL PROVINCES: POSSIBLE INTERPRETATIONS

I will use the term *local diversity* to refer to species richness on scales of 10 km^2 or less, and *regional diversity* for species richness of whole floras or faunas occupying regions large enough to have been separate from neighboring regions through at least some part of their evolutionary history. I will call the quantity that connects the two, the slope of a line between local and regional diversity on a log S–log A graph (fig. 15.1), *differentiation diversity,* following Whittaker (1977). Differentiation diversity describes species turnover along habitat (beta diversity) and geographical (delta diversity) gradients (see Cody, chap. 13).

An essential step in interpreting any particular difference in local diversity between regions is understanding the interrelations of local diversity with regional and differentiation diversity. My view of the possible interrelations is as follows: Regional diversity is the product of evolutionary history interacting with landscape history in speciation and extinction processes over millions to hundreds of millions of years. There is little reason to expect regional diversities to be constrained to similar values in Australia and elsewhere. In fact birds are the only case known to me in which Australian regional diversity is similar to that of another region (Australia 650, United

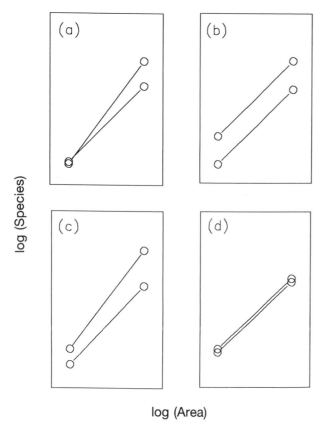

Figure 15.1 Schematic of four logically possible patterns relating species richness *S* to area *A*; two biogeographical regions are compared in each case. Points below and to left of graphs are local diversities, above and to right are regional diversities. Slopes of the lines measure differentiation diversities. See text for explanation of the four cases.

States 620, each about 8 million km²; Schall and Pianka 1978). I interpret this as an accidental similarity, and believe that in general, regional diversities will be different on different continents.

Local assemblages are drawn from regional faunas or floras over periods typically of hundreds to hundreds of thousands of years. Paleobiological evidence shows clearly that the species combinations that occurred in this relatively recent past were often different from those occurring today (e.g., M. B. Davis 1989; Graham 1986; Kershaw 1988; Webb 1987). This shuffling of community composition has occurred even though each species is presumably adapted to a limited range of temperature and rainfall conditions, which would tend to keep the same species combinations together through the considerable latitudinal or altitudinal migrations of the glacial-interglacial cycles. Thus a large proportion of a regional biota can be considered to have had the opportunity to reach any given site at some time during the last few hundred thousand years; the biota has been well mixed in this sense.

What rules might govern the size of the subset of a regional biota that is gathered into a local assemblage? Such rules might constrain local diversity directly, or might set differentiation diversity, that is, the proportion of regional

diversity that is subsampled in localities (the slopes in fig. 15.1). Differentiation diversity is not a property of any one locality, so the possibility of its being convergent between two continents by reason of similar physical environments does not arise. The possible rules are:

1. Local diversity might be constrained by some local ecological process operating similarly on both continents, and might therefore be convergent between Australia and elsewhere (in similar physical environments). Differentiation diversity would then represent a differential between local and regional diversity, and assuming regional diversities were different, differentiation diversity would have to differ between Australia and elsewhere (fig. 15.1A).

2. Differentiation diversity might be set by some process or principle operating similarly on both continents (for a given group of organisms), and might therefore be similar between Australia and elsewhere. Local diversity would then be different in Australia, in a way that mirrored the difference between the continents in regional diversity (Fig. 15.1B).

If regional diversity were not different between Australia and elsewhere, the similarity of differentiation diversities in this situation would lead to local diversities also being similar (fig. 15.1D). I would interpret similar regional diversities as an outcome of chance (see above).

3. Differentiation diversity might be different for reasons specific to each biogeographical region, and not because of constraints on local diversity. Differences between Australia and elsewhere in local diversity would then be a combination of effects due to regional diversity differences and effects due to differentiation diversity differences (Fig. 15.1C).

These three alternative sets of rules are related as follows to interpretations of the causes of differences or similarities in diversity between regions at different scales: Under rule 2, different speciation histories would have given rise to different regional diversities, but the difference in speciation history would not be associated with different degrees of geographical differentiation. Local diversity (in a given environment) would be determined as a subsample of regional diversity, the size of the subsample depending on the differentiation diversity, which would be characteristic of the group of organisms, not of the region in question. Under rule 1, local ecological processes would control local diversity to similar levels (in similar physical environments) in different regions, overriding any effect of speciation history. (This overriding is the effect I am calling "convergence," a usage different from that of Schluter and Ricklefs [chap. 21]). Different speciation histories would have given rise to different regional diversities. The difference in speciation history would be associated with a different degree of geographical differentiation. Thus the differentiation diversities might have been affected by the environments of the different regions, in the sense that landscape properties might have interacted with the speciation process to affect present-day differentiation diversity. The physical layout of the region would be the aspect of the environment relevant to differentiation diversity, while the climatic and soil properties of a locality would be the aspects relevant to local diversity. Under rule 3, different speciation histories would give rise to different regional diversities, and also to different differentiation diversities. Local diversities would be determined as the result of regional and differentiation diversities, and would not be convergent even in similar physical environments.

SOME CASES

Vascular Plants

Most data on local vascular plant diversity consist of species richnesses measured on plots of 0.1 ha (50 × 20 m), following the lead of Whittaker (1977).

Whittaker himself discussed Australia in the course of comparing richnesses in Mediterranean climates. He drew the conclusion that Australia, California, and Israel showed failure of convergence. Rice and Westoby (1983), on the other hand, worked on the basis that in Australia, sclerophyllous shrublands were commonly delimited by very infertile soils rather than by Mediterranean climate (see for reviews Beadle 1966; Specht 1970; Beard 1983). On this basis, shrublands in Australia could be treated as being in a similar environment to fynbos on the very infertile sandstone of the Cape Province of South Africa, but should not be expected to be similar to California chaparral or to Israeli shrublands.

Species richness on 0.1-ha plots varies widely within Australia, from 140 species for permanently moist tropical rainforest, to about 70 for sclerophyllous shrubland, to 50 or less for temperate forests. This ranking is the same as has been observed on other continents (fig. 15.2; Rice 1985). Provided that Australian sclerophyllous shrublands are compared only with those on comparably infertile soils in South Africa, the available evidence does not suggest that local species richness has reached different levels in Australia than elsewhere.

Regional richness diverges quite strongly between Western Australia and South Africa (fig. 15.3; Westoby 1985). Thus vascular plants appear to fit the pattern in figure 15.1A. Local richness is less than 5% of regional richness, so differentiation diversity is high.

Bracken Arthropods

Bracken arthropods are discussed by Lawton, Lewinsohn, and Compton in chapter 16. Briefly, Shuter and Westoby (1992) have found that local assemblages in Australia are different in both species richness and niche occupancy from those in other regions. Nothing is known of regional richness in Australia, but evidence from other continents indicates that local assemblages contain 50% or more of the regional faunas (i.e., differentiation diversity is low), and that rich regional faunas have richer local assemblages. Thus bracken arthropods appear to fit the model in figure 15.1B.

To some extent, differences between the Australian and British bracken faunas reflect order- and family-level differences in the overall insect faunas, further indicating that the herbivorous arthropod faunas on particular host plants can be seen as sampled from the regional faunas over evolutionary time, and shaped by evolutionary history in that way (Shuter and Westoby 1992).

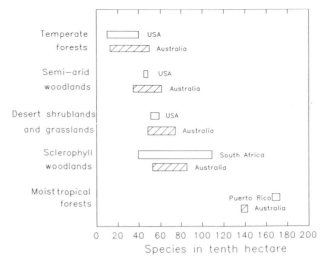

Figure 15.2 Range or mean of vascular plant species richness in 0.1-ha plots, comparing Australia (hatched bars) with vegetation of similar structure and in similar environments on other continents (open bars). Under "temperate forests," ranges are for most North American temperate forests versus many east Australian temperate forests. Under "semiarid woodlands," comparison is for mean of United States semiarid oak woodlands versus range for semiarid mallee woodland. Under "desert shrublands and grasslands," comparison is for means of Arizona desert shrublands and grasslands versus western New South Wales semiarid shrublands. Under "sclerophyllous woodlands," ranges are for South African fynbos versus coastal New South Wales sclerophyllous woodlands. Under "moist tropical forests," comparison is for a plot in Puerto Rico versus mean for Australian permanently humid tropical rainforest. (After Rice 1985; detailed sources are given therein.)

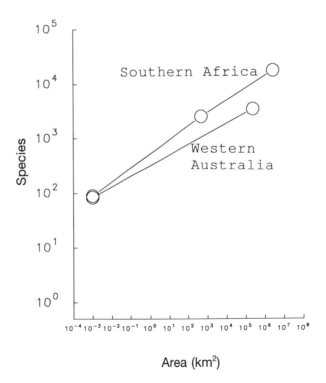

Figure 15.3 Species-area relations for vascular plants, comparing Western Australia with southern Africa. Local diversity is for sclerophyllous woodlands. (From Westoby 1985; detailed sources are given therein.)

Reef fishes

Coral reef fish species richness around an individual island is about twice as high on Australia's Great Barrier
Reef as in the Caribbean; this difference reflects the difference in regional faunas (fig 15.4; Westoby 1985). Thus
coral reef fishes appear to fit the pattern in figure 15.1B.
Individual islands sample about half the regional faunas,
so differentiation diversity is low.

Lizards

Morton (chap. 14) reviews the available information
about arid-zone lizard diversity. The regional fauna is
richer in Australia than in North America, whether the
regional fauna is taken to be that of the whole continent
(351 versus 99 species; Schall and Pianka 1978) or that of
the arid zone (170 versus 57 species). Local diversity
(mean 30 versus 7 species) is richer in Australia roughly
in proportion to regional richness (fig. 15.5), fitting the
model of figure 15.1B.

While differentiation diversity in the sense I am using—
the slopes of the lines in figure 15.5—is similar between
continents, it should be noted that Morton (chap. 14) and
Pianka (1986) have concluded that beta diversity is higher
in Australia.

Birds

Recher (1969) found no evidence that local bird species
diversity was different in Australian and North American
forests at a given level of foliage height diversity, though
response to foliage height diversity was strong and similar
on both continents. Similarly, Wiens (1991) found similar

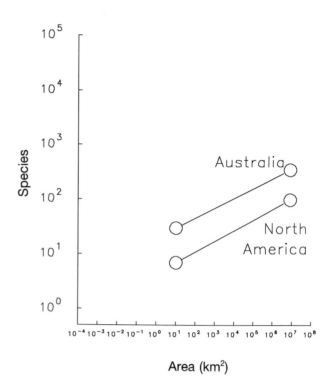

Figure 15.5   Species-area relations for lizards, comparing Australia
with North America. Local diversity is for arid-zone sites, from Pianka (1986; reviewed by Morton, Chap. 14), guessed to refer to
sample areas of about 10 km². Regional diversity from Schall and
Pianka (1978), areas each about 8 million km².

local species richness (ca. 6 species on average) in Australian and North American desert shrublands.

Regional bird species diversity is rather similar in Australia and North America (650 versus 620 species; Schall
and Pianka 1978). The bird situation (figure 15.6) therefore fits the pattern of figure 15.1D.

Summary of Evidence

The evidence shows different patterns in different groups
of organisms. Using some of the same evidence, I previously suggested that two distinct patterns could be seen
(Westoby 1985). These corresponded to the patterns of
figure 15.1A and 15.1B. In both patterns, regional diversity was not convergent, except by chance for birds; in
figure 15.1A local diversity was convergent and differentiation diversity was different, while in Figure 15.1B differentiation diversity was the same and local diversity differences reflected regional diversity. I also argued that the
pattern of figure 15.1A was associated with relatively high
differentiation—local diversity less than 5% of regional
diversity—while the pattern of figure 15.1B was associated with low differentiation diversity—local diversity
40% or more of regional diversity. I interpreted this difference as follows: Organisms conforming to the pattern
in figure 15.1B (reef fishes, arthropod herbivores of a particular host plant) have population dynamics involving
wide and partly stochastic dispersal. Considered over a
number of generations, genes of such organisms tend to
be mixed across the region, such that speciation along

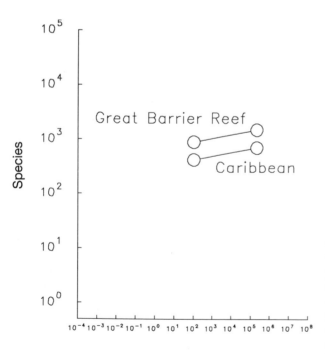

Figure 15.4   Species-area relations for coral reef fishes, comparing
the Great Barrier Reef (Australia) with the Caribbean. (After Westoby 1985; detailed sources are given therein.)

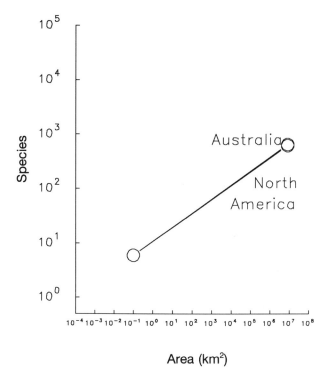

Figure 15.6 Species-area relations for birds, comparing Australia with North America. Local richness from desert shrublands, areas of about 10 ha (Wiens 1991); regional richness from Schall and Pianka (1978), areas of about 8 million km².

geographical gradients is restricted. Local diversity is then essentially a sample from regional diversity, with the percentage sampled being a constant for a particular group of organisms. In contrast, organisms conforming to the pattern in figure 15.1A (vascular plants) do not colonize so widely or with such a stochastic element. Differentiation along geographical gradients is steeper in such organisms. The extent of the differentiation may depend on the evolutionary history of the region, or on the layout of the landscape, or on the interaction of the two. Local diversity is a small percentage of regional diversity in such organisms, and the percentage may vary between regions.

For this chapter I have modified both treatment and data relative to my earlier study (Westoby 1985). The treatment of possible patterns has been expanded to include the possibilities in figure 15.1C and 15.1D. Figures 15.1A and 15.1B are no longer treated as being necessarily associated with a difference in differentiation diversity. To the data I have added lizards, but dropped harvester ants, because the data on ants seem to me too uncertain to allow any conclusion to be drawn. Morton and Davidson (1988) argued that local diversity in Australia was higher than in the United States considering all ants, similar considering only common harvester ants, but higher considering all harvester ant species. Further, there is great uncertainty about regional species richness for Australian ants (Morton, chap. 14).

The patterns described above show vascular plants conforming to figure 15.1A, reef fishes, bracken herbivores, and lizards conforming to figure 15.1B, and birds

conforming to figure 15.1D. For me, the important conclusion that emerges is that no known case conforms to figure 15.1C, that is, to a situation where *both* local diversity (in a similar physical environment) and differentiation diversity are different between Australia and elsewhere.

Lizards (the new case, compared with Westoby 1985) conform to figure 15.1B, that is, to a situation where local diversity is different between Australia and elsewhere in a way that reflects the difference in regional diversity. However, this case differs from reef fishes and bracken herbivores in that differentiation diversity is relatively greater, with local diversity 10% or less of regional diversity (though this percentage is still larger than for vascular plants or birds). It is far from clear why lizards should be in the category in which local diversity is essentially a given-proportion sample of regional diversity. They do not have in their life histories an essentially planktonic phase, like the larvae of reef fishes or the flying phase of herbivorous insects, that could account for their populations being well mixed across geographical space. Possibly it is relevant that the lizards live in the arid zone. Aridity in Australia developed comparatively recently, between about 6 million and 1 million years B.P. (Bowler 1982). The lizard assemblages have presumably been constructed from lineages moving into the developing arid core from surrounding higher-rainfall zones. In this sense, the arid-zone lizard assemblages could be interpreted as the product of a relatively recent mixing process from the regional lizard fauna. Under this interpretation, lizard assemblages in habitats with a longer Australian history, such as rainforest, might not be species-rich in proportion to the size of the regional fauna, and might conform to figure 15.1A rather than to figure 15.1B. Data are not available on this point.

Birds cannot be decisively classified as fitting either figure 15.1A or figure 15.1B, because their regional diversity happens to be the same in Australia and North America.

Thus, in the pattern represented by reef fishes, arthropod herbivores on bracken, and arid-zone lizards, differences in local diversity between Australia and elsewhere reflect differences in regional diversity. I interpret these cases as showing a significant influence of evolutionary history on local diversity, in addition to the influence of the local environment, which is always present. The influence is exerted via evolutionary history determining the size of the regional biota. In vascular plants and possibly birds, local diversity converges despite differences in regional diversity, suggesting that in these cases, evolutionary history need not be regarded as a significant influence on local diversity.

SOME IMPLICATIONS FOR CONSERVING BIODIVERSITY

Diversity is interesting not only as a property of assemblages, but as the repository of genetic resources. The genetic information in each species is very great, in that its DNA sequence is statistically highly improbable. Exceedingly large numbers of alternative sequences have been eliminated by natural selection in the course of evolution. Each extant sequence therefore embodies solutions to problems, though in most cases we do not yet fully understand the problems a species has faced nor why a particular DNA sequence has provided a solution. The DNA sequence of an extant species is potentially of great value to humans, because it has high information content and because it embodies solutions we may be able to use. Therefore, as a working policy, we should seek to conserve as many species as possible.

The patterns discussed in this chapter emphasize that conservation policy needs to be considered over wide areas, not as a problem of conserving locally diverse sites. For many organisms, differentiation diversity is what is important in contributing to Australia's overall biodiversity. Organisms with relatively slight geographical differentiation, such as reef fishes, may have population structures involving frequent exchange over geographical distances, so that conserving local assemblages will not be satisfactory even for them.

So far as we know at present, most terrestrial species biodiversity resides in angiosperms (about 25,000 species in Australia) and in insects. Taylor (1983) estimated 108,000 insect species in Australia. However, recent surveys of forest canopy insects have suggested much higher numbers. Calculations by Erwin (1982), May (1988), Stork (1988), and Thomas (1990) have estimated 10–80 million insect species worldwide; on this basis Australia might be expected to have 1–10 million. These high estimates are generated because large numbers of undescribed species are found in canopies of single tree species, and because a high degree of host plant specificity is assumed. In the previous section it was shown that angiosperms are biodiverse because of differentiation diversity, while insects do not have very high differentiation diversity across the range of any one host species (vide the bracken example discussed above), but have high differentiation diversity across the landscape in response to the distributions of the host species.

For any other group to be a major contributor to total biodiversity, it would have to share these traits of high local diversity plus high geographical differentiation, either directly or by host specificity on a group which is itself strongly differentiated. Present knowledge (Margulis and Schwartz 1988) indicates that vertebrates, crustacea, plants, mollusks, and annelids do not qualify; it remains uncertain whether mites, nematodes, protoctists, or prokaryotes might.

On Assessing Groups with Very High Biodiversity

There is a major practical and political problem in directing conservation policy at conserving biodiversity. Most biodiversity resides in the very speciose groups with high geographical differentiation. For groups such as forest canopy insects, fogging at one location may give rise to collections of many thousands of morphospecies, more than half of which may not previously have been collected, let alone described. This has led some to despair of assessing biodiversity or conserving such groups, because

complete cataloging and description would take several decades even given a crash program (R. L. Kitching, quoted in Monteith 1990).

Such despair is not called for. It is possible to assess biodiversity, even for very diverse groups, without cataloging the entire fauna. The principle is to measure dissimilarity between paired species lists and to relate that dissimilarity to measures of "distance" between the lists. For example, for lists of herbivorous arthropods on plants, the distance might be a phylogenetic distance between the two host species. From a set of such paired samples a function relating dissimilarity to distance can be empirically constructed. Then, given an estimated phylogeny for the set of hosts being considered, numerical methods can be used to estimate the size of the full set of herbivorous arthropods using that set of hosts (Ross Talent, Macquarie University, personal communication).

The implications for research policy are twofold. First, research needs to be concentrated on those groups that are known or suspected to be very biodiverse. Second, the research should characterize differentiation diversity by sampling to estimate the differentiation of species composition between pairs of sampling locations, or pairs of host species.

Australian Biodiversity in Relation to Global Change

Emissions of greenhouse gases into the atmosphere are modifying the world's energy balance with space at a rate that is expected to lead to a 2°–5° C warming by 2050. Increasing human food and natural resource demands are converting areas of natural vegetation into croplands or other intensively managed vegetation. Nothing in the future is certain, but an Australia considerably changed by 2050—warmer and with a larger area intensively managed by humans—is much more likely than an Australia unchanged from 1990. It would take considerable luck for the known processes driving change to be balanced out by countervailing effects.

Conservation policy still consists largely of identifying areas of land that currently contain biota of high conservation value and designating those areas for conservation as the primary land use. This approach is implicitly a response to the expectation that larger and larger land areas will be modified by intensive land use. However, it takes no account of the other major aspect of global change, expected climate warming. The basic effect of a global warming of 4° C will be a poleward shift of temperature zones by about 400 km, at temperate latitudes, or a shift to higher altitudes by about 400 m. Unless none of the biota to be conserved have current range limits set by temperature or temperature-related processes (which seems a large assumption), problems can be expected.

Thus, selecting conservation zones on the basis of high local diversity is inadequate, both because of expected climate zone shifts and because most diversity arises from differentiation across landscapes. Global change accentuates the necessity to approach conservation of biodiversity as a problem of the whole Australian region, a problem that requires much-improved geographical information and region-wide management. To limit the losses of biodiversity, we will be relying heavily on forests and rangelands occupied by native vegetation but managed for production. These lands occupy much wider areas than those reserved specifically for conservation, and must be correspondingly important, particularly in an era of shifting climate zones. There is an urgent need to develop objective data on the biodiversity contributions of managed forests and rangelands.

In highly diverse groups such as canopy arthropods, climate change could lead to losses of biodiversity in various ways. Some host plants may persist in their present locations but under a changed climate and pathogen regime, with consequences for their faunas. Some arthropod species may only be able to survive if they move their distributions to remain in a similar climate regime, but this might require them to operate within different food webs based on different host plants. If host plants disperse or are deliberately dispersed by humans to keep up with shifting climate zones, they nevertheless may be growing on different soils, or may take tens or hundreds of years to grow to canopy dominance, so they will not necessarily be capable of supporting a food web including the same set of species as currently.

How should we manage this complex of problems to restrict the loss of biodiversity as much as possible? In my view, the operating principle should be to give each species the greatest variety of opportunities to survive—first, by protecting its host plant within its present distribution, at least in part; second, by protecting native vegetation within areas to which its climate zone is expected to move; and third, by transplanting host plants to those areas. Such a risk-spreading policy would have two corollaries.

First, we would be accepting that host plant mixtures and food webs could not be sustained in their exact present configurations. I believe such loss of "community integrity" is not a serious problem. Where fossil records are available for plants and insects, it is clear that species have responded individualistically to past climate changes. Many species combinations have come into existence only recently, and the persistence of species has not been contingent on being embedded in a consistent food web.

Second, more information on the very biodiverse group such as canopy arthropods is urgently needed. Particularly important types of information would include the conservation value of native forests and rangelands managed for production, and degrees of host specificity and diversification along geographical gradients.

16

Patterns of Diversity for the Insect Herbivores on Bracken

John H. Lawton, Thomas M. Lewinsohn, and Stephen G. Compton

Bracken fern (*Pteridium aquilinum*) is a particularly useful model system for the study of herbivore diversity because it is a widespread and common native member of the flora of all the nonpolar continents (C. N. Page 1976). Fossil evidence indicates that *Pteridium* was already widespread by Tertiary times (Page 1976). The success of bracken is reflected in the fact that in countries such as Britain and Brazil it is rated as a serious weed, and the possibility of its biological control (e.g., Lawton 1988, 1990a; Lawton, Rashbrook, and Compton 1988) has provided one of the stimuli for extensive and intensive surveys of its insect faunas. In consequence, bracken is unique because its herbivorous insect communities have been studied in detail on several continents. Geographically, it may now be one of the best-studied ecological systems in the world.

After almost two decades of bracken-insect studies, we can now examine overall variation in the diversity of bracken-associated herbivores across a wide range of scales, and we can assess the importance of different factors involved in generating and maintaining this diversity. The results should prove useful for understanding the interplay of local and regional processes in determining the diversity of other assemblages of phytophagous insects (Strong, Lawton, and Southwood 1984; Cornell 1985a, 1985b) and for studying the diversity of all ecological communities (Westoby 1985; Ricklefs 1987).

BRACKEN HERBIVORES

Thorough surveys of bracken-feeding insects (and mites, here treated as "honorary" insects) have been conducted at sites in seven parts of the world: Hawaii, southwestern United States, Great Britain, South Africa, southeastern Brazil, Papua New Guinea, and Australia (Lawton 1982, 1984a, 1984b; Compton, Lawton, and Rashbrook 1989; R. P. Martins personal communication; Kirk 1982; Shuter 1990). As Figure 16.1 shows, the taxonomic composition of the assemblages feeding on bracken differs markedly between continents, and provides no obvious evidence that any one order is preadapted for feeding on bracken. Note, for example, the dominance of Lepidoptera in South Africa, of Coleoptera in Papua New Guinea, of Homoptera in Australia, and of Diptera and Hymenoptera in Britain, and the complete or virtual absence of these orders in other regions (there are no Coleoptera on

bracken in Britain, for instance, and no Hymenoptera in South Africa). These insects are a mixture of monophagous bracken specialists, oligophages exploiting bracken and other ferns, and broad polyphages for which bracken may be either an important or a minor foodplant (e.g., Lawton 1984a, 1984b).

There are few examples of congeneric bracken-feeding species on different continents or of individual species extending to more than one region. Insects appear to have independently colonized bracken in different parts of its range over evolutionary time. The taxonomic uniqueness of assemblages on different continents shows that no successful long-range colonization has occurred even over the considerable time span that bracken has existed in different parts of the world (C. N. Page 1990). Among the chewing insects that feed on the plant, the caterpillars of moths (Lepidoptera) predominate on the three southern continents, caterpillars of sawflies (Hymenoptera) in New Mexico and Great Britain, and adult and larval beetles (Coleoptera) in Papua New Guinea. No Diptera (fly larvae) exploit the plant in South Africa, but they occur in every other area surveyed so far; and so on. This taxonomic diversity is equally evident at lower levels, with, for example, each continent having its own blend of moth families contributing to the overall richness of the lepidopteran fauna (fig. 16.2). There is some evidence, at this finer level of taxonomic resolution, that certain groups of Lepidoptera have evolved to feed on bracken on several continents. Species within the Lithinini (Geometridae) are a good example, with related species exploiting bracken in Britain, South Africa, and Australia. It is not yet known whether this represents a single colonization and subsequent speciation, or independent colonization by related taxa on each continent (J. H. Lawton and M. J. Scoble, unpublished data).

REGIONAL SPECIES RICHNESS OF BRACKEN HERBIVORES

Thorough surveys of bracken herbivores have been carried out on a regional scale in five parts of the plant's range. Hawaii has no confirmed bracken herbivores, while Papua New Guinea has the richest recorded fauna, with 30 species. This variation in the total number of insect species exploiting bracken in different parts of the world is a function of how common and widespread

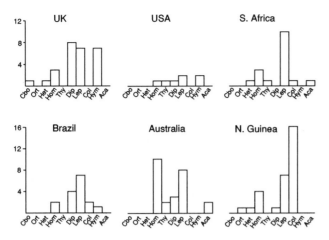

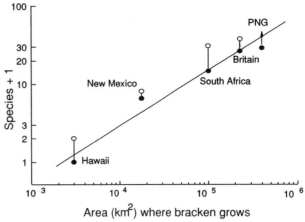

Figure 16.1 Taxonomic composition of the arthropod assemblages feeding on bracken in different parts of the world. Abbreviations: *Cbo,* Collembola; *Ort,* Orthoptera; *Het,* Heteroptera; *Hom,* Homoptera; *Thy,* Thysanoptera; *Dip,* Diptera; *Lep,* Lepidoptera; *Col,* Coleoptera; *Hym,* Hymenoptera; *Aca,* Acarina. Data from Lawton 1984b, (Britain and US); Compton, Lawton, and Rashbrook 1989, (South Africa); Martins, unpublished data (Brazil); Shuter 1990 (Australia), Kirk 1982 (Papua New Guinea).

Figure 16.3 Species-area relationship for the number of species of herbivores definitely feeding on bracken (•) in different parts of the world. Also shown (○) are total numbers of species feeding on bracken including possible, occasional, and uncertain records. The arrow in the data for Papua New Guinea indicates that the number may be an underestimate. Log (species + 1) has been used in the ordinate to allow the inclusion of a zero count for Hawaii. The regression line has been fitted to the definite records only, giving a conservative estimate for the intercept. The fitted regression is log 10(species + 1) = 0.70 log 10(area) − 2.35(r_2 = .97; *p* = .002). (After Compton, Lawton, and Rashbrook 1989.)

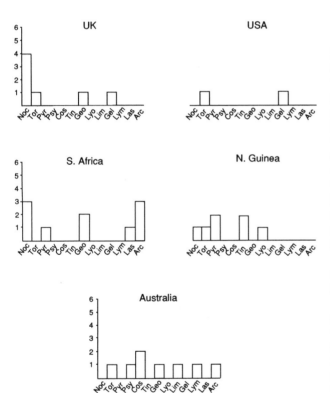

Figure 16.2 The families of Lepidoptera with species that feed on bracken in different parts of the world. Abbreviations: *Noc,* Noctuidae; *Tor,* Tortricidae; *Pyr,* Pyralidae; *Psy,* Psychidae; *Cos,* Cosmopterigidae; *Tin,* Tineidae; *Geo,* Geometridae; *Lyo,* Lyonetiidae; *Lim,* Limacodidae; *Gel,* Gelechiidae; *Lym,* Lymantriidae; *Las,* Lasiocampidae; *Arc,* Arctiidae. Data sources as in figure 16.1.

bracken is in each geographical region—there is a strong species-area effect (fig. 16.3). This relationship is well established in the literature on plants and their insect herbivores, although the correlation in figure 16.3 is unusually

high (Auerbach and Hendrix 1980; Strong, Lawton, and Southwood 1984; but see Zwölfer 1987; Lewinsohn 1991). The mechanisms generating species-area relationships, which are discussed by Strong et al., suggest that the very tight relationship depicted in figure 16.3 is probably fortuitous. Moreover, definition of the area over which bracken grows in each region is clearly prone to errors and uncertainties. In other words, addition of surveys from further geographical regions are likely to weaken rather than strengthen the correlation. Nevertheless, it will be surprising if additional studies obliterate the general pattern of more species of insect herbivores exploiting bracken in regions where it is more widespread.

We know little about how long ago, in evolutionary time, associations of particular taxa with bracken originated. Bracken from a Roman archeological site in Britain 1900 B.P. had galls of the cecidomyiid midge *Dasineura filicina,* which still exploits the plant in the same area (Seaward 1976), but two millenia are insignificant in a fossil history for *Pteridium* that extends back to the Tertiary (Page 1976). The leptosporangiate ferns, which include *Pteridium,* appear to have evolved simultaneously with the angiosperms during the Cretaceous (Page 1976). Certain phytophagous taxa (e.g., the selandriine sawflies and some of the anthomyiid flies) seem to have radiated and speciated predominantly on fern hosts, and we may therefore adduce that members of the bracken fauna in these groups have a long-standing association with the plants (Lawton 1984a). Other taxa (e.g., agromyzid flies and various Lepidoptera) evolved as herbivores in the Cretaceous, and most of them are predominantly associated with angiosperms (Lawton 1984a; Strong, Lawton, and Southwood 1984). Fern feeding must therefore have evolved subsequently in these taxa, but it is not necessarily recent. British bracken feeders include several species in monospecific genera that feed exclusively on bracken, although their nearest relatives exploit angiosperms (Lawton 1984a). Strictly monophagous, monospecific genera are unlikely to be recently evolved. Overall, we expect that most of the herbivores we are dealing with have been associated with bracken for many thousands of years, and in some cases, very much longer.

As a food resource, bracken cannot be easy to exploit, because it contains a wide variety of secondary plant compounds, as well as physical defenses and ant "guards" attracted to extrafloral nectaries (Lawton 1976; Jones and Firn 1978, 1979a, 1979b; Heads and Lawton 1984; Lawton and Heads 1984). We know very little about how the insect herbivores have adapted to circumvent or withstand the physical and biochemical defenses of the plant. On present evidence (e.g., Jones and Lawton 1991) allelochemicals are unlikely to reduce the number of species of insects exploiting the plant in different parts of the world, but they may modify the way in which insects are able to use the plants, and may influence their abundances. Ant–extrafloral nectary interactions have been better studied, and appear to play a role in structuring the herbivore community on bracken (Heads and Lawton 1985), favoring "internal feeders" (gall formers and leaf miners) over external chewers, for example.

REGIONAL AND LOCAL SPECIES RICHNESS

By "local," we mean patches of bracken of a few hundred to a few thousand square meters, within which species may potentially interact in ecological time.

The relationship between local and regional species richness of bracken herbivores could take several forms (Lawton 1982). First, all the species that occur regionally might coexist in every site; in this case there is no turnover of species across sites and beta diversity (Whittaker 1972) is zero. At the other extreme, local communities might become saturated with species, possibly because competition limits the number of coexisting species. Cornell (1985a) put forward an intermediate model, in which saturation is only manifested (and perceived) at higher values of regional richness. Another intermediate model might assume "proportional sampling," with not all of the species that occur regionally found at every local site, but with no tendency for local sites to become saturated with species. The proportional sampling model implies that species interactions do not limit the size of a local community: instead, since the total regional fauna is larger than the local assemblage, there must then be a measurable turnover of species across local sites, or, in other words, a substantial beta diversity (Whittaker 1972). Turnover of species across sites is also a corollary of the saturation model, with turnover increasing with greater regional diversity.

The local species richness of bracken-feeding insects depends weakly on the size of the censused bracken patches: larger patches have more species, even with constant sampling effort (Rigby and Lawton 1981; Compton, Lawton, and Rashbrook 1989). This is another manifestation of the species-area relationship, albeit at much smaller spatial scales (other studies of insects on local patches of host plants are discussed by Lawton 1978; Kareiva 1983; Strong, Lawton, and Southwood 1984). Strictly, therefore, data used to distinguish between the patterns listed above should be standardized to local patches of a fixed size. In practaice this is not possible, because some studies failed to record the size of the bracken patches. Fortunately, local species–patch size relationships appear to have such shallow slopes (Compton, Lawton, and Rashbrook 1989) that the error involved in comparing local patches of different sizes in different parts of the world is very small.

Additional problems result from seasonal variation in faunal composition in temperate regions (Lawton 1976, 1978), making single-season sampling inappropriate, and from the fact that more prolonged studies inevitably discover more species, particularly if they extend over several years and new species immigrate into a patch. Even if immigration is balanced by local extinctions of other taxa, the cumulative number of species recorded from the patch will increase over time. Our estimates of the local richness of bracken-feeding insects have therefore been based on annual figures for well-studied sites. These include detailed investigations lasting one year (e.g., Kirk 1982; Lawton 1982), and longer-running studies from which average annual data on local species richness can be extracted (e.g., Lawton and Gaston 1989; Gaston and Lawton 1989). Note, in passing, that the faunal composition and relative abundance of species at well-studied sites has remained very consistent over ten- to twenty-year periods (Lawton and Gaston 1989).

The relationship between local and regional species richness of bracken herbivores in different parts of the world is shown in figure 16.4. Given the small number of data points, the pattern should not be overinterpreted, but it appears that local richness is correlated with regional richness, with local patches in species-rich regions supporting larger numbers of herbivore species. The most economical description of the pattern in figure 16.4 is the simple proportional sampling model: during any one year, local communities contain roughly half the species in the regional pool. There is no indication of a limit to local assemblage size independent from regional richness—local communities do not become saturated with species. To date there are few other studies available for comparison, but Cornell (1985a, 1985b) found a similar pattern for cynipid gall wasps on different North American oak species. Indeed, some of the evidence for local saturation of communities with increasing size of the regional pool is less convincing when analyzed in alternative ways (e.g., Terborgh and Faaborg 1980 reanalyzed by Wiens 1989a), or is based on too few local-regional comparisons to allow definitive statements (e.g., Tonn et al. 1990). Thus, the bracken system may not be exceptional among a wider variety of ecological assemblages.

The proportional sampling model implies that local species richness is not constrained by interspecific competition, an assumption that is supported by population studies of bracken-feeding insects (Lawton 1982, 1984b). Furthermore, if interspecific competition were important in shaping bracken insect assemblages, apart from re-

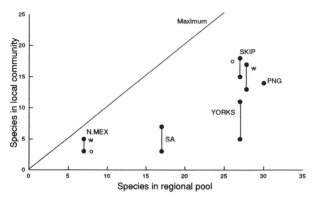

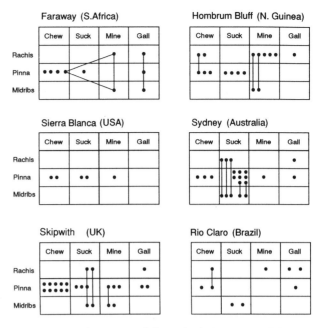

Figure 16.4 The relationship between local and regional species richness of herbivores feeding on bracken in different parts of the world. The "maximum" line has a slope of 1, corresponding to all species in a given regional pool occurring at every site. Date for local richness are from the following sources: New Mexico (NMEX), open (o) and woodland (w) sites, Lawton 1982; South Africa (SA), maximum and minimum species in patches of difference sizes in the eastern Cape, (Compton, Lawton, and Rashbrook 1989); Skipwith Common, England (SKIP), maximum and minimum numbers for open (o) and woodland (w) sites recorded in any year, Lawton and Gaston 1989; North Yorkshire Moors, England (YORKS), maximum and minimum numbers of species on bracken patches of different sizes, Rigby and Lawton 1981, amended in Compton, Lawton, and Rashbrook 1989; Hombrum Bluff, Papua New Guinea (PNG), numbers of species recorded by Kirk 1982.

Figure 16.5 Niche matrices defining the feeding sites and feeding methods of herbivorous arthropods on bracken at sites in different parts of the world. Feeding sites of species exploiting more than one part of the frond are joined by lines. "Midribs" is a preferable term (Thomson 1990) for "costa" and "costule" used previously (e.g., Lawton 1984b). Data from sources indicated in Figure 16.1

stricting species numbers, one might also expect it to produce convergence in the ways in which bracken is partitioned among the herbivores, given that structurally the plant is the same everywhere. What we see, however (fig. 16.5) is that, with the exception that all regions tend to have more species exploiting pinnae than other plant parts (see below), the distribution of species across resources on the plant is idiosyncratic from locality to locality, and region to region, with numerous vacant niches—ways of using the plant that are exploited in one part of the world, but which remain unoccupied elsewhere. The taxonomic distinctness of bracken-feeding assemblages on different continents (see figs. 16.1 and 16.2) is thus matched by idiosyncratic patterns of resource partitioning within local communities on different continents.

We cannot assert that contemporary bracken assemblages are not structurally more similar than ancestral assemblages. Convergence in community structure may be partial and cannot be ruled out merely because present-day assemblages are not identically structured (Schluter 1986; Schluter and Ricklefs, chap. 21). Indeed, Schluter has pointed out to us that all the bracken assemblages so far investigated tend to have proportionally more insect species exploiting pinnae ("leaves") than exploiting, say, the rachis ("stem"). To this extent the assemblages are convergent. However, in bracken as in most terrestrial plants, stems and petioles are tougher and more difficult to exploit than leaves. Resources that are difficult to exploit presumably have reduced chances of being colonized over evolutionary time, and hence support less diverse consumer assemblages. The relative prevalence of pinna-feeding bracken insects is most likely derived from the constraints on feeding on other plant parts rather than from interspecific interactions between herbivores.

LATITUDINAL AND HABITAT EFFECTS ON LOCAL SPECIES RICHNESS

Given that local assemblages of bracken herbivores are subsets of their respective regional pools, are they randomly sampled from those pools? Here, as in the continental comparisons, localities might show systematic differences either in species numbers or in species composition.

Data sets for Britain and South Africa are extensive enough to allow detailed analysis. In Britain, 216 sites were sampled over several years (Gaston and Lawton 1989) but, since sampling intensity affects the number of recorded species, only the 158 more thoroughly sampled sites were considered. In South Africa, samples for 80 different sites are available (Compton, Lawton, and Rashbrook 1989).

Covariance analyses were used to search for differences in local species richness among habitat categories: open versus woody sites, and, in Britain, pure bracken versus sites where it is interspersed with other vegetation. Site latitude and sampling date were included as (continuous) covariates (Lewinsohn, Compton, and Lawton, unpublished data).

In both Britain and South Africa, habitat categories have no measurable effect on species numbers, although on a more restricted geographical scale such effects are detectable. For example, in comparisons of sites close to one another, woodland sites had more species than open ones in New Mexico, and fewer species in Britain (Lawton 1982; MacGarvin, Lawton, and Heads 1986). In contrast, detailed studies of the effects of altitude on diversity again failed to detect any significant effects (Lawton, MacGarvin, and Heads 1987).

Local richness in both Britain and South Africa does, however, show a slight but significant latitudinal gradient. Local assemblages decrease in species richness toward higher latitudes. In Britain, assemblages change in size over the growing season as well (e.g., Lawton 1976), but the effect of latitude, although relatively smaller, is independent of sampling date. Thus, a separate analysis of the 51 British samples obtained in August (the peak of the bracken growing season) also showed a significant decrease in local species richness toward the north.

A latitudinal decrease in local richness can be generated in two ways (fig. 16.6). If it is due primarily to variation in species ranges, local and regional assemblages will be largest at lower latitudes and "lose" species gradually toward higher latitudes (fig. 16.6A). On the other hand, the same decrease in local richness can be generated even if all species span the entire region, provided that frequency of occurrence is reduced toward higher latitudes (fig. 16.6B). In this case, relative species turnover between sites will necessarily increase with latitude—or, in other words, similarity of species composition will decrease toward higher latitudes. Of course, the two schemes are not mutually exclusive.

The latitudinal ranges of bracken-feeding insects in the samples from Britain and South Africa are given in figure 16.7. Many of the British species are present throughout most of the latitudinal range of the samples, while others

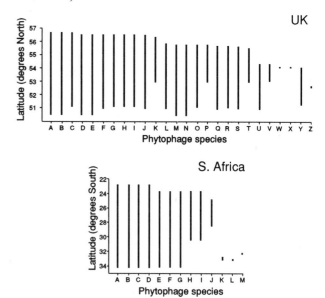

Figure 16.7   Latitudinal ranges of the arthropods found on bracken in Britain and South Africa. The vertical bars show the latitudinal extremes of the records for each species. Sampling did not extend beyond the ranges of the most widespread species.

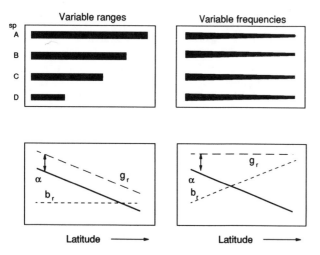

Figure 16.6   Variable range and variable frequency models for latitudinal changes in local herbivore richness (alpha-diversity). Variation in the ranges of herbivores (*Left*), with progressively fewer species represented at higher latitudes results in a decline in local richness, in parallel with the reduction of the regional pool (g_r), although the turnover (beta diversity) between sites (b_r), or relative similarity, may remain constant over the entire gradient. (*B*) The same decline in local species richness can occur with no decline in regional richness if the species occurrence frequencies are gradually reduced with increasing latitude (*Right*). The turnover will necessarily increase over the gradient in this case. The models are complementary and the distribution pattern of each component species may be independent of the rest. The vertical arrows show that the distance between the local and regional richness lines is an alternative representation of b_r.

are more restricted in their distribution. Although differences in species ranges contribute to turnover at different latitudes, they cannot account entirely for the higher richness of the more southern sites. However, the frequency of occurrence of most species within their geographical spans is not significantly influenced by site latitude (according to logistic regressions of individual species occurrences; Lewinsohn, Compton, and Lawton, unpublished data). Thus, the latitudinal decrease in local richness in Britain involves a combination of the two mechanisms.

Latitudinal trends in the species richness of bracken herbivores in South Africa seem to be easier to account for (fig. 16.7B). Most of the species were again effectively present throughout the latitudinal range of the samples, but the presence of three rare and localized polyphagous insects in samples from the south of the country failed to compensate for the addition of three more widespread species in the north. Range restriction rather than increased intersite turnover would seem to be primarily responsible for the observed pattern, although one species was more frequent in the lower-latitude samples within its range.

In figure 16.8 the geographical patterns are examined in a different way, showing species accumulation curves obtained by adding samples northwards and southwards. These are compared with the accumulation of species in a spatially unordered series of samples, represented in figure 16.8 by rarefaction curves derived from the species occurrence frequencies (Lewinsohn 1991). New species are initially encountered at a somewhat slower rate than in the unordered sequence in Britain, particularly when the samples are ordered from north to south. Conversely, in South Africa, new species accumulate extremely rapidly from the north, with most of the regional pool accounted for in the northernmost samples alone.

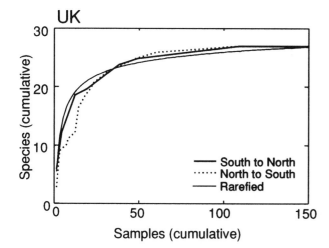

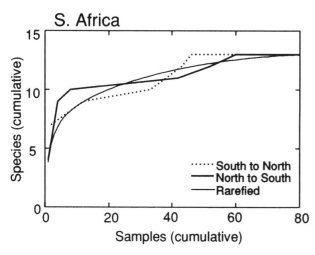

Figure 16.8 Species-site curves for British and South African samples. Sites were ordered latitudinally. The curves show the cumulative number of insect species obtained both from lower to higher latitudes (thick lines) and in the reverse direction (dotted lines). The thin lines represent rarefied frequency curves, obtained from the occurrence frequency of each herbivore in all samples (see Lewinsohn 1991 for formula and references). The rarefaction curve represents the average number of herbivores of all possible combinations for each number of samples on the abcissa; it gives the average expected accumulation of species in a random sequence, if herbivore distributions are independent from one another. Deviations from this curve indicate a spatial trend in herbivore occurrences.

DISCUSSION

Reasonably clear rules determining the species richness of insects feeding on bracken are slowly emerging (table 16.1). The general impression is that colonization of bracken by insects over ecological and evolutionary time in different parts of the world has been largely a stochastic sampling process. By this we mean that over evolutionary time, which taxa have colonized bracken, and which plant parts they have exploited, has not obviously been constrained by interspecific interactions between herbivores. Similarly, in ecological time we can find little or no evidence that local assemblages of bracken are sorted from the regional pool by significant interspecific interactions between herbivores. None of this is to say that some of the differences that exist between geographical regions, both in patterns of resource utilization by insects on the plant and in insect taxonomy, are entirely due to chance. For example, the resource pools of insects available for colonization of bracken differ between continents (Shuter 1990), affecting the likelihood that certain orders will colonize the plant on different continents over evolutionary time. Furthermore, bracken in different parts of the world (or even genotypically distinct clones within one region) may differ sufficiently as resources for herbivores to influence the structure of the insect community. Very little is currently known about direct host plant influences on local and regional species richness. On present evidence, however, these effects are small.

Different complexes of natural enemies in different parts of the world are another possible reason why patterns of resource exploitation and richness of herbivores differ between regions. The importance of parasitoids (e.g., Lawton 1986) remains unknown, but the effects of predatory ants (which are attracted by the extrafloral nectaries of the plant) have been studied in Britain and South Africa (Heads and Lawton 1984, 1985; Heads 1986; Rashbrook, Compton and Lawton, 1992). In both areas ants have very feeble (Heads 1986) or no measurable effect on populations of adapted bracken-feeding insects, although the possibility remains that certain taxa may be unable to exploit the plant, and that certain feeding niches may be unoccupied, because of ant predation (Heads and Lawton 1985). Although theoretically attractive, these arguments are very difficult to test rigorously in practice. The most parsimonious hypothesis for the rather idiosyncratic distribution of species across resources displayed in figure 16.5 is that colonization by insect herbivores over evolutionary time has been a largely stochastic process from taxonomically very different pools in each geographical region, weighted by the ease of exploitation of different plant parts, and that resulting local communities are little constrained by species interaction in contemporary ecological time.

In sum, geographical factors appear to provide the most deterministic influences acting on the species richness of bracken herbivores, in the form of the species-area relationship and latitudinal trends in species richness. The latter effects, however, are generated by the individualistic responses of different species. The greater local species richnesses seen in Britain and South Africa at lower latitudes, for example, are generated by rather different mechanisms. In southern Britain the addition of species absent from higher latitudes is not a significant contributory factor, while in South Africa both local and regional species richness increase toward the tropics. Different mechanisms therefore generate similar patterns in the two geographical regions, even for assemblages of insects feeding on the same plant.

Despite the fact that the latitudinal trend in diversity within regions may be generated by different mechanisms in Britain and in South Africa, the existence of this trend (albeit a slight one) is intriguing. It is, of course, commonplace to find trends of species richness with latitude, al-

Table 16.1.   Factors that Influence the Diversity of Insects Feeding on Bracken

Factor	Scale	Indicators	Importance
History	Continental	Faunal taxonomic composition	Strong
Plant range	Continental	Regional species-area effect	Strong
Latitude	Regional	Local assemblage size, turnover	Small but significant effects
Seasonality	Regional	Insect succession, dynamics	Variable
Habitat heterogeneity	Regional to local	Insect distribution	Variable
Patch size	Local	Local species-area effect	Weak
Community interactions	Local	Assemblage saturation, functional convergence	No detectable effects

though the underlying components (fig. 16.6) have rarely been examined. It has also proved difficult to factor out the effects of changing resource diversity from effects of resource predictability and stability. For bracken herbivores, at least, latitudinal trends in species richness cannot be generated by changes in resource diversity, and must be driven by changes in resource predictability and/or productivity. The problem deserves more attention. It would also be interesting to explore the contribution of latitudinal trends in diversity to the unexplained scatter in figure 16.4.

The insect herbivores of bracken have proved to be a rewarding system in which to explore the interplay of history and local and regional processes in determining patterns of species richness and community structure. The insights revealed by the system are very different from the ways in which ecologists have traditionally thought about species assemblages (table 16.1). Accidents of evolutionary history and large-scale biogeographical processes dominate local community structure; if species interactions are present, their effects are extremely feeble (Lawton 1982, 1984a, 1984b; Heads and Lawton 1984, 1985; Heads 1986; Rashbrook, Compton, and Lawton, 1992). Yet the system is *not* entirely unpredictable. It remains to be seen whether it is typical of other ecological assemblages. To date, too few have been studied in this way to tell.

17

Community Richness in Parasites of Some Freshwater Fishes from North America

John M. Aho and Albert O. Bush

Investigations attempting to identify *contemporary* determinants of species richness in natural communities are well represented in the ecological literature. In fact, during the past decade, there have been numerous edited volumes devoted solely, or largely, to addressing patterns and processes underlying community structure (e.g., Price, Slobodchikoff, and Gaud 1984; Strong et al. 1984; Diamond and Case 1986; Kikkawa and Anderson 1986; Esch, Bush, and Aho 1990). Much of the empirical data on community structure comes from studies on free-living organisms, but there is an increasing tendency to address symbiotic relationships. In this chapter, we provide a "parasite perspective." While "parasite" may imply different things to different people, we use *parasite* here to mean a parasitic helminth (i.e., the digenetic Trematoda, Cestoda, Nematoda, and Acanthocephala). As such, we are considering obligate endoparasites, incapable of emigration, many of which require intermediate hosts and most of which produce offspring that must physically leave the body of the definitive host prior to becoming infective. Thus, in contrast to some types of parasites (e.g., protozoans, viruses), the population in the definitive host is maintained through a balance of immigration and death.

A BRIEF HISTORY AND SOME CURRENT IDEAS

Perhaps surprisingly, questions related to species richness in parasitic communities have a lengthy history. Gregory, Keymer, and Harvey (1991) note that as early as 1920, Soviet scientists were attempting to delineate factors influencing the development of parasite communities. Much of that early work has been summarized and expanded by Dogiel (1964), who addressed both biotic and abiotic factors and their potential effects on parasite communities (his "parasitocoenoses"). Despite its age and lack of quantitative rigor, Dogiel's book, in our opinion, remains the best volume devoted to ecological parasitology.

Noble's (1960) paper on parasite communities (his "parasite-mix") in fishes, Holmes' (1961, 1962) papers on inter- and intraspecific competition in a parasitic guild, Schad's (1963) paper on niche diversification in a parasitic species flock, and Holmes' (1973) paper on site selection

The order of authorship for this chapter was determined by toss of coin and does not imply seniority.

are among the early papers that piqued interest in the study of parasitic communities in North America. Price's (1980) provocative book, *Evolutionary Biology of Parasites,* provided impetus for the growing interest in parasitic communities. Indeed, ecologists and parasitologists have begun to address such issues as temporal and spatial variation in parasite community richness (e.g., Rohde 1989; chapters in Esch, Bush, and Aho 1990).

Against this backdrop, synthetic theories of parasite community organization have emerged; their foundations are based largely on studies of free-living organisms. Holmes and Price (1986) identified a series of testable hypotheses that can be applied to parasite communities in host individuals, host populations, and host communities (which they consider to represent a hierarchical pattern of parasite communities). There have been explicit tests for some of these hypotheses, though they have been few in number. For example, Lotz and Font (1985), Bush and Holmes (1986b), Goater and Bush (1988), Stock and Holmes (1988), Rohde (1989 and references therein), and Moore and Simberloff (1990) have considered parasite communities with respect to an interactive versus isolationist continuum. Holmes and Price (1986) defined interactive communities as those comprised of species with high colonizing potential, leading to large populations and resulting in a high probability of biotic interactions; isolationist communities have the opposite attributes. Bush, Aho, and Kennedy (1990) have evaluated the hypothesis that diversity increases through evolutionary time, and Brooks (1988a and references therein) has pioneered the study of historical events in host-parasite systems. Wanntorp et al. (1990) emphasize the need to consider both ecological and phylogenetic analyses in interpreting community patterns.

In contemporary time, ecological processes can exert a major influence on patterns of parasite species richness. All else being equal, theory would predict that simple availability (the "supply-side ecology" of Lewin [1986]) would play a major role in the determination of richness. For example, we would expect aquatic-associated hosts to have more parasites (both species and individuals) because the majority of parasites require an aquatic environment for transmission. We would expect endotherms to have more parasites than ectotherms because the former must feed more frequently and intensively (and most parasites exploit food webs for transmission). We would expect migratory hosts, or those that are vagile (e.g., hosts move over small distances that encompass distinct habitats) to have more parasites simply because they are exposed to a greater diversity of habitats and therefore potential parasites. Kennedy, Bush, and Aho (1986) elaborate on these predictions. Alternatives to "supply-side ecology" are also known. For example, although endotherms may be expected to have more parasites for the reasons identified above, they also have a higher energy (and nutrient) budget and hence may be capable of supporting more parasites.

We acknowledge that most data are largely empirical, descriptive, scattered, and are not linked conceptually to higher-order processes (e.g., are without regard to the interaction of local and regional processes). However, a few very broad generalizations seem valid. It would appear that parasites of many invertebrates, birds, and marine fishes exhibit characteristics of interactive communities, whereas those of many freshwater fishes, amphibians and reptiles, and mammals show characteristics of isolationist communities (see chapters in Esch, Bush, and Aho 1990). Nonetheless, we wish to stress that to think one might identify a central, unifying theme for explaining patterns in parasite communities is unrealistic.

PARASITES AND COMMUNITY ECOLOGY

In many respects we believe that parasites provide powerful comparative systems for community analyses, largely because of their hierarchical nature. Here, we provide a simple overview of the hierarchy and conclude by showing where historical phylogenetic analyses rely on similarly nested levels. Holmes and Price (1986) and several chapters in Esch, Bush, and Aho (1990) expand on the hierarchical nature of parasite communities and elaborate on the utility of parasite communities for addressing questions in community ecology.

All parasites in a host individual, intermediate or final, define an *infracommunity* (Holmes and Price 1986). If interactions occur, this is where the scenario is played out. Even the presence of, and possible interactions between, adult parasites in definitive hosts may be mediated by events that transpire in infracommunities within intermediate hosts. Individuals of the same host species contain infracommunities that form natural replicates for comparative analyses. Infracommunities are unambiguous; they are uniquely defined by the host itself. We would emphasize, however, that they are incapable of self-perpetuation. As noted earlier, most parasites disperse their propagules into the free-living environment to undergo some development prior to exploiting trophic links to reach a definitive host and becoming adults. Whereas predatory effects are often implicated as being important in determining community patterns in many free-living systems, predation on adult parasites is unknown. (But we caution that the host's immunological response might provide a physiological analogue.)

All of the parasites in a given host species' population define the *component community* (Root 1973; Holmes and Price 1986). Whereas infracommunities have unambiguous boundaries, boundaries for the component community are less stringent and depend almost entirely on the spatial scale being considered. For example, one might consider the component parasite community of some species of fish as being all of the parasites in all of those individual fish throughout their geographical distribution. Alternatively, for migratory fishes, one might consider all of the parasites in all of the individual fish on their breeding grounds (or nonbreeding grounds). At a still finer scale, one might consider all of the parasites in all of the individuals of a species in a specific water body (e.g., a lake). Such a range of spatial scales makes words such as "local" and "regional" meaningless without further definition. For the purpose of the data that we present in this chapter, we consider *local* to mean the component community from a discrete body of water (or from a specific collection site of a large body of water), whereas we use *regional* to refer to all component communities throughout the host's geographical distribution (or at least that portion of its range for which we could obtain data). Despite these vastly different spatial scales, the idea of the component community is the same: it includes all of the parasites in a particular host species' population or some subset thereof. It is at this level that most investigations of parasite communities are conducted and that different spatial scales, including local and regional, can be investigated. It is also at this level that intentional dichotomies can be examined (e.g., parasites in young versus old, males versus females, etc.).

The final level is the *compound community* (Root 1973; Holmes and Price 1986), which includes all parasites, in all of their hosts, from a particular area. It suffers from the same potential ambiguity of spatial scales as does the component community. Because of the complexity of this level, it is seldom investigated. However, it is at this level that some important questions, such as to what extent parasite exchange and host specificity are important in structuring parasite communities, can be investigated.

It is at the component and compound community levels that historical questions have been meaningfully addressed. At the component level, host and parasite phylogenies can be compared in attempting to resolve cospeciation and host-switching events. By considering phylogenetic components with respect to both host and geographical distributions, the origin of compound communities can be investigated by performing historical biogeographical analysis. Both levels of historical analysis do require, however, that realistic phylogenies for hosts and parasites exist or can be derived. Brooks and McLennan (1991; chap. 24) elaborate on "cospeciation analysis" and provide a number of references to studies on historical analyses of parasite communities. For the remainder of this chapter, we will focus largely on contemporary patterns and proximate causes.

DETERMINANTS OF SPECIES RICHNESS

The search for patterns in species distribution is an important first step toward understanding community structure (e.g., MacArthur 1972; Brown and Gibson 1983). Nevertheless it is clear that such information, by itself, is insufficient to reassemble and interpret community dynamics, be they at the infracommunity, component, or compound level. Detailed local dynamics of species may be very complex. The interplay of local and regional dynamics can further exacerbate the problem. Though recent studies on species richness in parasite communities do exist, most address specific questions at local levels with no regard to putting the communities into perspective. Our aim will be to broaden this perspective by examining the relative importance of local versus regional processes in determining species richness in component communities of parasites. The patterns we address have been well documented for other organisms but have rarely been considered for parasites.

We begin with an overview by evaluating biogeographical patterns of parasite species richness in North American fishes. Our goal is to relate variation in the number of parasite species to the geographical range of the host species in order to evaluate whether increasing component community richness is simply a function of larger host geographical range, as might be expected if regional processes are more important than local processes. We then direct our analyses to specific hosts to determine whether local richness increases with regional richness, as would be expected in unsaturated communities (proportional sampling model of Cornell [chap. 22]), or whether local richness becomes asymptotic, indicative of saturated communities. We examine these patterns using original surveys on three species of *Micropterus* (Centrarchidae) and nine species of *Lepomis* (Centrarchidae) from North America.

Patterns in North American Fishes

Community ecologists must often explain why a particular site has more, or fewer, species than another. For symbiotic relationships, comparisons of the diversity of parasites can be made on a continuum of spatial scales ranging from local habitat patches to the geographical range occupied by the host species. It seems intuitive that hosts with large geographical ranges should support more parasites throughout the entire distributions than hosts with small isolated distributions. (By "more parasites" we mean regional richness, not local richness.) Many studies, particularly those on plant/herbivore systems (e.g., Lawton and Schroder 1977; Lawton and Price 1979; Strong 1979; Kennedy and Southwood 1984; Leather 1986) implicate the geographical area occupied by the host as being a prime determinant of richness on both local and regional spatial scales. On a regional level, determinants of richness have also been found to include such features as plant architectural complexity, taxonomic isolation, and the evolutionary age of the host plant (Strong, Lawton, and Southwood 1984).

Studies of the determinants of parasite diversity have lagged behind those focused on free-living systems, but biogeographical analyses have been receiving greater attention. Several recent studies on parasite component communities employ a biogeographical approach to evaluate the primacy of area in determining species richness. In a study of component communities in brown trout (*Salmo trutta*) from Britain, Kennedy (1978a) found lake area to be a major determinant of richness. In contrast, in a study of parasites of charr, *Salvelinus alpinus*, on Arctic islands (1978b), he found that island size (presumably larger islands would have more lakes) was unimportant in explaining species richness. Price and Clancy (1983) employed a regional scale in their investigation of why different species of British freshwater fish have different patterns of parasite richness. As in the studies in plant/herbivore systems, Price and Clancy found that some measure of host geographical range strongly influenced the total number of species per host (but see Guégan and Kennedy [in press] for a reanalysis and reinterpretation of these data). They also found that the feeding guild of the host was important. Gregory (1990) provided observations of parasite species richness in Holarctic waterfowl. Unlike previous studies, his study evaluated the effect of sample size and differences in taxonomic association on the species-area relationship. He emphasized the need to correct for different sample sizes and, when doing so, he found a robust relationship between the number of parasite species and host geographical range.

How general these trends are for other host/parasite systems is not known. In this section, we make an initial attempt to examine biogeographical patterns of parasite richness in North American freshwater fishes.

Parasite lists for different hosts were compiled from primary literature sources as well as from two checklists (Hoffman 1967; Margolis and Arthur 1979). From these sources, we found sufficient data to consider 87 species of fish representing six different families. (Our original data and references, including selection criteria and a list of species used, are available from the authors.) We examined three variables as a priori determinants of species richness: geographical area occupied by the host, adult host size, and host feeding guild (assigning fishes to one of five guilds using the criteria provided by Schlosser 1982). Data on feeding guilds were entered as numeric variables ranging from 1 to 5 (see Price and Clancy [1983] and Guégan and Kennedy [in press] for similar analyses). As others have suggested (e.g., Sugihara 1981) we found that a log/log or log/linear model provided the best fit to our data.

As in most previous studies on parasites (and other symbionts), host geographical range has a positive effect on regional species richness (fig. 17.1A). Host size and host feeding guild were also contributing factors in explaining parasite species richness (figs. 17.1B and 17.3). Comparing these single-factor analyses, geographical area accounts for a greater proportion of the variance than either size or feeding guild (table 17.1). Because some authors (e.g., Dogiel 1964; Esch et al. 1988) have argued that parasites with different colonizing strategies can influence species richness within communities, we also applied the same analyses to only the adult parasites (e.g., those that mature in fish). Again, all three variables show positive relationships with parasite species richness (figs. 17.2 and 17.3). In these analyses, however, host size explains a greater proportion of the variance than either area or feeding guild (table 17.1).

To determine the comparative contribution of the variables, we entered them into a stepwise multiple linear re-

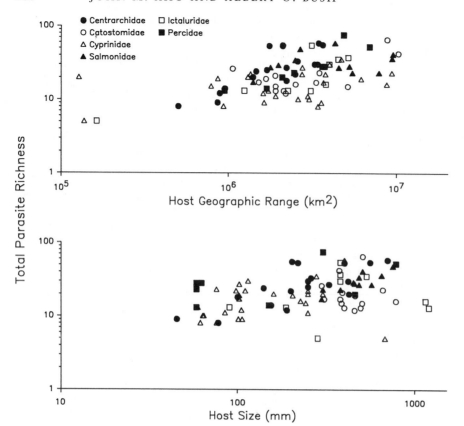

Figure 17.1 Relationships between total parasite richness (top) and geographical range and total parasite richness (bottom) and host size. Each point represents an individual species from one of six families of North American fishes.

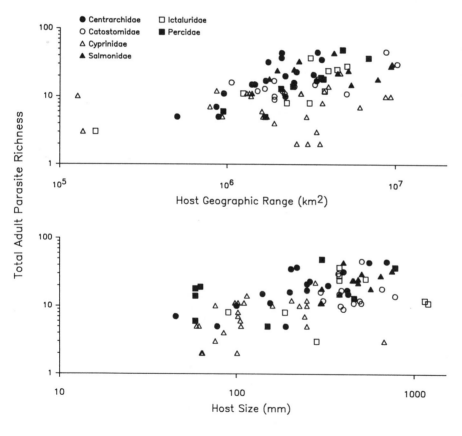

Figure 17.2 Relationships between adult parasite richness (top) and geographical range and adult parasite richness (bottom) and host size. Each point represents an individual species from one of six families of North American fishes.

gression. Considering all parasites, the three variables account for about 45% of the variance, with area being more important than size and feeding guild being least important (table 17.2). For adult parasites, the variables explain over 50% of the variance, and size explains more than twice the variance that area does (table 17.2). We conclude (as do Guégan and Kennedy [in press]), unlike other studies on parasites (e.g., Kennedy 1978a [who considered actual lake size as area] and Gregory 1990 [who considered host range as area]), that area alone is insufficient to account for different levels of species richness.

Our models explained only about half of the variance. With so much variance unexplained, we felt that the taxonomic relationships among hosts might be important in influencing species-area relationships. We thus tested for potential differences among the six families of North American freshwater fishes. Comparisons among fishes were made at the familial level because we had suitable data and because fundamental differences in life history characteristics can be recognized at that level. Using ANCOVA, we found significant differences in slopes among families, and in four families the richness/area regression had a coefficient of determination of at least 0.45 (table 17.3). For the Cyprinidae and Salmonidae there was no statistical significance for the species-area relationship, but they were retained in the covariate model for comparison with other families. We would emphasize that

Table 17.1. Percentage of Variance (r^2) for Linear Regression of Parasite Richness Using Host Geographical Area, Host Size, and Feeding Guild as Independent Variables

	Area	Size	Guild
Total richness	28.9[a]	15.4[a]	7.8[b]
Adult richness	20.5[a]	31.0[a]	10.9[b]

[a] $p < .001$
[b] $p < .01$

Table 17.2. Stepwise Regression Analyses of Determinants of Total and Adult Parasite Richness in North American Freshwater Fishes

	Total richness		Adult richness	
	Variance (%)	F-value	Variance (%)	F-value
Area	28.9	34.6[a]	15.4	24.1[a]
Size	11.1	15.6[a]	31.0	38.2[a]
Guild	4.7	7.1[b]	5.4	9.3[b]

[a] $p < .001$
[b] $p < .01$

Table 17.3. Analysis of Covariance Using Geographical Area and Familial Association as Independent Variables and Total Parasite Richness and Total Adult Richness as Dependent Variables

Total parasite richness[a]

Source of variation	df	SS	F-value	p
Family	5	0.69	4.43	.0014
Log Area	1	1.62	51.86	.0001
Log Area *Family	5	0.75	4.79	.0007
Catostomidae:	log richness = −1.61 + 0.45(area), r^2 = .39			
Centrarchidae:	log richness = −4.19 + 0.90(area), r^2 = .64			
Cyprinidae:	log richness = 0.21 + 0.15(area), r^2 = .11			
Ictaluridae:	log richness = −2.23 + 0.55(area), r^2 = .65			
Percidae:	log richness = −4.10 + 0.86(area), r^2 = .81			
Salmonidae:	log richness = 4.69 + 0.14(area), r^2 = .07			

Adult parasite richness[b]

Source of variation	df	SS	F-value	p
Family	5	1.06	4.49	.0012
Log Area	1	2.02	42.73	.0001
Log Area *Family	5	1.21	5.12	.0004
Catostomidae:	log richness = −1.84 + 0.47(area), r^2 = .49			
Centrarchidae:	log richness = −4.65 + 0.94(area), r^2 = .59			
Cyprinidae:	log richness = 0.39 + 0.06(area), r^2 = .01			
Ictaluridae:	log richness = −2.68 + 0.60(area), r^2 = .65			
Percidae:	log richness = −6.11 + 1.13(area), r^2 = .83			
Salmonidae:	log richness = 0.09 + 0.19(area), r^2 = .12			

[a] Overall model: df = 17, SS = 3.54, F = 4.4, p < .0001, and r^2 = .64.
[b] Overall model: df = 17, SS = 6.43, F = 9.4, p < .0001, and r^2 = .68.

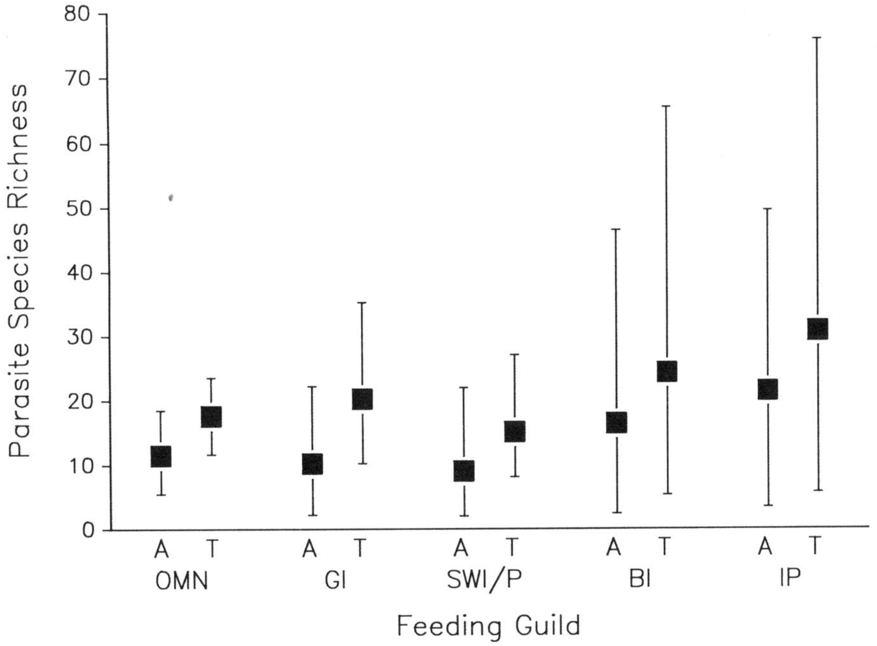

Figure 17.3 Relationship between host feeding guild and mean parasite richness in North American fishes. Vertical bars indicate the range of values for adult (*A*) and total (*T*) parasite richness in each guild. Fishes were assigned to guilds using the criteria of Schlosser (1982). *OMN*, omnivore; *GI*, generalist invertivore; *SWI/P*, surface water invertivore or planktivore; *BI*, benthic invertivore; *IP*, invertivore-piscivore.

these analyses are not independent of our earlier analyses and thus should be interpreted with caution. They do suggest that some measure(s) associated with taxonomic affiliation might bear consideration.

To make direct comparisons with patterns observed for parasites in British freshwater fishes, we reclassified the North American fishes into the same feeding guilds used by Price and Clancy (1983) and, following their example, applied linear rather than log/log regression. As in our previous results, area (for total richness) and both area and size (for adult richness) explained the greatest proportion of variance (table 17.4, columns 1 and 2). When we modified the data for British freshwater fishes (table 17.4, column 4) by eliminating monogenetic trematodes and lampreys, area still explained the greatest amount of variation, and guild was still significant, but host size also became significant. These differences between British and North American fishes may simply reflect a difference in the spatial scale over which the relationship is analyzed. Britain is small, and Price and Clancy found that a linear regression gave the best fit in 19 of 20 cases; when we used linear analyses on the North American data, explained variance for area decreased by 11%. There also may be variation due to differences in the diversity of the fish faunas. The British fauna is dominated by large-bodied cyprinids and salmonids (21 and 34 species selected from the British fauna by Price and Clancy) and is substantially less diverse than the North American fauna (we used data on 87 species).

Local versus Regional Richness in Black Bass and Lepomid Sunfishes

A major question arising from considering historical and geographical determinants of species richness is the importance of local versus regional processes. Terborgh and Faaborg (1980), Cornell (1985a, 1985b), and Ricklefs (1989b) argued that if local processes are important, local species richness should be asymptotic, whereas if regional processes are important, local richness should increase with regional richness. Studies invoking the importance of local processes in structuring lizard, bird, and mammal communities are common. The opposite seems true for most studies on herbivorous insects. For example, Lawton (1984b) and Lawton, Lewinsohn, and Compton (chap. 16) examined herbivores of bracken (*Pteridium aquilinum*) on several continents and argued that empty niches

were common and that local processes were unimportant. For parasite communities, there appears to be support for both conclusions. Aho (1990) examined the question using comparative data from parasite communities in five genera of amphibians. He observed either no significant change in local richness (four genera) or no consistent pattern of change (one genus) with increasing regional richness. While these findings may be taken as evidence for saturation of communities, he interpreted them as evidence for a pool exhaustion model (Lawton and Strong 1981). In contrast, Bush (1990) concluded that parasite communities in willets, *Catoptrophorus semipalmatus*, collected from several different wintering and breeding populations in North America were saturated and that local processes were important. He found essentially constant species richness in the different component communities (despite compositional differences); where significant differences in richness did exist, they could be attributed to parasite availability in specific habitats.

We began our investigation of the relationship between local and regional richness by using published reports on parasites in three species of black bass: Smallmouth (*Micropterus dolomieu*), Spotted (*M. punctulatus*), and largemouth (*M. salmoides*). We felt that bass would provide an appropriate test because they are very similar in ecology, they are relatively well studied (i.e., we could obtain a large number of studies from throughout their range), they have different but overlapping distributions, and their geographical ranges differ markedly in size (fig. 17.4). Following Ricklefs (1989b), we will accept the importance of regional processes if local richness increases in constant proportion with regional richness. Because we use original data and, for largemouth and smallmouth, have studies conducted throughout their original geographical ranges, we are able to describe their continental richness (rather than a potentially ambiguous "regional" richness).

Regional parasite species richness increased with increasing host range. Regional richness for largemouth bass was highest with 49 parasites, based on 24 studies conducted in 12 states or provinces; smallmouth bass had a total of 45 parasites, based on 17 studies conducted in 11 states or provinces; and spotted bass had a total of 24 parasites, based on 5 studies conducted in 3 states. Despite high regional richness, local richness for all three species never exceeded 50% of the regional pool (largemouth, 18 species; smallmouth, 21 species; and spotted, 12 species). Average parasite richness for all three host species was similar (largemouth, 9.1 ± 4.4; smallmouth, 11.3 ± 4.9; spotted, 10.2 ± 2.4). In figure 17.5, we plot local richness for each independent study against regional richness for the three host species. (In such a plot, a line with a slope of 1 would indicate the upper boundary for local richness.) ANOVA showed no significant difference among the host groups (total richness: $F_{2,43} = 1.21$, $p > .25$). Similar, but lower, values were found when only adult parasites were considered ($F_{2,43} = 2.88$, $p > .05$). However, as Cornell (1985a; chap. 22) notes, habitat specialization coupled with pool exhaustion could result in an asymptotic pattern, the reason being that be-

Table 17.4. Explained Variance (Percentage) for Determinants of Parasite Community Richness in North American and British Freshwater Fishes

	North America		Britain	
	Total Richness[a]	Adult Richness[a]	Price and Clancy	Price and Clancy[a]
Area	18[b]	17[b]	68[c]	66[c]
Size	8[c]	17[c]	8	14[d]
Guild	6[d]	9[c]	15[d]	17[d]

[a]Original data have been recalculated to make data sets as comparable as possible.
[b]$p < .001$
[c]$p < .01$
[d]$p < .05$

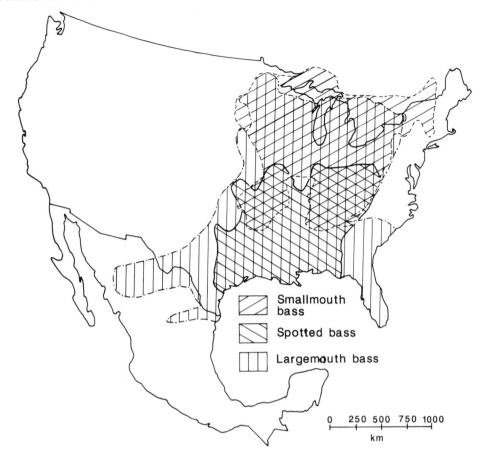

Figure 17.4 Native distributional ranges for three species of black bass. Geographical locations were determined from information provided in the *Atlas of North American Fishes* (Lee et al. 1980).

cause of specialization, regional richness may not reflect the true species pool. For our data, the parasites are generalists to centrarchid fishes, not specialists to any host species, and we therefore believe we have an accurate measure of the true species pool. We conclude from these data that local, and not regional, processes are most important in determining species richness in the component communities of bass.

Recognizing that these data reflect only three points, we extended our analyses to lepomid sunfishes to provide supporting evidence. We consider the lepomid data to be supporting rather than primary data because we could not obtain the breadth of coverage that we did for bass, and because the ecology of the different lepomid species is considerably more variable than is the ecology of the different species of bass. We obtained data on nine species of *Lepomis* (*L. auritus, L. cyanellus, L. gibbosus, L. gulosus, L. humilis, L. macrochirus, L. megalotis, L. microlophus,* and *L. punctatus*).

In contrast to the bass, for sunfishes there were apparent differences among host groups for both total parasite richness and total adult richness (fig. 17.6). We evaluated these plots using ANOVA with linear regression (more than one *y* for each *x*; Sokal and Rohlf 1981; 480). Although there were significant differences between host groups (which might be expected based on the different ecologies of the host species), the linear regressions were

not significant, and the deviation from regression was significant (table 17.5). Therefore, alpha diversity increases initially with gamma diversity, but then reaches a plateau and does not increase in constant proportion with regional richness.

CONCLUSIONS

Two caveats are important to our data sets. In attempting to explain species richness patterns in North American fishes (i.e., our analysis of parasite community richness and host geographical range for 87 species of fishes), we were forced to rely largely on data provided by checklists. There are two problems associated with using such data. First, checklists equate rare and common species, thus overemphasizing the potential importance of rare species at the regional level. Second, Gregory (1990; see also Kuris, Blaustein, and Alio 1980 for a similar argument) has recently contended that most studies ignore the importance of sampling effort. We attempted to minimize such effects by using only those species for which more than ten infracommunities had been examined in at least two different studies (species with small geographical ranges) and usually more than five studies (those with larger geographical ranges). Thus, we feel we have achieved a reasonable sampling effort and measure of re-

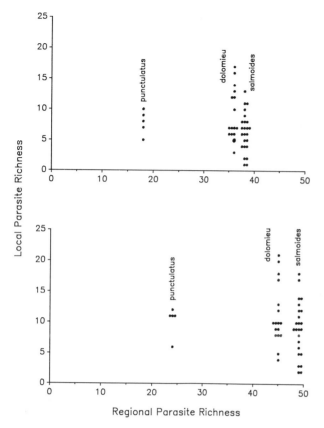

Figure 17.5   Plots of local richness versus regional richness in three species of bass. Plots are for adult parasites (top) and for all parasites (bottom). The upper boundary for local richness would have a slope = 1, indicating that local richness cannot exceed regional richness.

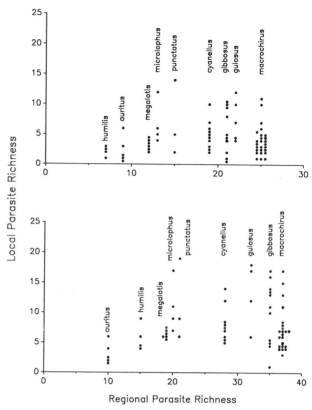

Figure 17.6   Plots of local richness versus regional richness in nine species of lepomid sunfish. Plots are for adult parasites (top) and for all parasites (bottom).

Table 17.5.   Analysis of Variance with Linear Regression on Local versus Regional Species Richness in Lepomid Sunfishes

Source of variation	df	SS	F-value
Total parasite richness			
Among groups	8	393.5	2.9[a]
Linear regression	1	43.0	0.9[b]
Deviation from regression	7	350.5	2.9[a]
Within groups	62	1039.0	

Source of variation	df	SS	F-value
Total adult richness			
Among groups	8	156.8	2.1[a]
Linear regression	1	8.4	0.4[b]
Deviation from regression	7	148.4	2.3[a]
Within groups	66	602.9	

[a]$p < .05$
[b]Not significant

gional diversity. Nevertheless, the caveat bears consideration.

A second caveat, one that transcends all our analyses, is that we relied on the original author's taxonomy. We do not perceive this to be a serious problem where we are simply concerned with numbers of species at the local level (we assume that, whether identifying them or not, authors can at least *differentiate* taxa), but it may be a more serious concern when the identity of the parasites is necessary. An associated problem might be that of "lump-

ers" and "splitters." Such a problem would affect all our analyses even when specific identities were not required. Because all groups have been examined by a variety of investigators, we assume that we have not introduced a directional bias into our analyses.

Given these caveats, we recognize our conclusions to be provisional and subject to change when more suitable data can be acquired (e.g., when more detailed systematic studies, more autecological studies, and more studies on parasitic communities are available). In particular, we need more studies that will allow us to meaningfully compare saturation versus proportional sampling. Such studies have been pioneered by students of phytophagous insects (e.g., Cornell 1985a, 1985b) and provide powerful insight into understanding community richness in such organisms (Cornell, chap. 22). Nonetheless, the general patterns we identify seem valid and should provide an appropriate departure point for future studies on parasitic communities.

Area (of either a physically discrete entity such as a lake or the geographical range of a species) cannot be disputed as an important variate in determining regional species richness, whether the communities are those of parasites, plant herbivores, or free-living organisms. Still, our data suggest that primacy should not be assigned to area, as we found host size and feeding guild to be significant factors. Such associations of variables are likely to be complex. Our interpretation is that both size and feeding guild are transmission phenomena, more so than is simple area. Large fishes are not born large. Many large species exhibit an ontogenetic change in food habits during development. Consequently, they have a greater opportunity to accumulate parasites than do smaller fishes, which often are restricted to a single feeding guild throughout their lives. When we considered only those parasites that mature in fishes, host size was much more important than host area. Why this is so is not clear. It may again be related to greater exposure (since larger fishes tend to pass through more feeding guilds and to ingest greater quantities of potential intermediate hosts), or it may be a measure of available size (resource base?), or another, as yet unexamined, variable. Our data do not allow us to address this further.

We also found a relationship associated with the taxonomic affiliation of the hosts. In our covariate analysis, we noted that four of the six host families included in our data exhibited significant and positive relationships with parasite species richness and area. Many parasites exhibit familial specificity (e.g., caecincolid trematodes in centrarchids, caryophyllaeid cestodes in catostomids), whereas others are generalists and may occur in several families. Different host families may have different regional pools of parasites (specialists and generalists) available. Taxonomic isolation, in conjuction with host distribution, may therefore be an important index of parasite species richness for any given host species. Exactly what might contribute to this pattern is unclear, but it probably has ecological (e.g., host capture), evolutionary (e.g., phylogenetic specificity), and coevolutionary (e.g., cospeciation) explanations.

When we took a more detailed look at three species of bass (using original rather than checklist data), we found no evidence that local richness responded to regional richness. Regional richness was high, but each local component community had only a small subset of that regional richness. Each local component community nevertheless had similar numbers of species. Asymptotic patterns suggesting community saturation have been observed in a variety of terrestrial (Cody 1966a; Terborgh and Faaborg 1980) and marine (Bohnsack and Talbot 1980; Abele 1984) systems. Many explanations have stressed the importance of interspecific interactions and/or limits to the size of niche space in a habitat. We do not interpret our data as supporting such a conclusion. A saturation pattern is insufficient evidence that species interactions limit local richness within a given habitat. We see only two common ($> 67\%$ prevalence) parasite species at the component community level in bass; the identity of the other species varies considerably. Beta diversity is high, but since parasite species composition is highly unpredictable, it is difficult to engender interactive processes as an explanation of constancy in local richness. We do see local processes (e.g., environmental factors) being more important than regional processes, and would suggest that bass have a pool of parasites from which they can draw. Just what

they do acquire is likely a function of who is present at the local level (e.g., supply-side ecology). Supply can be influenced by both biotic (e.g., a definitive host defecating and depositing parasite eggs in one pond but not in an adjacent pond) and/or abiotic (e.g., the lack of a molluscan fauna [and therefore digenetic trematodes] in a body of water due to acid rain) processes. Esch et al. (1988) have used such an argument in trying to explain the distribution of adult and larval stages of parasites in fishes from Great Britain.

For sunfishes, we see an overall pattern similar to that in bass (i.e., species richness becomes asymptotic despite increasing regional richness), and our interpretation of the pattern is the same. The parasites found in the sunfishes are largely the same centrarchid generalists that infect bass. However, unlike the bass, the sunfishes vary considerably in their ecology and morphology, and this alone may explain the significant differences in species richness between the groups.

In summary, we would argue that parasite component communities in the fishes that we examined can simultaneously provide evidence for the importance of both contemporary and historical events. We see no consistent evidence supporting either regional or local processes as being of overriding importance in determining species richness. For example, the contemporary patterns are superimposed on a historical component because of glaciation and reinvasion. The importance of the regional component is probably influenced by differential dispersal abilities of fish hosts, intermediate hosts, and parasites as well as by the degree of isolation of different water bodies. What is most evident is a need for considerably more data on the *autecology* of the individual parasites if we are to interpret the contributions made by historical and ecological factors in contemporary patterns of community richness. Brooks (1985; see also Brooks and McLennan, chap. 24) provides a similar view for parasites in freshwater stingrays from six river areas in South America. He not only found evidence to suggest that historical events could explain species richness in some areas, but also provides evidence that richness in at least one area was a function of local processes. Though few parasitologists have addressed their data with specific reference to local versus regional processes, a number have nonetheless done so in principle. For example, there are many studies that address component communities in oligotrophic and eutrophic aquatic systems (e.g., Wisniewski 1958; Chubb 1970; Esch 1971; Leong and Holmes 1981) that result from the interaction of regional and local processes. Unfortunately, such papers are published in journals largely peripheral to community ecology; their existence does imply that many students of free-living systems might benefit from considering parasitic systems. Finally, our inability to support consistently either local or regional processes as the primary determinant of species richness is not unexpected—we argued earlier that pluralistic views of parasitic community structure are required.

18

Evidence for the Influence of Historical Processes in Co-occurrence and Diversity of Tiger Beetle Species

David L. Pearson and Steven A. Juliano

Species co-occurrence and diversity of species are likely to be influenced by history (Endler 1982; Hilborn and Stearns 1982; Quinn and Dunham 1983; Brooks 1985; Dunham and Miles 1985; Dobson 1985; Pearson 1986; Ricklefs 1987; Jackson and Harvey 1989; Hughes 1989), but ecologists have largely ignored history in testing hypotheses. We define history as the integration of events over a long period that results in the accumulation of inherited species traits that affect co-occurrence and diversity. Contemporary environmental factors alone cannot always adequately explain the presence of patterns in these traits. Ideally the influences of history and ecology can be distinguished by tests that can account for traits inherited from ancestral populations (Mooi et al. 1989; Gould and Woodruff 1990). A more realistic likelihood is that history and current ecological factors will often interact to cause patterns of co-occurrence and diversity (Pearson 1977, 1982; Pearson et al. 1988; Altaba, 1991).

The goal of this chapter is to evaluate the relative importance of ecological and historical components in patterns of co-occurrence and biodiversity. Hypotheses will be tested by comparing species assemblages within and between geographical areas and habitat types.

TIGER BEETLES AS A TEST SYSTEM

An important first step in resolving the relative importance of ecology and history is the selection of an appropriate group of organisms to test pertinent hypotheses. The ideal test group should: (1) be well known taxonomically; (2) exhibit intersite variation in diversity, but have a total number of species small enough to make relatively short-term research projects feasible; (3) be readily observable; and (4) occupy relatively simple habitats that can be easily compared in different parts of the world.

Tiger beetles (Coleoptera: Cicindelidae) have many characteristics that meet these requirements for an ideal test organism. Taxonomically they are well known, with over 2000 species worldwide. All are predaceous on small arthropods, and all share a similar larval and adult body form. They are generally diurnal and are found primarily on soil surfaces from alpine meadows to tropical forest floors (Pearson 1988). An intensive survey of a single site,

even in tropical rainforests, can reliably find more than 85% of the species in 30–40 person-hours. Finally, many of the habitats in which tiger beetles occur are physically and ecologically uncomplicated. Saline flats, sand beaches, water edges, and alpine meadows each often share similar thermal properties, sight distances for predators, and prey abundances no matter where they occur in the world (Willis 1972; Hori 1982; Ganeshaiah and Belavadi 1986; Pearson 1988).

TESTS OF DIVERSITY HYPOTHESES POSSIBLE WITH TIGER BEETLES

Discerning the causes of patterns of co-occurrence is essential for a full understanding of biodiversity. The presence of a limiting resource may be one factor important for determining how many species can co-occur on a site. Mean mandible length (chord) is directly related to mean size of prey captured by tiger beetles throughout the world (Pearson and Mury 1979; Pearson 1980; Roer 1984; Ganeshaiah and Belavadi 1986). Usually, prey are either a limiting resource (Hori 1982; Pearson and Knisley 1985), or prey availability interacts with predator pressures (Pearson 1985), energetic needs (May, Pearson, and Casey 1986), and physical factors (Schultz and Hadley 1987; Pearson and Lederhouse 1987) to determine where each species can occur.

We address three specific hypotheses using tiger beetles. First, if historical (phylogenetic) effects on diversity are important, and if typical sizes of species within lineages differ on different continents, then assemblages in the same habitat on different continents should have different size distributions. In the most extreme scenario, typical sizes of species in all habitats may differ among continents due to limited availability of some size classes in the fauna of each continent. Alternatively, if local ecological processes have important effects on diversity, some degree of intercontinental convergence within habitats is predicted, and there should be little difference among continents in sizes of species in each habitat. A corollary is that if history (phylogeny) has important effects on co-occurrence and diversity, size of species within each lineage should be a conservative character. If, however, species sizes within a lineage are frequently and readily modified by selection, local ecological processes are likely more important.

Second, because soil desiccation and its association with temperature are critical for survival in larval and adult stages of these ectothermic beetles (Pearson and Lederhouse 1987; Hadley et al. 1990), annual precipitation is one readily measured ecological factor that may influence regional species distribution and in turn local diversity patterns (Terborgh and Faaborg 1980).

Finally, because different regions may contain different types and numbers of habitats, comparisons of relations between local diversity and regional diversity may only demonstrate that some parts of geographical areas have more species than others due to habitat or environmental differences. If local diversity within the same habitat on different continents is similar regardless of the continental diversity of species using that habitat, ecological factors are likely to be important. However, if the local diversity in the same habitat varies from continent to continent and is related to the continental diversity of species in that habitat, historical factors are indicated.

METHODS AND MATERIALS

Size Patterns for Different Continents and Lineages

We tested the prediction of strong intercontinental differences in size using data on mandible sizes for 17 well-studied assemblages from the Indian subcontinent and North America (Pearson and Juliano 1991). Four distinctive habitat types (open forest, pond edge, grassland, and sand) were all represented on both continents. Mandible lengths and lineage assignments were available for 53 species in these habitats (see Pearson and Juliano 1991). We performed two-way analysis of variance on mandible lengths and tested for effects of continent, habitat, and interaction. We were primarily interested in the effect of continent and whether any differences in mandible length between continents were consistent within habitat types (the interaction effect).

To test for body/mandible size constancy within phylogenetic lineages of tiger beetle species, we expanded the data base and examined all the species (a total of 413 species) known from each of the 29 lineages (genera and subgenera) of phylogenetically related species represented by the 53 species (Pearson and Juliano 1991) from which the calculations for the previous analysis of variance were made. These lineages included species from Neotropical, Australian, Nearctic, Oriental, Palearctic, and Ethiopian regions and additional habitat types. The lineages are per Rivalier (1954, 1963, 1969), Freitag (1979), Wiesner (1988), and Acciavatti and Pearson (1989). We placed each of these species within one of three size categories based on adult mandible length (chord): small (< 2 mm), medium (2–3 mm), or large (> 3 mm)(see table 18.2). These categories made allowance for variation caused by sexual dimorphism, and minor geographical differences, as well as local variation in adult size caused by differential feeding levels as larvae (Pearson and Knisley 1985). We then tested the null hypothesis that size class was independent of lineage using a likelihood ratio X^2 test. If size class is strongly associated with lineage, then size is an evolutionarily conservative character, and local selective forces have caused little modification of mandible length. Any constraints on sizes of co-occurring species, such as that due to competitive exclusion, are expressed by selective elimination of invaders that are the "wrong" size rather than by directional selection for evolution of size divergence. Our frequency data included many zero frequencies, and these were replaced for analysis by small positive values (1×10^{-20}) as recommended by SAS Institute, Inc. (1987) and Bishop, Fienberg, and Holland (1975).

Local and Grid Diversity versus Precipitation

Tests of the second hypothesis were based on regional surveys of terrestrial, diurnally active tiger beetle species from North America (Cazier 1954; Wallis 1961; Willis 1972), Australia (Freitag 1979; Sumlin 1984), and the Indian subcontinent (Acciavatti and Pearson 1989; Pearson and Ghorpade 1989). To quantify regional distributions and range comparisons, a grid of squares, each 275×275 km (350×350 km for Australia because of less intensive collections), was laid out randomly on a map of each of these areas (figs. 18.1–18.4). If a species was recorded in two noncontiguous squares, and if the lack of records from the intervening areas was most likely due to the absence of collections in appropriate habitat, the species was considered to be present in the intervening area as long as the disruption was no more than a single square. The number of species occupying each square was then tallied to determine a grid diversity. Local diversity was estimated by averaging the number of co-occurring species across all of the known assemblages of tiger beetle species within each square. Any grid square with three or fewer species was excluded from this analysis, as local diversity and grid diversity will be highly dependent in such squares.

To incorporate the relationship of an obvious abiotic ecological factor, precipitation, and species diversity (Gentry 1982; Pearson and Ghorpade 1989), we used general linear models. Local and grid diversity were the dependent variables, annual precipitation was the independent variable, and continent was a categorical variable. We tested for equality of slopes and for significant differences from zero for individual slopes. We then used partial correlation analysis to determine whether any relationship between pairs of variables from among local diversity, grid diversity, and precipitation was independent of the third variable. We also used the least squares means of precipitation in the three continental areas to provide a common and meaningful precipitation level against which the local and grid diversities could then be tested.

Cross-Continental Comparisons of Habitats

Three habitats commonly used by diurnal, terrestrial tiger beetles are floor of open forest (open canopy in deciduous to semideciduous forest), pond edge (muddy areas near standing water), and sandy river banks (moist sandy areas on the edge of or near running water courses). These three habitats occur within virtually all of the grid squares de-

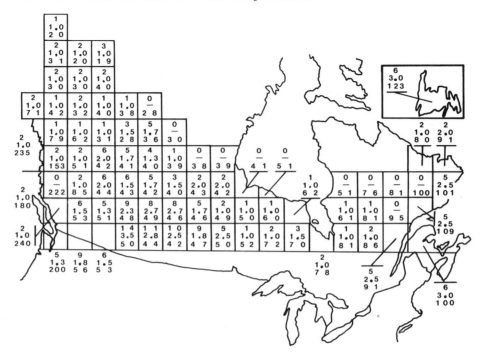

Figure 18.1 The number of terrestrial, diurnally active tiger beetle species and mean annual precipitation found within each square (275 × 275 km) of a grid covering the northern portion of the North American continent. For each square the top number is the total number of species known to occur (grid diversity), the middle number is the mean number of species found per assemblage (local diversity), and the bottom number is the mean annual precipitation (cm).

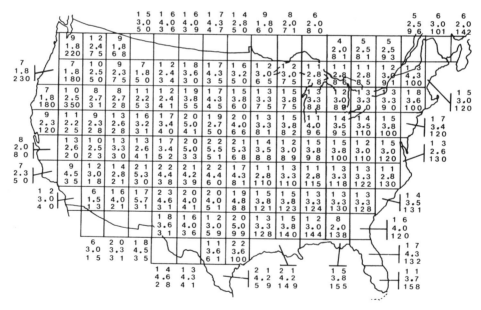

Figure 18.2 The number of terrestrial, diurnally active tiger beetle species and mean annual precipitation found within each square (275 × 275 km) of a grid covering the southern portion of the North American continent. For each square the top number is the total number of species known to occur (grid diversity), the middle number is the mean number of species found per assemblage (local diversity), and the bottom number is the mean annual precipitation (cm).

scribed for southern North America (fig. 18.2) and the Indian subcontinent (fig. 18.3).

We counted the number of co-occurring species found in each of these three habitats within each of the 131 grid squares for southern North America and the 61 squares for the Indian subcontinent. A *t*-test with unpooled standard deviations was used to compare the mean number of species per assemblage for each habitat within a square for each of the two continental areas. The total number of species found regularly in each of these habitats was then determined for each region and compared with the differences and similarities of mean number of species per habitat.

If history is a significant factor in determining local di-

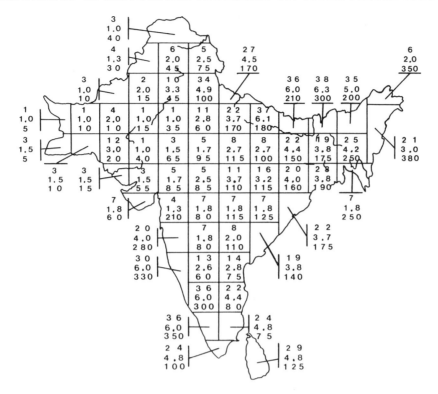

Figure 18.3 The number of terrestrial, diurnally active tiger beetle species and mean annual precipitation found within each square (275 × 275 km) of a grid covering the Indian subcontinent. For each square the top number is the total number of species known to occur (grid diversity), the middle number is the mean number of species found per assemblage (local diversity), and the bottom number is the mean annual precipitation (cm).

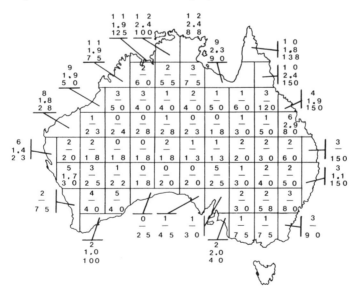

Figure 18.4 The number of terrestrial, diurnally active tiger beetle species and mean annual precipitation found within each square (350 × 350 km) of a grid covering the Australian continent. For each square the top number is the total number of species known to occur (grid diversity), the middle number is the mean number of species found per assemblage (local diversity), and the bottom number is the mean annual precipitation (cm). Because of the paucity of detailed data for much of the interior of the continent, local diversity figures are not available for many of the squares. The grid diversity figures, however, are considered reliable. No species of tiger beetles have been found on Tasmania.

versity, the region with the higher number of total species should also have a significantly higher mean number of species per assemblage in that habitat. If ecological factors are more important, the mean number of species within a habitat should be independent of the continental pool of species found in that habitat.

RESULTS

Intercontinental Size Differences and Lineage Effects on Size

Analysis of variance comparing mandible lengths of 53 Indian and North American species revealed significant effects of continent ($F_{1,45} = 12.52$, $P = .0009$) and habitat ($F_{3,45} = 5.36$, $P = .0031$), but no significant interaction effect ($F_{3,45} = 0.33$, $P = .8050$). Average mandible lengths thus differed between continents and among habitats, and the difference between continents was consistent. Across all four habitats, mandible lengths were longer for species from North America than for those from the Indian subcontinent (table 18.1). For both continents, least squares mean mandible length ($\pm$ SE) for open forest habitats (2.95 ± 0.15 mm) was significantly different from that for either grassland (2.12 ± 0.21) or pond edge (2.13 ± 0.19) habitats, which in turn did not differ from each other (overall $P = .05$). The least squares mean

for sand habitats (2.61 ± 0.15) was not significantly different from any of the other least squares means.

The species in this sample represented 20 lineages, with 9 lineages from North America and 13 lineages from the Indian subcontinent. Only one of the lineages (*Cylindera*) appeared on both continents. Thus, the species represented on each continent were primarily products of independent phylogenetic processes, and the differences in size are most likely a product of the different typical sizes from the available lineages.

When the sample size of species is expanded to include 29 lineages and additional habitats such as rainforest (table 18.2), the lineages represented from the Indian sub-

Table 18.1. Mandible Lengths of Tiger Beetle Species from Four Habitats in Two Regions

Continental region	Habitat				Overall least squares mean
	Grassland	Open forest	Pond edge	Sand	
India	1.90 ± 0.26	2.70 ± 0.39	1.79 ± 0.34	2.19 ± 0.15	2.14 ± 0.14
North America	2.30 ± 0.36	3.19 ± 0.06	2.48 ± 0.06	3.03 ± 0.25	2.76 ± 0.10

Note: Overall least squares means $\pm$ SE and within-habitat means (SAS Institute, Inc. 1987) of mandible lengths (mm) for representative tiger beetle species from the Indian subcontinent and North America occurring in each of four distinctive habitats.

Table 18.2. Numbers of Tiger Beetle Species in Three Mandible Length Categories

Genus (Subgenus)	No. species per mandible length class[a]			Habitat	Biogeographical range
	Small	Medium	Large		
Odontocheila (I)	5	2	1	Rainforest	Neotropical
Odontocheila (III)	—	4	—	Rainforest	Neotropical
Odontocheila (IV)	—	7	—	Rainforest	Neotropical
Odontocheila (V)	1	5	2	Rainforest	Neotropical
Pentacomia (II)	5	1	—	Rainforest	Neotropical
Pentacomia (III)	7	—	—	Rainforest	Neotropical
Therates (IV)	2	—	—	Rainforest	Australian
Therates (VIII)	—	2	—	Rainforest	Oriental
Therates (X)	—	6	—	Rainforest	Australian
Cicindela (III)	—	5	—	Sandy areas	Nearctic
Cicindela (VII)	—	12	2	Grassland to forest	Nearctic
(*Ancylia*)	—	—	5	Scrub forest	Oriental
(*Pancallia*)	—	—	4	Scrub forest	Oriental
(*Habroscelimorpha*)	1	13	—	Saline areas	Nearctic
(*Eunota*)	—	1	—	Salt flat	Nearctic
(*Pachydela*)	—	2	—	Sandy areas	Nearctic
(*Tribonia*)	—	5	—	Sandy areas	Nearctic
(*Cicindelidia*)	6	44	3	Mud flat to grassland	Nearctic
(*Cylindera*)[b]	38	2	2	Grassland to scrub	Palearctic, Ethiopian
	8	—	1		Nearctic, Neotropical
(*Cicindina*)	14	6	—	Mud flat to grassland	Palearctic, Oriental
(*Myriochile*)	—	17	—	Mud flat	Palearctic, Oriental, Ethiopian
(*Ellipsoptera*)	—	13	—	Water edge	Nearctic
(*Ifasina*)	62	2	—	Forest floor	Oriental, Ethiopian
(*Eriodera*)	—	1	—	Rocky streams	Oriental
(*Lophyridia*)	—	22	—	Water edge	Oriental, Ethiopian, Palearctic
(*Lophyra*)	—	23	2	Sandy areas	Oriental, Ethiopian, Palearctic
(*Calochroa*)	—	—	18	Moist forest	Oriental
(*Cosmodela*)	—	—	15	Moist forest	Oriental
(*Jansenia*-II)	14	—	—	Forest floor	Oriental
Total	163	195	55		
GRAND TOTAL = 413					

Note: Total number of tiger beetle species in each of three mandible length categories, habitat type, and biogeographical distribution for each of 29 genera/subgenera included in this study. Higher taxa and lineages based primarily on genitalia (Rivalier 1954, 1963, 1969; Freitag 1979; Wiesner 1988; Acciavatti and Pearson 1989.
[a]Small, < 2mm; medium, 2–3 mm; large, > 3 mm
[b]Lineage with representative species in both the Old and New World

Table 18.3. Results of Regressions for Local Diversity and Grid Diversity versus Precipitation in Three Geographical Areas

Area	n	Intercept	Slope ± SE		P	Mean ± SE	
Local diversity							
North America	197	2.7729	a	−0.0001 ± 0.0002	n.s.	a	3.0699 ± 0.0850
India	60	1.6293	b	0.0012 ± 0.0002	0.0001	a	2.9883 ± 0.1702
Australia	16	1.8473	a	0.0001 ± 0.0007	n.s.	b	2.0464 ± 0.2902
Grid diversity							
North America	197	10.7539	a	−0.0006 ± 0.0009	n.s.	a	12.1563 ± 0.4841
India	60	3.5264	b	0.0088 ± 0.0008	0.0001	a	13.1608 ± 0.9790
Australia	66	0.6350	b	0.0045 ± 0.0019	0.0210	b	8.3114 ± 1.5105

Note: Slopes significantly different from 0 are indicated by P values when $P < 0.05$. Slopes preceded by letters (a, b) are significantly different at an overall alpha = 0.05.

continent are still predominantly composed of small and medium-sized species, whereas the lineages represented in North America are predominantly composed of medium-sized and large species (table 18.2) regardless of habitat. Mandible length is significantly heterogeneous across lineages ($X^2 = 609.26$, df = 56, $P < .0001$), indicating that within lineages (across all continents) size is a highly conservative character. Length of mandible for species in assemblages on different continents can thus be interpreted as a highly conservative character, and mandible length does not regularly change in response to local ecological selective pressures.

Local and Grid Diversity versus Precipitation

For both local diversity ($F_{2,211} = 9.82$, $P < .0001$) and grid diversity ($F_{2,215} = 19.89$, $P < .0001$), there were significant interactions between continental region and precipitation, indicating that the slopes of the regressions within regions differed (table 18.3). For both local diversity and grid diversity, only the Indian subcontinent showed a significant positive relationship with precipitation. For North America, local diversity and grid diversity were unrelated to precipitation. For both local diversity and grid diversity, the least squares means for Australia was significantly less than for North America and the Indian subcontinent, and the latter two did not differ significantly from each other (table 18.3). Thus at average precipitation across all three regions, local and grid diversities are significantly lower in Australia. For pairs with indistinguishable slopes (North America and Australia for local diversity, Indian subcontinent and Australia for grid diversity) a strong regional and presumably historical effect on species diversity that manifests itself at grid and local levels is indicated.

Partial correlations among all three variables (local diversity, grid diversity, precipitation) for the three continents revealed that observed relationships between local and grid diversity were independent of precipitation. However, the relationship between local diversity and precipitation for the Indian subcontinent disappeared when the mutual relationship of these variables to grid diversity was taken into account (table 18.4).

If all species of tiger beetles, including nocturnal and arboreal species, are included in the analysis, relative species diversity patterns are similar to those that include

Table 18.4. Correlations among Local Diversity, Grid Diversity, and Precipitation for Three Geographical Areas

	Local diversity	Grid diversity	Precipitation
North America			
Local diversity	—	.8241 (158, .0001)	−.0889 (158, .2267)
Grid diversity	.8228 (158, .0001)	—	−.1344 (158, .0922)
Precipitation	.0390 (158, .6280)	−.1084 (158, .1766)	—
India			
Local diversity	—	.9553 (47, .0001)	.5586 (47, .0001)
Grid diversity	.9355 (47, .0001)	—	.6197 (47, .0001)
Precipitation	−.1433 (47, .3420)	.3508 (47, .0168)	—
Australia			
Local diversity	—	.2371 (12, .4581)	.4636 (12, .1290)
Grid diversity	.1170 (12, .7319)	—	.4481 (12, .0817)
Precipitation	.4238 (12, .1940)	.2185 (12, .5187)	—

Note: Grid squares with three or fewer species are not included. For each correlation matrix, simple correlations are given above the diagonal, and partial correlations, removing the effect of the third variable, are given below the diagonal. Sample sizes and P values are given in parentheses (n, P).

only diurnally active terrestrial species across all three continental areas. We connected the squares with similar numbers of species to delineate isoclines of 5, 10, 15, 20, 30, 40, and 50 species of tiger beetles per square (figs. 18.5–18.7). The greatest absolute differences between numbers of species before and after inclusion of nocturnal and arboreal species was on the Indian subcontinent. Here species of the arboreal genera *Collyris*, *Neocollyris*, *Tricondyla*, and *Derocrania* made up most of the difference.

On the Australian continent the difference is made up largely by nocturnal species of the genus *Megacephala*. Species of this genus make up over 30% of the Australian fauna. These nocturnal species outnumber diurnal tiger beetle species in 77% of the squares and are the only tiger beetles present in 11% of the squares (W. D. Sumlin III, personal communication). In none of the squares for

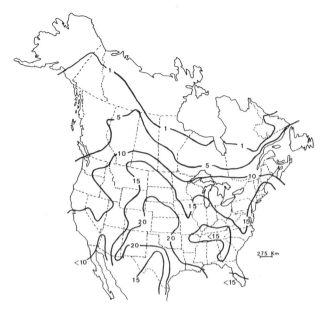

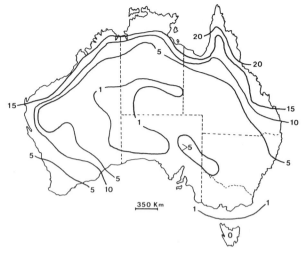

Figure 18.5 Isoclines drawn through the contiguous grid squares with an equal number of species of tiger beetles (including nocturnal, diurnal, terrestrial, and arboreal species for the North American continent. (From Pearson and Cassola, 1992; reprinted by permission of the Society for Conservation Biology and Blackwell Scientific Publication, Inc.)

Figure 18.7 Isoclines drawn through the contiguous grid squares with an equal number of species of tiger beetles (including nocturnal, diurnal, terrestrial, and arboreal species) for the Australian continent. (From Pearson and Cassola, 1992; reprinted by permission of the Society for Conservation Biology and Blackwell Scientific Publications, Inc.)

Cross-Continental Comparisons of Habitats

Of the 93 diurnal, terrestrial species occurring in North America, 45 (48%) were found regularly in one of the three habitats tested (open forest, pond edge, sandy river bank). Of the 161 species occurring on the Indian subcontinent, 71 (44%) were found in one of these three habitats.

For open forest and pond edge habitats, there was no significant difference in the mean number of co-occurring species between North America and the subcontinent (table 18.5). For the sandy river bank habitat, however, there was a significantly higher mean number of species on the subcontinent than on the North American continent. The continental pool of open forest species and sandy river bank species were both about twice that for the corresponding North American pools. The pool of pond edge species was about the same (15 and 12 species) for both continental regions.

Discussion

As tends to be the case in such comparisons (Clarke and Johnson 1990), few of our tests showed that either history or ecology alone was involved in the composition of co-occurrence and diversity. From our analyses, we concluded that for North America and the Indian subcontinent, historical processes are important determinants of co-occurrence and diversity that cannot be ignored. For tiger beetle systematists, the observation that size is a conservative character is not a revelation. Size has been considered a character state consistent within many tiger beetle species groups (Freitag and Barnes 1989). This is a pattern that ecologists have largely ignored. On the other hand, there were numerous exceptions to a relation between size and lineage that could be an indication of ecological processes.

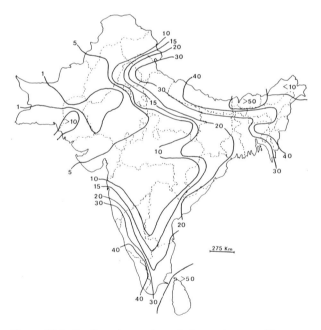

Figure 18.6 Isoclines drawn through the contiguous grid squares with an equal number of species of tiger beetles (including nocturnal, diurnal, terrestrial, and arboreal species) for the Indian subcontinent. (From Pearson and Cassola, 1992; reprinted by permission of the Society for Conservation Biology and Blackwell Scientific Publications, Inc.)

North America and the Indian subcontinent do nocturnal species approach this diversity. Nocturnal species make up less than 1% of the tiger beetle fauna of the Indian subcontinent and less than 6% of the North American fauna.

Table 18.5. Comparison of Local Diversity and Continental Species Pool of Diurnal, Terrestrial Tiger Beetles in Three Habitats in Two Geographical Areas

	Southern North America		India			
	Total no. species	$\bar{x}$ no. spp. /assemblage	Total no. species	$\bar{x}$ no. spp. /assemblage	t	P
Forest	18	2.4 (1.8)	37	2.9 (2.7)	−1.46	.15
Pond	15	2.3 (2.1)	12	2.6 (1.2)	−1.47	.14
River	12	2.1 (1.1)	22	3.7 (2.1)	−5.35	< .001

Note: Local diversity data are calculated from the number of co-occurring species in each habitat for 131 grid squares for North America and 61 for the Indian subcontinent (each square 275 × 275 km). Unpooled standard deviations (in parentheses) were used to calculate t values.

For Australia, the small sample sizes make any interpretation questionable. Local ecological processes appear to be generally of more importance there than in the other continental regions. The extreme unpredictability of precipitation over much of Australia, together with the extreme desiccation typical of most of this continent, may be of such overriding importance (Pianka 1989a) that it is inordinately significant in the patterns of co-occurrence seen in tiger beetles.

These extreme physical factors may have caused a general inhibition of dispersal of diurnal, terrestrial species, upon which we based our primary analysis. Australia is unusual in its large proportion of nocturnal tiger beetle species. Even when arboreal and nocturnal species are included, however, mean annual precipitation below 15 cm is associated with the lowest numbers of species in Australia (fig. 18.7) and in the Indian subcontinent (fig. 18.6), and with dramatic reversals of species number trends in the North American Southwest (fig. 18.5).

Temperature is a factor closely associated with moisture and thermoregulation. However, extreme temperature is another abiotic factor that can directly affect species distribution. Because larval tiger beetles avoid desiccation and temperature extremes by living in tunnels constructed in the substrate (Hadley et al. 1990), permafrost apparently makes it impossible for larval tiger beetles to descend sufficiently deep into the substrate to pass the winter months. The northern limit for tiger beetles as a group in North America (fig. 18.5) coincides with the southern limit of the climatic zone in which all otherwise appropriate habitats have shallow and continuous permafrost (Washburn 1980; Williams and Smith 1989).

It appears that in the Indian subcontinent, grid diversity is related to precipitation. Grid diversity is also related to local diversity. Because we used correlation analysis, we cannot determine whether these variables have a cause-and-effect relationship. However, if there is a causal connection among these three variables, it seems most logical that precipitation determines grid diversity, which in turn determines local diversity.

This interpretation suggests that at least on the subcontinent, historical processes, such as large-scale climatic patterns, influence grid diversity by influencing the ability of species to invade a region (Pearson and Ghorpade 1989). Grid diversity, in turn, influences local diversity by constraining the species available for colonization of a lo-

cal site. This pattern may generally apply to other regions, but perhaps because of the distinct geological history of an independent tectonic plate drifting across the Tethys Sea to collide with the Laurasian land mass (Pearson and Ghorpade 1989), the data are less ambiguous for the Indian subcontinent.

The relationship between continental diversity and the mean number of species co-occurring in each of three habitats suggests a mixture of ecological and historical influences. The sandy river bank habitat showed a positive relation between the continental species pool and the mean number of co-occurring species within a square—a result consistent with the interpretation that historical factors predominated. In the case of pond edge habitat, similar numbers of continental pools are associated with similar mean assemblage numbers, and these results are indeterminate. Contrastingly, the open forest species pool of the subcontinent was more than twice that of North America, but there was no significant difference in the mean number of co-occurring species within a square—a result consistent with an interpretation that ecological factors predominated.

As indicated by Pearson and Ghorpade (1989), this open forest habitat, especially in peninsular India, is dominated by a single lineage (*Jansenia*) within the genus *Cicindela*. Twenty-five of the 37 species included in the continental pool for this habitat are of this lineage, and all but 6 of its species occur in open forest. Phylogenetic, geological, and biogeographical evidence indicates that this lineage is an ancient one on the subcontinent. It is the only one that did not recently invade the subcontinent from the Asian mainland after the Indian plate contacted Laurasia in the Oligocene (Pearson and Ghorpade 1989). Antecedents of this lineage apparently drifted with the plate from its origin near the Madagascar and African plates. In this time more than 80% of the species have developed endemic ranges, in which many species replace each other from valley to valley. Although considerable controversy exists about the coevolution of species and how they enhance co-occurrence once they come into contact (Connell 1980; Diamond et al. 1989), tiger beetles apparently evolve most of their characters relatively slowly (Pearson et al. 1988). One interpretation is that ecological factors have had sufficient time to influence co-occurrence within this lineage and its primary habitat. Other lineages using habitats that lend themselves more to frequent dispersal, such as sandy river banks, still appear to be primarily under the influence of historical events. No parallel difference between ancient and recent lineages is known for diurnal North American tiger beetles.

Ideally, to assess the importance of phylogenetic factors (inheritance from common ancestors as a cause of the distribution of character states distinct from adaptations specific to terminal taxa), instances of loss or gain of character states should be counted and then tested for independence of ecological correlates (Felsenstein 1985; Coddington 1988; Donoghue 1989). The number of changes in the most parsimonious solution of the distribution of characters can then be compared with a robust phylogenetic tree and with a set of ecological parameters such as habitat type distributed among these species. Inspection of table 18.2 indicates that habitat type is also a conservative character strongly associated with lineage.

More sophisticated sampling techniques and experimental designs must be developed to distinguish the characters that are affected interactively by historical and ecological events. Tests of such patterns as recent arrival versus ancient persistence and overlapping of biogeographical tracks to produce local areas of high diversity are of importance in interpreting contemporary patterns of co-occurrence and biodiversity. They cannot be adequately tested with only ecological data. They must be placed in a historical perspective, and a phylogenetic tree is likely to be the most versatile tool in which to quantitatively couch this perspective.

Likewise, patterns of co-occurrence affected by predators (Pearson et al. 1988; Parmenter and MacMahon 1988; Mooi et al. 1989; Altaba 1991) and physical factors (Schultz, Quinlan, and Hadley 1991) should be tested against phylogenetic trees to distinguish ecological and historical effects on species co-occurrence and diversity.

The future goals of studies to test the influence of historical and ecological factors among tiger beetles are clear and feasible. They include: (1) development of a robust phylogenetic tree; (2) a survey of the entire world in which both regional and local diversity can be quantitatively compared (these data are now being gathered in Amazonian South America, Eurasia, and Africa); (3) a comparison of the local and regional diversity in many more than three habitats; and (4) a broader comparison of the influence of extreme physical factors and recent dispersal versus moderate physical factors and long-term presence of lineages.

SUMMARY

Assemblages of predaceous, adult tiger beetles from five distinct habitats in North America, Latin America, Asia, and Africa were used to test the influence of historical processes on patterns of co-occurrence and diversity. Body and mandible size of co-occurring species were found to be conservative characters more associated with phylogenetic lineage than with local ecological processes of evolutionary divergence.

We compared the total number of species within each of a gird of squares across North America, the Indian subcontinent, and Australia (grid diversity), the mean number of species per assemblage in each square (local diversity), and mean annual precipitation within each square. These comparisons indicated that large-scale climatic patterns probably affected grid diversity by influencing the ability of tiger beetle species to invade a region. Grid diversity in turn may have influenced local diversity by constraining the species available for colonization of a local site.

To assess the relative importance of history and ecology in determining patterns of diversity and co-occurrence among these tiger beetle species, we compared the mean numbers of co-occurring species in the same habitat type in North America and the Indian subcontinent. This local diversity was then compared with the size of the continental species pool available for this habitat. Of the three habitats compared, sandy river bank was interpreted as strongly under the influence of history, pond edge indeterminate, and open forest floor strongly under the influence of ecology, at least in the Indian subcontinent, where an ancient and endemic lineage of tiger beetles has undergone adaptive radiation and predominates in this habitat.

19

Pelagic Diversity Patterns

John A. McGowan and Patricia W. Walker

Understanding the origins and the maintenance of organic diversity are two of the most challenging intellectual problems in all of science. How there came into being so many kinds of living things is a question that has inspired theoreticians, experimentalists, and field observers for generations. But with the recent development of competition theory and its prediction of seemingly inexorable competitive exclusion, the question of how diversity is maintained has become equally important. This question is particularly puzzling in the open ocean, a large, old, moving, mixing, three-dimensional environment where there are no fixed substrates, no sessile species, and, therefore, no space limitations.

Lately it has become apparent that the study of species maintenance is of more than evolutionary or theoretical interest. The potential for species extinction rates far exceeding the natural ones is now a real threat due to human activities. One of the more serious results of these has been a change in the atmosphere itself due to human introduction of radiatively active gases. There are good reasons to believe that this atmospheric change will alter global climate patterns. Ecological theory predicts that disturbances such as these climatic anomalies will have consequences for the maintenance of diversity. However, the magnitude, rates, and even direction of diversity change remain unclear, as does the type or class of disturbance necessary to trigger the change. This state of ignorance about our natural world is not only unfortunate but potentially dangerous. To a large degree the structure and functions of pelagic ecosystems are dependent on the ocean's circulation. Changing climatic patterns will clearly affect this circulation.

In the surface waters dissolved CO_2 is generally in equilibrium with that in the atmosphere, but photosynthesis fixes much of it into plant carbon. The zooplankton grazers on the resulting fields of phytoplankton produce large quantities of particulate organic detritus, much of which falls out of the upper lighted zone and is consumed by microorganisms in the intermediate and deep waters. The amount of this respiration is so vast that subsurface waters, which are separated from the surface waters by sharp density gradients, are greatly oversaturated with CO_2 and undersaturated with O_2. When water from these depths is stirred up or upwells, it gives up much of its CO_2 to the atmosphere. If the upwelling is anomalous, as during a climatic perturbation, there could be many consequences for the carbon cycle. For example, CO_2 could be added to the atmosphere, thus enhancing the present anthropogenic excess (Keeling et al. 1989).

What has this carbon cycle to do with community diversity patterns? There are cogent and sound arguments that lead to the conclusion that the stability and efficient functioning of pelagic ecosystems, such as the phytoplankton-zooplankton-nekton-bacteria web, are tied to their species structure and are the product of a very long evolutionary history. If species loss and subsequent rearrangement of dominance hierarchies occurs in systems like these, the carbon cycle should, to some extent, be affected. Thus we are not only concerned with the maintenance of diversity as a concept, but also with the effects of loss of species on the stability of community structure and on functions such as the cycling of chemical compounds.

Pelagic communities are complex, and the processes that take place in them, such as the flow of energy and materials through webs, the cycling of essential plant nutrients, the sequestering of trace metals, and the flux of dissolved gases such as O_2 and CO_2 are clearly related to the kinds of different species present. For example, highly diverse phytoplankton communities are thought to be over ten times as efficient in recycling nutrients within the photic zone as are simpler assemblages (Eppley and Peterson 1979). We do not know if the high diversity is somehow the "cause" of this efficiency, but it seems to be clearly associated with it. Further, Michaels and Silver (1988) have shown in a simple model of oceanic food webs that "both the quantity and composition of sinking material are strongly determined by the community structure of the consumers."

Thus it seems prudent, given the nature of the threats to diversity, to review what we know (or rather what we don't know) about the mechanisms that maintain it in the world's largest ecosystem—and the world's largest reservoir of CO_2 and sink for pollutants—the open ocean.

SPATIAL PATTERN

The ocean's pelagic realm is vast, and it is appropriate to ask whether the study of community diversity problems can even be approached in such a huge area. Fortunately, through a combination of luck and forsight, a large number of appropriate samples of zooplankton and (later) phytoplankton were collected on far-ranging oceanographic expeditions in the 1950s and 1960s in the Pacific.

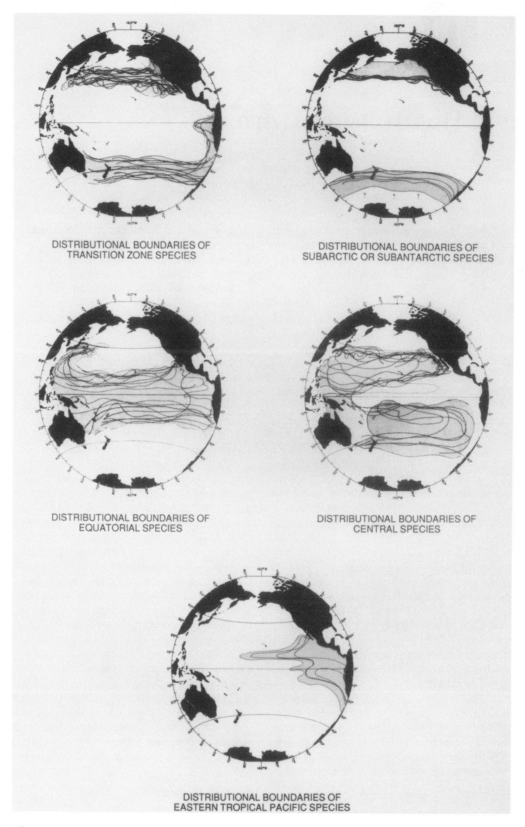

Figure 19.1 The major oceanic species provinces of the Pacific. The lines are species range boundaries for a variety of taxa of pelagic macrozooplankton. The shaded areas are the ranges. Phytoplankton and nekton species also fit this scheme. (From McGowan 1974.)

Further, most of the these were quantitative so that relative abundances could be determined. Taxonomic analysis was an early task so the question could be asked, how many species are present in the upper 200 meters of the Pacific Ocean? We can answer this question quite well for four major taxa, *Euphausiacea, Chaetognatha, Pteropoda,* and *Copepoda,* and reasonably well for quite a few others (McGowan 1974). The answer is that there are not very many species as compared to those on land. Although these four major taxa dominate the biomass of macrozooplankton everywhere, there are only about 80 species of euphausiids, 50 of chaetognaths, 40 or so of pteropods, and the most diverse group, calanoid copepods, number perhaps less than 2,000. Compare this with Amazonian beetles! It would be easy to dismiss this observation as due to lack of sampling (over 20,000 large, 300-m^3 net tows) or poor taxonomic effort, but there is evidence against this. The rate of new species descriptions has not increased with vastly increased sampling (McGowan 1974). Thus we see that, quantitatively at least, the oceans' diversity problems are of a different order than those on land.

Species range show repeatable spatial patterns (fig. 19.1). There are copepods whose range boundaries in the open ocean are sharp, and their ranges resemble those of euphasiids, chaetognaths, pteropods, thaliaceans, squid, and fishes. In other words, there are clearly evident assemblages or faunal groups of disparate species all of which are consistently present in large samples from the shaded areas, and consistently absent in the unshaded areas, of figure 19.1. These species ranges are all very large, so there are far fewer assemblage patterns than there are species. These patterns imply that there is a great deal of species co-occurrence, and such has been shown to be the

case. There are recurrent groups of species, the members of which occur together in the same sample more often than can be attributed to chance (Fager and McGowan 1963). These groups occupy (not surprisingly) the same large ranges that the composite maps of individual species show. Further, within the range of any one group there is a significant spatial concordance of abundance among species of that group (Fager and McGowan 1963). That is to say, there is an agreement among species on where to be abundant and where to be rare within their ranges. These and subsequent studies show that these large oceanic biogeographical provinces are complete, in situ, functional communities with all trophic levels present, in which immigration of allocthonous components is not an important factor (on this scale) contributing to structure and function.

There are distinct oceanic spatial patterns of species diversity, and these areas are large and few in number (fig. 19.2). However, these oceanic species gradients do not quite follow those established on land. The number of species per taxon is low at high latitudes, but rather than a regular, systematic increase equatorward, there is a sharp gradient at about 40° to 42° N. Diversity is high at mid-latitudes, but in the central and eastern Pacific it drops to intermediate levels in the equatorial zone at about 15° N to 15° S. Diversity increases once again at mid-latitudes in the South Pacific. There is a much broader gradient from the South Pacific maximum to the minimum near Antarctica than at the equivalent latitudes in the north. These patterns are similar for each of the taxa studied (euphausiids, pteropods, and chaetognaths). The lack of correspondence to the land-derived paradigm of steadily increasing diversity with decreasing latitude needs further documentation, but if found to be generally true for all or

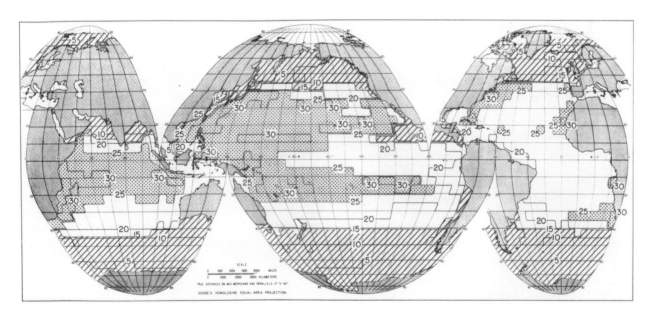

Figure 19.2   The number of species of pelagic euphausiids by five degree squares. North of 40° N in the Pacific there are only 5 or so species regularly present. Between 35° and 40° N, diversity increases rapidly until 30 or more species are found in the central gyres. In the eastern half of the tropical Pacific, especially near Central America,

diversity decreases. This meridional shift in diversity in the tropics and along the equator from low in the eastern half to higher in the west is due to the crossing of this zone by central species (see fig. 19.1). The South Pacific central gyre is about as diverse as the North. Other taxa follow these patterns (After Reid et al. 1978.)

most taxa in the pelagic realm, it complicates our theories of how species diversity arose and how it is maintained.

Why are these patterns the way they are? The correspondence between the shapes and sizes of these biotic patterns and those of the great gyrelike oceanic circulation and recirculation systems is unmistakable (fig. 19.3). These large cyclonic and anticyclonic gyres, as well as the circulation-recirculation system of the North Equatorial Current, the Equatorial Countercurrent, and the South Equatorial Current, have the unique and persistent physical-chemical properties necessary for the evolutionary development of true community ecosystems. In these vast ecosystems, in situ regulation (on a grand scale) is responsible for the state of each system, rather than the advective input or egress (immigration or emigration) of water with allocthonous properties, flora, or fauna. For example, the subarctic cyclonic gyre is a generalized upwelling system because of the direction of its rotation. It is therefore well supplied with nutrients from below. Because precipitation exceeds evaporation here, there is a low-salinity upper layer with a strong halocline. The weather is generally foggy and overcast, and winter storms are severe and frequent. The southern oceanic boundary of these conditions is about 40° N. South of this lies the great anticyclonic central gyre. Because of the central gyre's anticyclonic direction of rotation, it is a generalized downwelling system and it is therefore oligotrophic. Evaporation exceeds precipitation, so it is quite salty. Occasional storms occur in winters, but generally the weather is sunny and the winters are moderate. These two provinces have very different standing crops, productivities, and nutrient recycling regimes. The differences between them must have existed for a long time. As long as the poles have been relatively cooler than the equator, and as long as the earth has rotated in the same direction, and as long as there has been a Pacific basin, these large cyclonic and anticyclonic systems have existed; that is, there has been plenty of time for community evolution to have occurred.

There are three other recurrent groups of species with concordant ranges. South of the anticyclonic central gyre, between about 20° N and 20° S, lies the enormous equatorial circulation system. Although this is not a gyrelike system, there is continuity of flow because of the circulation-recirculation of the North Equatorial Current, the Equatorial Countercurrent, and the South Equatorial Current (fig. 19.3). The gross circulation in the equatorial province is, of course, quite different than that of the gyres in many ways—the band of equatorial upwelling being one of them. The eastern tropical Pacific also has its own assemblage (fig. 19.1). Here the Equatorial Countercurrent turns north and to the west, joining the North Equatorial Current. In doing so it creates a large cyclonic eddy: the Costa Rica Dome (fig. 19.3). There is intense oceanic upwelling here, as evidenced by the greatly undersaturated O_2 concentrations in the surface water, the cool euphotic zone temperatures, and the high nutrient concentration. The position and intensity of this large feature varies depending on the strength of the trade winds that drive the system. It is one of the few equatorial zone areas

with this intensity of upwelling and, consequently, sustains a high biomass (McGowan 1986).

Finally, there is a group of species that seem to be generalists in the warmer waters between 40° N and 40° S. These species are very seldom local dominants but rather seem to be opportunists, for on rare occasions they do bloom. Intense blooms occur over spatially limited areas, especially near the boundaries of their warm-water range.

The main gyrelike patterns, the subarctic gyre and the central gyre, are repeated in the South Pacific, where they share most of their species with their analogues in the north (fig. 19.1). The Southern Ocean with its circumglobal circulation has its own endemic low-diversity flora and fauna.

Thus oceanic species and diversity patterns are on a very large scale. They are clearly related to the large general features of circulation, which depend on climate. Smaller-scale circulation features, such as mesoscale eddies, rings, and small ocean basins, do not seem to have provided the degree of persistence of habitat features required for the coevolution of the diversity of species necessary for the establishment of functional ecosystems where the species structure and diversity are due to predictable in situ processes rather than advection, immigration, and emigration.

Most zooplankters are not strong swimmers, and their populations can be moved about by large-scale advective processes such as continental boundary currents or the stirring and mixing that occurs at the boundaries between biogeographical provinces. Such areas, or ecotones, have highly diverse flora and fauna that are mixtures of their source regions and are easily identifiable. The entire California Current is one such region where the diversity regimes seem to be a result of the physical movement of water rather than of evolution and species interaction (McGowan 1990; McGowan and Miller 1980).

DIVERSITY MAINTENANCE: NICHE SEPARATION

The central gyres of the North and South Pacific are the most diverse provinces of that great ocean. While temporal and spatial diversity patterns have been studied in both, it is the northern central gyre for which we have the most information. This province is an ideal system for such a study, for there is no evidence that allocthonous flora and fauna intrude into its interior. The spatial and temporal patterns of heterogeneity in density structures, plant nutrients, integrated water column plant biomass, integrated water column primary production, zooplankton biomass, and copepod species abundance and species rank order of abundance have been studied on spatial scales of sample separation ranging from less than 1 km to over 1,000 km and on temporal scales of from 1 day to over 12 years. Heterogeneity in environmental properties was low on all scales up to and including the mesoscale (about 500 km) as compared with other areas (Hayward, Venrick, and McGowan 1983). Even seasonal changes, except in water temperature, were difficult to detect. Because the gyre is so diverse, and because of the evidence for its being less patchy than other oceanic sys-

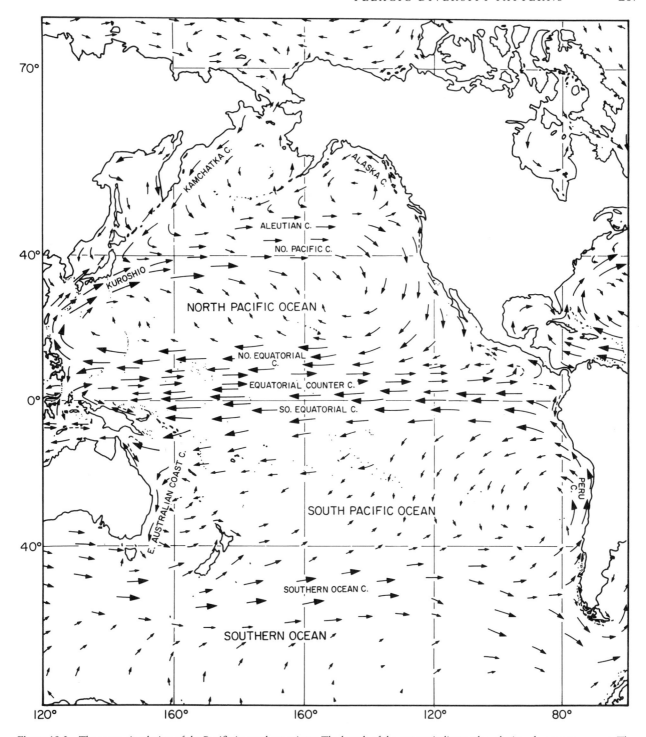

Figure 19.3 The mean circulation of the Pacific in northern winter. The length of the arrows indicates the velocity of water movement. The great cyclonic and anticyclonic gyre systems are clearly evident, as is the current-countercurrent recirculation system of the equator.

tems, the mechanisms for the maintenance of species diversity are not obvious. We have conducted studies of the two most promising theories of diversity maintenance here because of the strong evidence that in situ processes, rather than advective ones, are the main regulators of the state of the system, and because samples taken at the central region are likely to be representative of a much larger area due to the low level of spatial/temporal heterogeneity in all properties. The two theories tested were resource partitioning or niche separation and disturbance-perturbation or dispersal-reaction (McGowan and Walker 1979, 1985).

Although there is relatively little spatial heterogeneity, and although there are no obvious large-scale horizontal gradients, there are strong vertical gradients of almost all environmental properties, including temperature, light, plant nutrients, plant biomass, primary production, and zooplankton biomass. These vertical gradients are remarkably consistent over the seasons (except for temperature) and interannually. This situation provides an opportunity to test the theory that taxonomically or trophically related species partition resources by dividing "niche space" along resource gradients with a minimum amount of overlap between them and are thus able to coexist at the "same" geographical locale. That is, each tends to specialize on a fraction of the available resources, and because the resources vary vertically, the species mean abundances should be arrayed out in the vertical along the gradients (Whittaker 1975; May 1973).

To test this theory in the north central gyre, longreplicated series of vertically stratified samples of both phytoplankton and zooplankton were taken in different seasons and years. The intent was to find evidence for or against the idea that species abundance centers tend to be separated by depth. Such was not the case. The 317 species of phytoplankton present could be divided into only two distinct recurring associations, separated by a region of rapid transition near 100 m. There were 178 species in the shallow stratum and 139 in the deep stratum (Venrick 1982). Within strata they varied in their mean abundance by at least four orders of magnitude, with most of the species being rare. "Groups within each stratum show numerous intergroup affinities and considerable overlap of their vertical ranges" (Venrick 1982) (fig. 19.4). This spatial structure of species abundances occurred where there were steep vertical gradients of limiting resources—nitrate, phosphate, and light— in this oligotrophic province (Eppley et al. 1973; Hayward and McGowan 1985). The vertical distributions of 173 species of copepods were also determined over time and space, and on one occasion by following some drifting, current-following drogues for 10 days of continuous sampling at six depth ranges with paired nets. Here also the species fell into only a few groups in which each member had strongly overlapping vertical ranges with each other member (fig. 19.5). In this case there were seven separate categories, two of which were present only at night because their members were diel migrators (not shown in fig. 19.5). There were strong statistical concordances of abundance among the members within the five nonmigratory groups. This pattern held for both summer and winter data over the 3 years these vertical patterns were studied. There were strong gradients in resources over the depth range 0–600 m: plant biomass, microzooplankton, primary production, temperature, and density. Although entire recurrent groups of species were arrayed out along gradients, species within groups were not (McGowan and Walker 1979). Hayward (1980) examined the gut contents and times and places of feeding of a number of the more abundant co-occurring species. Using a "gut fullness" index, he found that there was agreement among species on what were "good" places and times to fill their guts. While some of the 173 species consistently present do seem to be specialized feeders judging from their mouthparts,

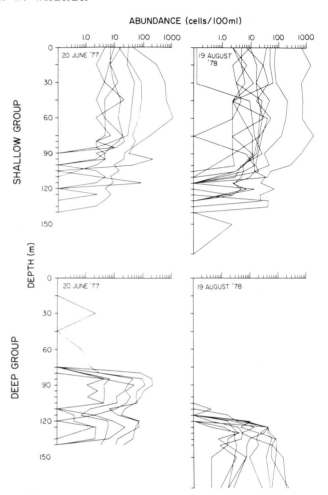

Figure 19.4 The vertical variations in abundance of phytoplankton species in two different years at the same locale in the north central gyre. There are only two associations, shallow and deep, and the same species occurred in the same associations in both years. There is strong overlap in vertical pattern within associations or groups and orders of magnitude differences in abundance within groups (From Venrick 1982.)

most are not. They appear to be omnivores taking small particles within a certain size range.

Thus the notion that there is a species equilibrium mediated by resource partitioning that can be detected by observing the manner in which species occupy spatial positions along multiple resource gradients is not supported by these studies, even though this ocean is quite oligotrophic and nutrients and food are limiting. Therefore we should expect intense competition among these very similar, co-occurring plant and animal species.

DIVERSITY MAINTENANCE: DISTURBANCES

Disturbance-perturbation, dispersal-reaction, and contemporaneous disequilibrium are similar theories used to explain the maintenance of diversity in communities where partitioning seems unlikely (Holling 1973; Levin 1974; Connell 1975, 1978; Yodzis 1978; Hubbell 1979; Paine and Levin 1981). This body of theory, observation,

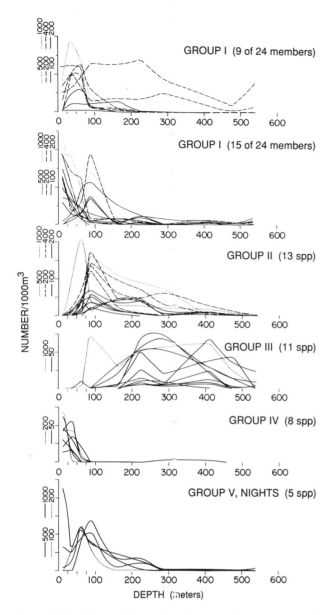

Figure 19.5 The vertical variations in mean abundance of north central gyre copepod species within recurrent groups. These patterns persisted, intact, over several years. There is strong overlap in mean abundance of copepod species within recurrent groups along vertical gradients of environmental properties (From McGowan and Walker 1979.)

All of these variants predict (some explicitly) that on some scale of patchiness there should be pronounced shifts or reshuffles of dominance structure, or, as Richerson, Armstrong, and Goldman (1970) put it, "at any one time many patches of water exist in which one species is at a competitive advantage relative to others." Therefore, by examining dominance hierarchies or the constancy of species' relative abundance on the appropriate scales of patchiness, we should be able to see these shifts in species structures. But what are the appropriate space and time scales? Since there were no good clues as to what these scales should be, a strategy was employed in which the species dominance structure of phytoplankton and (separately) copepods was studied in samples separated in time by about a half hour to over a decade and separated in space by a few hundred meters to over 800 km. The longest time scale represents well over 1000 generations of phytoplankton and many hundreds of copepods— enough time for competitive displacement and change of dominance structure to have occurred (McGowan and Walker 1985; Venrick 1990). In neither the phytoplankton nor the copepods were there significant changes in the rank order of species abundance (i.e., hierarchy) on temporal scales of sample separation up to and including 3 years for the shallow flora or on any up to 10 years for the copepods (figs. 19.6 and 19.7). Nor were copepod changes seen on spatial scales of meters to over 700 km. The overall mean levels of relative abundance in both phytoplankton and copepods showed that co-occurring species varied in abundance by at least five orders of magnitude (fig. 19.8). But over very many generations in both cases, the rare species stay rare and the abundant species stay abundant. There were overall statistical concordances of abundance among copepod species between cruises. That is, on a cruise-to-cruise and interannual basis, abundances of most species' populations tended to rise and fall in unison, thus preserving the rank order of abundance. This long-term persistence of rank order of abundance of species within two very different trophic levels on all relevant scales of sample separation in space and time is not what theory predicts.

Although we have not been able to validate the two most popular theories of diversity maintenance, we did acquire much descriptive information on species structure, and this information raises questions about why community structure and function are the way they are. These questions are especially acute as they relate to the role of persistently rare species. In both the phytoplankton and the copepods, most of the high level of diversity found in the central gyre is due to the presence of many rare or very rare species (fig. 19.6–19.8). Other studies, based on somewhat fewer samples, show this to be also true of adult meso-pelagic fishes and their larvae, chaetognaths, and amphipods (Barnett 1983; Loeb 1979, 1980; Lyons 1976; Shulenberger 1977). In the case of the phytoplankton and copepods, the rare and very rare species show no special characteristics (as, for example, spatial position or diel migration) that set them apart morphologically or behaviorally from the others. Some of the very rare species are congeners with the most abun-

and experiment basically states that although there is competition for resources, and although there are species present capable of outcompeting their neighbors, they are prevented from doing so. Thus competition does not lead to competitive exclusion and a reduction of diversity in the system. The mechanisms behind this result are patchy, episodic environmental disturbances that cause enough density-independent mortality to prevent the competitive dominants from winning out in general. Levin (1974), Yodzis (1978), and Paine and Levin (1981) have introduced a variant of this theory in which migration from patch to patch of propagules of varying competitive abilities can alter dominance structures and maintain diversity.

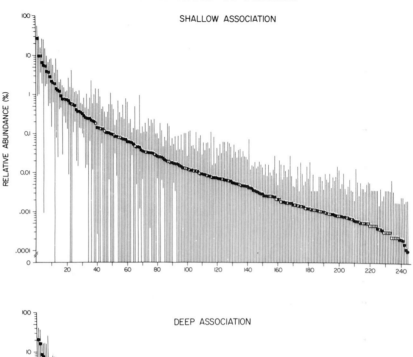

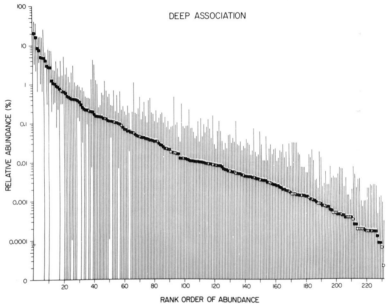

RANK ORDER OF ABUNDANCE

Figure 19.6 The mean rank order of north central gyre phyto-plankton species abundance (dots) over 1973 to 1985. The vertical bars are the overall range in abundance based on samples of 265 ml. Some species were missing from at least one sample, but there are orders of magnitude differences in mean abundance; most of the "rare" species (those making up less than 0.1% of all individuals) never became more abundant, and most of the "abundant" species (those making up more than 0.5%) never became "rare". (From Venrick 1990.)

dant ones and co-occur with them within the same recurrent group. Further, these rare species do not appear to be more "fragile" or subject to extinction due to environmental change than the numerical dominants. For example, as our sample separations approached 500–600 km distant from the center of study area (28° N, 155° W) there tended to be gradual environmental changes from the mean hydrographic conditions found at the centrum, until finally at some 1400 km distant, the major water mass and circulation outer boundary of the central anticyclonic gyre was reached (about 41° N). Our samples showed that as we approached this outer periphery of the gyre, some (central gyre) species tended to drop out of

the system. But there was no trend for these to be either the rare or the abundant ones, nor was there evidence of significant shifts in rank orders of abundance. The surviving dominants were in about the same hierarchical position as they were farther south, and the rare species for the most part maintained their status as well (McGowan and Walker 1985). Perhaps with further study of the life histories and physiology of the dropouts we can gain some idea of the attributes that make a species more vulnerable to extinction with changing environment, but as yet there appears to be no particular pattern related to rarity or abundance.

The central gyre (and perhaps other provinces) is

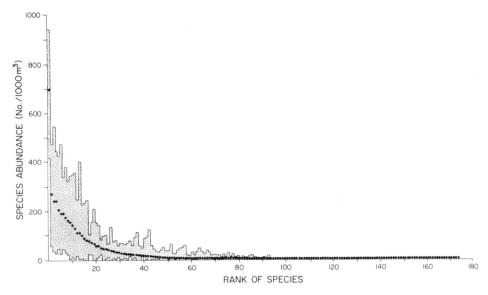

Figure 19.7 The mean rank order of abundance of north central gyre copepod species abundance (dots) over 7 years of observation. The vertical bars are the ranges of individual cruise mean abundances per species. Only 20 or so species were ever more than 1% of the total abundance of all copepods. There are many persistently rare and very rare species in this area, where there are about 175 species regularly present (Spearman rank correlation coefficient of order, positive $P < .001$ for all). (From McGowan and Walker 1985.)

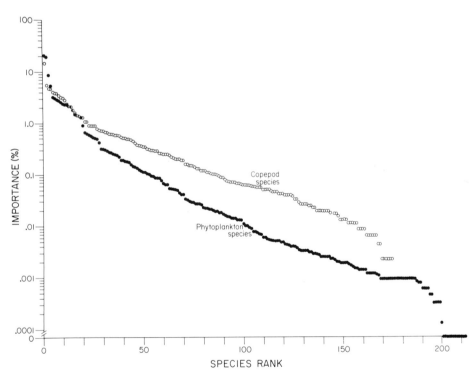

Figure 19.8 Species ranked by percentage importance where the mean abundance for each species is shown as a percentage of the total number of individuals collected of copepods or phytoplankton. In both cases, all species were found on all cruises (but not necessarily in all samples) and were therefore persistently present.

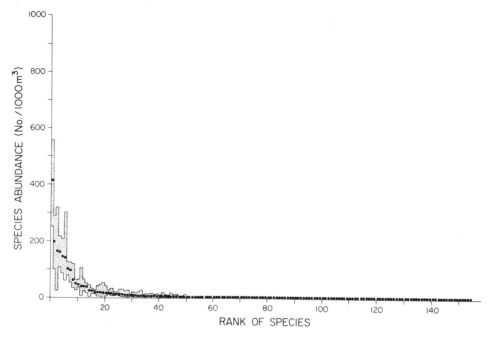

Figure 19.9 The mean rank order of copepod species abundance from the South Pacific central gyre (dots). The data are from three different years. The vertical bars are as in figure 19.7. This picture is very similar to that in the North Pacific, with only a few species ever being numerically dominant and most species being persistently rare or very rare, and with few of the dominants ever becoming rare.

highly diverse mainly because of the presence of many rare species. If we are concerned with the consequences of diversity loss, then the role of these rare species in the organization of the ecosystem must be understood. They cannot be very important predators or prey, and because of their rarity they certainly do not appear to be effective competitors with the dominants or with one another since there is an overall concordance of abundance. However, it should be noted that these species are so rare that a population doubling or halving would be difficult to detect, so there may be important interactions occurring that are below the level of sensitivity of our measurements. But this interaction, if it occurs, seems to be unrelated to processes affecting dominance structure on the left-hand side of the lognormal species curve where organisms are, on average, orders of magnitude more abundant (fig. 19.8). Here some shifts, or reshuffles, appear to occur among the top twenty species, but they are not frequent enough from time to time or large enough from place to place to render the overall Spearman rank correlation coefficients nonsignificant (McGowan and Walker 1985). Although the central gyre is calmer than other parts of the North Pacific, there have been storms and anomalies in wind speed and temperature that were large enough to affect the field of biomass (Venrick et al. 1987; McGowan and Hayward 1978). None of these perturbations have affected species dominance structure as theory predicts.

The South Pacific anticyclonic central gyre is an oligotrophic environment very similar to that in the North Pacific. It is a warm, saline downwelling system with a species list and dominance-diversity curve much like the north's (fig. 19.9). These two provinces are separated by the enormous equatorial current system. While a few species manage to "cross over" from South Pacific to North Pacific, most do not (see fig. 19.1). In spite of this substan-

tial barrier, both areas have very similar species lists. But the South Pacific gyre has about 30 fewer species of copepods than the north (out of 173 species). This apparent "loss" of species may be due to the fact that we have fewer samples and cruises in the South Pacific and therefore may be a sampling artifact. But there are other differences that cannot be explained away so easily. Some 10 rare species found in the south are not present in the much more numerous northern samples. Some other species, found in both gyres, have shifted position in the dominance hierarchy of the south, while most species have maintained the same relative positions in both gyres. These losses and/or rearrangements in the two gyres have not affected the long-term species dominance structure (i.e., the persistence of rank order of abundance) in either gyre; the dominants stay dominant and the rares stay rare on all scales of sample separation in both provinces (McGowan and Walker 1985). Using the rank order of abundance as a measure of stability, both gyres are remarkably stable as compared with other areas of the ocean (see below). Further, the temporal/spatial variations in population sizes, plant and animal biomass, productivity, and nutrient concentrations are, if anything, even less variable in the south than in the north.

Why are the rare phytoplankton and copepod species so persistent? What if anything is their role in the system? Many of the rare copepods are similar in body form, including mouthparts, to the dominant species, and congeners among them may or may not occupy the same spatial position in the water column. The plants are even more of a puzzle. There are many more species of them; their resource requirements are, presumably, more simple than those of the copepods; and greater numbers of them co-occur in the same vertical strata (Venrick 1982, 1990).

Bormann (1990) has suggested that parts of forest ecosystems have the ability to "perform functions beyond immediate needs" of the system and that this serves as a buffering capacity in times of stress. He calls this property "redundancy" and has given a number of examples from forest ecosystems, particularly in the case of nutrient cycling. The profusion of rare species in the central gyres may well be a form of "redundancy" where one or more of them may take over the function of an abundant, functionally important species should its population be somehow diminished.

THE CALIFORNIA CURRENT

The California Current is another system whose plankton have been studied for a long time. Conditions in this North Pacific eastern boundary current are very different from those in the two anticyclonic gyres. Water enters from the north and northwest and exits to the south. The area is rich in biomass, with high productivity and vigorous nutrient input. There are strong horizontal gradients of temperature, salinity, and biomass, as well as large seasonal and interannual changes. Plankton is very patchy. There is a strong inflow of cold low-salinity subarctic water from the north and of warm salty water from the south, and temperate waters are stirred in along the entire outer periphery in a series of quasi-permanent meanders and mesoscale eddies. These different inflows of water come from different biogeographical provinces (see fig. 19.1) and have different flora and fauna, all of which are swept into the area and stirred and mixed together on a massive scale. As a consequence, the California Current, especially its central sector, has one of the most diverse zooplankton assemblages in the Pacific. But here, as opposed to both gyres, there are large and frequent shifts or reshuffles in rank order of abundance on virtually all scales of spatial-temporal sample separation (fig. 19.10; McGowan and Miller 1980; McGowan 1990). There also have been very large (in terms of standard deviations from the mean) environmental perturbations recorded over the past 40 years. For example, both the 1958–59 and 1983–84 El Niños resulted in a decline by factors of ten in plankton biomass. However, these were only transient changes; "recovery" was very rapid. No permanent species losses or other structural or functional changes could be detected in this very noisy system after those or other events. That is, the perturbations or "stress" caused by climatic variations had no lasting ecosystem effects. The reason for this remarkable resiliency is no doubt the massive input of allocthonous populations from outside the systems. Local birth rate/death rate changes seem to be overwhelmed by immigration/emigration, or, in the lexicon of oceanographers, in situ events contribute little to the local variance compared with advective ones. Diversity here is therefore maintained by physics, not biology.

SUMMARY AND CONCLUSIONS

The pelagic biogeographical studies done in the Pacific during the 1950s and early 1960s showed that most oceanic zooplankton have very large populations ranging over huge areas. Both visual inspection of range maps and statistical studies of species presence and absence showed

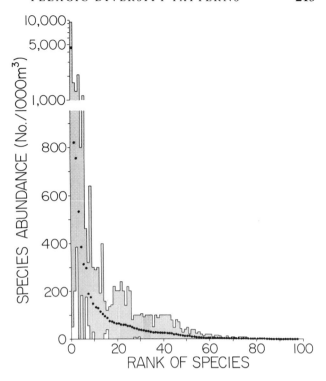

Figure 19.10 The mean rank order of abundance of copepod species from the California Current in a set of samples comparable in spatial/temporal array to those from the north central gyre. This pattern differs strongly from the dominance hierarchies shown in figures 19.6–19.8 in that on occasion dominance is much more pronounced and that on many occasions the species whose mean abundance was high became rare or absent in individual samples. Therefore, while many of the rare species stayed rare, the abundant species did not stay abundant (Spearman rank correlation coefficient nonsignificant). (From McGowan 1990.)

there was a large amount of co-occurrence. Other studies of phytoplankton, fishes, and squid tended to confirm these results. The patterns formed by these species assemblages resembled the great gyrelike circulation systems and the circulation-recirculation equatorial system. Because these circulation systems have considerable continuity and unique climatic regimes, and because their sense of circulation differs (viz., cyclonic versus anticyclonic), they differ strongly environmentally (viz., upwelling versus downwelling). These differences in circulation and climate have led to differing density structures and therefore differing physical stability—some of them mix vertically more easily than others. Therefore physical/climatic perturbations that affect one might not affect another, or might even affect it in an opposite way. Only very large and therefore infrequent physical perturbations affect entire circulation systems or ecosystems. Smaller perturbations, such as those caused by local winter storms, are stirred out of existence very quickly by the horizontal movement of water and its contents (i.e., nutrients and plankton).

For reasons we do not understand, some of these gigantic community ecosystems are more diverse than others. But these large-scale diversity patterns do not follow the classic terrestrial smooth latitudinal gradients of diversity, but rather show sharp changes in gradient at about 40° N and 50° S. Throughout the mid-latitudes they are rather flat, then they decrease in the equatorial zone east of the 180° meridian. This is not the canonical precept, and any theory to account for large-scale diversity patterns should take account of these observations.

Because the north central Pacific gyre is so highly diverse, and because it has far less heterogeneity in terms of biomass patchiness than other systems (except the south central gyre), it was selected for a detailed study of its ecological "structure." A high level of both phytoplankton and zooplankton species co-occurrence was found even on relatively fine scales. Many species abundance curves overlapped strongly in the vertical dimension, where there were strong resource gradients of temperature, light, density, nutrients, productivity, and biomass. These vertical patterns persisted over seasons, years, and decades. Species did not array themselves out over resource gradients, nor did the copepods avoid competition by eating at different times or places. Because of this lack of support for resource partitioning, a time series of species proportion was followed by an examination of the rank orders of abundance between samples separated in time by minutes to decades. The disturbance theories predict that on some scale of patchiness there will be changes in dominance hierarchies. We found no such changes in dominance on any scale: the numerical dominants stayed dominant, the common stayed common, and the rare stayed rare. While there were some shifts within abundance categories, there were orders of magnitude separating those categories. This same result was obtained when samples separated in space were examined. During this time and over this space there were storms and significant anomalies in climatic variables such as temperature and wind speed: that is, physical disturbances. A less extensive set of samples from the south central gyre, a very similar system, showed that while most of the same species were present, there were some changes. There was a strong and persistent dominance hierarchy here as well, and much small-scale (and large-scale) species co-occurrence. That is, neither the absence of species nor the changes in relative positions were reflected in a more variable species abundance or dominance structure or in a less structured community.

These two sets of observations indicate that pelagic communities are quite resilient to external disturbances and do not seem to be very dependent on a particular species diversity for maintenance of stable structure. Both the North and South Pacific central gyre communities are highly efficient at recycling nutrients within the photic zone. Although both theories used to explain the maintenance of diversity seem to have failed reasonable tests of their validity, something clearly does maintain it, for many phytoplankton and zooplankton generations have passed during our study and the species dominance structure has persisted.

The studies in the gyre have helped us to understand another system, the California Current. Here, most of the species have ranges outside of and far larger than even its 10^6 km^2. The central sector of this area is highly diverse, but as opposed to the two gyres, large reshuffles of species dominance take place from sample to sample and place to place on even very small time/space scales. It seems quite evident, in view of the strong horizontal physical gradients and opulent mesoscale eddy activity, that this diversity is almost entirely due to advective stirring and mixing of all allochthonous fauna from the north, west, and south.

The presence of large numbers of rare species is the chief determinant of the high diversity of the two gyres. Thus theory to account for diversity patterns, their persistence, and perhaps their origin must also account for the occurrence and persistence of rarity. The idea that communities, especially resilient ones, have a certain amount of redundancy of function and structure may be supported by these observations, even though we do not know how the rare species function, or their roles in the community.

Understanding why these community ecosystems are structured the way they are and what it will take to change them or disrupt their function is clearly tied to the problem of scale. Speaking of the ecological community, Ricklefs (1987) says that "structures will generally match the scales of processes responsible for them." That certainly seems to be true in the case of the oceans, where these processes seem to be of very large dimensions. There is a close relationship between the temporal and spatial scales of ecological events in the ocean (Haury, McGowan, and Wiebe 1978). Since large volumes of seawater take a lot of energy to warm or cool or overturn, such events must take a long time and therefore be of low frequency, and therefore community "perturbations" on this scale are rare. It must be that it is these larger-sale, large-amplitude events that are important ecologically and evolutionarily. It seems unlikely that short-term, small-scale, local studies of the structure and function of ecological systems will be adequate in the resolution of the problems in understanding the maintenance of structure of oceanic systems and how environmental changes will affect their structure and therefore their function.

20

Global Patterns of Diversity in Mangrove Floras

Robert E. Ricklefs and Roger Earl Latham

During the past three decades, ecologists have developed a body of theory that attempts to explain global patterns of species diversity in terms of interactions that limit the local coexistence of species (Pianka 1966; MacArthur 1972; Connell 1978; Tilman and Pacala, chap. 2). The outcome of these interactions is thought to depend on local conditions of the environment; as a result, patterns of diversity would parallel patterns of climate and other features of the physical world. Theories of diversity based upon local interactions predict that similar habitats in different parts of the world, in which biological communities have developed independently, should support similar numbers of species (Recher 1969; Cody 1975; Orians and Paine 1983; Ricklefs 1987). This is the principle of convergence. Although species richness generally does reflect physical factors in the environment (Rosenzweig and Abramsky, chap. 5; Wright, Currie, and Maurer, chap. 6), many comparisons of similar environments in different parts of the world have revealed strikingly different numbers of species (Orians and Paine 1983; Ricklefs 1987; Latham and Ricklefs, chap. 26; Morton, chap. 14; Schluter and Ricklefs, chap. 21; Westoby, chap. 15). Such "diversity anomalies" challenge our understanding of the origin and maintenance of biodiversity.

Diversity anomalies may result from local factors other than competition, including the effects of predation and disturbance (Orians and Paine 1983). However, these causes, to the extent that their influence is governed by local physical factors, should produce convergent diversity in similar habitats. Ecologists have also rightly pointed out that what appear to be similar habitats may, in fact, differ in fundamental attributes that affect species richness (Morton and James 1988). At the same time, however, ecologists must entertain the idea that differences in local species richness might arise from the particular history and biogeographical circumstances of each region, quite apart from the contemporary local environment.

Of the many examples of diversity anomalies, mangrove floras are one of the most enigmatic. Mangroves of the Indo–West Pacific (IWP) region have several times the species richness of comparable associations in the Atlantic-Caribbean-East Pacific (ACEP) region (Chapman 1976; Hadač 1976; Barth 1982; Tomlinson 1986; Duke 1993; see fig. 20.1). This anomaly parallels similar differences in the diversity of other associations in shallow tropical seas, notably seagrasses, reef-building corals, and their associated faunas (McCoy and Heck 1976; Rosen 1988; Woodroffe and Grindrod 1991). In both the ACEP and IWP regions, one may find mangrove[1] associations in the deltas of large rivers or along protected coasts, in areas with both wide and narrow tidal ranges, and bordering upon both arid and wet terrestrial environments. However, mangrove habitats in the IWP region consistently support more species than similar habitats in the ACEP region. On a global scale, the IWP has four times the number of genera (17 versus 4) and about six times the number of species (40 versus 7) as the ACEP (Saenger, Hegerl, and Davie 1983; Tomlinson 1986; table 20.1). Furthermore, the present ACEP flora is, with the exception of a single endemic genus, a subset of the IWP flora at the genus level. In the absence of obvious differences in the physical habitat, the difference in mangrove diversity between the IWP and ACEP regions may require an explanation based on regional processes or unique history. In this chapter, we address the mangrove diversity anomaly by considering the taxonomic positions of mangrove plants, their geographical distribution and fossil record, and the paleogeography and paleoclimatology of the mangrove habitat.

Taxonomic affinity provides clues to the historical development of present-day mangrove floras, suggesting the number of independent origins of modern mangrove taxa and, to the extent that their terrestrial sister taxa are geographically restricted, the region of their origin. The fossil record of mangroves is incomplete, but can help to distinguish regional differences in extinction and origination that may be causes of global patterns in species richness (Latham and Ricklefs, chap. 26; Van Valkenburgh and Janis, chap. 28; Valentine and Jablonski, chap. 29). Paleoclimatological and paleogeographical information may contribute to our understanding of the geography of origination and dispersal of mangrove taxa.

The considerations outlined below suggest that

1. Tomlinson (1986) suggests that "mangal" be used as a term for the community, and reserves "mangrove" for the plants themselves (Macnae 1968). Accordingly, the environment and habitat can be referred to by either term. Mepham and Mepham (1985) and Duke (1993) do not favor "mangal" because it is not commonly used in the English-language literature. In this chapter, we use "mangrove" as both noun and adjective in reference to both habitat and individual plants or taxa.

Table 20.1. Taxonomy and Biogeography of Exclusive Mangrove Trees and Shrubs

Subclass[a]	Order	Family	Genus	Total spp.[b]	Distribution[c] 2	1	6	5	4	3	A[d]	V[e]	I[f]	F[g]	Life form
Arecidae	Arecales	Arecaceae	*Nypa*	1*	1	1					−	+	Sf	*	Palm
Hamamelidae	Plumbaginales	Plumbaginaceae	*Aegialitis*	2*	1	1					−	+	G		Shrub
Dilleniidae	Malvales	Bombacaceae	*Camptostemon*	2+	1	2					+	−	G	+	Tree
		Sterculiaceae	*Heritiera*	2[b]+	2	1	1				±	−	S		Tree
	Primulales	Myrsinaceae	*Aegiceras*	1[i]+	1	1					−	+	G	+	Shrub
	Theales	Theaceae[j]	*Pelliciera*[k]	1+						1	−	+	G	*	Tree
Rosidae	Euphorbiales	Euphorbiaceae	*Excoecaria*	1[l]+	1	1					−	−	S		Tree
	Myrtales	Combretaceae	*Laguncularia*	1*				1	1	1	+	−	Tr	*	Sh/tr
			Lumnitzera	2*	2	2	1				+	−	G	*	Sh/tr
		Lythraceae	*Pemphis*	1+	1	1	1				−	−	S		Sh/tr
		Myrtaceae	*Osbornia*	1+	1	1					−	−	G		Shrub
		Rhizophoraceae[m]	*Bruguiera*	6*	6	5	1				++	++	Tr	*	Tree
			Ceriops	2[b]*	2	2					++	++	Tr	*	Tree
			Kandelia	1*		1					−	++	Tr		Tree
			Rhizophora	8[n]*	6	3	1	3	3	2	++	++	Tr	*	Tree
		Sonneratiaceae	*Sonneratia*	5*	3	5	1				++	−	G	+	Tree
	Sapindales	Meliaceae	*Xylocarpus*	5[o]+	3	5	2				++	−	S		Tree
Asteridae	Lamiales	Avicenniaceae	*Avicennia*	11[p]*	5	5	1	1	2	3	++	+	F	*	Tree
	Rubiales	Rubiaceae	*Scyphiphora*	1+	1	1					−	−	G		Shrub

Note: Table includes only exclusive mangrove species recognized by Tomlinson (1986). Saenger, Hegerl, and Davie (1983) additionally recognize *Conocarpus* (Combretaceae, 1 species), *Cynometra* (Caesalpiniaceae, 2 species), *Acanthus* (Acanthaceae, 3 species), and *Phoenix* (Arecaceae, 1 species); Saenger et al. do not include *Pemphis* (Lythraceae, 1 species). Duke (1993) includes the fern *Acrostichum* (Pteridaceae), *Diospyros* (Ebenaceae, 1 sp.), *Cynometra*, *Mora* (Caesalpiniaceae, 1 sp.), *Conocarpus*, *Pemphis*, *Aglaia* (Meliaceae, 1 sp.), *Acanthus*, and *Dolichandrone* (Bignoniaceae, 1 sp.).
[a] Higher taxonomy largely after Cronquist (1981).
[b] Number of species according to Saenger, Hegerl, and Davie (1983). *, major elements of the mangrove flora; +, minor elements (Tomlinson 1986, table 2.1).
[c] Regions described by Saenger, Hegerl, and Davie (1983): 2, Australia and New Guinea; 1, Asia and Indonesia; 6, East Africa and Madagascar (western Indian Ocean); 5, West Africa (eastern Atlantic Ocean); 4, western Atlantic Ocean and Caribbean Sea; 3, western Central and South America (eastern Pacific Ocean).
[d] Aerial roots or pneumatophores well developed (++), present (+), absent (−) (Hutchings and Saenger 1987).
[e] Viviparity well developed (++), present (+), absent (−).
[f] Taxonomic isolation at the level of species (S), genus (G), subfamily (Sf), tribe (Tr), or family (F).
[g] *indicates Paleogene fossil record; + indicates earliest fossils in Miocene.
[h] Duke (1993) recognizes 3 species.
[i] Tomlinson (1986) and Duke (1993) recognize 2 species.
[j] Airy Shaw (Willis, 1966) and Tomlinson (1986) place *Pelliciera* in a separate family, the Pellicieraceae.
[k] Genus name is *Pelliceria* according to Airy Shaw (Willis, 1966).
[l] Duke (1993) recognizes 2 species.
[m] The Rhizophoraceae are sometimes placed in a separate order, Rhizophorales (Tomlinson 1986). On the basis of a very thorough examination, Dahlgren (1988) considers the Rhizophoraceae as belonging to the order Celastrales and separates several terrestrial genera, including *Combretocarpus*, into the Anisophyllaceae (Rosales).
[n] Duke (1993) recognizes 6 species and 3 hybrids.
[o] Tomlinson (1986) recognizes only 1 species; Duke (1993) recognizes 2 species.
[p] Tomlinson (1986) and Duke (1993) recognize only 8 species.

throughout most of the Tertiary, conditions for the invasion of mangrove habitat by terrestrial taxa and their specialization as mangroves occurred primarily in Southeast Asia/Malaysia and, to lesser extent, East Africa/Madagascar. These conditions probably included the presence of a diverse terrestrial flora adjacent to mangrove habitat in areas of high, relatively aseasonal rainfall. Strong evidence supports the origin of only a single mangrove taxon in the Western Hemisphere, sometime prior to the early Eocene. The restriction of most mangrove taxa to the Indo–West Pacific region may have resulted from poor dispersal and from closure of the Tethys connection to the Atlantic Ocean in the middle of the Tertiary.

THE MANGROVE SYSTEM

Mangroves are defined as halophytic, generally woody plants that inhabit the upper intertidal zones of saltwater areas, primarily within tropical and subtropical regions (Tomlinson 1986; Hutchings and Saenger 1987; see Mepham and Mepham 1985 for a more detailed, critical evaluation). Mangrove vegetation usually occurs on soft sediments protected from extreme wave action, although many taxa may establish themselves on protected rocky shores (Thom 1982). Within the mangrove habitat, taxa are specialized and segregated (zoned) with respect to tidal height, salinity of the water, range of salinity of the soil, and aeration of the soil (Watson 1928; Macnae 1968; Chapman 1976; Oliver 1982; Snedaker 1982; Hutchings and Saenger 1987; Bunt et al. 1991; Duke 1993). Macnae (1966) recognized six zones in IWP mangrove habitat, distinguished by the dominant genus of mangrove tree: the landward fringe, *Ceriops* thickets, *Bruguiera* forests, *Rhizophora* forests, the seaward *Avicennia* zone, and the *Sonneratia* zone at the lowest level. The landward fringe is the most variable zone. Its floristic composition depends on the climate and vegetation of adjacent terrestrial habitats. In arid climates, owing to the evaporation of water between infrequent tidal coverage, the highest zone in the mangrove habitat can become so salty as to exclude woody plants, or even all vegetation (Walter 1985). In wet climates, salt concentration in the soil decreases landward, and terrestrial species may intermingle with mangrove species in the highest zone. In estuaries, saline conditions grade continuously into brackish and then fresh water. The palm *Nypa* and representatives of the genus

Acanthus, which some regard as mangrove taxa, occupy brackish-water zones. Because woody plants of freshwater swamp communities have no floristic relation to mangroves, they appear to have evolved independently.

Mangrove plants exhibit a variety of striking adaptations to salt stress and anoxic soils (Saenger 1982; Tomlinson 1986; Hutchings and Saenger 1987). Principal among these are various mechanisms of salt exclusion from roots or salt excretion from leaves; aerial roots and pneumatophores with openings (lenticels) to admit air; and viviparity. In the Rhizophoraceae, the seed germinates while still attached to the parent plant and is dispersed as a seedling with developing shoot and root axes. In several other mangrove taxa, embryonic development commences prior to dispersal, but the embryo does not break the pericarp of the seed. In all species of mangroves, propagules float and are dispersed by marine currents.

Mangrove vegetation consists of "exclusive" species that are limited to mangrove habitat and "nonexclusive" species that are distributed widely in terrestrial habitats but which also occur in the upper mangrove zones. Authors generally agree on which taxa belong in which group (but see Mepham and Mepham 1985 for a different, more inclusive, viewpoint): in a global mangrove flora of perhaps 20 genera and 50 species of exclusive mangrove taxa, Saenger, Hegerl, and Davie (1983) include 4 genera (7 species) that Tomlinson (1986) omits, but Tomlinson includes 1 genus (*Pemphis*: one mangrove species) not considered by Saenger et al. Chapman (1976) lists 16 genera and 55 species (10 are *Xylocarpus*). Duke (1993) includes 27 genera and 62 species plus 7 hybrids. In this chapter, we recognize 19 genera and 54 species or hybrids of exclusive mangroves (table 20.1); although mangroves have been studied intensively, their taxonomy is subject to revision (e.g., Duke 1991b; Duke and Jackes 1987; Juncosa and Tomlinson 1988a), and geographical ranges are imperfectly known in some areas (e.g., Bunt, Williams, and Duke, 1982).

Botanists also distinguish "major" elements of the flora, which are trees capable of forming dense stands (principally Rhizophoraceae, Avicenniaceae, Combretaceae, and Sonneratiaceae), from "minor" elements. Finally, mangroves harbor a diverse associated biota of marine and terrestrial plants and animals that may or may not be exclusive to the habitat but frequently occur there. Among plants, these include numerous epiphytes, parasites, climbers, and other herbaceous species that avoid salt stress by using mangrove plants as substrates for growth.

In general, mangrove species are widely and continuously distributed in suitable habitat within the region of their occurrence (Tomlinson 1986). Disjunctions (subpopulations separated by unoccupied suitable habitat) between populations within a species have been reported only in *Bruguiera hainsii* and *Pemphis acidula* (IWP region); disjunctions between sister taxa occur in the genera *Aegialitis* and *Camptostemon* (also IWP). However, Duke (1993) has called attention to the distinct mangrove floras of the northern and southern coasts of New Guinea, suggesting that "New Guinea marks a fusion boundary between two previously isolated and different mangrove

floras." Duke further makes the point that the distributional limits of many individual mangrove taxa are not well understood in terms of dispersal and ecological factors. In addition, Woodroffe and Grindrod (1991) emphasize the role of Pleistocene climate and sea level changes in modifying the present distributions of taxa.

TAXONOMIC AFFINITIES OF MANGROVES

Mangrove attributes appear to have evolved independently at least 15 times in 9 orders and 15 families (see table 20.1). Taxonomic isolation of mangrove species from non-mangrove sister taxa varies from the level of species within genera (*Heritiera*, *Excoecaria*, *Pemphis*, *Xylocarpus*) to the level of tribe (Rhizophoreae), or family if one separates the Avicenniaceae from the Verbenaceae (Tomlinson 1986). Taxa endemic to mangrove habitat at the species level generally are represented by single mangrove species, particularly if one accepts that all the mangrove taxa of *Xylocarpus* belong to a single species (Tomlinson 1986). Taxa endemic at the genus level also are represented mostly by single species, or by allopatric species (*Aegialitis* and *Camptostemon*). The exceptions are *Lumnitzera* (2 species) and *Sonneratia* (5 species); according to Cronquist (1981), the latter constitutes a family with only one other genus (*Duabanga*: 2 species of lowland rainforest trees), although Dahlgren and Thorne (1984) place *Sonneratia* in the Lythraceae. More than half the worldwide diversity of mangrove species is included within the Avicenniaceae and the Rhizophoreae, which evidently have undergone substantial diversification after invading the mangrove habitat (see below).

THE DIVERSITY OF MANGROVE TAXA

Saenger, Hegerl, and Davie (1983) tabulated the occurrence of species within six areas, three within the IWP region and three within the ACEP region (fig. 20.1). Although the IWP and ACEP regions presently have roughly equivalent total areas of suitable habitat (table 20.2), mangrove diversity differs between the regions by a factor of four for genera and about six for species. Within the ACEP region, most mangrove taxa, with the exception of *Pelliciera*, are widespread, and diversity does not vary markedly from site to site (Tomlinson 1986).[2] On the

2. Gentry (1982) lists 13 species of exclusive mangrove trees on the Pacific coast of Central and South America from southern Mexico to northern Peru, belonging to the genera *Avecennia*, *Crenea** (Lythraceae), *Laguncularia*, *Mora** (Leguminosae), *Pavonia** (Malvaceae), *Pelliciera*, *Phryganocydia** (Bignoniaceae), *Rhizophora*, *Tabebuia** (Bignoniaceae), and *Tuberostylis** (Compositae) (asterisks indicate taxa not found on the Caribbean side of the Isthmus). This area includes the Choco region of Columbia, one of the wettest on earth, where salt stress may be minimized and where uplifted mangrove habitats may permit intermixing of more terrestrial, floodtolerant (as opposed to salt-tolerant) vegetation. Such conditions may represent evolutionary entryways of terrestrial taxa into mangrove habitat. *Crenea* pollen appears in the Caribbean region in the Upper Eocene and Lower Miocene in association with *Rhizophora* pollen in coastal sediments, suggesting the presence then of a mangrove taxon within the genus (Germeraad, Hopping, and Muller 1968).

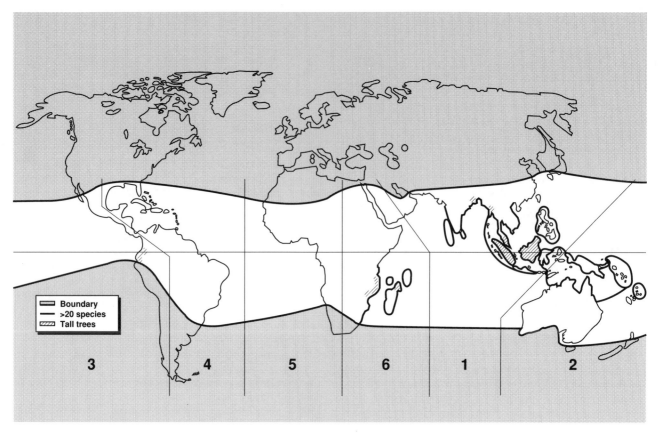

Figure 20.1 Worldwide distribution of mangroves. Coasts with more than 20 species are indicated by heavy lines; areas supporting very tall trees are indicated by hatching. (Distributions from Chapman 1970.) Vertical lines separate geographical areas used by Saenger, Hegerl, and Davie (1983) to tabulate regional diversity.

Table 20.2 Taxonomic Diversity of Mangrove Taxa in Different Biogeographical Regions

Region and subregion	Area of mangrove habitat (km²)	Number of genera	Exclusive species
IWP			
2. Australia/New Guinea	17,000	16	35
1. Asia/Indonesia	52,000	17	39
6. East Africa/Madagascar	5,000	8	9
ACEP			
5. West Africa	27,000	3	5
4. Western Atlantic/ Caribbean	48,000	3	6
3. Eastern Pacific	19,000	4	7

Source: Data from Saenger, Hegerl, and Davie 1983 and table 20.1.

ably as a result of the difficulty of long dispersal distance against prevailing ocean currents (Jokiel and Martinelli 1992). Species of *Bruguiera* and *Rhizophora* have been introduced successfully to Hawaii, which lacks native mangroves (Wester 1981). This suggests that diversity on the Pacific islands is indeed limited by colonization.

Local (i.e., hectare scale) diversity in mangrove habitat parallels regional diversity. Within the ACEP region, local diversity generally is 3–4 species, half the total number present in the region, but usually including all the species that co-occur geographically (Davis 1940; Chapman 1970, 1976). Within the IWP region, local diversity is more difficult to ascertain from published accounts, which tend to present floristic maps and idealized transects within localities. One such representation of a typical area on the Malayan west coast includes 6 genera and 11 species of predominant mangroves (Watson 1928); Macnae (1966) and Elsol and Saenger (1983) similarly depict several areas on the Queensland coast of Australia with 5–8 genera of exclusive mangrove taxa. Tomlinson (1986) indicates that certain localities on the coast of Queensland, Australia, may harbor up to 30 species of mangroves, most of them exclusive species. Transects from low to high water at several localities along the Endeavour River estuary in northeastern Australia revealed 7–15 species of mangroves per transect and a total of 25 species (Bunt et al. 1991). Thus, species richness in the

western edge of the IWP region, the coasts of East Africa and Madagascar (area 6) support restricted areas of mangrove habitat and relatively low diversities of taxa. This low diversity may be related to the small area of suitable habitat and to local environmental conditions. More than half the mangrove habitat in area 6 is on the island of Madagascar. The eastern coast of Africa lacks large rivers with well-developed deltas, and much of the coast is arid and unsuitable for mangrove genera that occupy the upper zones in wetter climates. Diversity also decreases eastward from New Guinea into the Pacific Islands, presum-

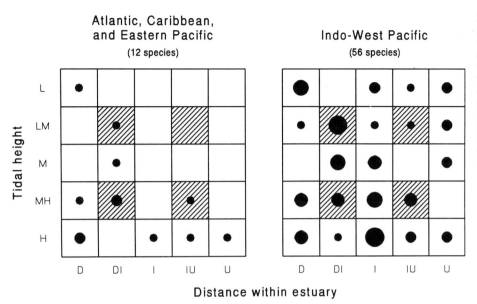

Figure 20.2 Distribution of mangrove taxa among categories of intertidal zone height (L, low; M, mid; H, high) and position within an estuary (D, downstream; I, intermediate; U, upstream) in the ACEP and IWP regions. Numbers of taxa per cell (1–6) are indicated by the sizes of the dots. Generalized taxa distributed in more than one category within each ecological axis are indicated by hatching. (Data from Duke 1993.)

IWP region exceeds that of the ACEP region by factors of 6 at the regional level and probably at least 2–3 at the local level.

Whether the greater diversity of IWP mangroves is accompanied by greater ecological specialization of taxa and greater zonation with respect to salinity and height in the tidal zone has not been well resolved. Zonation within the mangrove habitat and the tendency of stands to be dominated by a single species result in local coexistence of species being perceived over distances of tens or even hundreds of meters rather than on the scale of neighboring individuals (Macnae 1968; Chapman 1976; Hutchings and Saenger 1987). This dimension of heterogeneity appears to characterize the lower-diversity ACEP mangroves as well as the higher-diversity IWP mangroves (Davis 1940). Nonetheless, some zones of the IWP mangrove habitat may be occupied by mixed-species stands, particularly at mid-tidal levels (Macnae 1968; Elsol and Saenger 1983).

Duke (1993) characterized exclusive mangrove species and their hybrids on the basis of intertidal position and estuarine location. His tabulation allows one to determine whether ACEP and IWP mangroves occupy the same total niche space, and whether differences occur in the degree of specialization of taxa in the two regions. The intertidal zone was divided into three zones: L (lower, inundated more than 45 times per month), M (mid, 20–45 times), and H (high, fewer than 20 times). Distance up an estuary was divided arbitrarily into thirds: D (downstream), I (intermediate), and U (upstream). Each species of mangrove was cross-classified according to these two axes of the ecological space; each species was placed in one (e.g., L, M, or H) or two subdivisions (e.g., LM or MH) of each ecological axis. The number of species in each of the cross-tabulated categories in the ACEP and IWP regions is presented in figure 20.2.

The ACEP region lacks mangrove taxa in the lower intertidal upstream region of the ecological space (L-U, LM-U, L-IU, LM-IU). In the IWP region, this space is occupied by *Nypa fruticans*, *Rhizophora mucronata*, *Sonneratia*

apetala, *S. caseolaris*, *S. lanceolata*, and *Aegiceras corniculatum*. Suitable habitat presumably exists in the ACEP region; it is unclear why it is not filled. Evidently, the ACEP mangrove species have not expanded their niches to occupy this portion of the habitat, and terrestrial forms have not invaded.

In regions of low species richness, individual species often exhibit expanded ecological distributions, a phenomenon known as ecological release (MacArthur, Recher, and Cody 1966; Cox and Ricklefs 1977). Accordingly, one would predict that a larger proportion of species would occupy two, rather than one, of the habitat categories in the ACEP region compared with the IWP region. The proportions of species that Duke (1993) classified as occupying two categories of both tidal height and distance upriver (ACEP: 4 of 12; IWP: 13 of 56) do not differ significantly between the regions ($\chi^2 = 0.54$, $P > .10$). The same is true if one considers taxa specialized to just one category on both ecological axes. Thus, ACEP mangrove species do not exhibit increased niche breadth. This is consistent with the greater local diversity of mangrove species in the IWP region. Evidently, species have been added to the mangrove association in part by the invasion of niche space already occupied by other taxa. That is, the mangrove habitat does not appear to be saturated by species.

Regional species richness of other taxonomic groups ecologically associated with, but not exclusive to, mangrove vegetation also tends to be higher in the IWP region than in the ACEP region. Some of these associated taxa are marine, including numerous invertebrates and fishes, while others, particularly various plant taxa, have terrestrial affinities. Saenger, Hegerl, and Davie (1983) tabulated the number of species of mangrove-associated biota in several regions. Their results are excerpted in table 20.3 for the three areas that have been well surveyed: 1 (Asia/Indonesia), 2 (Australia/New Guinea), and 4 (western Atlantic/Caribbean). The details of this table undoubtedly reflect the varying degrees of attention that specialists have paid to each taxonomic group in each region. Of the

Table 20.3. Species Diversity of Plants and Animals Associated with Mangrove Vegetation

Taxonomic group	Area 1 (IWP)	Area 2 (IWP)	Area 4 (ACEP)
Monocotyledons	73	42	20
Dicotyledons	110	80	28
Total flowering plants	183	122	48
Algae	65	93	105
Non-polychaete worms	13	74	13
Polychaetes	11	35	33
Crustaceans	229	128	87
Mollusks	211	145	124
Echinoderms	1	10	29
Ascidians	0	8	30
Fishes	283	156	212
Total marine animals	748	556	528

Source: Data from Saenger, Hegerl, and Davie (1983).

Table 20.4. Geographical Distribution of Mangrove Genera

	Atlantic/Caribbean/East Pacific	Indo–West Pacific
Avicennia	3 species	4–6 species
Rhizophora	2 + 1 hybrid	3 + 2 hybrids
Laguncularia/Lumnitzera	1 species	2 species
Nypa	Paleogene fossil	1 species
Wetherellia	Paleogene fossil	
Pelliciera	1 species	
Indo–West Pacific endemics		14 genera
		32 species
Total	4 genera	17 genera
	7 species	40–42 species

three areas, the Asian/Indonesian part of the IWP region (area 1) is probably the most poorly known. Indeed, although more species have been reported from Asian/Indonesian mangroves (area 1) than from Australian mangroves (area 2) among herbaceous plants, crustaceans, mollusks, and fish, several invertebrate groups appear to be underreported. The data in table 20.3 reveal a nearly threefold or fourfold greater species richness of higher plants in the IWP region than in the ACEP region (122 or 183 versus 48). Among marine animals and marine algae, numbers of species in regions 2 and 4 are similar (556 versus 528 and 93 versus 105, respectively). Thus, the global diversity anomaly in mangrove habitat applies quite generally to elements of the flora derived from terrestrial habitats, but perhaps not so strikingly to the marine biota.

GEOGRAPHY OF MANGROVES

Mangroves are confined primarily to tropical latitudes (see fig. 20.1), although *Avicennia* occurs at high latitudes in some temperate areas of moderate climate, such as southern Australia and northern New Zealand (Chapman 1976; Wells 1983; Woodroffe and Grindrod 1991). The distributional limits of mangrove vegetation coincide approximately with the 24° C isotherm of mean sea surface temperature during the warmest month (Hutchings and Saenger 1987), the 15° C isotherm in the coldest month (Woodroffe and Grindrod 1991), and a variety of other climate indices. Regardless of the particular criterion, barriers to dispersal between the ACEP and IWP regions are presently maintained by the cool coastal environment of southern and southwestern Africa (Briggs 1974); within tropical latitudes across the central Pacific Ocean, dispersal is limited by distance (Chapman 1975). Within the ACEP region, the eastern Pacific Ocean (western coasts of Central and South America) has been isolated from the Caribbean and tropical Atlantic Ocean by the Panamanian Isthmus for the past three million years (Saito 1976; Keigwin 1978). Tomlinson (1986) does not recognize any taxonomic differences between the Pacific and Caribbean

sides of the Isthmus, although *Avicennia bicolor* is found only on the Pacific coast and *A. schaueriana* is restricted to the Lesser Antilles and Atlantic coast of South America. Duke (1993) discusses in more detail distribution anomalies within the IWP region, which indicate biogeographical subdivision.

The present distributions of mangrove taxa can be divided broadly into three types with respect to their occurrence in the ACEP and IWP regions: cosmopolitan, endemic to ACEP, and endemic to IWP (table 20.4). Only two genera are cosmopolitan: *Avicennia* and *Rhizophora*. The mangrove palm *Nypa*, currently widely distributed in the IWP region, is known from the fossil record of the early Tertiary in the ACEP region, including western Europe (fig. 20.3; see below), and has been reintroduced to Panama (Duke 1991a). According to Tomlinson (1986), *Laguncularia* (ACEP) and *Lumnitzera* (IWP) are sister taxa, suggesting a cosmopolitan distribution of their common ancestor. Endemic IWP genera can be further subdivided according to presence (6 genera) or absence (6 genera) in Madagascar and East Africa.

Within the cosmopolitan genera *Avicennia* and *Rhizophora*, each species is endemic to either the IWP or the ACEP region, although Tomlinson (1986) and Ellison (1991) suggest that the IWP species *Rhizophora samoensis* is the same as the widespread ACEP species *R. mangle,* perhaps having colonized the IWP region from the eastern Pacific.

Only one mangrove clade is, at present, restricted to the ACEP region, the monotypic family Pellicieraceae[3] (*Pelliciera rhizophorae*), although Gentry (1982) would include six others with distributions restricted, like that of *Pelliciera,* to the Pacific coast of the Americas.

In contrast to the low level of endemism of mangroves in the ACEP region (2 genera of exclusive mangroves including 2 species), 13 genera in 11 families, including 31 species, are restricted to the IWP region; these endemic IWP taxa represent 72% of the entire IWP exclusive mangrove flora at the genus level. With the exception of *Ceriops* and *Bruguiera* (*Paleobruguiera*) in the Eocene of England, none of these genera is known from the fossil record of the ACEP region (see below).

---

3. The name is given as Pelliceriaceae (genus *Pelliceria*) in Willis (1966).

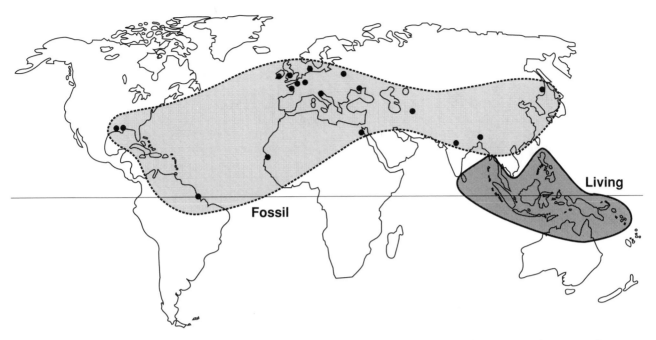

Figure 20.3   Fossil (symbols with dashed line) and contemporary (continuous line) distributions of the mangrove palm *Nypa*. (After Croizat 1968.)

THE FOSSIL RECORD OF MANGROVES

The mangrove habitat—protected, intertidal soft sediments within frost-free regions—presumably has existed for a very long time. According to Tomlinson (1986), fossil evidence suggests the continuous presence of mangrove vegetation since the end of the Paleozoic and certainly predating the origin of angiosperms. Croizat (1964) and others have even suggested that mangrove taxa may have left terrestrial descendants following uplift of mangrove habitat; certainly *Sonneratia* and *Pemphis* have sister taxa with restricted distributions at high elevations in Indomalaysia and Madagascar, respectively. Mepham and Mepham (1985) report the presence of many "exclusive" mangrove taxa at upland sites in various parts of the IWP region, including an inland stand of *Bruguiera* on Christmas Island in the Indian Ocean, which lacks shoreward stands of mangrove (Woodroffe and Grindrod 1991).

The presence of modern mangrove taxa in the fossil record dates back to the late Cretaceous (*Nypa*; Muller 1964), with dicotyledonous taxa (*Rhizophora, Pelliciera,* Sonneratiaceae) known from the early Eocene, at least 30 mya (Muller 1981). In general, the record is not particularly good. Mangroves are a restricted vegetation form difficult to recognize in the fossil record, except where remains of modern mangrove taxa are present; reproductive structures do not distinguish modern mangrove taxa from those of terrestrial habitats. Even when flowers, fruits, pollen, or wood can be assigned to a modern, exclusive mangrove genus, one cannot assume that the genus has been an exclusive mangrove taxon lacking terrestrial representatives in the past: the contemporary genera *Heritiera, Excoecaria, Pemphis,* and *Xylocarpus* include both terrestrial and mangrove species. However, a convincing

case for an extinct mangrove taxon has been made for *Wetherellia* and *Paleowetherellia* (possibly Euphorbiaceae), which are recorded from marine deposits in the Eocene of Maryland, Germany, and England (in association with *Nypa*) and the late Cretaceous and Paleocene of Egypt (Mazer and Tiffney 1982). The distribution of *Wetherellia* and *Paleowetherellia* is clearly Tethyan, but the absence of associations with the contemporaneous *Pelliciera* and *Rhizophora* is puzzling.

Many mangrove genera (e.g., *Aegiceras, Heritiera, Excoecaria, Osbornia*) apparently have not been reported from the fossil record because of poor sampling, difficulty of recognizing their fossil remains, or absence of the taxa from areas of fossilization. Considerable confusion has arisen because of the misidentification of some fossil remains, notably, claims of *Rhizophora* and *Bruguiera* pollen in the Paleogene of Europe (Muller 1981; but see Wilkinson 1981, 1983 for macrofossil evidence of *Ceriops* and *Paleobruguiera* [Rhizophoraceae]). Finally, many interpretations of the present distribution of mangroves incorporate tectonic conditions or events that predate the oldest fossils. Among these conditions is the availability of the Tethys connection between the IWP and ACEP regions, which closed 30–35 mya. We shall discuss implications of the fossil record below.

THE ORIGIN OF THE IWP-ACEP DIVERSITY ANOMALY

Diversity anomalies may arise historically from region-specific differences in the origin of clades, rates of diversification within clades, propensity for dispersal between regions, extinction, or some combination of these.

Extinction and Range Contractions

Differential extinction may cause disparity in species richness between regions, as it has contributed to the depauperization of the woody flora of temperate Europe (Latham and Ricklefs, chap. 26). However, the fossil record provides little evidence that extinction has been responsible for the low species richness of ACEP mangroves. *Nypa*, and possibly the Paleogene *Wetherellia*, are the only mangrove taxa that have disappeared from the ACEP region, although the distribution of the ACEP genus *Pelliciera* has been severely restricted as well. Furthermore, *Ceriops* and *Paleobruguiera* (Rhizophoreae) are known from the Eocene of southern England (Tethyan), but not from the Caribbean. Possibly these arrived at the doorway to the ACEP region but did not extend their ranges fully into it. *Wetherellia* appeared along the eastern coast of North American but apparently failed to enter the Caribbean.

No extinctions of mangrove taxa have been recorded from the IWP region. However, regional extinctions of mangrove taxa cannot be identified as such without contemporary representation of the taxa elsewhere. Thus, regional (= global) extinctions may have occurred of endemic taxa that were either not recorded in the fossil record or not recognized as mangroves.

Fossils do reveal substantial range contractions in the past in a few cases: (1) the disappearance of *Nypa* from an extensive distribution in the ACEP region in the early Tertiary (Croizat 1964, 1968; Daghlian 1981; Tralau 1964); (2) the contraction of mangrove vegetation that once occurred at high latitudes in central Europe (*Nypa*, *Ceriops*, *Paleobruguiera*: Daghlian 1981; Wilkinson 1981, 1983), in the Sahara of Egypt (Kräusel 1939), in southeastern Australia (*Nypa*, *Sonneratia*, Rhizophoraceae: Churchill 1973), on the Gulf Coast of North America (*Nypa*: Westgate and Gee 1990), and on the east coast of North America (*Wetherellia*: Mazer and Tiffney 1982); (3) contraction of the ACEP genus *Pelliciera* from its formerly widespread distribution to relictual populations (Graham 1977). The latitudinal contraction of mangroves may be explained by the general cooling of the earth, particularly at high latitudes, since the mid-Tertiary (Savin 1977; Keigwin 1980).

Pelliciera was widespread in the ACEP region from the Eocene (Panama, Greater Antilles) to the early Miocene, at which time it apparently disappeared from the northern parts of its distribution in the Greater Antilles and the Isthmus of Tehuantepec in Mexico (Graham 1977). It persisted on the northern coast of South America and Brazil through the Miocene, but is not reported thereafter, possibly owing to a paucity of suitable sediments (Graham 1977). The palm *Nypa* similarly disappeared from the Caribbean fossil record at the end of the Miocene. According to Graham (1977), *Rhizophora* pollen constitutes a minor part of *Pelliciera* associations in the Eocene of Panama and Jamaica, becoming abundant only during the Oligocene/Miocene time.

At present, *Pelliciera* occurs in wet refuges on the west coast of Central America and northern South America from Costa Rica to Ecuador, with a few relictual populations on the Caribbean coasts of Colombia, Panama, and Nicaragua (Calderon 1983; Winograd 1983; Jiménez 1984; Roth and Grijalva 1991). Graham (1977) suggested that the range contraction of *Pelliciera* may have resulted from a variety of factors, including sea level fluctuations, cooling climates, and competition from *Rhizophora*. Jiménez (1984) emphasized *Pelliciera's* intolerance of hypersaline conditions (soil salinity greater >37%, i.e., approximately that of seawater) and related its decline to drying conditions beginning in the Caribbean basin during the Miocene (references in Jiménez 1984). Today, *Pelliciera* occurs only under wet climate regimes (where abundant rainfall prevents mangrove soils from drying and prevents salinities from rising about the level of seawater) and in estuaries (e.g., the Caribbean coast of northern Colombia). The idea that drying climates caused the range contraction of *Pelliciera* is consistent with the disappearance from the region of *Nypa*, a species that occurs primarily in brackish estuaries. The distribution of *Pelliciera* prior to the Neogene also emphasizes the widespread presence of mangrove habitat adjacent to areas of wet tropical forest during the Paleogene, as well as the apparent absence of other mangrove taxa from the fossil record of the ACEP region.

Dispersal

Contemporary exclusive mangrove taxa disperse via floating propagules, whether fruits, seeds, or precociously germinated seedlings (Saenger 1982). Three connections between the ACEP and IWP regions may have been available at various times during the Cenozoic period: the Pacific Ocean, the Tethys Sea, and southern Africa. The IWP has experienced a change in configuration during the last 100 million years owing to the northward movements of the Indian and Australian landmasses, but the region has had a geographical continuity throughout the Phanerozoic. In contrast, the ACEP region as we know it today developed with the breakup of Pangaea beginning during the late Mesozoic. The opening of the Atlantic Ocean undoubtedly predates the origin of modern taxa (genera and species) of mangroves, but earliest appearances in the fossil record are relatively uninformative about the origins of most taxa. Prior to the development of the Atlantic Ocean, the ACEP region was restricted to the eastern Pacific Ocean, that is, the western coast of Pangaea, which presently supports a depauperate sample of IWP taxa of corals, mollusks, and fishes that apparently dispersed eastward across the Pacific Ocean (Newton 1988). Because of this Pacific component to the ACEP region, however, its biota potentially is as old as that of the IWP region. This fact takes on significance in light of Gentry's (1982) suggestion of high mangrove diversity in the Choco region of western Columbia.

Dispersal has always been possible between the IWP and ACEP regions across the Pacific Ocean, but movement of propagules between the west coast of tropical America and the nearest large islands in the western Pacific Ocean requires several months at present-day current velocities of about 1 knot. Probably only *Rhizophora* propagules remain viable long enough to make this journey (Rabinowitz 1978; Clarke and Myerscough 1991). Throughout the Tertiary, currents have been favorable

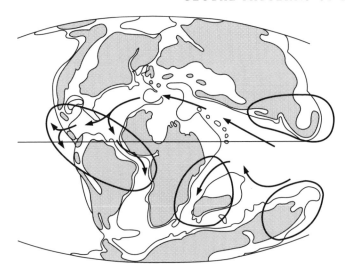

Figure 20.4 Map of the earth during the Paleocene–early Eocene showing the distribution of present-day mangroves. Arrows indicate ocean currents. Shaded areas indicate land masses; lines, continental shelves.

only for westward dispersal across the Pacific within tropical latitudes (Haq 1981; but see Newton 1988). The absence of localized eastern Pacific populations of IWP taxa suggests that mangrove taxa have not dispersed from east (IWP) to west; the presence of *Rhizophora samoensis* (= *R. mangle*) on New Caledonia, the new Hebrides, Tonga, and Samoa provides the only mangrove example of dispersal from west to east (Ellison 1991). We tentatively conclude, in agreement with Tomlinson (1986), Mepham (1983b), and others, that the Pacific is not now, and has not been in the past, an important dispersal route or center of evolution (in the sense of Croizat) for mangrove vegetation. However, Germeraad, Hopping, and Muller (1968) favored a trans-Pacific dispersal route for *Rhizophora* between the IWP and ACEP regions based on the appearance of *Rhizophora*-type pollen in the Caribbean during the late Eocene, possibly after the closure of the Tethys connection, and its later (Miocene) appearance in western Africa (Nigeria).

The second potential route of exchange between the ACEP and IWP regions was through the Tethys Sea, which connected the Indian Ocean to the developing Atlantic Ocean through the Mediterranean region during the Cretaceous and early Tertiary (fig. 20.4). The connection was closed off about 30–35 mya, during the Eocene. The presence of tropical elements in the European Paleogene fossil record (e.g., Reid and Chandler 1933; Chandler 1961), and the presence of mangrove genera in Tethyan and European deposits of Eocene age (Kräusel 1939; Prakash 1960; Tralau 1964; Haseldonckx 1972; Wilkinson 1981, 1983), suggest that the Tethyan region contained suitable mangrove habitat.

The only other marine connection between the IWP and ACEP regions is around the southern end of the African continent. Mangrove habitat is abundant on Madagascar and extends sporadically down the eastern (Indian Ocean) coast of Africa to Natal (Palmer and Pitman 1972; Moll and Werger 1978; Ward and Steinke 1982; Woodroffe and Grindrod 1991). At present, however, the south-

ern and western coasts of southern Africa are dry, cold (due to upwelling currents: Schulze and McGee 1978), and devoid of extensive, gently sloping coasts. Mangroves currently extend down the Atlantic coast of Africa, which is part of the ACEP region, only to a latitude of 12° S (Chapman 1976). The present situation does not represent conditions during the early Tertiary, when the oceans surrounding southern Africa were much warmer and may have provided a dispersal route for mangrove vegetation; Mepham (1983a) has argued persuasively for a southern, high-latitude dispersal route for mangrove taxa between southern Africa and Australia prior to the Eocene.

Regardless of the relative likelihood of any particular corridor, the continent of Africa provides a prominent biogeographical connection between ACEP and IWP mangroves, a connection that is also shared by the distributions of many terrestrial taxa (Croizat 1964). The apparent connection between the ACEP and IWP regions may be referable to the northern African shores of the Tethys Sea, as the continent of Africa has been intact through the whole of the evolution of contemporary mangrove taxa and an overland connection is out of the question.

With respect to the possibility of mangrove dispersal, we conclude that the distribution of mangrove habitat could have been continuous, or nearly so, on the scale of dispersal distances of propagules, between the ACEP and IWP regions during most of the Paleogene. Furthermore, dispersal through the Tethyan region by ocean currents was predominantly from east to west (Specht 1981a). Finally, the two regions have been isolated throughout the past 30–35 million years, as both the Tethyan connection and southern Africa would have been unavailable for mangrove dispersal.

Origin of Mangrove Clades

Clades occupying the mangrove habitat arose independently in at least 15 plant families. This count is based upon the premise that the ancestral habitat for modern mangrove clades was terrestrial, which is supported by the fact that most mangrove taxa exist within larger, otherwise terrestrial clades. Four of the 15 mangrove clades are or were cosmopolitan, 1 is restricted to the ACEP region, and 10 are restricted to the IWP region.

The fossil record often reveals that contemporary distribution may not coincide with or even include past geographical distribution. For example, the earliest Bombacaceae appear in the late Cretaceous (80 mya) fossil record of New Jersey, but the family presently occurs only in tropical regions of South and Central America, Africa, and Southeast Asia; similarly, *Symplocos* (family Symplocaceae) was first recorded from the late Cretaceous of western North America, and later was widespread across all of Laurasia, but presently occurs only in southeastern North America, South America, and eastern Asia to Australia (Krutzsch 1989). Similar examples occur in the Juglandaceae (Manchester 1989), and also include *Nypa* (see fig. 20.3).

In the case of clades in mangrove habitat, the geographical distributions of fossils support the generalization that present distributions include the area of origin.[4] When sister taxa are restricted to the same region, i.e., have shared geographical distributions, one can surmise either that the ancestral taxon had a similar distribution or that the ranges of the sister taxa have changed in parallel, perhaps even moving from the place of their origin. The latter scenario is unlikely in the case of mangroves, whose sister taxa now occupy a substantially different, terrestrial habitat, which may be influenced by different factors in the physical environment. Parsimony would lead one to conclude that the common geographical distribution of sister taxa includes, in a very general sense, the place of the origin of the derived clade.

The distributions of mangrove taxa and their terrestrial relatives clearly indicate the place of mangrove origins in many cases, including *Camptostemon*, *Osbornia*, and *Sonneratia*, which are discussed below. For several taxa, however, one cannot identify sister taxa of comparable taxonomic level owing to the distinctiveness of the mangrove taxon (e.g., *Nypa*, *Aegiceras*, *Aegialitis*, *Avicennia*). Additionally, when a sister taxon has a cosmopolitan distribution, one cannot identify the place of origin without additional historical or phylogenetic information. Of the three cosmopolitan mangrove clades, the subfamily Rhizophoreae belongs to a larger cosmopolitan family (Rhizophoraceae: 15 genera, 135 species; Juncosa and Tomlinson 1988a) with primarily Old World genera (including non-mangrove forms), but also including the cosmopolitan terrestrial genus *Cassipourea* (55 spp.), which is widespread in tropical America and the West Indies as well as in southern Africa, Madagascar, and Sri Lanka. The history of the Rhizophoraceae will be discussed at greater length below.

The *Laguncularia/Lumnitzera/Macropteranthes* clade is part of a larger, pantropical, woody family (Combretaceae) having 20 genera and over 500 species (Tomlinson 1986). The largest genera in the family (*Terminalia*, *Combretum*, *Quisqualis*) are pantropical in distribution. The terrestrial *Macropteranthes* (4 spp.) is restricted to tropical northern Australia. The ACEP nonexclusive mangrove *Conocarpus* also belongs to the Combretaceae. Thus, the origins of these mangrove taxa cannot be placed unambiguously.

The cosmopolitan *Avicennia* is usually placed in its own family (Avicenniaceae), although it has been included in the Verbenaceae. Willis (1966) suggests a relationship to the Salvadoraceae, whose members inhabit hot, dry regions, often coastal or saline, and whose largest genera (*Salvadora*, *Azima*, *Dobera*) are Old World groups centered around the Indian Ocean. The earliest fossil pollen records of *Avicennia* are from the IWP: the Eocene of Australia (probable), the lower Miocene of the Marshall Islands, and the upper Miocene of northwestern Borneo. The earliest ACEP record is from the Pliocene of Guyana

(Muller 1981); such a late arrival in the West is inconsistent with the much earlier closure of the Tethys corridor between the IWP and ACEP regions, but supports a trans-Pacific connection or southern African connection.

Most of the IWP endemics are closely allied to taxa having IWP-East African-Madagascaran distributions. The three genera having a single mangrove species (*Exocoecaria*, *Xylocarpus*, *Heritiera*) are presently confined to the IWP region. Seven exclusive mangrove genera restricted to the IWP region show varying degrees of affinity to IWP terrestrial taxa, as illustrated below.

Camptostemon (Bombacaceae, cosmopolitan) is placed by Tomlinson (1986) in the tribe Durioneae, whose principal genus *Durio* consists of twenty-seven species restricted to Burma and western Malaysia. Pollen of *Durio* is first recorded from the Oligocene-Miocene of northwestern Borneo, and that of *Camptostemon* from the lower Miocene of Borneo and possibly the upper Miocene of Papua (Muller 1981).

Osbornia (Myrtaceae, cosmopolitan) is quite isolated taxonomically but probably belongs to the Old World subfamily Leptospermoideae (Tomlinson 1986).

Sonneratia (Sonneratiaceae) has affinities with the cosmopolitan family Lythraceae. The only other genus in the Sonneratiaceae is *Duabanga*, which contains two allopatric terrestrial species restricted to Indomalaysia (Mahabale and Deshpande 1959; Willis 1966; Croizat 1968). *Sonneratia* is the only mangrove genus for which the pollen record reveals an intermediate form relating it to a taxon that is presently restricted to terrestrial habitats. The earliest pollen records of *Sonneratia* are from the lower and middle Miocene of Borneo (Muller 1981). A Lythraceae/Sonneratiaceae type of pollen (*Florschuetzia trilobata*) is known from Oligocene and Miocene coastal sediments of northwestern Borneo and possibly the Eocene of western Malaysia. This pollen is regarded as ancestral to that of Sonneratiaceae (Germeraad, Hopping, and Muller 1968) and disappears from the fossil record by the end of the Miocene, suggesting that it may be a transitional form connecting terrestrial and mangrove taxa.

Aegialitis belongs to the cosmopolitan family Plumbaginaceae and is probably closest to the cosmopolitan *Plumbago* (Willis 1966), which typically inhabits salt steppes and seacoasts. *Aegialitis* is first recorded in the pollen record from the middle Miocene of Borneo, where it no longer occurs.

Aegiceras (Myrsinaceae, cosmopolitan) is placed in the subfamily Myrsinoideae, tribe Ardisiëae, which includes the genera *Ardisia* (400 tropical species, cosmopolitan, including one mangrove associate in Southeast Asia) and *Tapeinosperma* (40 species in New Guinea, Queensland, New Hebrides, New Caledonia, and Fiji).

Scyphiphora (Rubiaceae, cosmopolitan) probably belongs to the tribe Gardeniëae, which contains both Old World and New World genera. The largest (*Gardenia*) is restricted to the Paleotropics.

Pemphis acidula (Lythraceae, cosmopolitan) shows no clear relationships within the family. A non-mangrove congener, *P. madagascariensis*, is restricted to the mountains of southwestern Madagascar.

---

4. However, Mepham (1983a) emphasizes the observation that several mangrove genera distributed widely during the Eocene (*Avicennia*, *Rhizophora*, *Sonneratia*) did not appear in the fossil record of Borneo until the Oligocene or Miocene (Muller 1964), and suggests that the present concentration of mangrove taxa in Southeast Asia/Malaysia is relictual.

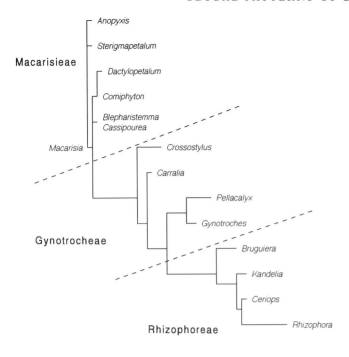

Figure 20.5 Cladistic phylogenetic tree of the Rhizophoraceae. Branch lengths are proportional to the number of character stage changes. The three tribes of the Rhizophoraceae are separated by dashed lines. (After Juncosa and Tomlinson 1988b.)

In summary, distributions of related taxa are either consistent with the origin of IWP endemic mangrove taxa within the IWP region (*Exocoecaria, Xylocarpus, Heritiera, Camptostemon, Osbornia, Sonneratia*) or uninformative (*Aegialitis, Aegiceras, Scyphiphora, Pemphis*). The pollen records of *Camptostemon, Sonneratia,* and *Aegialitis* are also consistent with IWP origins.

History of the Rhizophoraceae

Muller (1981) summarized the earliest records of *Rhizophora*-type fossil pollen: upper Eocene of the Caribbean, Brazil, and India; lower Oligocene of Australia (Queensland); Oligocene/Miocene of Mexico; Miocene of Borneo, Nigeria, and Senegal. Modern species of *Rhizophora* apparently were differentiated by the end of the Miocene. Pollen records of the IWP genera of Rhizophoraceae, *Bruguiera* and *Ceriops,* from the upper Miocene of the Marshall Islands require confirmation.[5] Muller (1981) rejected claims of *Bruguiera* pollen from the Oligocene of England, and of *Rhizophora* pollen from the Paleocene and early Eocene of western Europe. The evidence for mangrove vegetation in western Europe consists of Eocene records of fruits of *Nypa* and hypocotyls of *Ceriops* and *Paleobruguiera*.

Juncosa and Tomlinson (1988b) have recently produced a cladogram of the genera of Rhizophoraceae, based on 45 characters, which is reproduced in fig. 20.5.

5. According to Tomlinson, Primack, and Bunt (1979), other than *Rhizophora,* members of the Rhizophoreae are animal-pollinated and do not produce abundant wind-borne pollen.

The subfamilies Macarisieae, Gynotrocheae, and Rhizophoreae are well differentiated but are paraphyletic. The basal taxa of the Macarisieae have undergone relatively little evolutionary change with respect to the characters used. This cladogram is supported by another based on 16 seed characters (Tobe and Raven 1988), except that *Crossostylis* is located closer to the base of the Gynotrocheae, and the genera of the Macarisieae are arranged differently. A phenetic analysis based on pollen characters places, incorrectly it would seem, the Macarisieae between the Gynotrocheae and Rhizophoreae (Vezey et al. 1988).

Of the terrestrial taxa of the Rhizophoraceae, the basal taxa of the Macarisieae occur in Madagascar (*Macarisia,* 7 species), West Africa (*Anopyxis,* 2 species), and northern South America (*Sterigmapetalum,* 7 species). Generic diversity in the group is concentrated in West Africa and Madagascar. The Gynotrocheae occupy primarily Southeast Asia and islands to the east, as far as Fiji. The basal taxon *Carallia* extends west to India and Madagascar, providing the only contemporary geographical connection between the Gynotrocheae and the ancestral Macarisieae (fig. 20.6). The Rhizophoreae overlap the distribution of *Carallia* extensively, with the most highly derived genus, *Rhizophora,* extending into the ACEP region.

The cladogram and modern geographical distributions of the Rhizophoraceae suggest a scenario for the history of the family, which begins with the distribution of the subfamily Macarisieae in the upper Cretaceous in a broad belt extending across much of Gondwanaland (fig. 20.7). The Gynotrocheae, represented by the basal genus *Carallia,* possibly originated from Madagascan-Indian components of the Macarisieae (presently *Macarisia, Dactylopetalum,* and *Cassipourea,* with *Macarisia* being closest to the base). The connection between the Gondwanan distribution of the Gynotrocheae may have been provided by the drifting of the Indian subcontinent to its present position, leaving a remnant of *Carallia* behind on Madagascar. The origin of the Rhizophoreae from the Gynotrocheae might have occurred anywhere within the range of the basal *Carallia,* but Indochina/Malesia is strongly implicated by the high diversity of Rhizophoreae there. Subsequent spread of genera to the western part of the Indian Ocean, and of *Rhizophora* to the ACEP region, would have been accomplished by aquatic dispersal, either through the Tethys connection or eastward across the Pacific, prior to the end of the Eocene, by which time *Rhizophora*-type pollen appears in the ACEP region.

Proliferation of Mangrove Taxa

Once taxa have become established within mangrove habitat, they may speciate and diversify. The total number of species per clade is directly related to the taxonomic level of endemism of the clade in mangrove habitat (fig. 20.8). Similarly, the number of species per genus is related to the extent of geographical distribution (fig. 20.9). The latter relationship arises both from the presence of allopatric populations within a broad geographical distribution and from the presence of sympatric species. For example, four species of *Avicennia* co-occur in Malaysia, two in various parts of the ACEP region. Considerable

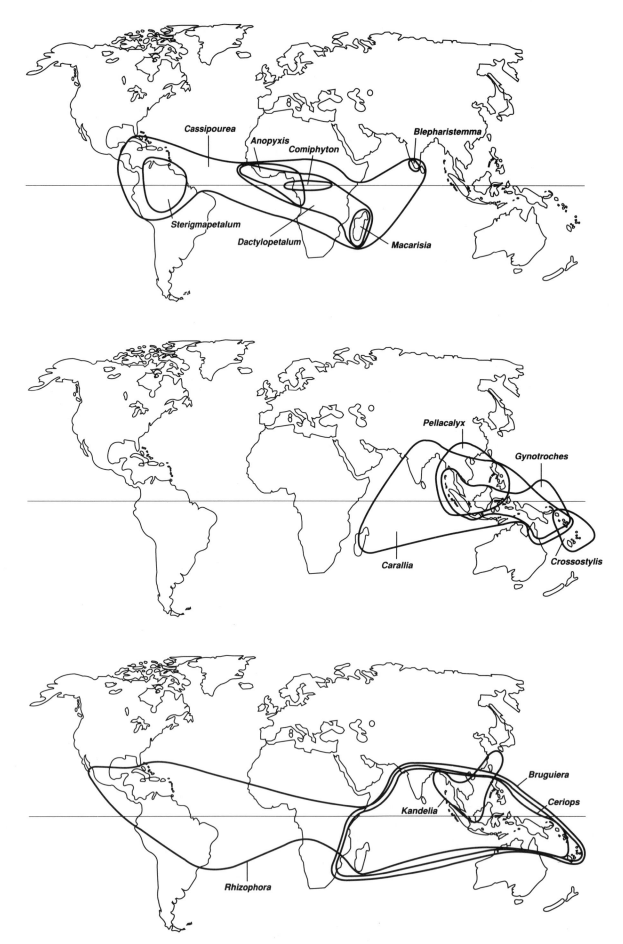

Figure 20.6　Distributions of the three tribes of the Rhizophoraceae: the Macarisieae (*top*), the Gynotrocheae (*center*), and the Rhizophoreae (*bottom*). The Macarisieae are the most primitive and the Rhizophoreae the most derived. (After Juncosa and Tomlinson 1988a.)

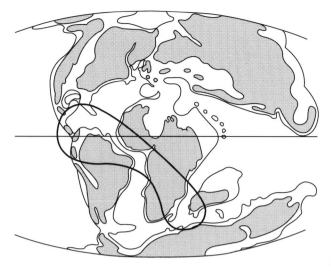

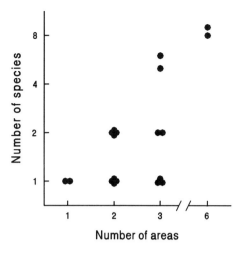

Figure 20.7 Contemporary distribution of the tribe Macarisieae superimposed on a map of the earth during the Paleocene–early Eocene. Shaded areas represent land masses; lines, continental shelves.

Figure 20.9 Number of species per genus of mangrove trees as a function of the number of areas (see Fig. 20.1) occupied. Also included are symbols representing the number of species of *Rhizophora* and *Avicennia* solely within the ACEP and IWP regions. (Data from table 20.1.)

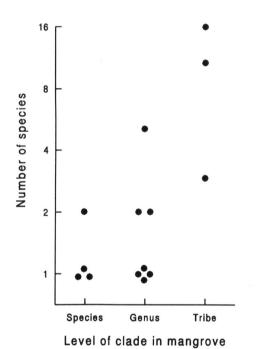

Figure 20.8 Number of species per clade of mangrove trees as a function of the taxonomic level of the clade. Each triangle represents a single clade. (Data from table 20.1.) The Rhizophoreae are considered a single mangrove clade; *Laguncularia* and *Lumnitzera* are placed in a single clade at the tribe level; *Xylocarpus* is considered to be a single species. (Tomlinson 1986.)

sympatry of species also exists within *Rhizophora, Bruguiera, Ceriops,* and *Sonneratia.* For example, *Sonneratia* contains two wide-ranging IWP species that include the ranges of three allopatric Austro-Malaysian species (fig. 20.10).

We presume that new taxa arise by allopatric speciation within mangrove habitat because of the unlikelihood

of sister taxa arising independently by parallel evolution from a more distantly related ancestor. Geographically separated sister taxa are known from *Bruguiera hainsii, Pemphis acidula, Aegialitis,* and *Camptostemon,* with most of the disjunctions occurring within Malaysia. These may represent cases of incipient or completed allopatric speciation. Allopatric models of isolation followed by secondary sympatry suggest reversible tectonic movements or cycles in climate that caused distributions to expand and contract, as might have occurred in the late Pliocene and Pleistocene (e.g., Prance 1982; Woodroffe and Grindrod 1991). Distributions of species boundaries of mangrove taxa within Malesia present a very complex picture (fig. 20.11), which suggests considerable independence of the factors affecting the distributions of each species.

Both *Avicennia* and *Rhizophora* have many species in the ACEP and IWP regions, but whether sister taxa within either genus occur within or between regions is not clear. Neither genus appears to have differentiated across the Panamanian Isthmus; *Avicennia schaueriana* and *A. bicolor* occur on opposite sides of the isthmus, but these probably are not sister taxa. *Rhizophora mangle* and *R. × harrisonii* are undifferentiated across the isthmus (Tomlinson 1986). Until phylogenetic relationships have been worked out in more detail, the geography of speciation in mangrove taxa will remain poorly understood.

CONCLUSION

We envision the origin of the ACEP-IWP diversity anomaly as the result of differences between the regions in the origin of mangrove clades and their subsequent diversification within the mangrove habitat. Extinction does not appear to have played a major role in the generation of the diversity anomaly, although the fossil record documents contractions of the ranges of taxa within each region, the extinction of *Nypa* and possibly *Wetherellia* in

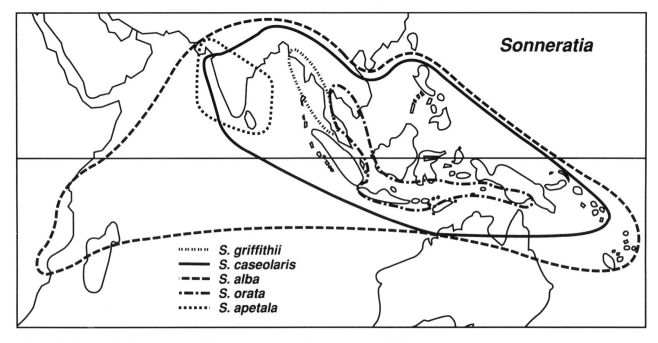

Figure 20.10 Distributions of species of the genus *Sonneratia*. (After Chapman 1970.)

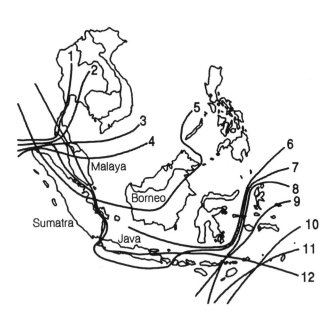

Figure 20.11 Limits to distributions of several mangrove species in Malesia: 1, *Aegialites rotundifolia* (Van Steenis 1949); 2, *Sonneratia griffithi* and *S. ovata* (Chapman 1970); 3, *Rhizophora stylosa* (Chapman 1970); 4, *Lumnitzera* spp. (Excell 1954); 5, *Kandelia candel* (Ding Hou 1958); 6, *Camptostemon philippinensis* and *C. schultzii* (Chapman 1976); 7, *Ceriops decandra* (Ding Hou 1958); 8, *Avicennia eucalyptifolia* (Chapman 1976); 9, *Aegialites annulata* (Van Steenis 1949); 10, *Rhizophora apiculata* (Chapman 1970); 11, *Lumnitzera* spp. (Excell 1954); 12, *Sonneratia ovata* (Chapman 1970).

the ACEP region, and the possibility that Southeast Asia/Malaysia is more a refugium than a center of origin. Although one cannot pinpoint the location of origin of cosmopolitan mangrove taxa (Mepham 1983a), origins since the closure of the Tethys Sea appear to have been restricted to the Indo–West Pacific region and more specifically to southeastern Asia. This raises the question of why entry into the mangrove zone and subsequent diversification did not happen more frequently in the ACEP region. It also raises the question of why clades arising prior to the closure of the Tethys connection, such as *Ceriops* and *Bruguiera*, did not extend into the heart of the ACEP region, or, if they did, why they disappeared without leaving a fossil record.

Presumably the terrestrial-mangrove transition is most likely in wet climates where a gradual transition in soil salinity exists between terrestrial and mangrove habitats, without the high salinity of the upper intertidal zone typical of arid regions (Hutchings and Saenger 1987). Southeast Asia and Malaysia are unique within the tropics in the continuous presence of extensive wet habitat since the end of the Cretaceous. Palynological and geological evidence suggests that most of the African and New World tropics were dry during the late Cretaceous (see Horrell 1990). Tropical latitudes became wetter during the early Tertiary, but the extinction of *Nypa* and the range contraction of *Pelliciera* since the Miocene suggest a marked drying trend in the ACEP region during the Neogene, at which time the ACEP and IWP regions were isolated from each other. Perhaps, then, contact between wet terrestrial vegetation and mangrove habitat was much reduced during this period, at which time diversity continued to increase in the Asian mangroves through invasion of new clades and autochthonous production of new taxa. Predictably, many more nonexclusive taxa of trees invade mangrove habitat from terrestrial habitats in the IWP region, especially Malaysia, than in the ACEP region (Saenger, Hegerl, and Davie 1983; Mepham 1983b; Mepham and Mepham 1985).

Mangrove taxa suggest a scenario for the development of diversity patterns in which elements of a diverse biota invade a stressful environment and require substantial evolutionary modification to make the transition. Farrell and Mitter (chap. 23) and Latham and Ricklefs (chap. 26) suggest a similar scenario for latitudinal gradients of diversity in terrestrial insects and forest trees. We suggest that invasion of novel environments by clades may be responsible for a component of the general relationship observed between diversity and habitat or other environmental characteristics.

21

Convergence and the Regional Component of Species Diversity

Dolph Schluter and Robert E. Ricklefs

Naturalists have long wondered whether independently evolved assemblages of species inhabiting similar environments might come to resemble one another (Cody and Mooney 1978; Orians and Paine 1983). Interest was provoked by early observations that the vegetation formations of climatically similar regions around the globe were superficially alike. A second stimulus came from systematists, who described an often astonishing degree of convergence in morphology among distantly related organisms exploiting similar resources. If individual species converged, might not communities of species also converge?

New interest in convergence at the community level rose with the development of theoretical ecology in the 1960s. Simple models suggested that local competition between species exploiting a given resource base would restrict the number and types of species that could coexist (MacArthur and Levins 1967; May 1973). It followed that if similar resources were present in two geographically isolated localities, analogous species sets would eventually result regardless of their origins. Initial work by Recher (1969), who found virtually identical relations between bird species diversity and foliage height diversity in North America and Australia, suggested that distant communities in similar environments were indeed convergent (MacArthur 1972).

Subsequent work has revised many ideas about the mechanisms maintaining community diversity, especially those concerning the omnipotence of competition (Connor and Simberloff 1979; Connell 1978, 1980; Wiens 1977; Tilman and Pacala, chap. 2), but the view that community pattern is governed by local processes still underlies the majority of theoretical and empirical studies (Ricklefs 1987; Schluter and Ricklefs, chap. 1). If this view is correct, then distant communities in similar environments are still expected to converge. Tests of convergence are therefore among the strongest available for evaluating the premise that the number and types of coexisting species are locally determined.

Orians and Paine (1983) drew attention to many examples in which diversity differed markedly between similar locations around the world. To a selection of cases from their list we have added several others (table 21.1), along with some examples of especially similar species numbers (table 21.2). Cases of the latter clearly exist, but counterexamples are no less spectacular or abundant. We may

conclude that within a taxon or guild, near-identical local diversities do not generally result when environmental conditions are judged to be the same.

However, the prevalence of convergence is difficult to establish from comparisons of diversity between matched sites in different parts of the world (Schluter 1986). The foremost problem is that similarity between communities in like habitats (e.g., table 21.2) need not imply convergence; nor need differences in community diversity preclude convergence. Two communities may be highly dissimilar (e.g., table 21.1) and yet have converged when compared with the differences that existed between their ancestors, or that exist between the regional species pools from which they are drawn. A second problem is that convergence tends to be viewed as an all-or-none phenomenon: communities must possess virtually identical sets of species to be considered convergent. Yet, convergence in species diversity is surely more widespread than indicated by the frequency of identical species sets. Finally, the unique history and geography of regions is often used to interpret dissimilar species diversities, but such effects are rarely measured or tested. Indeed, ecological research programs appear generally to lack a well-defined course of action once community dissimilarity is established.

We approach convergence in the number of coexisting species from a quantitative perspective. In particular, we view convergence as a component of variation between communities, whose importance can be measured and compared with that of other components including those reflecting properties of the surrounding region. We outline these ideas below, and apply them to a survey of data from the literature. The result is a preliminary estimate of the extent of convergence and regional effect in natural communities. We also identify limitations of the data for quantifying effects (especially regional) and suggest how future studies might minimize them. We conclude with some general points on mechanisms, and on the utility of partitioning community attributes into local and regional components.

THE COMPONENTS OF DIVERSITY

We illustrate the reasoning behind our analytical approach with an example. In an extensive series of field studies of desert lizards, Pianka (1986) counted 61 species in nine Western Australian sites, compared with only 22

230

Table 21.1   Examples of Highly Dissimilar Species Diversities in Similar Habitats Around the Globe

Organism	Habitat type[a]	No. species	No. species	Source
Mangroves	Mangal	Malaysia (40)	W. Africa (3)	Ricklefs (1987)
Algae	Rocky shore	Washington (17)	S. Africa (3)	Orians and Paine (1983)
Chitons	Rocky shore	Washington (10)	S. Africa (3)	Orians and Paine (1983)
Insects	Streams	Australia (60)	N. America (26)	Lake et al. (1985)
Insects	Bracken	England (21)	N. America (5)	Lawton, Lewinsohn, and Compton (chap. 16)
Bees	Desert	Argentina (188)	Arizona (116)	Orians and Solbrig (1977)
Bees	Mediterrean scrub	California (171)	Chile (116)	Mooney (1977)
Ants	Desert	Australia (37)	N. America (16)	Morton and Davidson (1988)
Ants	Mediterranean scrub	California (23)	Chile (14)	Mooney (1977)
Amphibians	Wetlands	Zambia (22)	Australia (14)	Simbotwe and Friend (1985)
Lizards	Desert	Australia (27)	N. America (7)	Pianka (1986)
Birds	Peatlands	Finland (33)	Minnesota (18)	Niemi et al. (1983)
Rodents	Desert	Arizona (16)	Argentina (5)	Mares (1980)

[a]Locations range in size from tree-holes to several km².

Table 21.2.   Examples of Nearly Equal Species Diversities in Similar Habitats Around the Globe

Group	Habitat	No. species	No. species	Source
Plants	Desert	Arizona (250)	Argentina (250)	Orians and Solbrig (1977)
Plants	Semiarid	N. America (70)	Australia (65)	Rice and Westoby (1985)
Sea anemones	Rocky shore	Washinton (11)	S. Africa (11)	Orians and Paine (1983)
Ants	Desert	Arizona (25)	Argentina (25)	Orians and Solbrig (1977)
Sapr. insects	Tree hole	N. America (6)	Ausralia (6)	Kitching and Pimm (1985)
Fishes	Forest lakes	Wisconsin (4)	Finland (4)	Tonn et al. (1990)
Lizards	Mediterranean scrub	California (9)	Chile (8)	Fuentes (1981)
Birds	Desert	Arizona (57)	Argentina (61)	Orians and Solbrig (1977)
Birds	Mediterranean scrub	California (30)	S. Africa (28)	Cody and Mooney (1978)
Birds	Shrub desert	Australia (5.5)	N. America (6.3)	Wiens (1991)
Small mammals	Shrubland	California (7)	S. Africa (6)	Fox, Quinn, and Breytenbach (1985)

species in the Kalahari Desert sites in southern Africa, and 14 species in twelve North American sites. Within-site diversities were also different among the three continents: an average of 27, 14.6, and 7.4 lizard species, respectively. Such differences in diversity were unexpected in environments apparently so similar, and their possible causes have provoked considerable discussion (Milewski 1981a; Morton and James 1988; Morton, chap. 14, Pianka 1986, 1989a, Westoby, chap. 15).

The significance of differences in species number between regions within a habitat type is best evaluated by additionally examining differences between habitats within each region (Cody and Mooney 1978; Blondel et al. 1984; Schluter 1986). For example, desert lizards are far more speciose in Australia than in southern Africa, but counts of lizards in tropical wetlands on the same two continents (fig. 21.1) add an additional perspective: local diversity varies with habitat in the same way in both places, despite large regional differences. That is, habitat type predicts lizard species diversity over and above the peculiarities of individual continents. Because prediction is a fundamental objective of convergence studies (see also Orians and Paine 1983), it makes sense to measure convergence by the degree to which diversity can be predicted from environment (e.g., habitat).

A second justification for assessing convergence by comparing within-habitat and between-habitat variation derives from the fact that convergence is a historical concept: it identifies differences among contemporary communities relative to differences among their ancestors

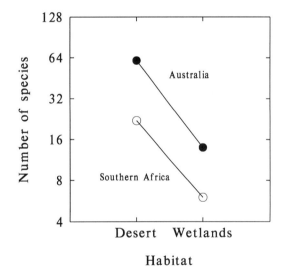

Figure 21.1 The number of lizard species in deserts and tropical wetlands of Australia and southern Africa. Desert data are from Pianka (1986), and combine species from all his study sites. Wetland data are from Simbotwe and Friend (1985), and represent complete lists from a national park on each continent.

(Cody and Mooney 1978; Orians and Paine 1983). Convergence at the level of individual species is best dealt with in a phylogenetic framework, where the characteristics of ancestors can be estimated (Wiley 1981). However, different communities within a geographical region do not often have a branching phylogeny in the sense that species do, because communities may exchange organisms at any point in their histories. Mixing can be observed directly in the fossil record (Valentine and Jablonski, chap. 29), and is also evident in the mixed taxonomic composition of communities, and in the broad habitat distributions of taxa within each region (Cody and Mooney 1978; Blondel et al. 1984; Schluter 1986; Morton and Davidson 1988; Cornell, chap. 22, Lawton, Lewinsohn, and Compton, chap. 16; but see Farrell and Mitter, chap. 23; Brooks and McLennan, chap. 24). Thus, communities in the same region are not fully isolated units, and their long-term histories are consequently blended. In the extreme, the best estimate of the ancestral state for any one community is the average of communities within the region. In this case, detecting convergence is equivalent to finding differences between habitats, consistent across regions, in the numbers of species present (Schluter 1986).

Comparison of diversities within and between habitats accomplishes an additional objective: the identification of consistent differences in diversity between regions. For example, lizard diversities in Australia parallel those found in southern Africa, but at a higher overall mean (fig. 21.1). Unique historical or geographical attributes of continents may underlie such differences (Pianka 1986; Ricklefs 1987).

The above arguments suggest that the number of species in a community may be profitably described as having at least two components: one attributable to the local habitat, and another to the region in which the community is embedded. The roles of these components may be estimated in a collection of communities by the fraction of the overall variance in diversity accounted for, as in a typical analysis of variance (ANOVA; Schluter 1986). We define convergence as the habitat component and seek to identify its importance.

METHODS

Data

Few studies were available for tests of convergence. The most extensive intercontinental comparisons are from the IBP investigations of mediterranean (Mooney 1977; Cody and Mooney 1978) and warm desert ecosystems (Orians and Solbrig 1977). Several papers were also found in a special Proceedings of the Ecological Society of Australia (vol. 14, 1985), devoted to the comparison of Australian communities with those of other regions. Other data sets are scattered throughout the ecological literature. Our final list (Appendix 21.1) is certainly incomplete, and we would appreciate being made aware of additional examples.

Field studies were invariably restricted in scope. The broadest comparisons in the literature were of diversities across habitats within a single climate zone (e.g., coastal scrub versus montane forest in mediterranean climates; rocky slope versus creosote flat in warm deserts). Still finer comparisons involved counts of species exploiting different "microhabitats" within a habitat type, such as substrates (e.g., insects located on different parts of plants; arboreal versus terrestrial lizards) and foods (e.g.,

nectar-feeding versus seed-eating birds). The third and broadest level of comparison, between habitats in different climate zones (e.g., mediterranean versus tropical wet forest), required that we combine data compiled by different researchers. In a few cases it was possible to simultaneously compare communities at more than one level of resolution (see Appendix 21.1).

Studies were also limited in the types of species investigated. All restricted study to specific taxonomic groups, even though many examined ecological guilds within these groups. A variety of taxa are represented, although terrestrial vertebrates predominate. Most studies were carried out on mainlands, although mediterranean studies by Cody and Mooney (1978; Mooney 1977) included the island of Sardinia. The island of New Guinea is represented in two other comparisons, Beehler's (1981) birds and Lawton, Lewinsohn, and Compton's (chap. 16) insect herbivores.

We restricted our regional comparisons to different continents in order to maximize the evolutionary independence of sample units. We were also interested mainly in "local" (alpha) diversities, the scale to which ecological theories apply. *Local* is difficult to define, however, and the most appropriate size of plot surely varies with taxon. We therefore relied mainly on each author's own judgement as to whether lists should be considered local or not. We included in a second category several studies in which lists were based on relatively large plots (e.g., 5–300 km²), and in which lists from several plots had been combined. Though somewhat arbitrary, the two categories permitted us a preliminary look at the effect of scale on patterns of convergence. No pattern was found, and we do not present these analyses.

A variety of methodological difficulties are inherent in any effort to compare species diversity, particularly when data from different researchers are combined. Not the least of these is the problem of how similar are matched sites from different continents. Another problem is that the number of species sampled may underestimate the number present. Orians and Paine (1983) discussed many of the problems of convergence studies, and we return to these and others in a separate section following the results.

Components of Diversity

Variation among localities in the number of coexisting species was partitioned into the following components:

$$(21.1) \qquad \sigma^2_{total} = \sigma^2_H + \sigma^2_{R,main} + \sigma^2_{R \times H} + \sigma^2_e$$

The term σ^2_H is the portion of total variance in species richness attributable to effects of local habitat (or microhabitat). In the example of figure 21.1, σ^2_H measures the variance between mean desert and wetland lizard diversities. The term includes habitat differences that are consistent across regions. For example, were lizard diversity highest in desert in one region, and highest in wetland in a second region (the opposite pattern to that shown in figure 21.1), then σ^2_H would be small. Therefore, σ^2_H is a natural measure of convergence because it represents the degree to which similar habitats yield similar diversities regardless of region.

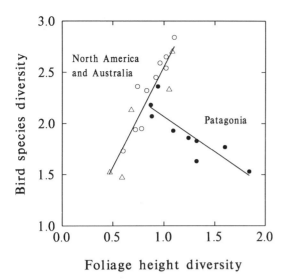

Figure 21.2 Bird species diversity and foliage height diversity in non-grassland temperate climate sites in Australia (triangles), North America (open circles), and South America (solid circles). (Data from Recher 1969; MacArthur 1964; Ralph 1985.)

The term $\sigma^2_{R,main}$ is the main effect of region on local species diversity, whereas $\sigma^2_{R\times H}$ represents the interaction between habitat and region. The main effect $\sigma^2_{R,main}$ is most simply thought of as representing differences among regions that are consistent across habitat types. For example, if one region has three times the number of species as another, averaged over all its habitats (e.g., fig. 21.1), then this will be evident as a large value of $\sigma^2_{R,main}$. In contrast, the interaction term measures the degree to which a region may have few species in some habitats compared with another region, yet have relatively more species in other habitats. An example is seen in figure 21.2, in which bird diversity declines with increasing vegetation diversity in South America, but increases with vegetation diversity in North America and Australia. This pattern might result if the habitat extremes are, for example, of opposite age or geographical extent in the different regions. In cases where the interaction is large (e.g., fig. 21.2), a main habitat effect $\sigma^2_{R,main}$ may also be present—for example, one region may have more species than the other when averaged over all habitats, even though an interaction is present. We therefore consider the total effect of region to include both $\sigma^2_{R,main}$ and $\sigma^2_{R\times H}$. The final term σ^2_e is the random error component.

Variance components were estimated from the data using the two-factor ANOVA in a general linear model (Neter and Wasserman 1974). Convergence (the habitat component) was estimated as

$$(21.2) \qquad V_C = \frac{(MS_H - MS_e)(h - 1)}{nrh},$$

where MS is the mean square from the ANOVA, h is the number of habitat types, r is the number of regions, and n is the number of sites within each combination of habitat and region. This is the typical variance component multiplied by $(h - 1)/h$, a modification to account for the fact that habitat is a fixed effect rather than a random

effect. V_C is an unbiased estimate of the "variance" among habitat means, whereas the usual variance component is not (e.g., Neter and Wasserman 1974).

The regional component was computed as

$$(21.3) \qquad V_R = V_{R,main} + V_{R\times H}$$

$$(21.4) \qquad = \frac{(MS_{R,main} - MS_e)(r - 1)}{nhr}$$

$$+ \frac{(MS_{R\times H} - MS_e)(h - 1)(r - 1)}{nhr}.$$

This is the typical variance component multiplied by $(r - 1)/r$, and provides an unbiased estimate of the "variance" among regional mean diversities.

Finally, we also computed the fraction of total variance in species diversity attributable to habitat and region. The convergence fraction was computed as the index

$$(21.5) \qquad I_C = \frac{V_C}{V_{total}},$$

where $V_{total} = V_C + V_R + MS_e$ is the total variance. Similarly, the fraction attributable to region was

$$(21.6) \qquad I_R = \frac{V_R}{V_{total}}.$$

I_C and I_R correspond to the quantities C and H in Schluter (1986).

Lack of replication in most published data sets emerged as a significant obstacle to estimating the full regional component V_R. Most studies provided only one value of species diversity for each combination of habitat and region (i.e., $n = 1$), and this prevented us from including the interaction term in the ANOVA, and hence from calculating the interaction component of variance $V_{R\times H}$ (eq. 21.3). Instead, the interaction is lumped with the error term in MS_e and cannot be extracted. Most of our measures of regional effect are thus underestimates because they include only differences between regions that are consistent over habitats ($V_{R,main}$).

A second problem, limited to data sets combined from those of different researchers (see Appendix 21.1), was that diversity values for certain combinations of habitat and region were missing. Such data can be analyzed with an additive model (no interaction term between habitat and region) provided that certain design criteria are met (Milliken and Johnson 1984; Dodge 1985). However, the variance components V_C and $V_{R,main}$ are computed differently from equations 21.2 and 21.4 when data are missing. The components can be calculated directly from the coefficients of the linear model (e.g., Dodge 1985), but an equivalent procedure when $n = 1$ is as follows: First, carry out the ANOVA and save the predicted values for each missing combination. Replace the missing values with their predicted values and recompute MS_H and $MS_{R,main}$ (not MS_e) in a second ANOVA that includes these new values. Finally, use equations 21.2 and 21.4 to calculate V_C and $V_{R,main}$ as before.

These methods can be extended to include more habitat and regional variables. For example, we analyzed some data sets by comparing the number of species in habitats, foraging microhabitats, and regions simultaneously. The calculations were similar to those for the two-way case, except that V_C included both habitat and micro-habitat components.

Finally, we used a similar, regression-based approach for data sets in which species diversity had been measured along a continuous habitat gradient (e.g., foliage height diversity; fig. 21.2). Such data sets have the advantage of replication (since only one degree of freedom is lost by fitting each line, assuming linearity). The habitat component of variation (i.e., convergence) is the fraction attributable to the independent variable. Differences between regions in the elevation of regressions measure the main regional effect. Variation in slopes represents the interaction between habitat and region, which we included in the total regional component.

All species counts were ln-transformed prior to analysis (we used ln (species + 1) when zeros were present). The ANOVAs required this transformation because variation among sites generally increased with the mean number of species present.

RESULTS

In this section we first illustrate convergence and regional effects in selected comparisons. We then summarize the general patterns shown by the complete set of available data.

Lizards

Pianka (1986) drew attention to the large differences between continents in the diversity of desert lizards (fig. 21.1; table 21.3). He suggested that historical differences, such as the great age of Australian deserts, may be responsible for this variation. Morton and James (1988) posited instead that Australia's particularly high lizard diversity is due in part to its poor soils, which engender a unique vegetation formation (*Spinifex* grass) that in turn favors a high density of termites, an important food for lizards (see also Morton, chap. 14). They raise the crucial issue that deserts on different continents may not be comparable ecologically, a possibility that is difficult to confirm or reject. However, high Australian lizard diversity appears not to be restricted to deserts; relatively high lizard diversities are also found in wetlands (fig. 21.1), where *Spinifex* is absent. This suggests that the unique histories or geographies of regions may indeed have affected diversity patterns over and above possible habitat differences.

We compared regional differences in lizard diversity with differences between habitat types, assuming for now

that habitats on different continents are ecologically similar. The data should be viewed cautiously, as they were gathered by several researchers employing different methods and are highly incomplete (table 21.3). Interestingly, habitat (convergence) and regional effects account for roughly equal amounts of recorded variation in lizard diversity ($I_C = 0.48$, $P < .05$; $I_R = 0.50$, $P < .05$).

Bird Species Diversity and Foliage Height Diversity

Perhaps the first, and certainly the most influential, quantitative study of convergence was Recher's (1969) comparison of bird species diversity (BSD) and foliage height diversity (FHD) in Australia with similar data collected in temperate North America by MacArthur (1964). Similar FHDs in sites from these two continents yielded similar bird species diversities (fig. 21.2). More recently, Vuilleumier (1972) and Ralph (1985) discovered a reversed pattern for non-grassland sites in temperate South America (fig. 21.2). Here, sites with the most complex foliage profiles (*Nothofagus* beech forest) had lower bird diversities than cedar and scrub sites with lower FHD values. One cannot rule out the possibility that unmeasured habitat differences between South America and the other continents have produced this surprising pattern. However, the low BSDs of *Nothofagus* forest probably result from its very restricted geographical distribution in South America, in contrast to the wide geographical extent of structurally less complex scrub-cedar habitats (Ralph 1985). This is a convincing demonstration of how effects of local habitat on species diversity may be superseded by regional ones.

Bird diversity was measured at several places along the FHD gradient within each continent, and three components of variation in BSD can therefore be extracted: habitat, region, and the interaction between habitat and region (assuming that the sample points within each region were independent). All three components were significant ($P < .001$). The interaction between habitat and region predominated, as reflected by the very different slopes in figure 21.2. The total regional component ($V_R = V_{R \times H} + V_{R,main} = 0.34$) was among the largest we encountered in the literature, accounting for nearly 80% of total variation in BSD among sample units (case 2 in appendix 21.1).

The similarity between Australia and North America is all the more extraordinary when viewed against the trend in South America (fig. 21.2; see Cody, chap. 13, for further corroboration). However, the total number of bird species in Australia is nearly identical to that in North America (Westoby, chap. 15), and the local pattern (fig. 21.2) may be in part a reflection of the (possibly coincidental) larger-scale resemblance.

More Birds

Birds provide one of the most extreme examples of regional effect on diversity (fig. 21.2), but they also provide some of the most detailed and extraordinary cases of convergence. Mediterranean habitats have been surveyed the most extensively, in California, Chile, South Africa, and Sardinia, by Cody (1975; Cody and Mooney 1978). Blondel et al. (1984) reanalyzed some of Cody's data and

Table 21.3. Number of Lizard Species in Different Habitats and Climate Zones

	South America	North America	South Africa	Australia
Warm desert	15[a]	13[b]	22[b]	61[b]
Tropical wetlands	—	—	6[c]	14[c]
Mediterranean scrub	8[d]	9[d]	—	—

Sources: [a]Orians and Solbrig (1977); [b]Pianka (1986); [c]Simbotwe and Friend (1985); [d]Fuentes (1981).

Note: Each value is a total species count across a variable number of sites. Dashes indicate missing values.

added their own observations from Provence in France. By mixing results from the two sets of researchers, we risk the possibility that sites are imprecisely matched or that censusing methodologies and criteria differed between Provence and the other sites. Nevertheless, virtually all analyses of all or parts of these data revealed a high degree of convergence in diversity (only a few examples are presented in appendix 21.1 to minimize redundancy). For example, Blondel et al. (1984) showed roughly parallel species counts in different diet categories in California, Chile, and Provence ($I_C = 0.74$, $P < .001$; case 26 in appendix 21.1). Regional effects consistent across habitats or diet categories were weak ($V_R \leq 0.02$). The communities were not identical: for example, Provence lacked nectar-feeding birds and had many fewer species of granivores than California (4 versus 14). Nevertheless, bird species diversity overall was highly predictable from habitat and diet category.

More recently, Erard (1989), building on earlier work by Karr (1976), compared bird communities of tropical wet forest in French Guiana and Gabon. He examined one habitat type only, but measures of convergence may be obtained for different diet groups within this habitat. Overall, the number of species exploiting different foods was strongly convergent between the two regions ($I_C = 0.80$, $P < .01$; case 13 in appendix 21.1). The largest discrepancy was in the number of nectarivores: thirteen in Guiana but only one in Gabon.

Erard (1989) also divided his birds into finer ecological subgroups, which allowed us to compare his findings with those of Cody and Mooney (1978) on the same two continents (fig. 21.3), after combining certain of the categories and deleting aquatic and ant-following categories. Several notable differences notwithstanding (e.g., in the diversity of nectarivores), the faunas are remarkably alike across habitats and foraging categories. Analysis revealed that most variation in species number was accounted for by habitat and diet ($I_C = 0.64$, $P < .01$); this fraction rises to $I_C = 0.82$ when the interaction between diet and habitat is also included in the convergence component ($P < .01$). The latter quantity allows for the possibility that the magnitude of the difference between two foraging categories (e.g., ground feeders and foliage insectivores; fig. 21.3) may differ predictably between tropical forest and mediterranean scrub.

We combined these mediterranean and tropical forest lists with those from other regions, and added paramo and temperate forest lists in a single overall analysis (case 19 in appendix 21.1). Ignoring interaction terms, we estimated that about 77% of variance in diversity among sample units could be explained by ecological factors (habitat and foraging category). This result is tentative since there are so many missing habitat/region combinations (see appendix 21.1), but the overall pattern was similar when some habitats or regions were deleted in order to minimize the fraction of missing entries. The regional effect was small ($I_R = 0.02$). Note that since a range of climates is included, the observed degree of convergence in this combined analysis partly reflects the familiar latitudinal and elevational gradients in bird species diversity. However, our analysis goes further by revealing (1) the remarkable consistency of diversity gradients in different parts of the world (e.g., fig. 21.3), and (2) that the numbers of species exploiting different foods in the same site (e.g., seeds versus arthropods; fig. 21.3) are often as different as the numbers in different climate zones (e.g., mediterranean versus topical forest; fig. 21.3). Clearly, resources may have large and consistent effects on species diversity.

Plant Diversity at 0.1 Ha

Rice and Westoby (1985) summarized the numbers of plant species in sites 0.1 ha in size, emphasizing the difference (or lack of it) between Australia and other continents. They contributed new data from Australia, and identified the most appropriate sites on other continents for comparison with these results (table 21.4). They argued that many previous comparisons involving some of these data did not carefully match sites either by physiognomy, level of human disturbance, or richness of soils. Thus the comparison raises once again the difficult question of how to find ecologically similar sites on different continents. The solution in this case might be to measure relevant soil properties of sites as well as plant diversities, as was done in comparisons of plant diversity in warm deserts of North and South America (Orians and Solbrig 1977; case 12 in appendix 21.1).

We based measurements of convergence and regional effect on the data in Rice and Westoby (1985; table 21.4). Overall, 80% of the variation among diversity values could be attributed to habitat type, indicating a strong level of convergence in diversity between Australia and elsewhere in similar habitats ($I_C = 0.80$). There was no consistent difference between Australia and other continents in diversity ($I_R = 0.03$), a result in agreement with

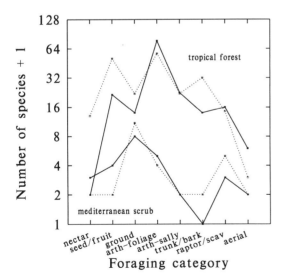

Figure 21.3 Bird diversity of different foraging guilds in mediterranean scrub and tropical wet forests of Africa (solid line) and South America (dotted line). Foraging categories are: nectar feeding; seed or fruit taken directly from plant; ground feeding; arthropods from foliage; arthropods from sallies; arthropods from trunk, bark, lichens or vines; raptors and scavangers; crepuscular and aerial insectivores. (Mediterranean data from Cody and Mooney 1978; forest data from Erard 1989.)

Table 21.4. Numbers of Plant Species in 0.1-Ha Sites

Habitat	Other continents	Australia
Moist temperate forest[a]	33	36
Temperate forest on fertile soils	78[b]	35[c]
Humid rainforest	219[d]	140[e]
Semiarid grassland and scrub[f]	70	65
Closed maquis shrubland/chaparral	38[g]	—
Mediterranean woodlands/fynbos[h]	85	77
Mallee or open oak woodland[i]	62	49

Note: Except where indicated, values are approximate midpoints of ranges summarized from various sources by Rice and Westoby (1985) in their fig. 3, or averages of values listed in their text.
[a]Types 1 (USA) and 2 in Rice and Westoby 1985 (R&W).
[b]Type 3 in R&W (USA).
[c]Temperate forests on fertile soils of the West Head.
[d]Mean of moist and wet forest values in Gentry and Dodson (1987) and Hall and Swaine (1976; cited in Gentry and Dodson).
[e]Type 9 in R&W.
[f]Types 4 (USA) and 5 in R&W.
[g]Type 10 in R&W (California and Israel).
[h]Types 12 (South Africa) and 13 in R&W.
[i]Types 6 (USA) and 7 in R&W.

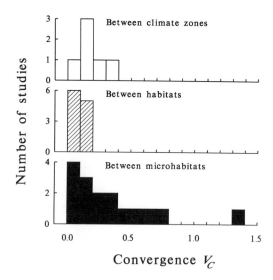

Figure 21.4 Convergence in relation to level of habitat or microhabitat resolution. The three categories refer to comparisons made between species diversities in different habitats in different climate zones (e.g., mediterranean versus tropical forest), different habitats within the same climate zone (e.g., points along a foliage diversity gradient in temperate forests), and different microhabitats or food categories within a habitat type.

Rice and Westoby's (1985) conclusion. Note that most values in our analysis are average or midpoints of diversities in 0.1-ha sites in habitats and regions and are less variable in species diversity than are individual site values. This analysis therefore overestimates the predictability of diversity in individual sites of 0.1 ha.

General Findings

The convergence component of variance in species diversity is summarized in figure 21.4. We caution that the values are not independent, since only a small number of habitat types are represented and since separate comparisons are often based on different taxa from the same study sites. Despite these problems, it is clear that different degrees of convergence were manifested in different situations. There was little indication that any taxon (e.g., birds) showed a higher degree of convergence than others. However, there was an effect of habitat resolution (fig. 21.4): the most impressive examples of convergence were in the numbers of species found in the same "microhabitats," or foraging categories within a habitat, on different continents. Similar patterns were evident in the convergence fraction, I_C, whose overall distribution was broad and roughly uniform between zero and one.

Convergence was strongly and positively related to the total amount of variance in diversity among sample units (fig. 21.5; this was true also of the index I_C). In other words, habitat or microhabitat differences were usually found to play a large role whenever sample units (sites, etc.) differed greatly in the numbers of species present. Conversely, chance and regional effects assumed the greatest relative importance when variation among sites in diversity was not large. A rough rule was that $V_C = V_{total} - 0.27$; on average, all but a constant portion of total variance in species diversity could be attributed to ecological differences among sites. Residual effects of habitat resolution on degree of convergence (fig. 21.4) disappeared once total variance among sampling units was accounted for. This suggests that the degree of convergence in similar habitats on different continents does not depend critically on how coarsely or finely the habitat cat-

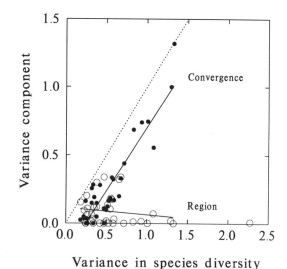

Figure 21.5 Variance components for convergence (V_C; solid circles) and regional effect (V_R; open circles) as a function of the total variance in species diversity among habitats and regions, V_{total}. The point $V_C = 1.82$; $V_{total} = 2.27$ is off the scale and not shown. Linear regressions of V_C and V_R on V_{total} exclude the points corresponding to $V_{total} = 2.27$. Dashed line corresponds to $Y = X$.

egories are defined, provided that comparisons are scaled by the total variance in diversity.

In contrast, the regional effect comprised roughly the same average portion of total variance in species diversity whatever the magnitude of V_{total} (fig. 21.5; V_R may actually decline slightly with increasing V_{total}). Thus, regional differences in diversity are a steady background feature of intercontinental community comparison. This suggests that one could estimate a regional effect using only a single habitat type, provided that replicate sites are available. Such data would not identify patterns consistent across habitat types, or allow comparison with effects of habitat, but they would nevertheless be informative and could be added to our results.

The average effect of region was found to be $\bar{V}_R = 0.09$, which should be considered a lower bound since V_R rarely included the full regional component (see "Methods" above). The upper bound to average $\bar{V}_R$ is 0.27, that portion of variance remaining after habitat effects are removed (see above). These bounds on regional variance of ln (number of species) correspond to standard deviations of 0.30 and 0.52. One may therefore expect the local number of species to vary between regions by a factor of about 1.4 to 1.7 (i.e., $e^{0.30}$ to $e^{0.52}$) when habitats are judged to be the same. We emphasize that these are averages, and that the regional effect may be larger or smaller in specific taxa or habitat types.

METHODOLOGICAL REMARKS

Up to now, we have ignored many possible problems in the data used to estimate diversity components. Here we discuss some of these problems and their possible effects on our estimates.

Nonidentical Sites. All our comparisons have assumed that sites on different continents matched by habitat and microhabitat are identical. However, perfectly matched sites are impossible to achieve, a problem that lies at the heart of current arguments over the causes of diversity differences between continents in some groups (e.g., Rice and Westoby 1985; Morton and James 1988; Pianka 1989a). Such debates will never be fully resolved because one cannot measure everything, and the ultimate importance of observed ecological differences is difficult to evaluate. The best that one can do is to match sites as carefully as possible and admit that the true magnitude of the habitat component may be underestimated somewhat because of residual ecological differences between sites on different continents. Such residual differences are not necessarily consistent between regions across all habitat types, and hence they may be more likely to inflate the error and interaction components of diversity than the main regional component.

No Replication. Most comparisons of diversity have not included replication, which means that only convergence and the main effect of region can be estimated. Future studies will need to incorporate more sites within a region to allow estimation of the interaction between region and habitat. It is not necessary that replicates be evolutionarily independent (one would not expect them to be, since they are from the same region), but only that their local fluctuations in diversity are uncorrelated. Replication and habitat variation among sites can often be achieved simultaneously in sites ordered along similar habitat gradients on different continents, such as foliage height diversity (fig. 21.2) or soil texture (Orians and Solbrig 1977; case 12 in appendix 21.1).

Nonindependence of Regions. Convergence measures the extent to which habitat predicts diversity across independent regions. "Independence" does not require that regions possess entirely different forms of life, but only that dispersal between regions, or recent common ancestry of their faunas, is not the present cause of any tendency for diversity to vary with habitat in the same way in different regions. For example, a manyfold difference in the number of species between tropical forest and grassland in both of two regions would not be considered remarkable if the regions had become isolated from each other only recently. In this case, the originally greater diversity in tropical forest might be an effect of habitat, but the repetition of the pattern in two regions would not represent convergence. In general, dispersal and common ancestry would tend to inflate the apparent habitat component of diversity. They would also reduce the regional component, but this would accurately reflect the histories of the regions and would not be an artifact.

We compared similar habitats on different continents rather than at opposite ends of the same continent in order to maximize the degree of independence of sites, but common history may nevertheless taint some of the comparisons. For example, the number of bird species feeding on nectar was low in mediterranean habitats of both North and South America (three and two, respectively; Cody and Mooney 1978). Is this consistently low diversity a direct consequence of the resource (i.e., convergence) or does it reflect some intrinsic property of the Trochilidae, to which all five species belong (i.e., common history)? To answer this we must examine the taxon in other habits (e.g., Erard [1989] documents thirteen nectar-feeding species in the tropical forest of South America, indicating that local diversity is not inherently low in Trochilidae), and look at other nectar-feeding taxa in the same habitat (e.g., Cody and Mooney [1978] also recorded three nectarivore species in mediterranean South Africa, but none are Trochilidae, suggesting that nectarivore diversity is low in mediterranean habitat regardless of taxon). Many intercontinental comparisons of diversity have examined the possible effect of taxonomy on degree of resemblance (e.g., Cody and Mooney 1978; Blondel et al. 1984; Lawton, Lewinsohn, and Compton, chap. 16). Future approaches might directly incorporate phylogenetic methods into the comparative framework (see Brooks and McLennan, chap. 24; Farrell and Mitter, chap. 23).

Measurement Accuracy. Problems of measurement accuracy arise in two forms. First, the number of species recorded tends to underestimate the number present (i.e., counts are biased estimates of diversity). Alternative measures of diversity such as Simpson's index are less biased, but only because they discount the rarest species, which is not always desirable. Second, different researchers use different plot sizes and survey techniques, making it difficult to compare their findings. Adopting a fixed plot size is a helpful way to standardize counts and minimize variation between studies (e.g., plants in 0.1-ha plots; Rice and Westoby 1985), although the result is that different numbers of individual plants are counted, confounding density and diversity. The best approach may be to count a fixed number of individuals, or to use a standard plot size and adjust for different densities using rarefaction. The quantitative consequences of bias and variability among studies for estimates of convergence and regional effect are unpredictable.

DISCUSSION

Despite possible problems with the data, it is clear that convergence in species diversity is not a rare or peculiar phenomenon in nature. Indeed, our survey suggests that at least some convergence is the rule rather than the exception. This does not imply that identical diversities are commonly the outcome of species assembly and evolution, but it shows that habitats exert repeatable effects on the number of coexisting species. Convergence was most noticible when the overall variance in species number among samples was high. On average, consistent habitat effects were uncovered whenever diversities among samples differed by more than a factor of about 1.7 (i.e., when the standard deviation of ln (diversity) exceeded 0.27).

The habitat component of diversity ranged between low and very high levels depending on circumstance, and communities could not be simply classified into convergent and nonconvergent groups. Rather, it is more realistic to view the number of coexisting species as having several components, of which common ecological environment is but one. The goal should be to better quantify the effects of habitat, and then to test alternative hypotheses giving rise to them. Interestingly, many of the most spectacular cases of convergence involved gradients in diversity of species on resources that exist side by side within the same habitat type (see appendix 21.1). Because of their small scale, such gradients may prove to be very useful in tests of hypotheses for convergence.

Alternative explanations for convergence fall into three classes. The first includes all local ecological processes favoring the coexistence of more species in some habitats than in others (see Underwood and Petraitis, chap. 4; Tilman and Pacala, chap. 2; Wright, Currie, and Maurer, chap. 6; and Rosenzweig and Abramsky, chap. 5). By itself, the presence of convergence does not imply that local diversities ever reach a ceiling. Rather, processes tending to reduce diversity, such as competition, predation, and environmental variability, and processes tending to increase it, such as mutualism, resource productivity, and resource diversity, need only depend on habitat type. The outcome is that diversity varies between habitats in similar ways on different continents, even though local diversity may currently be higher on one continent than another (e.g., fig. 21.1).

The second class includes all ecological processes consistently favoring a higher rate of species formation and spread in some habitats than in others. Speciation is not strictly a local process, but the mechanisms promoting it in this case must have their origins in features of the habitat in order to yield convergence. Features of a habitat type that might promote speciation include its typical geographical distribution. For example, montane habitats usually occur as fragmented archipelagoes of relatively small islands, whose distribution may promote speciation (but also impede subsequent spread, and perhaps promote local extinction). Habitats may also differ in the extent to which they promote speciation through ecological specialization, a factor proposed to explain the greater diversity of phytophagous insect taxa than sister taxa exploiting other resources (see Farrell and Mitter, chap. 23).

The third class includes the possibility that habitats in different regions have parallel histories. For example, as global climate cools, new habitats may form on mountaintops, which then begin to receive immigrants from older habitats at lower elevations. In this case, continents will exhibit similar relative numbers of species in lowland and highland habitats purely as a consequence of habitat age. Alternatively, a taxon that evolved in one habitat may spread very slowly to other habitats, owing to the difficulty of acquiring certain required adaptations (e.g., freezing tolerance). The process would then result in parallel gradients of diversity on different continents where the taxon is present. This argument has been used to explain latitudinal diversity gradients in phytophagous insects (Farrell and Mitter, chap. 23) and trees (Latham and Ricklefs, chap. 26), and tree diversity differences between terrestrial and mangrove habitats worldwide (Ricklefs and Latham, chap. 20). Diversity gradients of this third class are effects of habitat in the statistical sense, but the cause (e.g., habitat age) is not a measurable feature of present-day sites. The processes should probably be regarded as regional ones.

Regional effects were also present in many cases, although we were prevented from fully quantifying the regional component except when replication allowed (appendix 21.1). Localities in some regions possess consistently more species than those in others, even though habitats appear to be the same (see also Ricklefs and Schluter, chap. 30; Ricklefs and Latham, chap. 20; Westoby, chap. 15; Lawton, Lewinsohn, and Compton, chap. 16). Our results indicate that localities in separate regions differ on average by a factor of about 1.4 when habitats are similar. This estimate may be too low if the literature is biased toward particularly clear examples of convergence with low regional effect. We hope that the present results will stimulate others to survey more extensively both components of variation in diversity.

These differences between regions in the number of locally occurring species seem to have as their basis differences in regional diversity (see Ricklefs and Schluter, chap. 30; Westoby, chap. 15). Their causes may be varied, and are discussed in more detail in other chapters (e.g., Morton, chap. 14; Ricklefs and Schluter, chap. 30; Blondel and Vigne, chap. 12). Differences between regions that are consistent over several habitat types (i.e., the main regional effect) might stem from processes of a very general nature, including regional differences in age, rates of speciation and immigration, or history of extinctions. One must also consider the possibility that such effects are caused by consistent ecological differences between sites in separate regions (e.g., the poor soils of Australia). Habitat gradients in diversity that are not parallel between regions (e.g., fig. 21.2) represent the interaction component of the regional effect. These effects may stem from variation between regions in the available proportions of different habitats (an area effect), from differences in the taxa present, or from differences in the habitats through which species were able to enter a region. An ambitious future task will be to develop and test regional hypotheses based on these alternative mechanisms. The effort may be no less challenging than testing alternative hypotheses on convergence, especially because each region may require its own unique explanation.

APPENDIX 21.1

Components of total variance in species diversity (V_{total}): convergence (V_C) and regional effect (V_R). Scale refers to a comparison between local sites (1), or between larger areas and combinations of sites (2). Niche indicates whether diversity is compared between habitat types (1) or microhabitats (2) (food type, foraging substrate, or plant life form). Climate zone is left blank when more than one zone is involved. V_R includes the interaction between habitat and region only in cases marked by an asterisk. Literature sources and other details are provided below.

Organism	Scale	Niche	Climate zone	V_{total}	V_C	V_R	Source
Bird	1	1	Medit. scrub	0.24	0.16	0.00	1
Bird	1	1	Temp. forest	0.47	0.11	0.34*	2
Herp	1	1	Medit. scrub	0.38	0.05	0.30	3
Invert	1	1	—	0.66	0.20	0.32	4
Invert	1	1	Medit. scrub	0.32	0.15	0.06	5
Invert	1	1	Medit. scrub	0.35	0.19	0.11	6
Invert	1	1	Warm desert	0.18	0.02	0.15	7
Mammal	1	1	—	0.59	0.17	0.03	8
Mammal	1	1	Medit. scrub	0.22	0.05	0.02	9
Mammal	1	1	Warm desert	0.26	0.03	0.20	10
Plant	1	1	—	0.32	0.25	0.01	11
Plant	1	1	Warm desert	0.50	0.10	0.02*	12
Bird	1	2	Trop. forest	2.27	1.81	0.01	13
Herp	1	2	Warm desert	0.57	0.33	0.17	14
Herp	1	2	Warm desert	0.71	0.44	0.00	15
Invert	1	2	Bracken	0.55	0.17	0.18	16
Invert	1	2	Warm desert	0.23	0.04	0.09*	17
Plant	1	2	Medit. scrub	1.08	0.56	0.07	18
Bird	1	1+2	—	1.30	1.00	0.02	19
Bird	1	1+2	Medit. scrub	0.93	0.74	0.01	20
Bird	2	1	Peatlands	0.26	0.00	0.06	21
Herp	2	1	—	0.68	0.33	0.34	22
Invert	2	1	—	0.24	0.00	0.00	23
Invert	2	1	Temp. lake	0.54	0.18	0.10	24
Bird	2	2	Medit. scrub	0.59	0.32	0.00	25
Bird	2	2	Medit. scrub	1.00	0.74	0.00	26
Bird	2	2	Paramo	0.34	0.00	0.00	27
Bird	2	2	Warm desert	0.47	0.00	0.00	28
Herp	2	2	Swamp	0.27	0.00	0.10	29
Herp	2	2	Swamp	1.33	1.31	0.00	30
Herp	2	2	Swamp	0.33	0.28	0.03	31
Herp	2	2	Warm desert	0.42	0.28	0.00	32
Herp	2	2	Warm desert	0.83	0.69	0.01	33
Herp	2	2	Warm desert	0.52	0.16	0.00	34
Mammal	2	2	Warm desert	0.51	0.12	0.13	35
Bird	1+2	1	—	0.38	0.14	0.11	36

Sources and comments

1. Blondel et al. 1984, from their table 1. No. bird species in 3 mediterranean regions, by 4 habitats.

2. MacArthur 1964; Recher 1969; Ralph 1985. Bird diversity along a continuous foliage diversity gradient, including non-grassland sites only. Analysis is based on the diversity metrics (untransformed) rather than on log species counts, as the latter were not available.

3. Fuentes 1981. No. lizard species in mediterranean Chile, California, and Sardinia, by 3 habitats.

4. Lake et al. 1985, from their table 8. Projected no. stream insect species on 100 stones in North America vs.

Australia, temperate vs. tropics (latter missing in Australia; 2–6 replicate streams/category).

5. Mooney 1977. No. bee species in mediterranean Chile vs. California, by 3 habitats (coastal scrub to desert scrub).

6. Mooney 1977. No. ant species in mediterranean Chile vs. California, by 4 habitats (coastal scrub to montane forest).

7. Orians and Solbrig 1977, from their table 5-15. Ant species in sonoran vs. monte deserts, by habitat (*Larrea* flat vs. desert wash).

8. No. small mammal species in sites in Australia,

South Africa, and North America, by 2 habitats. Shrublands (4.3, 6.0, and 6.7 spp.; Fox, Quinn, and Breytenbach 1985) and desert (1.5, 0, and 5.3; Pianka 1986).

9. Mooney 1977. No. small mammal species in mediterranean Chile vs. California, by 4 habitats.

10. Mares 1980. No. small mammal species in Sonoran and Monte Deserts, by 3 habitat categories.

11. See table 21.4.

12. Orians and Solbrig 1977, from their tables 4-9 and 4-10. No. perennial plant species in Sonoran vs. Monte Deserts, along a continuous soil texture gradient (% rock + gravel).

13. Erard 1989, from his table 4.2. No. bird species in rainforest of Gabon vs. French Guiana, by 7 feeding categories.

14. Pianka 1986. Mean no. lizard species in desert sites of North America, Kalahari, and South America, by 5 microhabitats (ground vs. arboreal, diurnal vs. nocturnal, and subterranean).

15. Orians and Solbrig 1977, from their table 7-6. No. lizard species in Sonoran Vs. Monte Desert sites, by 5 foraging categories. Sonoran data taken from Pianka (1973). Monte list is from Andalgala.

16. Lawton, Lewinsohn, and Compton, chap. 16, from their fig. 16.5. No. herbivores on bracken in sites of 6 regions (South Africa, New Guinea, UK, USA, Australia, and Brazil), by 3 plant parts (rachis, pinna, and midribs).

17. Morton and Davidson 1988, from their tables 1 and 2. No. ant species in desert sites of USA vs. Australia, as a function of precipitation, a continuous "habitat" variable. Used *Acacia* shrubland sites in Australia, as they were most similar to USA sites in vegetation.

18. Cody and Mooney 1978, from their table 2. No. plant species in 4 mediterranean regions, by 9 life forms.

19. No. bird species in 8 foraging categories (see fig. 21.3), 4 habitats (tropical and temperate forest [tr & te], mediterranean [m], paramo [p], and 5 regions. The no. species in the 8 categories is listed by region and habitat: Africa ([tr: 1, 21, 13, 76, 22, 13, 15, 4; Erard 1989], [m: 3, 4, 8, 5, 2, 1, 3, 2; Cody and Mooney 1978]), South America ([tr: 13, 51, 22, 57, 22, 32, 15, 3; Erard 1989], [m: 2, 2, 11, 4, 2, 2, 5, 2; Cody and Mooney 1978], [p: 1, 1, 3, 3, 0, 0, 2, 0; Dorst and Vuilleumier 1986]), Malaysia-New Guinea ([tr: 12, 56, 21, 32, 13, 4, 19, 5; Beehler 1981, cited in Erard 1989], [p: 0, 0, 1, 4, 0, 0, 0, 1; Dorst and Viulleumier 1986]), Europe ([m: 0, 3, 9, 1, 0, 3, 2; Cody and Mooney 1978], [te: 0, 6, 11, 19, 5, 7, 9, 8; Tomialojc, Wesolowski, and Walankiewicz 1984, cited in Erard 1989]), North America ([m: 0, 3, 7, 7, 2, 2, 5, 2; Cody and Mooney 1978], [p: 0, 2, 0, 1, 0, 0, 1, 0; Dorst and Viulleumier 1986]). There are many missing values in this design (data are available for few habitats within each region), but combining regions to minimize this (e.g., Europe and Southeast Asia) gave similar results.

20. Blondel et al. 1984, from their table 1. No. bird species in 3 mediterranean regions, by 4 habitats, and additionally by 4 diet categories (G, F, N, & I; Fo and O were excluded; species in H were divided between I and G, those in J were split among I and F, and species in K were divided between F and G.)

21. Niemi et al. 1983. No. bird species in peatlands of Minnesota and Finland, by 3 habitats (open habitats excluded, semi-open missing from Finland).

22. See table 21.3.

23. No. ant species in study areas, summed across sites, in USA, South America, and Australia, by 2 habitats: warm desert (16, 28, and 37 spp.: Orians and Solbrig 1977; Morton and Davidson 1988) and mediterranean scrub (23, 10, and — [missing]; Mooney 1977).

24. Timms 1985. No. species of macrobenthos in small lakes (< 2 km²) of Australia vs. other continents (Europe and USA), by three water types (acid, saline, fresh).

25. Blondel et al. 1984, from their table 2. No. bird species in mediterranean areas of 3 regions, by 7 foraging substrates.

26. Blondel et al. 1984, from their table 3. No. bird species in 3 mediterranean regions, by 9 diet categories.

27. Dorst and Vuilleumier 1986, from their table 6-4. No. bird species in paramo areas of South America and Africa (both ≈ 300 km²), by 7 foraging categories (combining predators and scavengers, "other guilds" excluded).

28. Orians and Solbrig 1977, from their table 5-13. No. bird species in Sonoran vs. Monte Deserts, by 6 foraging categories.

29. Simbotwe and Friend 1985. No. lizard species in tropical wetlands of Australia vs. Zambia, by 3 microhabitats (arboreal, semiarboreal, and terrestrial/aquatic).

30. Simbotwe and Friend 1985. No. lizard species in tropical wetlands of Australia vs. Zambia, by 2 microhabitats (arboreal and terrestrial/aquatic).

31. Simbotwe and Friend 1985. No. amphibian species in tropical wetlands of Australia vs. Zambia, by 2 microhabitats (arboreal and terrestrial/aquatic).

32. Orians and Solbrig 1977, from their table 7-6. Total lizard species counts in Sonoran vs. Monte Desert areas, by 5 foraging categories.

33. Orians and Solbrig 1977, from their table 5-7. Snake species counts in Sonoran vs. Monte Deserts, by 4 foraging microhabitats.

34. Orians and Solbrig 1977, from their table 5-9. Anuran species in Sonoran vs. Monte Deserts, by 2 tadpole diet groups.

35. Orians and Solbrig 1977, from their table 5-5. Mammal species counts (excl. bats) in Sonoran vs. Monte Deserts, by 6 diet categories.

36. Schluter 1986. No. of finch species in 9 habitats and 4 regions.

PART FOUR

Historical and Phylogenetic Perspectives

Communities have histories of development. Their present properties reflect the phylogenetic origins of the taxa they contain, as well as the unique events and geographical circumstances that occurred during their formation. The effects of history are most apparent at the regional level, but the connections between regional and local levels transmit this historical influence to the local community.

History raises many questions concerning local diversity: What is the history of the association between the members of a local assemblage? Do coexisting species coevolve, or is their association brief and changeable? How does the phylogenetic origin of available taxa influence the patterns of resource use, the rate of origination of new forms, and the strength of interactions between species? What are the relative influences of invasion by clades from nearby communities (adaptive shifts) and of the in situ proliferation of clades (adaptive radiation) on community diversity? To what degree do assemblages at different localities within a region share a common history? Do communities have time to achieve equilibria between speciation and extinction rates, in the way that island communities may achieve an equilibrium between colonization and extinction?

Answering these questions requires historical reconstruction. The chapters in this part of the book approach this problem from three directions: the fossil record, systematics, and phylogenetics. Each of these approaches has different advantages and disadvantages and problems of interpretation and completeness of data. Nevertheless, together they begin to form a picture of the historical development of regional biotas and the effect of that history on species diversity within local communities.

Cornell finds little evidence for ecological saturation of phytophagous insect communities. Phytophagous insect species are heterogeneously distributed among species of trees in the British flora with respect to insect taxonomy and guild membership (chewers, miners, sap feeders, gall formers). Patterns of community resource use are strongly influenced by taxonomic proliferation within guilds. Furthermore, patterns of diversity cannot be related to the properties of the trees as hosts. Cornell suggests that historical accidents and biogeographical circumstances account for most of the variation in taxonomic composition of the community and guild structure.

Farrell and Mitter address diversity of phytophagous insects and their host plants from a phylogenetic perspective. This approach is particularly successful because host use by many phytophagous insects is a relatively conservative trait, and in extreme cases host and insect phylogenies are tightly concordant. The authors focus on evidence bearing on Ehrlich and Raven's "escape and radiation" hypothesis, which postulates that insect and host lineages have evolved successive adaptations to one another (new defenses by plants, and new counteradaptations by insects), and that each group experienced bursts of diversification while it (temporarily) had the upper hand. Among the most striking finds is that one plant defense, secretory canals, is consistently associated with higher plant diversity within clades possessing this trait. Interestingly, comparisons show no difference in rates of proliferation of sister clades present in tropical and in temperate regions. Hence speciation rate is not higher in the tropics; rather, the latitudinal gradient in species diversity may result because insect clades originated in the older tropics, and only later colonized more recent temperate environments.

Brooks and McLennan search for coevolution and host switching using a phylogenetic analysis of coevolved ecological associations, particularly those of hosts and parasites. They assess history by genealogical relationships among species, then use these data to determine how species came to jointly inhabit a region and how they interact with each other. Their methodology requires a phylogeny, but shows great power in distinguishing historical (genealogical) and nonhistorical contributions to the composition of ecological communities.

Cadle and Greene use phylogenetic analysis to demonstrate the effect of genealogical history on the composition and niche occupation of Neotropical snake assemblages. North and South American communities contain different proportions of different clades of snakes, which exhibit disparate morphological and ecological characteristics. Thus, whether or not a particular guild of snake is present in a local community depends greatly on the availability of suitable clades within the region. Cadle and Greene's analysis implies that ecological niches are often unfilled, and that the occupation of empty niche space by particular clades (and hence overall diversity) is limited by evolutionary barriers. This recalls a theme developed by Farrell and Mitter, Ricklefs and Latham, and others in this volume.

Latham and Ricklefs examine the well-known anomaly in temperate-zone tree diversity between North America, Europe, and Asia. Angiosperm trees constitute an excellent system for historical and ecological analysis because they fill virtually an entire trophic level in forested habitats, they are well represented in the fossil record, and most Tertiary genera persist to the present. The fossil record shows clearly that at all taxonomic levels tree diversity in Asia has persistently exceeded that in North America. Hence the diversity anomaly is very old. European diversity was reduced from levels now found in Asia by extinction during the Neogene. Systematic and fossil data indicate that the North American and European floras are mostly subsets of the Asian flora. Latham and Ricklefs emphasize the contribution of the unique biogeography of each region to the development of diversity in forest vegetation.

Kauffman and Fagerstrom use the excellent fossil record of reef-building organisms to illustrate the tremendous variation in reef communities over time in response to changes in physical characteristics of the marine environment (such as ocean chemistry affecting precipitation of calcium carbonate, sedimentation, extent of shallow marine platforms, and climates), and in the aftermath of catastrophic extinction events. The evidence suggests that mass extinctions have been a primary force regulating reef diversity through time, periodically reducing diversity to near zero and sometimes resulting in long periods of negligible reef development. Such extinctions also apparently allow the reef habitat to become dominated by taxa much less prevalent prior to the extinction event.

Van Valkenburgh and Janis demonstrate the power of fossil data to resolve historical, phylogenetic, and geographical components of community organization. Local and regional diversity of large herbivorous mammals declined in North America since the mid-Miocene, and varied from a high of 98 species to the present low of 12. Lower diversity was accompanied by lower spatial turnover (beta and gamma diversity) and larger geographical ranges for the remaining species. Carnivore diversity varied less through time than herbivore diversity, remaining relatively constant; predator/prey ratios have therefore fluctuated greatly. In contrast to results presented in several other chapters, waves of immigration of species from Eurasia and South America did not cause a noticeable rise in regional or local diversity in North America, suggesting saturation.

Valentine and Jablonski use fossil data from the Pleistocene to examine the response of communities of marine invertebrates to rapid environmental changes and to immigration of new species from outside. Their analysis suggests that species associations are loosely formed and lack long duration. The invasion of communities by species from outside brought about little extinction, suggesting that ecological interactions within communities do not tightly regulate diversity. Valentine and Jablonski emphasize a constant flux in community organization and changing distributions over geographical areas within evolutionary (geological) time. Accordingly, the fossil record provides little evidence for the conditions of long-term association that would promote formation of coadapted species complexes.

The historical record, whether it is inferred from fossil or from systematic and geographical evidence, presents a world in constant flux and out of equilibrium with the contemporary ecological scene when viewed on large scales. Adaptive shifts and adaptive radiation appear to be important components of the filling of ecological niche space, even on the local level. Community ecology clearly exhibits a tension between the tendency of species diversity to conform to local ecological conditions, on one hand, and the disruptive force of regional processes, such as species production, and of unique historical events, such as tectonic movements and bolide impacts, on the other. Ecologists have tended to emphasize the former, systematists and paleontologists the latter. The truth lies somewhere in the middle ground, which can be cultivated only by integrating the insights and approaches of these disciplines.

22

Unsaturated Patterns in Species Assemblages: The Role of Regional Processes in Setting Local Species Richness

Howard V. Cornell

WHAT IS SATURATION?

The principal concern of this chapter will be to discuss how one might go about answering the question, are ecological communities saturated with species?, that is, do they exhibit upper limits to species richness? The term *saturation* has had a variety of meanings in the past (see Terborgh and Faaborg 1980 for a short review), but its usage here will be restricted to an upper limit to local richness that is independent of the size of the regional colonization pool in the sense that further increases in regional richness will have little influence on local richness. Such a limit must result from species interactions in local habitats. Accordingly, coexistence depends upon resource partitioning or other mechanisms to avoid such strong competition as to cause the local exclusion of interacting populations. Normally, species composition will be relatively stable after an initial period of assortative shuffling. Saturation in this sense then does not mean a limit to richness set by the size of the colonization pool, sometimes called "pool exhaustion." It also does not apply to a stochastic equilibrium of independent colonization and extinction rates in the habitat, even though the resulting turnover can produce a richness ceiling (an upper limit to richness that can occur for any reason). However, if turnover is driven by successive invasion of competitively superior species, then interactions are clearly restricting local richness, and the resulting richness ceiling will fulfill the definition.

SATURATION IN THEORETICAL COMMUNITIES

The first and perhaps most obvious question to ask is, does theory predict that saturation should occur? Examinations of several model systems have addressed the maintenance of richness in ecological time.

Noninteractive Communities

Theoretical communities can be classified as interactive or noninteractive depending upon whether strong biotic interactions take place among the residents of a local habi-tat. Strong, Lawton, and Southwood (1984, 112–113) present two verbal models for noninteractive communities. The first is based on strong density-independent fluctuations in the abiotic environment, whereas the second is based on the limited ability of colonists to invade a habitat (see Strong, Lawton, and Southwood 1984 for detailed descriptions). The habitat occupied by the community may thus be thought of as a population "sink" (Pulliam 1988) for a large proportion of the constituent species. In "sink" habitats, populations undergo frequent extinctions and must be rescued by recolonization from "source" habitats where they are persistent. If local extinction in the sink is more a function of abiotic fluctuation than of competition or predation, and if local reproduction rates are too low to balance local mortality rates, then local richness will depend more upon extrinsic recolonization than upon intrinsic interspecific interactions (see also Holt, chapter 7, for a more detailed exploration). Such conditions are believed to be characteristic of many insect communities on host plants (Strong, Lawton, and Southwood 1984). Noninteractive communities of this sort are by definition unsaturated, and are likely to be common where abiotically generated "sink" habitats prevail, for example, in extreme environments, or at the edges of species' ranges.

Caswell (1976) developed what appear to have been the first noninteractive community models to predict the structure of noninteractive assemblages of species (see also Caswell and Cohen, chap. 9). Caswell's Model I is the most purely noninteractive; it assumes that all species respond to abiotic conditions identically, and that all biotic interactions among species are absent. Colonization by new species is a random process that is independent of the number of species already present. Once a new species enters the community, its birth rate and death rate are linear and equal, and the resulting population dynamics of each species is completely independent of all others. Niche space is always open since there is no limit on the number of individuals summed over all species that can coexist in the community.

As expected, Caswell's model predicts that given an infinitely large regional pool, there is no upper limit on the

Table 22.1. Saturation or Lack of Saturation In Community Models

Model Type	Source	Saturation
Noninteractive	Caswell 1976	No
Interactive		
Niche heterogeneity		
Classical niche	MacArthur 1972	Yes
Resource ratio	Tilman 1985	Yes
Temporal niche/	Chesson and Huntly 1989	Yes
subadditivity		
Spatio-temporal heterogeneity		
Lottery	Sale 1977	No
Random walk	Hubbell and Foster 1986	No
Aggregation	Shorrocks and Rosewell 1986	Yes
Disturbance	Huston 1979	Yes
Specialist predator	Janzen 1970; Armstrong 1989	No

richness of the local community. Local richness depends only upon the length of time the community has existed and over which colonization has occurred (Caswell 1976).

Interactive Communities

Interactive communities can be of many types, but generally fall into two categories, niche heterogeneity types and spatio-temporal heterogeneity types (table 22.1). This grouping is unconventional and somewhat oversimplified relative to other classification attempts (cf. Schoener 1986), but the models in each category have sufficient features in common to make the grouping logical and convenient for the discussion here.

Niche Heterogeneity Models. Niche heterogeneity models are simply variations on "Gause's paradigm," which holds that coexistence is favored by species differences in one or more important niche dimensions. Each species does best in one part of the niche space, thereby maintaining its place in the assemblage. In niche heterogeneity models coexistence depends upon classic assumptions about niche packing, niche partitioning, and limiting similarity, either in a spatial context (MacArthur 1972; Tilman 1985) or in a temporal one (Chesson and Huntly 1989), and, as such, all of these models predict saturation resulting from interactions in limited niche space.

Spatio-Temporal Heterogeneity Models. In spatio-temporal heterogeneity models, environmental heterogeneity or uncertainty is an important condition for coexistence, and the niche space is often assumed to be uniform and unpartitioned. In spatio-temporal heterogeneity models, there are four ways species can coexist: (1) by having roughly the same competitive abilities, coupled with unpredictable recruitment via chance disturbance (Sale 1977; Hubbell and Foster 1986); (2) by reducing the overall intensity of competition via aggregated utilization of fragmented resources (Shorrocks and Rosewell 1986); (3) by periodic disturbance to a single habitat (i.e., without a spatial dimension) coupled with slow population growth rates (Huston 1979; see also Caswell and Cohen, chap. 9);

and (4) by spatial variation in the risk to new population recruits of attack by specialist predators (Janzen 1970; Armstrong 1989). The specific details of how local richness is maintained in each of these models are treated in each individual study, but as regards saturation, the results are mixed (see also Cornell and Lawton 1992).

The aggregation and disturbance models predict ceilings on local richness for a given species pool, but for somewhat different reasons than in niche heterogeneity models. The aggregation model is formally related to the patch models of Skellam (1951) and Horn and MacArthur (1972); species can always be added to these models if they have the right combination of competition coefficients, dispersal ability, and extinction rates. But Shorrocks and Rosewell (1986) have shown that communities drawn from collections of species with randomly assigned competitive abilities and aggregation parameters always reach a richness value (roughly seven species) where new invasions are not possible. Communities represented by both aggregation and disturbance models will collapse to single-species monocultures in the absence of mitigating aggregation or disturbance. Richness is maintained, or at least exclusion is slowed, by a dynamic balance between disturbance or dispersal among fragmented resources on the one hand and elimination of species by superior competitors by means of population growth on the other. Niche space need not be partitioned in order for coexistence to occur (although "meta-niche" space can be; Tilman and Pacala, chap. 2).

The lottery and random walk models are fundamentally stochastic, so that species composition and relative abundance change temporally. Niche space is unpartitioned, but interactions among species can be intense. However, species interactions play no role in competitively driven turnover, as defined in the introduction, because all species have equal competitive abilities and therefore cannot eliminate one another. At the extreme, ceilings to richness can arise, but not because of saturation; they exist only because habitat space is finite, placing restrictions on the total number of individuals that can occur there. As richness increases, average population size decreases until local extinction matches the rate of appearance of new species at the site. This stochastic equilibrium is scale-dependent because larger habitats will have lower average extinction rates, all else being equal. In a simple model, Hubbell and Foster (1986) have shown that average extinction probabilities are likely to be quite low even in moderate-sized habitats, suggesting that such stochastic communities are unlikely to reach a ceiling in moderate- to large-sized habitats at realistic levels of richness.

The specialist predator model also predicts no saturation. A simple patch model of Janzen's hypothesis developed by Armstrong (1989) has shown that specialists can stabilize prey relative abundances in a frequency-dependent manner for any number of prey species. Regulation becomes weaker as the diversity of the prey community increases, so that the model approaches the behavior of the random walk models at very high diversities.

Mixed Predictions. So, to return to the original question, does theory predict that saturation should occur?, the answer must be yes and no. If Caswell's model of a noninteractive community is included, nearly as many models predict no saturation as predict saturation, and saturation may be absent even in strongly interactive communities. Since real communities are probably complex mixtures of processes drawn from several models, caution must be exercised in drawing lessons for the real world from a survey such as this. But it should be emphasized that the theoretical position is at best ambiguous on this very important point.

SATURATION IN REAL COMMUNITIES

Given this theoretical background, the second obvious question is, can we observe saturation in natural species assemblages? In order to answer this question a method is needed that will allow us to determine whether there are ceilings to species richness in a local habitat. One approach has been to compare the local richness in a series of roughly uniform habitats that are subject to colonization from larger or smaller species pools (Terborgh and Faaborg 1980; Cornell 1985a, 1985b).

A Comparative Method

To compare local species richness, standardized samples are taken in comparable habitats from different defined geographical areas. The defined geographical areas can be islands, or they can be areas of uniform habitat. For phytophagous insects, one system of interest here, they can be geographical ranges of host plants. The salient point is that the defined geographical areas must contain different numbers of species acting as species pools for each local habitat. Host plants supporting phytophagous insects can differ in several respects that influence the size of the regional pool, including the diversity of habitats in which the host occurs and taxonomic propinquity to other hosts in the region. Often there is a strong relationship between the regional richness of phytophages and the host's geographical range; larger and smaller regional pools can thus be attained by sampling closely related host species with different geographical ranges, or the same host species in different regions where its geographical range differs.

The number of species in the standardized sample (or more accurately, the average number of species over several standard samples) represents local richness for a habitat. Regional richness is the total number of species supported by the host over its entire geographical range in the region. Regional richness is then treated as an independent variable and correlated with local richness among host species or regions.

Saturation or Proportional Sampling? Two generalized results are possible. In communities with no ceiling, similar habitats with richer regional colonization pools can exhibit proportionally richer local assemblages ("proportional sampling"; fig. 22.1). Proportional sampling means that local richness is dependent upon the richness of the

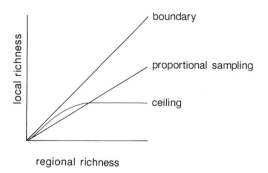

Figure 22.1 Theoretical relationships between local and regional richness in communities exhibiting proportional sampling and in those exhibiting richness ceilings. The relationship labeled "boundary" represents complete regional pool exhaustion and is the upper boundary on local richness.

regional pool and independent of biotic interactions that may be occurring in the habitat. In the extreme, all of the species in the regional pool occur in a single local habitat, and the slope of the line takes its maximum value of one ("boundary"; fig. 22.1). This pattern represents complete pool exhaustion and is probably unlikely for the vast majority of real communities. More realistically the slope will be somewhat less than one, either because species associated with the host are specialized to other habitats, resulting in high beta diversity, or because periodic extinctions occur and are balanced by stochastic colonization from the regional pool.

It must be emphasized that a proportional sampling pattern does not necessarily mean that the community is noninteractive. Although the absence of a ceiling is characteristic of and strongly suggestive of a noninteractive community, interactive communities can show the same pattern, as was suggested in the theoretical discussion above. All that proportional sampling indicates is that regardless of the nature of any local interactions that may be occurring, they are not sufficient to limit local diversity. Supporting evidence on lack of abundance compensation or other evidence from manipulations is required to confirm that a community is noninteractive.

Given that an assemblage without a ceiling can be interactive, it may be desirable to refine the search for limits on local richness by focusing on species groups most likely to interact. These may, although not necessarily, show a differential tendency to level off. Hanski (1982) has suggested that the species in a community will generally fall into two rather distinct groups: core species, which are widely distributed, locally abundant, and are likely to interact, and satellite species, which are narrowly distributed, locally rare, and are less likely to interact. Not all communities support the predictions of the core-satellite hypothesis (see Gaston and Lawton 1989 for review), but for those that do, proportional sampling patterns will be more persuasive if the community is subdivided into core and satellite species and each group is tested independently for richness ceilings (e.g., Cornell 1985a). In this way, subtle tendencies toward saturation may be more clearly revealed.

If the assemblage reaches a ceiling, then local richness will not increase as a constant proportion of regional richness. It may at first increase with increases in regional richness, but will eventually become constant or at least independent of regional richness ("ceiling"; fig. 22.1). Data that exhibit proportional reductions of local richness with increasing regional richness thus point to a saturated community.

If a ceiling is found, then additional observations and/or experiments will be required to distinguish true saturation from stochastic equilibrium or pool exhaustion. A stochastic equilibrium is simply a balance of colonization and extinction rates that is independent of species interactions (i.e., a noninteractive equilibrium; Wilson 1969). Pool exhaustion results when all or nearly all of the species from a given regional pool that can arrive at and survive in a given habitat have done so. As defined geographical areas get larger, habitat diversity and, in turn, regional richness and beta or between-habitat richness can increase, but local richness may remain constant due to pool exhaustion. In such a case, regional richness overestimates the species pool of the habitat in a nonlinear pattern, and produces a ceiling on local richness.

Interactions among the constituent species are a central requirement for saturation as defined in this chapter; thus evidence for density compensation, niche shifting, the absence of identifiable empty niches, and/or high ratios of core versus satellite species would support saturation. High ratios of satellite species and/or rapid species turnover independent of observable interaction among newly arriving and disappearing species would support stochastic equilibrium, whereas low turnover coupled with lack of niche shifting, abundance compensation, and obvious empty niches would support pool exhaustion.

Tests for Saturation

Vertebrates. An early test for saturation by this method was performed on bird communities occupying West Indian islands (Terborgh and Faaborg 1980). When the numbers of species in standard samples from matched habitats were plotted against the number of species on each island, a ceiling was apparent, suggesting saturation. Both species interactions and pool exhaustion were explored as possible explanations, but they could not be readily distinguished, making interpretation of the ceiling ambiguous at best. Moreover, Wiens (1989a) replotted the data from this study and showed pictorially and statistically that proportional sampling explains the pattern nearly as well as the apparent ceiling. Finally, Ricklefs (1987) tested assemblages of songbirds in the same archipelago and observed a strong dependence of local richness on regional richness with no hint of a richness ceiling. The weight of evidence thus favors proportional sampling rather than saturation as the most plausible description of this assemblage.

In another study involving vertebrates, Tonn et. al. (1990) purported to demonstrate a ceiling in small-lake fish assemblages. However, their conclusions were based on only two data points; an argument for proportional sampling could almost as convincingly be made, leaving the issue ambiguous for this particular assemblage.

Phytophagous Insects. Saturation in phytophagous communities was first tested in assemblages of cynipid gall wasps associated with California oaks (Cornell 1985a, 1985b). The cynipids are a very diverse group with over 600 species in North America (Krombein, Hurd, and Smith 1979) but are virtually restricted to the oak genus. Distributional ranges of the oaks vary widely, and cynipid regional richness correlates strongly with oak distribution in California (Cornell and Washburn 1979; Cornell 1985a). Saturation can be evaluated by correlating local richness with regional richness on oaks with various ranges. When this is done, a strong proportional sampling pattern emerges (Cornell 1985a, 1985b) indicating that cynipid assemblages are not saturated.

In addition, the expected correlation between oak geographical range and local richness is strong (Cornell 1985b), indicating that the correlation between regional richness and oak range is not due exclusively to geographical turnover of cynipid species among local sites. In sharp contrast, Stevens (1986) found no correlation between local richness of wood-boring scolytid beetles and the geographical range of their host plants. However, he did report a strong correlation between regional richness and host range, and concluded that richer regional faunas on widely distributed hosts were mainly due to habitat heterogeneity (i.e., geographical turnover). This result implies a richness ceiling; however, factors other than host geographical range can influence regional richness, and thus the best test for a ceiling is a direct correlation between local and regional richness. When this is performed (after removing seven host species that were sampled at only one site), a weak but significant correlation emerges ($Y = 0.39X + 0.33$; $R^2 = .26$; $P < .001$; $n = 40$). The Y-intercept was not significant ($t = 0.86$; $P = .39$), and moreover, a test for curvilinearity in the data revealed none; the quadratic term (and all higher-order terms) of a curvilinear regression was not significant ($t = -0.7$; $P > .4$), indicating that a linear relationship between local and regional richness was the most parsimonious description of the pattern. Thus, although there is considerable scatter in the correlation, the results are more suggestive of proportional sampling than of saturation.

More recently, John Lawton (1990b, 1990c) used this method to test for a ceiling in herbivore assemblages associated with bracken. Bracken has a worldwide distribution, but varies from region to region in the extent of its geographical range. Furthermore, those regions where bracken has a wider distribution also have greater numbers of herbivores associated with it. Species pools thus vary from place to place.

When the number of bracken herbivores in local samples is correlated with the size of the regional pool, again one observes strong proportional sampling, indicating no saturation. This is a particularly satisfying result, because previously gathered information about the system, such as the prevalence of empty niches and the lack of abundance compensation during population fluctuations (Lawton 1982, 1984b; Lawton and Gaston 1989; Compton, Lawton, and Rashbrook 1989), had already suggested that bracken assemblages were noninteractive, and as such they would have been predicted to be unsaturated.

Hawkins and Compton (1992) plotted local against regional richness for South African fig wasps and their parasitoids and demonstrated a striking proportional sampling relationship in both groups. This study is notable for the quantity of data collected, and is the first rigorous test for saturation in the third trophic level.

Finally, Zwölfer (1987) and Lewinsohn (1991) correlated the number of insect species on flower heads of temperate and tropical Asteraceae, respectively, with several host attributes. Using path analysis, they demonstrated a very strong correlation between local richness on flower heads and the size of the regional pool, again suggesting that such assemblages were unsaturated. No other tested variable (head size, host taxonomic isolation, host distribution) significantly influenced local richness. As in the case of wood borers, host distribution made virtually no contribution to regional richness, effectively decoupling these two variables in these systems.

Data for parasite communities in amphibians (Aho 1990) and fishes (Aho and Bush, chap. 17) do not support proportional sampling, and appear more consistent with saturation. These data on parasite assemblages from vertebrate hosts provide the best evidence currently available for an asymptotic relationship between local and regional richness.

As far as I am aware, these are the only data sets complete enough to test for saturation by the method advocated in figure 22.1. Some studies with fewer data may support the saturation model: e.g., ants in sclerophyllous vegetation (Westoby 1985); invertebrates in grass beds (Heck 1979); crustaceans on coral heads (Abele 1984); and fishes on coral heads (Bohnsack and Talbot 1980). Others suggest proportional sampling: e.g., freshwater ciliates (Taylor 1979); leaf miners on oaks (Opler 1974); coral reef fishes (Westoby 1985). However, none of these results are sufficiently rigorous to allow unambiguous rejection of either model.

Implications

With the exception of parasites on vertebrate hosts, all of the examples reported in the foregoing section either supported proportional sampling or were ambiguous. Without a doubt, cases of saturation will be discovered as more systems are rigorously examined. However, because these are the only data sets that are complete enough to test for saturation by this method, nascent rules governing the relationship between regional and local richness for different types of assemblages are not yet forthcoming. Predictable rules will be necessary to fuel progress in community theory, and they also will be useful to conservationists who need to evaluate the biodiversity of particular segments of a biota over regional spatial scales. The shape and slope of the sampling curve may vary in predictable ways among different taxa, feeding guilds, and organisms of differing body size. Accurate estimates of regional diversity may thus be secured from small local samples such as the recently popular tree canopy censuses in tropical forests (Gagne 1979; Erwin 1982, 1983; Adis, Lubin, and Montgomery 1984; Morse, Stork, and Lawton 1988) coupled with the appropriate sampling rule. It is unfortunate that we have so little

information on so fundamental an ecological question, given its practical utility.

Despite the absence of fixed rules, clearly defined unsaturated patterns have been documented in at least five phytophagous assemblages. To people who have worked extensively with insects on host plants, it is probably not surprising that they are unsaturated, given that herbivores generally (although not always) show weak interspecific interactions (Hairston, Smith, and Slobodkin 1960; Slobodkin, Smith, and Hairston 1967; Lawton and Strong 1981; Price 1983; Schoener 1983; Connell 1983; Strong, Lawton, and Southwood 1984; Fritz, Sacchi, and Price 1986; Hairston 1989; Evans 1989; but see Stiling and Strong 1984; Karban 1989; other references cited in Strong, Lawton, and Southwood 1984). But the implications for community theory in general are more profound. Unsaturated patterns suggest that the determinants of local richness cannot be discovered by studying local species assemblages in isolation. Since local richness is largely predictable from regional richness, we need to know what determines the size of the species pool from which the local assemblage is drawn.

A CHICKEN AND EGG PROBLEM

The origins of species richness at different spatial scales are obviously linked, but offer something of a chicken and egg problem in that it is unclear whether local richness derives mainly from broad-scale processes underlying regional richness or vice versa. Niche packing or other locally operating mechanisms may influence regional richness in a reversal of cause and effect. Put another way, is local richness the dependent variable, as indicated in figure 22.1, or should the axes be reversed, making regional richness the dependent variable?

Local Processes Limit Regional Richness

The answer to this question will most certainly depend upon whether local richness is saturated. A strong local-regional correlation may result simply from larger niche space on some host populations than on others, and all assemblages may be saturated within their respective niche spaces, some large, some small. If true, then the egg comes before the chicken, and regional diversity may in some sense be limited by local process.

The limit would be most obvious if habitats from place to place in the region are essentially uniform, with the same set of species adapted to live in each. Beta or between-habitat diversity would then be low (at the extreme it would be zero), and local and regional diversity would converge on a line of equality with a slope of one (fig. 22.2). However, the line in this case represents saturation, rather than pool exhaustion as it did in figure 22.1. Of course, this outcome is an improbable extreme. More likely, habitats in the region will be subtly different, and the impact of local process on regional diversity will be softer and less obvious. Each habitat will have a different species composition, between-habitat diversity will be larger, and limits to regional richness will be set by the number of distinct habitats, each supporting a saturated assemblage (Cornell and Lawton 1992; fig. 22.2).

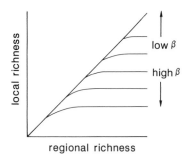

Figure 22.2 Expected relationship between fully saturated local communities and regional richness at different levels of habitat heterogeneity in a defined geographical region. The straight line with a slope of one represents a boundary where all habitats in the region are uniform and beta diversity is zero. (From Cornell and Lawton, 1992; by permission of Blackwell Scientific Publications, Ltd.).

Regional Processes Limit Local Richness

If local diversity is not often saturated, as is likely if the arguments in this chapter are correct, then the chicken comes before the egg, and other limits on regional diversity, which in turn limit local diversity, must be sought. Questions often asked by systematists, such as, where do species come from? and, "why do certain taxa radiate extensively? suggest profitable approaches to the problem. Conditions that encourage rapid phylogenetic diversification in lineages over evolutionary time scales will result in increased regional richness, and may ultimately set its limit (fig. 22.3; proportional sampling).

REGIONAL RICHNESS AND THE ADAPTIVE ZONE

In this section I wish to explore alternative mechanisms that can lead to phylogenetic diversification. In the classic view (Simpson 1953; Hutchinson 1959), rapid diversification is thought to be encouraged by the creation of a new adaptive zone where evolving species can exploit previously unutilized resources or escape mortality from natural enemies. New adaptive zones are typically thought to encompass large amounts of empty heterogeneous niche space; species can thus proliferate into the new zone via niche differentiation, that is, by evolving specializations that allow coexistence with each species occupying a small subset of the total space. Diversification can thus be related back in some sense to local ecology, but in this case in evolutionary time.

There have been some very interesting recent attempts to test the adaptive zone hypothesis (that diversification is accelerated in new adaptive zones) by comparing the species richness of extant, equal-aged sister groups, one of which occupies the fully utilized ancestral adaptive zone and the other of which has developed key innovations that have allowed it to exploit a new adaptive zone. If the adaptive zone hypothesis is correct, the latter should be more species-rich than the former, subject to the constraints that the new zone is sufficiently large to allow extensive proliferation and that the sister groups of a given lineage are dominant clades in both zones. The method is improved by examining multiple lineages with independent evolutionary origins to control for the effect of common phylogenetic history.

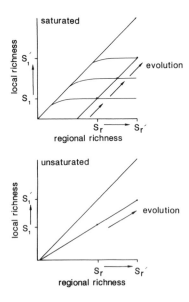

Figure 22.3 Possible changes in local richness in saturated and unsaturated communities as regional richness increases via phylogenetic diversification over evolutionary time scales. Local richness can increase (although it need not be proportionate) with regional richness in unsaturated communities because limits on local richness are ultimately set by regional pools. In saturated communities, limits are set by ecological processes, but can still increase through evolutionary time if the average characteristics of the constituent species are altered, allowing more to be packed in.

A Test of the Adaptive Zone Hypothesis

Charles Mitter, Brian Farrell, and Brian Wiegmann (1988) tested the adaptive zone hypothesis for the origins of phytophagy in insects. Only nine of the thirty orders of insects feed on plants, suggesting that plant feeding presents a formidable evolutionary barrier to insects. But these researchers have shown that in the majority of cases, once the barrier is breached, radiation is extensive, and that phytophages (when they represent the derived condition) from multiple lineages are almost always more diverse than their sister groups. Farrell, Dussourd, and Mitter (1991) have shown the same pattern for plants. Those that contain latex/resin canels are presumably better defended than those that do not. Latex/resin canels are a derived trait with multiple origins, and those taxa that possess canels are generally more diverse than those that do not. Both patterns support the original predictions of coevolutionary theory as proposed by Ehrlich and Raven (1964).

Sister group comparisons of this type provide the most rigorous test of the adaptive zone hypothesis that we have to date. However, the classic model of adaptive zone diversification via niche differentiation need not be correct in order for such radiations to occur. More recent models have suggested that phylogenetic diversification into new adaptive zones does not require niche differentiation. It may also occur through decreased rates of competitive extinctions of new reproductive isolates or through increased opportunities for geographical speciation arising from expanded species ranges or tectonic activity, neither of which necessarily involves diversifying selection into new adaptive zones (Mitter, Farrell, and Wiegmann 1988; Slowinski and Guyer 1989).

Alternatives to the Adaptive Zone Hypothesis

Moreover, not all sister group comparisons exhibit higher diversification in the new adaptive zone (Mitter, Farrell, and Wiegmann 1988), suggesting that alternative processes may be at work. There are at least two alternatives to the adaptive zone hypothesis per se that do not require open niche space as a prerequisite to diversification. The first is that diversification is promoted by habitat specialization, which encourages isolation and divergent selection imposed by each habitat (Eldridge 1976). The second is that diversification is promoted by "intrinsic" properties of certain taxa that predispose them to increased rates of speciation or decreased rates of extinction. For example, the propensity to be organized into social groups may increase speciation rate by reducing the size of the breeding population, consequently increasing interdemic isolation (Bush et al. 1977). Other intrinsic constraints might include particular chromosomal arrangements, as in the case of the genus *Clarkia* (Lewis and Lewis 1955), or the prevalence of sexual selection and founder effects, as in the case of the Hawaiian drosophilids (Carson and Templeton 1984; Dominey 1984) and the cichlid fishes of the great African lakes (Fryer and Iles 1972; Dominey 1984). All of these can encourage rapid speciation independently of a species' niche.

It is likely that all of these processes contribute to diversification to some extent, depending upon the phylogenetic group. But the arguments suggest that open niche space and/or niche heterogeneity are neither necessary nor sufficient for species diversification. Instead, the tendency to proliferate may be an intrinsic historical property of specific taxa, based on their genetic and biogeographical attributes, and may be effectively independent of local ecology and niche complexity. Local ecological opportunity may thus have little impact on regional richness, and ultimately, on richness levels in local assemblages, over both ecological and evolutionary time (Cornell and Lawton 1992).

PHYLOGENY AND THE STRUCTURE OF REGIONAL POOLS

Diversification

The specific conditions that encourage rapid diversification of herbivores on host plants are not well understood, but several lines of evidence suggest that phylogenetic history of hosts plays a role. For example, average herbivore richness per host species often decreases with decreasing numbers of closely related host species in a given flora, an effect that is sometimes called taxonomic isolation (Kennedy and Southwood 1984; Strong, Lawton, and Southwood, 1984; Cornell 1985a). The relationship between taxonomic isolation and decreased herbivore richness per host species has frequently been attributed to the presence of biochemical, physical, and phenological barriers to host sharing among extant herbivore species (Strong, Lawton, and Southwood 1984; Cornell 1985a). However, taxonomic isolation may also depress the richness of highly host-specific herbivore taxa in which the vast ma-

jority of species are restricted to single hosts. For example, leaf miners associated with the genus *Quercus* (oaks) endemic to the Pacific slope of California normally feed on only a single species of tree (Opler 1974). But the residuals of a regression of leaf miner richness on *Quercus* geographical range exhibit a pattern consistent with the effects of taxonomic isolation; they fall above the regression line for white oaks (the richest oak subgenus), and they generally fall on or below the line for the black and primitive oaks (the more taxonomically isolated species). Thus, white oak species have richer miner faunas, on average, than the other subgenera even though host sharing is rare. This pattern suggests that diversification may also play a role in setting regional richness levels, either by means of cospeciation due to shared history of host species or by host shifting. Host shifting by herbivores is generally easier among hosts that are close relatives, and such shifting may result in the isolation of subpopulations and subsequent speciation (Bush 1975; Wood and Guttman 1981; Guttman, Wood, and Karlin 1981; Feder, Chilcote, and Bush 1988). The fact that closely related herbivore species are often found on closely related hosts (Price 1977; Cornell and Washburn 1979) supports either mechanism. Low levels of taxonomic isolation among hosts may thus drive herbivore diversification over evolutionary time, in addition to facilitating simple host sharing among herbivore species.

Taxonomic Composition and Guild Structure

Other features of regional herbivore pools may also be influenced by phylogenetic history. For example, historical differences in the extent of proliferation of specific herbivore taxa can have a strong effect on the regional pool's taxonomic composition and guild structure. The point is illustrated by an analysis of regional richness among guilds of the herbivorous arthropods associated with the British trees (Cornell and Kahn 1989). The list of British arboreal arthropods was recently revised (Kennedy and Southwood 1984) and now represents the most complete and accurately documented phytophagous arthropod fauna currently available. This makes it particularly well suited for an investigation of the effect of herbivore phylogenetic history on regional species pools.

The herbivores on 28 British tree species were cross-classified into 4 guild and 15 taxonomic categories using the separate breakdowns given in Kennedy and Southwood (1984). The relative proportions of herbivores simultaneously falling into each guild and taxonomic category are portrayed in separate spectra for each tree species in figure 22.4. It is apparent from the figure that guild spectra and taxonomic spectra (simultaneously classified by guild) both vary widely among the tree species. This variation is statistically confirmed by chi-square contingency analysis of a guild × tree species cross-classification table in the case of guild structure, and a chi-square goodness-of-fit test of each individual tree spectrum against a standard spectrum representing the entire fauna in the case of taxonomic structure (simultaneously classified by guild; Cornell and Kahn 1989). Moreover, guilds and taxa that are particularly outsized (again, using a simple statistical criterion) are coded in figure 22.4; they seem

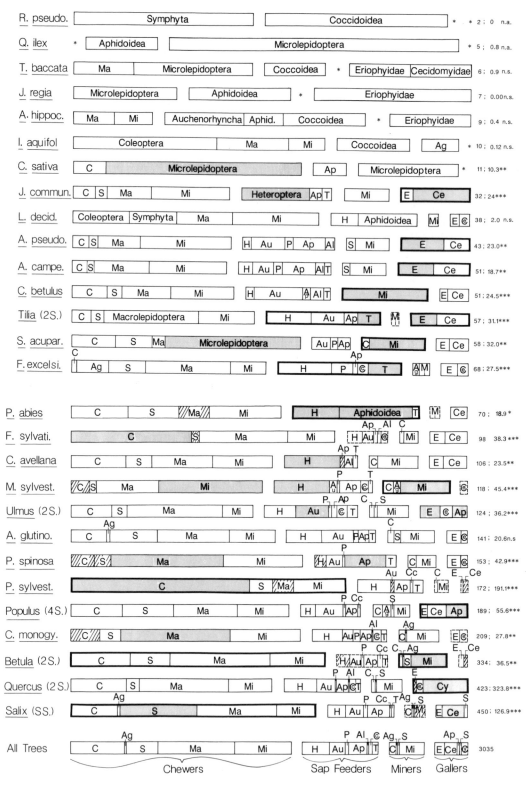

Figure 22.4 Guild and taxonomic spectra for the herbivorous arthropods on British tree species. Each guild spectrum is represented by four bars separated by spaces; the lengths of the bars indicate the proportion of each guild on each tree. Asterisks indicate guilds missing on a tree. The bars are subdivided, and each subdivision indicates the proportion of each guild represented by a particular arthropod taxon. *Ag,* Agromyzidae; *Al,* Aleyrodoidea; *Ap,* Aphidoidea; *Au,* Auchenorhyncha; *C,* Coleoptera; *Cc,* Coccidoidea; *Ce,* Cecidomyidae; *Cy,* Cynipoidea; *E,* Eriophyidae; *H,* Heteroptera; *Ma,* Macrolepidoptera; *Mi,* Microlepidoptera; *P,* Psylloidea; *S,* Symphyta; *T,* Thysanoptera. Thick-walled and dash-walled bars are over- and undersized guilds respectively. Shaded and cross-hatched subdivisions are over- and undersized taxa respectively. Numbers at right margin are the total number of herbivores species on each tree and the chi-square values for the goodness-of-fit tests of each individual tree spectrum to the standard spectrum (all trees). *$P < .05$, **$P < .01$, ***$P < .001$. (From Cornell and Kahn 1989; by permission of Blackwell Scientific Publications, Ltd.)

to be scattered haphazardly and idiosyncratically across all tree species.

Attempts to explain the wide variation in guild structure have largely failed. In multiple regressions, the proportion of herbivore species in the chewers guild increased slightly on more abundant, more closely related tree species, with 46% of guild variation explained by these tree attributes. However, other tested attributes (time since last colonization by tree in Britain, evergreenness, palatability) had no effect on this guild, and sap feeders, leaf miners, and gall formers were little influenced by any tree attribute. Moreover, there was no evidence of compensatory changes in species richness between different guilds that might offer clues to negative species interactions between them. Guild richnesses either increased or decreased independently of one another, or in one case, increased in parallel (Cornell and Kahn 1989). In summary, regional pools vary widely in guild structure and taxonomic composition among tree species, but there is little evidence that this fauna is highly organized at the functional level by conventional ecological mechanisms.

Instead, guild structure is strongly influenced by taxonomic proliferation. Part of the reason is that for herbivores, guild membership is strongly constrained by taxonomic membership. Virtually all Thysanoptera are sap feeders, all aphidoids are sap feeders or sap feeder/gall formers, all Macrolepidoptera are chewers, and all cecidomyiids, cynipids, and eriophyids are gall formers. Thus, if a guild includes a taxon that has proliferated extensively for unknown historical reasons, that guild is likely to be overrepresented in the fauna of a particular tree species. Conversely, if particular taxa are depauperate, the guild will likely be underrepresented (fig. 22.5A; table 22.2). Some outsized taxa are particularly prominent. For example, *Pinus sylvestris* has an excess of chewing beetles (largely due to a proliferation of bark beetles) but a dearth of chewing caterpillars, whereas *Salix* is overrepresented by chewing sawflies. *Tilia* and *Fraxinus excelsior* have in-

Table 22.2. Chi-Square Test for Association between Outsized Guilds and the Occurrence of at Least One Outsized Taxon in a Guild

	Contingency coefficient[b]	Outsized taxa[a]	Outsized guilds[a]		
			Yes	No	Total
Chewers	0.10	Yes	2	6	8
		No	1	12	13
		Total	3	18	21
Sapfeeders	0.19	Yes	4	7	11
		No	1	9	10
		Total	5	16	21
Leaf miners	0.51	Yes	5	0	5
		No	3	13	16
		Total	8	13	21
Gall formers	0.46	Yes	8	2	10
		No	2	9	11
		Total	10	11	21
All guilds	0.39				
	$\chi^2 = 16.61^c$	Yes	19	15	34
		No	7	43	50
		Total	26	58	84

Source: Cornell and Kahn 1989.
Note: Only 21 of the 28 tree species had sufficient numbers of herbivores to be included in this analysis.
[a]Coded in figure 22.4.
[b]Indicates relative contributions of each guild to the association.
[c]All guilds combined to attain sufficiently large cell sizes for test; $P < .001$

flated thysanopteran sap feeder faunas, and *Picea abies* has more than its share of sap-feeding Heteroptera. In the gall former guild, *Juniperus communis* is overrepresented by gall midges (Cecidomyiidae), whereas *Populus* and *Quercus* have excessively high numbers of aphid and cynipid gall formers respectively. The proliferation of cynipids on *Quercus* is particularly noteworthy in that the species richness of this group is very high, but it is virtually restricted to the oak genus. Oak also shows a corresponding drop in the proportion of other gall-forming taxa, presumably because of the amazing proliferation of cynipids on this genus.

Guild structures among British tree species are neither stable nor predictable, suggesting that organizational rules that might govern the functional structure of phytophage assemblages in contemporary time are absent in this system. Instead, the idiosyncratic distribution of individual taxa among tree species suggests that historical influences having little to do with attributes of contemporary tree species probably contribute strongly to the diversity of specific herbivore taxa and thus to guild representation in the British arboreal arthropods. Chance colonization events, host shifting and adaptation, host specialization, and differential rates of extinction, colonization, and adaptive radiation intrinsic to each systematic group all are likely to contribute to the unique taxonomic composition and richness of each guild on each tree species. The reasons for such differences in taxonomic distribution are obscure, but they surely reflect differences in the evolutionary history of herbivore accumulation on these trees, which may have little to do with present-day tree attributes.

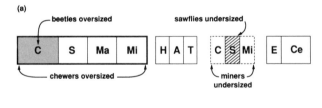

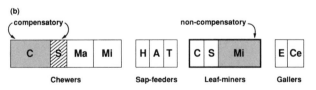

Figure 22.5 Schematic representations of guild and taxonomic spectra of individual tree species, demonstrating (a) that when guilds are outsized they often include at least one outsized taxon, and (b) that there is less association between outsized taxa and outsized guilds in some guilds because of compensatory reductions in some taxa as others increase, resulting in guilds that are not outsized. Abbreviations as in figure 22.4.

Saturation

Such historical processes may have as significant an effect on the structure of local communities as they do on regional diversity for the obvious reason that regional and local diversity are often linked. However, even if local communities are saturated, history may still exert an influence (fig. 22.B; ceiling). This is simply because species from regional pools composed of different taxonomic and guild spectra may differ in their abilities to coexist together in a rich local assemblage. Some species may be better adapted to local environmental conditions, or have, on average, narrower niche requirements, allowing richer local assemblages to stably persist (fig. 22.3; ceiling). In herbivore assemblages, gall formers and leaf miners generally have more specialized feeding habits than chewers (Strong, Lawton, and Southwood 1984), and are also more sedentary, suggesting that assemblages drawn from regional pools historically composed largely of gallers and miners should on average have more invasible niche space, given equivalent numbers of species (Cornell and Kahn 1989).

I am not aware of any data suitable for testing this proposition for local assemblages, but patterns of association between outsized taxa and guilds in the regional British insect fauna hint that miner and galler assemblages ought to be more invasible than chewer and sap feeder assemblages. Specifically, the association between outsized taxa and outsized guilds in table 22.2 is stronger for the leaf miners and gall formers than for the chewers and sap feeders. The weaker association in the latter two guilds is mainly due to compensatory reductions in some taxa as others increase, resulting in little net change in guild size (fig. 22.5B). In guilds that are less invasible, such compensatory shrinkage of some taxa as others increase would be expected, whereas in guilds with more empty niche space, taxa can proliferate without displacing other taxa in the guild.

Conclusions

Several conclusions can be drawn from the foregoing discussion. First, the theory is ambivalent regarding its predictions that communities should or should not be saturated, and the meager information available indicates that phytophagous species assemblages are not saturated. We need a lot more data on this fundamental point to make any generalizations. At this point I think proportional sampling ought to be viewed as a null hypothesis since it represents the absence of any effects of species interactions on the structure of the local community. Second, if communities are generally unsaturated in ecological time, the direction of richness control will be from regional to local, not vice versa. Establishing the direction of control will rely on a combination of evidence on saturation (or lack thereof) and on the role of niche diversification in the proliferation of regional biotas. At this time, such data are lacking. Finally, the generation of regional richness and other structural characteristics of regional pools in evolutionary time also may not rely on local ecology, but may be heavily influenced by phylogenetic history. Studies of phytophagous species assemblages suggest an important role for regional-historical processes in setting the structure of local communities. The key to community structure may thus be found in evolutionary history and extrinsic biogeography rather than in intrinsic local processes, making community ecology a more historical science.

23

Phylogenetic Determinants of Insect/Plant Community Diversity

Brian D. Farrell and Charles Mitter

UNDERSTANDING INSECT/PLANT DIVERSITY

Most recent reviews have concluded that communities of herbivorous insects on plants are seldom governed by competition and rarely converge in either diversity or guild structure (Price 1983; Lawton 1984b; Strong, Lawton, and Southwood 1984; Cornell, chap. 22; but see Zwölfer 1988; Jaenike 1990). Diversity in such plant faunal communities may thus be set by the balance among independent rates of colonization, speciation, and extinction, which in turn might strongly reflect the geographical and phylogenetic history of the faunal community and its constituent species. In this chapter we review the ways in which explicit study of that history, barely begun, has contributed to, or might contribute to, our understanding of insect/plant diversity.

Two approaches to assessing the nature and magnitude of historical effects on insect/plant communities can be recognized. The first involves island-biogeography-like studies of the regional phytophage faunas associated with individual plant taxa (reviewed in Strong, Lawton, and Southwood 1984; Cornell, chap. 22). Much of the variation in faunal diversity appears explainable by the size of the area occupied by the plant species, and by the size and complexity of the plant itself (Strong, Lawton, and Southwood 1984). These factors might be provisionally classified as ahistorical: they are not obviously related to long-term plant history (but see below), and they probably reflect simple mechanisms, such as encounter frequency, which should apply similarly to all insect-plant interactions. Long-term processes generally enter the models through secondary factors added to explain residual variance from these main effects. Thus, characteristics such as time since entry into the regional flora, taxonomic or biochemical "isolation" (i.e., distinctiveness), and clade membership (Kennedy and Southwood 1984) are clearly tied to plant history, and the response to them seems likely to vary strongly among herbivore clades. However, our current ignorance of the processes underlying these "historical" effects leaves unclear whether these can be adequately rendered by such simple models (Strong, Lawton, and Southwood 1984; Cornell, chap. 22).

A complementary approach, then, would begin with

explicit consideration of how interactions evolve. This approach is exemplified by Ehrlich and Raven's (1964) model of coevolution, under which successive evolutionary innovations in plant defenses, and in their circumvention, permit alternate episodes of plant and insect radiation. Under this model, contemporary dominance, diversity, and trophic relationships of different insect and plant clades are ascribed to their position in the historical sequence of "escape and radiation" (Thompson 1989). Advances in phylogeny reconstruction are now allowing explicit examination of the macroevolutionary questions implicit in Ehrlich and Raven's thesis, which until recently have been mostly neglected in the vast ecological literature their essay inspired.

Clearly, a full understanding of insect/plant community diversity will require a synthesis of evolutionary and ecological approaches. In this chapter, we first attempt to catalog and review the evidence for those properties of the phylogenesis of insect/plant associations that might be expected to influence their contemporary diversity. It will be convenient to organize this survey around several broad questions arising from Ehrlich and Raven's model. We then attempt to place these issues in the context of ecological diversity models. At what scales do plant faunal communities reflect the evolutionary forces we review? Have these in fact influenced their diversity? How does the answer vary among communities, and how exactly can we find it?

Any theory ascribing community structure to evolutionary history, as opposed to, say, local partitioning of resources, presupposes genetic limits on the evolution of species' interactions. In Ehrlich and Raven's model, for example, novel defenses would be unlikely to permit plant radiation unless evolution of the corresponding insect countermeasure were a rare event occurring only in preadapted lineages. The notion of constraint plays a similar role in colonization-rate models of insect diversity. For example, the question of how to model plant "isolation" (from the pool of potentially colonizing herbivores on other plants: see Strong, Lawton, and Southwood 1984) is essentially that of modeling constraints on the evolution of host use by herbivorous insects. Thus we first survey the variety and intensity of possible limitations on the evolution of host selection, and the kinds of evidence that might bear on them.

An obvious (if little studied) premise of Ehrlich and Ra-

This chapter is, in part, contribution no. 8690 from the Maryland Agricultural Experiment Station.

253

ven's model is that, if currently affiliated insect and plant lineages have affected each other's evolution, they or their ancestors must have interacted before the recent past. Thus, we next review evidence on the timing of insect versus plant evolution. If interactions are so conservative as to persist over geological time, contemporary community structure could reflect just the relative ages of associated insects and host plants.

Third, we consider the evolutionary consequences of long-term interaction. Can we demonstrate, as Ehrlich and Raven predict, that insects and plants have evolved successive adaptations to each other, and that these escalations drive diversification? More generally, if insect/plant diversification may be related to their interactions, how might this affect the diversity of particular insect/plant assemblages?

Finally, we consider how all these lines of evidence might be synthesized in the attempt to explain faunal diversity, with particular focus on one major insect/plant community pattern, latitudinal gradients of species diversity.

Evolutionary Conservatism in Host Plant Use

Evolutionary lability of host use by insects, which could erase the imprint of history from contemporary associations, might be predicted from the frequent occurrence of heritable variation for host use traits and from the rapid accumulation of herbivores on some introduced plant species (Bernays and Graham 1988). However, genetic theory and experimental evidence are also compatible with strong constraint on diet evolution; change in diet probably requires concerted change in multiple, genetically independent traits (Futuyma and Moreno 1988; Gould 1991; Jaenike 1990). Thus, to judge the importance of such constraints for the assembly of particular plant faunas, we need evidence on the major patterns, if any, of long-term diet evolution. Given that interacting insects and plants rarely fossilize together, the major source of such evidence will be phylogenetic comparisons among extant species. The discussion that follows will presume that we have superimposed host associations on herbivore cladograms derived from other evidence, thereby reconstructing most parsimonious histories of change in host use (details in Mitter and Farrell 1991).

Given such reconstructions, there are two types of phylogenetic patterns that would suggest genetic limits on diet evolution, pending corroboration from parallel studies of its genetic and adaptive bases (Courtney and Kibota 1990; Futuyma and McCafferty 1990; Miller and Feeny 1989). First, constraint on the sequence or direction of evolution—on what changes can occur and in which direction—is suggested when the evolutionary sequence of host colonizations implied by the insect cladogram is predictable from the likely genetic barriers to transition between hosts that are in some way most similar (e.g., only between chemically similar or confamiliar host species, or only between host species in a single habitat), then the direction of transfer (i.e., the range of potential hosts) may be inferred to be under constraint. Second, "constraint" of host use evolution is implied if change occurs rarely or slowly (e.g., only with some small frac-

tion of all speciation events) whatever its sequence or direction.

Given the genetic complexity underlying host use, the variety of potential "constrained" patterns is large (Mitter and Farrell 1991). We focus here on the major prediction of Ehrlich and Raven's model: that host use should be conservative with respect to plant secondary chemistry or other supposed defenses. These characteristics will generally be correlated with plant taxonomy; indeed, because so many plant characteristics are correlated with it, taxonomy offers a convenient if coarse measure of the overall lability of host associations.

The extent of that lability has been debated. Jermy (1984) pointed out that to judge the applicability of Ehrlich and Raven's model, we need an idea of the frequency distribution of evolutionary patterns of host use. He erected four types of such patterns and argued, contra Ehrlich and Raven, that while most insect species are monophagous or oligophagous, closely related insects most commonly feed on only distantly related sets of plants. However, the evidence on this question has been almost entirely anecdotal. In a preliminary quantitative study (Mitter and Farrell 1991), we compiled rates of change in plant family association, based on the relatively few (25) phytophage groups for which both a substantial number of host records and a species-level cladogram were available. Among these, the majority class was that in which 20% or fewer of speciation events were associated with change in host family (fig. 23.1), a common indicator of "major" diet differences (Jermy 1984) and frequently the limit of host record precision. Thus, there is quantitative support for the generalization that related insects most commonly use related, hence probably chemically similar, plants.

While host genus or family affiliation appears relatively conserved in the evolution of most herbivore groups (compared with traits such as body size, for example), some clades shift more freely among disparate plants. For example, in the moth genus *Yponomeuta* (Menken 1982),

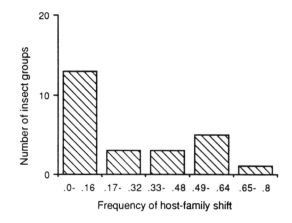

Figure 23.1 Frequency distribution of shifts in host family in 25 herbivore groups. The largest category is of insect clades for which fewer than 20% of speciation events are associated with host transfer between plant families. (Data from Mitter and Farrell 1991).

about half the speciation events are associated with host family shift. Some clades, such as the swallowtail butterfly subgenus *Graphium* (Saigusa et al. 1982), appear associated with particular secondary compounds that occur in sometimes distantly related plant families; thus, host family affiliations may underestimate conservatism with respect to host chemistry. Other clades appear to be restricted to hosts sharing (through convergence) some other property, such as occurrence in habitats like deserts or acid bogs whose abiotic rigor demands specialized insect and plant adaptations (Mitter and Farrell 1991). Many hypotheses to explain variation in the degree and kind of phytophage conservatism have been proposed, invoking insect life history (mode of feeding, plant tissue attacked, defense strategy, etc.), plant defense strategy, and plant community structure. However, statistical tests of these hypotheses, based on the rates of evolution of diet in multiple independent insect clades, are presently lacking.

Diet Breadth

Although most phytophage species are oligophagous, feeding on several related plant species or genera, some are strictly monophagous, and some are polyphagous, feeding on multiple, unrelated plant families. Variation in diet breadth can affect both the specificity of evolutionary interaction between particular insect and plant clades and the trophic structure of contemporary insect/plant communities. Clearly, then, we need to understand both the adaptive significance of that variation and the possible evolutionary constraints on it.

The selective forces potentially governing diet breadth, reviewed by several authors (Futuyma and Moreno 1988; Bernays and Chapman 1987; Jaenike 1990), are still problematic. Among the ecological correlates of diet breadth demonstrated in various groups (see review in Jaenike 1990) are adult longevity (Janzen 1984) and size (Wasserman and Mitter 1978; Neimela, Hanhimaki, and Mannila 1981), larval phenology (Schneider 1980), and host chemistry (Janzen 1984; Futuyma 1976; Feeny 1977). For example, within the coleopteran superfamily Chrysomeloidea, members of primitive groups, which mostly attack dead and decomposing trees, tend to be polyphagous, whereas species in more derived groups, which attack healthy trees or herbaceous plants, are much more often host specific (Linsley 1961).

Because insect diet breadth often seems to involve multiple, concerted morphological and physiological adaptations, it might be expected to evolve less predictably or frequently in response to selection than do life history features with simpler genetic bases, such as dispersal ability (Denno, Olmstead, and McCloud 1989). Average breadth of diet varies strikingly among herbivore clades, though there is some evidence that it can evolve rapidly without change in the characters (e.g., larval phenology or adult size) correlated with it on a broad scale (Mitter and Farrell 1991). The distribution of diet breadths could also be influenced by a correlation between specialization and diversification rate, but as yet there is no explicit evidence for this (Futuyma and Moreno 1988).

Origins of Herbivore Faunas

If the phylogenetic patterns just reviewed do reflect underlying genetic constraint, then this might affect the diversity of the herbivore communities on various host plants in several ways. In the context of island-colonization models, it could reduce invasion rates by preventing most lineages from becoming established on (or adapting to) the locally available host plants. One expectation, if diet conservatism is in fact so pervasive as to limit plant faunal diversity, is that the herbivores on a given plant fauna should be derived from a restricted, closely preadapted subset of ancestors, as implied by the restriction of their near relatives to a taxonomically or otherwise limited set of plants. Such patterns would also result from the sequential circumvention of successive plant defense innovations expected under Ehrlich and Raven's model. A comprehensive accounting of phylogenetic connections among plant faunas, analogous to the cladistic relationships among areas sought by historical biogeographers (Nelson and Platnick 1981), would enhance our ability to model the effects of plant "isolation" on diversity and to detect diffuse or guild coevolution.

There have been few attempts to characterize the phylogenetic origins of entire faunas, although some insights have resulted from the related approach of clustering plants by their shared herbivore species (Futuyma and Gould 1979; Cornell and Washburn 1979; Holloway and Hebert 1979; Berenbaum 1981; Claridge and Wilson 1981; Neuvonen and Neimela 1983). We have only begun to identify the major phylogenetic connections among plant faunas. Striking, previously unsuspected ones, such as those based on convergent plant defenses such as furanocoumarins (e.g., in umbellifers and composites, which share related herbivores: Berenbaum 1983) or secretory canals (Farrell, Dussourd, and Mitter 1991; Dussourd and Denno 1991) are still being discovered. We do not yet know how many groups of related insects use plants that share by convergence a particular defense or habitat or other ecological characteristic, as opposed to groups in which near relatives use taxonomically related host plants.

Evidence for marked restriction of phylogenetic origins, at least on a broad scale, exists for a few plant faunas. For example, although explicit phylogenetic information on the direction of host shift is missing, the near relatives of herbivores that attack the umbellifer tribes containing angular furanocoumarins are almost always herbivores adapted to linear furanocoumarins (Berenbaum 1983).

A similar pattern emerges in the world fauna of the clade that comprises the families Asclepiadaceae and Apocynaceae. These plants, which we shall loosely term "milkweeds," contain defenses consisting of latex canals combined with cardenolides and (in more primitive members) iridoid glycosides. Their herbivores are mostly aposematic, often sequestering plant toxins for defense against predators, and are both specialized and conservative in their feeding habits. For example, with one exception, the herbivores attacking *Asclepias syriaca* at one site in Ohio (Dailey, Graves, and Kingsolver 1978) all belong

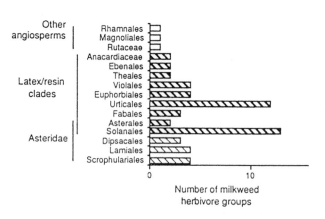

Figure 23.2 Frequency distribution of affiliations with other plant orders among fifty independent taxa containing herbivores of Asclepiadaceae/Apocynaceae. Most records are from the subclass Asteridae or from other latex-canal-bearing groups. Each plant order is presented on the vertical axis (with ordinal names following Cronquist 1981).

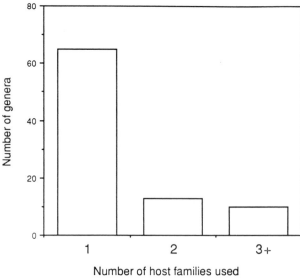

Figure 23.3 Frequency distribution of number of host families used by species in 107 leaf beetle genera found in Missouri (Riley and Enns 1979). Most species are restricted to the same host genus or family used by their typically neotropical congeners.

to very broadly distributed groups invariably affiliated with milkweeds.

Worldwide, milkweed feeding has arisen independently approximately a hundred times (Mitter and Farrell, unpublished data). Whether this number is unusually low remains to be determined. There are almost no cladograms for insect groups containing milkweed herbivores, but feeding habits are known for other members of the same higher taxon in about fifty cases, of which some probably represent ancestral habits and others secondary shifts. About half of these records (fig. 23.2) are from a set of just ten families in five orders, all belonging, like milkweeds, to the subclass Asteridae (Cronquist 1981). For example, the introduced weevil species *Gymnetron tetrum* attacks *Asclepias* in Ohio, while European populations (and the rest of the tribe) attack the iridoid-containing asterid orders Scrophulariales and Oleales. As asterids make up only about 15% of dicot families, and 25% of the species, this concentration is unlikely to be random. The underlying mechanism merits investigation, because the asterid families are characterized by highly disparate "major" classes of secondary compounds (Cronquist 1981).

Nearly all the other affinities of the relatives of milkweed herbivores, similar in number to the asterid connections, are with just seven plant families (in seven orders) which, though unrelated, share with milkweed the possession of latex and/or resin secretory canals (Farrell, Dussourd, and Mitter 1991). Thus, the broad affiliations of milkweed insects seem to represent an intersection of two larger, phylogenetically coherent faunas: asterid-associated insects, and those specialized on latex/resin plants (Farrell, Dussourd, and Mitter 1991).

Similar analyses of other asterid and other canal-bearing plant faunas might show similar patterns. However, as illustrated by the best-studied asterid fauna, the specialized insects of thistles (Asteraceae: Cynarioideae: reviewed by Zwölfer 1988; Zwölfer and Herbst 1988), the degree and kind of restriction on faunal sources is highly variable, and is dependent in part on the scale of comparison. Thistles seem to have been originally colonized by insects preadapted by occupying similar plant structures and xeric habitats, often from chemically and taxonomically unrelated plants, such as Chenopodiaceae and Brassicaceae. Subsequent evolution of thistle herbivores, however, has been conservative: about 80% of insect species restricted to Cynarioideae appear to have been derived most immediately from progenitors attacking plants in the same genus or tribe (Zwölfer 1987). Indeed, some of these herbivore lineages may have undergone early diversification in parallel with the Cynarioideae (see below).

While we have concentrated above on the faunas of plant taxa, these may not be entirely natural units for studying regulation of local or regional phytophage diversity, because the faunas of co-occurring plants can show substantial overlap (Futuyma and Gould 1979), and because different guilds within the same fauna may show very different patterns (Cornell and Kahn 1989; Cornell, chap. 22). Thus, studies of the origins of broad herbivore guilds across co-occurring plant species provide a necessary complementary perspective.

We have begun an analysis of the chrysomelid beetles of the midwestern United States as represented in the fauna of Missouri (Riley and Enns 1979). Together these represent 336 species in 107 genera and 13 subfamilies. Although explicit phylogenies for these predominantly New World genera are lacking, nearly half can be characterized as invariably restricted to one host genus or family (fig. 23.3), while the others show broader but still definite host restrictions (e.g., to asterid families). Thus, the host affiliations of nearly half of the species in this region are predictable from those of their relatives elsewhere. The broader geographical distributions of nearly all of these oligophagous genera correspond to those of their host

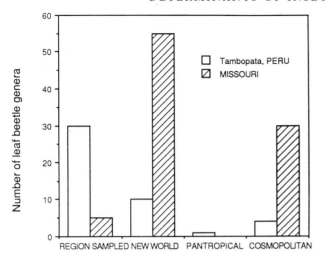

Figure 23.4 Frequency distribution of biogeographical regions occupied by leaf beetle genera found in Missouri contrasted with those occupied by genera sampled in a Peruvian rainforest. Peruvian leaf beetle genera are significantly more often restricted to the region sampled (i.e., Amazonas) than are Missouri genera (i.e., to North America), reflecting the largely Neotropical origins of New World leaf beetles (chi-square = 65.36; $P < .001$; df = 3).

plants, further suggesting an influence of host plant taxonomic isolation on local beetle diversity. Nearly three-fourths of Missouri leaf beetle genera are predominantly Neotropical, affiliated with largely tropical plant groups, and many have presumably only recently entered the midwestern United States (fig. 23.4; see below). For example, *Monomacra* flea beetles probably occur in Missouri only because two species of *Passiflora*, their hosts throughout the Neotropics, occur there as well. Similarly, species of *Chelymorpha* and *Typophorus* attacking *Ipomoea* are presumably derived from the constant but very diverse association of these beetles with *Ipomoea* in the tropics. The few strictly temperate groups in the Missouri fauna, such as the Holarctic *Syneta*, which is affiliated with various northern hardwoods and conifers, tend to have broader host ranges, but even these are often conservative. For example, *Chrysomela* occurs with its hosts *Salix* and *Populus* in both the North and South temperate zones (Brown 1956). In sum, from an evolutionary perspective, Missouri chrysomelid/host plant communities seem to represent in large part a subset of widespread, long-standing associations.

AGE AND PERSISTENCE OF INSECT/ PLANT ASSOCIATIONS

We have seen that a preponderance of insect groups are evolutionarily conservative in their host taxon affiliation. Sufficiently conservative associations, once established, could persist through subsequent evolution of the interacting lineages. Such parallel diversification should increase the likelihood of long-term coevolution between particular sets of lineages, though it could also occur without adaptive reciprocity. More generally, parallel diversification would imply a historical cause for current associations, and the structure and diversity of those associ-

ations might reflect in part just the sequence of origin of the respective plant and insect groups.

A spectrum of mechanisms can be envisioned for synchronous diversification in the broad sense. In extremely specific interactions, new associations might arise only through speciation in previously associated lineages, leading to a detailed match of insect and plant phylogenies. Ehrlich and Raven's model, in contrast, predicts not exactly parallel insect/plant speciation, but a coarser match, between the evolutionary sequence of plant defense types and the ages of origin of the successive herbivore groups that overtake and radiate on them. Even if newly evolved hosts were colonized at random, the ages of their present-day associates might be correlated if change in host use were slower than host diversification.

Most insect phytophage species belong to clades that have been phytophagous since the Cretaceous, in many cases much longer (Zwölfer 1978). Thus, phytophage diversification has been broadly contemporaneous with that of higher plants, especially of angiosperms. The question we address in this section is how common the traces of this shared history are, on what scale of comparison they are most evident, and how their presence can be rigorously demonstrated, given that many associations, in contrast, undoubtedly reflect recent host shifts. Subsequently, we consider the possible coevolutionary consequence of such enduring interactions.

Concordance of Insect and Plant Cladograms

The most obvious expectation, if particular insect and plant lineages have diversified in association, is that the order of divergence among host taxa should correspond to that among their specific herbivores. The problem of measuring and statistically assessing the resulting concordance of plant and insect cladograms is analogous to that of vicariance biogeography, and is not fully solved; recent advances are reviewed by R. D. M. Page (1988, 1990) and by Farrell and Mitter (1990).

There have been few explicit studies of insect/plant phylogeny concordance. Among fourteen assemblages in which at least some phylogenetic evidence is available (Mitter and Farrell 1991; plus *Tetraopes*, introduced here), nearly two-thirds showed some suggestion of cladogram concordance (fig. 23.5). However, even when cladogram concordance is documented, it might not represent parallel cladogenesis: it could instead arise entirely subsequent to host diversification, through herbivore colonization and speciation constrained by features correlated with plant phylogeny. Thus, multiple lines of evidence must be brought together to assess parallel diversification. For some cases where host and herbivore phylogenies show partial correspondence (e.g., *Rhagoletis* fruit flies), evidence on relative ages and the number of unoccupied hosts rules out diversification in parallel, though such cases would still suggest constraint on the direction of host shifts. For some other groups, cladogram correspondence is also partial (e.g., heliconiine butterflies: K. S. Brown 1981; *Larinus* weevils: Zwölfer and Herbst 1988) but information on insect and plant ages favors the hypothesis of synchronous diversification.

The two strongest, most statistically defensible cases

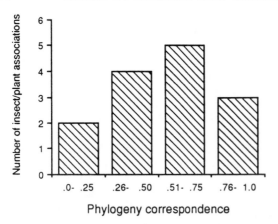

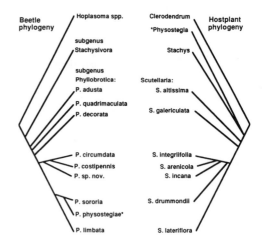

Figure 23.5 Frequency distribution of correspondence between herbivore and host phylogenies for fourteen groups (data from Mitter and Farrell 1991). The horizontal axis is a quantitative measure of cladogram correspondence (see Farrell and Mitter 1990). Two-thirds of the insect clades surveyed show some correspondence to host phylogeny.

Figure 23.6 Phylogeny estimate of *Phyllobrotica* leaf beetles, compared with host Lamiales phylogeny synthesized from literature (Farrell and Mitter 1990). Cladogram correspondence is significant or nearly so under several randomization models. The exceptional association of *P. physostegiae* with the perennial mint *Physostegia* (indicated by the asterisks) probably represents recent colonization from an annual, xeric-adapted ancestral host in the same habitat.

of parallel diversification are those of the chrysomeloid beetle genera *Phyllobrotica* and *Tetraopes* and their relatives (Farrell and Mitter 1990; Farrell 1985, 1991). Review of the biologically similar *Phyllobrotica* and *Tetraopes* systems will help illustrate the issues. With few exceptions, each individual species or subspecies in both these beetle genera attacks a single, different species of the same herbaceous plant genus. The nearest relatives of these two beetle genera, *Hoplasoma* and *Phaea* respectively, attack more primitive, woody members of the respective host groups. The adults feed on the flowers and leaves, while the larvae attack roots (Farrell and Mitter 1990; Farrell 1985, 1991).

A phylogeny estimate for the *Phyllobrotica* species with known host plants is shown in figure 23.6, with a literature-based estimate of host relationships. The beetle host plants belong to the asterid order Lamiales. *Phyllobrotica* and relatives are absent from two large, advanced lamialean clades, the lamiaceous Nepetoideae and the verbenaceous Verbenoideae. They are restricted to the subfamilies Viticoideae (Verbenaceae) and Lamioideae (Lamiaceae), basal (probably paraphyletic) elements of their respective families (Cantino 1982; Cantino and Sanders 1986).

The overall match of cladograms (figure 23.6) is significant under some (though not all) reasonable randomization approaches (Farrell and Mitter 1990) despite one major disagreement, the placement of the mint *Physostegia;* the adjacent *Scutellaria* sections Lateriflorae and Annulatae (which respectively include *S. lateriflora* and *S. integrifolia* and relatives) are also transposed between the cladograms. This pattern is consistent with largely parallel phylogenesis and only occasional host transfers.

A recent phylogeny estimate of the *Tetraopes* and *Phaea* species with known host plants (Farrell 1991) also shows significant concordance with a literature-based estimate of host phylogeny (figure 23.7), the two major exceptions being two apparent colonizations of advanced milkweeds by the relatively primitive beetles *T. pilosus*

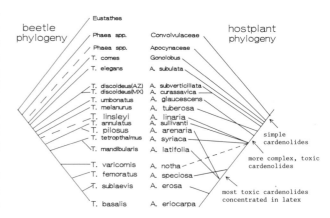

Figure 23.7 Phylogeny estimate of *Tetraopes* beetles based on morphology and allozymes, compared with literature-based relationships of host plants (Farrell 1991). Concordance is significant ($P < .01$) under several randomization models. Host *Asclepias* shows apparent progression toward increased complexity and toxicity of cardenolides, perhaps representing escape and radiation. Distributions of beetle and host plant clades suggest Oligocene origins.

and *T. mandibularis*. Within *Asclepias,* however, there is an apparent phylogenetic progression in the complexity, toxicity, and tissue distribution of cardenolides (Nelson, Seiber, and Brower 1981; Woodson 1954), which could have guided beetle colonization entirely subsequent to host diversification. (The possibility of analogous chemical variation among the hosts of *Phyllobrotica* has not been studied.) Clearly, additional evidence is needed to decide whether diversification in these assemblages has been contemporaneous.

Temporal Duration of Insect/Plant Affiliations

A critical question in judging whether current associations reflect a joint evolutionary history is whether the associated lineages are of equivalent ages. If the hosts are much older, parallel cladogenesis cannot have occurred. At present, estimates of the age of associations come mostly from fossils (limited by the generally poor record for insects) and from biogeography. However, rates of molecular divergence may prove to be sufficiently clocklike within narrowly circumscribed clades to provide an important future source of datings (DeSalle and Templeton 1988; Hafner and Nadler 1988; R. D. M. Page 1990).

Because evidence on the interactions themselves is rarely fossilized, most paleontological evidence on the age of associations is somewhat indirect. On the broadest scale, Zwölfer (1978) presented evidence suggesting that the major independent phytophage clades originating in successive geological eras are currently associated on average with successively younger plant taxa. Further, the basal divergences within such clades are often roughly concordant with host relationships. For example, the oldest (Mesozoic) fossils of Cerambycidae and Scolytidae represent cladistically basal subgroups associated at present with conifers, whereas advanced members of these families mostly attack flowering plants (Linsley 1961; Wood 1982). The most plausible interpretation is that the older insect groups have retained associations established before the younger plant groups appeared.

Similar patterns at lower levels probably await discovery, as illustrated by Tertiary beetle herbivores (fig. 23.8; Farrell and Mitter, unpublished data). The host taxa used by extant genera in the related superfamilies Chrysomeloidea and Curculionoidea that are also found in Paleocene/Eocene fossils are significantly older than those used by extant genera found only in Oligocene/Miocene fossils. For example, the conifer-feeding scolytids, platypodids, and cerambycids are common in Eocene ambers (including coniferous Baltic amber) and shales, while many genera of these groups preserved in the fabaceous *Hymenea*-derived, Oligocene/Miocene Dominican amber attack this or other legumes today. The primitive chrysomelid genus *Donacia* is reported from the same Paleocene shales as its present-day host *Nymphaea* (Crowson 1981), whereas chrysomelid genera currently affiliated with composites and Convolvulaceae appear only in fossils of Oligocene and Miocene age. The older of these associations are thus likely to have persisted for 55–65 million years (see also Eastop 1973; Hickey and Hodges 1975; Opler 1973; Moran 1989).

Many apparently relictual insect/plant distributions also suggest long-continued interactions. Among the oldest of these are the primitive cerambycid beetle and homopteran herbivores of *Nothofagus*, whose Gondwanian distributions suggest persistent affiliation since the Cretaceous (Linsley 1942, 1961, 1963; Humphries, Cox, and Nielsen 1986). The flea-beetle genus *Monomacra* and its host *Passiflora* are similarly distributed throughout much of the Neotropics, North America, and Australia, a disjunction that might also suggest Cretaceous origins for both groups. A more recent pattern is that much of the disjunct, North American/Southeast Asian "Arcto-Tertiary" flora is host to similarly distributed cerambycid and chrysomelid beetle genera (Tiffney 1985b; Linsley

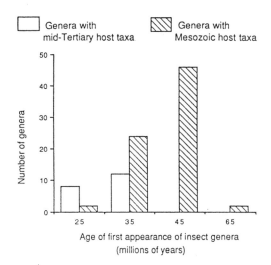

Figure 23.8 Paleontological ages of some extant beetle/host plant affiliations, compiled from Heer (1865), Handlirsch (1904), Klebs (1910), Statz (1938), Linsley (1942, 1961, 1963), Cherepanov (1988), Larsson (1978), Wood (1982) and our own studies of Dominican amber (Farrell and Mitter, unpublished data). Phytophagous beetle genera presently affiliated with older plant groups appear significantly earlier in the fossil record than beetles presently affiliated with younger groups (chi-square = 36.7; $P < .001$; df = 4).

1961, 1963); Moran (1989) gives an example from aphids.

These broad compilations suggest that many patterns of plant/herbivore associations have persisted for much of the Tertiary or longer. Because these plants and insects have surely speciated during that interval, they have in a broad sense diversified contemporaneously. However, the best evidence on the relative contribution of parallel diversification to extant associations should come from those relatively few cases in which cladogram concordance has also been quantified.

Although there are no fossils of the *Phyllobrotica/Hoplasoma* clade reviewed earlier, these beetles' "Arcto-Tertiary" distribution would date them to the mid-Tertiary, in agreement with fossil pollen dating for the origin of the Lamiales (Muller 1984). Similarly, the scant tetraopine and asclepiad fossils suggest that *Tetraopes* and *Asclepias* are both of mid-Oligocene age (Statz 1938; Tiffney 1984). While additional, molecular datings are needed, the statistically supported cladogram concordance in these cases is thus plausibly attributable to parallel diversification.

Evidence on age can also be decisive when cladogram concordance is ambiguous. Thus the weevil genus *Larinus* and its host thistles show only weak, higher-level cladogram concordance; but biogeography, fossils, and allozyme divergence levels indicate a late Oligocene origin for both groups, supporting partial parallel phylogenesis (Zwölfer and Herbst 1988). A similar interpretation is plausible for heliconiine and related butterflies (K. S. Brown 1981) and carsidarid psyllids (Hollis 1987), which despite only partial phylogeny correspondence with their respective passifloraceous and malvalean hosts, show striking geographical disjunctions, suggesting old associations. Conversely, the low level of allozyme divergence in *Ophraella* leaf beetles supports the hypothesis of recent colonization of its much older asteraceous hosts, consistent with the lack of phylogeny concordance (Futuyma and McCafferty 1990). A similar argument was applied to *Rhagoletis* flies by Berlocher and Bush (1982).

Evolutionary Mechanisms

As we have seen, many assemblages may reflect complex mixtures of parallel phylogenesis and host transfer. Apportionment of these would be more credible if we could independently predict (or at least rationalize) where parallel phylogenesis and departures from it should occur. In *Phyllobrotica*, for example, the unusual natural history of the "aberrant" association of *P. physostegiae* with *Physostegia* (see fig. 23.6) supports its interpretation as an isolated instance of host transfer, from an unusual annual, xeric-adapted prairie *Scutellaria* (the host of its sister species) to a related, chemically similar perennial host plant in the same habitat (Farrell and Mitter 1990). Host transfer from another unpredictably available, annual host to a sympatric confamilial perennial has also been reported in the *Ophraella* leaf beetles (Futuyma 1991). Whether the faunas of annual plants are particularly prone to shift, however, is not yet known.

One possible mechanism for parallel phylogenesis is joint geographical isolation, as suggested by the distributions of some *Phyllobrotica* and *Tetraopes* host pairs. Thus, *P. costipennis* and its host *Scutellaria arenicola* are both endemic to the biotically distinctive Lake Wales Ridge region of central Florida. They were probably isolated together by early Pleistocene (or late Pliocene) sea level fluctuations from their respective sister species to the northwest (Ward 1979). Similarly, the apparently young species *Tetraopes pilosus* and *A. arenaria,* both of which show the white pubescence characteristic of dune inhabitants, were probably jointly isolated as their midwestern United States sandhill habitat formed in the early Quaternary. An alternative model was presented by Zwölfer (1988), who suggested that partial, higher-level congruence between the phylogenies of *Larinus* weevils and their host thistle tribes (see below) may result simply from gradual restriction of inter- versus intratribal host shifts as these plant tribes diverged.

While the mechanisms just discussed carry no necessary implication of coevolution (sensu Janzen 1980), there exists some evidence consistent with a more coevolutionary model in the *Asclepias* system. Preliminary field and laboratory studies show that the highly diverse, more advanced, toxic milkweeds are free from the milkweed generalist arctiid moths and chrysomelid beetles that attack more primitive congeners, suggesting possible escape and radiation (Farrell 1991). *Tetraopes* beetles are the only folivores other than monarch butterflies (*Danaus* spp.) that attack these most derived milkweeds.

Variation in the degree of parallel phylogenesis seems related to the intricacy of the interaction. Thus, *Tetraopes* and *Phyllobrotica* show several features thought to promote specificity and conservatism of feeding habits, including larval endophagy and adult feeding on the larval host. Moreover, both have "toxic" hosts on whose chemicals the apparently aposematic adult beetles may be dependent for defense. *Tetraopes* sequesters *Asclepias* cardenolides (Nishio, Blum, and Takahashi 1983); for example, the advanced Sonoran Desert species, *Tetraopes sublaevis*, sequesters both primitive and advanced cardenolides from its host *Asclepias erosa* in very high concentrations (see fig. 23.7; Roeske et al. 1976; R. Martin and S. Lynch, unpublished data). Studies of sequestration by *Phyllobrotica* are currently under way. Many of its hosts share the widespread iridoid glycosides catalpol and aucubin, important to other herbivores for predator defense (M. D. Bowers 1988; personal communication). Such defense systems may promote evolutionary conservatism (Feeny 1987). Patchy host distribution and sedentary adult behavior, which at least in *Tetraopes* has been shown to be linked to genetic population subdivision (McCauley and Eanes 1987), may further increase the chance for joint geographical speciation. Closely parallel phylogenesis of insects and host plants seems most likely to be found in other assemblages with a similar natural history, which is common in other leaf beetles, long-horned beetles, and their relatives.

ESCALATION, DIVERSIFICATION, AND COEVOLUTION

If, as we suggest above, insects and plants have frequently had the opportunity to affect each other's phylogenesis, how have they done so? Clear cases of adaptive reciprocity between particular insect and plant lineages are few, and close pairwise coevolution may be rare (see review in Futuyma and Keese 1992). However, as we have seen, fossils and distributions indicate that phytophagous insects and higher plants have diversified in at least broad synchrony, and that many associations are conservative and old. Long-associated insect clades have certainly formed part of the community of enemies with which plant evolution has had to contend, and should have participated in such diffuse or guild coevolution as these two trophic levels have undergone, whether or not they are the dominant selective force on plants.

Ehrlich and Raven's "escape and radiation" model embodies a cardinal theme of the New Synthesis (Simpson 1953; Futuyma 1986): that diversification is driven by improved adaptation to biotic interaction. More recently, however, there has been skepticism that interactions, including those of insects and plants, are intense or consistent enough to be major evolutionary forces (Strong, Lawton, and Southwood 1984; but see V. K. Brown 1990). Moreover, advocates of a "hierarchical" theory of evolution have argued that evolutionary trends might be governed by species selection or sorting unrelated to increased individual adaptation, or might even be random (Raup et al. 1973; Gould 1985, 1989).

A forceful reassertion of the classic view, which he terms the hypothesis of escalation, was provided by Vermeij (1987), who cited support from marine paleontology. Confirmation of this hypothesis, presently lacking for insect/plant interactions, has at least three requirements: evolutionary advances must have in fact occurred in traits affecting the interaction; these changes must have been adaptations to the interaction; and they must have led to increased diversification.

Plant Defense Escalation and Radiation

Because they are rarely fossilized, temporal sequences of escalation, for example, in plant defense, must be inferred from comparison among extant species. Potential difficulties are that species exhibiting the earlier stages in such a sequence might be extinct and unavailable for study, and that progression in "diffuse" adaptation of an entire assemblage might be only weakly mirrored in the phylogenies of individual lineages.

Comparison of present-day plant taxa nevertheless reveals a number of potential escalation sequences (Mitter and Farrell 1991). For example, Berenbaum (1983) postulates an evolutionary sequence in umbellifers and other groups, supported by taxonomic distributions and biosynthetic pathways though not yet by explicit phylogenies, of ever more toxic and complex coumarin compounds, each reducing attack by enemies adapted to the antecedent defense. As we noted earlier, in *Asclepias* milkweeds there is an apparently similar phylogenetic progression toward more toxic and complex cardenolides, which moreover become increasingly concentrated in the latex, where their effects against herbivores should be maximal (Nelson, Seiber, and Brower 1981).

The issue of whether such traits actually evolved as defenses has proven controversial. Ideally, one would show that plants bearing the advanced trait suffer less fitness reduction from herbivores than do relatives lacking it. Related, less direct evidence would be reduced herbivore species diversity on such plants or higher plant population size. Such comparisons are essentially nonexistent, although consistently elevated plant population density has been suggested for one form of defense, secretory canals (Farrell, Dussourd, and Mitter 1991; see below).

A potential difficulty with contemporary fitness comparisons is that the original selective advantage may have been lost (Futuyma 1983). For example, the current diversity and abundance of insects attacking milkweeds might suggest that those plants' latex canals and other defenses have brought little protection from herbivores. However, the phylogenetic origins of such faunas could provide indirect evidence for an initial fitness differential. That is, if milkweed herbivores could be shown to represent radiations from a relatively small number of initial colonists among a potential, preadapted pool of insects, we could infer that the milkweeds' defenses had at first permitted them to shed many of their herbivores. As yet, this approach has been little exploited (but see Price and Pschorn-Walcher 1988).

How might we test Ehrlich and Raven's postulate that novel defenses foster plant radiation? One might simply compare species counts in groups bearing and lacking a defense. For example, Berenbaum (1983) showed that among the Apiaceae, genera with both angular and linear furanocoumarins have markedly higher average numbers of species than genera bearing linear furanocoumarins only, and that the latter in turn are more diverse than gen-

era that lack coumarins. However, two difficulties with this analysis are that genera may be of very different ages, and that genera bearing similar coumarins are probably not phylogenetically independent. Angular furanocoumarins in Apiaceae, for example, are restricted to two adjacent tribes and may thus characterize a single lineage (Nielsen 1971). Their effect on diversification is thus confounded with that of all the other features that define that lineage.

Both these problems may be overcome if we can identify multiple lineages bearing independent origins of the same defense and the sister groups of those lineages (Farrell, Dussourd, and Mitter 1991). Since sister groups are by definition of equal age, differences in their diversities must reflect different rates of diversification (origination minus extinction: Stanley 1979; Mitter, Farrell, and Wiegmann 1988). Ehrlich and Raven's hypothesis predicts that lineages bearing the new defense should be consistently more diverse than their sister groups when the latter retain the more primitive defense.

Application of this approach to another broad class of apparent defenses, secretory canals bearing latex or resin, provides statistical evidence that defense escalation can promote plant radiation (Farrell, Dussourd, and Mitter 1991). Such canals have arisen at least forty times, ranging across all the angiosperm subclasses in addition to ferns and gymnosperms, and show great variation in the details of their anatomy of the chemistry of their contents. However, they have such strong functional similarities that they can be considered a distinct syndrome. Secretory canals are typically tubes, consisting either of living cells (laticifers) or of intercellular spaces (resin canals), which ramify extensively through the plant. They are filled with viscous fluids that ooze or spurt out when the canals are severed. Many nondefensive functions have been suggested for secretory canals, but evidence for these is lacking. In contrast, there is much support for a defensive function (Dussourd and Eisner 1987; Dussourd and Denno 1991).

Sister group comparisons are available for only sixteen origins of canal systems, but the canal-bearing lineage is more diverse than its presumed sister group in thirteen of these. This statistically significant trend strongly suggests that secretory canals have promoted radiation in the plant groups that have evolved them. Moreover, the diversities of canal-bearers relative to their sister groups were found to be independent of geological age or latitude, suggesting that the advantages of canals may still be evident, and do not reflect the often-supposed higher rates of diversification in the tropics (see below). Canal-bearing taxa dominate communities in particularly stressful habitats, such as boreal forests and African and American deserts (Langenheim 1973), that may largely preclude entry to plant groups lacking such especially effective defense systems.

Recent inventories of tropical forests have reported that individual species bearing secretory canals seem unusually abundant (Boom 1986; Prance, Rodriguez, and DaSilva 1976), a conclusion confirmed by our finding that, within a Peruvian rainforest, individual species representing eight lineages with independently acquired canals have consistently higher mean population densities than their near relatives (Farrell and Mitter, unpublished data). Reports of the avoidance of canal-bearing plants by leaf-cutting ants (Stradling 1978) may help explain this observation. Thus, the local abundance and diversity of canal-bearing species may reflect their evolutionary success on a global scale, and may also bear on the unsolved problem of how increased individual adaptedness (i.e., escalation) is translated into increased diversification. Analogous observations on other taxa (Kochmer and Handel 1986; Ricklefs 1989b; Jablonski 1987, Lidgard and Jackson 1989) suggest that local demographic characteristics may often persist through geological time, and that they reflect long-term evolutionary trends.

Escalation and Diversification in Insects

Ehrlich and Raven suggested that radiation onto the diverse resources offered by plants played a major role in insect diversification. A basic prediction we might make, then, is that plant feeders as a whole should have diversified relatively rapidly, as compared with groups retaining older trophic habits (Southwood 1973; Strong, Lawton, and Southwood 1984). Sister group analyses as described above, applied to independent origins of phytophagy from saprophagy or predation, statistically support this prediction (Mitter, Farrell, and Wiegmann 1988).

This finding is consistent with Ehrlich and Raven's model, but there is an important alternative (though complementary) explanation. Price (1980) and others have argued for a broad concept of "parasitic" organisms—including most phytophages—and suggested that these might speciate more rapidly than predators or saprophages as a result of their typically extreme ecological specialization. That is, the discrete, patchy distribution of the hosts they live on should impose unusually fragmented population structure and disruptive selection. An analogous hypothesis relating niche width to diversification rate has been advanced in paleontology (Lidgard & Jackson 1989), but has had essentially no rigorous tests. This hypothesis predicts, first, that host differences might drive insect reproductive isolation. Evidence that such differences (e.g., in host phenology or physiology) are the primary factors in mediating reproductive isolation in some flies and homopterans have been recently reported (Wood, Olmstead, and Guttman 1990; Feder, Chilcote, and Bush 1990), but it is still not at all clear that they are general (Mitter and Futuyma 1983).

A second prediction of the specialization hypothesis is that other types of parasitism, such as parasitism on animals, should also accelerate diversification. Carnivorous parasitism has arisen from saprophagy or predation at least several dozen times among insects (Wiegmann, Mitter, and Farrell, 1993). Sister group comparisons are available for fourteen such lineages of parasites. No trend in relative diversities was apparent, either in the entire sample of comparisons or when these were variously subdivided as to the mode of parasitism or the host group attacked. This finding suggests that the "parasitic" lifestyle and its genetic consequences alone do not always result in enhanced diversification. Elevated phy-

tophage diversity, and the lack thereof for animal parasites, are probably related instead, or in addition, to the differing resource bases available to different trophic groups.

If resource diversity underlies phytophagous insect diversity, shifts within phytophagous lineages to new resource "adaptive zones," such as new plant tissues or new groups of plants as defined by taxonomy or secondary chemistry, should also enhance diversification. This proposition seems almost self-evident, yet it might be false: the diversity of feeding habits could be incidental to, not a cause of, the diversity of phytophage species. Tests of this hypothesis are essentially nonexistent. A logical place to start would be the conservative dimensions of resource use discussed earlier, such as those reflecting plant defense types, since if "adaptive zones" exist and have figured in radiation, then they should characterize groups of related species. There are a number of likely candidates (see review in Mitter and Farrell 1991). The strongest argument to date for insect radiations following colonization of plant groups bearing successively more advanced defenses is Berenbaum's (1983; Berenbaum and Feeny 1981) analysis of coumarin plant faunas, but sister group comparisons for these are lacking.

Finally, as stressed by the biogeographers of the New Synthesis, there is likely to be an important spatial component to escalation and diversification (Darlington 1957; Vermeij 1987). Plants and insects can escape their predators or competitors either by evolving a new adaptation or by invading a biota that lacks countermeasures to the ones they already have. We return to this theme below.

EVOLUTIONARY EFFECTS ON DIVERSITY

In the foregoing sections we have considered a number of evolutionary properties of insect/plant associations which might affect their contemporary diversity. A more difficult question is whether they have done so, and how exactly one would find this out. A satisfying synthesis of the ecological and evolutionary views of insect/plant diversity is not at hand. In this section we review some aspects of previous diversity studies that suggest a contribution from evolutionary history. We then focus on the possible historical mechanisms underlying one striking pattern of variation in diversity, that associated with latitude.

Strong, Lawton, and Southwood (1984) point out that the residuals from regression of plant faunal diversity on geographical area and plant size are often large. There are several reasons to think that much of this "unexplained" variation is in the broad sense due to history. Inspection of Strong et al.'s table 3.1 suggests that the residuals are larger in those studies that artificially classify plants as "trees," "shrubs," and "herbs" than in more taxonomically controlled studies of the faunas on particular plant taxa such as oaks or bracken. The positive correlation between residuals for different groups of plant enemies, such as fungi versus insects, further suggests consistent differences among plant taxa (Strong, Lawton, and Southwood 1984). In a number of cases, variables implicitly or explicitly invoking plant history, such as time since introduction, taxonomic "isolation," or even taxon membership per se (Kennedy and Southwood 1984), have explained a significant fraction of the residuals. The explanatory power of "isolation" in particular should increase as phylogenetic studies improve our understanding of what constrains host colonization.

The combined effects of isolation and age are illustrated in some studies of introduced plants (Strong, Lawton, and Southwood 1984). It is not surprising that faunas on sugarcane, a widely introduced grass, rapidly equilibrate in size (Strong, Lawton, and Southwood 1984): the cosmopolitan, chemically relatively homogeneous Poaceae support a large, relatively broadly oligophagous fauna of potential colonists (Jermy 1984). Other, more taxonomically and biochemically isolated introduced plants, such as *Eucalyptus* in North America or *Opuntia* in Australia, have accumulated extremely few insect enemies. *Asclepias syriaca*, introduced into milkweed-poor Europe in the seventeenth century, has yet to be colonized (Roeske et al. 1976); we might predict that it will eventually draw herbivores from other asterid and other latex-bearing families. Similar effects can be seen on an evolutionary time scale. For example, the Miocene-age Sonoran Desert flora supports a distinctive complement of endemic herbivores, nonexistent in the much more recent Sahara (Linsley 1961). The thistle tribe Cardueae, host to a diverse, specialized, and conservative fauna in its Palearctic center of origin, still has a depauperate fauna in the New World, despite the proliferation of over a hundred endemic species since its invasion in the Miocene (Zwölfer 1988).

While it is thus likely that evolutionary history plays a significant role in faunal diversity, there has been little attempt to explicitly reconcile historical processes with diversity models. An example of the kind of connection that needs to be made is the relationship between models of plant "isolation" and "escape and radiation" coevolution. Under Ehrlich and Raven's model, biochemically and taxonomically isolated plants might actually be the remnants of groups that are declining because their outmoded defenses no longer deter attack by enemies. An example is *Gingko*, the sole survivor of an old clade that has been eclipsed by its resin-canal-bearing sister group, the conifers. Such groups might tend to have relatively higher herbivore loads than their more advanced, more recently diverse relatives. Conversely, "isolated" plants could represent clades whose novel defenses have just begun to promote radiation. Possible examples are the relatively young crucifer genera *Erysimum* and *Cheiranthus*, which differ from the family ground plan in possessing cardenolides (Feeny 1977). Given the strong initial evidence for a causal connection between chemical innovation, enemy attack, and plant diversification (Farrell, Dussourd, and Mitter 1991), we cannot hope to fully understand how "isolation" figures in plant faunal diversity without distinguishing among its historical origins. As yet, there has been no attempt to do this.

In sum, our understanding of the evolutionary sources of diversity variation among co-occurring plant faunas must be judged rudimentary. There is, however, another major axis of variation in diversity to which a form of coevolutionary model may be more directly applicable.

Diversity Differences across Latitudes

Plants and their insect herbivores provide some of the most striking examples of decreasing species diversity with increasing latitude. It seems possible that aspects of the escape and radiation model will help explain such gradients.

Extensive regions of temperate climate are a relatively recent occurrence in earth history, reflecting a global cooling and drying trend beginning in the early Tertiary. Thus, most occupants of temperate regions derive from lineages that evolved in more tropical climates. Harsh climate has surely acted as a filter in the differential invasion of, or survival in, Tertiary temperate regions by plant groups, as evidenced by the special adaptations it evokes. For example, the north temperate representatives of many primitively woody, tropical lineages are herbaceous, an apparent adaptation for overwintering that permits escape underground during harsh winter seasons (Wolfe 1978; Judd, Sanders, and Donoghue 1993). In turn, the recently evolved temperate flora may have a depauperate insect fauna because the same climatic barriers are only slowly being overcome by primitively tropical herbivores, a form of (climatic) enemy escape analogous to that permitted by novel defenses. This climatic "defense" is overcome only by the evolution of migration or diapause, which allows insects to avoid harsh winter seasons. Insect diapause consists of a very complex, polygenic suite of concerted neurological and physiological traits that permit early detection and avoidance of severe seasons. Diapause for overwintering by newly temperate insects appears to have arisen only in tropical lineages "preadapted" via traits permitting avoidance of seasonal periods of low resource availability or inundation (Tauber, Tauber, and Masaki 1986). There are apparently no known cases of diapause arising de novo in newly temperate insect groups whose tropical relatives do not show some form of seasonal quiescence (Tauber, Tauber, and Masaki 1986). Because host use is so often taxonomically conservative, herbivores might thus be expected to show even more pronounced latitudinal gradients than insects of more generalized trophic habits, since they face not only the barrier of overwintering but also the additional barrier of finding suitable hosts in the depauperate temperate flora.

We have explored these hypotheses in a comparison of the history and biogeography of two New World beetle taxa of similar age, one herbivorous and one predaceous (Farrell et al., unpublished data). In studies of the insect fauna of tropical forest canopies at the Tambopata Reserved Zone, a site in Amazonian Peru (Farrell and Erwin 1988), the largest group of herbivores has proven to be the leaf beetles (Chrysomelidae), whereas the largest group of generalist predators is the rove beetles (Staphylinidae). All the major subfamilies of these two families are represented in these samples.

The local distribution patterns of leaf versus rove beetles at Tambopata correspond to the differences in their trophic habits (Farrell and Erwin 1988). Rove beetle species are distributed evenly across habitats of different floristic composition, with species diversity and abundance predictable from the total canopy foliage volume. In con-

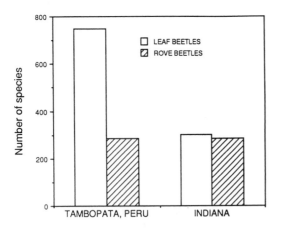

Figure 23.9 Diversity of leaf and rove beetle species represented in samples from Tambopata, Peru, and in the fauna of Indiana. Rove beetle diversity is similar, but leaf beetles are far more diverse in the much smaller tropical sample area (chi-square = 75.8; P < .001; df = 1). (Note: Both estimates of rove beetle diversity are probably low, though the comparisons with leaf beetles will not likely change [J. S. Ashe, personal communication].)

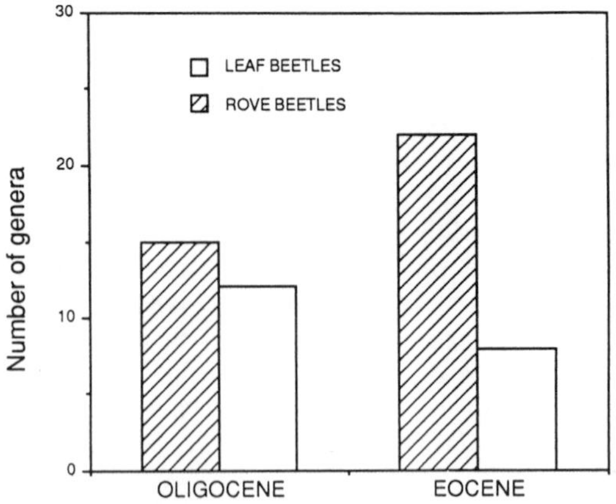

Figure 23.10 Ages of first appearance of extant leaf and rove beetle genera. Although both families arose in the Jurassic, there is an apparent trend for leaf beetle genera to appear later in the fossil record, which would suggest more recent diversification (chi-square = 1.73; P = .18; df = 1).

trast, most leaf beetle species are restricted to a single canopy type, with diversity and abundance little related to foliar volume, as might be expected of specialized herbivores (Farrell and Erwin 1988).

These ecological differences are mirrored in strikingly different gradients with latitude when the Tambopata samples are contrasted with the well-documented fauna of Indiana (Blatchley 1910). For leaf beetles, the number of species found just at Tambopata exceeds that from the entire state of Indiana, a much larger area (fig. 23.9). Conversely, rove beetles are about equally diverse in the two localities. This difference does not simply reflect host plant diversity, because there are far fewer species of canopy trees at Tambopata than there are plant species in Indiana.

Several lines of evidence support the hypothesis that the steeper gradient in leaf beetles reflects coevolution in a broad sense with their host taxa. Both beetle families appear to have arisen in the Jurassic (Crowson 1981), but the major diversification of leaf beetles apparently occurred later (fig. 23.10). Thus, extant rove beetle genera are often present in Eocene deposits, whereas extant leaf beetle genera more often first appear during the Oligocene, coincident with the generic-level diversification of their main host groups: Asteridae, advanced Rosidae, and monocots (Jolivet 1988; Muller 1984). Present distributions also support the idea that rove beetle genera are older (fig. 23.11): among the genera in the Tambopata samples, leaf beetles are significantly more often restricted to South America than are the comparatively cosmopolitan rove beetles.

In sum, we suggest that New World leaf beetles, at least, have diversified to some extent concurrently with tropical host plant groups, and that their conserved affiliation with those plants, themselves only infrequently able to invade or adapt to harsh temperate climates, is one reason why there are fewer leaf beetles in temperate than in

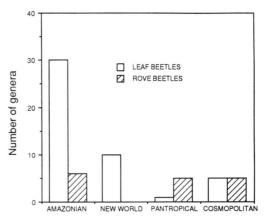

Figure 23.11 Biogeographical distributions of leaf and rove beetle genera from rainforest canopy samples at Tambopata, Amazonian Peru. The possibly more recently diversified leaf beetles are significantly more often restricted to Amazonia (chi-square = 14.5; P = .0023; df = 3).

tropical areas (Farrell et al., unpublished data). We might further speculate that among the herbivores that do reach the temperate zone, the greatest radiations may be among those able to shift onto the characteristic dominant temperate plant groups. Thus, for several largely tropical asterid-specialist leaf beetle (e.g., *Capraita*, *Longitarsis*, *Trichaltica*, and *Octotoma*) and sphingid moth (e.g., *Manduca*) genera, an apparent evolutionary pathway to association with the temperate forest flora has been through their colonization of *Fraxinus*, one of the very few temperate woody asterids. Similarly, while the "reverse" latitudinal diversity gradient in aphids and parasitic Hymenoptera has been said to reflect difficulties in host-finding by these host-specific insects in diverse tropical communities (Janzen 1981; Gauld 1986; Dixon et al. 1987), it may instead reflect radiations in the temperate zone, spurred by adaptation to temperate climate and host groups.

While the narrative above seems plausible, the hypotheses it embodies will clearly need rigorous, replicated testing against competing ones. In recent decades, reflecting the climate in evolutionary biology generally, such historical explanations have been eclipsed by those based on inherent differences between temperate and tropical regions. "Equilibrial" hypotheses have invoked (among other factors) differential climatic stability, productivity, effective spatial heterogeneity, and frequency of disturbance (see reviews in MacArthur 1972; Brown and Gibson 1983; Stevens 1989) in explaining why tropical regions should foster either increased species coexistence or greater diversification rates. Thus, temperate species are said to have broader latitudinal ranges (a possible consequence of adaptation to greater climatic variation at temperate sites: Janzen 1967; Stevens 1989) or expanded niche breadths (Pianka 1978), and these are suggested to constrain opportunities for successful invasion or diversification in temperate areas.

Hypotheses implying higher diversification rates in the tropics should be amenable to explicit phylogenetic tests. Simple counts of relative diversities do not suffice for this purpose because they do not control for the possibility that there has been more time for diversification in the tropics; temperate zones, as we have seen, are of relatively recent origin. In the absence of fossil or other datings, we can achieve such control by restricting comparison to sister groups, which by definition are of equal age (fig. 23.12). Among the five temperate/tropical sister-group comparisons that we have so far identified among herbivorous insects (table 23.1), there is no obvious trend toward faster diversification in the tropics (Farrell et al., unpublished data). This result is at least consistent with a historical explanation for depauperate temperate faunas, though there is little evidence on whether niche breadth or other dimensions of resource use vary consistently between temperate and tropical sister groups. Much of the greater diversity in the tropics thus lies at higher taxonomic levels (i.e., many tropical lineages are absent from temperate areas: Ricklefs 1989b) and seems much older than could be plausibly attributed to vicariance between Pleistocene refugia (e.g., Prance 1982).

SUMMARY AND CONCLUSIONS

1. Because phytophagous insect communities show little evidence of saturation or convergence, their phylogenetic history may be essential to understanding their diversity. We review the growing evidence on major evolutionary features of insect/plant affiliations, implicit in Ehrlich and Raven's model of coevolution, that could influence diversity.

2. Accumulating evidence suggests that insect host use is typically conservative, a prerequisite for long-term coevolution; the range of evolutionarily accessible hosts seems strongly constrained, limiting the potential for local optimization and community convergence.

3. Aggregate phylogenetic histories of plant-faunal or regional herbivore communities have rarely been exam-

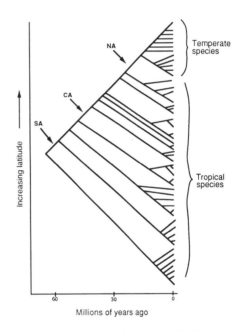

Figure 23.12 Hypothetical example of a primitively tropical clade (found in South and Central America: SA and CA, respectively) with recently evolved temperate (i.e., North American: NA) representatives. The tropical element has more species only because it is older; the rate of diversification is higher in temperate zone.

Table 23.1. Temperate/Tropical Diversity Comparisons for Five Sister Group Pairs of Phytophagous Insects

Tropical group	Probable ancestral distribution	No. of spp.	Temperate group	Probable ancestral distribution	No. of spp.	Sign of difference
Coleoptera						
1. *Hoplasoma* + *Hoplasomoides* (Farrell and Mitter 1990)	Southeast Asia	40	*Phyllobrotica*	Holarctic	22	+
2. *Tetraopes* (subgroup)	Central America	≤3	*Tetraopes* (subgroup)	Nearctic	13	−
3. *Sibinia* (subgroup) (Clark 1978)	Neotropics	≤11	*Sibinia* subgroup *Sibinia*	Nearctic → Palearctic	≈125	−
Lepidoptera						
4. *Lamproptera* + *Graphium* (Miller 1987)	Paleotropics	91	*Iphiclides*	Palearctic	2	+
5. *Audea* complex (Mitter and Silverfine 1988)	Paleotropics	≤38	*Catocala*	Palearctic, Nearctic	≈240	−

ined. Several examples suggest that such assemblages are typically drawn from a restricted pool of preadapted lineages, implying that local diversity could be limited by herbivore conservatism, particularly for taxonomically "isolated" plants.

4. While phylogenetic studies of associated insect and plant lineages are still very few, prolonged evolution in close association, indicated by concordant species phylogenies, seems uncommon. However, close parallel phylogenesis is apparent in two beetle groups with life histories tied exceptionally closely to their hosts. Moreover, fossils and biogeography suggest that many broader insect and plant clades have remained associated over geological time, thus diversifying in part together, with consequent opportunity for coevolution. Current association patterns reflect in part just the relative ages of the interactants.

5. If current diversity differences among insect and plant groups are due in part to "escape and radiation" coevolution, phylogenetic comparisons among extant species should suggest historical sequences of defenses and counteradaptations. Evidence is scant, but several likely examples of plant defense escalations exist.

6. One form of plant defense innovation, secretory canals, has been associated with consistently elevated diversification rates, further supporting the Ehrlich and Raven model; more studies are needed. Canal-bearing species may also have elevated population densities, suggesting that both dominance and diversity reflect long-term evolutionary success.

7. Sister group comparisons support the postulate that attacking higher plants (in contrast to carnivorous parasitism) has promoted insect diversification, but radiation attendant on colonization of successively escalated plant groups has not yet been rigorously sought.

8. Evolutionary constraints on insect/plant associations appear widespread, but their effects on local diversity are unclear. History-related variables may explain considerable variation in faunal diversity among plant species, but ecological and coevolutionary approaches to diversity need explicit reconciliation.

9. An evolutionary approach may be especially applicable to diversity variation with latitude. Greater reduction of diversity in the geologically young temperate zone in some phytophagous, as contrasted with predatory, beetles may reflect the twin barriers of climate and of coevolved, conserved affiliations with tropical host plant groups. Preliminary sister group comparisons provide no evidence for consistent differences in diversification rate between temperate and tropical clades, consistent with this historical explanation of diversity gradients. Whether the shift to temperate resources accompanying some temperate invasions spurs radiation awaits testing.

24

Historical Ecology: Examining Phylogenetic Components of Community Evolution

Daniel R. Brooks and Deborah A. McLennan

RESEARCH IN COMMUNITY STRUCTURE

Ecologists have striven for decades to untangle and delineate the forces involved in assembling biological communities. This quest has produced both a wealth of information and a wealth of controversy surrounding the interpretation of that information. For example, some ecologists have argued that community structure is an emergent property of generalized energy flow patterns, rather than of specific interactions among particular organisms or species (Lindeman 1942; Odum 1969; Patten and Odum 1981; Wright 1983; Glazier 1987). By moving the selection arena from the individual to the ecosystem (Oksanen 1988), these researchers are able to examine communities without "detailed information about individual species" (Loehle and Pechmann 1988). Other ecologists have sought general assembly rules for community structure at the level of interspecific interactions such as competition (Dobzhansky 1951; Williams 1964; see reviews in Strong et al. 1984) and predation (Hairston, Smith, and Slobodkin 1960; Paine 1966; Parrish and Saila 1970; for additional references see Oksanen 1988). These researchers view community structure through a window framed by individual selection, fitness, and optimization arguments. Because community systems are complex, however, there has been a tendency to base models of community species composition and structure on one particular process, and to examine that process within a population context (J. H. Brown 1981; McIntosh 1987; but see Karban 1989; Menge and Olson 1990). This purely populational view equates evolution with changes in gene frequencies in populations, and this, in turn, construes evolution as a reversible phenomenon, in spite of all of our empirical evidence to the contrary.

Both systems ecologists and evolutionary ecologists have discovered valuable clues in their quest to solve the mysteries of community assembly and evolution. We believe, however, that the mysteries will remain unsolved if we do not incorporate the vital information contained at the species level into our explanatory framework. In the following pages, we will present ideas about how we can uncover that information and, working with a coalition of ecologists and systematists, solve the one mystery and begin on the next.

The Importance of Species

Speciation is a temporally irreversible and historically unique phenomenon. It is a property of collections of demes that is not manifested by the demes themselves. Although evolutionary changes are always occurring within populations, the coherent structure of a species is not affected unless gene flow between populations is severed. Not surprisingly, the collection of demes construed as representing a species often exhibits more geographical and ecological stability than the demes themselves (i.e., demes can disappear and re-form without destroying the species). The most important characteristic of a species is that its members are bound together by unique common ancestry (the *evolutionary species concept*: Wiley 1978, 1980; see also Endler 1989; Templeton 1989). Over time, distinct historical trajectories emerge from the speciation process, with each species differing to some degree from its ancestor and closest relatives, but retaining some of its ancestry in the form of shared, derived traits (synapomorphies). Species, then, are vessels of future potential, living legacies of past modifications and stasis shaped by millennia of biotic and abiotic interactions. They are history embodied. This is the essence of evolution.

The preceding discussion leads us into the murky realm of ambiguous terminology. In this case, the word "history" is the source of confusion. *History* has been variously used to mean (1) environmental changes in the past (e.g., Hughes 1989; Facelli and Pickett 1990; Foster, Schoonmaker, and Pickett 1990); (2) individual experience (e.g., McCrea and Abrahamson 1987; Peterson and Black 1988); (3) changes in populations (e.g., Buss and Yund 1988); and (4) the order of species invasion into an area (e.g., Robinson and Dickerson 1987 and references therein; Drake 1990a). We will use the word *history* in its evolutionary sense; that is, we will equate history with the genealogical relationships among species. This definition hearkens back to Darwin's original insight that all organisms are connected by common genealogy (Darwin 1872, 346): "The characters which naturalists consider as showing true affinity between any two or more species, are those which have been inherited from a common parent, all true classification being genealogical."

Although evolutionary ecologists have examined experimental data within a comparative (historical) framework on a regular basis, few researchers have incorporated phylogenetic information into their evolutionary explanations. For example, suppose you are interested in the question of species coexistence. As MacArthur (1972) noted, the best place to look for the factors involved in species coexistence is among sympatric populations of congeners. The assumption behind this recommendation is a historical one: members of the same genus should theoretically share a number of ecological, morphological, and behavioral characters because they are all descended from a common ancestor. The recognition that the genealogical relationships among species may influence the outcome of an experimental investigation provides the historical context in any evolutionary ecological study. Having discovered an appropriate group of sympatric congeners, you may set about collecting a wealth of data concerning feeding behavior, habitat preference, and breeding cycles in order to identify the way(s) in which the species are partitioning their environment, and use that information to explain the evolution of the diversity you find. This second step is primarily nonhistorical because it requires assumptions about the evolutionary *past* of the species' interactions, based upon characters and interactions observed in the *present* environment. What is missing here is information about the evolutionary origin and elaboration of the characters and of the associations themselves. Macroevolutionary patterns uncovered by phylogenetic analysis can provide this information, because a phylogenetic tree is at once a hypothesis of genealogical relationships (species origin) and of structural and functional relationships (character origin).

Historical Ecology

Biologists began to focus their attention on macroevolutionary patterns of diversity in the early 1970s (see, e.g., Ross 1972a, 1972b; Boucot 1975a, 1975b, 1978, 1981, 1982, 1983). Brooks (1985) consolidated the research of these authors, as well as the results from his own studies of parasitic organisms, into a discipline that he called *historical ecology*. Initially, historical ecology was concerned with studying the macroevolutionary components of tightly coevolved ecological associations such as host-parasite or herbivore-plant systems. Recently, the boundaries of historical ecology have been expanded to include two general evolutionary processes, speciation and adaptation, and the macroevolutionary effects of these processes in the production of both evolutionary groups of organisms and multispecies ecological associations (Brooks and McLennan 1991).

How would a historical ecologist approach the question of community evolution? Let us answer this question by considering a hypothetical example (elaborated from Mayden 1987): Imagine discovering that two fish species in a large lake (area 1, fig. 24.1A) do not overlap ecologically; say, for example, one is a benthic forager (species D) and one is limnetic (species Z). A possible explanation for such habitat separation is that it represents the effects of competition between these two species in the past. Is this a reasonable hypothesis? The short answer to this question is, "we don't know." A longer, more informative answer requires additional information. Without a phy-

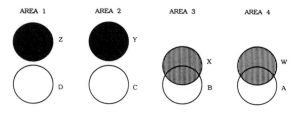

(a) DISTRIBUTION OF FISH SPECIES IN WATER COLUMN

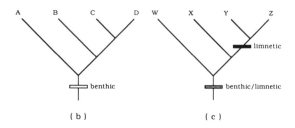

Figure 24.1 The influence of history at the community level. (*A*) Distribution of fish species in the water column by foraging preferences. *White circles*, species foraging in the benthos; *striped circles*, species foraging in both the benthos and the limnos; *black circles*, species foraging in the limnos. (*B*) Phylogenetic relationships for clade A–D based on nonecological data. (*C*) Phylogenetic relationships for clade W–Z based on nonecological data.

logeny for the fishes and a record of their relatives' interactions with each other, it is impossible to ascertain whether the association in our research lake is a result of interactions between Z and D or a historical legacy of interactions between their ancestors. So, after extensive fieldwork, we uncover more co-occurring species from the same two clades represented by Z and D in other lakes: species C (benthic) and Y (limnetic) in area 2, species X (demonstrates both foraging modes) and B (benthic) in area 3, and species W (demonstrates both foraging modes) and A (benthic) in area 4 (fig. 24.1A). As luck would have it, phylogenies for the two clades, based on morphological data, exist. When the foraging modes are optimized on the trees, we discover that foraging on the benthos is plesiomorphic for all members of the A + B + C + D clade (fig. 24.1B). These species have not changed their foraging habits, interactions with members of the other clade notwithstanding. Conversely, foraging on both benthic and limnetic prey was primitive for the W + X + Y + Z clade (fig. 24.1C) but something happened during the interaction between the ancestor of Y and Z and the ancestor of C and D, and the former moved out of the benthic into the limnetic realm. So, while this does rule out a role for interspecific competition between past populations of species D and Z in shaping the current foraging modes in these fishes, it does not rule out the possibilities that competition was either involved in the habitat shift in the appropriate ancestors or is maintaining the divergent foraging habits today.

In order to examine community assembly within a historical (phylogenetic) framework, we must ask and answer two questions, the first being, How did the species come to be in this area/association? Within any group of geographically associated species there will be nonrandom ecological interactions, ranging from casual to obligatory, among some of those species. The problem lies in determining which components, if any, of the associations among extant organisms can be traced through a history of association between the ancestors of those organisms. This is the central question in historical or comparative biogeography. The second question is, How are the members of an association interacting with one another? This question focuses our attention on the details of specific interactions between organisms as we examine information about the origins and the modifications of these interactions within a phylogenetic framework. In the remainder of this chapter, we will discuss the methods available for answering these questions and the results obtained by researchers who study communities from a phylogenetic perspective.

COSPECIATION

How Did the Species Come to Be in This Area/Association?

One of the fundamental advances in evolutionary biology over the past 25 years has been the formal articulation of two different perspectives on the manner by which species achieve their geographical distributions. These perspectives may be categorized loosely as *island biogeography* (MacArthur and Wilson 1963, 1967), which calls attention to the propensity for organisms to move about, and *vicariance biogeography* (Croizat, Rosen, and Nelson 1974; Rosen 1975, 1985; Platnick and Nelson 1978; Humphries and Parenti 1986; Wiley 1981, 1988a, 1988b), which reminds us that those movements may not be unconstrained. Both research programs have contributed valuable insights into the problem of species cooccurrence: first, two or more species may be associated today because at least one of the species evolved elsewhere and subsequently became involved in the interaction by colonizing a host or dispersing into a geographical area. This is referred to as *association by colonization*. Second, two or more species may be associated today because their ancestors were associated with each other in the past. For example, suppose that after extensive collecting, you discover two distinct fish communities on either side of a mountain range. Community B comprises representatives from three different families: species 2, 7, and 12. Amazingly, community C comprises different members from the same three families: species 3, 8, and 13. How were these communities assembled? In order to answer this question, you need information about the evolutionary history of the species. Fortunately, well-supported phylogenetic trees exist for all the fishes you have collected (fig. 24.2A). Examination of these trees in light of the geological history of the area leads you to hypothesize that: sister species 2 and 3, sister species 7 and 8, and sister species 12 and 13 were all formed when their ancestors were subdivided by the uplifting of the mountain range. The associations among species 2, 7, and 12 in community B and the associations among species 3, 8, and 13 in community C are thus a historical reflection of

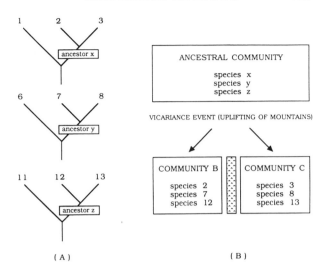

Figure 24.2 Association by descent. *Left:* Phylogenetic trees for the three fish families collected by an intrepid historical ecologist on two sides of a mountain range. *Right:* Hypothesis of community assembly based upon information from speciation studies. The ancestral community is hypothesized to have been fragmented into two new communities, B and C, by the uplifting of the mountains. All the species in the ancestral community speciated as a result of this range division.

the association between their ancestors, species x, y, and z, in the ancestral community that existed before the uplifting of the mountains (fig. 24.2B). In this case, the contemporaneous biogeographical relationship is a persistent ancestral component of the biotic structure within which the interacting species reside. This is referred to as *association by descent* because each of the species has inherited the association. It is likely that many, if not most, communities contain both vicariant and dispersalist elements; therefore, it is important to have a method that elucidates the relative roles that geological changes and colonization have played in determining the patterns of species origin and geographical occurrence.

The first methods developed in this area were designed to provide qualitative documentation of general biogeographical distribution patterns based upon the sister group relationships of members of different clades. Rosen (1975) presented an approach using *reduced area cladograms* in which distributional elements not common to all clades were eliminated from the data base, resulting in a simplified area cladogram depicting the general pattern. The Platnick and Nelson (1978) and Nelson and Platnick (1981) approach, called *component analysis* (see also Humphries and Parenti 1986), relied on the use of *consensus trees* (Adams 1972; Nelson 1979, 1983) to summarize the common biogeographical elements (for a discussion of the limitations of consensus trees, see Miyamoto 1985; Wiley et al. 1990). In this method, elements that depart from the general pattern are depicted as ambiguities. The possibility that these ambiguous elements actually represent instances agreeing with the general pattern is then investigated by invoking one of two assumptions.

Both the reduced area cladogram and the component analysis approaches have been criticized by ecologists and by systematists. Simberloff (1987, 1988) and Page (1987, 1988) pointed out that removal of ambiguous or conflicting data might make it appear that there is more evidence for general patterns than the data actually support. These authors also objected to the lack of statistical significance tests (but see Simberloff 1987) to determine the probability that apparent general patterns are due to a common cause (vicariant speciation) or to a series of unrelated parallel speciation events (Endler 1982).

Wiley (1988a, 1988b) and Zandee and Roos (1987) also criticized component analysis for obscuring evolutionarily relevant aspects of biogeographical patterns. They called for a methodology that would allow an integration of the exceptions with the more general patterns. The new methodological approach (Brooks 1981; Wiley 1988a, 1988b; Zandee and Roos 1987; Brooks 1990; Brooks and McLennan 1991) is not designed to ask, either qualitatively or quantitatively, whether there is a single general distribution pattern that might explain all of the data. Rather, it is based on the assumption that *any given biota is likely to be a combination of species that evolved where they now occur and those that evolved elsewhere and dispersed into the area where they now occur.*

This method (Brooks parsimony analysis: Wiley 1988a) has five basic steps: (1) reconstruct the phylogenetic relationships of the organisms (fig. 24.3A); (2) designate the areas in which the species occur as if they were taxa; (3) treat the phylogenetic relationships of the species *as if they were a completely polarized multistate transformation series,* in which each taxon and each internal branch of the tree is numbered (fig. 24.3B); (4) construct an area cladogram based on the phylogenetic relationships of the species (fig. 24.3C); and (5) compare this cladogram with a cladogram for the areas based upon geological evidence (fig. 24.3D). This comparison allows us to identify the species in any biota that co-occur due to a common history of allopatric speciation, and those that are present due to dispersal from another area. The historical component of the community comprises species whose phylogenetic history corresponds with the geological history of the areas in which the communities are found. We can identify these species by searching for points of congruence between the two area cladograms (association by descent). The nonhistorical components of the community are highlighted by the incongruencies between the two cladograms because there is no reason to expect a species' history to be congruent with the history of an area into which it dispersed (association by colonization). Of course, any species that originally appears as a colonizing influence in a community can become phylogenetically associated with it in the subsequent evolutionary history of that community.

It is important to determine which components of an ecological association are linked with either vicariance or dispersal events in order to assess the temporal and environmental context in which particular interactions emerged. However, the possibility of shared phylogenetic histories among geographically co-occurring species can be investigated without making assumptions about the type or extent of their ecological interactions. In fact, beyond occupying the same general area, the study species need not interact at all. In a simple sense, we are interested in asking to what extent speciation has been promoted by the active movements of organisms and to what extent it has been promoted by the active movements of the earth. The robustness of such studies depends on whether we have an explicit independent evolutionary history for the areas and how many clades we are able to analyze phylogenetically.

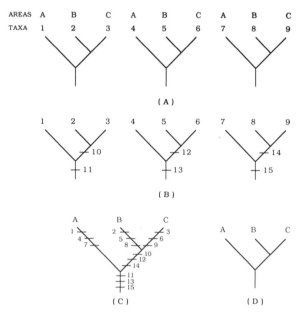

Figure 24.3 Essentials of Brooks parsimony analysis. Numbers = species; letters = areas. (A) Phylogenetic trees for three clades, members of which are found in three different communities. Areas in which the species reside are mapped above the trees. (B) Phylogenetic trees numbered for analysis. All the species, including the ancestors, receive a number. (C) Area cladogram reconstructed using phylogenetic relationships of the species. (D) Cladogram for the areas based upon geological data. Points of congruence between (C) and (D) indicate episodes of vicariant speciation; incongruence highlights episodes of dispersal by a species into an aresa. In this example, the two area cladograms are 100% congruent.

North American Freshwater Fishes: Pre- or Post-Pleistocene Origins? North America contains approximately 1,000 described and undescribed species of freshwater fishes, compared with 230 in Australia, 250 in Europe, 1,500 in Asia, 1,800 for Africa, and 2,200 in South America (Gilbert 1976). This diversity has traditionally been attributed to the dispersal of fishes into North America from Europe, Asia, and South America. If this is true, then North American fish communities are younger than their counterparts elsewhere. The patterns of this diversity also vary across the continent. In general, the drainages east of the Rocky Mountains are species-rich and dominated by the members of only a few genera in a small number of families (Cyprinidae, Percidae, Centrarchidae, and Ictaluridae). By contrast, the western fauna is somewhat depauperate, and is composed of a different assortment of fishes (distinctive cyprinid and catostomid genera, Salmonidae, Cyprinodontidae, and Cottidae). Researchers have sought explanations for these patterns by identifying the putative center of origin for a particular group, then invoking a variety of dispersals and extinctions (see Mayden 1988 for a detailed discussion). Such explanations were often predicated on the hypothesis that most species had dispersed into the continent from elsewhere.

Mayden (1988) examined the historical biogeography of fishes in seven different clades: the *Notropis leuciodus*, *Luxilus zonatus-coccogenis*, *Etheostoma variatum*, *Nocomis biguttatus*, and *Fundulus catenatus* species groups, and the subgenera *Etheostoma* (*Ozarka*) and *Percina* (*Imostoma*). These fishes inhabit drainage systems within the Central Highland region. The Central Highland consists of three currently disjunct regions. To the west of the Mississippi River lie the Ozark and the Ouachita highlands, separated by the floodplain of the Arkansas River. The western highlands, in turn, are separated from their expansive counterpart, the eastern highlands, by the floodplain of the Mississippi River. Prior to the disruptive influences of Pleistocene glaciation, these regions were continuous (see references in Mayden 1988).

Two hypotheses have been proposed to explain the diversity patterns of the freshwater fish fauna in this region. The first hypothesis is based upon observations that the same or related species often occur on the eastern and western sides of the Mississippi. Given this, several researchers proposed that much of the current diversity was produced by the fragmentation and isolation of populations during Pleistocene glaciation. According to this scenario, speciation occurred subsequent to this isolation, and has been accompanied by widespread dispersal of the new species following the retreating glaciers. The second hypothesis postulates that current diversity existed before the Pleistocene glaciation. According to this scenario, the glaciers fragmented the freshwater fauna existing in the expansive and continuous highland province. While this does not rule out a role for glaciation in the production of the current disjunct distribution patterns of freshwater fishes in the Central Highland region, it minimizes glaciation's role as the stimulus for widespread and recent speciation by these organisms. Mayden (1988) focus his attention on the Central Highland drainages in an attempt to shed further light on the problem:

> For Central Highland fishes one may examine the origin of the fauna by comparing the history of the drainage basins involved and the history of the fishes, inferred from geologic data and phylogenetic relationships, respectively. If congruence is obtained between the phylogenetic relationships and drainage relationships existing prior to the Pleistocene then one may predict that the fish groups existed prior to glaciation and the vicariance hypothesis would be supported. However, if relationships of fishes are congruent with drainage patterns developed after glaciation, then an explanation of dispersal during and after the Pleistocene glaciation may be appropriate.

Mayden performed a phylogenetic analysis using 34 river drainages as area taxa, and using the phylogenetic

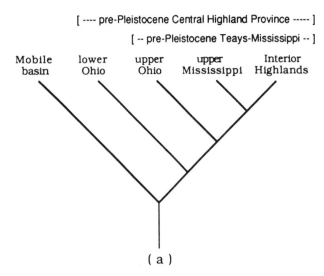

Figure 24.4 Area cladogram for the Central Highland drainage systems reconstructed from the phylogenetic relationships of 40 species of freshwater fishes. The original area cladogram has been simplified to highlight five groups of rivers that indicate preglacial origins of the fishes that inhabit them: Mobile basin; lower Ohio; upper Ohio (part of the old Teays River drainage); upper Mississippi; and Interior Highlands. (After Mayden 1988.)

relationships from the seven different clades as characters (in all, 37 characters). This analysis produced 33 equally parsimonious area cladograms. We will base the following discussion upon a simplified version of Mayden's consensus tree for those cladograms (fig. 24.4). This area cladogram highlights four points of agreement among all the hypotheses. First, all area cladograms place the rivers of the Interior Highlands in a single clade. That clade, in turn, is the sister group to the drainages of the upper Mississippi River and the old Teays River. Second, prior to the Nebraskan and Kansan glaciations, the Teays River flowed west into the upper Mississippi. Following these glacial advances, the course of the Teays was shifted southward to flow into the Ohio River, which is the present-day (post-glaciation) configuration. All 33 area cladograms place the rivers of the old Teays system with the upper Mississippi drainages; the Teays system never clusters with drainages forming part of the lower Ohio River system. Third, the Mobile basin, which was separate from the Mississippi drainage long before the Pleistocene, retains its independent status in this analysis, clustering at the base of the area cladogram. And finally, one clade encompasses all the Mississippi offshoots thought to have formed the drainages for the pre-Pleistocene Central Highland, while yet another clade reconstructs the pre-Pleistocene Mississippi–Teays River system.

All of these distributions coincide with *pre-Pleistocene*, rather than post-Pleistocene or contemporary, drainage patterns (for a detailed discussion of individual rivers, see Mayden 1988). This suggests that there was a diverse and widespread Central Highland ichthyofauna prior to the Pleistocene glaciation, corroborating the studies of speciation patterns reported by Wiley and Mayden (1985). This does not mean that dispersal and glaciation have been unimportant in this system. For example, in seven cases, Mayden was able to demonstrate that specific instances of geographical homoplasy coincided with episodes of Pleistocene glacial alterations in river flow patterns that apparently resulted in some faunal mixing. Not surprisingly, then, current distributions reflect an interaction between the relatively ancient origins and diversification of the fauna and the recent effects of large-scale environmental changes.

The Freshwater Stingrays of South America. Members of the stingray family Potamotrygonidae are the only elasmobranchs that are permanently adapted to freshwater habitats. Because potamotrygonids are restricted to rivers that empty into the Atlantic Ocean, ichthyologists have tended to assume that the ancestor of the potamotrygonids was an Atlantic marine or euryhaline stingray that dispersed into fresh water, adapted to the new surroundings, and then dispersed throughout eastern South America, speciating along the way. The most common evolutionary scenario postulated that the potamotrygonid ancestor moved from the Atlantic Ocean into the Amazon Basin during the Pliocene marine ingression. Subsequent to this invasion, a population was isolated from the ancestor, progressively adapting to fresh water and spreading throughout South America by stream capture during the past 3–5 million years.

Brooks, Thorson, and Mayes (1981) uncovered a new perspective on the origins of this enigmatic group of stingrays in their phylogenetic investigations of the ecological association between potamotrygonids and their helminth parasites. They began with a historical biogeographical analysis to discover how long the potamotrygonids had been in fresh water. If they had arrived relatively recently and had speciated as a result of independent dispersal, then potamotrygonids, and their parasites, *should not show correlated patterns of speciation with organisms that evolved in the freshwater habitats.* The geographical distribution patterns for the parasites of potamotrygonids are complex: some of the parasite species appear to be restricted to single river systems, while others are more widespread (table 24.1).

The area cladogram for these groups (fig. 24.5) highlights four evolutionary components that have contributed to the helminth community composition in the Neotropical freshwater stingrays (see Brooks and McLennan 1991 for details of the analysis). The first component is the historical geological, or vicariant, backbone linking the Paraná, upper Amazon, Orinoco, and Magdalena areas. These areas all contain species whose phylogenetic relationships correspond to the geological history of the regions: these are the species that evolved in the communities where they are now found. The remaining components of the parasite distribution patterns involve three sequences of dispersal into these areas along the following routes: (1) from the Paraná to the mid-Amazon to the Orinoco; (2) from the upper Amazon into the Orinoco, and (3) from the upper Amazon, the Orinoco, and the Magdalena to form the Maracaibo fauna.

The congruence between the area cladogram based on the phylogenetic relationships of the parasitic worms in-

Table 24.1. Geographical Distribution of 23 Species of Helminth Parasites Inhabiting South American Freshwater Stingrays.

Parasite species[a]	Locality[b]					
	1	2	3	4	5	6
1. *Acanthobothrium quinonesi*	0	0	0	0	+	+
2. *Acanthobothrium regoi*	0	0	0	+	0	0
3. *Acanthobothrium amazonensis*	0	0	+	0	0	0
4. *Acanthobothrium terezae*	+	0	0	0	0	0
8. *Potamotrygonocestus magdalenensis*	0	0	0	0	0	+
9. *Potamotrygonocestus orinocoensis*	0	0	0	+	0	0
10. *Potamotrygonocestus amazonensis*	0	0	+	+	+	0
13. *Rhinebothroides moralarai*	0	0	0	0	0	+
14. *Rhinebothroides venezuelensis*	0	0	0	+	+	0
15. *Rhinebothroides circularisi*	0	0	+	0	0	0
16. *Rhinebothroides scorzai*	+	0	0	+	0	0
17. *Rhinebothroides freitasi*	0	+	0	0	0	0
18. *Rhinebothroides glandularis*	0	0	0	+	0	0
24. *Eutetrarhynchus araya*	+	+	0	+	0	0
25. *Rhinebothrium paratrygoni*	+	+	0	+	0	0
26. *Paraheteronchocotyle tsalickisi*	0	0	+	0	0	0
27. *Potamotrygonocotyle amazonensis*	0	0	+	0	0	0
28. *Echinocephalus daileyi*	0	0	+	0	0	0
29. *Paravitellotrema overstreeti*	0	0	0	0	0	+
30. *Terranova edcaballeroi*	0	0	0	+	0	0
31. *Megapriapus ungriai*	0	0	0	+	0	0
32. *Leiperia gracile*	+	0	0	0	0	0
33. *Brevimulticaecum* sp.	+	0	0	0	0	0

[a]Species are numbered for biogeographical analysis.
[b]1, upper Paraná, including the lower Mato Grosso; 2, mid-Amazon, near Manaus; 3, upper Amazon, near Leticia; 4, delta of the Orinoco; 5, Lake Maracaibo tributaries; 6, mid- to lower Magdalena.

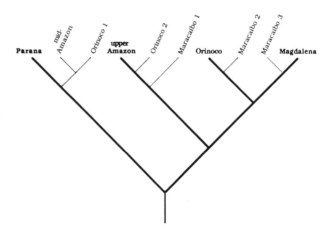

Figure 24.5   Area cladogram for six river systems in eastern South America, based on phylogenetic relationships of the helminth parasites that inhabit stingrays living in those rivers. Bold lines show the origin of rivers based on phylogenetic data that corresponds with the geological cladogram for these rivers. Species whose phylogenetic history supports these branches represent the historical component of communities in those regions. Dotted lines show the placement of rivers based on phylogenetic data that disagrees with the geological cladogram (dispersalist elements). Species whose phylogenetic history supports these branches represent the dispersalist component of communities in those regions.

habiting potamotrygonids, the areas of endemism for ostariophysan fishes inhabiting the Neotropics, and the geological history of the river systems in which they occur, pushes the origin of the parasite communities back to at least the mid-Miocene. Given that the parasite fauna is an old one, there are two alternative explanations for the origins of the parasite/stingray associations. The first is that the parasitic fauna that now inhabits the potamotrygonids occurred in South America prior to the appearance of the stingrays. In this case, the parasites or their closest relatives should be found in freshwater organisms, such as ostariophysan fishes, that were living in South America during the mid-Miocene or earlier. This explanation supports the proposal that the potamotrygonids are recently derived. Alternately, the freshwater stingrays may be older than biologists once believed, and their parasites reflect that ancient ancestry. If this is the case, the parasites inhabiting potamotrygonids, or their closest relatives, should inhabit marine stingrays whose geographical distribution is consistent with a hypothesis that the group originated as a result of marine invasion of South America no later than the mid-Miocene.

In order to examine these alternatives, we must expand the scope of the study to include the closest relatives of the parasites found in the freshwater stingrays. Four of the 23 parasite species listed in table 24.1 inhabit either teleosts or crocodilians in particular areas, where local potamotrygonids have picked them up. The remaining 19 species of parasites are restricted to stingray hosts. The closest relatives of these species inhabit marine stingrays (with the exception of *Megapriapus ungriai*, the only acanthocephalan known to inhabit elasmobranchs, and whose relationship to other acanthocephalans is uncertain). Hence, it seems likely that most of the parasite groups inhabiting potamotrygonids were brought into Neotropical freshwater habitats with the ancestor of the stingrays themselves.

If we entertain the possibility that the ancestor of the potamotrygonids (and its parasites) arrived in Neotropical freshwater habitats no later than the mid-Miocene, we must reevaluate our ideas about the source of those marine ancestors. The geography of South American prior to the mid-Miocene differed in three significant ways from what we see today: (1) Africa and South America were joined (i.e., there was no Atlantic Ocean at the mouth of the Amazon); (2) the Andes began sweeping upward from the south beginning in the early Cretaceous and moved northward; (3) the Amazon River flowed into the Pacific Ocean until the mid-Miocene, when it was blocked by Andean orogeny, becoming an inland sea and eventually opening to the Atlantic Ocean. This led Brooks, Thorson, and Mayes (1981) to conclude that potamotrygonids must have come from an ancestor in the Pacific Ocean.

Enlarging the spatial scale of this study to include the geographical distribution of the marine relatives of the parasites inhabiting potamotrygonids brings to light support for the hypothesis that these stingrays and their parasites originated from marine ancestors that were isolated in South America from the Pacific Ocean by Andean orogeny. The closest relatives of the parasites inhabiting potamotrygonids occur in Pacific marine stingrays (Brooks and Deardorff 1988). A similar origin has been suggested for Amazonian freshwater anchovies (Nelson 1984) and possibly for Neotropical freshwater needlefish (Collette 1982). In addition, each of the parasite species inhabiting potamotrygonids requires a mollusk or arthropod intermediate host, so it seems likely that mollusk and arthropod species derived from marine ancestors also moved into Neotropical freshwater habitats along with the ancestor of the potamotrygonids. Hence, a sizeable component of current Neotropical freshwater diversity seems to be derived from Pacific marine ancestors.

COADAPTATION

How Are the Members of an Association Interacting?

Cospeciation analysis does not incorporate assumptions concerning particular adaptive processes, so such analyses will not provide any information about the ways in which associated lineages may have influenced each other's evolution. This question falls into the domain of coadaptation studies, which are designed to uncover the adaptive components within the macroevolutionary patterns of association between and among clades. Studies of coadaptation must begin with a cospeciation analysis to provide the phylogenetic background against which episodes of mutual modification can be highlighted. Without this information, it is impossible to objectively differentiate scenarios based on the assumption that current associations reflect historical associations from other scenarios which presuppose little, or no, history of interaction.

There is a marked similarity in many ecological associations, especially of specialized species, around the world, which some researchers believe is evidence of phylogenetic influences in interaction structure as well as species composition. For example, McCoy and Heck (1976) examined the communities of corals, seagrasses, and mangroves throughout the tropics of the world, and concluded that these ecological associations all had a common origin. Hill and Smith (1984) discovered similar roosting patterns among six species of bats in a Tanzanian cave and fourteen species of bats in a cave in New Ireland Island off the northeastern coast of New Guinea. In both communities, species of the genus *Hipposideros* lived near the rear of the cave in association with species of *Rhinolophus*, while species of *Rousettus* roosted just inside the first major overhang. Ross (1986) reported a high degree of phylogenetic constraint in the structuring of contemporaneous reef fish communities. Boucot (1982, 1983) concluded that the fossil record demonstrates the conservative nature of community structure throughout evolutionary history. On the parasitological side, Benz (cited in O'Grady 1989) examined the distribution of copepods living on sharks' gills. He discovered substantial diversification in habitat preference (site selection) within these copepod communities. Interestingly, the phylogenetic structure of these gill niches was retained even though the specific shark and copeped species varied from community to community. In almost every case, species of *Pandarus* inhabit the gill arch; *Eudactylinodes* and *Gangliopis* attach themselves to the secondary lamellae; *Nemesis* burrow into the efferent arterioles; *Phyllothyreus* inhabit the superficial portions of the interbranchial septum; *Paeon* embed in the interbranchial septum; and *Kroyeria* are found in the water channels of the secondary lamellae or embedded in the interbranchial septum. To other re-

Table 24.2. Historical and Ecological Influences on Community Structure

Species	Species interactions	
occurrence	Ancestral	Derived
Ancestral (resident)	I Historically constrained residents	II Residents change stochastically
Derived (colonizer)	III Non-competitive colonization	IV Colonization by competitors

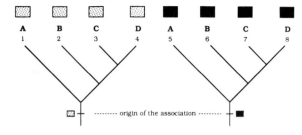

Figure 24.6 Phylogenetic effects on community structure in area D. Letters = area; numbers = species. Species 1 and 5 occur together in area A, species 2 and 6 occur in area B, species 3 and 7 occur in area C, and species 4 and 8 occur in area D. The state of a particular character involved in the interaction between these community members is depicted by the boxes. In this situation, both the origins of the traits and the origins of the association between community members are old. Extant community structure in area D thus represents *the persistence of an ancestral association, coupled with the persistence of ancestral interaction traits in both members of the community.*

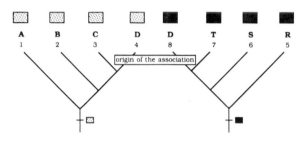

Figure 24.7 Effects of colonization by a "preadapted specialist" on community structure in area D. Letters = areas; numbers = species. The state of a particular character involved in the interaction between these community members is depicted by the boxes. Species 8 has colonized area D and is now interacting with species 4. In this situation, the origins of the traits are plesiomorphic while the origins of the association between community members are apomorphic. Extant community structure in area D thus represents *the appearance of a colonizing species, coupled with the persistence of plesiomorphic interaction traits in both members of the community.*

searchers, however, such similarities indicate widespread convergence in community structure rather than common origins.

We think it likely that any given multispecies association may be characterized by the interaction of a variety of influences. Studies of such systems can be further enhanced by incorporating concepts of spatial and resource allocation among co-occurring species (J. H. Brown 1981, 1984; Brown and Maurer 1987, 1989). For example, all species within a community contain information about their origin with respect to the biota (spatial allocation) and about the origin of traits relevant to the association, i.e., traits that characterize a species' interactions with other species and with the environment (resource allocation). The occurrence of a given species in a community may be due to either phylogenetic association (its ancestor was associated with the ancestors of other community members), in which case we refer to it as a resident species, or to colonization, in which case, not surprisingly, the taxon is termed a colonizing species. Similarly, the ecology of any given species in a community may reflect the presence of persistent ancestral traits or of recently evolved, autapomorphic, traits. Brown and Zeng (1989) recently suggested that communities should be considered mosaics of all four of these types of historical and ecological influences. The combinations of species occurrence (speciation processes) and interactions (adaptation processes) that influence community composition are depicted in table 24.2.

PHYLOGENETIC HISTORY (table 24.2, type I). The conservative, homeostatic portion of any community is composed of species that evolved in situ through the persistence of an ancestral association (congruent portions of phylogenies in a cospeciation analysis). Such species display the plesiomorphic condition for characters involved in interactions with other community members and with the environment (fig. 24.6). Since this section of the community is characterized by a stable relationship across evolutionary time, it may act as a stabilizing selection force on other members of the community by resisting the colonization of competing species.

COLONIZATION BY PRE-ADAPTED SPECIES (table 24.2, type III). This portion of the community contains species that have been added by colonization. Such species can be recognized in part because their phylogenetic history is incongruent with the histories of other community members. In this context, the term "preadapted" implies only

that these individuals are able to colonize the area because they already possess traits that do not conflict with the existing community structure (fig. 24.7). This scenario postulates that, initially at least, there is no competition between colonizing individuals and established (resident) members of the community. If the appearance of these species reduces the possibilities for the subsequent addition of species to the community, then this type of macroevolutionary pattern corresponds to the asymptotic equilibrium model of MacArthur and Wilson (1967). If the rates of colonization are low enough, the community may persist below expected equilibrium numbers. Finally, if the colonizers are so specialized ecologically that they do not affect other members of the community or preexisting potential niche space, community diversity may increase without approaching an apparent equilibrium.

COLONIZATION BY COMPETING SPECIES (table 24.2, type IV). All species that colonize a community will exhibit in-

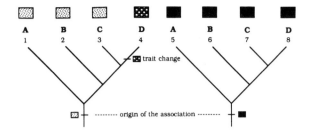

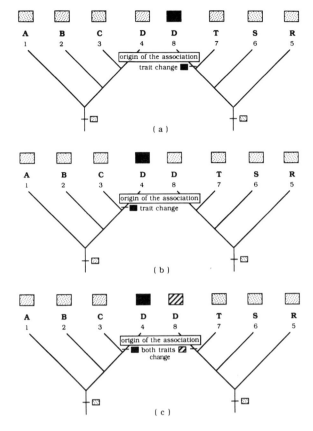

Figure 24.8 Effects of colonization by a competing species on community structure in area D. Letters = areas; numbers = species. The state of a particular character involved in the interaction between these community members is depicted by the boxes. Species 8 has colonized area D and is now interacting with species 4; therefore the origins of the association between community members are apomorphic. (A) The origin of the trait is plesiomorphic in the resident species and apomorphic in the colonizer. (B) The origin of the trait is plesiomorphic in the colonizer and apomorphic in the resident species. (C) The origin of the trait is apomorphic in both the colonizing and resident species. Extant community structure in area D thus represents *the appearance of a colonizing species followed by modifications in the trait(s) involved in the interaction between resident and colonizer.*

Figure 24.9 Effects of stochastic or nonequilibrium changes on community structure in area D. Letters = areas; numbers = species. Species 1 and 5 occur together in area A, species 2 and 6 occur in area B, species 3 and 7 occur in area C, and species 4 and 8 occur in area D. The state of a particular character involved in the interaction between these community members is depicted by the boxes. In this situation the origin of the association between community members is plesiomorphic; however, while species 8 retains the ancestral interaction character (plesiomorphic), the character has been modified in species 4 (apomorphic). Extant community structure in area D thus represents *the persistence of an ancestral association, coupled with the appearance of an autapomorphic trait in one of the members of the community.*

spond most closely with the asymptotic equilibrium model of MacArthur and Wilson (1967).

STOCHASTIC (NONEQUILIBRIUM) EFFECTS (table 24.2, type II). If there are unused types of resources in a community over extended periods of time, stochastic evolutionary changes, operating on resident species, may result in the use of some of that previously unoccupied space. In this scenario, evolutionary changes in ecological characters occur within a cospeciation framework. Species contributing to this portion of the community structure can be recognized by their historical congruence with other community members, coupled with the presence of apomorphic traits characterizing their interactions with other species and with the environment (fig. 24.9). Since these changes do not affect the evolution of other community members, they may represent a type of stochastic wandering through modifications allowed by the existing community structure. Such species appear to diverge in ecological traits for no apparent reason, although care must be taken to rule out the effects of previous competition. The longer a community exists below equilibrium numbers of species, through any of the processes described under colonization by preadapted species, the greater the possibility that resident species will experience these sorts of evolutionary changes.

Neotropical Stingrays and Their Helminth Parasites Again. The relationship between Neotropical freshwater stingrays and their helminth parasites clearly illustrates one aspect of the historical ecological approach to studying community structure. The patterns of spatial allocation (historical biogeography) for the various host and parasite species were discussed previously. Here we use the patterns of host utilization by each helminth species in each area as an indication of resource allocation.

The area cladogram (fig. 24.5) and the host cladogram (based on the phylogenetic relationships of the parasites:

congruence in a cospeciation analysis. However, unlike the conservative situation depicted for preadapted species, varying patterns of character evolution will be traced upon this phylogenetic framework if colonizing individuals compete with resident species. In this situation, at least one of three things must happen in order for the colonizer to become established: (1) the colonizing species will change (fig. 24.8A). (2) the resident species competing with the colonizer will change (fig. 24.8B), or (3) both the resident and the colonizer will change (fig. 24.8C). The changes will produce a pattern in which the colonizer, the resident, or both exhibit an apomorphic condition of the traits relevant to the competitive interaction. Replacement of the resident by the colonizer is indicated on a cospeciation analysis if the extinction event is coupled with the colonization event and if other members of the extinct species' clade have resource requirements similar to those of the colonizer. These macroevolutionary patterns corre-

Table 24.3. Matrix Listing 23 Species of Helminth Parasites Inhabiting Neotropical Freshwater Stingrays and the Parasite's Contribution to Community Structure in Each of Six Communities.

Parasite species	Locality[a]								
	1	2	3a	3b	4	5	6a	6b	6c
Acanthobothrium quinonesi	I	III	0	0	0	0	0	0	0
Acanthobothrium regoi	0	0	I	0	0	0	0	0	0
Acanthobothrium amazonensis	0	0	0	0	0	I	0	0	0
Acanthobothrium terezae	0	0	0	0	0	0	I	0	0
Potamotrygonocestus magdalenensis	I	0	0	0	0	0	0	0	0
Potamotrygonocestus orinocoensis	0	0	I	0	0	0	0	0	0
Potamotrygonocestus amazonensis	0	III	IV	0	0	I	0	0	0
Rhinebothroides moralarai	I	0	0	0	0	0	0	0	0
Rhinebothroides venezuelensis	0	III	I	0	0	0	0	0	0
Rhinebothroides circularisi	0	0	0	0	0	I	0	0	0
Rhinebothroides scorzai	0	0	IV	III	0	0	I	0	0
Rhinebothroides freitasi	0	0	0	0	III	0	0	0	0
Rhinebothroides glandularis	0	0	IV	0	0	0	0	0	0
Eutetrarhynchus araya	0	0	III	0	III	0	I	I	0
Rhinebothrium paratrygoni	0	0	III	0	III	0	I	I	I
Paraheteronchocotyle tsalickisi	0	0	0	0	0	I	0	0	0
Potamotrygonocotyle amazonensis	0	0	0	0	0	I	0	0	0
Echinocephalus daileyi	0	0	III	0	0	I	0	0	0
Paravitellotrema overstreeti	II	0	0	0	0	0	0	0	0
Terranova edcaballeroi	0	0	II	0	0	0	0	0	0
Megapriapus ungriai	0	0	II	0	0	0	0	0	0
Leiperia gracile	0	0	0	0	0	0	II	0	0
Brevimulticaecum sp.	0	0	0	0	0	0	II	0	0

[a]1, Magdalena River area in host *Potamotrygon magdalenae*; 2, Maracaibo area in host *Potamotrygon yepezi*; 3a, Orinoco delta in host *Potamotrygon orbignyi*; 3b, Orinoco delta in host *Paratrygon aiereba*; 4, mid-Amazon in host *Potamotrygon motoro*; 5, upper Amazon in host *Potamotrygon constellata*; 6a, upper Paraná in host *Potamotrygon motoro*; 6b, upper Parana in host *Potamotrygon falkneri*; 6c, upper Parana in host *Paratrygon aiereba*. 0, parasite is not known to occur in that host in that area; I, phylogenetic component in terms of both spatial (biogeographical) and resource (host) allocation; II, nonequilibrium or stochastic effects (mostly colonization by parasites that normally inhabit teleost fishes or crocodilians but use intermediate hosts that are ingested by stingrays); III, colonization by "preadapted specialists" (i.e., geographical dispersal without host switching, or with host switching when there is no other congener in the colonized host); IV, colonization by potentially competing species (indicated by the presence of another congener in the host).

see Brooks and McLennan 1991 for a discussion of methodology) provide evolutionary information about the occurrence of each helminth species in a given area and in a particular host. This information, in turn, can be used to identify the portion of the community structure that is associated with each of the four influences discussed in the preceding paragraphs (table 24.3). The effects of these various influences are not evenly distributed among the six communities. If we examine each community graphically (fig. 24.10), we find three general pictures of community structure. Three communities are dominated by type I influences: the Upper Paraná (seven type I and two type II influences), Upper Amazon (six type I influences), and Magdalena (three type I and one type II influences). The structure of these communities is essentially determined by historical factors. Similarities in structure among these communities are due to common history and not to convergence. Two communities are dominated by dispersal and host switching by preadapted species from other source areas: the mid-Amazon (three type III influences) and Maracaibo (three type III influences). In these cases, the parasite species have moved into hitherto untapped resources in colonizing the new host. No resident species are displaced; the colonizers simply increase the community's diversity. Finally, one community, in the stingrays of the Orinoco delta, has been assembled according to a variety of influences (three type I, two type II, four type III, and three type IV influences). The structure of this community is therefore the end product of a complex interaction between history, host switching into

unfilled resource space, host switching by potential competitors, and stochastic effects.

Category "0" in table 24.3 indicates the number of parasite-host associations that could possibly exist if all parasite species inhabited all host species in each area, less the number of associations actually discovered. This may represent an estimate of the amount of open niche space available for future evolutionary change within each community, or the degree to which communities of specialists exist as nonequilibrium systems, or it may simply be an artifact arising from the occurrence of some ecological specialists in any community, depending on one's initial assumptions about the underlying causes of community evolution. From the historical ecological perspective, there is little to say about category "0" observations, because they summarize what was not found, something phylogenetic analysis has not been designed to explain.

COMBINING COSPECIATION AND COADAPTATION: THE FLAGSHIP STUDY

Gorman (1992) studied some North American freshwater fish communities to discern the relative role of historical constraints on the species composition and the species associations within those communities. He used three rivers in the extensive Central Highland Mississippi drainage system: the White, the Gasconade, and the Wisconsin driftless. The White and the Gasconade are both part of the Ozark drainage, and are closer together and more similar to each other ecologically than either is to the Wiscon-

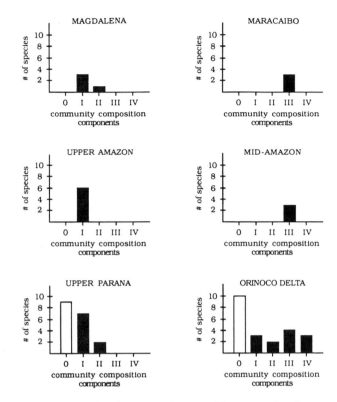

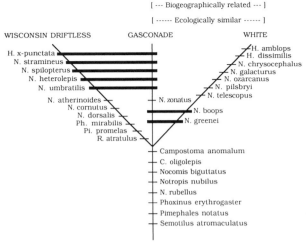

Figure 24.11 Distribution of twenty-nine North American freshwater fishes in three river drainages. *H. = Hybopsis; N. = Notropis; R. = Rhinichthys; Ph. = Phoxinus; Pi. = Pimephales; C. = Campostoma.*

Figure 24.10 The relative contributions of four types of evolutionary influences on the organization of six communities of helminth parasites inhabiting neotropical freshwater stingrays. *O,* parasite is not known to occur in that host in that area ("unoccupied niche space"); *I,* phylogenetic component in terms of both spatial (biogeographical) and resource (host) allocation; *II,* nonequilibrium or stochastic effects (mostly colonization by parasites that normally inhabit teleost fishes or crocodilians but utilize intermediate hosts that are ingested by the stingrays); *III,* colonization by "preadapted specialists" (i.e., geographical dispersal without host switching, or with host switching when there is no other congener in the colonized host); *IV,* colonization by competing species (indicated by the presence of another congener in the host).

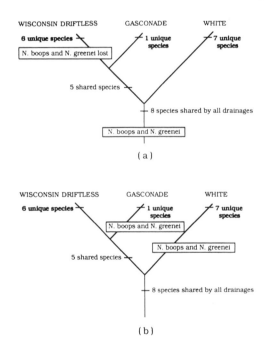

Figure 24.12 Relationships among the three river drainages, based upon shared species. This is a *phenetic* analysis of these relationships: drainages are grouped according to raw similarity (presence/absence of species). The presence of *Notropis boops* and *N. greenei* in the Gasconade and White rivers is unresolved; either these species are primitively present for all drainages and have become extinct in the Wisconsin (*A*) or they originated in one of the Ozark rivers and dispersed into the other (*B*).

sin; therefore, it is not untoward to predict that the communities in these two rivers should show a closer affinity. Gorman tested this ecological prediction by examining distribution patterns for twenty-nine fish species at two levels of analysis: the rivers, and specific habitats within the rivers.

Comparisons among River Systems: Species Composition on a Large Spatial Scale

Figure 24.11 provides a picture of the distribution of the twenty-nine species among the river drainages. If we construct relationships among the river drainages by clustering according to the presence/absence of species (fig. 24.12), the Gasconade occurs with the Wisconsin driftless drainage. The results of this analysis thus do not support the ecological expectation that the two Ozark drainages should display a more similar community composition. Mayden (1988) reconstructed the historical relationships among the rivers of the Mississippi drainage based upon comparing information from two sources: the geological relationships of the river drainages *and* the phylogenetic

relationships of fishes living in those drainages. According to this analysis, the three rivers of interest to us are associated in the manner shown in figure 24.13.

The phylogenetic analysis supports the intuitive expectations of community ecologists: the two Ozark drainages are more closely related to each other than either is to

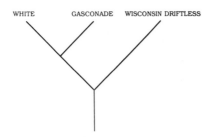

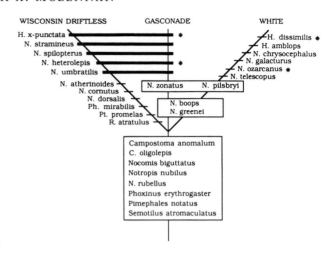

Figure 24.13 Mayden's (1988) consensus cladogram for the three river drainages, based on an analysis of forty fish species. This cladogram indicates that *the White and Gasconade are more closely related to each other than either is to the Wisconsin.*

Figure 24.14 Distribution of fishes among the three river drainages (ecological data) examined from a phylogenetic perspective. Historical components of community structure are enclosed within *boxes.* * = putative members of the same clade. *H. = Hybopsis; N. = Notropis; R. = Rhinichthys.*

the Wisconsin driftless system. How can we explain the anomalous results of the nonhistorical analysis? On the surface, the Wisconsin and the Gasconade drainages appear to have a more similar community structure because they share five species in common, whereas the Gasconade only shares two species with the White River (fig. 24.12). Since this conflicts with the phylogenetic hypothesis (fig. 24.13), we would have to conclude that history is a poor predictor of community composition at this level of investigation. However, there are two types of similarities in community structure, based upon the presence of (1) shared common species and (2) *sister species.*

A nonhistorical approach paints an incomplete picture because it does not examine the entire community; only the shared species are investigated, and the endemic or unique species are disregarded (fig. 24.12). By contrast, a phylogenetic analysis incorporates both historical and nonhistorical information from all species within the community. Bearing this in mind, let us reexamine the distribution of these fishes (fig. 24.14). With an eye to history (i.e., the Ozark drainages are more closely related), we can say that, of the twenty-nine species:

We can predict the distribution patterns for twelve species on the basis of history: *C. anomalum, C. oligolepis, N. biguttatus, N. nubilus, N. rubellus, N. boops, N. greenei, P. erythrogaster, P. notatus,* and *S. atromaculatus. N. zonatus* and *N. pilsbryi* also represent a historical component of the community structure because they are sister species (Mayden 1988) in sister river systems.

We cannot predict the distribution patterns for the five species shared between the Gasconade and Wisconsin driftless system (*H. x-punctata, N. stramineus, N. spilopterus, N. heterolepis,* and *N. umbratilis*) on the basis of the historical relationships of the drainages or their current proximity. These species are assumed to be where they are because of dispersal, but, in the absence of phylogenies for the fishes, we do not know whether they dispersed from the Wisconsin into the Gasconade or vice versa. For example, a member of both the *H. x-punctata* (*H. dissimilis:* Wiley and Mayden 1985) and *N. heterolepis* (*N. ozarcanus:* Mayden 1989) clades is found in the White River. If these pairs of relatives are sister species, their presence in the White and Gasconade Rivers is explained by common history (predictable), followed by the dispersal of *H. x-punctata* and *N. heterolepis* from the Gasconade into the Wisconsin. Once we have identified the existence and direction of dispersal events, we can begin

to investigate the environmental variables in common between the two river systems and the effect of dispersing species on an established community.

There is not enough information about the phylogenetic relationships of the remaining endemic species to determine how many are present due to dispersal and how many are present due to common speciation patterns. In order to resolve this problem we need phylogenies for the problematic groups.

Comparisons among Habitats within the Rivers: Species Composition on a Small Spatial Scale

Gorman next examined fish communities inhabiting pools and slow raceways of third- and fourth-order streams in the White, Gasconade, and Wisconsin driftless river systems. Since the presence or absence of rare fishes may be strongly affected by sampling errors, we will base this analysis upon the predominant species in these habitats. We begin by mapping the distributions of predominant species onto an unresolved diagram for the creeks (fig. 24.15). Next, we reexamine these distributions in light of the historical relationships of the river (creek) systems (fig. 24.16).

The distribution of species in these headwater communities indicates that Roubidoux Creek (Gasconade River drainage) and the Norfork River (White River drainage) are more closely related to each other than either is to Nippersink Creek (Wisconsin driftless drainage). Since this *agrees* with Mayden's phylogenetic analysis, it appears that *the effects of historical constraints on community structure can be detected at both the large spatial scale of river drainages and the small spatial scale of individual habitats within those drainages.* The incorporation of additional phylogenetic information into this study uncovers additional aspects of the evolution of community structure. For example, the relationships depicted in figure 24.14 indicate that *N. nubilus* and *C. oligolepis* are primitively present in all three drainages, while *N. boops* is shared between the Gasconade and White Rivers. Based

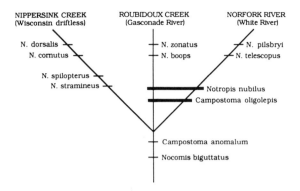

Figure 24.15 Distribution of predominant species (> 5% of all individuals) collected in pool and slow raceway habitat of third- and fourth-order streams from the three river drainages. *N.* = *Notropis*.

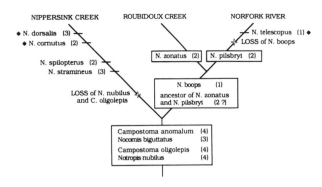

Figure 24.16 Analysis of data within the phylogenetic framework presented by Mayden (1988) and incorporating additional information from the drainage-level analysis (figure 24.14). Ecological profiles of dominant species are mapped onto the cladogram for the river systems. Foraging categories: *1*, upper pelagic; *2*, mid-pelagic; *3*, lower pelagic; *4*, near-benthic. * = endemic species.

upon this we can see that thirteen of the eighteen predominant species in these pool and slow raceway habitats represent historical constraints on species composition in the communities:

(1) Roubidoux Creek community species composition is completely predicted by history (6/6 species): four species are primitively present in all these river drainages (*N. biguttatus, C. anomalum, N. nubilus, C. oligolepis*), one species is shared with its sister system, the White River (*N. boops*), and one species, *N. zonatus*, occurs due to a shared speciation event with the White River (remember, *N. zonatus* and *N. pilsbryi* are sister species).

(2) Norfork River community species composition is almost completely predicted by history (5/6 species): four species are primitively present in all these river drainages (*N. biguttatus, C. anomalum, N. nubilus, C. oligolepis*), and one species, *N. pilsbryi,* occurs due to a shared speciation event with the Gasconade River. The remaining species, *N. telescopus,* is endemic to the White River drainage, so we would have expected to find it based upon our analysis at the drainage level. However, this does not tell us where that species came from originally (see discussion of endemics in the preceding section).

(3) Nippersink Creek community species composition is only weakly predicted by history (2/6 species): two species are primitively present (*N. biguttatus, C. anomalum*), and two species are endemics (*N. dorsalis* and *N. cornutus*). The presence of *N. spilopterus* and *N. stramineus* is problematic; as discussed in the previous section, these species may have dispersed into the area.

This portion of Gorman's study raises the following questions: Why is *N. boops* absent in the Norfork River? Why are *N. nubilus* and *C. oligolepis* absent in Nippersink Creek? Why are *N. spilopterus* and *N. stramineus* absent in Roubidoux Creek? The answers to these questions might be found by examining the ecological interactions among species within these communities.

Analysis Based upon Categories of Ecological Interactions: Community Structure in General

The addition of ecological profiles for the species in these communities (fig. 24.16) allows us to reconstruct the evolutionary sequence of assemblage of community structure. The oldest component of these creek communities is

the near-benthic forager. In the Ozark systems, the upper and mid-pelagic species may have been added next if the ancestor of *N. zonatus* and *N. pilsbryi* was a mid-pelagic forager. If not, the upper pelagic forager was added to the community first, and the mid-pelagic species, second. In addition, both the Ozark communities have the same functional structure: three near-benthic species, one lower pelagic species, one mid-pelagic species, and one upper pelagic species. This agrees with both the historical relationships of the areas and the similar environmental parameters found in the creeks.

Nippersink Creek has a radically different ecological composition from the Ozark creeks: one near-benthic species, three lower pelagic species, and two mid-pelagic species. The phylogenetic relationships for the four most recent additions to this community must be resolved in order to determine which, if any, of them are present because they evolved there (perhaps the endemics *N. dorsalis* and *N. cornutus*) or because they dispersed in (perhaps *N. spilopterus* and *N. stramineus*). If we can find historical evidence for dispersal, then the phylogenies can be used to determine whether the current interactions among the residents and colonizers are ancestral or derived. This, in turn, will tell us where to look for interspecific competition in this community.

Why is *N. boops* absent in the Norfork? The macroevolutionary patterns indicate that *N. boops* was potentially replaced by its ecological equivalent *N. telescopus*. Is there any way to move from a hypothesis based upon patterns to an experiment designed to test processes? In this case, the answer to this question is definitely affirmative. Although rare, *N. boops* is in fact present in the Norfork. Given this, we need to collect further information about the history of *N. telescopus* (i.e., what is its sister species, and did it evolve in the Norfork or disperse into that area?). If it dispersed into the Norfork, and if experiments reveal that it is currently outcompeting *N. boops* in that river, then we will have convincing evidence for competitive exclusion *without having to invoke the ghost of competition past.*

Why are *N. nubilus* and *C. oligolepis* absent in Nippersink Creek? Contrary to the *N. boops* example, these species have not been replaced with ecological equivalents. The answer to this may lie in the fact that the substrate in Nippersink Creek is more sandy than in the Ozark streams and that both "missing" species are near-benthic foragers. If so, habitat and food preference experiments should help resolve this problem.

Analysis Based upon a Specific Ecological Interaction: Community Structure in Particular

In the final section of his study, Gorman performed a series of experiments investigating the role of habitat preference in the assemblage of the Roubidoux Creek community. Preferred vertical distributions were determined for *N. biguttatus, N. nubilus, C. anomalum, C. oligolepis, N. zonatus,* and *N. boops* by observing isolated individuals in the laboratory. These patterns were compared with distributions observed in the field and with those obtained in the laboratory using mixed species groups. The fishes that demonstrate the same distributions both in laboratory experiments and in the field are the four oldest members of the communities: *N. biguttatus, N. nubilus,* and the two *Campostoma* species. Additionally, the overlap in preferred habitats among these species is low. On the other hand, the more recently derived members of the community (*N. zonatus* and *N. boops*) show evidence of interspecific interactions; they are displaced from their preferred position in the field and in the mixed species groups. Evidence of competitive interactions should be sought in two places within the community: between an older member of the community and its function replacement, in this case *N. boops* and *N. telescopus,* or between the most recently derived member of the community, in this case *N. zonatus,* and the original residents.

Three generalizations about the evolution of these freshwater fish communities can be drawn from this study: (1) communities are composites of both historical and nonhistorical components; (2) this pattern can be detected on both large and small spatial scales; and (3) historical ecological analysis can provide a framework for experimental investigations of the impact of ecological interactions on community assemblages.

SUMMARY

The macroevolutionary patterns uncovered by historical ecological research provide information about *the origins of species* in a community and *the origins of characters* involved in the interactions among those species. Essentially we are on a quest to answer two questions: How did the species come to be in this community? (speciation studies), and, "When did the traits that characterize the interactions among those species originate? (adaptation studies). Incorporating the answers to these questions into our evolutionary framework will allow us to pursue three old problems from a new perspective. First, we can construct community profiles based upon the relative composition of phylogenetic (historical) and nonphylogenetic components. We can use these profiles to focus our search for the processes underlying community evolution. For example, consider the helminth parasites of freshwater

stingrays. If you were interested in studying the effect of colonization on community structure, you would conduct your fieldwork in the Maracaibo region. Conversely, the upper Paraná and upper Amazon would yield valuable information about the effects of long-term historical factors on community structure and stability. We can also use these profiles to ask questions about the relative contributions of historical and nonhistorical factors to community assembly. Are there a limited number of community types, or does each community represent a unique combination of these influences?

Second, phylogenetic systematics provides a robust methodology for distinguishing similarities among communities that are due to common history from similarities that are due to convergence. Convergent evolution of similar traits in different lineages, or of unrelated species fulfilling ecologically equivalent roles in different communities, is considered to be strong evidence for adaptation; however, convergence is often asserted without its demonstration by phylogenetic analysis. Phylogenetic systematics provides a strong test of convergence because putative homoplasies are identified a posteriori from analyses based on a set of characters for which no postulate of homoplasy was proposed a priori. Once convergence has been identified, adaptive hypotheses can be constructed by looking for similarities in environments inhabited by taxa exhibiting the convergent traits or positions within the community.

Third, we can distinguish species that exhibit adaptive shifts (apomorphic or derived characters) in the community from species that exhibit ancestral (plesiomorphic) characteristics. Given this, we can begin to ask questions about the evolutionary influence of residents on colonizers and colonizers on residents. For example, if resident species tend to retain their ancestral ecological characteristics and colonizers tend to change, we could say that phylogenetic history exerts a cohesive and conservative influence on the evolution of community structure. Combining the results from all these studies will provide us with a more direct estimate of the relationship among the processes underlying the origin, divergence, and maintenance of biological communities.

Ricklefs (1987, 1990) and Brown and Maurer (1989) have recently discussed the significance of scaling effects in explaining the structure and evolution of biotas. Brooks (1988b) suggested that phylogenetic studies in biogeography could be related to the biological relationships between phylogenetic (temporal) and spatial scaling effects, leading to two general expectations: (1) the greater the temporal scale, the larger the spatial scale on which the relevant evolutionary patterns are manifested, and (2) the greater the spatial scale, the greater the influence of phylogeny on shaping the patterns we see. Hence, we would expect phylogenetic analysis to be more useful as we expand the temporal and spatial scale of our studies. However, as the study by Gorman (1992) indicates, there may be phylogenetic influences even on small spatial scales; and, as the freshwater stingray study indicates, there may be departures from phylogenetic influences on large spatial scales.

25

Phylogenetic Patterns, Biogeography, and the Ecological Structure of Neotropical Snake Assemblages

John E. Cadle and Harry W. Greene

Although the potential role of history is often acknowledged in passing, much of modern community ecology is operationally predicated on the notion that patterns of occurrence and interaction among species result from contemporary phenomena (see Pianka 1989b; Winemiller and Pianka 1990, and citations therein). Thus, most studies of latitudinal diversity gradients look first to contemporary abiotic and biotic factors to explain variation in local patterns of diversity. Similarly, null models for community composition and organization typically are constructed as if potential participants are randomly or ubiquitously available. However, as we will show with examples from snakes, historical contingencies can affect contemporary local patterns in fundamental respects; moreover, those effects can be addressed, given sufficient information on phylogenetic relationships.

Historical contingency enters into community organization in several ways. The unique history of each of the clades that make up a community imposes an array of constraints (morphological, physiological, behavioral) influencing the potential ecological roles of clade members, and speciation events increase the available species pool for community composition. Through evolutionary time local ecological interactions may cause the extinction or exclusion of particular clades from communities, thus altering the "starting rules" for further community change. Finally biogeographical events influence community composition by changing the array of potential interactors at particular points in time and space. All of these factors restrict the degree to which the composition and organization of contemporary communities are predictable solely from contemporary factors.

We will demonstrate regional differences in the character of Neotropical snake assemblages, which have a historical basis in the relative representation of alternative clades. Because the clades have different biogeographical histories, compelling arguments favor the primacy of history in determining some aspects of snake community organization in the Neotropics. We analyze general evolutionary trends in Neotropical snakes with respect to body size, habitat, and diet, and we relate these trends to present-day structure in rainforest snake assemblages. In essence, we pose the inverse of the question asked by

Brown and Maurer (1989), who sought to identify ecological processes affecting the diversification of clades on large land masses. We ask whether there are evolutionary patterns and processes within clades that affect continental patterns of community organization. These, obviously, are two perspectives on the same problem: What are the determinants of community organization in space and time? We argue that consideration of phylogeny and biogeography should be a major concern of researchers studying comparative community ecology. Similar views are expressed by Brooks (1985) and Brooks and McLennan (1991; chap. 24). We begin by summarizing the phylogenetic and biogeographical aspects of Neotropical snake assemblages, then consider general evolutionary trends in body sizes, macrohabitats, activity times, and diets for each lineage. Finally, we assess the implications of these patterns for community composition.

PHYLOGENTIC AND BIOGEOGRAPHICAL BACKGROUND

Our research concerns aspects of the systematics, behavior, and natural history of snakes in the cosmopolitan family Colubridae. We base many of our observations herein on our field experience with these snakes throughout much of the Neotropics. As a detailed review of Neotropical snake ecology is beyond the scope of this chapter, we provide only summary references where possible.

Much uncertainty exists concerning the phylogenetic relationships among major lineages of the family Colubridae. However, the colubrid lineages with representatives in the Neotropical fauna are few, and some aspects of their radiation within the region are becoming apparent (Cadle 1984a, 1984b, 1984c). Nevertheless, our exploration of patterns and our inferences concerning phylogenetic and historical aspects of Neotropical colubrid assemblages do not hinge on particular phylogenetic details, but are apparent when the phylogeny is considered at a general level. In those cases in which we have considered more detailed phylogenetic information, we have used conservative estimates of major groupings (fig. 25.1, appendix 25.1).

Nearly all colubrids in present-day Neotropical com-

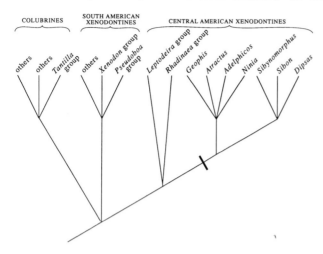

Figure 25.1 Inferred relationships among the major lineages of Neotropical snakes and some generic groups considered in this chapter. The genera belonging to each group that were included in our survey of rainforest sites are listed in appendix 25.1; additional genera not encountered in that survey pertain to most of the groups. The bar denotes a clade (genera to the right of the bar) referred to in the text as "Goo-eaters." The *Xenodon, Pseudoboa*, and *Tantilla* groups are frequently given formal status (as discussed by Myers [1986], Bailey [1967], and Savitzky [1983], respectively). Otherwise, the groups are for convenience in our discussion and are not given formal taxonomic status here. *Adelphicos* and *Atractus* are included in the *Geophis* group according to work in progress by F. J. Irish (contra Cadle 1984b). (After Cadle 1984a, 1984b, 1984c, 1988.)

munities belong to one of three major historical lineages (Cadle 1984a, 1984b, 1984c, 1985; fig. 25.1), to which we refer by informal names, as the taxonomy is still unsettled. One lineage, the *Colubrines*, is a cosmopolitan group whose distribution approximates that of the entire family Colubridae (all continents except Antarctica). The other two, *Central American Xenodontines* and *South American Xenodontines* (Cadle 1985), apparently are endemic to the New World. Although the members of the two xenodontine lineages were once considered parts of a larger group ("Xenodontines" sensu lato), there is no compelling evidence that the larger unit is monophyletic (Cadle 1984a, 1984b, 1984c) and we treat them separately in our analyses. It is important to stress that the names Central American Xenodontines and South American Xenodontines in the sense used here refer to evolutionary lineages (clades), and not strictly to the areas of occurrence of snakes within those clades. The reason for these appellations will become apparent below. Within the Neotropics the two xenodontine lineages account for most of the colubrid diversity, with about 90 genera and over 500 species. Colubrines, by contrast, are less diverse in the Neotropics (25 genera and about 100 species south of Mexico), but are diverse in North America.

The distributions of the two xenodontine clades overlap broadly in Central and South America (fig. 25.2A), but each is thought to have diversified in situ in Central America and South America, respectively (see Savage 1982; Cadle 1985). Presently, the Central American Xen-

odontine clade has its greatest diversity in Central America and in the northern Andean region of South America, whereas South American Xenodontines are, with few exceptions, restricted to South America (fig. 25.2B; Cadle 1985). Moreover, each of these lineages includes many species whose geographical ranges are quite localized, resulting in many areas of endemism (fig. 25.2B). Central American Xenodontines extend well into South America and are represented there by numerous species in a variety of genera. South American Xenodontines, however, are poorly represented outside South America; only four genera occur north of Costa Rica. Biogeographical and phylogenetic data suggest that dispersal of xenodontines between Central and South America occurred through much of the Tertiary and was not a direct response to the completion of the Panamanian Isthmus during the Pliocene (Cadle 1985; Vanzolini and Heyer 1985). Hence, mixing of the faunas is potentially old and complex.

Colubrines are distributed essentially like the Central American Xenodontine lineage in the Neotropics (fig. 25.2A), and although they are diverse throughout Central America, most species have broad ranges, resulting in low endemicity in the Neotropics (fig. 25.2B). Biochemical data suggest that the radiation of Colubrines into all continents is more recent than the separation between the two xenodontine lineages (Cadle 1984c). Presently, no evidence suggests that Neotropical or New World Colubrines are monophyletic relative to Colubrines elsewhere.

Although the only Tertiary colubrid fossil known from the Neotropics (Colombia) is 12 to 15 million years old (Estes and Baez 1985), some biogeographical considerations and molecular comparisons suggest that the three lineages we consider are at least twice that age (Savage 1982; Cadle 1984c, 1988). The Central and South American Xenodontine clades are inferred to be early to mid-Tertiary components of the Neotropical fauna. On the other hand, the pattern of differentiation of Colubrines suggests that they may be more recent components of the Neotropical fauna (Cadle 1984c; Cadle, unpublished data).

Thus, contemporary assemblages of Neotropical colubrids comprise members of two ancient lineages who radiations were centered in Central and in South America, respectively, and a third lineage whose radiation in the region is perhaps more recent. The geographical deployment of these lineages (fig. 25.2) results in a greater proportional representation of Central American Xenodontines in South American communities than vice versa. Because of the broader geographical ranges of Colubrine species, there should be more similarity in species composition among assemblages with respect to this lineage than for either of the xenodontine clades. Each of the three clades also differs with respect to ecologically relevant variables, which results in marked differences in assemblage composition as a function of their differing geographical deployment. We now explore these variables (body size, macrohabitat, and diet) as a prelude to examining their effects on contemporary communities.

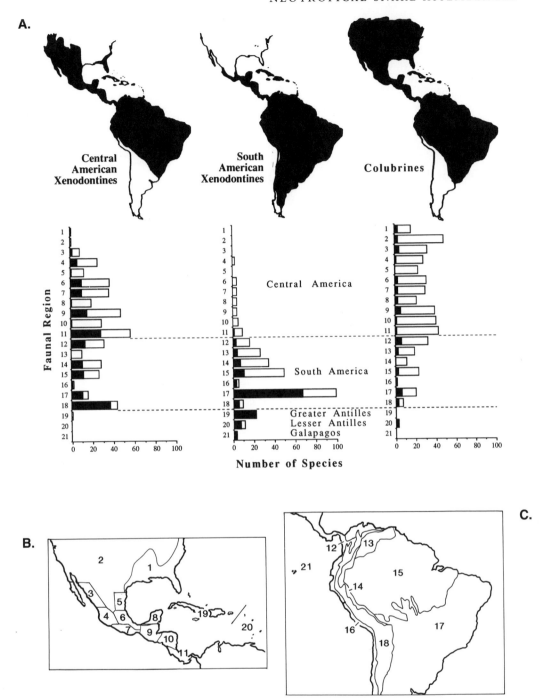

Figure 25.2 Geographical distributions of three clades of Neotropical colubrids. Maps at the top show the overall distributions of each clade. Histograms show, for each clade, its species richness according to faunal regions shown in the numbered maps at the bottom. Faunal regions are listed roughly north to south along the vertical axis of each histogram. Each bar represents the total species occurring in a region; shaded portions indicate the proportion of those species that are endemic to a given region. (After Cadle 1985; details concerning the faunal regions and the general approach are given therein.)

METHODOLOGICAL BACKGROUND

Body Size

We considered two aspects of adult body size: length and mass. Body lengths (snout to vent; SVL) were taken from the literature and from museum specimens; body masses were measured on specimens in the National Museum of Natural History, the Museum of Vertebrate Zoology at the University of California, Berkeley, and the Academy of Natural Sciences of Philadelphia. Significant geographical size variation exists for many widespread snake species, but the nature of that variation is almost completely undocumented (an exception being that of *Leptophis ahaetulla;* Oliver 1948). We attempted to use measurements from the literature for those assemblages we examined or from geographically close regions; each species is given a single maximum length value. The difficulty of obtaining reliable estimates of mean adult sizes for these species (many of which are uncommon in collections) presently precludes a more rigorous statistical examination of geographical size variation. The general trends we discuss here will be robust to such considerations.

Macrohabitat and Time of Activity

In considering the general environment in which given species perform their activities, we consider the structural habitat and the temporal pattern of use—that is, whether the species is nocturnal, diurnal, or both. We distinguish between "macrohabitat," an organism's general position in the environment (aquatic, fossorial [capable of burrowing], cryptozoic [often hidden in leaf litter], terrestrial, or arboreal); and "microhabitat," an organism's precise position in a macrohabitat, such as within the axils of bromeliads. The former has been referred to as an "adaptive zone" by some researchers (e.g., Henderson, Dixon, and Soini 1979; Henderson and Crother 1989), but we prefer to retain that term for its more widespread and inclusive use in the evolutionary literature (see Baum and Larson 1991). In snakes, use of a particular macrohabitat is usually associated with modifications of body form involving aspects of length, mass, and anatomical structures (table 25.1).

In practice, assigning particular tropical snake taxa to a macrohabitat category can be complex for two reasons. First, some snakes are eurytopic while active: although often on the ground, species of *Urotheca* and *Oxyrhopus* sometimes climb in vegetation; conversely, some species of *Imantodes* and *Dipsas* are dramatically specialized for arboreality, yet frequently crawl on the ground. Second, some Colubrines (e.g., species of *Chironius* and *Mastigodryas*) routinely forage diurnally on the ground but sleep several meters above ground on tree limbs at night. In characterizing taxa we have generally used the macrohabitats for *active* snakes, but we realize that important features of life history may eventually be understood only by

considering the total environmental space used by particular species.

Similar problems arise in considering activity periods. Some species of tropical snakes are strictly nocturnal (e.g., all species of *Imantodes, Leptodeira, Dipsas*), whereas others are exclusively diurnal (e.g., all *Chironius, Rhadinaea*). Time of activity, however, can vary geographically (e.g., species of *Tantilla* in the southwestern United States are typically nocturnal, but *T. melanocephala* is diurnal in Costa Rica), or even seasonally within a population (e.g., *Philodryas* in Argentina: Thomas 1976). Some species appear to be both diurnal and nocturnal within populations (e.g., some species of *Clelia* and *Oxyrhopus:* Vitt and Vangilder 1983). We have relied on our observations and substantiated literature reports for characterizing activity periods.

Diet

Dietary variation in Neotropical rainforest colubrids is summarized in appendix 25.1. Although there is substantial unstudied variation (geographical, ontogenetic, seasonal: Greene 1989) in the diets of Neotropical snakes, this is unlikely to affect our broad conclusions in this chapter.

Fifteen Rainforest Assemblages

Patterns of assemblage structure in Neotropical snakes were examined for fifteen rainforest communities whose species composition has been well characterized (table 25.2; fig. 25.3). We restrict our attention to rainforests because these contain the most diverse assemblages of Neotropical snakes. We also use "rainforest" in its broadest sense to include tropical and subtropical, lowland and montane forests. Although the patterns we demonstrate

Table 25.1. Aspects of Body Form and Structure Associated with Specific Macrohabitats in Snakes

Macrohabitat	Features[a]
Fossorial	Small body size (length), reduced head width, scale reductions, small eyes, inferior mouth, skull reinforcement, narrow snout, short tail
Cryptozoic	Small size, generalized morphology
Terrestrial	Medium to large body size, generalized morphology
Arboreal	Small body mass (high length/mass ratio), compressed body, relatively long tail (sometimes prehensile), relatively large eyes, enlarged vertebral scale row, center of gravity shifted posteriorly
Aquatic	Dorsal and terminal displacement of eyes and nosrils, valvular closure of nasal cavity and/or mouth

Sources: Dixon, Thomas, and Greene 1976; Greene 1989; Marx and Rabb 1972; Peters 1960; Savitzky 1983; and references cited therein.
[a]Features under one macrohabitat are not necessarily independent.

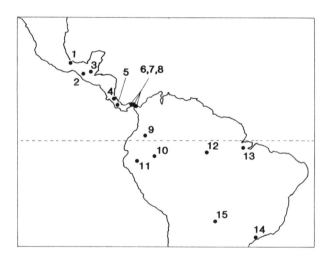

Figure 25.3 Locations of fifteen sites used for analyzing properties of Neotropical rainforest snake assemblages (see table 25.2), numbered roughly north to south. 1, Los Tuxtlas, Veracruz, Mexico; 2, Finca Santa Julia, San Marcos, Guatemala; 3, El Petén, Guatemala; 4, La Selva, Heredia, Costa Rica; 5, Rincón, Puntarenas, Costa Rica; 6, Barro Colorado Island, Canal Zone, Panama; 7, Agua Clara, Colón, Panama; 8, Yavisa, Darién, Panama; 9, Santa Cecilia, Napo, Ecuador; 10, Iquitos, Loreto, Peru; 11, Rio Cenepa, Amazonas, Peru; 12, Manaus, Amazonas, Brazil; 13, Eastern Pará, Pará, Brazil; 14, Atlantic Forest, São Paulo, Brazil; 15, Misiones, Argentina.

Table 25.2. Fifteen Rainforest Assemblages Used for Analysis of Assemblage Composition

Locality[a]	N[b]	% of fauna[c] CA	SA	CO	Reference
1. Los Tuxtlas Veracruz, Mexico	28	36	11	53	Pérez Higareda 1978
2. Finca Santa Julia San Marcos, Guatemala	18	67	5	28	MVZ[d]
3. El Petén El Petén, Guatemala[e,f]	32	47	9	44	Duellman 1963
4. La Selva Heredia, Costa Rica	46	43	11	43	Guyer, in press
5. Rincón Puntarenas, Costa Rica	35	38	15	47	R. W. McDiarmid, personal communication
6. Barro Colorado Island Canal Zone, Panama	37	38	19	38	Rand and Myers 1990
7. Agua Clara Colón, Panama	37	38	19	38	Dunn 1949
8. Yavisa Darién, Panama	29	31	28	41	Dunn 1949
9. Santa Cecilia Napo, Ecuador	38	26	45	29	Duellman 1978
10. Iquitos Loreto, Peru[f,g]	65	28	36	36	Dixon and Soini 1977
11. Rio Cenepa Amazonas, Peru	36	29	37	34	J. E. Cadle and R. W. McDiarmid, unpublished data
12. Manaus Amazonas, Brazil	47	19	45	36	Zimmerman and Rodrigues 1990
13. Eastern Pará Pará, Brazil[f]	58	16	53	31	Cunha and Nascimento 1978
14. Atlantic Forest São Paulo, Brazil[f]	42	10	69	21	I. Sazima, unpublished data
15. Misiones Misiones, Argentina[f]	33	15	73	12	Gallardo 1986

[a]See Figure 25.3 for locations.
[b]Total number of colubrid species at each site.
[c]Percentage of the colubrid fauna comprising each of three lineages: CA (Central American Xenodontines), SA (South American Xenodontines, and CO (Colubrines). Some percentages do not add to 100% because these assemblages include snakes whose lineage membership is uncertain (see Appendix 25.1).
[d]Fauna based on extensive collections deposited in the Museum of Vertebrate Zoology, University of California, Berkeley.
[e]Includes species probably in the area but undocumented by collections (see Duellman 1963).
[f]Samples from these localities are regional samples, as compared with point samples from the other localities.
[g]Sixty-five species of colubrids are known from the Iquitos region. The single site with the greatest number of species, Moropon, had 45 species. Percentages are based on the regional sample.

are most apparent with reference to rainforest assemblages because of their diversity, the imprint of history is evident in non-rainforest Neotropical assemblages as well (see below).

EVOLUTIONARY PATTERNS IN BODY SIZE AND RESOURCE USE IN NEOTROPICAL COLUBRIDS

Ecological Consequences of Body Size in Snakes

We first contrast the three lineages in terms of body size distributions. Body size is an important variable that influences many aspects of an organism's life history, energy requirements, and interactions with other components (biotic and abiotic) of its environment (Calder 1984). The lengths and masses of snakes relate to two aspects of their natural history that strongly influence the ecological roles of particular species. First, body size is a major (but not the only) determinant of the types of prey that a given species can subdue (see Greene 1983, 1989; Pough and Groves 1983; Emerson, Greene, and Charnov 1993). Second, use of certain macrohabitats is correlated with body size in snakes. For example, most fossorial or cryptozoic

Neotropical snakes have body sizes less than 40 cm standard length; most arboreal snakes have high length/mass ratios compared with snakes of comparable body lengths using other habitats (table 25.1; see Vitt and Vangilder 1983; Guyer and Donnelly 1990). Each major clade of Neotropical colubrids differs from the others in body size distribution and, therefore, in terms of diversification with respect to the two ecologically relevant variables influenced by body size: prey and habitat utilization.

Body Size Distributions in Neotropical Colubrid Lineages

As an initial appraisal of differences in body size among the three lineages, we compared the distributions of body lengths among rainforest members of each clade (fig. 25.4). This comparison demonstrates two points relevant to further considerations of the composition of assemblages of these snakes. First, each distribution is significantly different from the others (Kolmogorov-Smirnov two-sample test; $p < .01$ for each pairwise comparison). Two of the clades, Central and South American Xenodontines, show rather narrow size ranges, whereas Colubrines span and exceed the size distributions attained by the

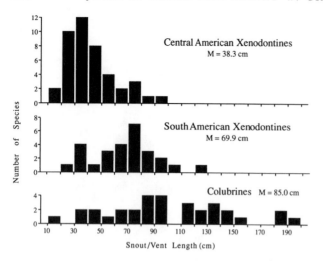

Figure 25.4 Frequency distributions of body size in Neotropical rainforest snakes belonging to three lineages. *M* = median value of each distribution.

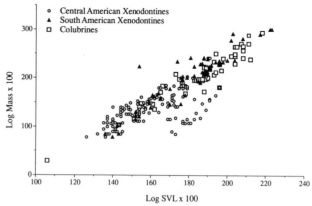

Figure 25.5 Relationship between log (body mass) and log (body length) for three Neotropical snake lineages. Each point represents one individual. A given species may be represented by multiple points.

other two clades. Second, median lengths differ substantially among the three clades. Note especially that the median length of South American Xenodontines (70 cm) is nearly twice that of Central American Xenodontines (38 cm).

Despite the significant differences in size distribution among clades, there is considerable overlap. However, consideration of body mass as an additional measure of size (fig. 25.5) highlights an important difference among the lineages, particularly the two xenodontine clades. Most Central American Xenodontines above 60 cm SVL (100 log SVL > 175), and some between 50 and 60 cm SVL (100 log SVL 170–175) have lesser body masses than similar-sized South American Xenodontines and most Colubrines (fig. 25.5). This difference is a reflection of body forms associated with arboreality (attenuation and compression; table 25.1). A few Colubrines and South American Xenodontines also show similar tendencies in length/mass relationships reflecting arboreality, but in general such forms are not developed in either of these lineages to the extent that they are in Central American Xenodontines.

The distributions shown in figures 25.4 and 25.5 reflect fundamental differences in the patterns of evolutionary radiation in these clades, and they are independent of local community effects (see below)—they are manifestations of evolutionary processes affecting length and mass within each lineage. These phylogenetically based differences in size distributions bear on the *potential* biological roles of these snakes in their communities, and some ecological roles are precluded for particular snakes purely as a result of lineage-specific size parameters. Next, we explore these ecological ramifications with respect to macrohabitat associations and predatory characteristics of rainforest snakes.

Resource Utilization: Habitat

We characterized the frequency (in terms of number of genera and species) with which taxa of each lineage use five macrohabitats. Although this survey was somewhat

subjective, and was hampered by lack of detailed observations on many species, some trends in macrohabitat use are evident.

Macrohabitat use by the three clades overlaps greatly, but differences are apparent (table 25.3) with reference to the fossorial, arboreal, and cryptozoic macrohabitats (mainly used by Central American Xenodontines), and the aquatic macrohabitat (mainly used by South American Xenodontines). Most South American Xenodontines are generalized terrestrial snakes, and are usually diurnal, but some are diurnal and nocturnal or strictly nocturnal (especially members of the *Pseudoboa* group). Few Central American Xenodontines are generalized terrestrial or aquatic snakes, and none are terrestrial/arboreal, but there are diverse radiations of cryptozoic, fossorial, and arboreal snakes in this clade. Although some cryptozoic Central American Xenodontines are diurnal (e.g., *Rhadinaea*), all arboreal members of this lineage are nocturnal (e.g., *Dipsas, Imantodes, Leptodeira*). With the exception of cryptozoic and fossorial species of the *Tantilla* group (the Sonorini of Savitzky [1983]), and nocturnal species of *Elaphe, Trimorphodon*, and (sometimes) *Lampropeltis*, Neotropical Colubrines are mainly diurnal terrestrial and/or arboreal snakes.

If we assume that the terrestrial macrohabitat is the probable primitive (ancestral) association, then departures from this mode of life (and accompanying modifications of body form [table 25.1]) clearly have evolved repeatedly in colubrids (Marx and Rabb 1972; Peters 1960; Savitzky 1983). For the xenodontines, sufficient phylogenetic data are available (Cadle 1984a, 1984b, 1984c) to indicate that, even within each of these clades, all nonterrestrial macrohabitat associations (and their morphological correlates) have evolved repeatedly. A similar analysis for Colubrines is not presently possible because of the lack of a detailed phylogeny for that clade, but we suspect this to be the case in that lineage also.

Resource Utilization: Diet

The correlation of particular prey types with particular clades in snakes ranges from the highly specific (all species of particular genera consume only one kind of prey, e.g.,

Table 25.3. Use of Macrohabitats and Some Prey Classes by Three Lineages of Neotropical Snakes

	South American Xenodontines		Central American Xenodontines		Colubrines	
	G	S	G	S	G	S
Macrohabitats						
Fossorial	2	23	3	107	1	1
Cryptozoic	3	6	6	66	5	33
Terrestrial	22	122	4	17	9	31
Terrestrial/Arboreal[a]	1	3	0	0	7	30
Arboreal	4	4	4	63	0	0
Aquatic	5	23	2	5	0	0
Prey						
Earthworms/mollusks	1	2	7	150	0	0
Arthropods	0	0	0	0	4	28
Snakes	2	12	0	0	0	0

Note: The numbers of genera (G) and species (S) using a macrohabitat or prey class as a *major* resource are indicated. For the xenodontine lineages, only genera for which some information was available on intragroup relationships were included (references in Cadle 1984a, 1984b, 1984c); for Colubrines, only genera and species occurring south of Mexico were considered. For both variables we assumed that species of a genus were similar except for specific cases in which we knew otherwise (e.g., *Coniophanes bipunctatus* is the only aquatic species of its genus, so it is counted under "aquatic," whereas all other *Coniophanes* spp. are counted as cryptozoic).

[a]Includes a large number of Colubrines that routinely use both arboreal and terrestrial habitats when active; this is qualitatively different to us than either macrohabitat as a separate entity.

Dipsas) to the very general (individuals, species, or genera showing a wide range of prey types, e.g., *Drymarchon*). The association of prey types with particular clades of Neotropical colubrids is best illustrated by consideration of specific prey classes. At the most general level, all three clades of Neotropical colubrids include many species that consume frogs and lizards, perhaps the primitive diet. Within each clade, however, particular subclades (sometimes individual genera) have diets restricted to other prey classes; conversely, some prey classes are strongly associated with particular clades.

Most Central American Xenodontines prey upon a variety of poikilothermic vertebrates, including lizards, salamanders, fish (rarely), and frogs and their eggs; one speciose subclade feeds on earthworms and gastropods (see below). No Central American Xenodontines prey significantly upon birds, mammals, or snakes. Among South American Xenodontines, snakes preying mostly upon invertebrates are rare (only members of the genus *Tomodon*, which feed on slugs). Moreover, numerous species of South American Xenodontines prey on mammals (*Oxyrhopus, Clelia, Philodryas, Pseudoboa, Tropidodryas*), birds (*Philodryas, Oxyrhopus*), and snakes (*Clelia, Erythrolamprus, Philodryas*), all of which are portions of the resource space not used by any Central American Xenodontines. As in the Central American clade, many South American Xenodontines prey upon lizards, amphibians, and fish. Most Neotropical Colubrines eat frogs and/or lizards, although a few genera (*Lampropeltis, Spilotes, Elaphe*) frequently take endothermic prey. Only members of the *Tantilla* group feed on invertebrates (arthropods only), and no Neotropical Colubrines feed on mollusks or fish. Some dietary differences among the lineages that reflect use of unusual food resources are summarized in table 25.3; these lineage effects may well reflect special handling requirements for these types of prey (see Peters 1960; Savitzky 1983; Sazima 1989; and below).

Perhaps because it is an unusual feeding strategy in snakes, the existence of a putative clade of Central American Xenodontines feeding on soft-bodied invertebrates—gastropod mollusks (slugs and snails) and annelids (earthworms)—is noteworthy. This clade comprises 7 genera and > 150 species; we informally refer to its members as the Goo-eaters (see fig. 25.1; *Tomodon*, a South American Xenodontine, also feeds largely on slugs and is the only other Neotropical colubrid with such feeding habits). Among Goo-eaters, members of one putative subclade (F. J. Irish, personal communication) feed almost exclusively on earthworms (*Geophis, Adelphicos*, and *Atractus*) or earthworms and mollusks (*Ninia*); members of their presumed sister group (*Sibon, Dipsas*, and *Sibynomorphus*; Cadle 1984b; F. J. Irish, personal communication) are almost exclusively mollusk feeders. Given the unusual nature of these feeding habits among snakes generally (Mushinsky 1987), we find the virtual restriction of them to one clade of Central American Xenodontines significant, and propose that there may be phylogenetic constraints upon the feeding response in this clade based on either the morphology of the feeding or the chemosensory apparatus, or both (Peters [1960] and Downs [1967] discuss some aspects of morphology). That the constraint may be chemosensory in nature is suggested by the fact that some species of *Atractus* are robust snakes in excess of one meter in length and seem physically capable of subduing other kinds of prey, yet they feed exclusively on giant earthworms!

Behavioral genetic studies of the chemoreceptive responses of snakes suggest a mechanism by which constraints on prey selection, such as we propose for the Goo-eaters, could occur. Responses to prey odors in snakes are known to show heritable variation (Arnold 1980, 1981). With reference to the Goo-eaters, it is interesting that in a North American natricine colubrid, *Thamnophis elegans*, a significant phenotypic and genetic correlation between chemoreceptive responses to leeches (annelids) and slugs (mollusks) exists, perhaps as a result of shared surface chemistries (Arnold 1980, 1981). A common basis for these behaviors in phylogenetically distant snakes, such as natricines and the Goo-eaters, would not be surprising if the shared feeding responses are based on simple chemical cues. If these correlated responses remained intact within a diversifying clade, they might result in the fixation of specialized feeding behaviors within the lineage. Moreover, correlated feeding responses that have a genetic basis could have important consequences for the ecology of particular clades because selection to recognize or avoid particular prey could affect responses to other prey species (cf. Arnold 1980, 1981).

In other snakes, for example, those preying on mammals and certain hard-bodied arthropods (Savitzky 1983), physical constraints on feeding ecology might be operative, and might explain why few Neotropical snakes feed on mammals. Most mammals are heavier than many amphibians and lizards, and their mass-specific abilities for sustained struggling with a snake predator are enhanced by endothermy. For these and perhaps other reasons (e.g., formidable dentition), mammals are only available to snake taxa with characteristics (large size, constricting ability, and/or venom) that can overcome their defenses (see Greene and Burghardt 1978; Reynolds and Scott 1982; Greene 1983; Pough and Groves 1983). Among Neotropical colubrids, venom injection mechanisms are found in most Central American and South American Xenodontines (they are evidently absent in Goo-eaters), and a few Colubrines (e.g., *Oxybelis*, *Tantilla*). Constriction is apparently restricted to one genus of Central American Xenodontines (*Imantodes*), some South American Xenodontines (*Philodryas*, *Tropidodryas*, and the *Pseudoboa* group), and a few Colubrines (*Elaphe*, *Spilotes*, *Lampropeltis*). The fact that no Central American Xenodontines feed on mammals is very likely a consequence of their small size (see above) and their lack of constricting capability as a result of either behavioral or morphological limitations.

COMMUNITY ASSEMBLY OF RAINFOREST ASSEMBLAGES

Differential body size patterns emerge as primary characteristics of the three Neotropical snake lineages, and these patterns affect two other aspects of their radiation, the exploitation of prey and of habitat resources. That particular diets and body sizes are seen in members of each clade independent of major habitat types (and, therefore, of particular combinations of biotic and abiotic interactions) suggests that these features are lineage-specific characteristics. Moreover, these differential patterns are not likely to be the result of long-term evolutionary interactions between the lineages if each has evolved in separate biogeographical centers (see "Phylogenetic and Biogeographical Background", above). We now explore the implications of these findings for interpreting the composition of particular rainforest assemblages.

First, we should ask whether the lineage-specific size distribution patterns translate into real differences in size distributions among snake assemblages, given the pronounced differences in geographical deployment of the lineages. If the lineage-specific patterns were not reflected in real assemblages based on their geographical location (and, hence, subject to the vagaries of biogeographical history), then there would be a stronger case for arguing

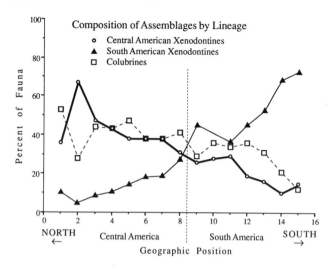

Figure 25.6 Proportional representation of species belonging to three lineages in fifteen Neotropical rainforest assemblages, arranged north to south. Each set of points in a vertical line represents one assemblage, numbered as in figure 25.3.

that local interactions "adjust" assemblage compositions to achieve some balance of sizes and size-related parameters. On the other hand, if the size distributions and ecological characteristics of individual assemblages are largely a function of the proportional representation of each lineage, then the imprint of historical differences among the lineages emerges as a primary force molding the composition and ecological characteristics of any given assemblage.

The proportional representation of each clade in the fifteen rainforest assemblages follows a pattern closely paralleling the pattern of biogeographical deployment of the clades as a whole (fig. 25.6; compare fig. 25.2): South American Xenodontines are poorly represented in assemblages outside South America, whereas the proportions of both Central American Xenodontines and Colubrines show a more gradual decline in assemblages from north to south. Moreover, body size distributions of snakes *within* the rainforest assemblages show patterns that are direct consequences of the geographical deployment and size distributions of the clades (figs. 25.4 and 25.5). Figure 25.7 shows the size distributions of snakes in six rainforest assemblages, and indicates the lineages making up each size class. Note that the Central American assemblages have a high proportion of species at the lower end of the size distribution (< 50 cm SVL). One of the three South American assemblages, Santa Cecilia, has some species within this range, but there is a relatively greater proportion of species with body lengths greater than 50 cm in the South American assemblages (fig. 25.7). The long right-hand tail of the distributions (> 100 cm SVL) is mainly due to the Colubrines present in these assemblages (compare fig. 25.4). Statistical comparisons of size distributions among assemblages within either Central America or South America are nonsignificant ($p > .05$, Kolmogorov-Smirnov two-tailed test), whereas all pairwise tests between Central and South American assemblages are significant ($p < .02$ for the Los Tuxtlas–Santa Cecilia comparison, $p < .01$ for all others).

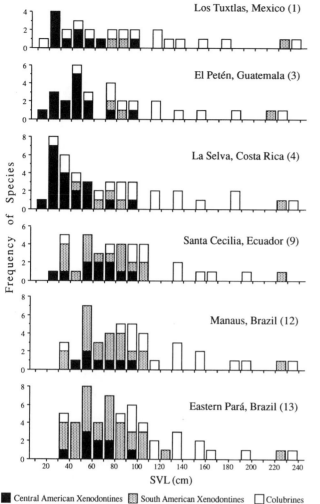

Figure 25.7 Frequency distributions of body sizes of colubrids in three Central American rainforest assemblages (Los Tuxtlas, El Petén, La Selva) and three South American assemblages (Santa Cecilia, Manaus, Eastern Pará). The number adjacent to the name of each assemblage corresponds to its location, shown in figure 25.3. The composition of each size class by lineage is indicated by shading. Several colubrids for which adequate measurements were unavailable were omitted, the maximum being four species in the Manaus assemblage. These omissions would not significantly alter the distributions.

Figure 25.7 shows that snakes within Central American rainforest assemblages are significantly smaller than those within South American assemblages. Correspondingly, South American rainforests often lack diverse guilds associated with the lower end of the size distribution (see below). This trend reflects the differing geographical deployments and size distributions of the lineages. Unfortunately, a full exploration of the implications of this trend is not currently feasible because our understanding of snake ecology in tropical forests is inadequate. Nevertheless, we can suggest several ways in which the phylogenetic and biogeographical patterns we have identified underlie particular aspects of community organization. Several examples from rainforest snake assemblages emphasize the history-dependent nature of particular

components of the predator communities in these systems:

First, in most South American rainforest assemblages, a guild of nocturnal, arboreal predators on frogs and lizards is composed almost entirely of snakes whose phylogenetic affinities are with Central American Xenodontines (*Imantodes, Leptodeira*). Among South American Xenodontines, only *Tripanurgos compressus* and perhaps species of *Siphlophis* approach this pattern of resource use, and they appear less commonly in South American rainforests than do species of *Imantodes* and *Leptodeira*.

Second, predators on soft-bodied invertebrates are represented in many South American rainforest communities by two divergent, species-rich groups of Central American origin: a guild of fossorial earthworm eaters (*Atractus*) and a guild of nocturnal, arboreal gastropod eaters (*Dipsas*). These guilds sometimes make up a substantial portion of the colubrid fauna at a site (e.g., seven species of *Atractus* constitute 18% of the colubrids at Moropon in Amazonian Peru: Dixon and Soini 1977), and would be entirely absent in South American rainforest communities were it not for the historical events (dispersal and subsequent speciation) that resulted in mixing of the biotas of Central and South America. Indeed, walking through a South American rainforest at night would be a very different experience without the commonly observed arboreal snakes of Central American origin (*Dipsas, Imantodes,* and *Leptodeira*). Similarly, fossorial snakes whose phylogenetic affinities are with South American Xenodontines are uncommon; in some South American rainforest communities the fossorial macrohabitat is used by species of *Apostolepis* (a South American Xenodontine; never more than two species in known communities) but, unlike *Atractus*, they feed on snakes and amphisbaenians.

Third, a diverse guild of aquatic snakes is largely absent from Central American rainforests (usually represented by no more than a single species of either *Tretanorhinus* or *Hydromorphus; Coniophanes bipunctatus*, the only aquatic member of its genus, occurs with *Tretanorhinus* in one assemblage). Yet in South America, aquatic snakes in the South American Xenodontine genera *Helicops, Hydrodynastes, Pseudoeryx, Hydrops,* and some species of *Liophis* are conspicuous elements in many lowland rainforest communities. These stem from several subclades within the South American Xenodontine radiation.

Finally, the cryptozoic, or "leaf litter," macrohabitat in Central American rainforest communities contains a diverse array of snakes in the Central American Xenodontine genera *Rhadinaea, Coniophanes, Urotheca, Trimetopon, Ninia, Amastridium,* and some species of *Sibon;* and the Colubrine genera *Tantilla, Stenorrhina,* and *Scaphiodontophis*. At the Central American rainforest sites we examined, these snakes make up 24% to 44% of the colubrid fauna. In South America this guild comprises fewer species (the South American Xenodontines "*Rhadinaea*" *brevirostris, Xenopholis,* and some species of *Liophis*).

These examples illustrate some of the ways in which the assembly of Neotropical colubrid communities is, to a significant extent, a result of admixture of three evolutionarily independent lineages, each bringing to contemporary communities sets of species bearing different morphological and ecological syndromes. Without the perspective of phylogeny, the underlying historical basis of many aspects of community assembly would go unnoticed. For Neotropical snakes, we suggest that many features of present ecological relationships among species can only be understood in terms of these historical patterns.

DISCUSSION

A Phylogenetic Perspective on Another Type of Assemblage

We illustrate the important perspective that phylogenetic information can offer community studies by examining the Brazilian caatinga snake assemblage studied by Vitt and Vangilder (1983), and contrasting our viewpoint with theirs on some aspects of community organization. We do not mean to single out this study for criticism—in fact, it is the best available analysis of a tropical snake assemblage. Rather, we emphasize how phylogenetic information can augment community studies by giving insight into the historical basis of certain patterns.

Vitt and Vangilder (1983) attempted to identify the causative agents for particular patterns of resource use in the caatinga community. Specifically, they offered proximate explanations for three patterns: (1) the absence of snakes feeding on invertebrates, which they attributed to competition from insectivorous mammals, the presence of predators on small snakes, and lack of suitable microhabitats; (2) the large number of snakes feeding on anurans, attributed to convergence and the year-round abundance of frogs at the study site; and (3) the lack of snakes specializing on mammals, attributed to the unpredictability of mammals as a resource. Although these explanations are plausible, our phylogenetic perspective on the snakes presents a different view. Only two of the major lineages here under consideration made up the colubrid assemblage studied by Vitt and Vangilder: South American Xenodontines (80%) and Colubrines (20%), with a total of 15 colubrids at the site. With this in mind, we consider the three aspects of community organization discussed by Vitt and Vangilder.

First, snakes that feed primarily on invertebrates, although speciose, are phylogenetically restricted among New World colubrids (and among colubrids generally; see Savitzky 1983). Arthropod eaters, in particular, are found only among a putative clade of mainly North American colubrines, the Sonorini (Savitzky 1983). Of this group, *Tantilla* has a substantial diversity in the Neotropics, but only *T. melanocephala* is widely distributed east of the Andes. Although this species is known from the caatingas (Vanzolini, Ramos-Costa, and Vitt 1980), most of its distribution is in more mesic habitats (L. D. Wilson 1987). Given the restricted phylogenetic distribution of arthropod feeding among colubrids, its restriction in the New World to a primarily North American clade, and the occurrence of only a single species of this clade in eastern South America, the absence of a diverse guild of arthropod feeders in this caatinga community is not especially

surprising, regardless of the abundance and diversity of arthropods—few Neotropical communities have a diversity of arthropod-eating colubrids, simply because appropriate lineages are absent from the Neotropics. Similarly, the prevalence of arthropod eaters in Sonoran Desert communities (as contrasted by Vitt and Vangilder 1983) reflects, to us, the historical association (and possible in situ evolution) of this particular clade with North American arid environments, where eight genera and more than thirty species occur.

Second, six species of three genera (*Liophis, Waglerophis, Leptophis*) in the caatinga community fed exclusively on frogs, and two species of *Philodryas* had frogs as a minor dietary component. *Liophis* and *Waglerophis* belong to a clade within the South American Xenodontines, the Xenodontini (Myers 1986); species of both of these genera are primarily frog feeders, as are all members of the Xenodontini with the exception of *Erythrolamprus* and perhaps *Umbrivaga* (Appendix 25.1). Similarly, frogs are the main dietary component of all species of *Leptophis*, and form a minor dietary component in some species of *Philodryas*. Hence, we do not view the incidence of frog eaters in this community as an example of convergence due to the abundance of frogs at the particular study site, but rather as a simple reflection of dietary habits primitive for or fixed in the groups under consideration.

Finally, the lack of mammal specialists in the caatinga assemblage is explained more simply by phylogenetic considerations than by fluctuations in prey availability. In general, Neotropical snakes rarely feed on mammals. Among colubrids, only the Colubrines *Elaphe triaspis, Lampropeltis triangulum,* and *Spilotes pullatus* eat rodents frequently enough to be termed mammal specialists. The mammal specialists in Vitt and Vangilder's (1983) assemblage were moderate to large constrictors (*Spilotes, Boa,* and *Epicrates*), venomous constricting South American Xenodontines (*Clelia, Oxyrhopus, Philodryas*), or venomous pitvipers (*Crotalus*)—a pattern suggesting that large size and a derived method for subduing prey are required for specializing on mammals (see above). Furthermore, to argue that mammal specialists are absent from the caatinga assemblage because prey populations are unpredictable is to argue that they are unpredictable throughout the Neotropics, since few Neotropical snake assemblages thus far studied show a high frequency of mammal specialists (see Emmons 1984 for discussion of variation in mammal densities in Neotropical rainforests).

Furthermore, we reject the assumption that proximate resource bases will determine, in more than a trivial sense, the use made of them by particular community members. At most, resources *permit* the presence of particular kinds of organisms in the community, but the morphological and behavioral characteristics (size, venom, constriction, chemical perception) that control use of a resource often have deeper phylogenetic roots that transcend present-day communities. Biogeographical events *may* result in the distribution of species possessing these characteristics to particular communities, but that is a historical contingency (Henderson and Crother [1989; 483] allude to a similar explanation for West Indian snake diets). Simple availability of a resource or niche space does not create selective pressures to "fill" the space with an appropriate organism (cf. Lewontin 1977, 1987).

Ecomorphs and Convergence in Assemblage Organization

A recurrent theme in community ecology is the occurrence of "ecological equivalents" in communities otherwise unrelated by either history or phylogenetic composition. Such equivalents are often characterized by sets of co-occurring morphological features that have some discernible relationship to ecological attributes: the "ecomorphs" of some authors (Williams 1972; Karr and James 1975; Losos 1990). Many comparative studies of community organization (e.g., Diamond 1975; Pianka 1986; Sazima 1986) are initiated with the expectation that communities of particular organisms (e.g., birds or lizards) will show repeated properties based on the existence of such ecological equivalents.

We see few examples of true ecological equivalents in Neotropical rainforest snake assemblages. Most of the more derived ecomorphs (e.g., nocturnal arboreal lizard and frog eaters such as *Imantodes,* or fossorial earthworm eaters such as *Atractus* and *Geophis*) are the only Neotropical snakes with such ecological roles. If these clades happen to be absent from a particular assemblage, then these roles simply do not exist in the assemblage. If ecological equivalents are truly uncommon (see also Pianka and Huey 1978; Pianka 1986), then the extent to which community convergence occurs also is questionable.

Although some studies support the view that communities with differing evolutionary backgrounds converge to similar structures (e.g., Williams 1972; Emmons, Gautier-Hion, and Dubost 1983; Van Valkenburgh 1988; Losos 1990), others, including our own, have failed to demonstrate such convergence, even with attempts to control the phylogenetic background by limiting consideration to particular lineages (Williams 1983; Roughgarden, Heckel, and Fuentes 1983; Pianka 1986; Schluter 1988b). We suspect, as did Pianka (1986), that the phylogenetic histories of the clades making up these communities force different organizational properties on contemporary communities. In effect, every lineage carries some evolutionary baggage that permits only certain options in terms of resource use and ecological interactions (see Lewontin 1987 for extended discussion). For our snake communities, this evolutionary baggage comprises a suite of features related to body size and prey characteristics.

Clearly, the search for convergence in community organization must assume that the communities being compared are in equilibrium. For our snake assemblages, in other words, are the differences we perceive among the assemblages and that we attribute to historical differences among their component clades merely a result of failure of these communities to reach equilibrium? Given sufficient evolutionary time, would the rainforest snake assemblages attain similar organizational properties in terms of diversity with respect to the macrohabitat, size, and prey variables we have discussed? Are there *assembly rules* (Diamond 1975; Fox 1987) for community organization? These questions are not answerable for the complex assemblages we have discussed. However, we have demonstrated the virtual restriction of certain ecological roles to particular clades, and there is good reason to believe that the vagaries of biogeographical history largely determine the distribution of clades among communities. Biogeography (and, therefore, history) plays a key role in structuring these contemporary communities.

Moreover, community equilibrium is, necessarily, ephemeral in evolutionary time due to the differential extinction, speciation, and colonization of different clades constituting a community. Yet, even given much evolutionary time and stable ecosystems, the extent to which we should expect communities to converge is not clear. For example, Pianka (1986) failed to discover convergence in lizard communities occupying ancient and apparently stable desert ecosystems and concluded that the effect of history on those communities was considerable. The age of the rainforest assemblages we considered is controversial (see Prance 1982), although their component clades are apparently old (see "Phylogenetic and Biogeographical Background", above). Yet we do not see extensive convergence even at the level of particular ecomorphs. This suggests that the evolutionary history of these clades (including intrinsic morphological attributes and extrinsic biogeographical events) exerts a major influence on the assemblage structures of these snakes. If the effects of history on community organization are profound and widespread, then formulating theory to account for historical contingency will be a major challenge to ecologists (cf. Hubbell and Foster 1986).

Beyond Ecomorphology: Evolutionary Ecomorphology

Our study raises many questions concerning the structural and functional basis of ecological differences among Neotropical snake lineages—questions usually addressed within the context of functional or ecological morphology. We do not answer those questions here, but outline what we consider to be some of the necessary components leading toward their resolution.

First, body size is clearly a major factor influencing the pattern of radiation of these clades, but how prey and macrohabitat attributes are mediated through size is virtually unexplored. Structural and functional studies could specify more precisely the relationship between size and each of these parameters. Such studies would not answer questions concerning the basis for the lineage-specific size distributions. However, when viewed in the context of a phylogeny, functional morphological studies might yield clues about constraints and patterns of morphological transformations in snakes (cf. Schaefer and Lauder 1986), and about why certain ecological patterns are not evolutionary options for particular clades.

As one example, consider the prevalence of arboreal and fossorial snakes in the Central American Xenodontine clade, and, more intriguingly, their phylogenetic juxtaposition as sister taxa within one subclade, the Goo-eaters. At first the juxtaposition seems odd, since arboreal and fossorial snakes are modified in such different ways (table 25.1). Does it, in fact, reflect the necessity of concertina locomotion (as opposed to lateral undulation) in both macrohabitats and, thus, shared functional morphological syndromes in both burrowing and climbing (cf. Gans 1974)? To answer these sorts of questions, phylogenetic, functional, and ecological perspectives need to be integrated so that the effect of history *on* ecology can be assessed.

We applaud the burgeoning interest in "ecomorphology" as a subdiscipline of both ecology and morphology, but until phylogenetic perspectives also become part of the discipline, it will result, at best, in an incomplete description of present ecological/morphological associations. Certainly, without the addition of phylogenetic perspectives, questions about adaptation and the evolution of particular ecomorphs cannot be addressed (Baum and Larson 1991; Donoghue 1989; Losos 1990). Although the mechanisms are not yet clear, morphologists recognize that there are intrinsic phylogenetic components to structure that channel the directions of morphological and functional evolution (see, for examples, Liem and Wake 1985). These conceptual developments need to be assimilated into ecomorphology if it is to address questions such as why certain clades and not others perform particular ecological roles. By evaluating the effect of phylogenetic history on relationships between contemporary morphologies and ecologies, ecomorphology itself is transformed into a more dynamic (and demanding) discipline, evolutionary ecomorphology.

CONCLUSIONS AND RECOMMENDATIONS

In seeking explanations for patterns in community organization, the roles of phylogeny and historical events are often overlooked or, more commonly, considered unavailable for study. Our survey shows that phylogenetic trends are relevant to community ecology, and that recognition of this fact may suggest the level at which general explanations for particular patterns should be sought. Thus, for example, we do not need to "explain" the absence of invertebrate-eating snakes from a particular community in terms of present-day ecological factors if historical events resulted in the absence of appropriate lineages from the community. The question of why are there more species of snakes in some assemblages, or more particular kinds of snakes, changes from consideration of niche metrics to questions about rates of speciation in those taxa contributing to high diversity (e.g., *Atractus*, *Dipsas* in the communities we considered). These findings shift the explanatory level to questions concerning evolutionary opportunities and constraints within lineages, origination and extinction of lineages, patterns of regional diversity, and historical biogeography (see also Hubbell and Foster 1986; Graham 1986). History and phylogeny, therefore, should be given more attention by ecologists, and systematists and evolutionary biologists should strive to integrate their findings into ecology.

In taking a strong position on the importance of historical events for community organization, we do not deny the contribution of local interactions to structuring Neotropical snake communities (reviewed by Vitt 1987). However, those contributions are poorly understood

(though often asserted). We have documented the influence of historical contingencies on some aspects of local community organization. The ramifications of those contingencies are unexplored, but might extend to both lower and higher trophic levels (Greene 1988). Just how the differences in size distributions among the rainforest communities that we documented influence community organization is not known. Comparative studies of Neotropical snake communities might begin with an in-depth assessment of these differences. History will then be brought fully into the realm of contemporary community ecology, and the effects of historical differences among lineages on community structure can then be more cogently evaluated.

APPENDIX 25.1. NEOTROPICAL RAINFOREST SNAKES AND THEIR DIETS

The following genera occur in the fifteen rainforest assemblages we consider. Under each of the xenodontine lineages some generic groupings to which we refer in the text are indicated (see fig. 25.1). They are *not* intended to have formal taxonomic status; "unplaced" genera are not collectively monophyletic relative to the named generic groups. Abbreviations for prey types: ar, arthropods; ew, earthworms; cr, crustaceans; mo, gastropod mollusks; fi, fishes; ca, caecilians; fr, frogs; sa, salamanders; tr, turtles; li, lizards; am, amphisbaenians; sn, snakes; bi, birds; ma, mammals. Frog, reptile, and bird eggs are lumped with their parent taxon. Prey types known to occur frequently in the diet of one or more species are in capitals; lowercase letters indicate a food generalist (e.g., *Drymarchon*) or insufficient data to determine whether one or more prey types predominate. For *Elaphe*, a mainly north temperate genus, we used diet information only for the species with a primarily tropical distribution (*E. triaspis*).

Colubrines

Tantilla group (Sonorini of Savitzky [1983]): *Ficimia* (AR), *Stenorrhina* (AR), *Tantilla* (AR)

Unplaced: *Chironius* (FR, sa, li, bi, ma), *Dendrophidion* (FR), *Drymarchon* (fi, fr, tr, li, sn, ma), *Drymobius* (FR, li, ma), *Drymoluber* (fr, li), *Elaphe* (MA), *Lampropeltis* (sn, bi, MA), *Leptophis* (FR, li, bi, sn), *Masticophis* (ar, li, sn, bi, ma), *Mastigodryas* (fr, LI, sn, bi, ma), *Oxybelis* (FR, LI, BI, ma), *Pseustes* (BI, ma), *Scaphiodontophis* (fr, LI), *Simophis* (?), *Spilotes* (li, bi, MA)

Central American Xenodontines

Geophis group: *Atractus* (EW), *Adelphicos* (EW), *Geophis* (ar, EW), *Ninia* (ar, EW, MO, ca)

Dipsas group: *Dipsas* (ar, MO), *Sibon* (MO), *Sibynomorphus* (MO, ar)

Leptodeira group: *Imantodes* (FR, LI), *Leptodeira* (FR, li)

Rhadinaea group: *Coniophanes* (ar, ew, fi, FR, sa, li, sn), *Rhadinaea* (ew, FR, sa, LI, sn), *Urotheca* (fi, FR, SA)

Unplaced: *Amastridium* (fr), *Hydromorphus* (?), *Nothopsis* (fr, sa), *Tretanorhinus* (fi), *Trimetopon* (li)

South American Xenodontines

Pseudoboa group (see Bailey 1967): *Clelia* (LI, SN, MA), *Drepanoides* (li), *Oxyrhopus* (fr, LI, am, MA), *Phimophis* (li), *Pseudoboa* (LI, bi, ma), *Siphlophis* (li, bi), *Tripanurgos* (LI)

Xenodon group (see Myers 1986): *Erythrolamprus* (fi, li, SN), *Liophis* (ar, ew, FI, FR, sa, li, am, bi, ma), *Lystrophis* (FR, li), *Umbrivaga* (?), *Waglerophis* (FR), *Xenodon* (FR)

Unplaced: *Apostolepis* (SN, AM), *Elapomorphus* (am), *Helicops* (FI, fr, li), *Hydrodynastes* (cr, fi, fr), *Hydrops* (ca, FI), *Philodryas* (fr, LI, sn, bi, MA), *Pseudoeryx* (FI), *Pseudotomodon* (li)

"*Rhadinaea*" *brevirostris* group (FR, LI; see Cadle 1984a for phyletic placement of this group), *Thamnodynastes* (FR, li), *Tomodon* (mo), *Tropidodryas* (fr, LI, MA), *Xenopholis* (FR)

Unassigned genera (lineage uncertain or data contradictory)

Enulius (li), *Rhinobothryum* (li), *Sordellina* (FI, fr)

Sources

Beebe 1946; J. E. Cadle and H. W. Greene, unpublished data; Cunha and Nascimento 1978; Dixon and Soini 1977; Dixon, Thomas, and Greene 1976; Duellman 1958, 1978; Gallardo 1970; Gertler and Morales 1980; Greene 1975; Henderson 1982, 1984; Henderson and Binder 1980; Henderson and Crother 1989; Henderson and Hoevers 1977; Hoogmoed 1980; Martin 1958; Michaud and Dixon 1989; Mole 1924; Myers 1974, 1982; Oliver 1948; Sazima 1989, unpublished data; Scott 1969; Seib 1984, 1985a, 1985b; Serie 1920; Sexton and Heatwole 1965; Swanson 1945; Test, Sexton, and Heatwole 1966; Thomas 1976; Vitt 1980, 1983; Vitt and Vangilder 1983.

26

Continental Comparisons of Temperate-Zone Tree Species Diversity

Roger Earl Latham and Robert E. Ricklefs

Biogeographers and ecologists have sought for nearly two centuries to make sense of the striking differences in tree species diversity among the major forest regions of the earth. Alexander von Humboldt and Aimé Bonpland may have been the first to comment on these differences (1807) when they remarked on the far greater species diversity among tropical tree species than in the temperate zone. With some hyperbole, they also claimed that North America had three times as many species of oaks alone as Europe had tree species. Moist tropical forests have since been found to have at least ten times as many tree species as moist forests of the northern and southern temperate zones, at several spatial scales. Biologists have proposed various explanations for the temperate-tropical disparity in tree species diversity (Pianka 1966, 1989b; Hubbell and Foster 1986; Stevens 1989). No one mechanism has found general acceptance, and several may contribute to the pattern. Within temperate latitudes, the mesic forests of eastern Asia have three times more tree species than forests in eastern North America and six times more than those in Europe. Although this pattern has been recognized for more than a century (Gray 1878), it has attracted far less attention from theorists than the temperate-tropical disparity.

Suggesting an explanation for regional disparities in temperate tree species diversity, Gray (1878) and other early authors pointed to probable differences in extinction rates due to differences between the continents in the severity of climatic cooling during the Quaternary Ice Ages. Continental ice sheets covered much of Europe's present-day temperate forest region and advanced deeply into eastern North America, but they never reached the mid-latitudes of eastern Asia (fig. 26.1). Fossil floras show that forests in Europe were far more diverse in the Tertiary than at present (Reid and Chandler 1933; Mai 1960; Kilpper 1969; van der Hammen, Wijmstra, and Zagwijn 1971; Mai 1971a, 1971b; Takhtajan 1974; Łańcucka-Środoniowa 1975; Collinson and Crane 1978; Friis 1979; Mai 1980, 1981; Gregor 1982; Friis 1985; Kvaček and Walther 1987; Mai 1987a, 1987b; Mai and Walther 1988; Sauer 1988; Kvaček, Walther, and Bužek 1989). Most, if not all, of the genera lost from Europe in the late Tertiary continue to inhabit forests of eastern Asia or North America. Plant distributions underwent severe contraction and confinement in Europe, according to some views, because most plants failed to migrate southward beyond 40° to 45° north latitude over or around the east-west–trending mountain ranges and the Mediterranean basin (e.g., Gray 1878). The effects were less severe in eastern North America, where the barriers—the Gulf of Mexico and the Mexican highlands—lay south of 30° north latitude. For 1,200 km inland across continental eastern Asia from the eastern coast, no physical barriers would have impeded southward migration by plants well beyond the Tropic of Cancer.

Ecologists generally have invoked geographical variation in the outcome of small-scale deterministic processes to explain global-scale diversity patterns (Connell and Orias 1964; MacArthur 1972; Ricklefs 1977, 1987; Huston 1979). These hypotheses describe a world in equilibrium. Even when so-called nonequilibrial mechanisms such as natural disturbance are called into play (Huston 1979), it is assumed that disturbance and interactions between populations, such as competitive exclusion and colonization of disturbed patches, would show balance within a landscape if it were sampled at large enough spatial and temporal scales (Petraitis, Latham, and Niesenbaum 1989).

A major problem with invoking competition and other interactions to explain global diversity patterns is that the scale of the patterns is grossly mismatched to the scale of the putative causal process. It is a large step from local-scale processes to predictions of regional-scale species richness. Such predictions would require plausible hypotheses linking the expression of local processes to regional variation in the physical environment (Ricklefs 1987). Furthermore, the fossil record gives no reason to expect equilibrium in regional species richness. The global number of vascular plant species apparently has risen steadily throughout the Phanerozoic, most sharply since the radiation of angiosperms in the late Cretaceous (Crepet 1984; Niklas, Tiffney, and Knoll 1985), while the area of land in terrestrial plant communities has perhaps decreased, at least since the Oligocene, due to the growth of continental ice sheets. Local processes could determine regional species richness if the region were merely a collection of habitats and localities within which diversity were regulated. However, differences in diversity between regions with similar climate suggest that local and regional processes contribute separately to regional species richness.

In this chapter we attempt to understand geographical

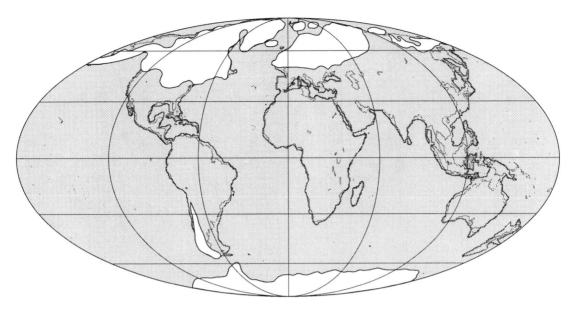

Figure 26.1 Maximum Quaternary advance of continental ice sheets (white areas) superimposed on present-day coastlines (after Nilsson 1983). Maximum areal coverage has been estimated for North America (excluding Greenland), Europe (including western Siberia), and eastern Asia at approximately 16, 9.4, and 1.2×10^6 km², respectively (Nilsson 1983).

patterns in the contemporary diversity of regional tree floras in the temperate Northern Hemisphere by examining both historical and ongoing processes. We begin by comparing diversity in contemporary regional floras at several taxonomic levels. Next, we compare regional fossil floras beginning in the early Tertiary, when forests of modern aspect first appeared, with present-day regional floras, seeking patterns in the survival and extirpation (regional extinction) of genera. We then review tests of equilibrial hypotheses developed to explain local diversity. Finally, we suggest a scenario for the establishment of global patterns in tree species diversity. We infer that contemporary patterns of tree species diversity owe much to historical and evolutionary contingency, in contrast with some other authors who have interpreted the patterns as arising from, and maintained in equilibrium by, ecological interactions. We propose ancient roots for contemporary diversity anomalies, considerably older than the Pliocene and Quaternary climate cooling and resulting widespread extirpations. We conclude by discussing testable predictions suggested by our interpretation.

CONTEMPORARY GEOGRAPHICAL PATTERNS IN TEMPERATE TREE SPECIES DIVERSITY

Species Diversity in the Four Regions

In order to compare taxonomic diversities among the major moist temperate forest regions of the Northern Hemisphere, we defined the areas covered by moist temperate forest and compiled lists of all of the characteristic tree species. The regions are the four warm-temperate humid and temperate-nemoral climate biomes of Walter (1979). These mid-latitude regions extend varying distances toward mid-continent from the east and west coasts of Eurasia and North America (fig. 26.2). Appendix 26.1 pres-

ents the criteria used in compiling the tree floras and an abridged version of the floras themselves, with numbers of species tabulated by genus and region, information about the contemporary distributions of the families and genera, and Tertiary fossil data on the genera. Table 26.1 summarizes the total flora by the number of species, genera, families, orders, and subclasses occurring in each region. A total of 1,166 species make up the characteristic north temperate tree flora. Species are distributed among Europe (including the Caucasus), eastern Asia, North America's Pacific slope, and eastern North America, respectively, approximately in the ratio 2:12:1:4.

Adjoining Subtropical Forest

Of the four moist temperate forest regions, only the one in eastern Asia shares a long, common border with moist subtropical forest. Thus, it is reasonable to conjecture that the high diversity of temperate eastern Asia's tree flora may be due to the incursion of subtropical elements. Many tree species occurring in the eastern Asian temperate zone that have mainly subtropical ranges were already excluded from our list. We retallied the list also excluding those species that have mainly temperate ranges but that, nonetheless, belong to genera with predominantly tropical distributions (table 26.1). This did change the eastern Asian term in the ratio of species numbers among regions, but a large disparity between regions remained. When predominantly tropical genera were excluded, the ratio became approximately 2:9:1:4.

Area

The disparities in tree diversity among the four moist temperate forest regions cannot be explained as an area effect, because three of the regions (excluding the much smaller temperate forest zone of North America's Pacific slope) cover similar areas. The areas of the three larger regions

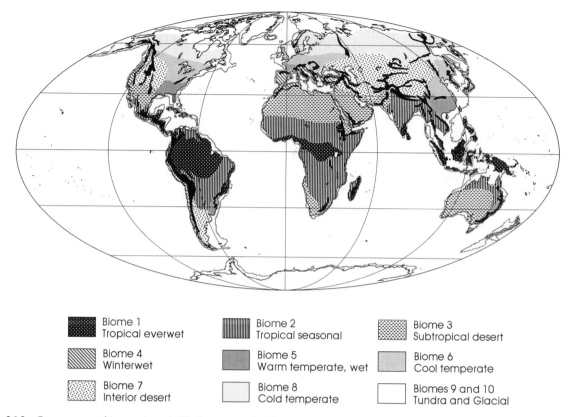

Figure 26.2 Contemporary biomes. Areas in black are major highlands. The study focuses on biomes 5 and 6. (After Walter 1979.)

Table 26.1. Summary by Taxonomic Level and Region of Moist Temperate Forest Trees in the Northern Hemisphere

Taxonomic level	Number of tree taxa characteristic of moist temperate forests in:				
	Northern, central, & eastern Europe	East-central Asia	Pacific slope of North America	Eastern North America	Northern Hemisphere (total)
Subclasses	5	9	6	9	10
Orders	16	37	14	26	39
Families	21	67	19	46	74
Genera	43	177	37	90	213
Species	124	729	68	253	1,166
Families excluding those of predominantly tropical distribution (% of total)	18 (86%)	37 (55%)	18 (95%)	29 (63%)	41 (55%)
Genera excluding those of predominantly tropical distribution (% of total)	41 (95%)	121 (68%)	35 (95%)	77 (86%)	149 (70%)
Species exclusive of predominantly tropical genera (% of total)	122 (98%)	570 (78%)	66 (97%)	236 (93%)	987 (85%)

were estimated by transferring Walter's (1979) biome delineations to 1:12,000,000-scale Miller oblated stereographic projection maps (Rand McNally 1969) and planimetering. In Europe, eastern Asia, and eastern North America, moist temperate forest biomes were estimated by this method to cover approximately 1.2, 1.2, and 1.8 × 10⁶ km² respectively. These estimates are generally lower than those from sources that present more detailed surveys of potential natural vegetation (roughly, preagri-

cultural vegetation) in only one region. For example, Wolfe's (1979) map of eastern Asian mid-latitude forests shows approximately 1.8 × 10⁶ km² in forest types dominated by broad-leaved deciduous trees. Braun's (1950) map of deciduous forest associations in eastern North America indicates approximately 2.4 × 10⁶ km² in the entire temperate forest region. Despite the discrepancies in magnitude, the ratio of these estimates—3:4—is similar to the ratio of the areas defined by Walter—2:3. (Potential

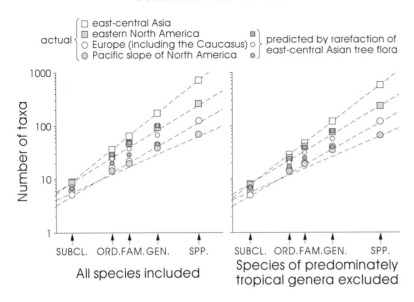

Figure 26.3 Numbers of tree taxa of the four moist temperate forest regions in the Northern Hemisphere and results of simulated rarefaction of the east-central Asian tree flora to the species numbers of the other tree floras (see table 26.1).

natural vegetation maps are common for countries or regions within Europe but are virtually nonexistent, [except for Walter 1979] for the entire temperate forest region of Europe including the Caucasus, even in the most geographically comprehensive treatments [Rubner and Reinhold 1953; Mayer 1984; Jahn 1991].) By area alone, one would expect North America's temperate forest zone to have the highest tree species diversity.

Taxonomic Diversity Patterns

There are many other means of detecting patterns in taxonomic diversity among regions with similar climate and vegetation besides simply comparing the numbers of species. We explored several, including comparing numbers of higher taxa (table 26.1; fig. 26.3), comparing numbers of species per genus and other ratios of lower to higher taxa, comparing numbers of genera and families consisting of only one or two species, and examining patterns in the overlap of taxa among regions. It is implicit that higher taxa are older than lower taxa. Thus, contemporary distributions of tree genera, families, and orders among regions may offer clues about the historical relationships among the regions' tree floras.

Simulated Rarefaction. Whether or not regions differ in their distributions of species among higher taxa could have broad implications for interpreting historical relationships among regional floras and possible causes of regional diversity differences. For example, suites of physiological or anatomical traits associated with plant families may have been crucial to species' regional survival or extinction during episodes of climate change that affected the regions differently. In this case, we would expect regions to differ significantly in taxonomic structure; that is, the flora of one region should diverge significantly from a random subset, containing the same number of species, of a more diverse flora in another region, in the number of families represented and in the frequency distribution of species per family.

We compared the taxonomic structures of the four re-

gions' tree floras by simulating the rarefaction (Simberloff 1979) of the most diverse floras to match the species numbers in other, less diverse floras. We performed the rarefaction tests by computer, picking randomly from the species pool of one temperate tree flora until the number present in a less diverse flora was reached. For example, in simulating the rarefaction of the 729-species eastern Asian temperate tree flora to 124—the number of species that belong to the European temperate tree flora—the program randomly picked 124 species from the eastern Asian list and tallied the genera, families, orders, and subclasses represented by those species. Each simulation was repeated 1,000 times. Rarefaction was simulated initially using the entire tree list and again using only species that do not belong to genera with predominantly tropical distributions.

The actual numbers of genera, families, and orders in the temperate tree floras of Europe and extreme western North America differ substantially from the mean numbers obtained by rarefaction of the eastern Asian temperate tree flora (tables 26.2 and 26.3; figs. 26.3 and 26.4). Statistical analysis of the rarefaction results shows that the European temperate tree flora is consistently poorer in genera, families, and orders than the eastern Asian temperate tree flora, whether or not the predominantly tropical genera are omitted. Rarefaction of the eastern North American tree flora to the numbers of species in Europe yielded similar differences, which are also highly significant. Rarefaction revealed almost no significant differences in hierarchical patterns of tree diversity between east-central Asia and eastern North America or between Europe and the Pacific slope of North America.

We assume, for the moment, that frost tolerance is a characteristic of higher taxa (we will return to this assumption later) and that frost tolerance was the key to a taxon's regional survival through Quaternary cooling. It follows that regional extinction would have been nonrandom among higher taxa. Based on these assumptions, the rarefaction results support the hypothesized link between extirpation and low tree species richness in Europe and

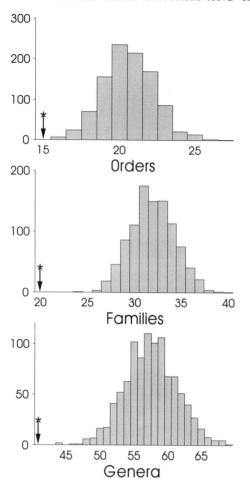

Figure 26.4 Frequency distributions of higher taxa resulting from simulated rarefaction of the east-central Asian moist temperate tree flora to the species number of the European moist temperate tree flora, with genera of predominantly tropical distribution omitted (see table 26.3). Numbers of taxa actually present in the European moist temperate tree flora are marked by asterisks (*).

western North America, since the distributions of species among higher taxa in these regions differ sharply from random subsets of the east-central Asian and eastern North American forests. However, given the same assumptions, the rarefaction results suggest that something other than differences in regional extinction rates may be responsible for the lower tree species richness in eastern North America relative to east-central Asia. To interpret these results, we seek a factor that is unbiased toward or against particular genera, families, or orders, in contrast to regional extinction based on frost tolerance or intolerance, which we assume to be phylogenetically selective.

Species per Higher Taxon. The temperate tree floras of Europe and the Pacific slope of North America clearly are depauperate at the higher taxonomic levels relative to those of eastern Asia and eastern North America. Furthermore, their higher taxa, on average, have fewer species, even though many genera and families are represented in the eastern Asian temperate forest region by only one or a few species (fig. 26.5A). While *numbers* of genera represented by one or a few species are far higher in eastern Asia than in Europe, eastern Asia and Europe have almost identically low *proportions* of these regionally low-diversity genera, lower than the corresponding proportions in North America (fig. 26.5B). In genera common to each pair of regions, Asian temperate forests are two to five times more speciose, on average, than forests in the other regions (table 26.4 columns A and B).

Globally Depauperate Genera. We examined the distribution of globally monotypic or ditypic (comprising only one or two species) tree genera among the regions (table 26.5). We were interested in the biogeography of these very low diversity genera because presumably some are relics of formerly diverse lineages, and some are autochthonous and perhaps new taxa that have not diversified. In either case they probably are more likely to go extinct

Table 26.2. Results of Simulated Rarefactions of Contemporary Tree Species

Regions compared	Taxon	Taxa present in region 1	Mean taxa in 1,000 floras drawn randomly from region 2	$t_{(df = 999)}$	Significance
1 Europe	Order	16	25.2	−4.80	**
2 East-central Asia	Family	21	38.5	−6.22	**
	Genus	43	70.2	−6.46	**
1 Pacific slope of North America	Order	14	20.7	−3.11	*
2 East-central Asia	Family	19	29.1	−3.65	**
	Genus	37	47.0	−3.16	**
1 Eastern North America	Order	26	30.4	−2.42	(NS)
2 East-central Asia	Family	46	50.4	−1.63	(NS)
	Genus	90	105.5	−3.34	**
1 Europe	Order	16	22.4	−4.58	**
2 Eastern North America	Family	21	36.5	−7.13	**
	Genus	43	58.8	−4.73	**
1 Pacific slope of North America	Order	14	18.8	−2.99	*
2 Eastern North America	Family	19	28.1	−3.80	**
	Genus	37	39.8	−0.90	(NS)
1 Pacific slope of North America	Order	14	13.5	0.46	(NS)
2 Europe	Family	19	16.8	1.51	(NS)
	Genus	37	32.6	2.01	(NS)

Note: We used *t*-tests to compare the actual numbers of taxa present with the means from 1,000 randomly generated floras in which the moist temperate tree flora of region 2 is rarified to the number of species in the moist temperate tree flora of region 1. The Type I error rate was adjusted using the Dunn-Šidák method (Sokal and Rohlf 1981): for $\alpha = .05$, $\alpha' = .0085$ and for $\alpha = .01$, $\alpha' = .0017$ (*$P < .05$, **$P < .01$).

Table 26.3. Results of Simulated Rarefactions of Contemporary Tree Species

Regions compared	Taxon	Taxa present in region 1	Mean taxa in 1,000 floras drawn randomly from region 2	$t_{(df = 999)}$	Significance
1 Europe	Order	15	20.6	−3.30	**
2 East-central Asia	Family	20	32.0	−5.07	**
	Genus	41	57.5	−4.30	**
1 Pacific slope of North America	Order	14	17.3	−1.95	(NS)
2 East-central Asia	Family	19	24.9	−2.45	(NS)
	Genus	35	40.0	−1.56	(NS)
1 Eastern North America	Order	25	23.8	0.75	(NS)
2 East-central Asia	Family	41	39.0	0.97	(NS)
	Genus	77	80.6	−0.92	(NS)
1 Europe	Order	15	21.3	−4.49	**
2 Eastern North America	Family	20	32.9	−6.41	**
	Genus	41	53.2	−4.01	**
1 Pacific slope of North America	Order	14	17.8	−2.42	(NS)
2 Eastern North America	Family	19	25.2	−2.78	*
	Genus	35	36.4	−0.47	(NS)
1 Pacific slope of North America	Order	14	12.9	0.97	(NS)
2 Europe	Family	19	16.1	2.00	(NS)
	Genus	35	31.2	1.83	(NS)

Note: Simulations were performed and tested for significance as described in table 26.2.

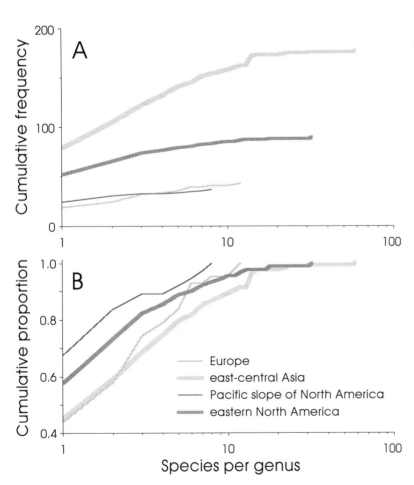

Figure 26.5 Frequency distributions of species per genus in the four regions. (*A*) Cumulative frequency; (*B*) cumulative proportion of total frequency.

Table 26.4. Comparison of Species Diversity of Tree Genera among Regions

Regions compared	(A) Mean species/genus in region 1 of genera common to both regions	Significance of Wilcoxon T_s (H_0: $\bar{y}_A = \bar{y}_B$)	(B) Mean species/genus in region 2 of genera common to both regions	Significance of Mann-Whitney U (H_0: $\bar{y}_B = \bar{y}_C$)	(C) Mean species/genus in region 2 of genera extirpated from region 1
1 Europe 2 East-central Asia	3.03 $n = 39$	$P < .000001$	10.1 $n = 39$	$P < .000001$	2.69 $n = 74$
1 Pacific slope of North America 2 East-central Asia	2.00 $n = 31$	$P < .00001$	10.0 $n = 31$	$P < .0005$	3.54 $n = 35$
1 Eastern North America 2 East-central Asia	3.57 $n = 61$	$P < .000005$	7.59 $n = 61$	$P < .01$	2.00 $n = 8$
1 Europe 2 Eastern North America	3.43 $n = 30$	(NS)	5.20 $n = 30$	$P < .0005$	1.82 $n = 34$
1 Pacific slope of North America 2 Eastern North America	2.08 $n = 26$	$P < .005$	5.19 $n = 26$	(NS)	2.67 $n = 21$
1 Pacific slope of North America 2 Europe	2.27 $n = 22$	(NS)	3.64 $n = 22$	(NS)	2.50 $n = 8$

Note: We compared species diversities of tree genera currently inhabiting regions of higher overall diversity (B) with the same genera persisting in regions of lower overall diversity (A), and with genera extirpated from regions of lower overall diversity during the mid- to late Tertiary and Quaternary (C). Nonparametric methods were used to test the statistical significance of differences between groups in species numbers per genus: the Wilcoxon signed-rank test for paired samples and the Mann-Whitney test for unpaired samples. Adjusting the Type I error rate by the Dunn-Šidák method (Sokal and Rohlf 1981), for $\alpha = .05$, $\alpha' = .0085$ and for $\alpha = .01$, $\alpha' = .0017$.

Table 26.5. Geographical Distribution of Globally Mono- and Ditypic Genera of Moist Temperate Forest Trees in the Northern Hemisphere.

	Northern, central, & eastern Europe	East-central Asia	Pacific slope of North America	Eastern North America	Entire Northern Hemisphere (total)
Genera of > 2 species worldwide	41	148	34	78	168
Genera of ≤ 2 species worldwide (% of total)	2 (4.7%)	29 (16%)	3 (8.1%)	12 (13%)	45 (21%)

Note: One ditypic tree genus, *Liriodendron*, lives in two of the regions.

than are multispecies genera. Their distribution may reflect differences among regions in rates of extirpation or production of new taxa.

There are 45 globally monotypic or ditypic tree genera native to the north temperate forest regions. Six make up globally mono- or ditypic families: four in eastern Asia (Ginkgoaceae, Eucommiaceae, Cercidiphyllaceae, Tetracentraceae) and two in eastern North America (Leitneriaceae, Cyrillaceae). The globally mono- and ditypic genera are distributed among Europe, east-central Asia, North America's Pacific slope, and eastern North America, respectively, at a ratio of approximately 1:12:1:5, compared with the ratio of total numbers of genera of approximately 1:4:1:2. We used the G-test to compare the distribution among regions of genera consisting of only one or two species worldwide with the distribution of genera with more than two species (table 26.5). The test showed the two distributions to be marginally significantly different ($G = 6.7$, df $= 3$, $.05 < P < .1$). Thus, globally mono- and ditypic genera may be overrepresented in eastern Asia and eastern North America relative to Europe and North America's Pacific slope.

The fossil record shows 25 globally mono- and ditypic temperate tree genera to have relict distributions; that is, they formerly ranged across at least one more of the four regions than they do currently. These genera (table 26.6)

include *Ginkgo*, 7 conifers (including *Glyptostrobus*, which ranges in eastern Asia's temperate zone but occurs primarily southward), and 11 hamamelids. The known relicts thus belong disproportionately to the older classes and subclasses. *Ginkgo* and conifers belong to the oldest surviving lineages of temperate trees, and hamamelids include the oldest known angiosperm temperate trees, members of the formerly diverse and now depauperate Platanaceae (Schwarzwalder 1986).

Of the 21 globally mono- and ditypic genera that appear to be endemic to a single region, 13 (62%) are absent from the fossil record in any of the four regions. They may have occurred sparsely in Tertiary forests, their Tertiary ranges may have been small, they may have first appeared only recently, or they may be cryptic in the fossil record owing to low pollen output, non-wind-dispersed pollen, or restriction to habitats not conducive to fossilization of leaves, flowers, fruits, or seeds. Most are probably within-region relicts or lineages that never were diverse or abundant.

Global Distribution of Genera. We tallied genera that occur in either two or three of the regions, that is, those that are neither endemic nor cosmopolitan (fig. 26.6). Of the 63 genera so distributed, 59 (94%) are present in eastern Asia. The largest tally—20 genera common only to east-

Table 26.6. Taxonomic Distribution of Globally Mono– and Ditypic Tree Genera with Relict Distributions

Class	Subclass	Total genera	Mono- and ditypic genera	
			Fossils in other regions	No fossils in other regions
Ginkgoopsida		1	1	0
Pinopsida		27	7	5
Magnoliopsida	Magnoliidae	18	1	1
	Hamamelidae	45	11	2
	Dilleniidae	32	2	5
	Rosidae	67	1	7
	Asteridae	18	1	1
Liliopsida	Arecidae	3	1	0
	Commelinidae	3	0	0
	Liliidae	1	0	0

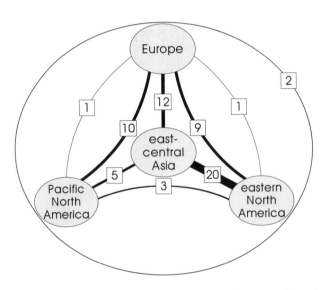

Figure 26.6 Numbers of tree genera native to either two or three of the four north temperate forest regions. Straight lines and the middle curved lines (bowing out) indicate genera that occur in two regions. The inner curved lines (bowing in) indicate genera that occur in three regions including east-central Asia. The outer ellipse indicates genera that occur in the three regions excluding east-central Asia.

ern Asia and eastern North America—reflects the well-known range disjunction displayed by many moist temperate forest plants that inhabit both regions (Li 1952; Graham 1972; Boufford and Spongberg 1983). Three tallies nearly tie for second rank: Europe and eastern Asia; Europe, eastern Asia, and Pacific North America; and Europe, eastern Asia, and eastern North America. Despite their proximity, the temperate forests at the two ends of North America share no genera uniquely, and they share the fewest genera as members of three-region groups. Eastern Asia emerges strikingly and overwhelmingly as the core area of Northern Hemisphere temperate tree genus distributions.

Exceptional Genera. A few genera run counter to the general trend of greatest diversity in eastern Asia followed, in sequence, by eastern North America, Europe, and Pacific North America. Most notable are *Carya,* the hickories, with 13 species in eastern North America and one in east-

ern Asia, and *Crataegus,* the hawthorns, with approximately 18 tree-sized species (≥ 8 m maximum height) in eastern North America, 2 each in eastern Asia and Europe, and 1 on North America's Pacific slope. Most *Crataegus* species are shrubs, but the trend in total species distribution parallels that of the few tree-sized members of the genus: approximately 220 of the global total of 306 species in a recent reexamination of *Crataegus* taxonomy (Phipps et al. 1990) are centered in the eastern North American moist temperate zone. Another genus that strikingly defies the trend is *Quercus,* the oaks, with tree-sized species numbering 32 in eastern North America, 21 in eastern Asia, 11 in Europe, and 5 in Pacific North America.

Juglandaceae, the family to which *Carya* belongs, is among the very few angiosperm tree lineages distributed widely across the north temperate zone for which there is fossil evidence that early diversification took place outside of eastern Asia, in this case in eastern North America and Europe (Manchester 1989). The tribe Querceae of the Fagaceae, including *Quercus,* may also have originated in eastern North America and Europe (Crepet and Nixon 1989). *Crataegus,* on the other hand, may have originated in southern China in the early Tertiary (Phipps 1983) despite its current locus of highest diversity in eastern North America.

CLUES FROM THE FOSSIL RECORD

In examining fossil distributions, we focused on the Paleogene, over 40 million years of warm, relatively stable climate during which forests spanned most of the present-day Arctic and covered nearly the entire breadths of Eurasia and North America, but were interrupted intermittently on both continents by north–south trending shallow seas in mid-continent (Figure 26.7). We used two different estimates of the fossil tree floras for the four contemporary north temperate forest regions: (a) genera actually represented in the fossil record of each region (Reid and Chandler 1933; Hu and Chaney 1940; Traverse 1955; Mai 1960; Kilpper 1969; van der Hammen, Wijmstra, and Zagwijn 1971; Mai 1971a, 1971b; Tanai 1972; Takhtajan 1974; Łańcucka-Środoniowa 1975; Rachele 1976; Collinson and Crane 1978; Friis 1979; Mai 1980; Potter and Dilcher 1980; Mai 1981; Gregor 1982; Freder-

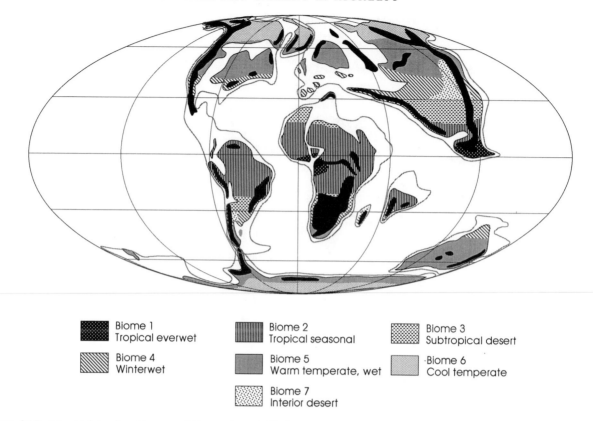

	Biome 1 Tropical everwet		Biome 2 Tropical seasonal		Biome 3 Subtropical desert
	Biome 4 Winterwet		Biome 5 Warm temperate, wet		Biome 6 Cool temperate
			Biome 7 Interior desert		

Figure 26.7 Maestrichtian (late Cretaceous) biomes. Areas in black are major highland regions. (Redrawn from Horrell 1991.)

iksen 1984a, 1984b; Wing and Hickey 1984; Friis 1985; Kvaček and Walther 1987; Mai 1987a, 1987b; Mai and Walther 1988; Sauer 1988; Kvaček, Walther, and Bužek 1989; Guo 1990; McCartan et al. 1990; Manchester, unpublished data; Friis, personal communication) and (2) an expanded list of genera also including those that are absent from the Tertiary fossil record of the region but present in its contemporary flora *and* in the Tertiary fossil record of at least one other region. The latter we term the "inferred" Paleogene tree flora of each region. We acknowledge that these are imperfect estimates of the actual regional Paleogene tree floras for many reasons, including the infrequency and nonrandom distribution of fossilization events, the scarcity of Tertiary sedimentary deposits, and the rarity of paleontologists with the skills, specific curiosity, and time to analyze Paleogene tree floras.

In comparing fossil and contemporary floras, we included some present-day occurrences of genera not included in considerations of contemporary floras alone. All such occurrences (marked (P) in Appendix 26.1) fall into one of two categories: certain trees at the edges of their ranges, and certain shrubs. We included a genus in a region's contemporary flora even if represented there only by shrubby species if the genus also includes tree-sized species in another region, because trees may not be distinguishable from non-tree congeners in fossils. For such comparisons we also included a genus in a region's contemporary flora even if its range is peripheral to the region if it is also a characteristic member of a contemporary moist temperate tree flora in another region. For example, *Larix* is a member of the moist temperate tree floras of

Europe and east-central Asia, but its occurrences in the temperate zone of eastern North America are peripheral or disjunct from the main North American range of the genus, which is boreal. For another example, *Platanus* species belong to the moist temperate tree floras of eastern and Pacific slope North America, but in Eurasia, native stands of the genus are confined mainly to seasonally arid (eastern Mediterranean), montane (the Himalayas), and tropical (Southeast Asia) regions. Such genera may have contributed to moist temperate forest fossil assemblages both in regions where they were characteristic of the moist temperate flora and in those where their presence was infrequent, signaling spillover from a neighboring biome.

Our compiled fossil data show a tally of 107 tree genera that have disappeared since the Oligocene from some, but not all, of the present-day north temperate regions (table 26.7). We sought patterns in the division of tree genera in each region into three categories: present as Paleogene fossils but extirpated, present as fossils and persisting in the contemporary flora, and known only from the contemporary flora (see Appendix 26.1).

Comparability of Fossil Data among Regions

First, we compared total extant genera in each region with extant genera that are also represented in the fossil record for that region. This comparison provides a rough estimate of the consistency among regions of the fossil data reliability (table 26.8), assuming that most genera occurring in each of the four regions now also occurred somewhere within that region during the Tertiary. The

Table 26.7. Genera Extirpated Regionally during the Late Tertiary and Quaternary

Genera	Europe	Northern & east-central Asia	Western North America	Eastern North America
Acanthopanax	×	●		
Ailanthus[E]	×	●	×	
Alangium[E]	×	●	×	×
Albizia[E]		●	×	
Aphananthe	×	●	×	●
Aralia	×	●		●
Asimina[N]	×			●
Broussonetia	×	●		
Bumelia[E]			×	●
Calocedrus	×	(P)	●	
Camellia[E]	×	●		
Carpinus[N]	●	●	×	●
Carya[N]	×	●	×	●
Castanea[N]	●	●	×	●
Castanopsis[E]	×	●	●	
Catalpa	×	●	×	●
Cedrela[E]	×	●	×	
Cephalanthus[N]	×			●
Cephalotaxus[N]	×	●		
Cercidiphyllum[N]	×	●	×	
Chamaecyparis[N]	×	●	●	●
Chionanthus[N]	×	●		●
Cinnamomum[E]	×	●		
Clerodendrum[E]	×(?)	●		
Clethra	×	●		(P)
Cunninghamia[N]	×	●		
Cyclobalanopsis[N]	×	●		
Cyclocarya[N]	×	●		
Cyrilla[N]	×			●
Dendropanax[E]		●		×
Diospyros[E]	×	●	×	●
Disanthus[N]	×	●		
Distylium[N]	×	●		
Emmenopterys[N]		●	×	
Engelhardtia[E]	×	(P)		×
Enkianthus[N]	×(?)	●		
Eucommia[N]	×	●	×	×(?)
Euptelea[N]	×	●		
Evodia[E]	×	●		
Fagus[N]	●	●	×	●
Fortunearia[N]	×	●		
Ginkgo[N]	×	●	×	
Glyptostrobus[E]	×	(P)	×	×
Gordonia[E]	×	(P)		●
Halesia[N]	×			●
Hamamelis[N]	×	●		●
Hemiptelea[N]	×	●		
Hydrangea	×	●		(P)
Illicium[E]	×	●		
Kalmia[N]	×			●
Keteleeria[E]	×	●	×	
Koelreuteria	×	●		
Lagerstroemia[E]	×	●		
Leitneria[N]	×	×		●
Lindera[E]	×	●	×	(P)
Liquidambar[N]	×	●	×	●
Liriodendron[N]	×	●	×	●
Lithocarpus[E]	×	●	●	
Litsea[E]	×	●		(P)
Lyonia	×	(P)		●
Magnolia	×	●	×	●
Mallotus[E]	×	●	×(?)	
Manglietia[E]	×	●		
Meliosma[E]	×	●		
Metasequoia[N]	×	●	×	
Michelia[E]	×	●		
Neolitsea[E]	×	●		
Nyssa	×	●	×	●
Osmanthus[E]	×	●		●
Ostrya[N]	●	●	×	●
Paulownia	×	●	×	
Persea[E]	×	●	×	
Phellodendron	×	●		
Phoebe[E]	×	●		
Planera[N]			×	●
Platycarya[N]	×	●	×	×
Poliothyrsis[N]	×	●		
Pseudolarix[N]	×	●		
Pseudotsuga[N]	×	●	●	
Pterocarya[N]	●	●	×	●
Pteroceltis[N]	×	●		
Rhus	(P)	●	×(?)	●
Robinia	×	×		●
Sabal[E]	×	●	×	●
Sapindus[E]	×	●	×	●
Sapium[E]	×	●		
Sassafras[N]	×	●	×	●
Schefflera[E]	×	●		
Sciadopitys[N]	×	●		×
Sequoia[N]	×	×	●	
Serenoa[N]	×			●
Sinowilsonia[N]		●		●
Staphylea[N]	×	●		●
Stewartia[N]	×	●		●
Symplocos[E]	×	●	×	●
Tapiscia[N]	×	●	×	
Taxodium[N]	×	●	×	×
Ternstroemia[E]	×	●		
Tetracentron[N]	×(?)	●		
Thuja[N]	×	●	●	●
Tilia[N]	●	●	×	●
Torreya[N]	×	●	●	●
Tsuga[N]	×	●	●	●
Turpinia[N]	×	●		
Ulmus[N]	●	●	×	●
Zanthoxylum[E]	×	●		
Zelkova[N]	●	●	×	×(?)

Note: (×) indicates extirpation from a region, (●) indicates contemporary tree species existing in a region. Contemporary genera are marked (P) if they do not attain tree height or if they rarely occur in the region's flora but do inhabit an adjoining biome and are characteristic members of a moist temperate tree flora in another region. Fossil genera are marked (?) if identified only tentatively by recent authorities (cited in this chapter). Present distributions of genera are indicated by superscripts: E, predominantly tropical; N, predominantly temperate or extending into tropical latitudes mainly at high elevations. (See Appendix 26.1 for sources of fossil and distributional data.)

completeness of the fossil record for extant genera does not differ significantly among the regions ($G = 6.21$, df = 3, $P > .1$).

Persistence/Extirpation Rates among Regions

Next, we compared total fossil genera in each region with fossil genera that are also extant in that region, as an estimate of the relative survival of genera among regions (ta-

ble 26.9). The four regions differ significantly in tree genus survival rate ($G = 53.9$, df = 3, $P \ll 0.001$). Extirpation rates of tree genera were radically unequal among the four regions. Europe was especially hard hit.

Extirpated Genera

Next, we looked at contemporary floras for differences between the genera that died out regionally and those that

Table 26.8. Relative Index of Completeness of Tertiary Fossil Record, by Region, of Moist Temperate Forest Tree Genera in the Northern Hemisphere

	Europe	Northern & east-central Asia	Western North America	Eastern North America
Total extant genera[a]	53	185	42	99
Represented in fossil record[b]	38 (72%)	117 (63%)	35 (83%)	49 (49%)

[a]Includes extant genera that do not attain tree height or that rarely occur in the region's flora but do inhabit adjoining biomes and are characteristic members of a moist temperate tree flora in another region.
[b]Includes a few fossil genera identified only tentatively by recent authorities (cited in this chapter).

Table 26.9. Survival since Mid-Tertiary, by Region, of Moist Temperate Forest Tree Genera in the Northern Hemisphere

	Europe	Northern & east-central Asia	Western North America	Eastern North America
Total fossil genera[a]	130	122	75	60
Surviving[b]	38 (29%)	117 (96%)	35 (47%)	49 (82%)

[a]Includes a few fossil genera identified only tentatively by recent authorities (cited in this chapter).
[b]Includes extant genera that do not attain tree height or that rarely occur in the region's flora but do inhabit adjoining biomes and are characteristic members of a moist temperate tree flora in another region.

survived. The mean numbers of species per genus in the eastern Asian temperate tree flora are significantly greater in the genera that persist in the other regions than in the genera that were extirpated from them. A similar relationship holds between the eastern North American temperate tree flora and that of Europe (table 26.4, columns B and C; fig. 26.8). In other words, genera that are currently more speciose in eastern Asia (or eastern North America) are more likely to have survived in other regions than genera that are currently less speciose.

For random extinction to have produced this effect, species diversity would have to be a distinctive property of individual genera that endures across many millions of years and among continents. The data suggest that this is unlikely. We compared numbers of species in genera common to each pair of the six possible pairs of continental areas using the G-test (adjusting Type I error rate as in table 26.2), testing genus-by-genus whether the species numbers in the less diverse region of the pair differ significantly from expected values generated by proportionally reducing the species numbers in the more diverse region. Two of the pairs show significant differences: eastern North America and eastern Asia ($G = 255$, df = 60, $P < .01$) and eastern North America and Europe ($G = 61.1$, df = 29, $P < .01$); eastern North America and the Pacific slope of North America differ marginally significantly ($G = 49.4$, df = 26, $.05 < P < .1$).

Furthermore, extirpated genera are nonrandomly distributed among higher taxa (table 26.10). For example, in Europe all tree genera of the now mostly tropical subclasses Magnoliidae and Arecidae died out. In the subclasses Hamamelidae and Rosidae, which have radiated widely in temperate, boreal, and high-altitude habitats, nearly half of the tree genera persisted. Counter to the trend of species dropping out that belong to subclasses of mainly tropical distribution, two-thirds of the genera of conifers, which inhabit mostly temperate and boreal regions, disappeared. However, over half of these genera now are monotypic and most have very small global ranges, indicating their relictual status. In contrast, more

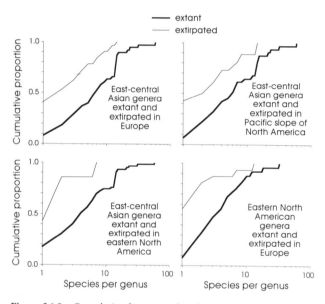

Figure 26.8 Cumulative frequency distributions (as proportions of total frequency) of species per genus in high-diversity regional tree floras for genera that are also extant in less diverse regional tree floras versus genera extirpated from less diverse regions.

than three-fourths of the conifer genera persisted on the Pacific slope of North America, the other region hit hard by extirpation. There, also, as in Europe, the magnoliids experienced the greatest attrition.

Global Distribution of Paleogene Genera

We tallied fossil genera that occurred in either two or three of the regions defined earlier for contemporary floras (fig. 26.9). In parallel with the contemporary pattern, 88 of the 95 genera in the inferred Tertiary floras that were so distributed (78 of 81 genera in the fossil-only floras) were present in eastern Asia. Of the 170 genera in our inferred Tertiary flora, 19 (11%) have been found in only one region, 95 (56%) are represented in two or three regions, and 56 (33%) are represented in all four regions.

Table 26.10. Taxonomic and Regional Distributions of Extant and Extirpated Genera of North Temperate Forest Trees Belonging to the Conifer and Dicot Classes

Class	Subclass	Europe		Northern & east-central Asia		Western North America		Eastern North America	
		Extant	Extirpated	Extant	Extirpated	Extant	Extirpated	Extant	Extirpated
Pinopsida		7	15	24	2	14	4	11	2
Magnoliopsida	Magnoliidae	0	14	15	0	1	5	7	0
	Hamamelidae	16	21	42	2	9	13	20	6
	Dilleniidae	7	13	23	0	6	4	20	0
	Rosidae	19	16	61	1	10	8	28	2
	Asteridae	4	6	15	0	2	3	9	0

Note: Includes a few fossil genera identified only tentatively by recent authorities (cited in this chapter) and extant genera that do not attain tree height or that rarely occur in the region's flora but do inhabit adjoining biomes and are characteristic members of a moist temperate tree flora in another region.

Of the 213 genera in our extant flora, 122 (57%) are endemic to one region, 63 (30%) are present in two or three regions, and 28 (13%) are in all four regions. The Tertiary and extant distributions differ highly significantly ($G = 240$, df = 2, $P \ll .001$). The low numbers of endemic fossil genera may be due in part to the lower likelihood of discovering fossil genera that were present in only one region. However, the pattern of decline in cosmopolitan distributions is striking. The well-known concurrence between the temperate floras of eastern Asia and eastern North America (Li 1952: Graham 1972; Boufford and Spongberg 1983) appears to be merely a vestige of the formerly even stronger affinity among the floras of these two regions and that of Europe. Most of the changes from figure 26.9 to figure 26.6 are due to extirpations in Europe.

Rarefaction of Paleogene Genera

We compared the higher taxonomic structures of the inferred Paleogene tree floras (table 26.11) by simulating the rarefaction of the eastern Asian flora to match the numbers of genera in the other regions' floras (method given above). Unlike the simulated rarefaction of contemporary floras, the analysis of Paleogene floras does not compare regions in numbers of tree *species* among genera, families, and orders because species number cannot be reliably estimated from fossil remains. Rarefaction of the eastern Asian fossil tree *genera* indicated that the distribution of fossil genera among families and orders in Europe, the Pacific slope of North America, and eastern North America did not differ from random samples of the Asian fossil genera (table 26.12). We also simulated rarefaction of the contemporary tree genera of temperate east-central Asia to contemporary numbers of tree genera in the other three regions, for comparison with the rarefactions of fossil genera and of contemporary species (table 26.13). Rarefying contemporary floras by genera paralleled the results of rarefying contemporary floras by species (see table 26.2).

CLIMATE AND TREE SPECIES DIVERSITY

Several comparative studies have revealed a direct relationship between species diversity and various climate variables, particularly precipitation or estimates of actual evapotranspiration (AET) (Richerson and Lum 1980;

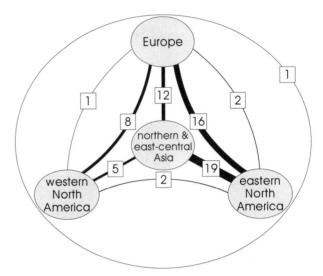

Figure 26.9 Numbers of tree genera occurring in either two or three of the four contemporary north temperate forest regions during the Paleogene (genera represented by fossil remains plus those inferred as present during the Tertiary because they are present in the region's contemporary flora *and* in the Tertiary fossil record of at least one other region). See figure 26.6 for an explanation of the lines.

Wright 1983; Turner, Lennon, and Lawrenson 1988). Patterns of this type have been demonstrated for trees in both temperate (Currie and Paquin 1987; Adams and Woodward 1989) and tropical (Gentry 1988a) regions, using sampling areas ranging from grid blocks as large as 100,000 km² to small plots of 0.1 to 1 hectare. Furthermore, several authors have cited the consistency of these relationships between regions (e.g., Adams and Woodward 1989) as evidence of convergence and determination of species richness by local physical factors.

In general, tree species richness increases in direct relation to precipitation and AET, suggesting that a positive relationship exists between diversity and productivity of the habitat (but see Tilman and Pacala, chap. 2, and Rosenzweig and Abramsky, chap. 5 for evidence that diversity declines at high habitat productivity). This relationship forms the basis of "species-energy theory" or energy-diversity theory (Wright, Currie, and Maurer, chap. 6), which relates diversity to energy flux by means of several possible mechanisms. In general, high produc-

Table 26.11. Summary by Taxonomic Level and Region of Forest Trees in "Inferred" Tertiary Floras

Taxonomic level	Europe	Northern & east-central Asia	Western North America	Eastern North America	Total
Subclasses	8	7	8	7	8
Orders	33	33	24	26	34
Families	61	63	43	49	67
Genera	140	156	81	98	170

Table 26.12. Results of Simulated Rarefactions of Tertiary Tree Genera

Region	Taxon	Taxa in early Tertiary tree flora of region	Mean taxa in 1,000 floras drawn randomly by genus	$t_{(df = 999)}$	Significance
Europe	Order	33	31.2	2.25	(NS)
	Family	61	59.4	1.04	(NS)
Western North America	Order	24	26.5	−1.45	(NS)
	Family	43	44.1	−0.39	(NS)
Eastern North America	Order	26	28.2	−1.50	(NS)
	Family	49	49.0	0.02	(NS)

Note: We used *t*-tests to compare the numbers of tree taxa inferred to have inhabited high-latitude forest regions during the Tertiary with the means from 1,000 randomly generated floras in which the Tertiary tree flora of northern and east-central Asia is rarified to the number of genera in the Tertiary tree flora of each of the other regions (see text for method of inferring paleofloras). The Type I error rate was adjusted using the Dunn-Šidák method (Sokal and Rohlf 1981): for α = .05, α′ = .017 and for α = .01, α′ = .0033 (*P < .05, **P < .01).

Table 26.13. Results of Simulated Rarefactions of Contemporary Tree Genera

Region	Taxon	Taxa in present-day tree flora of region	Mean taxa in 1,000 floras drawn randomly by genus	$t_{(df = 999)}$	Significance
Europe	Order	16	21.6	−2.78	*
	Family	21	30.5	−4.14	**
Pacific slope of North America	Order	14	19.9	−3.00	**
	Family	19	27.3	−4.00	**
Eastern North America	Order	26	30.0	−2.16	(NS)
	Family	46	48.2	−0.82	(NS)

Note: We used *t*-tests to compare the actual numbers of taxa present with the means from 1,000 randomly generated floras in which the moist temperate tree flora of contemporary east-central Asia is rarified to the number of genera in the moist temperate tree flora of the other regions. Simulations were performed and tested for significance as described in table 26.12.

tivity maintains larger numbers of individuals per species and thus reduces the probability of stochastic extinction. High production in habitats with little stress may also increase the total variety of microhabitats and permit greater microhabitat specialization. For certain types of organisms, notably trees, high precipitation and temperature may be associated with the occupation of a greater variety of habitats, thereby increasing sample diversity through increased habitat heterogeneity on a regional scale and through spillover or mass effects on a habitat scale. Regardless of the mechanism, the correlation between diversity and physical factors suggests that the outcome of species interactions depends on the physical conditions of the environment.

Alternative viewpoints must be entertained when one identifies diversity anomalies, in which habitats with similar physical conditions are occupied by different numbers of species in different regions (Schluter and Ricklefs, chap. 21). Such is the case in both local and regional comparisons of mangrove species between the depauperate Caribbean and species-rich Indo-Pacific regions (Ricklefs and Latham, chap. 20). It is also certainly the case in the regional comparisons of species richness in north temperate deciduous broad-leaved forests presented here. The evi-

dence for a parallel diversity anomaly at the local level (1- to 10-ha plots) is weak due to inadequate sampling in Asia, but appears to be consistent with the regional trend between Europe and eastern North America (Latham and Ricklefs 1993). Lacking contrary evidence, we accept the possibility that eastern Asian temperate forests contain markedly more species at the local level than do temperate forests elsewhere.

Both Currie and Paquin (1987) and Adams and Woodward (1989) claimed general similarity among eastern Asia, Europe, eastern North America, and temperate regions of the Southern Hemisphere in species richness of temperate forests. We have reviewed and criticized these conclusions in detail elsewhere (Latham and Ricklefs 1993). Briefly, in these comparisons Asian data cited by Adams and Woodward (1989) were restricted to boreal and island localities. Moist temperate continental Asia, where regional tree species diversity is higher by far than anywhere else in the earth's temperate zones, was not sampled. Although diversity in Europe was claimed to be comparable to that in eastern North America, seven out of eight European sampling areas fell below the North American regression of species richness on AET. In these studies, species richness was tabulated for large grid blocks (51,000 to 100,000 km²), which introduces unspecified contributions of habitat heterogeneity to species diversity. In arid regions, for example, in which average climate conditions do not support forest vegetation, tree species were recorded only from riparian habitats. Furthermore, both Currie and Paquin (1987) and Adams and Woodward (1989) mixed broad-leaved and needle-leaved (including boreal) forests, perhaps making direct comparisons inappropriate. Latham and Ricklefs (1993) found that tree species richness in 0.5- to 10-ha samples of temperate broad-leaved forests was unrelated to AET. Because the outcome of any ecological interaction that may restrict local coexistence of species is determined at the local scale, we feel that our analysis more directly tests the relationship between species richness and energy flux within the temperate broad-leaved deciduous forest biome of eastern North America.

The increase in tree species richness from temperate to tropical latitudes is generally thought to reflect parallel gradients of physical conditions (Ricklefs 1977; Gentry 1988a), with temperature and moisture generally higher toward the equator at low elevation. While species richness of trees, large shrubs, and lianas on 0.1-ha plots in the tropics appears to increase with annual precipitation up to 300 to 500 cm (Gentry 1988a), Latham and Ricklefs (1993) failed to find a relationship between species number of trees alone and AET on 1-ha plots in tropical forests. A latitudinal gap of at least 15° separates broad-leaved forests in the Neotropics and in temperate North America. As noted by Latham and Ricklefs (1993), a corresponding discontinuity exists in tree species richness. In an analysis of covariance of tree species richness and AET on 0.5- to 10-ha plots, temperate and tropical plots differed significantly between each other (by an order of magnitude) but independently of AET, even though the ranges of temperate and tropical AET values overlapped. Thus, the latitudinal difference in tree species diversity is not a direct consequence of a latitudinal difference in physical conditions, because tree species diversity is statistically unrelated to AET within latitudinal belts.

In general, we conclude that regional effects influence tree species richness independently of, and in addition to, local effects of climate. Diversity-climate correlations among large sampling blocks may reflect increased variety of habitats suitable for trees as productivity increases (increased beta diversity). Temperate-tropical differences in diversity in the Americas and Europe/Africa may represent discontinuities in diversity along continuous environmental gradients. Greater sampling, particularly of temperate and subtropical forests in eastern Asia, will be required to clarify these relationships. But for the present, we feel that a simple, continuous relationship between diversity and local climate does not provide an adequate description of contemporary patterns of tree species diversity.

THE HISTORICAL DEVELOPMENT OF TEMPERATE TREE FLORAS

North temperate broad-leaved deciduous tree floras present the following patterns. First, a large proportion of these floras, particularly in the northerly parts of the bi- omes, belong to families that are characteristic of north temperate regions, primarily Betulaceae, Fagaceae, Hamamelidaceae, Juglandaceae, Salicaceae, Cornaceae, Rosaceae, and Aceraceae. Toward the southern parts of the biomes, representatives of more tropical families appear, but generally not in large numbers.

Second, the difference in diversity between tropical and temperate floras, and between temperate floras in different regions, resides at high taxonomic levels. There are roughly 11 species per family in temperate eastern Asia and half that number in temperate eastern North America, which has about a third the total number of species (see table 26.1). Rarefaction of the Asian species indicates that the distribution of North American species among higher taxa does not differ from that in a random sample of the Asian taxa. On 0.1-ha plots, temperate floras exhibit about 1.4 to 2.3 species per family, while lowland tropical floras, having up to 10 times as many species, exhibit species/family ratios of between 2 and 4, with as many as 58 families represented at a single site (Gentry 1988a). Thus, patterns of diversity are expressed at a high taxonomic level (Ricklefs 1989b). This suggests that contemporary patterns were established long ago by colonization and cladogenesis, which played roles at least as important as that of extinction.

Third, temperate flora diversity and the proportion of species belonging to predominantly tropical families and genera are both highest in eastern Asia, where there is and perhaps has been since before the Tertiary a continuous corridor of mesic forest connecting tropical and temperate latitudes. Colonization of temperate biomes in Asia from the tropics over long time periods has probably played an important role in the development of temperate forest communities there.

Fourth, although both European and western North American temperate floras suffered extinctions during the mid- to late Tertiary, the primary temperate diversity anomaly, that distinguishing eastern Asia from other temperate regions, is old and probably was established primarily by regional differences in colonization and autochthonous production of new taxa.

We propose that the differences in diversity of temperate tree floras among continents reflect the history of colonization of temperate biomes, which appears to have occurred more frequently in Asia, and the subsequent production of new autochthonous taxa and their geographical spread within temperate biomes. Furthermore, we propose, as have Farrell, Mitter, and Futuyma (1992) for insects, that differences between temperate and tropical floras reflect a physiological barrier to colonization of temperate zones that can be crossed only by the evolution of freezing tolerance mechanisms. Thus, the relatively low diversity of angiosperm trees in temperate areas arises because of the difficulty of colonizing temperate regions, rather than, or in addition to, any intrinsic limits either on species production or on coexistence of species within temperate areas. Accordingly, explanations for latitudinal gradients (actually disjunctions) in diversity can be traced to historical and evolutionary factors rather than to contemporary ecological interactions. We discuss these ideas in more detail below. It is not our purpose here to provide a balanced evaluation of alternative models. Rather, we advocate a particular model that must be properly evaluated in the future.

Most angiosperm families arose during the late Cretaceous and Paleogene. During this time, frost-free climates covered much of the world's land surface (see fig. 26.7). The oldest fossils of contemporary moist temperate zone tree families in the Northern Hemisphere date from over 100 mya to less than 15 mya, with most falling within the range 30 to 90 mya (fig. 26.10). During the early part of this period, eastern Asia was the only region in the Northern Hemisphere where a more or less continuous zone of forest vegetation might have existed between the tropics and high latitudes (see fig. 26.7). Fossil data from the late Cretaceous to the mid-Tertiary indicate an arid zone covering most inland fossil collection sites between subtropical moist forests in northeastern Asia and tropical moist forests in southeastern Asia (see figure 26.7; Song, Li, and He 1983; Horrell 1991). However, moist conditions are likely to have existed near the coast throughout this period, as they do today (A. M. Ziegler, personal communication). Europe was isolated from tropical Africa by the Tethys Sea and from eastern Asia by large inland seas during much of this period. North America was isolated by water from extensive tropical areas in South America. The southern portions of the north temperate regions were separated from each other by two oceans and two shallow mid-continental seas, but their northern portions were at least intermittently connected via Greenland, Ural, Bering, and mid-Canadian land bridges (Tiffney 1985a). Because several of the most prominent temperate families of trees have fossil records dating back to the late Cretaceous (Betulaceae, Fagaceae, Juglandaceae) or early Paleogene (Hamamelideceae, Nyssaceae), we presume that the development of temperate floras occurred at this time.

In our scenario, the development of north temperate forests involved the crossing of a major physiological boundary—the evolution of freezing tolerance—and reflected routes of colonization and dispersal from frost-free areas into various areas north of the frost boundary. Initially, areas of the Northern Hemisphere exposed to freezing were very restricted and distributed far to the north. Most angiosperm families, including those restricted to frost-free areas at present, inhabited what are now mid-temperate latitudes. Palynological data demonstrate that the replacement of gymnosperms by angiosperms during the Cretaceous began in equatorial latitudes but quickly spread far to the north (Crane and Lidgard 1990). The broad latitudinal distribution of forest vegetation in eastern Asia would seem to have been especially conducive to the evolution and spread of tree taxa.

With cooling beginning in the Oligocene and the expansion of the frost zone, most angiosperm families vacated the high-latitude areas that now make up the temperate zone. At the same time, the separation of Eurasia from Africa and India by the Tethys Sea and a wide separation between North and South America continued to limit any possible connection between moist tropical and moist temperate regions solely to eastern Asia (see fig. 26.7). By the Miocene, the two biomes shared a common boundary or transition zone in east-central Asia ex-

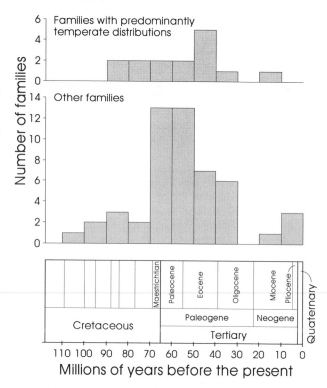

Figure 26.10 Frequency distribution of earliest known fossil ages of angiosperm plant families represented in contemporary moist temperate tree floras of the Northern Hemisphere. (Data from Cronquist 1981 and Muller 1981, using the greater age where the two sources disagree.)

tending more than 1,500 km westward from the coast (Song et al. 1981).

The present high diversity of the temperate Asian tree flora, particularly at the family and genus levels, combined with the strong representation of modern taxa in the fossil record of eastern Asia, indicates Asian origins for much of the north temperate flora. Contemporary distributions of cosmopolitan elements of this flora give no clues to the locations of their origins. Several cases are known in which the present distribution of a taxon does not include its fossil distribution. For example, *Platycarya* is restricted at present to temperate and subtropical eastern Asia (one species) but is known from the fossil record only in North America and Europe (Manchester 1987). Whether its presence in Asia is relictual (as in the case of *Cyclocarya*) rather than representing a recent colonization depends on whether it occurred historically in the region, which cannot be ruled out. Details of this sort cannot be adequately resolved, even for the best-known of families like the Juglandaceae, owing to the inadequacies of the fossil record.

By all measures, eastern Asia's forests claim most of the diversity of the north temperate tree flora. All but 4 families of the total flora—95%—live in temperate eastern Asia (all but 7 of the families include tree-sized species there); the exceptions are the primarily tropical Cyrillaceae and Sapotaceae, the Leitneriaceae (a family in southeastern North America without clear relationships), and the more widespread Platanaceae, which currently inhabits tropical Southeast Asia and occurs as fossils at higher latitudes in Asia. In total, these account for only 8 temperate species—less than 1% of the total flora. Of the genera, 87% live in eastern Asia (most include tree-sized species there). The 28 genera that are absent from eastern Asia account for only 3% of the total species. The pattern still holds when families or genera of predominantly tropical distributions are omitted: 95% of the nontropical families and 85% of the nontropical genera occur in eastern Asia.

We presume that the dispersal of cosmopolitan families and genera occurred early in the evolution of the north temperate tree flora, and that most higher taxa restricted to eastern Asia originated there at a later time. If exchanges between the regions were possible throughout the Tertiary, we would have expected less endemism in the eastern Asia temperate flora, particularly its southerly elements. By Manchester's (1987) account, the cosmopolitan distribution of the Juglandaceae, including most of its modern genera, was established by the mid- to late Eocene. This particular case would seem to go against the general rule. The fossil record indicates an origin for the Juglandaceae in North America in the Paleocene, with nearly contemporary distribution of most of the genera in Europe, followed by their somewhat later appearance in eastern Asia. The family is also unusual in having roughly equal numbers of species in eastern North America and eastern Asia, primarily owing to the thirteen species of tree-sized *Carya* (hickories) in North America.

While cosmopolitan north temperate genera might have originated anywhere in the Northern Hemisphere, the distributions of more restricted genera clearly place east-central Asia at the center of dispersal (see fig. 26.6). The corresponding distributions of Tertiary fossil genera (see fig. 26.9) give approximately equal weight to east-central Asia and Europe as possible centers of origin. These patterns, together with the presence of a high diversity of endemic Asian taxa, point strongly to a Eurasian—most likely an eastern Asian—origin for much of the temperate flora. The scarcity of evidence for extirpation of genera in eastern North America suggests that the Asia bias reflects origination and not contraction of global ranges to relictual distributions (although, admittedly, fewer Tertiary fossil assemblages have been discovered so far in eastern North America than elsewhere in the contemporary north temperate zone).

In addition to the larger number of higher taxa in eastern Asia, widespread genera also tend to be more species-rich in that region. Conspicuous examples include *Carpinus*, with 14 tree-sized species in eastern Asia and 1 in eastern North America, *Alnus* (12/2), *Populus* (14/3), *Malus* (15/5), *Prunus* (32/10), *Sorbus* (15/2), *Acer* (58/9), and *Fraxinus* (14/7). The only exceptions of note are *Quercus* (21/32), *Carya* (1/13), and *Crataegus* (2/18). These genera are typically northern in distribution, and we therefore assume that the diversity differences represent differential proliferation and perhaps survival of species within temperate clades, rather than a differential frequency of invasion of temperate biomes by groups of mainly southern distribution. We are not comfortable speculating on the conditions in eastern Asia that might promote cladogenesis, although these might be associated with the varied topography of the region. Nor are we

ready to conjecture about the possible role of the age of taxa within each of the regions, although we presume that genera present in more than one region have had relatively long independent histories in each. Based on paleobotanical data, Wolfe (1981) has suggested that the Asian–eastern North American connection within *Acer*'s range was severed by the end of the Eocene, with a connection between eastern Asia and western North America persisting into the early Miocene.

The diversity anomaly in temperate tree floras between eastern Asia and other temperate regions appears to have arisen in part from the more frequent invasion of temperate biomes by tropical and subtropical vegetation in that region. If this is true, it also sheds light on the origin of the diversity contrast between tropical and temperate tree floras. We presume that the invasion of contemporary temperate biomes required the acquisition of frost tolerance, which involves extensive (and presumably costly) elaboration of biochemical mechanisms to protect stems and dormant buds from freezing (Sakai and Larcher 1987). Thus, frost tolerance presents a physiological barrier to dispersal that precludes most higher taxa of tropical and subtropical plants from the frost zone and has resulted in their withdrawal from temperate latitudes with recent cooling and more pronounced latitudinal stratification of temperature. Several of the taxa that crossed the frost barrier early in the Tertiary or perhaps in the Cretaceous have proliferated tremendously and achieved family status. Many taxa in typically tropical and subtropical families and genera have also crossed the barrier to varying degrees, but have not proliferated to the same extent nor penetrated the temperate biomes so extensively. Of genera from typically tropical families, only *Magnolia*, several genera in the Lauraceae (*Cinnamomum*, *Lindera*, *Litsea*, *Machilus*), and *Tilia* have achieved even moderate levels of diversity in temperate zones, and then primarily in eastern Asia and southern portions of eastern North America.

As expected, high-latitude subtropical Tertiary floras show strong taxonomic affinities with modern low-latitude subtropical floras (Sharp 1951). However, many plant families represented in subtropical Tertiary floras unearthed in the present-day temperate zone still have species living there. One possible explanation is that most temperate taxa may have dispersed and diversified globally under frost-free conditions across northern (but subtropical) latitudes, and then invaded the frost zone within each continent. This scenario suggests that there may have been "preadaptation" among certain subtropical groups for frost tolerance when global cooling near the end of the Eocene expanded the area of temperate climate in the north. Alternatively, the temperate flora may have invaded the frost zone at relatively few times and places and then dispersed and diversified globally, mainly within the frost zone. This hypothesis receives support from the simulated rarefaction of the Paleogene fossil tree genera of eastern Asia to match the numbers of fossil tree genera found in the other three regions, which showed no significant differences between the smaller fossil floras and random samples of the Asian fossil flora.

Phylogenetic analysis would provide a test of these

hypotheses and might offer clues about the relative ease or difficulty of evolving frost tolerance. Such a test would focus on genera with species inhabiting temperate areas and species inhabiting frost-free areas on two or more continents, or families with genera similarly distributed. The aim would be to determine which are more closely related: frost-tolerant and frost-sensitive sister species or genera from a single continent (consistent with taxa dispersing globally across the subtropics, then invading the frost zone) or frost-tolerant sister species or genera from different continents (consistent with taxa invading the temperate zone, then dispersing globally across it).

Our hypothesis concerning the historical development of temperate tree floras can be applied more widely whenever a major physiological barrier must be crossed. Farrell, Mitter, and Futuyma (1992) make a similar case for the lower diversity of insects in temperate zones relative to the tropics. Certainly the invasion of mangrove habitat by angiosperms requires a comparable evolution of new physiological capabilities (Tomlinson 1986). The hypothesis is consistent with the distribution of diversity among habitat types within temperate biomes as well. If habitat shifts require any level of physiological adaptation, then the highest diversity will likely occur in the habitat where species invaded the temperate zone. This is likely to be the oldest habitat occupied and the closest to the habitat type of the external source of colonists. In the case of trees, if colonization of temperate habitats occurred from the wet tropics or seasonally moist subtropics, we might expect diversity in temperate biomes to be highest in those habitats with warmer and more moist conditions. Productivity in such habitats ranks among the highest in the temperate zone. Thus, the observed correlation between species diversity and energy flux in temperate biomes could as well represent the historical origins of the biota as it could variation in the outcome of local interactions under different physical conditions.

CONCLUSIONS

The diversity anomaly in temperate forest tree species between east-central Asia and other regions of the Northern Hemisphere appears to be ancient and to have arisen from differences between the regions in colonization history and perhaps in subsequent rates of proliferation of endemic taxa. The fossil record also supports the old hypothesis that the low diversity of temperate tree species in Europe and North America's Pacific slope (but not the intermediate diversity of tree species in eastern North America) resulted from extinctions during the Neogene period of cooling climate and glaciation. These extinctions were nonrandom, being centered on old, relictual taxa of gymnosperms and old, primarily tropical families of angiosperms.

Geographical distributions and the fossil record suggest that most cosmopolitan taxa of temperate trees originated in eastern Asia and dispersed to Europe and North America, with conspicuous exceptions in the Juglandaceae and, probably, the Fagaceae. Additional temperate taxa appeared in Asia after dispersal routes to other temperate regions were largely closed off, giving rise to a large

number of endemic temperate taxa there, many with tropical affinities.

This pattern of colonization of temperate regions suggests that the disjunction in diversity between temperate and tropical tree species may have arisen in part due to physiological constraints on crossing the freezing tolerance barrier. Thus, diversity patterns may have significant evolutionary as well as biogeographical and ecological bases.

Further resolution of the causes of diversity patterns will require new paleontological, biogeographical, and taxonomic data and synthesis. It must also be based on increased understanding of the physiological basis for the relationship of species' distributions to the physical environment.

APPENDIX 26.1
REGIONAL DISTRIBUTION OF TREE SPECIES INHABITING NORTHERN HEMISPHERE MOIST TEMPERATE FORESTS

Tree is defined as a self-standing woody perennial that reaches a maximum height of eight meters or more. The four regions are the Northern Hemisphere warm-temperate humid and temperate-nemoral climate biomes of Walter (1979). Taxonomy and range information are from Bailey and Bailey (1976), Braun (1950), Camus (1936–1938), Chi'ên (1921), Cronquist (1981), Elias (1980), Kartesz and Kartesz (1980), Krüssman (1979, 1984), Li (1935, 1973), Meyen (1987), Mirov (1967), Mitchell (1974), Ohwi (1965), Petrides (1972), Rehder (1940), Uphof (1968), Walter (1979), Wang (1961), and Zheng (1983, 1985). Taxonomy follows the most recent source for genera and species and Cronquist (1981) and Meyen (1987) for families and higher taxa of angiosperms and gymnosperms, respectively. Sources of fossil data are: Europe: Reid and Chandler (1933), Mai (1960), Kilpper (1969), van der Hammen, Wijmstra, and Zagwijn (1971), Mai (1971a, 1971b), Takhtajan (1974), Łańcucka-Środoniowa (1975), Collinson and Crane (1978), Friis (1979), Mai (1980, 1981), Gregor (1982), Friis (1985), Kvaček and Walther (1987), Mai (1987a, 1987b), Mai and Walther (1988), Sauer (1988), Kvaček, Walther, and Bužek (1989), and Friis (personal communication); eastern Asia: Hu and Chaney (1940), Tanai (1972), Takhtajan (1974), and Guo (1990); western North America: Wing and Hickey (1984), Sauer (1988), and Manchester (unpublished data); eastern North America: Traverse (1955), Rachele (1976), Potter and Dilcher (1980), Frederiksen (1984a, 1984b), and McCartan et al. (1990).

Northern tree species with pan-continental distributions and southern tree species with more than 50% of their ranges extending into subtropical or mediterranean areas were omitted from the list, even if they inhabit substantial fractions of the moist temperate forest regions. Omitted species that live in moist temperate forests only at high latitudes or high elevations often are more widespread in the biome poleward from the moist temperate forest zone which, in the Northern Hemisphere, also extends much farther east or west across the continent; examples include *Betula papyrifera* Marsh. (paper birch),

Picea mariana (Mill.) B.S.P. (black spruce), and _Populus tremuloides_ Michx. (quaking aspen) in North America and _Betula pubescens_ Ehrh. (downy birch), _Picea abies_ (L.) Karst. (Norway spruce), and _Populus tremula_ L. (aspen) in Eurasia. Omitted species that grow in moist temperate forests only at low latitudes, often on protected sites, represent incursions from the moist subtropics or from the mediterranean winter rain biome. Many subtropical tree species occur in scattered locations along the southern fringes of the moist temperate forest zone in China, including species of _Mangletia_ and _Michelia_ (Magnoliaceae), _Actinodaphne_, _Cinnamomum_, _Lindera_, _Litsea_, _Machilus_, _Neolitsea_, and _Phoebe_ (Lauraceae), _Cyclobalanopsis_, _Lithocarpus_, and _Quercus_ (Fagaceae), _Elaeocarpus_ and _Sloanea_ (Elaeocarpaceae), and numerous other genera (Wang 1961). Examples in the eastern North American moist temperate forest zone—far rarer than in eastern Asia—include _Pinus clausa_ (Chapm.) Vasey (sand pine) and _Quercus chapmanii_ Sarg. (Chapman oak).

Also omitted from the list were some cold-weather deciduous tree species in eastern Asia that occur mainly in the mountains in the subtropics and more sparsely northward into the temperate forest. Examples include _Bretschneidera sinensis_ Hemsl. (of the monotypic Bretschneideraceae), which occurs at elevation 800 to 1,500 m in China's Guizhou, Yunnan, and Hunan provinces (Li 1935); _Cathaya argyrophylla_ Chun & Kuang (Pinaceae), from elevation 920 to 1,800 m in Guangxi, Sichuan, Hunan, and Guizhou (Zheng 1983); and _Rhoiptelea chiliantha_ Diels & Hand.-Mazz. (of the monotypic Rhoipteleaceae), from elevation 500 to 1,400 m in Yunnan, Guangxi, Guizhou, and south beyond the borders of China (Li 1935). A large number of cold-hardy deciduous and evergreen species with ranges mainly in the mountains of China's western Sichuan and Yunnan provinces and the eastern Himalayas also were omitted, including scores of tree-size _Rhododendron_ species.

Tree species that commonly dominate the moist temperate forest canopy on North America's Pacific slope were included on the list even if their ranges lie mainly outside the moist temperate forest region. Examples include _Pseudotsuga menziesii_ (Mirbel) Franco (Douglas-fir), _Abies grandis_ (Dougl. es D.Don) Lindl. (grand fir), and _Populus balsamifera_ L. subsp. _trichocarpa_ (Torr. & Gray) Brayshaw (black cottonwood). Dominance of communities across several biomes is common among species inhabiting this smallest of the north temperate forest regions, as one might expect from the so-called "mass effect" of spillover among habitats and the larger ratio of this region's perimeter to its total area.

The filters applied to the total floras to derive our regional tree species lists result in underestimation of the true tree species richness in all four regions. However, we believe that our lists for Europe, the Pacific slope of North America, and eastern North America closely approximate the total numbers of native tree species actually present in those regions. Our list for east-central Asia, by contrast, substantially underestimates the total number of native tree species actually present due to spillover from subtropical, tropical, and montane forests along the southern and southwestern margins of the region.

| Class /subclass | Order | Family | Genus | Number of tree species | | | |
				Northern, central, & eastern Europe	East-central Asia	Pacific slope of North America	Eastern North America
Ginkgoopsida	Ginkgoales	_Ginkgoaceae_*[N]	_Ginkgo_*[N]	†	†	1	†
Pinopsida	Pinales	_Cephalotaxaceae_[N]	_Cephalotaxus_[N]	†	3		
		Cupressaceae[N]	_Calocedrus_	†	(P)	† 1	
			Chamaecyparis[N]	†	2	† 2	1
			Cupressus		1	2	
			Juniperus	3	† 4	† 1	3
			_Platycladus_*[N]		1		
			Thuja[N]	†	† 3	† 1	1
			_Thujopsis_*[N]		† 1		
		Pinaceae[N]	_Abies_[N]	† 4	14	† 6	† 2
			Keteleeria[E]	†	1	†	
			Larix[N]	† 1	† 5		† (P)
			Picea[N]	† 2	† 9	† 2	† 1
			Pinus	† 6	† 14	† 7	† 12
			_Pseudolarix_*[N]	†	† 1		
			Pseudotsuga[N]	†	† 3	† 1	
			Tsuga[N]	†	† 7	† 2	† 2
		Taxaceae[N]	_Taxus_[N]	† 1	† 3	† 1	(P)
			Torreya[N]	†	4	† 1	1
		Taxodiaceae[N]	_Cryptomeria_*[N]	†	1		
			_Cunninghamia_*[N]	†	† 1		
			_Glyptostrobus_*[E]	†	† (P)	†	†
			_Metasequoia_*[N]	†	† 1	†	

Codes: E (Equatorial), family or genus is predominantly tropical in distribution; N (Nemoral), family or genus is predominantly temperate in distribution or extends into tropical latitudes mainly at high elevations; *, globally monotypic or ditypic family or genus; †, genus is represented in the Tertiary fossil record of the region; (P), geographical range of the genus is peripheral to the region and/or the maximum height of the largest species in the genus occurring in the region is less than 8 m (given only for those genera for which Tertiary fossil information is included); (?), fossil genera identified only tentatively by recent authorities (cited above).

(continued)

				Number of tree species							
				Northern, central, & eastern Europe		East-central Asia		Pacific slope of North America		Eastern North America	
Class /subclass	Order	Family	Genus								
			Sciadopitys*^N	†			1			†	
			Sequoia*^N	†		†		†	1		
			Sequoiadendron*^N					†	1		
			Taiwania*^N			†	1				
			Taxodium*^N	†		†		†		†	2
Magnoliopsida /Magnoliidae	Illiciales	Illiciaceae	Illicium^E	†			3				
	Laurales	Lauraceae^E	Actinodaphne^E			†	3				
			Cinnamomum^E	†		†	11				
			Lindera^E	†		†	8	†			(P)
			Litsea^E	†		†	9				(P)
			Machilus^E			†	10				
			Neolitsea^E	†		†	5				
			Nothaphoebe^E				1				
			Persea^E	†				†			1
			Phoebe^E	†		†	11				
			Sassafras^N	†		†	1	†			1
			Umbellularia*^N					†	1		
	Magnoliales	Annonaceae^N	Asimina^N	†							1
		Magnoliaceae	Liriodendron*^N	†		†	1	†		†	1
			Magnolia	†		†	14	†			7
			Manglietia^E	†			4				
			Michelia^E	†			3				
	Ranunculales	Sabiaceae^E	Meliosma^E			†	5				
Magnoliopsida /Hamamelidae	Daphniphyllales	Daphniphyllaceae	Daphniphyllum^E				2				
	Eucommiales	Eucommiaceae*^N	Eucommia*^N	†		†	1	†		†(?)	
	Fagales	Betulaceae^N	Alnus^N	†	3	†	12	†	1	†	2
			Betula^N	†	3	†	14	†	(P)	†	5
			Carpinus^N	†	1	†	14	†		†	1
			Corylus^N	†	3	†	3			†	(P)
			Ostrya^N	†	1	†	4	†		†	1
		Fagaceae^N	Castanea^N	†	1	†	4	†		†	2
			Castanopsis^E	†		†	7	†	1		
			Cyclobalanopsis^E	†		†	4				
			Fagus^N	†	3	†	6	†		†	1
			Lithocarpus^E	†		†	5	†	1		
			Quercus^N	†	11	†	21	†	5	†	32
	Hamamelidales	Cercidiphyllaceae*^N	Cercidiphyllum*^N	†		†	1	†			
		Eupteleaceae^N	Euptelea^N	†			2				
		Hamamelidaceae^N	Altingia^E				1				
			Disanthus*^N	†		†	1				
			Distylium^N	†		†	1				
			Fortuneria*^N	†		†	1				
			Hamamelis^N	†		†	2			†	1
			Liquidambar^N	†		†	2	†		†	1
			Loropetalum*^N				1				
			Sinowilsonia*^N				1			†	
		Platanaceae	Platanus^N	†	(P)	†	(P)	†	1	†	1
	Juglandales	Juglandaceae^N	Carya^N	†		†	1	†		†	13
			Cyclocarya^N	†		†	1				
			Engelhardtia^E	†			(P)			†	
			Juglans	†	1	†	6	†	1	†	2
			Platycarya*^N	†		†	1	†		†	
			Pterocarya^N	†	1	†	6	†			
	Leitneriales	Leitneriaceae*^N	Leitneria*^N	†		†					1
	Myricales	Myricaceae	Myrica	†	(P)		1		1	†	1
	Trochodendrales	Tetracentraceae*^N	Tetracentron*^N	†(?)		†	1				
	Urticales	Moraceae^E	Broussonetia	†		†	1				
			Cudrania^E				1				
			Maclura*^N								1
			Morus	†	(P)		7			†	1
		Ulmaceae	Aphananthe	†			1	†			
			Celtis^N	†	3	†	8	†	(P)	†	4
			Hemiptelea*^N	†			1				
			Planera*^N			†				†	1
			Pteroceltis*^N	†		†	1				
			Ulmus^N	†	6	†	8	†		†	6
			Zelkova^N	†	2	†	2	†		†(?)	
		Urticaceae^E	Villebrunea^E				1				

(continued)

Class /subclass	Order	Family	Genus	Northern, central, & eastern Europe		East-central Asia		Pacific slope of North America		Eastern North America	
Magnoliopsida /Dilleniidae	Ebenales	Ebenaceae^E	Diospyros^E	†		†	3	†			1
		Sapotaceae^E	Bumelia^E					†(?)			3
		Styracaceae	Halesia^N	†							3
			Pterostyrax^N				2				
			Styrax^E	†	(P)	†	7	†	(P)	†	1
		Symplocaceae^E	Symplocos^E	†		†	2	†		†	1
	Ericales	Clethraceae^E	Clethra	†		†	1			†	(P)
		Cyrillaceae^E	Cliftonia*^N								1
			Cyrilla*^N	†						†	1
		Ericaceae	Arbutus		1			†	1		
			Elliottia*^N								1
			Enkianthus^N	†(?)			1				
			Kalmia^N	†		†					1
			Lyonia	†		†	(P)			†	1
			Oxydendrum*^N							†	1
			Rhododendron^N	†	(P)	†	6	†	1	†	2
			Vaccinium^N		(P)		(P)		(P)	†	1
	Malvales	Elaeocarpaceae^E	Sloanea^E				1				
		Sterculiaceae^E	Firmiana^E			†	1				
		Tiliaceae^E	Tilia^N	†	5	†	14	†		†	3
	Primulales	Myrsinaceae^E	Ardisia^E				1				
			Myrsine^E				1				
	Salicales	Salicaceae^N	Populus^N	†	2	†	14	†	1	†	3
			Salix^N	†	4	†	23	†	8	†	8
	Theales	Theaceae^E	Camellia^E	†		†	2				
			Franklinia*^N								1
			Gordonia^E	†		†	(P)			†	1
			Stewartia^N	†		†	4				1
			Ternstroemia^E	†		†	1				
	Violales	Flacourtiaceae^E	Idesia*^N			†	1				
			Poliothyrsis*^N	†		†	1				
			Xylosma^E			†	1				
Magnoliopsida /Rosidae	Apiales	Araliaceae	Acanthopanax	†			2				
			Aralia	†		†	2				1
			Dendropanax^E				2			†	
			Evodiopanax*^N				1				
			Kalopanax*^N			†	1				
			Schefflera^E	†			1				
	Celastrales	Aquifoliaceae	Ilex	†	1	†	14			†	5
		Celastraceae^E	Euonymus		1	†	6			†	1
	Cornales	Alangiaceae^E	Alangium^E	†		†	2	†		†	
		Cornaceae^N	Aucuba^N				1				
			Cornus^N	†	1	†	7	†	1	†	3
			Macrocarpium^N				2				
			Torricellia				1				
		Nyssaceae^N	Davidia*^N				1				
			Nyssa	†		†	1	†		†	3
	Euphorbiales	Buxaceae	Buxus	†	2	†	(P)				
		Euphorbiaceae^E	Mallotus^E	†		†	3	†(?)			
			Sapium^E	†		†	1				
	Fabales	Caesalpiniaceae^E	Cercis^N		1	†	3				1
			Gleditsia			†	5			†	2
			Glymnocladus^N			†	1				1
		Fabaceae	Cladrastis^N			†	5				1
			Dalbergia^E				1				
			Erythrina^E								1
			Laburnum*^N		2						
			Maackia^N			†	2				
			Ormosia^E				1				
			Robinia	†		†					2
			Sophora			†	1			†	(P)
		Mimosaceae^E	Albizia^E			†	3	†			
	Myrtales	Lythraceae^E	Lagerstroemia^E	†			2				
		Myrtaceae^E	Syzygium^E			†	1				
	Rhamnales	Rhamnaceae	Hovenia^N			†	2				
			Rhamnus^N		(P)		2	†	2	†	(P)
			Ziziphus^E		1	†	1				
	Rosales	Hydrangeaceae^N	Hydrangea	†		†	1				(P)

(continued)

Class /subclass	Order	Family	Genus	Northern, central, & eastern Europe	East-central Asia	Pacific slope of North America	Eastern North America
		Rosaceae[N]	Amelanchier[N]	1	1	1	2
			Chaenomeles[N]		1		
			Crataegus[N]	† 2	† 2	† 1	18
			Eriobotrya[E]		† 1		
			Malus[N]	3	† 15	1	5
			Mespilus*[N]	1			
			Photinia		3		
			Prunus[N]	† 6	† 32	† 2	† 10
			Pyrus[N]	4	10		
			Sorbus[N]	† 8	† 15	1	2
	Santalales	Olacaceae[E]	Schoepfia		1		
	Sapindales	Aceraceae[N]	Acer[N]	† 12	† 58	† 3	† 9
			Dipteronia*[N]		1		
		Anacardiaceae[E]	Choerospondias*[N]		1		
			Cotinus[N]				1
			Pistacia[E]	1	† 1		
			Rhus	† (P)	† 4	†(?)	† 3
			Toxicodendron[N]		2		† 1
		Hippocastanaceae	Aesculus[N]	† 1	† 3	† 1	4
		Meliaceae[E]	Cedrela[E]	†	† 1	†	
		Rutaceae[E]	Evodia[E]	†	† 5		
			Phellodendron	†	† 5		
			Ptelea			(P)	† 1
			Zanthoxylum[E]	†	† 3		2
		Sapindaceae[E]	Koelreuteria	†	† 2		
			Sapindus[E]	†	† 1	†	2
		Simaroubaceae[E]	Ailanthus[E]	†	† 3	†	
			Picrasma[E]		1		
		Staphylaceae	Staphylea[N]	†	1		1
			Tapiscia*[N]	†	1	†	
			Turpinia[E]	†	1		
Magnoliopsida /Asteridae	Dipsacales	Caprifoliaceae[N]	Sambucus	† 1	2	1	1
			Viburnum	† (P)	† 2		4
	Lamiales	Boraginaceae	Ehretia[E]		† 3		
		Verbenaceae[E]	Clerodendrum[E]	†(?)	1		
			Premna[E]		1		
	Rubiales	Rubiaceae[E]	Adina[E]		1		
			Cephalanthus	†			1
			Emmenopterys*[N]		1	†	
			Pinckneya*[N]				1
			Randia[E]		1		
	Scrophulariales	Bignoniaceae[E]	Catalpa	†	† 3	†	2
			Paulownia	†	4	†	
		Oleaceae	Chionanthus[N]	†	† 1		1
			Forestiera[E]				1
			Fraxinus[N]	† 6	† 14	† 1	† 7
			Ligustrum	(P)	† 1		
			Osmanthus[E]	†	† 4		1
			Syringa[N]		1		
Liliopsida /Arecidae	Arecales	Arecaceae[E]	Sabal[E]	†		†	1
			Serenoa*[N]	†			1
			Trachycarpus[E]		2		
Liliopsida /Commelinidae	Cyperales	Poaceae	Arundinaria		3		1
			Phyllostachys[N]		12		
			Semiarundinaria		1		
Liliopsida /Liliidae	Liliales	Agavaceae	Yucca[E]				1

27

The Phanerozoic Evolution of Reef Diversity

Erle G. Kauffman and J. A. Fagerstrom

DIVERSITY OF REEF COMMUNITIES

Biotic reefs are among the best preserved ancient ecosystems and thus can be compared with modern analogues to study the evolution of diversity within a single set of ecologically related communities. Ecologically simple stromatolitic reefs span the last 3.2 billion years of Earth history. Complexly structured reef ecosystems, characterized by high species and community diversity and extensive symbioses, are continuously represented in strata spanning the last 650 million years (my), apparently supporting time-stability evolutionary hypotheses. But the geological evolution of reefs was highly dynamic, and tropical/subtropical regions were frequently perturbed by large-scale changes in climate, oceanography, geology, and plate tectonic configuration that dramatically influenced the evolution and extinction of reef communities. Global mass extinctions in particular have set back the evolution of reef diversity for 3–10 my intervals, every 26–30 my during the last 250 my, and episodically before then. Modern reef ecosystems are a product of only the last 45–50 my of evolution, and even during this interval have experienced major ecological crises.

The high diversity of modern and fossil reef communities is so well established that it has been used as a standard for comparison with diversity in other communities. Yet the higher taxa characterizing reefs, their functional roles and relative importance in reef building, and overall diversity in reef communities have varied enormously through geological history.

Earlier in geological time, reef communities were ecologically and numerically dominated at various times by cyanobacteria, algae, archeocyathids, diverse spongiomorphs, primitive tabulate and rugosan corals, early scleractinian corals, bryozoans, brachiopods, and bivalves (especially rudistids). Changing biological composition within reef ecosystems through time was accompanied by dramatic changes in diversity of reef biotas within tropical-subtropical environments. These fluctuations reflect broad variations in global climates and marine systems, in evolution and competition, and the devastating effects of mass extinctions on reef communities. Nevertheless, various reef-building taxa have maintained similar characteristics, i.e., evolutionarily convergent skeletal form and rigidity, positive relief, rapid growth rates leading to large size, photosymbioses, etc., for about two billion years (Fagerstrom, 1987).

Diversity of reef communities can be described in at least five ways: (1) species diversity or richness, (2) number of successional stages, (3) number of ecological units (subcommunities or ecological zones), (4) number of guilds, and (5) species diversity within subcommunities, zones (Goreau 1959; Goreau and Goreau 1973), and guilds (Fagerstrom 1987). Measures 1 and 2 are the most common and are the only ones considered here. Most of our ancient examples reflect diversity among well-skeletonized taxa in photic-zone, non-cryptic communities, which exceed diversity levels found in other types of reef communities.

Because no comprehensive theory exists to explain why diversity is so variable among different communities in space and time, species diversity is of great interest to both ecologists and paleoecologists. Reviews of species and guild diversity in marine reef and non-reef communities through the Phanerozoic include those by Valentine (1970, 1971), Bambach (1977, 1983), Signor (1985), and Fagerstrom (1987). Comprehensive analyses of modern reef diversity are rare. An early estimate of species diversity in modern reef communities (i.e., 3,000 spp; Protozoa through fishes) was that of Wells (1957). Species diversity, excluding Mollusca, Crustacea, and Echinoidea, on a small Carribbean patch reef (Carrie Bow Cay, Belize) is 664 species (see fig. 27.5; Rützler and Macintyre 1982, 153–416). By contrast, there are nearly 1,300 benthic species in the Capricorn and Bunker Groups, Great Barrier Reef, Australia (Mather and Bennett 1984) and 2,300 benthic species in the reef communities of French Polynesia (Richard 1985). Planktic and nektic species have been omitted from these estimates because they migrate among both reef and open ocean communities. The above discrepancies in diversity probably reflect variable sample sizes, habitat diversity, the species-area effect, and/or biogeographical history, all of uncertain relative importance.

In addition to their high diversity, reef communities are also characterized by intense spatial competition, and high ecological/trophic complexity including symbioses among algae, corals, protozoans, and sponges, and commensalism among corals, crustaceans, and fishes (Fagerstrom 1987). Thus, high diversity and ecological complexity appear to be interrelated within reef communities, especially when compared with other benthic community types.

Historical Determinants of Quaternary Reef Diversity

Complexly structured reef ecosystems are delicately perched in tropical environments. They are frequently perturbed by physical, thermal, chemical, and biological disturbance events causing drastic reductions in diversity, with recovery times of decades to centuries. Such disturbances have local to regional effects and do not necessarily lead to global species extinction or to significant changes in reef community composition through geological time. Recovery from physical disturbance is more rapid (with repopulation mainly from propagule fragments) than recovery from chemical or biological disturbance (with repopulation mainly by larval recruitment). Correlations between the frequency of disturbance events and diversity increases in coral reefs and rainforests support the "Intermediate Disturbance Hypothesis" of Connell (1978), which may explain high local species diversity during ecological time.

Grassle (1973) explained high reef species diversity as the result of high trophic specialization, reflecting the general stability of reef environments and absence of stochastic disturbance events. Even during the Late Pleistocene and Holocene, the tropics were characterized by stability and predictability of factors controlling reef growth, such as temperature, water chemistry, substrate, and trophic resources. The most important exception was the rapid rise in sea level (5–10 mm/yr between 18,000 and 6,000 B.P.: Fagerstrom 1987, 74) that drowned many deeper-water reefs having upward accretion rates less than the rate of sea level rise. However, for shallower reefs that kept pace with sea level rise, high reef diversity was maintained, or perhaps even increased, which appears to support the "Time-Stability Hypothesis" of Sanders (1969) rather than the Intermediate Disturbance Hypothesis. Previous interpretations of historical and geographical variations in reef diversity have commonly utilized different ways of measuring species diversity as well as geologically short-term variation of presumed causal factors for diversity differences.

Geographical Determinants of Quarternary Reef Diversity

With both increased distance from the equator and greater water depth, there is a progressive decrease in species diversity among reef communities (Stehli and Wells 1971; Wells 1955, 1957; Rosen 1977). Variations within these general patterns may reflect differences in habitat diversity and temporal stability. Meaningful comparisons of geographical diversity differences must consider problems of: (1) the interrelations among diversity, relative species abundance, and the size of the sampled area, i.e., the species-area effect; (2) the nature, variety, longevity, and abundance of suitable reef habitats; and (3) diversity differences between biogeographical provinces reflecting their long-term evolutionary histories.

The species-area effect is illustrated by the data for Recent reefs of the Pacific and Caribbean (see fig. 27.5). The Pacific data (1,268 total species; 850 shelled species) are from Heron Island, Great Barrier Reef, Australia, which

has an area of 36 km², whereas those for Carrie Bow Cay in the Caribbean (664 total species; 200 shelled species: see fig. 27.5) are from a 0.065 km² transect. The species-area effect is also shown by comparison of only the coral species diversity data from the following four well-studied reef locations (indicative of alpha diversity for locations 1, 2, and 3, and gamma diversity from location 4). (1) At Carrie Bow Cay, Rützler and Macintyre (1982, 13–36) identified 23 coral species on patch reefs 1–30 m deep. (2) At Ocho Rios, Jamaica, Goreau (1959, 72–80) identified 40 coral species in an area of approximately 0.142 km² on a reef flat 0–15 m deep. (3) At Heron Island, Endean (1982, 223) reported 107 coral species in 36 km². (4) For the lagoonal and shallow seaward slope reefs of French Polynesia covering an area of about 7,500 km², Pichon (1985) listed 168 coral species.

Reef substrate cover by live corals may reach 60%–70% (Laxton and Stablum 1974, 7), and only a few coral species with large colonies provide most of the coverage. Thus, most of the species diversity of reef communities is among rare, small, symbiotic/commensal (sensu Gotto 1969), endolithic, epibiont and free-living organisms (Endean and Cameron 1990) located in cryptic reef framework habitats and adjacent sedimentary (level-bottom) substrates. For example, in the Capricorn and Bunker Groups there are about 175 gastropod, 45 bivalve, and 169 coral species (Mather and Bennett 1984), and in French Polynesia there are 1,004 gastropod and 118 bivalve species (Richard 1985).

Historical biogeography also affects reef diversity. Species diversity in the geologically older and larger Indo-Pacific Province is much greater for most higher taxa than in the younger and smaller Atlantic Province (Newell 1971, 25–26; Fagerstrom 1987, 77–81). Within the Indo-Pacific Province, reef diversity is highest around Indonesia and the Malay Peninsula (Indo–West Pacific and Malay Regions/Subregions; Ekman 1953). Whereas convergence of increasing species diversity gradients among several higher taxa toward one geographical area (Ekman 1953; Stehli and Wells 1971) has classically been cited as evidence for a center of origin, Ekman (1953, 78–79) ruled out the Center Of Origin Hypothesis for the Malay Region. In the Caribbean, the Carrie Bow Cay (see fig. 27.5; Recent Caribbean) and Ocho Rios reefs are in the lower-diversity Caribbean Province, whereas the Great Barrier Reef (see fig. 27.5; Recent Pacific) and French Polynesia are in the higher-diversity Indo-Pacific Province, suggesting that biogeographical history (e.g., the longer evolutionary and biogeographical history of the Indo-Pacific) may be a major factor in explaining these diversity differences. Conversely, higher diversity in the Indo-Pacific may also be due to its greater area and habitat diversity. We conclude that diversity in modern reef communities is regulated by several interrelated environmental factors.

Diversity in Fossil Reef Communities

The composition and structure of fossil reef communities is generally better preserved than that of coexisting level-bottom or pelagic communities. Thus, there is a rich and relatively continuous record of reef ecosystems and reef-associated facies in the Phanerozoic strata (figs. 27.1A

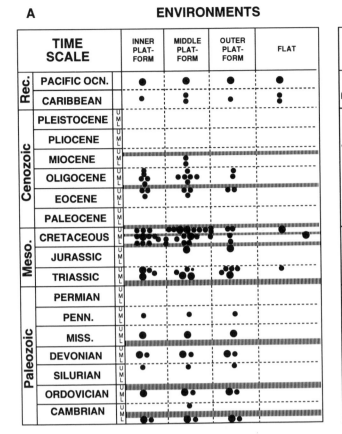

Figure 27.1 The data base for this study is shown by the stratigraphic (temporal) and habitat distributions of well-known fossil reefs through time along an onshore (left) to offshore (right) transect. Reef occurrences are represented by black dots keyed in size to the taxonomic diversity (completeness) and quality of the original data for each citation. Only selected high-quality data points were used in making final compilations (e.g., Figs. 27.5, 27.6); the largest dots represent the best data. The patterned horizontal lines indicate major mass extinction events. (The authors will provide a list of references used in assembling the data base upon written request.)

and B); this record constitutes the data base for the present study. In reefs with abundant binding organisms (e.g., laterally expanded algae, spongiomorphs, Bryozoa, and corals), the skeletonized elements of ancient communities may be essentially frozen in time and space. However, there are fundamental differences between modern and fossil reef communities regarding their preservation and the methods of data gathering and analysis used that make them difficult to compare.

Determination of the postmortem, preburial, and early postburial histories of dead skeletons (i.e., taphonomy; Bates and Jackson 1980, 638) is of great importance in the paleoecological reconstruction of fossil communities. Only skeletons in growth position or toppled in situ and their associated epibionts and endobionts are used to reconstruct fossil communities; these skeletons must be distinguished from those of exotic taxa that have undergone postmortem current transport to the final burial site. In fossil reef communities, such skeletal transportation and accompanying fragmentation is reduced by the binding and mutual support provided to both live and dead skeletons by encrusting organisms. Entrapment of resident and introduced bioclastic sediment fragments is enhanced by baffling in loosely constructed reefs having low framework/internal sediment volumetric ratios.

Taphonomic processes are followed by diagenetic processes (i.e., postburial compaction, alteration, replacement, and dissolution of skeletons) that further reduce the preservation potential of reef communities. The preservation potential of soft-bodied organisms on reefs is very low, whereas it is very high for thick, calcitic invertebrate shells; preservation potentials of porous aragonite skeletons among many corals and other mollusks are intermediate.

Frost (1977, 101–102) estimated the percentage of skeletal organisms with moderate to high preservation potential in each of seven "benthic trophic and functional categories" for four "ecozones" in modern Caribbean reef communities (fig. 27.2). The highest preservation potential is for "frame hermatypes" (about 40%–70% are skeletal); the next highest is for "microherbivores" (about 20%–60%), and the lowest is for "tentacle feeders" (about 0%–20%). Gaillard (1983) estimated the amount of biomass lost by higher taxa and trophic categories during fossilization of Jurassic reefs in France. Here the greatest loss during transition from living to fossil reef communities was among algal producers, but virtually all of the consumers are represented as fossils.

Species diversity data in modern reefs are usually derived from a thin, subhorizontal surficial veneer that represents a single moment in time; comparably exposed time planes at outcrops of ancient reefs are rare. Diversity data

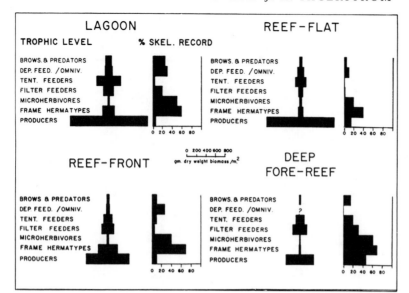

Figure 27.2 Percentage of skeletal organisms with moderate to high preservation potential in each of seven benthic trophic and functional categories, for four ecozones in modern Caribbean reef communities. Notice that the highest preservation potential is among frame hermatypes (40%–70%) and microherbivores (20%–70%). Left-side histograms record biomass for each trophic level; biomass scale is in center of figure. (From Frost 1977, 101–102; reprinted by permission.)

from ancient reefs are usually derived from a near-vertical outcrop surface consisting of an extended time dimension of variable duration and continuity. Because of the vagaries of orientation, size, shape (Nelson, Brown, and Brineman 1962), weathering, community composition, and location of collecting sites on both living and fossil reef surfaces, field data from originally similar reefs may differ significantly.

The same species-area effect, biogeographical province, and water temperature, and depth gradients that shape modern reefs presumably also affect diversity comparisons among fossil reef communities. Estimates of paleolatitudes, original water depths, chemistry, and temperature can be made using a great variety of geological, geochemical, and paleobiological techniques and used to infer the influence of these factors on diversity. Comparisons of Mesozoic species diversities along a gradient from the equatorial Tethys Sea toward the poles show the same general patterns of decreasing species and community diversity as those for modern reef communities, but due to warmer global temperatures, Mesozoic diversity gradients are less steep.

The one aspect of species diversity that probably is better preserved in fossil than in modern reef communities is change during long-term ecological succession. Recovery of reef communities from major disturbance takes decades to centuries; recovery from ancient mass extinctions may take up to eight million years. But ecological studies of reef succession rarely last more than three to five years (cf. Connell 1988). By contrast, nearly complete temporal successions spanning tens to thousands of years are common in fossil reef communities.

Despite these problems and biases, the diversity data sets for modern and ancient reef communities used herein (see figs. 27.1, 27.5) are broadly comparable, suggesting compensating or canceling effects for these varied data sets. This fact is especially apparent when comparing only shelled taxa or single higher taxa of common reef builders. For example, Recent coral diversity for Carrie Bow Cay in the Caribbean (23 species) is about average for coral and other reef builders over the entire Phanerozoic

(see fig. 27.5), even though the total number of shelled species at this site is considerably greater (reflecting greater access to the entire surface of modern reefs).

GEOLOGICAL FACTORS AFFECTING THE EVOLUTION OF REEF ECOSYSTEMS

The geological history of reefs—their composition, size, structure, longevity, and biogeography—is one of considerable temporal change (Newell 1971; Heckel 1974; Toomey 1981; Boucot 1983; Fagerstrom 1987; Talent 1988). Reef ecosystems have diversified and waned in response to major fluctuations in Earth environments. Further, there is a positive correlation between the diversity of reef communities and the size, diversity, abundance, and distribution of favorable reef habitats. Six important geological factors have caused dramatic changes in reef habitats through time: (1) global climatic change; (2) global sea level changes and their effects on climate and habitat dispersion; (3) ocean chemistry, including nutrient flux; (4) global and regional tectonics and its effects on the distribution of shallow marine platforms relative to climate zones; (5) rates and character of sedimentation; and (6) mass extinction events or intervals.

Global Climates

Long-term (millions of years; my) temperature changes are driven by a variety of forces acting on climate and atmospheric circulation patterns. Among these are: (1) plate tectonics and the arrangement of continents and oceans; (2) continental topography, especially linear mountain ranges; (3) glacial versus interglacial stages; (4) oceanic and atmospheric circulation patterns; (5) global sea levels; and (6) solar radiation and the Earth's albedo. Short-term (20–400 ky) changes in temperature are mainly related to Milankovitch climate cyclicity.

Major changes in reef ecosystems reflect long-term fluctuations of Earth environments, especially geologically rapid changes in global mean ocean temperatures, which may exceed 5° C over less than 1 my (Hay 1988). Because of the presumed stenothermal nature of many reef-building organisms, relatively rapid changes in climate may greatly reduce reef diversity and even cause widespread extinction (see Stanley 1984). Seventy-seven percent of low-diversity troughs in reef ecosystems are correlated with high and low Phanerozoic temperature extremes. The highest diversity and development of reef communities occurred during periods of moderately elevated eustatic sea level and moderate thermal regimes (see fig. 27.6A and G) accompanied by expansion and long-term stability of shallow-water habitats in tropical/subtropical climate zones. These relationships may be moderated, however, by other factors such as biological competition, as in the Jurassic and Cretaceous.

Fischer (1984) proposed that the history of global temperature change was divisible into two alternating climatic regimes, which he called icehouse and greenhouse supercycles (see fig. 27.6B). Greenhouse worlds were thought to be characterized by global warming, increased atmospheric CO_2, and high sea level (see fig. 27.6E, H, and G), with resulting expansion of shallow epicontinental and coastal seas in tropical zones and widespread reef development. Conversely, icehouse worlds were characterized by lowered CO_2, temperature, and sea level and by glaciation, all of which restricted the reef belt and thus tropical diversity. On a global scale, these broad supercycles generally correlate positively with expansion or contraction of reef habitats, but they do not directly correlate with diversity trends (see figs. 27.6, 27.7), perhaps because thermal and salinity thresholds for reef organisms were exceeded during various greenhouse intervals and because of the various biases noted in the introduction.

Sea Level

Throughout geological history sea level has fluctuated from more than 300 m above (mid-Cretaceous) to more than 130 m below (Wisconsin glacial stage) the present level. Vail et al. (1977) and Haq, Hardenbol, and Vail (1987) summarized both relative and eustatic sea level changes (see fig. 27.6G) and recognized fluctuations of several magnitudes (first- to seventh-order cycles) with varying durations (30–50 my to 20 ky) and causes, both tectonic and climatic (e.g., glacioeustasy). These changes have had profound effects on reef development in three ways. First, sea level highstands result in prolonged global warming, stabilization of equable maritime climates, and up to 25% expansion of the tropical zone. Second, highstands usually open up marine connections between formerly isolated areas, enhancing gene flow and the possibility for taxonomic diversification through immigration and niche partitioning. Such connections may also act as corridors for new oceanic circulation patterns, changing regional/global climates, further expanding reef habitats, and enhancing community diversity; low sea levels have the reverse effects. Third, highstands commonly occur during periods of global warming (e.g., Cretaceous-Eocene; see fig. 27.6G and E) when there is no permanent ice at the poles, resulting in a warm, maritime-dominated world and thus expansion of tropical and subtropical climate zones and of reef habitats. Climate models for these conditions predict the presence of an additional near-equatorial "supertropical" climate zone (fig. 27.3; Barron and Washington 1984; "Supertethys" of Kauffman and Johnson 1988) characterized by waters slightly warmer (2°–5° C) and more saline (> 38–40 ppt) than those in the central marine tropics today. This may have had varied effects on reef diversity. For example, emplacement of a supertropical climate zone during the middle and Late Cretaceous seemingly retarded the development of diverse tropical coral-algal–dominated reef communities, which today have narrowly defined upper thermal limits, and

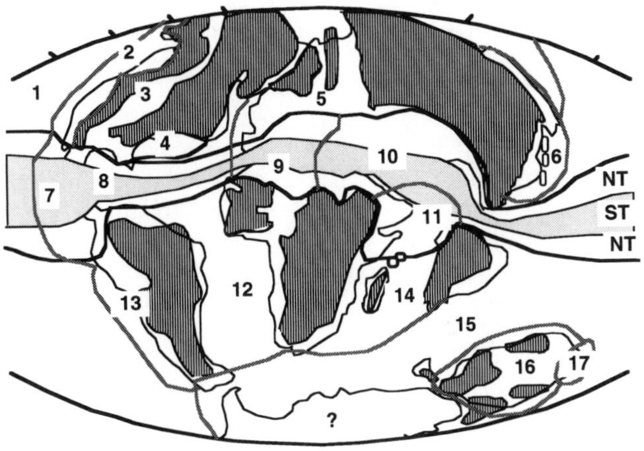

Figure 27.3 Paleogeography of the Cretaceous greenhouse interval, which was characterized by extensive reef building, though not necessarily by high diversity. This greenhouse world was characterized by: absence of permanent polar ice; high sea level with broadly flooded continental margins and interiors; warm equable moist climates; an expanded tropics (by 20%–25%) and subtropics (by 25%–30%) relative to modern climate zones; continental dispersion allowing development of a through-flowing circumglobal tropical sea (Tethys) and intermittent development of a warmer and more saline supertropical climate zone (shaded area: Supertethys of Kauffman and Johnson 1988). The Cretaceous is unusual in that the biological and climatic tropics lay predominantly north of the paleo-equator due to plate tectonic constraints. Bold black lines define the Cretaceous tropical boundaries. *NT*, normal tropics, as measured by modern reef-forming environments; *ST*, supertropics (Supertethys), which commonly had unique low-diversity reef biotas. Numbers encircled by gray lines refer to Cretaceous paleobiogeographical provinces and subprovinces (Kauffman, 1973). White areas on continental outlines are shallow warm epicontinental seas developed during eustatic highstand. (Plate reconstruction after Smith, Briden, and Drewry 1977, as modified by Kauffman and Johnson 1988.)

allowed their replacement by more high-temperature/hypersaline–adapted taxa, such as the rudistid bivalves. Even though rudistid diversity increased toward the core of Supertethys, overall, this change in reef communities resulted in lower diversity than had previously existed. Coral-algal communities continued to build reefs in normal tropical environments marginal to Supertethys. On the other hand, the ancient supertropical climate zone may have enhanced global diversity by acting as a barrier to north-south gene flow, giving rise to unique normal tropical biotas flanking it. Global cooling breaks down the supertropical zone, causing extinction of stenothermal taxa uniquely adapted to it and bringing the dissimilar northern and southern tropical biotas into direct competition. Despite the loss of some taxa, such climatic shifts could result in increased core tropical reef diversity by a kind of "diversity pumping" mechanism (Valentine 1968) associated with niche partitioning. Fourth, rising sea levels may drown carbonate platforms and reefs ("give-up" reefs; Neumann and Macintyre 1985; Schlager 1981). Reduced light intensity due to deeper reef submergence diminishes the efficiency of photosymbioses and leads to reef destruction (e.g., the Late Cenomanian-Early Turonian mass extinction event). Conversely, prolonged sea level fall may result in the draining of platform reef habitats, as in the Permian and latest Cretaceous, contributing to widespread mass extinction.

Ocean Chemistry

In modern reef habitats dominated by stenotopic taxa, rapidly induced and/or prolonged intervals of abnormal salinity produce major changes in composition, diversity, population, and guild structure among stenotopic taxa. The same ecological restrictions probably applied to ancient reef-building taxa, especially corals. During greenhouse intervals, there was an even greater potential for salinity fluctuations due to enhanced evaporation rates and monsoonal flooding compared with modern icehouse conditions. The hypersalinity of the Late Miocene (Messinian) Mediterranean Sea (Fagerstrom 1987, 234–235) and the Jurassic Gulf of Mexico (Scott 1984b) were both correlated with major reductions in reef diversity, as was Supertethys of Kauffman and Johnson (1988; see fig. 27.3).

Oxygen deficiency is another factor that may have lowered ancient reef diversity. During the Jurassic-Cretaceous greenhouse interval (see fig. 27.6B and I) deep to intermediate ocean circulation was largely driven by descending dense, hypersaline, tropical surface water and upwelling deep, nutrient-rich water near the poles and along continental shelves. Descending organic-rich surface waters used up much of the available oxygen, causing exceptional expansion of the oxygen minimum zone (OMZ) (e.g., Jenkyns 1980; Arthur, Schlanger, and Jenkyns 1987). During rapid sea level rise, the OMZ overstepped the continental platforms and epicontinental seas, causing widespread decimation of tropical reef biotas (e.g., Arthur and Schlanger 1979). For example, many of the Late Albian-Early Cenomanian (middle Cretaceous) platform and barrier reefs of eastern Mexico and Texas were probably killed abruptly by overstepping of the OMZ during

rapid sea level rise events (Scott 1984a, 1984b). No reef diversity peaks occur during peak oxygen depletion in our analysis; most diversity peaks are associated with fully oxygenated marine waters.

Nutrient supply and its regulation of trophic resources exercise a strong control on diversity in marine ecosystems (Valentine 1971, 1973; Hutchinson 1978; Dodd and Stanton 1981; Hallock et al. 1988). However, in the case of reefs dominated by photosymbiotic coral-algal communities, the relationship is inverse: high nutrient levels produce lower diversity (Hallock 1987; Hallock and Schlager 1986). Large-scale, long-term increases in nutrient concentrations across continental platforms are caused by inundation of nutrient-rich coastal areas with rising sea level and/or by changes in the location and intensity of coastal upwelling and of currents that redistribute upwelled nutrients across shallow platforms (Parrish 1982). Nutrient fluctuations in living reef ecosystems trigger changes in diversity, biomass, and productivity, and may be inferred in ancient reefs by proxies of those factors, such as diversity, population structure, skeleton and reef sizes, and by the amount of organic carbon stored in reefs and associated rocks. Modern reefs do not flourish in waters of excessive nutrient concentrations (e.g., Kaneohe Bay, Hawaii: Smith et al. 1981; see also Hallock et al. 1988) because of several factors. Hallock and Schlager (1986) have predicted that during periods of nutrient excess, plankton blooms will reduce light intensity enough to diminish the efficiency of the algal-coral symbiosis, causing reduced growth rates of corals. Furthermore, excess nutrients may produce an increase in competitive fleshy algal populations and in boring bivalves, annelids, and sponges. These then overgrow reefs and/or actually destroy reef frameworks, respectively (Hallock, 1988).

Plate Tectonics

Compilations of the Earth's plate tectonic history can be found in studies by Smith, Briden, and Drewry (1973), Smith, Hurley, and Briden (1981), Scotese and Sager (1988), Scotese and McKerrow (1990), and Ziegler, Scotese, and Barrett (1982), among others. During extended periods of Earth history characterized by rapid plate movement, dispersed continents, and topographical elevations of the seafloor, sea levels were relatively high, warm equable climates prevailed, and oceanic circulation commonly included through-flowing circumequatorial or subequatorial currents in latitudinally continuous tropical seas (e.g., fig. 27.3). Such circulation patterns, which characterized the early and middle Paleozoic and late Mesozoic–early Cenozoic, were also conducive to the widespread distribution of shallow reef habitats. Long intervals of continental aggregation, low sea levels, and colder, more seasonal climates, in some cases leading to continental glaciation, are characterized by diminished reef development and community diversity (see fig. 27.6). Continental dispersion associated with high sea level and the formation of shallow continental platforms and epicontinental seas marginal to circumequatorial oceans favor widespread development and diversification of reef communities. These factors increase the habitat area and heterogeneity available for reef builders sufficiently isolated from one another to develop endemic biotas, such as the Cretaceous eastern and western Indo-Mediterranean, Caribbean, and East Pacific Provinces (Kauffman 1973). The drift of continents from temperate to tropical locations also increased reef diversity by increasing tropical ecospace and providing new endemic centers. The northward drift of Australia during the late Mesozoic–Recent carried it from cool temperate into tropical climates by the Miocene, initiating development of the Great Barrier Reef (Davies 1988).

Volcanism associated with active plate spreading may cause short-term cooling and reef death by filtering solar radiation through suspended ash and/or, in the case of the Cretaceous Caribbean Province (Kauffman and Johnson 1988), by drowning of island arc and platform-associated reefs in volcaniclastic sediments.

Rates and Patterns of Sedimentation

Most reef-building organisms are highly sensitive to turbidity, reduced irradiance, which retards photosynthesis, and suffocation by rapid sediment accumulation on tissue surfaces. Extended periods of rapid sedimentation, due to increased erosion of adjacent land areas exposed by falling sea level or to global climate change that produces increased runoff, may broadly curtail or terminate reef growth by sediment drowning. Short-term storm-induced sedimentation interrupts but rarely terminates reef growth; recovery generally requires years to decades. Grain size also determines the effect of rapid sedimentation events on reefs. Most coarse sediment is carried along the adjacent substrate rather than across elevated reef surfaces and accumulates at rates slower than the upward growth rates of the reef builders. Influx of fine suspended sediment is more deleterious to reef growth because it can be deposited on the live reef surface, suffocating taxa such as sponges that lack sediment-shedding adaptations, and diminishing light penetration to coral-algal surfaces. Only taxa with well-developed sediment-shedding capabilities will survive long intervals of increased sedimentation rates over reef tracts.

Effects of Mass Extinction

Mass extinction events are characterized by loss of more than 50% of the world's species, representing ecologically and genetically diverse groups, within a short period of geological time (days/weeks to a few million years: Kauffman 1988). These events occurred periodically every 26–30 million years during the last 250 million years (Raup and Sepkoski 1984, 1986) and irregularly before then. Figure 27.4 shows the principal Phanerozoic mass extinction events as black horizontal lines. Kauffman (1988) recognized three patterns of mass extinction. Catastrophic mass extinctions are brief (days to thousands of years) global events resulting from extraordinary Earth perturbations such as large meteorite impacts and their secondary effects; an example is the Cretaceous-Tertiary mass extinction (Alvarez et al. 1980, 1984). These mass extinctions are not ecologically very selective. Stepwise mass extinctions are of intermediate duration (0.5–3 my) and consist of a succession of abrupt, ecologically graded extinctions, each usually less than 100,000 years' dura-

tion, between which extinction rates return to background levels. Stepwise mass extinctions may be caused by a variety of mechanisms (e.g., multiple extraterrestrial impacts, a series of abrupt temperature or chemical changes, multiple volcanic outgassing or oceanic trace element advection events; for examples see Kauffman 1988). Most involve widespread destruction of reef ecosystems early in the extinction interval. Graded mass extinctions are caused by acceleration of background extinction rates and their Earthbound environmental causes without compensatory increases in speciation. These extinctions are ecologically graded from (first) more stenotopic (e.g., reefs) to more eurytopic taxa; their duration is typically 1–5 million years or more. An example is the Permo-Triassic mass extinction (Teichert 1990). Reef ecosystems apparently were the first to be affected and most thoroughly destroyed in mass extinctions, and the last to recover. Reconstitution of shallow-water tropical reef ecosystems following mass extinctions characteristically took 1–8 million years, and newly recovered reefs were commonly dominated by very different taxa than those that formed pre-extinction reefs (fig. 27.4).

These factors result in a geological history of reef communities (see figs. 27.5, 27.6) characterized by intervals tens of millions of years long during which diversity gradually builds to a major evolutionary peak, alternating with shorter intervals, commonly following regional to global mass extinctions, in which reefs are absent or diversity is extremely low (see fig. 27.6; Boucot 1983; Sheehan 1985; Fagerstrom 1987). Most mass extinctions were multicausal and generally involved very rapid, large-scale shifts in geological factors (discussed above) within time frames that exceeded the evolutionary rates and adaptive ranges of the affected taxa, especially taxa with narrow adaptive ranges such as those found in tropical reef communities.

Figure 27.6A shows the generally good correlation between major Phanerozoic mass extinction events and sharp reductions in the species diversity of reef communities (see also Fagerstrom 1987, 444–456). However, there are some apparent exceptions: the peak rise in diversity during the Late Triassic actually precedes the mass extinction event, and many authors do not regard the Late Miocene extinction as a major event. Thus, regardless of the cause, tropical taxa, especially reef builders, are among the first to undergo mass extinction and the last to recover, suggesting that complex ecosystems cannot rapidly rebuild even though some reef-associated taxa may survive the extinction event. Reef communities with their complex biological interactions evolve over long time intervals; they are not inherited from surviving component lineages.

Mass extinction, recovery, and their associated environmental and biological changes have been the most powerful influences on the evolutionary history of reef communities, including species diversity. The repeated decimation and recovery of reef communities (Newell 1971; Boucot 1983; Sheehan 1985; Fagerstrom 1987) suggest that the high diversity of both modern and ancient reef communities is not the product of a geologically long and stable environmental history, but rather of shorter-

A

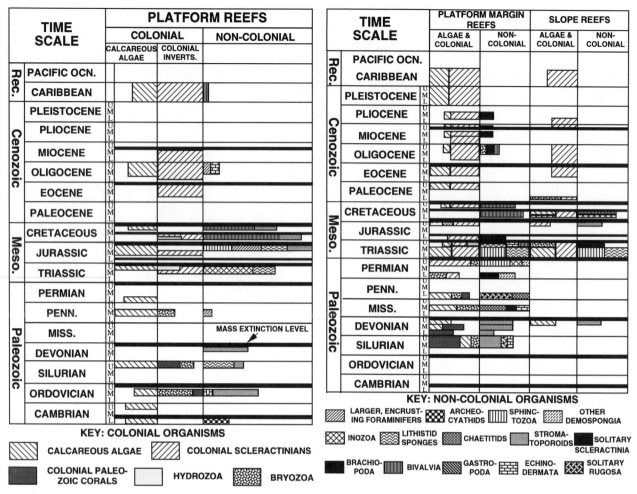

Figure 27.4 Dominant reef builders through time: temporal variation in the taxonomic composition of Phanerozoic reef communities based on the high-quality data points in figure 27.1; data are keyed to habitats and ecological zones across tropical platforms, platform margins, and slopes. The horizontal extent of each patterned bar represents the relative importance of each group in reef building for that interval of time; long bars represent the most dominant reef-building taxa. Keys to the major colonial and noncolonial taxa making up reef ecosystems through time are at the base of the figure. Note frequent changes in the composition of reefs, changes in dominance of colonial versus noncolonial reef builders (including all sponges) through geological time, and the relationship of these changes to mass extinction events (solid black horizontal bars).

term radiations between mass extinctions. Thus, the geological history of reef community diversity does not support the Intermediate Disturbance Hypothesis of Connell (1978), nor the Time-Stability Hypothesis of Sanders (1969) and Grassle (1973). The high diversity of modern reefs results mainly from the evolution of clades since the Oligocene, and by speciation during the Miocene-Holocene; that is, during the last 25 million years.

However, not all intervals lacking major reef development are entirely the result of mass extinction. For example, the long (65 my) Middle Cambrian–Middle Ordovician interval and the complete absence of reefs during the earliest Mississippian (Tournaisian) and Early Triassic (Scythian) reflect the paucity of skeletonized, colonial-pseudocolonial reef-building invertebrates at these times. In addition, the competitive displacement and/or ecological replacement of the diverse coral-dominated reef communities of the Early Cretaceous by the low-diversity,

rudistid-dominated communities of the middle to Late Cretaceous took nearly 50 my, and was largely unaffected by the Aptian mass extinction (well shown in fig. 27.6A; see also Kauffman and Johnson 1988; Johnson and Kauffman 1990).

COMPOSITION OF PHANEROZOIC REEF COMMUNITIES

The Phanerozoic history of compositional changes in reef communities (see fig. 27.4) consists of extended periods of uniformity in the higher taxa present, punctuated by brief episodes of rapid taxonomic turnover (Boucot 1983; Sheehan 1985). Fagerstrom (1987, 445–462) used these turnover events to subdivide the Phanerozoic history of reef community evolution into nine units (see fig. 27.6D). Each unit is based on its dominant reef-building higher taxa. These units and taxa, from oldest to youngest, are as follows

1. *Early Cambrian*. Problematic skeletal Cyanophyta (e.g., *Renalcis, Epiphyton, Sphaerocodium*) and Archaeocyatha (primitive nonspicular sponges?).

2. *Middle Cambrian-Early Ordovician*. Various nonskeletal (e.g., *Girvanella*) and skeletal (e.g., *Renalcis, Epiphyton, Calathium, Pulchrilamina*) Problematica, demosponges, and early corals.

3. *Middle Ordovician-early Late Devonian (Frasnian)*. Stromatoporoidea (primative nonspicular sponges?), tabulate and rugose corals, varied "algae," and Bryozoa.

4. *Late Late Devonian (Famennian)-Late Permian*. Exceedingly varied higher taxa including, at different times, Chaetetida, Sphinctozoa, Inozoa (calcareous sponges), demosponges, Bryozoa, Brachiopoda, Crinoidea, and the greatest diversity of higher "algal" taxa.

5. *Middle-Late Triassic (Carnian)*. Sphinctozoa, Inozoa, and "algae." Sponge families, genera, and species differ from those in unit 4.

6. *Late Triassic (Norian)-Early Cretaceous (Barremian)*. Scleractinia and some Hippuritacea (aberrant Bivalvia, "rudistids" of many authors; in shallow-water reefs) and a great variety of siliceous sponges and various "algae" (in deeper-water reefs).

7. *Early Cretaceous (Aptian)-Late Cretaceous*. Hippuritacea and some Scleractinia. "Algae" are commonly absent.

8. *Paleocene-Eocene*. Scleractinia (fewer families and genera than unit 7) and Corallinacea.

9. *Oligocene-Holocene*. Scleractinia (generally same families and genera but different species than unit 8), Milleporina, and crustose Corallinacea.

DIVERSITY EVOLUTION AMONG REEF-ASSOCIATED TAXA

Determining diversity trends through the last 570 my and comparing them with modern reef diversity is a difficult task. The potential for greater relative preservation of entire reef communities in progressively younger rocks suggests that reef diversity among shelled organisms should increase gradually through the Phanerozoic and culminate in a modern diversity peak—but this is far from being the case (see figs. 27.5, 27.6). Reef diversity has fluctuated markedly through geological time.

We have taken three steps in compiling the diversity data shown in figures 27.5 and 27.6 to try to equalize the obvious discrepancies between fossil and modern reef data. First, only taxa with preservable hard parts were used in calculating modern reef diversity (open ovals; fig. 27.5) for comparison with ancient counterparts: i.e., skeletonized algae, benthic protists, spongiomorphs, corals, bryozoa, brachiopods, tube-building annelids, mollusks, certain arthropods (e.g., ostracods), and echinoderms. Second, a conservative estimate of fossil reef diversity is utilized in this analysis. Factors of fossil preservation, variable degrees of discovery and documentation of taxa, and the fact that many reefs can only be studied in outcrop cross-section rather than in three dimensions, naturally deplete the paleontologist's record of diversity in reef communities. Further, many older reef studies only list

genera or, less commonly, families, orders, or classes. Whereas these single data entries may represent numerous species, we have either logged them as one species (if only genus is listed), or if comparable species-level data are available for the same higher taxon in reefs of the same general age, we have used the mean number of species reported elsewhere. Third, to equalize, in part, the loss of three-dimensional observations in fossil reefs, we have utilized time-averaged taxa lists for each fossil reef community.

Our data sets on reef diversity are documented in figure 27.1, and compared in Figure 27.5 for different reef-associated environments: lagoon or inner, middle, and outer platform, backreef, reef core, forereef, and forereef slope to basin facies. Whereas the data are uneven for the reasons listed above, they are the best data available to us from a thorough review of the literature and our collective decades of field observation. The data set represents virtually all major intervals of Phanerozoic time (570 my to present) (see figs. 27.1, 27.5). Data are resolved to the series, stage (average 4–6 my duration) or substage (average 1–3 my duration) level and provide an adequate proxy of reef diversity through time.

Patterns of Diversity Change through Time

Figures 27.5 and 27.6 do not show the pattern of diversity change in fossil reef ecosystems predicted by rarefaction statistics applied either to Phanerozoic diversity trends, reflecting increasing preservation potential through time (e.g., Raup 1972, 1975; but see also Tipper 1979), or to diversity trends within large, contemporaneous, geographically widespread sample sets (Koch 1987; Koch and Sohl 1983). Nor do the data support any broad geological application of the Time-Stability Hypothesis of Sanders (1969). These hypotheses would predict a general evolutionary increase in reef diversity through time, with the highest peak in the Neogene to Recent, but for quite different reasons.

Instead, figures 27.5 and 27.6A show dramatic changes in reef diversity through time, with some Phanerozoic peaks that equal or even exceed modern Caribbean diversity among shelled taxa, and troughs reflecting paucispecific reef communities, even in some cases during peak reef-building times (e.g., the Cretaceous period). These dynamic changes can be linked primarily to short-term geological and biological events, especially mass extinctions, and to longer-term changes in Earth climate (especially temperature) and ocean systems (especially sea level and chemistry).

Peaks in reef diversity through geological time can be divided into first-, second-, and third-order peaks based on the maximum number of shelled taxa recorded for each interval: first-order peaks have over 100 recorded taxa, second-order peaks 50–100 taxa, and third-order peaks 30–50 taxa (mainly recorded as species). To evaluate diversity for any time interval, we have selected the highest level recorded for any single reef of that age. Rarely is there sufficient data from multiple reefs to establish a statistical mean value. Further, shallow-water tropical reef biotas (inner platform to upper forereef slope habitats; solid line curve, figs. 27.5, 27.6A) have been

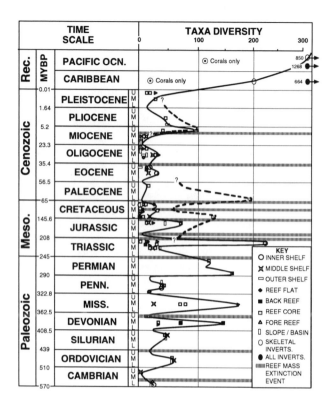

Figure 27.5   Temporal variation in taxonomic diversity of reef communities, based on the highest-quality data points in figure 27.1. The solid line represents shallow-water tropical reef diversity; the dashed line represents deep-water reef diversity. Where there are no reef data, the curves are smoothed between adjacent data points, except where we are certain that few, if any, reefs existed, in which case we have shown a trough between data points. Mass extinctions are represented by the patterned horizontal lines. Data for the Recent Caribbean reef are based on Carrie Bow Cay, Belize (Rützler and Macintyre 1982). Recent Pacific data are from Heron Island, Great Barrier Reef, Australia (Mather and Bennett 1984). The available Recent taxonomic data are not easily divisible into environmental zones of the reef, so that comparison with more specific fossil data is difficult in some cases. Key in lower right shows symbols representing different ecological zones of the reefs. Dark ovals represent non-skeletal plus skeletal taxa, i.e., all taxa recorded from this site. Open ovals represent skeletal taxa only, comparable to the fossil record. Time scale from Harland et al. 1989.

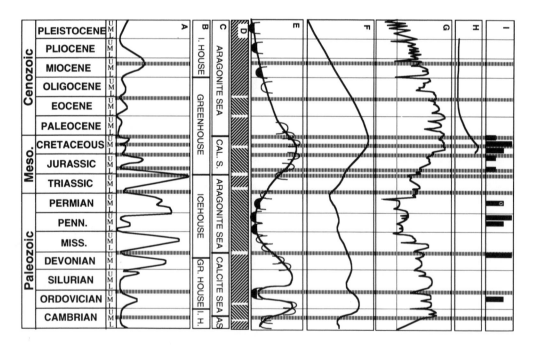

Figure 27.6   Long-term fluctuations in global environmental factors (B–I) that have influenced the species diversity of reef communities (A) during the Phanerozoic. Extended patterned horizontal lines indicate major mass extinction events. All curves directed toward the right of the diagram indicate increasing values. B, C: Intervals of icehouse and greenhouse climatic supercycles of Fischer (1984) and aragonite-calcite–facilitating seas of Sandberg (1983), respectively. D: Changes in major reef community composition, the so-called ecological-evolutionary units of Fagerstrom (1987). E: Climate cy-cles and general temperature fluctuations (after Fischer 1984); warm climates are indicated by open-left symbols (⊃) and cold climates by open-right symbols (⊂) with times of continental glaciation indicated by half-filling of the symbol. F: Intensity of volcanism (after Fischer 1984). G: Global sea level (after Vail et al. 1977; Haq, Hardenbol, and Vail 1987; Ross and Ross 1988). Thin vertical line represents the present sea level. H: Atmospheric CO_2 (after Berner, Lasaga, and Garrells 1983; Berner 1991). I: Oceanic anoxic events (dark bars) (after Hay 1988).

separated from deeper-water reef biotas dominated by sponges and ahermatypic corals (middle to lower slope and basin habitats; outer shelf of continental margins; dashed line curve, fig. 27.5) in the analysis.

First-order peaks in shallow-water reef diversity are as follows (figs. 27.5, 27.6A), from oldest to youngest: (1) Middle Devonian (Givetian) on shallow carbonate platforms, with diversity rise starting in the Middle Silurian and declining somewhat just prior to the Frasnian-Famennian (Late Devonian) mass extinction event, which devastated reef communities; (2) Late Early to Middle Mississippian (mainly reef core plus forereef data, including some transported taxa), beginning in the earliest Mississippian and subsequently declining to an Early Pennsylvanian low point; (3) Early Permian on shallow carbonate platforms, mainly among reef-building brachiopods, with the rise beginning in the Late Pennsylvanian and a subsequent fall to the Middle Permian diversity trough; (4) Late Permian, on shallow carbonate platforms, as a lower but significant diversity peak on the decline from the Early Permian diversity peak; the subsequent fall largely took place just below the Permo-Triassic boundary, marking the first extinction event (step) in the Earth's greatest mass extinction interval; (5) the Late Triassic of Europe, on shallow carbonate platforms, with initial rise in the Middle Triassic and subsequent fall just prior to the Triassic-Jurassic boundary, as a first phase of this mass extinction event; (6) the Pleistocene to Recent reef-building interval. Two first-order deep-water reef diversity peaks occur in the (7) Early Cretaceous in deeper-water slope-basin habitats, with the rise initiating in the Middle Jurassic and the subsequent fall terminating in the late Early Cretaceous (base of Aptian); and (8) the late Early Paleocene in deep-water shelf facies (European passive margin), mainly among reef-associated Bryozoa and ahermatypic corals, following the great Cretaceous-Tertiary mass extinction. Peaks 3 and 8 are mainly recorded from single localities (Glass Mountains of Texas and Denmark, respectively).

Second-order peaks in reef diversity (from oldest to youngest) occur in the Middle Ordovician, Late Jurassic, and Late Miocene. Important third-order diversity peaks occur in the Early Cambrian, Late Silurian, Middle Pennsylvanian, middle Cretaceous, Middle Eocene, and Middle Oligocene. Major intervals characterized by low reef diversity, or a general lack of reefs, occurred in the Precambrian, Late Cambrian to earliest Ordovician following the Lower Cambrian mass extinction, in the Late Ordovician to Middle Silurian, in the Late Devonian following the Frasnian-Famennian mass extinction, in the Late Mississippian-Early Pennsylvanian, in the Late Pennsylvanian, in the Early and Middle Triassic following the Permo-Triassic mass extinction, in the Early to Middle Jurassic following the Triassic-Jurassic mass extinction, in the earliest Cretaceous following the Jurassic-Cretaceous mass extinction, in the latest Cretaceous through Early Eocene associated with the great Cretaceous-Tertiary mass extinction, in the Early Oligocene following the Eocene-Oligocene mass extinction, in the latest Oligocene-Early Miocene (a major Caribbean Province extinction event), and Late Pliocene-earliest Pleistocene. Sixty-nine percent of these low-diversity intervals follow mass extinction events.

The average interval between first- and second-order diversity peaks is 52.7 my (range 3.6–147.8 my), and the average interval between all diversity peaks combined, regardless of order, is 36.5 my (range 3.6–96.5 my). The extremely broad range of durations between peaks would seem to rule out the possibility of cyclicity in the data set. Utilizing the Maximum Entropy Spectral Analysis (MESA) program, a time series analysis of these data was attempted. The frequencies for all high-diversity peaks were 0.0, 0.166, 0.33, and 0.5, but the analysis had no power. Frequencies for only first- and second-order diversity peaks were 0.0, 0.25, and 0.5, again without power. Frequencies for all low-diversity troughs were the same as for first- and second-order diversity peaks, and without power. These analyses suggest a lack of cyclicity in the data.

The average duration between all intervals of very low reef diversity is 43.2 my, predictably similar to that for all diversity peaks. The broad range of interval values (8–91.2 my), low correlation levels, and lack of power in the MESA analysis seemingly precludes cyclicity in the data set.

Mass extinction is the main forcing mechanism for changes in reef diversity during the last 570 my. Whereas a 26–30 my cyclicity has been demonstrated by Raup and Sepkoski (1984, 1986) and others for Mesozoic-Cenozoic mass extinction events, the actual record of peaks and troughs in reef diversity (fig. 27.5) is not cyclic, and the average time span between diversity troughs during this interval is 37.9 my. This suggests that other factors (physical, chemical, and biological) besides mass extinctions have also affected diversity trends in the evolution of reef communities (see below). It also suggests that some mass extinctions did not have a great effect on reef diversity. For example, reef diversity during the Early Cretaceous was already very low as a result of competitive displacement and/or ecological replacement of diverse coral-algal–dominated reef communities by paucispecific rudistid bivalve–dominated reef communities, so that the middle Cretaceous (Aptian) mass extinction event (see fig. 27.6A) did not have an important effect on reef diversity (Kauffman and Johnson 1988; Johnson and Kauffman 1990). Furthermore, the Eocene-Oligocene mass extinction event had a minimal effect on reef diversity, whereas the Oligocene-Miocene reef extinction was a major event, but not a global mass extinction in the Raup-Sepkoski data (1984, 1986).

Peaks and troughs in the deep-water reef diversity data operate independently of the shallow-water data (fig. 27.5). It is important to note, however, that diversity peaks in the deep-water data commonly follow mass extinction events, raising the possibility that deep-water reefs represent refugia for shallow-water reef builders during and just after mass extinctions.

Factors Affecting the History of Reef Diversity

Figure 27.6 compares the various geological and environmental factors that potentially affect reef diversity against the actual geological record of shallow-water diversity (curve derived from fig. 27.5 and smoothed). Regulatory factors include: global temperature (E) and sea level (G) changes; levels of global volcanism (F); marine water chemistry (C, I); atmospheric CO_2 levels (H), greenhousing and global warming (B), as opposed to cooler glacial "icehousing" intervals (Fischer 1984); and biological phenomena like mass extinction events (A) and major changes in reef ecosystems (D). From this critical comparison of the peaks and troughs of reef diversity (figs. 27.5 and 27.6A) with variations in environmental factors, the following controls on reef diversity through time are apparent, in order of importance:

Mass Extinction Events. Mass extinction events or intervals are the primary regulating force on reef diversity through time. Major global mass extinctions, shown as bold or patterned horizontal lines in figures 27.1 and 27.4–6, commonly incorporate in their early phases abrupt diversity drops in reef ecosystems and are commonly followed by long intervals without reefs, or with low reef diversity. The relationship between rapid decline in reef diversity and mass extinction events/intervals exists for all three types of mass extinctions (catastrophic, stepwise, and graded), and for all proposed causes of these mass extinctions, including associated extraterrestrial impact, global cooling, sea level fall, oceanic anoxia, acid rain, and/or enhanced global volcanic activity. These various phenomena associated with multicausal mass extinction events have, individually, also occurred many times in Earth history without significantly affecting biological diversity.

Figure 27.7 documents the relationship between major drops in reef diversity through time and various proposed causal mechanisms. Of the 12 major mass extinctions recorded during the Phanerozoic, 11 (92%) are associated with major decline in reef diversity either just before (as a first step) or at the main extinction boundary; the Miocene mass extinction, the weakest in the extinction data (Raup and Sepkoski 1984, 1986), is not characterized by a reef extinction, although a major extinction occurs about 10 my before this at the Oligocene-Miocene boundary. In the total record of peak lows in Phanerozoic reef diversity (fig. 27.7), 11 of 13 (85%) immediately follow mass extinction events.

Three important patterns emerge from our data set regarding reef diversity and mass extinction. First, it is apparent in almost all examples that reef ecosystems collapse, and diversity declines abruptly, early in the mass extinction intervals (Johnson and Kauffman 1990), even when they are characterized by some catastrophic event like the end-Cretaceous bolide impact (Kauffman 1979, 1984, 1988). Figure 27.6A generally shows this pattern, with a rapid diversity drop just prior to each mass extinction peak. This response apparently reflects the predominantly stenotopic nature of reef-dwelling taxa in the past, as is the case today. The rate and magnitude of environmental changes commonly associated with even the early phases of mass extinctions apparently exceeded the adaptive ranges and/or evolutionary rates of taxa within ancient tropical ecosystems well before global temperate ecosystems were affected.

Second, reef ecosystems were not only the first to be decimated during most mass extinction events, but they were commonly the last to recover following mass extinctions. Typically, a 1–8 my interval following a global mass

A — CORRELATION WITH ENVIRONMENTAL FACTORS

PEAK (HIGH DIVERSITY)	ORDER	CHEMISTRY			TEMPERATURE				CO2	SEALEVEL		VOLCANISM	
		AS	CS	OMZ/OAE	IH	GH	HI	LO	PEAK	HI	LO	HI	LO
HOLOCENE	1	●			●			●			●		●
LO. PLIOCENE	2	●			●			●			●		●
UP. MIOCENE	2	●			●			●	●				●
M. OLIGOCENE	3	●				●		●	●				●
M. EOCENE	3	●				●		●	●			●	
MIDDLE CRETACEOUS	3		●	●		●	●		●	●		●	
UP. JURASSIC	2		●	●		●	●		?	●		●	
UP. PERMIAN	1	●			●			●	↑		●		●
LO. PERMIAN	1	●			●			●			●		●
M. PENNSYL.	3	●		●	●			●	↓		●		
M. MISSISSIP.	1	●			●			●	?	●			●
M. DEVONIAN	1		●			●	●		↑	●		●	
UP. SILURIAN	2		●			●	●			●			●
M. ORDOVICIAN	2		●			●	●		?	●			●

B — CORRELATION WITH ENVIRONMENTAL FACTORS

TROUGH (LOW DIVERSITY)	CHEMISTRY			TEMPERATURE				CO2	SEALEVEL		VOLCANISM	
	AS	CS	OMZ/OAE	IH	GH	HI	LO	PEAK	HI	LO	HI	LO
UP. PLIOCENE	●			●			●*			●		●
UP. OLIGOCENE	●				●		●*?			●		●
LO. OLIGOCENE	●				●		●		●			●
LO. EOCENE	●				●		●		●		●	
L. CRETACEOUS		●	●		●	●		●	●		●	
MID. JURASSIC	●				●		●	?	●		●	
MID. TRIASSIC	●				●	●		↑		●		●
MID. PERMIAN	●		●		●		●*			●		●
UP. PENNSYLVAN.	●		●		●		●*	↓		●		●
LO. PENNSYLVAN.	●				●		●*	?		●		●
UP. DEVONIAN		●	●		●	●		↑		●	●	
LO.-MID SILURIAN		●			●	●				●	●	
UP. CAMBRIAN		●			●	●		?		●	●	

Figure 27.7 Correlation of reef diversity peaks (**A**) and troughs (**B**) with environmental factors. *AS,* aragonite seas; *CS,* calcite seas; *OMZ/OAE,* expanded oxygen minima zones/oceanic anoxic events; *IH,* icehouse world; *GH,* greenhouse world; *HI,* high values; *LO,* low values; * indicates continental glaciation. Attempts to apply nonparametric correlation methods to this data base have failed to yield significant results because it was not possible to rank the data, and the data set is too limited. Data summary: (**A**) AS/CS = 9/5; AS correlates with 64.3% of the diversity peaks. OMZ-OAE/Oxygenated = 3/11; oxygenated water correlates with 78.6% of the peaks. IH/GH = 7/7; IH/GH show no correlation. High temperature/moderately low temperature = 5/9; moderately low temperature correlates with 64.3% of the peaks. HI/LO sea level = 9/5; relatively high sea level correlates with 64.3% of the peaks. HI/LO volcanism = 6/8; low volcanism correlates with 57.2% of the peaks. (**B**) AS/CS = 9/4; AS correlate with 69.2% of the diversity troughs. OMZ-OAE/Oxygenated = 4/9; oxygenated water correlates with 69.2% of the troughs. IH/GH = 8/5; IH correlates with 61.5% of the troughs. High temperature/moderately low temperature = 5/8; low temperature correlates with 61.5% of the troughs. HI/LO sea level = 7/6; relatively high sea level correlates with 53.9% of the troughs. HI/LO volcanism = 4/7; low volcanism correlates with 63.6% of the troughs. Number of mass extinctions associated with diversity troughs = 11/12 (91.7%). Number of diversity troughs associated with mass extinctions = 11/13 (84.6%).

extinction either lacks significant reefs throughout the tropics, or is characterized by scattered, low-diversity reefs composed of more eurytopic organisms like sponges, crustose algae, or noncolonial invertebrates. The broad low troughs following many mass extinctions in figure 27.6A reflect this pattern.

Third, there seems to be a persistent pattern of recovery of reef diversity between mass extinctions. Following a post-extinction interval with little or no reef development, reef diversity slowly builds over the next several million years, not reaching its pre-extinction levels (if at all) until just prior to the next mass extinction event. Inasmuch as high reef diversity is largely a function of complex ecological structure within communities, including high levels of symbioses, niche partitioning, and trophic specialization, the slow recovery of reef diversity after a mass extinction probably reflects the time involved in evolving these relationships among component taxa of new reef ecosystems. The asymmetrical nature of the diversity peaks in figure 27.6A symbolically reflects this relationship, which is recorded in detailed stratigraphic data that cannot be shown at the broad scale used in this analysis.

Ocean Chemistry. The second most important correlation with changes in diversity among ancient reef ecosystems is with ocean chemistry (fig. 27.6C and I), and in particular, oxygen and pCO_2 levels. Sandberg (1983) noted broad fluctuations in the mineralogy of nonskeletal carbonate in ancient seas, reflecting plate tectonically influenced oscillations in pCO_2. He proposed a system of alternating calcite seas characterized by low Mg-calcite deposition and aragonite seas characterized by high Mg-calcite and aragonite deposition (fig. 27.6C); these phases also describe the principal materials available for skeletal mineralization. These phases correlate generally, but not precisely, with the greenhouse and icehouse intervals of Fischer (1984) (fig. 27.6B). Sixty-four percent of the diversity peaks in Phanerozoic reef ecosystems correlate with aragonite seas, but only 50% with either greenhouse or icehouse intervals. Sixty-nine percent of the diversity troughs also correlate with aragonite seas and 62% with icehouse intervals characterized by cool seasonal climates and low global sea level (fig. 27.7).

Many Phanerozoic intervals lacking permanent ice sheets at the poles and characterized by warm equable climates were also characterized by sluggish ocean circulation driven by warm, oxygen-depleted, hypersaline bottom water. These intervals were commonly associated with sea level rise, expanded oxygen minima zones (OMZs), and oceanic anoxic events (OAEs). Rising sea level episodically brought the upper portions of these oxygen-depleted water masses in contact with tropical carbonate platforms and reefs, effectively drowning them in dysoxic to anoxic water and killing off diverse organisms; OAEs should predictably be times of low reef diversity. Figure 27.6I shows this to be the case, where 79% of the OAEs are associated with low troughs in the reef diversity curve, exceptions being the Late Permian, Middle Jurassic, and certain middle Cretaceous OAEs or OMZ expansions. Major oceanic advection events of trace elements, especially trace metals, also kill off tropi-

cal reefs, as in the middle Cretaceous (Cenomanian-Turonian boundary; Johnson and Kauffman 1990).

Global Temperature Change. Global temperature change and its effect on climate and glacial history has long been thought to be a primary control on reef diversity. Figure 27.6 compares the historical development of reef diversity against the global marine temperature curve (Hay 1988); greenhouse versus icehouse intervals (Fischer 1984; Sheehan 1985); the global CO_2 curve reflecting greenhousing and global warming potential (Berner 1991; Berner, Lasaga, and Garrells 1983); and known intervals of extensive continental glaciation leading to development of large ice sheets (Fischer 1984). As would be predicted from the restriction of modern shallow-water reef development to water warmer than 70°F mean annual temperature, there exists a strong relationship between trends in Phanerozoic reef diversity and global temperature levels. Figures 27.6 and 27.7 show that 64% of the Phanerozoic peaks in reef diversity occur with moderate to relatively low or falling global temperatures, as do 62% of the low-diversity troughs. Only 50% of the diversity peaks are associated with warm greenhouse intervals. Surprisingly, the three major high-temperature peaks in figure 27.6E (Late Cambrian-Early Ordovician; Silurian; Jurassic-Cretaceous, during which times climates were highly equable and the tropics broadly expanded, are all associated with diversity minima, not maxima, as are the major intervals of global cooling with continental glaciation in the Precambrian, Late Ordovician, Permo-Pennsylvanian, and Pliocene.

Thus, major peaks in reef diversity during the Phanerozoic appear to correlate with a critical thermal window between peak high and low global temperatures, both of which exceeded the tolerance limits of most stenothermal reef-associated organisms. Small-scale diversity peaks in the Late Silurian, Jurassic, Cretaceous, and Paleogene that are associated with major thermal peaks are mainly based on algal, sponge, and rudistid bivalve reef communities, which have relatively low diversity compared with living coral-algal reef communities. Kauffman and Johnson (1988) attempted to explain the low Cretaceous reef diversity during an extreme thermal peak by suggesting that a significantly warmer, somewhat hypersaline supertropical water mass (Supertethys) that developed in the core tropics exceeded the adaptive ranges of many corals, leading to competitive displacement and/or ecological replacement of diverse Jurassic-Cretaceous coral-algal reef communities by paucispecific rudistid bivalve-dominated reef communities. The data presented in this chapter support this hypothesis and suggest that these factors may also have been operative during Paleozoic thermal peaks (fig. 27.6E) characterized by low reef diversity. The Cretaceous to Recent CO_2 curve (fig. 27.6H) shows a similar relationship: high CO_2 resulting in a major Cretaceous global warming is correlated with relatively low reef diversity, and low CO_2 during the Neogene is characterized by highly diverse reef ecosystems. These factors explain why peaks in reef diversity, which should predictably be associated with peaks of global warming, do not occur at the highest Phanerozoic temperature maxima.

There is a clearer correlation between low global temperature and diversity troughs, however; 62% of the minima in reef diversity also correlate with thermal lows, periods of continental glaciation, and icehouse intervals. The Miocene and Holocene peaks of diversity associated with a major glacial interval are an important exception. But the highest of these small diversity peaks tend to be associated with interglacial intervals and moderate climate warming, even though overall reef diversity does not show any major decline during this icehouse interval (fig. 27.6B) (Wise and Schopf 1981).

Sea Level. There is a general correlation between global reef diversity and eustatic sea level fluctuations. Major elevations of the global Phanerozoic sea level to as much as 300 m above the present stand produced warm, stable, ameliorated, maritime-dominated global climates, expansion of the tropical/subtropical climate zones by up to 25–30%, and possibly the episodic development of a warmer, more saline supertropical climate zone (Supertethys). Reef-forming environments expanded and diversified, producing a general correlation (64.3% of the data) of high sea level and reef diversification events (fig. 27.6G, 27.7). However, this correlation is not very strong because highly elevated global sea level results also in conditions that may be detrimental to diversification of reef organisms: (1) more widespread drowning of reef-forming sites to depths that will no longer support reef growth; (2) overstepping of reef-bearing tropical platforms with the dysoxic to anoxic waters of expanded OMZs and OAEs; and (3) increases in temperature and salinity levels in the core tropics during sea level highstands and development of supertropical climate zones, to the point where these levels exceed the adaptive ranges of diverse stenothermal/stenohaline reef-associated taxa. Thus, not all reef diversification peaks correlate with sea level highstands (e.g., the Permian and Pleistocene-Recent), and not all sea level highstands are intervals of high reef diversity (e.g., the Cambrian, Silurian, and Cretaceous), especially when they are associated with thermal maxima.

Conversely, we might expect a strong correlation of sea level lowstands, which reached as much as 130 m below present stand, broadly exposing tropical platforms, with diversity lows and even extinctions in reef biotas (Hallam 1989). This correlation does not exist; only 46% of the low reef diversity troughs correlate with sea level lowstands, probably reflecting the ability of reefs to migrate (prograde) downslope with at least slowly falling global sea level.

Plate Tectonics. Plate tectonic factors, as they affect continental and island platform dispersion, tropical oceanic circulation, and thermal history, also show a broad but not pervasive relationship to reef diversity through time. The principal relationship is the one between diversity peaks and times when continuous tropical/subtropical currents were active within a single Earth-encircling tropical sea made possible by continental dispersion on either side of Tethys. These conditions favored widespread dispersal and diversification of reef-building taxa. This may be the case for the Middle Ordovician, Late Silurian, and Cretaceous through Oligocene diversity peaks (fig. 27.6A), but not for others, including most first-order peaks of the late Paleozoic and Mesozoic.

Active plate tectonic intervals during Earth history are commonly associated with increased volcanic activity, especially explosive volcanism, which may result in the introduction and dispersal of harmful aerosols into the atmosphere, short-term fallout of thick ash deposits, and short-term global cooling—all detrimental to shallow-water reef ecosystems. Figure 27.6F shows a general curve for global volcanism during the Phanerozoic. Figure 27.7 shows no correlation between times of increased volcanism and low-diversity troughs in reef ecosystems; only 36.4% of the low-diversity intervals are correlated with global peaks in volcanic activity. Only 57% of the peaks in reef diversity are associated with periods of little or no global volcanism. We conclude that, in general, levels of global volcanism have little effect on reef diversity through time, although local to regional effects may be devastating (Kauffman and Johnson 1988).

Biological Factors. Biological factors have strongly influenced reef diversity in at least two major intervals, and very possibly in other cases as well. Competitive displacement and/or ecological replacement of dominant reef builders by newly evolved/introduced taxa, significantly changing reef diversity, has been documented for the Triassic and the Cretaceous periods. In the Middle to Late Triassic (Stanley 1981, 1988), scleractinian corals gradually displaced sphinctozoan and inozoan spongiomorphs as the dominant reef builders. This resulted in a marked increase in diversity of reef-associated taxa, leading to one of the highest first-order diversity peaks in Earth history (figs. 27.5, 27.6A). Modern sponges are highly competitive, especially biochemically, and tend to dominate low-diversity communities on reefs; they also provide poor substrates for attachment or boring by the secondary reef dwellers (certain corals, bryozoa, worms, mollusks, echinoderms, etc.) that form the bulk of modern reef diversity. Conversely, modern coral-algal–dominated reef communities are highly diverse, not just because of the complex ecological zonation and successional relationships between the major reef builders, but also because they provide favorable hard substrates, with many microhabitats for reef-associated taxa. Triassic reefs reflect these same relationships, as well as the competitive advantage of scleractinians over sponges as major framework builders.

During the middle to Late Cretaceous, moderately diverse scleractinian coral-algal–dominated communities were competitively displaced/replaced by low-diversity rudistid bivalve–dominated communities, at least in the Caribbean Province (see fig. 27.4). Changing tropical climates (especially rising temperatures) and ocean chemistry (especially oxygen, salinity, and dissolved nutrient levels) may have favored rudistids over corals in the core tropics (e.g., Scott, 1984a; Scott et al. 1990). The latest Cretaceous mass extinction decimated rudistid reef communities for another 7–10 my. Although noncolonial rudistids provided hard substrates suitable for attachment of diverse epibionts and endobionts, these smaller taxa did not effectively colonize rudistids until the latest Cretaceous, and then at low levels of diversity, possibly as a result of chemocompetitive strategies among rudistids (Kauffman and Sohl 1974, 1978; Kauffman and Johnson 1988). This abrupt decrease in reef diversity (figs. 27.5, 27.6) added to the effects of the middle and latest Cretaceous mass extinctions of reef ecosystems, which collectively resulted in the long terminal Cretaceous and Paleogene interval (7–10 my) during which no significant reefs formed on shallow tropical platforms worldwide (Kauffman 1984, 1988; Johnson and Kauffman 1990).

In our analysis, we also investigated the possibility that dominance of reef ecosystems by colonial versus acolonial taxa may have directly affected reef diversity, as in the case of the Triassic-Jurassic scleractinian coralalgal–dominated reef communities, versus the middle and Late Cretaceous rudistid bivalve–dominated communities (see fig. 27.4; Kauffman and Johnson, 1988), and in the case of modern reefs, where sponge-dominated, vermetid gastropod–dominated, bivalve-dominated, and wormdominated reefs have much lower diversity than coralalgal reef communities in the same areas. But no consistent relationships were found, and in the Early and Late Permian, some of the highest first-order diversity peaks were in reefs strongly dominated by brachiopods and spongiomorphs, respectively (in our analyses, spongiomorphs were accepted as noncolonial organisms, following the arguments of Fry 1979).

Other Factors. Other environmental fluctuations that were tested against the history of reef diversity in this study included the position of CCD (carbonate compensation depth) in the world's oceans, relative size of the tropics, and rates of plate spreading. None of these showed any correlation to, or even common association with, major fluctuations in reef diversity through time.

CONCLUSIONS

Phanerozoic reef diversity, as measured by shelled species, has shown dynamic fluctuations through time that do not reflect the expectations of broadly applied time-stability evolutionary theory, nor of rarefaction statistics. Instead, reef diversity shows a series of major peaks, some of which approach modern diversity levels among shelled taxa, and troughs, including long intervals of geological time without prominent reef development.

Several dynamic geological, climatic, and oceanographic factors correlate with a significant number of fluctuations in Phanerozoic reef diversity. Ocean chemistry, and especially oxygen and pCO_2 levels in ancient oceans is one such factor. Reefs thrived in oxygenated waters, but were broadly decimated when expanded oxygen minima zones (OMZs) or oceanic anoxic events (OAEs) inundated reef platforms with dysoxic to anoxic waters during rapid sea level rise. pCO_2 regulated skeletal mineralogy and sedimented carbonate minerals in Phanerozoic oceans. Most diversity peaks and troughs in reefs were associated with pCO_2 levels producing aragonitedominated seas. High dissolved nutrient levels did not fa-

vor reef diversification except among nonphotosymbiotic taxa, which were predominantly suspension feeders.

Temperature fluctuations in world oceans exercised a profound but unexpected control on reef diversity levels. Major thermal peaks and troughs during the Phanerozoic were both associated with low reef diversity. Both thermal extremes, associated on one hand with high ocean temperatures, low oxygen levels, and hypersalinity, and on the other with global cooling, higher seasonality, and episodic polar glaciation, apparently exceeded the adaptive ranges of largely stenotopic Phanerozoic reef-associated taxa. An intermediate temperature window, especially as developed during cooling trends, favored reef diversification.

Sea level history, which records global rise of more than 300 m and fall of more than 130 m through time, compared with the present sea level stand, has controlled the expansion and contraction of tropical reef ecospace, as well as global temperature and climate equitability. There is a significant correlation between elevated global sea level, warm equable climates, and reef diversification events, but not as high as expected. This is because of detrimental environmental effects of high sea level, such as drowning of reef tracts, overstepping by oxygen-depleted waters, and elevated temperature and salinity, which may exceed the adaptive ranges of diverse stenotopic reef builders.

Plate tectonic movements have affected the dispersion of continental margins, island platforms, and other favorable reef habitats around the ancient tropics, and episodically have created circumglobal tropical seas, which enhanced tropical reef ecospace, larval dispersal, and diversification. Closure of tropical seas, or establishment of dispersal barriers as a result of plate tectonics, is usually associated with low reef diversity.

But the major factors affecting reef diversity through time are those associated with mass extinction events/intervals. Stenotopic components of reef ecosystems were characteristically rapidly and broadly decimated early in mass extinction intervals. Mass extinctions are further followed by long intervals of time (1–8 my) without significant reefs, during which the basic reef ecosystem slowly restructures from surviving and newly evolved bioconstructional elements. Mass extinctions are the major driving force behind reef diversity, extinction, and evolution, removing established reef ecosystems and creating vacant ecospace into which new types of reef ecosystems can evolve. Another biological factor that has had an important positive or negative effect on reef diversity is competitive displacement of one dominant reef-building group by another.

The fluctuating history of reef diversity on Earth thus reflects two major forces: the dynamic history of the Earth, its atmosphere, and its oceans, which rarely permits environmentally sensitive tropical ecosystems and environments to stabilize for long intervals of time; and the effects of major "disturbance" events—mass extinctions, their extraterrestrial and terrestrial causes, and rapid destabilization of tropical ecosystems for long periods of time.

28

Historical Diversity Patterns in North American Large Herbivores and Carnivores

Blaire Van Valkenburgh and Christine M. Janis

The North American fossil record of mammals is exceptional and offers a unique opportunity to study historical trends and broad-scale patterns in diversity. Past reviews of diversification among Cenozoic mammals have utilized this resource to test hypotheses of extinction-origination equilibria and community stability at the continental level (Lillegraven 1972; Gingerich 1984a; Webb 1984, 1989; Stucky 1990). There has been less attention paid to issues of diversity control at the local level, such as the relationship between regional and local diversity, and the effect of historical events, such as immigration or extinction, on provincialism (but see Rose 1981). Nevertheless, these are issues that are accessible through the mammalian fossil record and of great interest to ecologists. Despite some shortcomings, studies of extinct communities offer some clear advantages over those of extant communities. First, extinct communities provide views of pre–Ice Age and pre-human communities, serving to highlight the effects of these two major perturbations on present-day ecosystems. Second, the development of ancient communities can be followed over extended time spans, revealing the effects of the addition and loss of species and providing a record of long-term significant trends in species and community evolution.

In this chapter we explore changes in the diversity of large mammalian predators and their presumed prey over the last 44 million years (late Middle Eocene-Recent) of North American history. North America was chosen over other continents because of its relatively rich and continuous Cenozoic record and the fact that it has existed as an island continent for much of its history. Evolution has proceeded largely in situ, and episodes of immigration from other continents are relatively discrete and well documented (Woodburne 1987; Webb 1989). The analysis begins in the late Middle Eocene, some 20 million years after the extinction of the dinosaurs, at a time when the mammals take on a more modern aspect. Below the Middle Eocene, there are abundant mammalian fossil remains and well-preserved faunas (e.g., Gingerich 1989; Rose 1990), but species are often members of extinct orders. The resulting morphological gap between ancient and modern forms makes it more difficult to infer ecological roles through comparisons with modern species. From the Late Eocene onward this situation improves as the living families becomes increasingly well represented.

Unlike previous studies of Cenozoic mammalian diversity, this study utilizes faunal lists from individual fossil localities to track changes in predator and prey diversity. In the past, most similar work has relied on continental lists of species occurrences (e.g., Romer 1966; Savage and Russell 1983), and analysis has been done at the level of mammalian genera or higher (families, orders). The use of species lists from localities chosen to represent distinct regions of the continent permitted us to explore questions of changing provincialism (beta and gamma diversity) and the association between regional and local diversity in North America. In particular, the data were applied to questions of continental diversity, local versus continental diversity, faunal provincialism, and predator and prey diversity. Throughout this chapter, *diversity* is equivalent to species richness without consideration of relative abundances.

Provincialism, local diversity, and continental diversity are three related issues, all of which are likely to have been affected by the profound climatic and topographical changes that occurred in North America over the last 44 million years. Since the Middle Eocene (ca. 50 mya), global climates have become significantly cooler, drier, and more seasonal, with dramatic effects on the distribution of vegetational habitats (Leopold 1967; Wolfe 1978; Singh 1988). In North America, the predominant change was a shift away from relatively closed, high-biomass, mesic forest toward more open, lower-biomass, and savanna-like environments, and ultimately deserts, although these are unknown in the North American fossil record before the Pleistocene (Wolfe 1985). This change in vegetational habitats is expected to have been associated with changes in herbivore diversity and perhaps provincialism. Over the same interval, the West Coast became more isolated from interior regions due to climatic and orogenic changes. Latitudinal temperature gradients increased across the continent, bringing seasonal droughts to the West Coast (Wolfe 1978). The Rocky Mountains uplifted, as did the Sierra Nevada, creating a major topographical barrier and rain shadow between coastal and interior areas (Dott and Batten 1976; King 1977). Thus, a rise in provincialism might be expected due to the greater isolation of coastal faunas and the effects of a rain shadow on interior faunas.

Continental diversity and provincialism are also likely to have been affected by episodes of increased extinction or immigration. Obviously, extinction reduces diversity at the continental level, but its effect on local or community diversity is less clear. For example, a reduction in the total number of herbivore species in North America might be accompanied by an increase in the number of widespread, as opposed to locally endemic, species as survivors expand their ranges in response to the removal of competitors. A relative increase in the number of widespread species will produce a decline in provincialism, and local diversity could remain unchanged. Similarly, a surge of immigrants from Eurasia or South America is expected to boost continental diversity but might or might not alter patterns of provincialism. One note of caution should be interjected at this point concerning matters of resolution and scale: provincialism is likely to rise initially, since immigrants will tend to appear as local endemics in and near the region of entry (e.g., Beringia or the Gulf Coast), but should drop as the immigrants colonize larger areas of the continent. However, if colonization has proceeded at a rate similar to that observed for range expansion in several extant large mammals (e.g., 1,000 km per 100 years; Kurtén 1957), our analysis will not detect the gradual spread of immigrants because communities are sampled at approximately 2-million-year intervals.

Predator and prey are expected to coevolve at some level, but the strength of the linkage between the two groups is unclear. Morphological analyses of presumed adaptations for prey capture and predator escape in North American Cenozoic mammals have revealed a surprising lack of evidence for a coevolutionary arms race between carnivores and ungulates (Bakker 1973; Janis 1993). Long-limbed, presumably fast ungulates appear in the Early Miocene, well before the appearance of long-limbed predators in the Pliocene (Janis 1993). Moreover, predators and prey differ significantly in general evolutionary pattern. The history of herbivores in North America is dominated by two morphological trends: increasing cursoriality and increasing dental specializations for grazing. Both can be explained as adaptations to the spread of grasslands (Janis 1988, 1993). The history of carnivores (inclusive of the Creodonta) is not so clearly directional; instead, it is characterized by the repeated evolution of particular ecomorphs, such as bone-cracking and catlike species (Martin 1989; Janis and Damuth 1990; Van Valkenburgh 1991). Nevertheless, there may still be a predictable relationship between the diversity of large mammalian predators and of their prey within a community, and it is this relationship that is explored here.

DATA BASE AND METHODS

The data base consists of numbers and names of large mammalian carnivore and herbivore species for a set of 115 fossil and 8 Recent localities, spanning the past 44 million years of North American history (late Middle Eocene to Recent; Table 28.1). The large herbivores are primarily ungulates (including proboscideans) and edentates; the carnivores include members of the orders Creodonta (family Hyaenodontidae) and Carnivora (families Canidae, Felidae, Nimravidae, Amphicyonidae, Ursidae, and Mustelidae). Excluded from our analysis are small

herbivores such as rodents and lagomorphs, and predators estimated to be smaller than the living gray fox (*Urocyon cinereoargenteus*). This is done for two reasons: to focus the study on large ungulates and their presumed predators, and to utilize the more extensive literature on the functional morphology and paleobiology of the larger species in the interpretation of results (e.g., Scott 1985; Janis 1984, 1990; Van Valkenburgh 1987, 1989). The species lists for the fossil localities are all from the literature and are amended with updates from various researchers (see Acknowledgements); a list of localities and citations is available from the authors on request. The data for the Recent communities are from Hall (1981).

The 44-million-year time span was divided into 25 time intervals, each of which corresponds to a subdivision of a North American land mammal age (e.g., early Chadronian, early Barstovian; table 28.1). Each time interval is represented by a set of two to eight localities, each of which samples a moderately to highly distinct zoogeographical area within North America, such as the California Coast, Central Great Plains, or Gulf Coast (fig. 28.1). (Unfortunately, localities east of the Mississippi River are extremely rare throughout most of this interval, so this region of the United States is not included in our analysis.) The eight regions are not equally represented in the fossil record over the last 44 million years, and consequently, all zoogeographical regions could not be sampled for each time interval (see "Regions Sampled," table 28.1). We included one locality per region per time interval, and the species list for each fossil locality can be considered a sample of the actual diversity that existed in the region. To minimize problems of missing species resulting from a lack of preservation, we chose fossil localities with the greatest number of included species of both carnivores and herbivores. Localities were only included if they appeared to be representative of all localities in a region. Within the entire set of 115 fossil localities, the number of preserved species per locality ranged from 9 to 54, with a mean of 21 (SD = 7.0).

For all these analyses, we calculated both minimum and maximum estimates of species richness (i.e., number of species, and equivalent to species diversity in this chapter) because of the presence of indefinite species assignments (e.g., *Equus* sp.) in the fossil faunas. The minimum estimate ("Total," table 28.1) was calculated under the assumption that all indefinitely assigned species within a particular genus during a time interval were identical to some other contemporaneous member of the same genus. For example, in calculating a minimum estimate of turnover between the Central Great Plains and Southern Great Plains regions in the early Clarendonian, *Calippus* sp. (listed for the Central Great Plains locality; Minnechaduza fauna; Ash Hollow Formation: Webb 1969a) was considered specifically identical to one of the three *Calippus* species that occur in the approximately coeval (i.e., contemporaneous) Southern Great Plains locality (Clarendon Local Fauna: Schultz 1977). The maximum estimate was calculated assuming all such species were distinct, and would count four, as opposed to three, species of *Calippus* in this example. With the exception of figure 28.1 and table 28.1, all of the results presented here

Table 28.1.   Data Used in the Analyses

Interval[a]	Epoch[b]	Mammal age[b]	mya[c]	Carnivore[d] species		Herbivore[d] species		Total[e]	Predator/prey ratio	Mean Turnover estimate	Regions[f] sampled
1	M. Eocene	L. Uintan	44	12	(14)	33	(49)	45	0.36	0.65	4
2	L. Eocene	Duchesnean	40	9	(9)	43	(50)	52	0.21	0.79	5
3	L. Eocene	E. Chadronian	36	12	(19)	37	(44)	49	0.32	0.75	3
4	L. Eocene	M. Chadronian	35	9	(12)	23	(27)	32	0.39	0.64	3
5	L. Eocene	L. Chadronian	34	8	(9)	37	(48)	45	0.22	0.59	3
6	Oligocene	Orellan	33	7	(9)	20	(32)	27	0.35	—	2
7	Oligocene	E. E. Arikareean	28	19	(21)	37	(43)	56	0.51	0.80	3
8	Oligocene	L. E. Arikareean	25	9	(13)	16	(28)	25	0.56	0.70	4
9	E. Miocene	E. L. Arikareean	22	13	(15)	25	(27)	38	0.52	0.88	3
10	E. Miocene	L. L. Arikareean	21	11	(13)	42	(55)	53	0.26	0.73	4
11	E. Miocene	E. Hemingfordian	19	15	(25)	42	(55)	57	0.36	0.80	4
12	E. Miocene	L. Hemingfordian	17	10	(14)	54	(79)	64	0.19	0.70	5
13	M. Miocene	E. Barstovian	16	22	(27)	98	(127)	120	0.22	0.74	6
14	M. Miocene	E. L. Barstovian	14	23	(36)	76	(111)	99	0.30	0.79	6
15	M. Miocene	L. L. Barstovian	12	14	(21)	78	(105)	92	0.18	0.72	5
16	L. Miocene	E. Clarendonian	11	16	(19)	63	(90)	79	0.25	0.74	6
17	L. Miocene	L. Clarendonian	9.5	15	(25)	71	(97)	86	0.21	0.68	6
18	L. Miocene	E. E. Hemphillian	8	9	(20)	38	(67)	47	0.24	0.65	6
19	L. Miocene	L. E. Hemphillian	6.5	13	(21)	50	(77)	63	0.26	0.53	6
20	L. Miocene	L. Hemphillian	5.5	13	(24)	46	(80)	59	0.28	0.58	7
21	Pliocene	E. Blancan	3.5	12	(19)	21	(32)	33	0.57	0.54	5
22	Pliocene	L. Blancan	2.5	14	(17)	32	(49)	46	0.44	0.46	6
23	Pleistocene	Irvingtonian	1	15	(16)	33	(41)	33	0.45	0.54	7
24	Pleistocene	Rancholabrean	0.3	17	(23)	35	(44)	35	0.49	0.62	6
25	Holocene	Recent	0	14	(14)	12	(12)	12	1.17	0.33	8

Note: Epoch dates from Woodburne 1987 and from Swisher and Prothero 1990.

[a]Time interval numbers used for figures 28.2–28.5.

[b]Abbreviations for epochs and mammal ages: E, early; M, middle; L, late.

[c]Approximate midpoint date of interval in millions of years.

[d]Total number of species in each category. Numbers in parentheses refer to estimated maximum diversity when all indefinitely assigned species in a given time interval are considered to be separate species.

[e]Total number of combined herbivore and carnivore species.

[f]Number of sampled geographical regions, as shown in figure 28.1

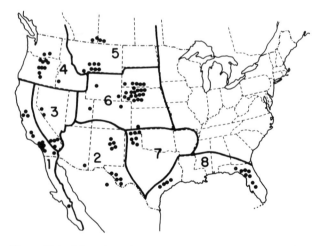

Figure 28.1   Map of North America with the eight regions utilized in this study outlined. Dots indicate approximate sites of fossil localities from each region. Regions are as follows: 1, California Central and Coast; 2, Southern Great Basin; 3, Northern Great Basin; 4, Pacific Northwest; 5, Northern Great Plains; 6, Central Great Plains; 7, Southern Great Plains; 8, Gulf Coast. The regions were modified from those used by Tedford et al. (1987) in their review of North American Miocene mammalian biostratigraphy.

are based on the minimum estimates of diversity. Trends with time in total, herbivore, and carnivore diversity levels were similar for both minimum and maximum estimates.

The analysis of provincialism was based solely on the

herbivores. Predators were excluded because carnivores tend to be more widespread and less closely tied to particular vegetation assemblages than herbivores, and therefore are less sensitive indicators of local endemism. Provincialism was estimated using the following index of species turnover between coeval localities:

$$1 - \frac{L_{12}\,(L_1 + L_2)}{2\,L_1\,L_2},$$

where L_1 and L_2 are the numbers of species in the two localities being compared, respectively, and L_{12} is the number of species shared by the two. The value of the index varies between 0 and 1, with 1 representing complete turnover, in which the localities have no species in common. This index has been used by Cody (chap. 13) to explore patterns of beta (between adjacent, dissimilar habitats) and gamma (between geographically distinct similar habitats) diversity in extant continental avifaunas. Because of insufficient data on the habitats represented by the fossil localities, it was not possible to separate these two components of species turnover in our analysis.

Indices of turnover were calculated for all pairwise comparisons of localities for each of the 25 time intervals, and then averaged to produce a mean turnover value for each time interval. In all cases, both minimum and maximum turnover values were calculated, as explained above. Only minimum estimates are presented here, as they are considered to be more realistic given that sampling biases, as described below, are likely to inflate turnover values for fossil time intervals.

All of the above measures of species diversity and geographical turnover were analyzed with respect to time. As such, they are time series and subject to statistical problems that are not present with data collected for a single point in time. Each datum in a time series cannot be considered to be independent of all others; instead, it is likely to be correlated with the points directly before and after it (McKinney 1990). This potential lack of independence can strongly bias the results of statistical analyses based on the assumption of independence, such as regression and correlation. The magnitude of this series correlation (or autocorrelation) can be examined by determination of the serial correlation coefficient (Madsen and Moeschenberger 1983). Serial correlation coefficients vary between 0 (no autocorrelation) and 1 (complete dependence), and their significance can be ascertained by calculating the Durbin-Watson statistic. For an excellent discussion of time series analysis and evolutionary trend data, see McKinney (1990). All of the time series data for this chapter were tested for serial correlation following the methodology outlined by McKinney (1990), and results indicated no significant autocorrelation.

There are, of course, biases inherent in estimating diversity and turnover indices with fossil data. There are at least three sources of variation in the fossil data that are relevant to this study: missing species; differences in the number or array of regions sampled per interval; and variation in the degree of synchroneity of localities per interval. Because of missing species, our estimates of species diversity are all minima, but we feel they are adequate for documenting relative trends through time. The proportion of species that existed but were not preserved is likely to decrease from the Eocene to the Recent because the quality of the fossil record (number of specimens, geographical coverage) improves. This tends to drive diversity estimates upward among younger intervals. However, despite this bias, our results suggest a decline in total diversity with time. The consequence of missing species for turnover indices (if the proportion missed is approximately the same in both regions) arises from underestimating the number of shared species, and thus overestimating turnover (M. Cody, personal communication). Because species sampling tends to improve toward the Recent, younger intervals may exhibit lower turnover values than older intervals.

If the localities sampled for a particular interval were not contemporaneous, both richness and turnover may be overestimated because faunas that were in fact sequential are lumped into a single time interval. The time intervals (represented in the analyses by an approximate midpoint date; table 28.1) range in duration from 0.01 million years (Recent) to 5 million years (Duchesnean) with a mean length of 1.73 million years (SD = 0.96; duration data from Woodburne 1987). Least squares linear regression of either diversity or turnover values against interval duration revealed no significant correlations, indicating that duration length is not a serious bias in our analysis. The final problem of variation among time intervals in the number of localities available for study is more difficult. An increase in the number of localities and, consequently,

the number of specimens, is likely to produce an increase in observed continental species diversity. Estimates of mean geographical turnover per interval are also likely to be affected by the number and array of regions sampled. This problem is particularly acute when comparing turnover values between early and late intervals within our sample because there are almost no Gulf Coast or California Coast localities prior to the early Miocene (late Arikareean), and the two regions are consistently sampled for only the last 15 million years. Thus, it is unreasonable to compare degrees of endemism between the Late Eocene and Pliocene when the former is represented by only three regions and the latter by seven. In the following discussions, we attempt to deal with all the above biases in at least a qualitative way and temper our conclusions accordingly.

RESULTS AND DISCUSSION

Continental Species Diversity

Continental species diversity is estimated as the sum of all species sampled in a given time interval ("Total," table 28.1). Because each region is represented by a single locality, our estimates of regional diversity are minima, but can be considered to represent approximately the same proportion of the actual regional diversity for each time interval. Similarly, continental diversity estimates are minima and represent some proportion of actual continental diversity, but this proportion will vary according to the number of regions sampled per interval. Given this caveat, the data presented here indicate that species diversity of large carnivores and herbivores in North America has not been stable over the past 44 million years. Instead, the number of large carnivore and herbivore species in North America has fluctuated widely, ranging from the present minimum of 26 species to a Middle Miocene (early Barstovian) maximum of at least 120 species (table 28.1; fig. 28.2). From the late Middle Eocene (late Uintan; interval 1) to the Late Oligocene (late early Arikareean; interval 8), species richness fluctuated and then rose steadily to a peak in the Middle Miocene (early Barstovian; interval 13). Subsequent to this peak, richness declined to the present (Recent; interval 25). This approximately 15-million-year decline was punctuated by three relatively steep drops in richness, in the early Late Miocene (late Clarendonian; interval 17), near the Miocene-Pliocene boundary (late Hemphillian; interval 20), and between the late Pleistocene (Rancholabrean; interval 24) and the Recent.

The late Middle Eocene to Middle Miocene variation in species diversity may be an artifact of sampling. As can be seen in figure 28.2, the number of geographical regions sampled follows roughly the same pattern as species richness up to the Middle Miocene. Previous researchers have noted the presence of a diversity peak for North American mammalian genera in the Middle Miocene (Webb 1983, 1989; Savage and Russell 1983), but Stucky (1990) has argued that the peak is a sampling artifact, reflecting the greater numbers of specimens available for study at this time. If Stucky is correct, and if greater geographical coverage necessarily results in a larger sample of species, then it is difficult to be confident of the reality of this Middle

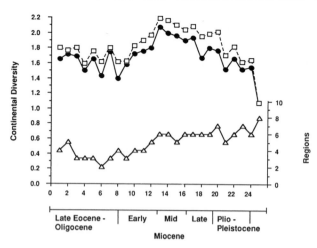

Figure 28.2 Log (10) species richness (left) and number of sampled regions (right) against time interval, oldest to the left. Solid circles, minimum estimate of diversity ("Total," table 28.1); open squares, maximum estimate of diversity; triangles, number of sampled regions ("Regions Sampled," table 28.1). The numbers for the time intervals correspond to the subdivisions of land mammal ages listed in table 28.1.

Miocene peak. However, the subsequent drop seems real, given that the number of sampled regions remains relatively large after the Middle Miocene, and that the number of specimens available for study increases as one approaches the Recent (Gingerich 1987; Stucky 1990). Notably, diversity declined over the past 15 million years, despite an overall increase in the number of immigrant species, particularly in the Middle Miocene and Pliocene (Webb 1977, 1989). Thus, immigration during these two intervals did not produce a positive pulse in continental species diversity of large herbivores and carnivores.

The probable causes of the Late Miocene to Recent drop in diversity are the changes in climate and habitat distributions mentioned earlier. In fact, the curve of decline in diversity is very reminiscent of that published for temperatures of the Northern Hemisphere (see Gingerich 1984b). Paleobotanical evidence suggests that desert and steppe environments expanded relative to forest and woodland over this interval, as the climate became more arid, seasonal, and temperate (Leopold 1967; Wolfe 1978; Singh 1988). Studies of large herbivore extinctions suggest similar changes: browsing ungulates declined first, followed by grazing species (Webb 1983; Janis 1989). Among smaller herbivores, several late Neogene rodent taxa evolved higher-crowned or ever-growing cheek teeth (hypsodonty and hypselodonty) adapted for the coarse and gritty vegetation typical of steppe and desert (Webb 1977). Moreover, these small herbivores diversified as the larger ungulates declined (Webb 1969b, 1984, 1989). This elevation in the relative diversity of small as opposed to large herbivores can perhaps be explained as a response to an increasingly continental climate with greater seasonal extremes of temperature and aridity. If such changes resulted in an absolute decrease in the length of the season favorable for reproduction (birth and growth) of herbivores, then very large species with long and intraspecifically variable gestation periods would suffer a decline in

reproductive fitness (Guthrie 1984; Kiltie 1984). Presumably, a drop in reproductive fitness would have been associated with a decline in abundance and therefore with increased vulnerability to extinction among large as opposed to small herbivores.

Local Versus Regional Diversity

Several studies of species diversity within comparable habitats in different regions have revealed marked differences in local diversity levels that do not appear explicable on the basis of local environmental conditions (Ricklefs 1987; Westoby, chap. 15; Ricklefs and Latham, chap. 20; Latham and Ricklefs, chap. 26). Rather, it appears that in some instances the number of species in a given habitat is partially determined by the number of species present within some larger geographical unit, such as a region or continent. Continental (or regional) diversity is likely to fluctuate due to historical phenomena, such as immigration and orogenic events, and thus is expected to track environmental conditions less closely than does local diversity. Ricklefs (1987) discussed the possible relationships between diversity at the local or community level and that at the regional or continental level and proposed two models of the relationship: a saturation model and a regional enrichment model. In the saturation model, species diversity within communities achieves a saturation level that is determined by local environmental conditions. Higher levels of regional diversity are not necessarily accompanied by higher levels of local diversity; instead, they reflect greater geographical turnover among species and/or the addition of new communities. In the regional enrichment model, species diversity within communities does not always reach saturation, and greater regional diversity is associated with greater local diversity. The increased local diversity is permitted by shifts in resource utilization among competitors.

Comparisons of local and regional diversity for modern communities are done under the assumption that the regions and localities being compared represent similar environments and therefore that differences in diversity are not due to environmental factors. However, in the present analysis, the relationship between local and regional (continental) diversity is analyzed over time within the same continent, and it is clear that the environment (climate, vegetation) is changing. This complicates the interpretation of results considerably, as the effects of these changes on local and continental diversity are difficult to predict. An increase in local diversity might reflect either an improved climate or greater regional diversity. Nevertheless, if local diversity was entirely dependent on regional diversity and local environmental conditions were without influence then a regression of local diversity (species per region) versus continental diversity for distinct time intervals would result in a line with a slope of one and a correlation coefficient of one; that is, every species added to the continent would appear in each of the local communities as well. In addition, this pattern would indicate that provincialism (beta and gamma diversity) did not contribute to continental diversity levels over the sampled time span. If, on the other hand, local diversity showed virtually no correlation with regional diversity, then this would favor the saturation model, in which local diversity is entirely determined by local conditions. Moreover, it would reveal a substantial role for provincialism in building up continental diversity levels. Of course, the truth probably lies somewhere between these two extremes. A least squares regression of local on continental diversity for each time interval reveals a significant, positive relationship ($p < .001$), with 57% of the variance in local diversity explained by continental diversity, but the slope is much less than one (slope = 0.17, $r = .76$; fig. 28.3A). A double logarithmic regression of the same data indicates that local diversity increases as approximately the 0.44 power of regional diversity (fig. 28.3B; slope = 0.438 [95% confidence interval = 0.189], $p < .01$, $r =$

.73). Given the lack of strict independence between the two diversity estimates, and the fact that both are estimated with error, the slope and significance values should be viewed with caution. Nevertheless, it is clear that average local diversity per time interval, defined as mean species richness per locality per interval, increases gradually with continental diversity. This suggests that regional enrichment occurs, but that it is constrained by effects of local saturation. Consequently, a buildup in continental diversity must reflect in part the addition of new communities or greater geographical turnover among contemporaneous communities. To examine this question of shifting levels of geographical turnover, it is useful to consider the history of provincialism on the continent.

Provincialism

Provincialism is a biogeographical concept that concerns the extent of local endemism within a continental fauna. A province is recognized as a restricted area that includes a set of locally endemic species (Brown and Gibson 1983). Thus, a continent characterized by a high degree of provincialism includes a number of sets of local endemics distributed in separate areas, whereas one characterized by a low degree of provincialism would include few such sets and many more widespread taxa. The number of provinces within a continent is likely to reflect environmental heterogeneity and the presence of major geographical barriers such as mountain ranges and deserts. The significant changes in climate and topography that occurred over the last 44 million years affected both environmental heterogeneity and the magnitude of geographical barriers within North America, and thus were likely to affect the prevalence of local endemism.

Mean turnover values between regions for the 24 fossil time intervals (fig. 28.4A) range from a low of 0.46 in the Pliocene (late Blancan; interval 22) to a high of 0.88 in the Early Miocene (early late Arikareean; interval 9), with a mean of 0.68 (SD = 0.11). As might be expected given the sampling problems, all the turnover values for fossil intervals exceed that calculated for Recent North American herbivores (0.33), for which data on species richness per region are complete and comparisons are certainly contemporaneous. The variation, as evidenced by the standard deviation bars in figure 28.4A, appears high, but this is expected as adjacent regions are likely to exhibit lower turnover values than are more distant regions. Nevertheless, the unusually high variance for the Recent is surprising, but is explained by a combination of relatively homogeneous interior faunas and more distinct coastal (Florida and California) faunas. The plot of turnover indices against time (fig. 28.4A) shows an initial pattern of fluctuations about a steady state up until the Middle Miocene (early Barstovian, interval 13), followed by a moderately steady decline to the present. With the exception of the notable low in the latest Eocene (late Chadronian, interval 5), the turnover values prior to the Middle Miocene do not differ significantly among themselves (Tukey test, $P > .10$). In part, this reflects the relatively small number of regions sampled for most of these older intervals (see table 28.1) and the disproportionately large effect of the addition or loss of a single region on mean turnover val-

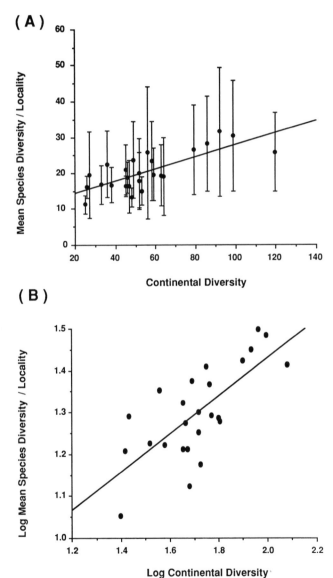

Figure 28.3 (A) Regression of mean species diversity per locality against estimated continental diversity (TOTAL, table 28.1) for each time interval. Error bars represent one standard deviation. Equation for least squares regression: $y = 0.168x + 11.2$; $r = .76$; 95% confidence interval for slope = 0.066. (B) Log/log regression of the same data as in (A). Equation for least squares regression: $y = 0.455x + 0.521$; $r = .73$; 95% confidence interval for slope = 0.189.

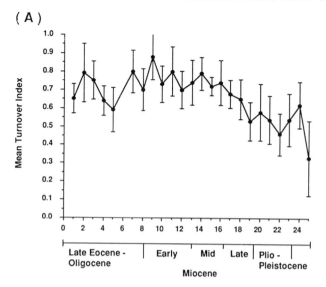

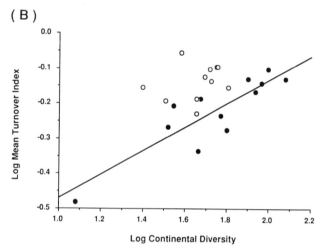

Figure 28.4 (*A*) Mean turnover index per interval ("Mean Turn-over estimate," table 28.1) plotted against time interval (as in fig. 28.2). Error bars represent one standard deviation of the mean. (*B*) Log (10) of mean turnover index plotted against estimated continental diversity ("TOTAL," table 28.1) for each time interval. Solid circles represent younger time intervals, from Barstovian to Recent; open circles represent older, pre-Barstovian intervals. Equation for least squares regression based on younger intervals: $y = 0.806 + 0.336x$; $r = .88$; $n = 12$. Note that removal of the value for the Recent represented by the low, outlying point lowers the r to .74.

ues. For example, the peak turnover value reached in the Early Miocene (early late Arikareean, interval 9) may be a sampling artifact, as this is the earliest time interval for which Gulf Coast localities are available. Not surprisingly, the Gulf Coast fauna shares relatively few species with the approximately coeval Southern Great Basin and Central Great Plains faunas, and the average turnover rate for the interval is high. Given the potentially large effects of sampling different regions on the turnover index, it is unwise to draw conclusions concerning trends in provincialism for time intervals represented by fewer than six regions (i.e., before the Middle Miocene [Barstovian]). All Barstovian and later faunas include samples from California, the Gulf Coast, and the interior regions, and thus can be compared with some confidence.

The trend of declining provincialism from the Middle Miocene to the Recent differs significantly from the near steady-state conditions that occurred prior to the Middle Miocene. This was determined by calculating the mean and standard deviation of the pre-Barstovian turnover values, and establishing that the post-Barstovian values drifted below two standard deviations of the mean (see McKinney 1990). This decline is interrupted by a single marked rise between late Pliocene and Pleistocene time (intervals 22 and 23). Sampling biases might be partially responsible for the post–Middle Miocene decline in turn-over values, since the proportion of missing species is expected to fall as the Recent is approached. Although it is not possible to accurately assess this bias at present, we doubt this proportion has changed dramatically for large herbivores between the Middle Miocene and Ranchola-brean. Most of the included herbivores were larger than 15 kg, and taphonomic studies indicate that mammals in this size range are likely to be represented in fossil assemblages (Behrensmeyer, Western, and Boaz 1979). Moreover, paleoecological analyses of several Middle and Late Miocene ungulate assemblages reveal structures (arrays of body sizes and dietary types) that are comparable to those of living assemblages, suggesting that missed species have not strongly distorted our view of the past (Janis 1982; Webb 1983).

The long-term decline in provincialism after the Middle Miocene was accompanied by a similar drop in continental species richness (see fig. 28.2; table 28.1). Thus, as the number of herbivore species on the continent fell, those that remained or appeared tended to be more widespread. For example, the number of equid species within our sample declined from twenty-two to three between the Late Miocene and Pliocene. Of the surviving three species, two were widely distributed, occurring in three to five of the five sampled regions, whereas only one of the Late Miocene species was similarly distributed. Preferential extinction of local endemics is expected if changes in climate reduce habitat diversity such that the evolution and survivorship of species with more generalized habitat and dietary requirements are favored.

The positive association between continental species richness and turnover values that characterizes the later Cenozoic (Middle Miocene-Recent) intervals is not apparent among earlier intervals (open versus solid circles, fig. 28.4B). As stated above, turnover values for earlier intervals should be viewed with caution because they are based on a smaller number of geographical regions than those for later intervals. Exclusion of these earlier intervals from the regression analysis increases the correlation coefficient (*r*) considerably, from .60 to .88 (.74 when the Recent is deleted).

The apparent presence of a long-term trend of declining provincialism is surprising, given that the time span encompasses two notable extinction events, significant fluctuations in the intensity of immigration, and major climatic and orogenic changes. Rapid and widespread extinction, such as occurred in the late Hemphillian (ca. 5 mya) and Rancholabrean (ca. 0.01 mya) could produce a sharp drop in local endemism if more ecologically restricted taxa were relatively susceptible. The turnover data suggest that such a drop occurred after the Rancholabrean extinctions, but not after those of the late Hemphillian, despite their greater magnitude (Webb 1984). To test whether the decline in provincialism was due to preferential extinction of localized species, it would be necessary to classify species prior to the extinction event as ecologically restricted or widespread and then compare their survivorship (see, for example, Jablonski 1986). Alternatively, the drop in endemism after the Hemphillian extinction event might simply reflect the fact that surviving species were able to expand their ranges because of the decline in total diversity, and might imply little concerning the relative specialization of survivors and victims.

Apparently, the number of Eurasian immigrants per interval had no noticeable effect on turnover values at the examined time scale. There are two major waves of Eurasian immigrants in the late Cenozoic (see fig. 28.4A), one during the late Early Miocene (Hemingfordian, intervals 11–12) and the other during the late Pliocene (Blancan, intervals 20–21) (Tedford et al. 1987; Webb 1989). Although both are associated with slight drops in the turnover index, these are not more conspicuous than drops at other times. By contrast, the invasion of South American mammals following the closure of the Panamanian isthmus does appear to have contributed to a rise in the mean turnover index (intervals 22–24). This presumably reflects the relatively restricted distribution of some South American invaders to the warmer southern regions of the United States. These southern immigrants suffered greatly during the Rancholabrean extinctions (Webb 1984), and their loss contributes to the sharp decline in provincialism between the Rancholabrean and the Recent.

It is perhaps most surprising that the climatic and orogenic changes of the last 15 million years are not associated with increasing provincialism in our analysis. As mentioned above, the Sierra Nevada underwent significant uplift over the last 9 million years, and this is likely to have isolated West Coast from interior faunas. Moreover, the increase in latitudinal temperature gradients in the Northern Hemisphere in the later Cenozoic is likely to have produced a continent with more distinct climatic and vegetational zones. Such increases in environmental heterogeneity are expected to be associated with greater geographical differentiation of the mammalian fauna (Cody 1975, 1986). In addition, a number of previous researchers (Webb 1977; Tedford et al. 1987; Stucky 1990) have stated there was a general pattern of increasing provincialism among North American mammals over the course of the Cenozoic. Detailed work on the transition from Pleistocene to modern distributions of living species in North America reveals numerous, significant range contractions such that formerly sympatric species are now allopatric (Graham 1976; Graham and Lundelius 1984; Guthrie 1984). This should be manifested as a rise rather than a decline in provincialism. The reasons for this discrepancy between our results and the previous work are not entirely clear, but probably relate to differences in data and approach. First, the present analysis is based solely on large herbivores, whereas the earlier studies considered a broader diversity of mammals, including small herbivores. Small herbivores, such as rodents, are likely to have smaller geographical ranges and exhibit greater degrees of local endemism than are large, wide-ranging herbivores, such as equids (Brown and Gibson 1983). Second, in the previous studies, conclusions concerning provincialism were based on qualitative assessments of geographical turnover among particular regions. In particular, the increasing isolation of the West Coast faunas has been emphasized (Webb 1977; Tedford et al. 1987). It may be that the quantitative approach taken here provides a more comprehensive view of total continental provincialism, rather than focusing on greater local endemism in one or two regions. Notably, however, a review of average turnover values per interval for comparisons between the California Coast region and all other regions reveals no significant trend in greater endemism in the California Coast region since the Middle Miocene.

Predator versus Prey

The number of large carnivore species appears to have roughly tracked the number of large herbivore prey over the last 44 million years, although after the Middle Miocene, herbivore diversity appears to have declined while carnivore diversity remained relatively stable (fig. 28.5A). A test of cross-correlation between successive changes in predator and prey richness after detrending each data set (performed with the SERIES module in SYSTAT:PC; Wilkinson, 1988) showed a significant association between contemporaneous shifts without any significant lags. Nevertheless, a log/log regression of predator on prey richness shows only a weak association ($r = .50$) and a general pattern of prey numbers increasing much more rapidly than predator numbers (fig. 28.5B; slope = 0.30, $p <$ .05). Estimates of continental prey diversity over the past 44 million years (see table 28.1) range from a low of 12 (Recent) to a high of 98 (Middle Miocene [early Barstovian]), whereas those for predators span a narrower range, from 7 (Early Oligocene [Orellan]) to 23 (Middle Miocene [early late Barstovian]). The Recent North American fauna is clearly unusual in that the number of carnivores exceeds the number of large herbivores. Previous work on modern carnivore guilds indicates that this is a consequence of the Pleistocene extinctions and of the prevalence of omnivory among North American mammalian carnivores (Van Valkenburgh 1989).

The ratio of predator to prey richness appears to decline rapidly as prey diversity rises (fig. 28.5B). This is shown more clearly in figure 28.6, where the ratio of predators to prey is plotted against total continental diversity. This intriguing result suggests that the ecological relationship between predators and prey has varied significantly over the Neogene (see table 28.1), with some intervals showing much higher ratios of predators to prey (e.g., 0.57 in the early Pliocene [Blancan]) than others (e.g., 0.19 in the Early Miocene [late Hemingfordian]). However, as with the Recent fauna, this might be partially a result of differences among intervals in the relative numbers of highly carnivorous as opposed to more omnivorous carnivores. Future work should focus on comparisons of dietary specialization among predators from time intervals with differing predator-prey ratios. It is predicted that intervals characterized by high predator-prey

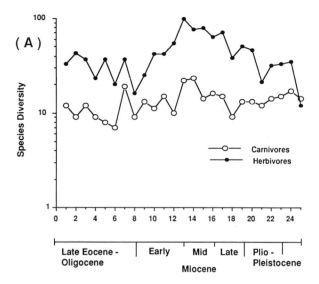

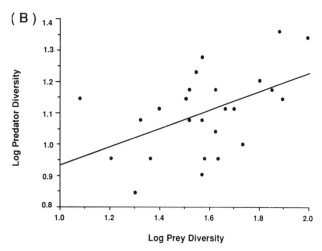

Figure 28.5 (*A*) Logarithmic (base 10) plot of estimated total number of large prey (solid circles; Herbivore species, table 28.1) and total number of large predators (open circles; Carnivore species, table 28.1) plotted against time interval (as in fig. 28.2) and (*B*) against each other. Equation for least squares regression: $y = 0.64 + 0.30x$; $r = .50$; $n = 25$.

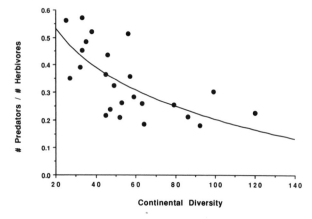

Figure 28.6 Plot of predator/prey ratio (see table 28.1) against estimated continental diversity ("Total," table 28.1) for all 24 fossil time intervals. The Recent value is not shown as it appears aberrantly large (1.17). The line was fit to the data by a logarithmic least squares regression: $y = 1.15 - 0.475\ln(x)$; $r = .68$; $n = 24$.

ratios will include relatively more omnivorous and generalized predators than those with low predator-prey ratios.

The pattern of declining predator-prey ratio with increasing total species richness (fig. 28.6) indicates that the relationship between species richness of predators and prey is not constant. During times of diversity buildup or decline (e.g., post–Middle Miocene to Recent), the numbers of herbivore species changed much faster than those of their presumed predators. A one-to-one relationship between predator and prey was not expected, but considerations of trophic energetics might have predicted a stable relationship such that a doubling or halving of prey species was always associated with a predictable response, such as a 10% to 20% shift, in the number of predator species. Although the slope of the log/log regression is significant, ($p < .05$) and suggests such a predictable relationship, the correlation coefficient is low (.50) and there is considerable scatter about the line. Certainly a tighter fit would be expected if predator and prey were considered as units of biomass (kg/km^2) rather than numbers of species. However, it is more troublesome to predict how the addition or loss of a new prey species to a community modifies the resource base from a predator's point of view. A new species might result in an increase in available biomass and, depending on its ecology, might represent a new type of prey that favors the evolution of a specialized predator. However, it might not. Resource partitionment among living sympatric carnivores is accomplished largely on the basis of prey size rather than specific prey identity (Rosenzweig 1966; Schaller 1972; Gittleman 1985), and it may be that the addition of a few herbivore species to a community contributes relatively little to the expansion of potential prey opportunities for predators.

The pattern of prey species diversity outpacing predator diversity suggests that large herbivore diversity is less tightly controlled by either predator diversity or resource levels than is carnivore diversity. Given their elevated position in the trophic chain and the costs of prey capture, carnivores are expected to compete more intensely for food than are large herbivores (Hairston, Smith, and Slobodkin 1960; Schoener 1974). Although competition is expected to drive diversification, it may be unable to do so among carnivores, for which it appears there are a limited number of ways to partition the resource (prey) successfully. Prey represent a very different type of resource than plants, all of which are composed of separate, renewable parts (e.g., fruits, leaves) that can often be removed without destroying the entire plant. Although sympatric herbivore species may be specialized to feed on different parts of the same plant, predators are unlikely to specialize on a portion of a prey animal, given that acquisition of a particular part (e.g., liver) requires killing the entire animal. Among terrestrial mammals, predation is often difficult and risky and the cost appears too great not to consume all or most of the prey. If the array of successful predatory strategies is constrained by characteristics of the resource, then diversification might be dampened despite competition. Again, it would be useful to study the variation in predator and prey numbers over time from a morphological viewpoint, with analyses of body size distributions and locomotor adaptations of predator and prey for each time interval.

CONCLUSIONS

The diversity of large North American herbivores and carnivores has declined since the Middle Miocene, both at the continental and the local level. Over the same interval, provincialism among the herbivores has become less pronounced. This pattern suggests that the decline in diversity lowered levels of competition and allowed the remaining species to expand their ranges, thus reducing geographical turnover. The data imply that a rise in diversity at the continental level is likely to be accompanied by greater geographical differentiation among regional faunas, perhaps as competition results in niche compression and range reduction. The decreased provincialism that accompanied reduced continental (and local) diversity during the later Cenozoic is consistent with this explanation. Alternatively, climatic change might have affected the distribution of habitats such that the continent became more homogeneous (resulting in less provincialism among herbivores) and species specialized for the diminishing habitats became extinct.

The paleoclimatic and paleobotanical records indicate that there has been substantial change in temperature, precipitation, and seasonality since the Middle Miocene. Since the Eocene, much of North America has shifted from predominantly wooded, closed habitats to more open, savannalike grasslands and then to steppe. Temperatures dropped, aridity increased, and seasonality became more pronounced. Aspects of this transition should have favored the diversification of large, terrestrial herbivores and their predators as low-stature, accessible vegetation such as grass replaced high-stature, relatively inaccessible vegetation such as trees (Guthrie 1984). Nevertheless, large herbivore diversity declines after the Barstovian, suggesting that conditions were deteriorating. Given the coincident rise in small herbivore diversity, it is suggested that increased seasonality and shorter summers may have selected against species such as large herbivores with relatively long gestation periods.

Here and in previous examples (Webb 1969b, 1983, 1989; Marshall et al. 1982), it appears that species diversity among North American large herbivores has at least approached, if not achieved, saturation levels set by current environmental conditions, at least over the examined time scale. If this were not the case, pulses of immigration, such as occurred in the Miocene and Pliocene, should have produced a noticeable rise in continental diversity. In fact, Webb (1989) has argued that the rate of immigration of species to North America has been controlled by continental diversity levels, such that immigrants are only successful when North American diversity levels have dropped sufficiently (for extrinsic, climatic reasons) to "make room" for new taxa.

Although the parallel declines in diversity, temperature, and precipitation may be understandable, the coincident drop in provincialism is more difficult to rationalize given that environmental heterogeneity appears to have increased over the same interval. Evidence for increased het-

erogeneity includes an enhancement of the latitudinal temperature gradient, which is likely to have narrowed latitudinal vegetational zones, and the continued uplift of the western mountain ranges, which probably further isolated western from interior faunas. Nevertheless, the turnover data indicate that geographical turnover among regions was higher in the Middle Miocene when the climate was milder and the mountains presented less of a barrier. It is possible that our interpretation of the degree of environmental homogeneity in the Middle Miocene is incorrect, and that there may actually have been a wide array of distinct habitats that encouraged the evolution of local endemics. The array of habitats was reduced after this time, despite the increased latitudinal temperature gradient, and provincialism among large herbivores declined. Clearly, a parallel analysis of the history of plant community diversity in North America would improve our understanding of changing patterns of herbivore diversity.

Additional support for a positive correlation between species diversity and provincialism can be found in studies of global taxonomic richness among Paleozoic marine invertebrates. For example, Sepkoski (1988) demonstrated that an increase in global generic diversity in the Ordovician was accompanied by both greater levels of alpha diversity within communities and greater geographical differentiation between communities. Similarly, Bambach (1985) argued that diversity increase in the Phanerozoic was accomplished by packing more species into communities and by increased specialization of species on previously underutilized aspects of the ecospace. Together, studies such as these and the present analysis, done at disparate time scales (hundreds versus tens of millions of years) and on very different communities (marine invertebrates versus terrestrial mammalian herbivores), provide strong evidence concerning processes of ecological diversification. In all cases, greater total diversity (global or continental) is achieved both through increases in local diversity and through greater geographical differentiation among communities; that is, through increases in both beta and gamma components of diversity.

Future studies of patterns of diversification and decline could benefit from ecomorphological analysis of species. During times of diversification, were the new species more likely to be added to communities at the periphery of occupied ecospace (as inferred from morphology)? During times of decline, did the victims share any set of features, such as large size or specialized dental or limb morphology, that might suggest why they were vulnerable? Established morphological indicators of body size and dietary and locomotor adaptations for large herbivores could be applied to these questions (e.g., Scott 1985; Janis 1984, 1988, 1990; Solounias and Dawson-Saunders 1988; Kappelman 1988). In addition, analysis of the geographical distributions of species that fail to survive an extinction event should reveal the relative susceptibility of stenotopic and eurytopic species.

The analysis of diversity trends in large predators and their presumed prey suggests a more complex relationship between the two than was expected. Although the number of predator species is related to that of prey species, the ratio of predator to prey declines as the number of herbivore species rises. It is difficult to discern why predators fail to diversify as much as their prey do, but it may reflect both the greater tendency toward provincialism among herbivores and the limited number of ways predators can subdivide their resource (prey). As mentioned above, carnivores are less likely than herbivores to be restricted to a particular vegetation assemblage; the quality of meat does not vary geographically to the same degree as that of vegetation, and although optimal predation strategies may differ among habitats (e.g., ambush in forest, pursuit in savanna), many predators are able to hunt successfully in a wide range of vegetation types (e.g., wolf in forest and prairie). If there are intrinsic limitations to diversification among predators, then it is not surprising to discover that repeated examples of parallel and convergent evolution characterize the history of mammalian carnivores. In North America, sabertoothed-cat-like predators have evolved independently at least three times (Felidae, Nimravidae, Creodonta); hyenalike beasts have appeared at least twice (Canidae, Oxyaenidae); and large doglike predators have appeared in four families (Canidae, Amphicyonidae, Ursidae, Hyaenidae) (Butler 1946; Savage 1977; Martin 1989). By contrast, North American herbivores exhibit much less iterative evolution of ecomorphs over the Cenozoic. Instead, as mentioned in the introduction, the history of large herbivores is dominated by directional trends in limb and dental morphology that can be explained as responses to environmental changes (Janis and Damuth 1990). Not surprisingly, morphological evolution among herbivores appears much more closely tied to changes in the physical environment than that among predators. Moreover, ecomorphological analyses suggest that coevolution among predators is at least as important as coevolution between predator and prey in explaining the adaptive morphology of carnivores (Bakker 1973; Van Valkenburgh 1985, 1988, 1991; Janis 1993).

The unique contributions of this chapter to the study of mammalian diversification patterns are its quantitative analysis of provincialism based on species-level data from selected localities and its examination of predator and prey richness through time. Although the above-discussed results are intriguing and gratifying in their confirmation of patterns predicted by previous work, this chapter must be considered a beginning and far from the final answer to questions of diversity controls at the local and regional level. The analysis of provincialism would be strengthened by the addition of more localities per region, and its conclusions should be tested with a parallel analysis of small herbivore diversity. In addition, the hypothesized process of increased diversification through greater specialization of species and endemism within communities should be explored with ecomorphological analysis. It is time we stopped simply counting taxa and tracking their numbers over time, and began looking at them, measuring them, and estimating their ecological roles. Without this added perspective, our conclusions concerning diversification over evolutionary time remain incomplete.

29

Fossil Communities: Compositional Variation at Many Time Scales

James W. Valentine and David Jablonski

The contrast between the apparent harmony of activity observed within biotic associations and the often highly distinctive, individualistic requirements and tolerances of different species has prompted very different views of the nature of biological communities. In one view, communities are visualized as little more than the coincident overlap of the ranges of species that happen to share certain tolerances (Gleason 1926). In other views, communities are lightly integrated through adaptations that might permit such processes as competitive exclusion of one community by another (Sepkoski and Sheehan 1983), or are so thoroughly integrated as to become virtual superorganisms (Clements 1936; Dunbar 1968). Our view, based chiefly on fossil evidence from marine communities, is that removal, addition, and substitution of species within marine community associations is common in nature and in fact is the rule over time. The data also suggest that resources within marine communities are not fully allocated, so that many are so "open" and loosely integrated that the fraction of resources actually exploited may vary importantly over time. Species immigration and local extinction may create demand for or release of local resources over scales of 10^3–10^5 years—one to three orders of magnitude less than average species durations. These inferences lead to our main conclusions, which are hypotheses to the effect that change rather than stability is the normal lot of communities over ecological as well as evolutionary time, and that their flexibility is a normal outcome of macroevolutionary processes and indeed permits ecosystem structure to survive environmental perturbations without undue effects on incumbent populations.

Although the fossil record cannot be used routinely to study details of population and community interactions and dynamics (as proposed by Caswell and Cohen, chap. 9), it can reveal patterns of association and change in community composition over time scales that are beyond the reach of neontology, and to which modern patterns cannot necessarily be extrapolated. Marine fossil data can usefully be brought to bear on some of the questions that are of interest in community ecology: for example, whether communities normally utilize all available resources to support a maximum number of species, and whether invasions create important effects among native populations. Accordingly, we shall emphasize these particular questions from the standpoint of the marine fossil record. It is nevertheless appropriate to examine briefly some studies of living communities that provoke us to ask questions of the fossil record.

LIVING COMMUNITIES

In a review of the effects of introduced species upon their new communities, nearly all terrestrial, Simberloff (1981) found that surprisingly little effect has been reported for most introductions: the effects of the human population itself aside, it is uncommon for the species that humans introduce to cause extinctions (only 71 out of 854 studied introductions are associated with extinctions). A difficulty in accepting these results at face value is that the time spans, both following the introductions and over which the studies were conducted, are quite short, and it is conceivable that as generations pass, some populations in the invaded communities may become extinct as a result of the introductions. Furthermore, some effects on the native populations short of extinctions were detected, and other effects may be difficult to detect, so that we cannot conclude that introductions cause *no* change in community structures beyond an increase in diversity (and see other criticisms in Herbold and Moyle 1986). Having said that, however, these studies do suggest that, while introductions may occasionally devastate portions of an ecosystem, creating local extinctions (Vitousek 1986; Atkinson 1989), they commonly have little effect, especially when they involve non-island settings. Just how often introduced species find hospitable communities cannot be specified, however, because the percentage of attempted introductions that are unsuccessful because of one or another form of resistance from native community members is not usually known (Simberloff 1981). However, for one data set of insects that does contain failure statistics, Simberloff (1986) tentatively concluded that the success of the introductions depends more on the habitat requirements of the introduced species than on the composition of the target community. Overall, Simberloff's reviews do suggest that communities are commonly invaded, implying that there are resources that are either unallocated or that can be released to support immigrant populations without undue effects on the ecosystems involved.

There have been few neontological studies that have dealt with introductions and invasions of marine communities. The notable exceptions are in bay and estuarine settings, where introductions have been extensive indeed.

Species such as oysters were introduced for commercial purposes, along with their associated parasites, and others arrived as larvae in ships' ballast (see Carlton 1987; Williams et al. 1988). In San Francisco Bay, for example, nearly 100 species in six phyla have been identified as having been introduced within the last 130 years or so, and probably many more are yet to be detected (Carlton 1979). Most of the introduced species that have been identified are conspicuous and at least locally abundant forms. In one well-studied natural arm of the Oakland Estuary, a brackish environment, about 80% of the invertebrate species are introduced forms. The introductions have come from numerous regions of the world, including New Zealand, Chile, the Mediterranean, Japan, and the North Atlantic. The species thus assembled evolved not only in different regions but of course in different biotic milieux, yet they display no obvious difficulty in exploiting their new environments: they assimilate the local production, channel energy along trophic chains and webs, and perform an array of community functions that seem completely ordinary. If one were not aware of their exotic and disparate provenances, these species associations would not be thought at all unusual (Carlton, personal communication.). This situation makes it difficult to detect invaders, which are sometimes formally described as if they were native species. The very ordinariness of these assemblages holds little support for those who suggest that evolution creates tightly coadapted communities.

Whether any of the invaders of San Francisco Bay caused local extirpations among native species is not certain, but no actual extinctions have been recorded. Competitive displacement of a native gastropod by an introduced species has been recorded on the New England coast (Brenchley and Carlton 1983); both species continue to coexist, but the native is now restricted to a portion of its niche that is unavailable to the introduced form. No extinctions, but possible niche adjustments, have been

suggested in the eastern Mediterranean, where over 200 Red Sea species have invaded via the Suez Canal (Por 1978). It is possible that some of the invaders in San Francisco Bay have caused similar displacements; for example, the Asian bivalve *Potamocorbula amurensis*, which arrived in 1986, has almost entirely preempted the return of the native dry-period community in the northern Bay (Nichols, Thompson, and Schemel 1990). Some of the invaders may simply have filled "empty niches," for the early records of the fauna in the invaded habitats indicate that it was relatively depauperate (quoted by Carlton 1979). Thus, to date, the invading species have served to enrich the local fauna, and significantly so. Northeastern Pacific invasions are not restricted to San Francisco Bay but are of widespread occurrence in bay to estuarine settings in the region (see Carlton 1989 for introductions into the National Estuarine Research Reserve, Coos Bay, Oregon), and are known from many parts of the world. Figure 29.1 shows the best-known invasion routes across the Pacific Ocean, involving hundreds of successful species introductions.

Some caveats are required before the San Francisco Bay examples can be generalized, to be sure. Those novel communities are composed of organisms that are adapted to estuarine conditions, one of the more variable of situations, and thus are perhaps more capable of successful adventitious integration than organisms that normally face fewer environmental perturbations. That the introduced species are hardy is indicated by their survival of transport and their colonization. Alas, we have no idea how many species have "tried" to join these communities and failed because they could not find a viable role within the new ecosystem. Similar wholesale invasions have not occurred in open ocean habitats such as exposed rocky shores, beaches, and shallow shelves. However, it may be that this habitat selectivity is due to selectivity in the sources and destinations of ballast water and of such commercial transplants as oysters, or that, as the bay habitats have

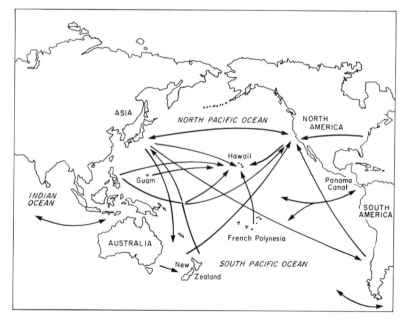

Figure 29.1 Some dispersal routes of introduced species across and into the Pacific Ocean. (After Carlton 1987.)

been anthropogenically altered, they may have become particularly receptive to some invaders. At any rate we now have some questions to ask of the fossil record: Do natural communities appear to have flexible compositions? Can nonanthropogenic biogeographical invasions from distant sources regularly succeed? Are some types of communities more prone to invasion than others? Are some community roles (herbivores, top predators, etc.) more easily replaced or partitioned than others?

MARINE PLEISTOCENE COMMUNITIES

The Pleistocene (1.6 to 0.012 million years ago) was a time of unusually rapid environmental change, geologically speaking, with a series of glaciations and associated climatic modifications and shifts, and swinging sea levels with their associated changes. Many of the Pleistocene marine invertebrate species are still extant, and their present distributions, associations, and in some cases general biologies are known. Thus the living biota may serve as a reference datum against which to measure changes in those features.

During the Pleistocene, glacial lowstands of the ocean alternated with interglacial highstands. Several coastal regions and islands of the world contain remarkably good fossil records of high interglacial sea stands on marine terraces, with rich shallow-water marine invertebrate faunas, chiefly mollusks. Low glacial oceans, the shorelines of which were well below the present sea level, have left less of a record (that is accessible, at any rate), although a few down-warped coastal basins captured glacial faunas. The extratropical northeastern Pacific molluscan fauna is well known both from Pleistocene and Recent studies. Here we use this fauna as an example. In the Californian Province, nearly 80% of the living molluscan spe-

cies are found as fossils, and no species that are common today or that are believed to be of special ecological importance are missing from the Pleistocene record (Valentine 1989a). However, many species that live today only in provinces to the north or south of the Californian region are found in Californian Pleistocene assemblages (17.4% of the fossil species). At many fossil localities, species living only to the north today co-occur with species living only to the south, all intermixed with Californian residents. Many other species that now end their ranges within the Californian region are found as fossils outside of their present living ranges. The fossil records of groups other than mollusks are not so well known, but they do display similar biogeographical patterns.

Typical interglacial Pleistocene assemblages, represented at fossil localities in Baja California, are summarized in figures 29.2 and 29.3. These localities probably represent oxygen isotopic state 5e and are about 125,000 years old. At Punta Baja (fig. 29.2), the assemblage is chiefly a rocky shore community; there are fourteen species present that live only to the north of Punta Baja today, and none that live only to the south. Thus these extralimital forms have shifted to the north since the assemblage accumulated, suggesting that temperatures are warmer today. The withdrawal of species northward has been variable in extent, however, with some forms retreating entirely into the Oregonian province, while others remain in the Californian province today but only to the north of the fossil locality. Other species may well have undergone range shifts as well, but unless their range endpoints pass the fossil locality they do not appear to shift in the figure. Some of the species have ranges that do not overlap today, thus the assemblage is anomalous by present biogeographical standards.

The other locality, at Bahía San Quíntín (fig. 29.3), is a

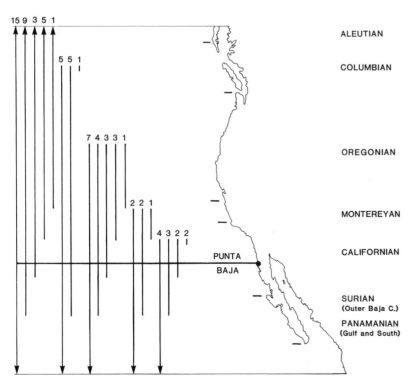

Figure 29.2 Living latitudinal ranges (vertical lines) of species occurring in the Pleistocene at Punta Baja, Baja California (at latitude of horizontal line), classed by province or subprovince and plotted to their centers. Numbers of species are indicated at the tops of the range lines. Within the Californian province, the segments north and south of the fossil locality are treated as if they were subprovinces. (Provincial regions after Valentine 1966; fauna from Emerson and Addicott 1958 and Valentine and Meade 1961.)

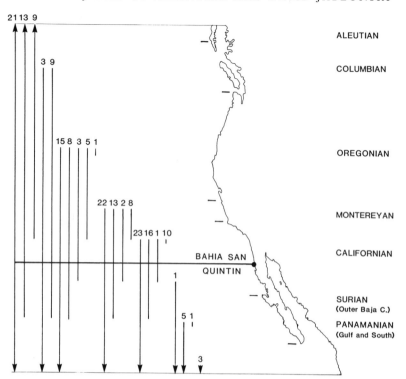

ALEUTIAN

COLUMBIAN

OREGONIAN

MONTEREYAN

CALIFORNIAN

BAHIA SAN
QUINTIN

SURIAN
(Outer Baja C.)

PANAMANIAN
(Gulf and South)

Figure 29.3 Living latitudinal ranges (vertical lines) of species occurring in the Pleistocene at Bahía San Quíntin, Baja California (at latitude of horizontal line), classed and plotted as in figure 29.2. (Fauna from Jordan 1926 and Valentine 1961.)

mixture of a shallow quiet-water assemblage with rocky shore forms (the site is in the lee of volcanic cones). As there are at least two major biotopes represented, all the species probably did not share the same food web. Of the 191 species for which data seem adequate, 32 live today only to the north, while 10 live only to the south. This mixture creates a number of anomalous species associations, not only among the extralimital species but between them and those forms that live at the locality today but do not range far enough northward or southward to share overlapping living ranges with them. The climatic conditions involved in creating this assemblage are uncertain.

Although some cases of species interactions in such assemblages can be worked out from taphonomic information in the shells (e.g., Walker 1989), details of interactions among the majority of the populations are lacking. Resolution of aspects of these communities is thus poor by neontological standards, but the consistent anomalous associations, found at scores of Pleistocene localities over thousands of kilometers in the Californian and adjacent regions, leave little doubt that the communities were compositionally distinct from those of today.

Clearly, the compositions of these local marine invertebrate communities have been changing significantly through time. As the known fossil ranges usually underestimate the actual ranges—there is little chance that we will find the most northern or most southern individual as a fossil—the number of range changes, and the consequent changes in species associations, must have been even greater than the data show. Terrace deposits that accumulated during interglacials that were relatively warm, (judging, for example, from oxygen isotope studies of deep-sea cores: Shackleton and Opdyke 1973), contain numbers of southern species. Deposits representing cooler

interglacial times may lack southern species altogether but contain species found in northern provinces today (see Wehmiller et al. 1977). Although there are no closely dated basinal deposits from glacial times, the presence of significant numbers of northern species at some horizons suggests glacial ages. In short, so far as we can tell, the range changes were driven by changes in the climatic regimes associated with glacial cycles.

The use of the present biogeographical pattern as a datum is not meant to imply that the present ranges and associations carry some special significance (Russell 1990). If we were to wait for the next interglacial age, we would surely find that the present fauna contains a goodly number of species whose ranges lie outside the (future) datum ranges. Which species these might be seems impossible to predict; communities are not precisely duplicated among interglacials, each of which may represent somewhat unique conditions. The Pleistocene picture, then, is one of species shifting independently up and down the shelf as sea level rises or falls, and northward or southward as temperatures warm or cool; the living biota is merely a "snapshot" of this process. So far as can be told, the shifts do not involve entire communities as such, but rather, species respond individually, some changing their ranges by hundreds of miles and some by tens.

The communities of the Californian shelf, then, have undergone multiple species invasions and departures, but the compositional changes do not seem to have had detectable effects upon the biota. By the standards of the Neogene (Stanley 1979), the extinction rate on the Pleistocene shelf is low, showing no particular "spikes" of acceleration despite the climatic swings with their associated range shifts and compositional changes within communities (Valentine and Jablonski 1991). Furthermore, although the Pleistocene fauna does include some estuarine communities, by far the greatest proportion of the fauna represents open marine habitats, many along what were clearly exposed rocky headlands and some from moderate depths, and these "normal" marine communities contain the vast bulk of mobile species. Thus, these invasions and departures were not restricted to communities in particularly variable habitats. Among the species that display range shifts are infaunal siphonate bivalves, nonsiphonate bivalves, epifaunal bivalves, grazing gastropods, browsing gastropods, predatory gastropods of several types, and ectoparasitic gastropods—essentially the entire molluscan spectrum of habitat types and trophic predilections is represented. It does not seem possible to say that any particular faunal element has been more or less prone to invasion than others.

The Pleistocene example supports those cases described among living biotas that suggest that communities may easily be altered in composition, and therefore are capable of maintaining trophic webs that function well despite important species turnovers and, in all probability, changes in diversity. It seems unlikely that all of the invaders are being extensively adapted by selection to form components in some closely integrated coadapted ecosystem. A more plausible explanation for these data is that the communities were rather open and that species could enter or leave with little effect on the ecosystem as a whole. In this case, energy pathways and other community-level features were in flux, and the "adaptive space" occupied by community member species varied through time, as might have also the resources potentially available to support additional species' populations.

The Pleistocene data do not permit us to rule out the possibility that some would-be invaders were excluded by native populations. Neither are the data good enough to permit the identification of any changes in native populations or any population interactions, short of extinction, that may have been associated with the invasions. One would expect such changes to have occurred. There is some experimental evidence (Goodnight 1990a, 1990b) suggesting that replacement or addition of prey, predators, or competitors may result not only in simple ecological repatterning within the community but also in genetic modifications of some of the species involved. It seems possible that if such evolution occurred during an interglacial and, especially, during the next glacial, it would present invaders in succeeding interglacials with novel biotic conditions, perhaps resulting in a new pattern of success and failure among invading populations. However, any such evolution does not appear to have lowered the survivorship of native species.

Pleistocene marine immigrations and emigrations are not restricted to the northeastern Pacific, but are recorded virtually wherever shelf-depth associations are described. In many regions the incursions of invading species are useful for local biostratigraphic correlation and sometimes for interregional correlation when they are associated with a widespread climatic event. Even tropical reef communities were subject to compositional changes during the Pleistocene, when sea level changes and oscillating temperatures caused extensive local extinction and recol-onization (for reviews see Paulay 1991a, 1991b). For example, three successive interglacial limestones on Aldabra Atoll, Indian Ocean, each contain very different coral and mollusk faunas, none of which resembles the present-day communities (Taylor 1978). Similar compositional shifts are documented by interglacial assemblages in Kenya (Crame 1986) and Polynesia (Grigg 1988; Paulay 1991a; Paulay and Spencer 1989). The marine species pool of the Indo–West Pacific was relatively unscathed by these range shifts—there is only one globally extinct gastropod species out of 162 in the Hawaiian Pleistocene record, for example (Kohn 1980)—but again associations at any point in geological time are fleeting and reflect available microhabitats and idiosyncratic range extensions and contractions. Finally, Lindberg (1991) has documented temperate marine species interchanges between the Northern and Southern Hemispheres that involved crossings of tropical latitudes and which apparently enriched both temperate biotas. The exchanges have been episodic, beginning well back in pre-Pleistocene times. The temperate communities seem not to have been saturated with species.

The Pleistocene biotic changes are certainly driven primarily by climatic changes associated with glacial cycles, which have been shown to correspond to astronomical Milankovitch cycles (fluctuations in seasonality resulting from the earth's precession, from variations in orbital eccentricity, from and "wobble" about the axis of rotation) (Hays, Imbrie, and Shackleton 1976; Shackleton, Imbrie, and Pisias 1988). The astronomical cycles have probably been present throughout the Phanerozoic, and indeed, cyclic patterns in ancient sediments have been interpreted as reflecting climatic effects entrained in Milankovitch cycles (see Fischer 1986). Bennett (1990) has suggested that the magnitude of the response to Pleistocene Milankovitch forcings can be taken as typical of the biotic response throughout earth history. The magnitude of the climatic variations, and therefore the biotic effects that derive from Milankovitch cycles, obviously varies with other (presumably earthbound) conditions. Ice ages are not caused by Milankovitch cycles, but if ice ages occur, their periodicity may be regulated thereby. Over much of Phanerozoic time, global climates seem not to have tracked the cycles with the extreme swings of conditions that characterized Pleistocene glaciations. The extent to which communities were invaded as a result of more moderate cyclic changes is not yet known from empirical studies. Although we expect that Bennett is correct that Milankovitch effects have been pervasive throughout the Phanerozoic, we doubt that associated biotic changes have approached the magnitude of the Pleistocene events, with the possible exception of other intensively glaciated times such as during the Ordovician and Carboniferous.

Many invasions are registered in the marine fossil record, ranging from intrusions involving a few lineages (such as invasions by Pacific mollusks into the North Atlantic during the Neogene; Durham and MacNeil 1967; Nelsen 1978; Vermeij 1989; Reid 1990) to the coalescence of entire provincial biotas (as during the assembly of Pangaea in the Paleozoic; see, for example, Sheehan 1988). These events certainly resulted in changes in community compositions. They were made possible by geographical changes resulting either from sea level changes or from global tectonic processes. Such cases are difficult or impossible to evaluate in terms of community integration at present. In the Neogene case, it is not now possible to determine whether or not the few invaders had any effect upon their new communities. In most of the Paleozoic cases there are important extinctions associated with the massive invasions. Many of the extinctions can be attributed to the physical changes that permitted the invasions, but it is not clear whether biotic interactions were also responsible for some of the losses.

The Pleistocene situation in the northeastern Pacific is unusual in the combination of high temporal resolution (geologically speaking), control over the distributions and community associations of the species today and in the past, the virtual lack of extrinsically imposed extinctions, and the presence of large-scale migrations. These factors permit the inference that the marine communities are "open" and that species are by and large acting individually. A similar combination of circumstances is still lacking in most fossil situations, although it is not impossible to find roughly analogous circumstances in older settings in principle.

The characterization of paleocommunities relies heavily upon the identification of "recurrent assemblages," which are widely described as occurring in pre-Pleistocene marine and terrestrial strata. These occurrences might suggest that community compositions were less ephemeral in pre-Pleistocene times, perhaps owing to a more subdued pace of environmental change. However, recurrent associations are recognized within the Pleistocene record (Valentine and Mallory 1965), indicating that many species shared patterns of occurrence in the fossil record despite the compositional changes. Furthermore, the more ancient associations are not without compositional variation. Increasingly detailed analysis of such assemblages has prompted either the recognition of a host of formally named but often intergrading assemblages (e.g., over 40 benthic associations in Fursich and Werner 1986) or a realization that gradient analysis may better characterize fossil distributions than does the forcing of taxa into discrete clusters (Cisne and Rabe 1978; Springer 1985; Springer and Miller 1990). The temporal and spatial covariation of some taxa within those gradients is at least partly due to their similar environmental requirements. We certainly do not deny that if particular environments do happen to persist over geological time scales (as Wing and Farley 1990, argue for early Cenozoic forests; see also DiMichele and Phillips, 1990), prolonged and strong biotic interactions might occur.

MARINE PHANEROZOIC
EVOLUTIONARY TRAJECTORIES

Community components also exhibit considerable independence over macroevolutionary time scales (that is, long enough that clades may wax and wane, becoming more or less dominant contributors to communities). The taxonomic turnover rates of many clades have character-

istic modes, as discussed by Simpson (1944, 1953) and treated at some length by Stanley (1979; see Holman 1989 for recent studies). Sepkoski (1981, 1984, 1987) found a large-scale pattern in the distribution of turnover rates among benthic marine taxa over the Phanerozoic (see also Van Valen 1985 and Valentine 1989b, for additional discussion). Taxa that dominate many early communities (i.e., trilobites during the Cambrian, Sepkoski's Fauna I) tend to have high turnover rates. They are succeeded as dominants in many communities by clades with slower turnover rates (i.e., crinoids and articulate brachiopods during the post-Cambrian Paleozoic; Sepkoski's Fauna II). These dominant taxa then fade in importance and are succeeded by taxa that turn over at yet slower rates (gastropods and bivalves coming to dominance during the Mesozoic and Cenozoic; Sepkoski's Fauna III, still dominant in marine communities today).

Although dominant at different times, all of the taxa typifying Sepkoski's three faunas overlap for hundreds of millions of years. Some decrease in turnover rates occurs within taxa through time (e.g., the gastropod and bivalve families of Fauna II are shorter-lived than those of Fauna III; Valentine, Tiffney, and Sepkoski 1991), but this is small relative to between-fauna differences. Across the Phanerozoic, faunal turnover rates vary by factors of 8 to 10 at the family level, while within major taxa such turnover rates vary by factors of 2 or 3. Thus, extinctions and originations of important community elements are not synchronized in any way that might suggest long-term cohesion. This is not itself strong proof that communities are not well integrated, for integrating adaptations within communities might be inherited by new species or at least accompany the speciation process, however long the species may endure thereafter. Still, the disparate turnover rates do suggest that local ecosystem structure or process has little effect on regulating turnover, which depends on speciation and extinction rates, and that communities can accommodate differential background levels of extinction or origination of some components without this affecting some others.

The individualistic ecological behavior of benthic clades over evolutionary time is corroborated when the data for faunal changes within and between Sepkoski's faunas are placed in an environmental framework. Analyzing Paleozoic marine communities, Sepkoski and Sheehan (1983) and Sepkoski and Miller (1985) showed that Faunas II and III each became dominant in onshore environments first and then expanded outward across the continental shelf. However, this pattern is underlain by idiosyncratic shifts of individual taxa and functional groups. Bretsky and Klofak (1986) found little evidence that the first occurrences of Fauna III species, and their subsequent shifts offshore, involved displacements, replacements, or reorganizations within the Fauna II–dominated communities they entered. Different bivalve life-habit groups (e.g., epibyssate suspension feeders, free-burrowing deposit feeders) also expanded across the shelf at different rates and at different times, giving rise along the way to unique, geologically transient associations (Miller 1988, 1989). Miller's discovery of contrasting bivalve patterns in terrigenous versus carbonate-rich environments may also indicate clade-specific rather than community-level controls on environmental shifts.

Jablonski and Bottjer (1990b, 1991b) document similar patterns within Fauna III. Eighteen of the twenty-five marine benthic orders that have originated since the beginning of the Mesozoic first appeared in onshore environments. As Miller observed for Paleozoic bivalves, the post-Paleozoic orders expanded across the continental shelf at different times and rates. Thus, a mid-shelf assemblage in the late Jurassic might contain stalked crinoids, epifaunal salenioid echinoids, and burrowing bivalves; a late Cretaceous one might lack crinoids, gain cheilostome bryozoans and spatangoid echinoids, and still include salenioids and burrowing bivalves; and a late Tertiary one would lack salenioids but contain cheilostome bryozoans, infaunal spatangoid echinoids, and a different suite of burrowing bivalves. Clades expand or shift their environmental preferences independently, and these changes are evidently neither driven nor accompanied by unusual extinction events (Jablonski and Bottjer 1990b).

This is not to say that there are no clade interactions over evolutionary time; some broad faunal and floral changes do appear to involve replacements, often mediated by mass extinctions (reviewed by Hallam 1987; Jablonski and Bottjer 1990a). Some potential marine clade interactions might involve the effects of durophagous marine predators and other agents of biotic disturbance. These have probably caused extinctions and elicited the evolution of morphological defenses (e.g., Thayer 1983; Vermeij 1987; Jackson and McKinney 1990), and might even have mediated clade retreats into exclusively offshore distributions (see discussion by Jablonski and Bottjer [1990b], who are doubtful). Our overall impression remains, however, that species and clades have originated and spread individualistically.

TERRESTRIAL COMMUNITIES

In the terrestrial realm, the migration of species and the reconstitution of communities occurred in ways rather analogous to the marine systems. Invasions of living species have been reviewed by Simberloff (1981, 1986); as remarked previously, there seem to be sufficient "openings" in communities that successful, extinction-free invasions are not uncommon, although many of the invasions occur in disturbed habitats. Here we also wish to review briefly the considerable plant and animal evidence that bears on terrestrial Pleistocene communities.

North American examples of changes within Pleistocene mammal associations have been summarized by Graham and Lundelius (1984), Graham (1985), Webb and Barnosky (1989) and in references therein. The latest Pleistocene records are within the range of radiocarbon dating, and thus the sequence of events during that interval, and during the succeeding Holocene (the last 12,000 years), is even better resolved than the marine terrace sequences. There are important Pleistocene extinctions among large mammals, chiefly at the close of the epoch. Small mammals suffered some extinction, but their chief

response to the changes of the times was migration. Those movements brought into association species that today live in separate areas and are said to be ecologically incompatible under present conditions, much as in the marine Pleistocene. Such associations are termed "disharmonious" in terrestrial literature and "anomalous" in marine studies. As the mammal faunas included species that are now allopatric, the Pleistocene communities may have been more species-rich. A difference between the range shifts in these two realms is that in the marine case cited, the shifts were along a narrow north-south-trending shelf, and the biogeographical changes were essentially onedimensional; whereas among the continental faunas some shifts have involved important east-west components as well. The terrestrial disjunctions in present ranges are thus two-dimensional. In both realms, the pattern of change between the Pleistocene and today indicates that, far from responding as units, communities became reconstituted as each individual species responded according to its unique tolerances and requirements (Graham 1985).

Graham and Lundelius (1984) speculated that some of the small mammal extinctions of Pleistocene time were associated with a breakdown in coevolved adaptations as climatic changes of the glacial cycles drove communities apart, causing a "coevolutionary disequilibrium;" competition sprang up between species brought into new associations, and specific plant resources shifted in response to the changes. Human agencies have also been implicated, especially for the large mammal extinctions. Rates have been compiled at the generic level by Gingerich (1984a, 1987) and were recently reviewed by Barnosky (1989). As was the case in the marine realm, the Pleistocene mammal extinction rates are not unusual for the Neogene and indeed lie on the low side. The late Pleistocene extinction peak, chiefly of large-bodied mammals, rises above the preceding background rate, but is matched or exceeded by several Cenozoic extinction events. Thus, although geographical range changes are likely to have been unusually high during the Pleistocene (though data at comparable resolution is in fact not available for earlier Neogene faunas), there is evidently little need to invoke unusual conditions to explain Pleistocene extinction rates (Barnosky 1989).

Other terrestrial groups besides the mammals display similar patterns of extensive Pleistocene range shifts, including reptiles and amphibians (Brewer 1985) and especially beetles, whose spectacular range shifts in Europe have been documented by Coope (1970, 1979), and which display changes in North America as well (Morgan 1987). Patterns of shifting distributions and associations of plants, including considerable reconstitution of some forests, have been well documented from megafloral and pollen evidence in Europe, North and South America, Australia, and Africa (see Livingstone 1982; Barnosky, Anderson, and Bartlein 1987; Colinvaux 1987; Walker and Chen 1987; Jacobson, Webb, and Grimm 1987; Huntley and Webb 1988, 1989). In the case of the plants it is particularly clear that the floristic changes were driven by glacial cycle climate changes; today's plant communities may be less than 6,000–8,000 years old, at least in the Northern Hemisphere where the data are most detailed.

As with marine faunas, the more ancient terrestrial biotas display unceasing compositional changes that are, however, difficult to assess with regard to their community-level consequences. The mammalian radiation in the wake of the end-Cretaceous extinction of dinosaurs presumably represents an instance in which removal of preemptive interactions permitted extensive diversification. At the clade level, such preemptive interactions are more commonly seen than possible displacement competition, although the rise of the angiosperms and the concomitant decline of gymnosperms may be an example of the latter (e.g., Knoll 1984, 1986). The paleobotanical data, however, suggest that other processes were also at work: conifers show little decline, and cycadophyte macrofloral diversity decline (but not the palynological decline) begins well in advance of the mid-Cretaceous angiosperm diversification (Lidgard and Crane 1990; Crane and Lidgard 1991).

Phanerozoic evolutionary trajectories among terrestrial clades have been discussed by Valentine, Tiffney, and Sepkoski (1991). Land plant clades display increasing turnover rates; that is, the groups dominating earliest have the lowest species extinction rates, while later dominants have higher rates, with angiosperms the highest of all. This situation is precisely the opposite of the marine pattern. However, *within* plant clades there is evidently a slowing of rates, as there is within marine invertebrate clades. Terrestrial tetrapods have a complicated pattern but, unlike the plants, tend toward a dominance of slower-evolving clades through time. Thus the terrestrial tetrapods and their plant resources display some discordance in their evolutionary trajectories, and extinctions and originations even of taxa that must interact are not closely correlated.

SUMMARY AND CONCLUSIONS

The fossil record of communities is incomplete at every level, and the temporal resolution of the record that we do have is quite coarse by neontological standards. Direct study of the processes that have occurred in ecological time is usually impossible. On the other hand, the results of ecological processes can commonly be inferred with some confidence from the fossil record, over time scales that are beyond the reach of neontological data. We have examined the migration of species into and out of communities over the last million years or so. It seems likely that most, perhaps all, of the migrations have been driven by climatic changes associated with the glacial-interglacial cycles. The migrations have resulted in important changes in terrestrial and benthic marine species, often yielding associations without present-day analogues. These are geologically temporary associations (like the present associations), and they reflect the tendency for climate cycles to generate singular combinations of temperature, equability, and moisture, rather than simple binary glacial-interglacial alternations or simple latitudinal shifts of parallel isotherms (see Huntley and Webb 1989 for discussion and references). Species'

differences in tolerances and migration rates will then inevitably combine to produce complex dynamic patterns such as those discussed here. Most authors suggest that reproduction and recruitment may be the crucial factors, which hints at potential modeling directions for past and future changes, but interactions of demography and climate can be complex. For example, Wethey's (1985) analysis of rocky intertidal communities of New England suggests that the persistence of species in the present-day climate depends at least partly on the relation of age at first reproduction to the frequency of sea-ice formation that can catastrophically scour the mid-intertidal zone (see also Perry et al. 1990 on the possible role of lags between tree migration and movement or stability of soil fungi and bacteria in the response of plant communities to climate change). Most details of autecology and of species interactions within fossil communities are lost, but by analogy with historical invasions, it is a viable hypothesis that at least some invaders were able to utilize resources that were either untapped or could be partitioned to permit the immigrant populations to survive without creating local extinctions among the natives. If exclusions accompanied the invasions they must have been chiefly local, for extinctions were not accelerated over pre-Quaternary Neogene rates.

Biotic invasions are quite common over geological time. Astronomically forced, periodic climatic changes (such as the Milankovitch cycles) have presumably pervaded the entire Phanerozoic, and even during long stretches of geological time when no extensive glaciations are recorded, climatic variations may have permitted invasions. Other invasions appear to be based on the creation or improvement of dispersal routes through geographical changes accompanying sea level change or tectonic displacements. Control over the details of compositional change within communities over ecological time is poor for most fossil cases, but the fact of such changes is clear.

Were communities bound in rigid webs of coadaptation, the pace of ecological change exhibited by Quaternary associations would create constant disruptions and certainly local and probably total extinctions (for virtually no latitude or locality is free from some invasions). It seems likely to us that the formation of even moderately diverse communities would be precluded. Instead, if communities contain sufficient unoccupied "adaptive space" with resources available to immigrants, then the coherent behavior of communities in the face of incessant change is explicable. Although examples of true coevolution doubtless exist (e.g., Kiester, Lande, and Schemske 1984), even tight associations of species in Recent communities—e.g., plants and their pollinators—presumably reflect preadaptation (with perhaps some short-term adjustments) more often than they do long-term reciprocal evolutionary accommodations (see discussions in Futuyma 1986; Westman 1990).

That there are indeed available resources and ecological opportunities within communities is evidently due to extinctions that occur at characteristic "background" rates within the various taxa (Walker and Valentine 1984). Speciation rates observed in the fossil record are too low to maintain diversity levels that "fill" the environment, considering the rates of species extinctions. Inspection of the lists of extinct marine species that are found in the Pleistocene of the Californian region indicates that, judging from their nearest living relatives, the extinct forms (like the immigrants) represent essentially the entire range of habitat and trophic requirements existing in benthic marine communities, with no obvious biases considering the nature of the data.

It seems to us difficult to support a case that the widespread changes in Pleistocene community composition, documented in every ecological and biogeographical realm and among all taxonomic groups for which data are available, have entailed important disruptions to community organization, for extinctions during these changes were unexceptional. Extinction of large-bodied mammals does represent a noticeable Quaternary bias, but such losses are a common theme in terrestrial animal extinctions, and furthermore may involve the consequences of human hunting rather than reflecting any community collapse. Earlier range changes cannot be documented now with the resolution available for the Quaternary record, but as the climatic drive was usually far less intense, species ranges must usually have been less labile and associations usually less variable before the Pleistocene. Extinction rates, however, were not correspondingly lower. If significant numbers of Pleistocene extinctions were due to disruption of coadapted associations, then they must have been matched by similar or at least corresponding extinctions in earlier times.

That some sorts of bonds help to structure communities is a plausible generality, for which experimental evidence exists (see the discussion in Pimm 1991, especially chap. 15), and in some circumstances coadaptation may forge obligatory interrelationships. However, the evidence that we have reviewed suggests that most such bonds are easily broken and reforged, a case of microevolutionary fine-tuning that is not necessarily associated either with extinction or with speciation. A dissociation of climatically driven changes from many macroevolutionary events has been argued by Barnosky (in press) for terrestrial situations and appears to be quite general. Climate changes certainly can cause extinction and certainly can create opportunities for speciation, but on the community level it would seem to be nearly every species for itself. Although introductions and exclusions within communities may be of great consequence for the individuals and, to a lesser extent, the populations that are directly affected, the communities appear to be quite resilient from a long-term perspective. Indeed, it may be the very flexibility and openness of communities that provides a buffer to dampen the ripple effects among species that might arise from climatic variation. It is our interpretation that such openness is characteristic of "natural" communities and is not restricted to habitats disturbed by humans.

30

Species Diversity: Regional and Historical Influences

Robert E. Ricklefs and Dolph Schluter

Chapters in this book provide convincing evidence that regional processes influence both regional and local diversity. For ecologists, this carries the important message that taxonomic components of community organization can be comprehended only by placing the local community in its historical and biogeographical context. To some extent, this may apply to functional components of community organization as well. Clearly, one cannot comprehend geographical variation in the species richness of local communities, including the pervasive increase in species diversity from the poles toward the equator, solely by studying dynamical interactions among species within local communities. Geographical, evolutionary, and historical factors play a strong role in determining a community's species composition, diversity, niche occupancy, and perhaps other attributes of its organization. Ecological insights gained from simple models, laboratory systems, and controlled experiments in nature, while valid in their own contexts, do not transfer well to natural systems, in which spatial heterogeneity over a variety of distances, historical development of species assemblages, and evolution enter the overall equation for coexistence.

The issue of species diversity has now come full circle. Early in this century, biologists had concluded that the diversity of a region reflected its age, and they did not clearly distinguish between regional and local diversity. Beginning in the late 1950s, and especially in the 1960s, many ecologists partitioned regional from local diversity and took on local diversity as an ecological problem that they could solve by resorting to the insights of population biology. This was a period of local determinism in ecology. With regard to explaining global patterns in species diversity, that paradigm has had limited success. To advance beyond contemporary understanding, ecologists must now reject the parochial view of local determinism and recognize that ecology, evolution, geography, and history are different facets of a single set of processes and the patterns they generate. One cannot isolate any one system of a particular dimension from processes and structures at a smaller scale embedded within it or from those at a larger scale containing it.

The issue of species diversity provides an excellent starting place for a major reappraisal of ecology. Its resolution will require a new synthesis of disciplines—ecology, evolution, biogeography, systematics, paleontology—that have drifted apart during this century. Because

the phenomena of species diversity span such large dimensions, the comprehensiveness of the research effort required to tackle the phenomenon must be equally great. Many of the chapters in this book present clear examples of this developing synthesis. In this concluding chapter, we shall discuss some of the major issues and suggest questions and approaches for the future.

PATTERNS AND ISSUES

The diversity of species included within a small area of relatively uniform habitat reflects a variety of processes either operating locally within the area or influencing the local association of species from without. In general, local processes reduce diversity through competitive exclusion, overexploitation, and stochastic variation; regional processes balance these, elevating diversity through movement of individuals between habitats and habitat patches, production of new species within regions, and exchange of species between regions. The relative importance of these processes to local species richness undoubtedly varies among regions and taxonomic groups; understanding how attributes of the environment and of organisms influence this balance is the key to understanding patterns of diversity.

Global comparisons make several strong and apparently generalizable points about diversity. First, within regions, local diversity correlates strongly with various aspects of ecology, including climate, habitat structure, and mesoscale (that is, landscape) heterogeneity (Gentry 1988a; Currie 1991; Wright, 1992; Rosenzweig and Abramsky, chap. 5; Tilman and Pacala, chap. 2; Wright, Currie, and Maurer, chap. 6). This ecological determinism of diversity is further supported by theory and simulations with model communities illustrating the resistance of communities to invasion by new species once an equilibrial (saturated) diversity has been achieved (Case 1990; Drake 1990a; Haydon, Radtkey, and Pianka, chap. 11). In addition, within habitats, species exhibit characteristic distributions among ecological roles. In many cases, these relationships are consistent between regions, producing a strong effect of local ecology on species diversity (the "convergence" effect of Schluter and Ricklefs, chap. 21); in some other cases, notably the array of herbivores on bracken, niche occupancy differs noticeably between regions (Lawton, Lewinsohn, and Compton, chap. 16; see also Aho and Bush, chap. 17, Cadle and Greene, chap. 25, and Ricklefs and Latham, chap. 20).

Second, the influence of regional and historical factors becomes apparent when local species richness in a given habitat type differs between regions. This disparity demonstrates that local assemblages are not isolated within the narrow context of contemporary and local processes. Local (alpha) diversity reacts to regional and historical processes and circumstances; regional diversity, local diversity, and turnover of species between habitats (beta diversity) covary.

Theories of local diversity based on the influence of regional processes and unique historical events, on one hand, and of local interactions among species, on the other, are not irreconcilable. To the contrary, both are essential terms of the diversity equation (fig. 30.1). The linking of patterns at local and regional scales suggests that local and regional diversity cannot be separated; instead, they must be considered together as different expressions of an integrated ecological system within each region.

Diversity, whether at the regional or local level, reflects the resolution of internal/external dichotomies. That is, the character of any given ecological unit—the assemblage of species occupying a local uniform habitat, a landscape of habitats, or a region of landscapes—is shaped both by internal processes and external influences. Depending on the unit defined, the character of the internal and external processes varies. For example, the behavioral versatility of individuals, the movement of individuals between habitats and areas, and the evolutionary invasion of ecological space may be either internal or external factors depending on the spatial and temporal extent of the ecological structure of interest.

We view the community as a hierarchy of structures with different sets of processes connecting each level of the hierarchy (fig. 30.2). Diversity may be described for each level of the hierarchy ranging from a small area of uniform habitat to the globe. Constraining influences on diversity travel from bottom to top through the hierarchy, that is, from the smallest unit to the largest; augmenting influences travel from top to bottom (Cornell and Lawton 1992). The level of diversity at any point in the continuum reflects the balance of processes acting at higher and lower levels in the hierarchy.

Each level in this ecological hierarchy is connected to the next higher and lower levels by a variety of processes whose influences touch more than one level. With regard to the problem of species diversity, we may be concerned with the movements of individuals within populations, the spread of populations geographically, the formation of species, and the shifting or expansion of habitats. Source-sink effects (Pulliam 1988), mass effects (Shmida and Wilson 1985), supply-side ecology (Roughgarden, Gaines, and Pacala 1987), landscape ecology (Turner 1989), and mesoscale models (Hanski, Kouki, and Halkka, chap. 10; Holt, chap. 7) all represent concepts or processes that link ecological structures of different dimensions. Species formation and habitat shifts occur over evolutionary periods. Although they lie outside the traditional scope of ecological study, the ecological context of these processes is critical to understanding how they work; complementarily, each has important consequences for ecological structures.

Contemporary patterns of diversity may originate in part from the unique history and biogeography of each region (Brooks and McLennan 1991, chap. 24; Latham and Ricklefs, chap. 26; Ricklefs and Latham, chap. 20). Historical reconstruction provides information about what did happen within continually evolving ecological settings, thus elucidating the dynamics of many large-scale processes. At times, processes reveal their strength only following perturbations; we can regard the expression of history in contemporary diversity as the record of response to perturbation (e.g., Kauffman and Fagerstrom, chap. 27). To the extent that we can identify historical events by their effects, we can judge the power of ongoing local and regional processes to restore an equilibrium condition. Thus, ecologists should consider historical reconstruction of paramount interest (Brooks and McLennan, chap. 24; Cadle and Greene, chap. 25).

The hierarchical scheme portrayed in figure 30.2 raises several issues. First, do opposing factors define an equilibrium level of diversity at each or any of the levels in the hierarchy? An equilibrium requires that rates of increase and decrease in number of species depend on the diversity of the system (MacArthur 1969; Rosenzweig 1975). Second, if an equilibrium exists, has the community actually achieved that point? We presume that answers to this question can come only from historical analysis; the fossil record is generally either mixed or equivocal on this point (e.g., Sepkoski 1981; Knoll 1986). Third, if the community has achieved an equilibrium at a particular level, how

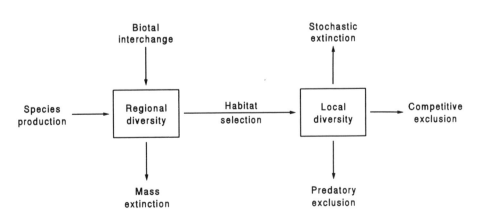

Figure 30.1 The influence of regional and local processes on regional and local diversity. The arrows point in the direction of augmentation of diversity. Habitat selection (the inverse of beta diversity) adjusts the relationship between regional and local diversity.

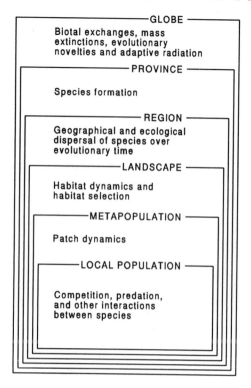

Figure 30.2 A hierarchical viewpoint of processes influencing species diversity that have different temporal and spatial dimensions. Each level includes all lower ones and exists within all higher ones. Unique (chance) events may occur at any level of the hierarchy.

do the physical characteristics of the local environment, the landscape arrangements of habitats, and the climate and biogeography of the region influence the position of the equilibrium?

For decades, ecologists have pursued a general theory relating diversity to the local physical environment with little consensus on the specific mechanisms responsible (Connell and Orias 1964; Pianka 1966; Connell 1978; see also Strong et al. 1984). Tilman and Pacala (chap. 2) review many plausible local explanations for the regulation of species diversity, several of them having expressions sensitive to the physical environment (e.g., Ricklefs 1977), all of them difficult to test. Explanations also have been produced for patterns in diversity observed at larger dimensions of the community. Theoretical treatments have explored the coexistence of species in metapopulation models (Levin 1974; Yodzis 1978; Shorrocks and Rosewell 1986; Ives 1991; Holt, chap. 7; McLaughlin and Roughgarden, chap. 8; Caswell and Cohen, chap. 9), and habitat selection in simple landscapes (Rosenzweig 1981, 1985, 1987; Morris 1987, 1988); modeling has led to promising experimental work as well (e.g., Rosenzweig and Abramsky 1985). Similar efforts should be directed toward understanding the influence of environmental factors on rates of species production and the origin and spread of evolutionary novelty.

To understand diversity we need to see the local community in its proper context: as a patch of habitat in a large region of patches among which there is perpetual exchange of individuals and species and production of new species; and as an assemblage whose properties re-

main affected by local or widespread events that occurred long ago. Thus, our goal is to understand how regional processes affect the regional species pool and, in turn, how they affect the local community. In this chapter we identify the regional and historical forces that shape local assemblages, summarize existing evidence of their effects, and outline the main questions that researchers of diversity face in the years ahead.

ASSESSING THE ROLES OF LOCAL, REGIONAL, AND HISTORICAL FACTORS

Historically, the development of modern ecology during the past fifty years brought the consideration of global patterns of species diversity into the realm of ecological investigation. In undertaking the study of the diversity issue, ecologists had to divorce local diversity from the influence of regional and historical factors (Kingsland 1985; Ricklefs 1987). R. H. MacArthur and some of his students and associates struggled with this problem because the impact of history and geography presented itself forcefully in the guise of differences in diversity between regions with similar ecology; this struggle was resolved by compartmentalizing local and regional diversity as separate phenomena (MacArthur 1965, 1969; Cody 1975).

Most ecologists now hold a more balanced view of the regulation of local diversity, in which local interactions among species constitute one set of terms in a larger equation. For the future, we must set out to determine the relative contributions of various mechanisms of exclusion and coexistence to the diversity balance.

A major hypothesis arising from the idea of local determinism of diversity is that of community saturation: local interactions establish rigid upper limits to diversity, whose values depend on local environmental conditions. This hypothesis predicts that local diversity should vary independently of regional diversity, which is clearly false for some groups of organisms (e.g., Lawton, Lewinsohn, and Compton, chap. 16; Westoby, chap. 15; Cornell, chap. 22; Latham and Ricklefs, chap. 26; Ricklefs and Latham, chap. 20; Schluter and Ricklefs, chap. 21). However, the hypothesis appears to be supported in some cases. For example, Van Valkenburgh and Janis (chap. 28) have shown that increased regional diversity of mammals following faunal mixing had no effect on the diversity of local fossil assemblages.

The dynamic balance between local and regional processes is revealed by various "natural" experiments in which regional diversity is perturbed. These experiments belong to two classes: First, migration between a local area and its surrounding region may be cut off by the formation of a barrier. Second, routes of exchange may be opened between regions previously separated by a barrier, elevating the total number of species that can potentially reach any locality. As summarized below, results of such experiments confirm that local interactions tend to reduce local diversity, whereas regional inputs augment it.

Community Enclosure

A strong test of whether diversity in a locality depends on that of the surrounding region is to place a barrier around it and see if the local species persist. If they do, then regional factors were unimportant in maintaining diversity; if species disappear, then regional factors were important (Brown and Gibson 1983). Rising sea levels coinciding with the retreat of glaciers during the past 18,000 years created islands from areas once continuous with the mainland along several coasts. This "manipulation" reduced or stopped the movement of individuals between the local area and its surroundings. Such events clearly were associated with decreases in diversity, demonstrating the role of migration in maintaining local populations and the efficacy of regional processes in maintaining local (and regional) diversity (Diamond 1972). If one assumes that isolation blocks migration thoroughly, the decline in number of species with time estimates the local extinction of species through competitive or predatory exclusion, stochastic extinction, or local calamitous events (Case, Bolger, and Richman 1992).

Wilcox (1978) demonstrated "relaxation" of diversity of lizards on landbridge islands in the vicinity of Baja California by relating diversity to time since isolation (fig. 30.3, left; see also Case, Bolger, and Richman 1992). Terborgh (1974) provided analogous data for the species richness of birds on landbridge islands, showing the general effect of island size (fig. 30.3, right). Birds apparently suffer higher rates of extinction than lizards do (see Case and Cody 1987). In the presence of land bridges, migration of individuals from other areas prevents the local extinction of populations resulting from local processes.

Ecological Invasions

The idea of saturation predicts that an alien species should not invade an intact community, or should do so only with the coupled competitive exclusion of a resident member. Thus, the fate of introduced and pest species has been of considerable interest to population biologists and community ecologists (Elton 1958). Disturbed habitats, including such naturally distressed habitats as riverbanks, appear to receive aliens most readily (e.g., MacDonald, Powrie, and Siegfried 1986; Crawley 1987). Disturbance presumably reduces competition from native plants and animals by reducing their populations from time to time, making way for aliens to invade. In general, species invasions of any habitat are rarely accompanied by the exclusion of native competitors (Por 1978; Carlton 1979; Simberloff 1981; Case and Bolger 1991; Case, Bolger, and Richman 1992; Valentine and Jablonski, chap. 29). Even though occasional instances result in dramatically altered local ecosystems (Vitousek 1986; Atkinson 1989), a general consensus is emerging that competitive interactions between species rarely determine the success or failure of colonization (Simberloff 1986).

Biotal Interchange and Ecological Mixing

Occasionally, climate change and geographical rearrangements bring about conditions for interchange of taxa between previously isolated regions. Among the best known of these conditions are the formation of a land connection across the Tethys Sea between Eurasia and Africa about 18 million years ago (mya), the contact of Australia and New Guinea with Southeast Asia about 15 mya, the formation of the Bering land bridge about 70 mya (broached during the mid-Pliocene, reforming a connection between the North Pacific and North Atlantic Oceans), and the formation of the Panamanian land bridge, first crossed by large numbers of terrestrial taxa about 6 mya and finally completed by 3 mya. A more recent anthropogenic example is the opening of the Suez Canal, which formed a marine connection between the Red and Mediterranean Seas. Several chapters of this volume, particularly those by Van Valkenburgh and Janis (chap. 28) and by McGowan and Walker (chap. 19), address the augmenting influence of biotal mixing on regional diversity. Vermeij (1991b) reviewed instances of in-

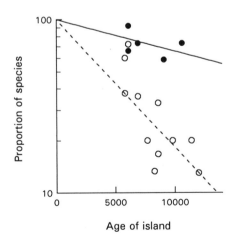

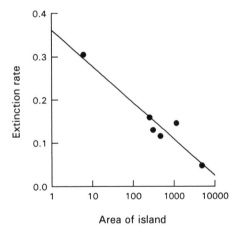

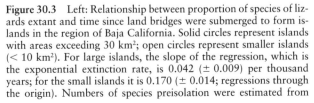

Figure 30.3   Left: Relationship between proportion of species of lizards extant and time since land bridges were submerged to form islands in the region of Baja California. Solid circles represent islands with areas exceeding 30 km²; open circles represent smaller islands (< 10 km²). For large islands, the slope of the regression, which is the exponential extinction rate, is 0.042 (± 0.009) per thousand years; for the small islands it is 0.170 (± 0.014; regressions through the origin). Numbers of species preisolation were estimated from faunas of nearby mainland areas. (Data from Wilcox 1978.) Right: Relationship between exponential extinction rate and area for terrestrial birds on landbridge islands off the coasts of Central and northern South America. Numbers of preisolation species were estimated from the species-area regression for mainland areas of similar environment. (Data from Terborgh 1974.) (For similar analyses see Case and Cody 1987; Case, Bolger, and Richman 1992; Richman, Case, and Schwaner, 1988.)

terchange during the Neogene, and came to several general conclusions.

First, the species that invade usually constitute a small percentage of the total available pool: 2%–11% of mammal genera were involved in exchanges across the Panamanian Isthmus; 4% of fish species from the Red Sea have invaded the Mediterranean; of 250 species of gastropods inhabiting the Line Islands in the Central Pacific, 13% have invaded offshore islands in the eastern Pacific, while only 4% have reached the coasts of the mainland.

Second, invaders constitute a variable percentage of the recipient pool: in the eastern Pacific, 100% of corals have Central/West Pacific origins, but only 5% of mollusks do; 90% of Central American lowland rainforest angiosperms have South American ancestry.

Third, interchanges usually exhibit considerable asymmetry. Transporting currents can explain this asymmetry in some marine cases, but not others. Invaders may have found some biotas weakened by prior episodes of extinction, as, for instance, in the marine faunas of the eastern Pacific and Mediterranean. More frequently, however, one biota exhibits competitive superiority over another. Such asymmetry appears to characterize the mammal interchange between North and South America, during which northern forms proliferated and diversified in South America, pushing out many of the old natives. Webb (1991) suggests, however, that the invasion from the north involved mostly temperate, xeric-adapted taxa that crossed the land bridge during dry interglacial periods; he attributed the greater success of northern forms to the fact that the area of the colonizing region exceeded by six times that of the receiving region, and immigrants from the north greatly outnumbered those from the south. Furthermore, many of the southern immigrants were tropical taxa whose spread into North America was severely limited by the availability of suitable habitat.

Spread of taxa from regions of high diversity to regions of low diversity appears to be a general feature of biotic interchanges, out of proportion to the relative numbers of taxa. Certainly exchange of taxa between islands and mainlands appears to be one-sided. Several authors have suggested that taxa from diverse biotas have a general competitive superiority, perhaps related to the evolution of defenses against a greater variety of predators and disease organisms (Brown 1957; Warner 1968; Ricklefs and Cox 1972; Van Riper et al. 1986).

Finally, as in the case of invasions by single species, interchanges rarely cause extinctions in recipient biotas, with the result that regional diversity generally increases, at least for a while (Barry et al. 1991; Flynn, Tedford, and Zhanxiang 1991; Vermeij 1991a; Lindberg 1991). Interchanges usually are transient events, running their course as barriers to dispersal re-form or as most potential colonists complete their spread. The biotas of some areas clearly have come from several sources and exhibit elevated diversity due to the mixing of taxa from two or more regions. Examples of such histories include the parasites of fishes in the Orinoco Basin of northern South America (Brooks and McLennan, chap. 24), zooplankton faunas at the confluence of marine currents (McGowan and Walker, chap. 19), and the coral reef fishes of Guam and Tahiti (Thresher 1991).

THE RELATIONSHIP BETWEEN LOCAL AND REGIONAL DIVERSITY

Because regional processes and unique historical events influence local diversity, the study of diversity must include an examination of the processes that connect ecological structures of different spatial and temporal dimensions. Metapopulation models and experimental studies of habitat fragmentation have already advanced our understanding of this issue considerably. Ecologists have begun to develop techniques for characterizing landscapes, and studies of the relationship between local and landscape diversity and landscape configuration will follow, building on the foundations laid by research on habitat selection. Ecologists have not, however, touched upon the problems of species production and adaptive radiation. And they have entrusted unique historical events, such as biotal interchange, the origin of novelty, and mass extinction, to paleontologists. We shall return to these issues below.

The interdependence of local diversity and regional diversity depends on habitat breadth, the beta component of diversity. In community saturation models, local processes determine local diversity independently of variation in regional diversity. This implies that habitat breadth accommodates variation in regional diversity and that habitat specialization varies independently of within-habitat specialization. Most empirical studies have shown that alpha and beta diversity covary in direct relation to regional diversity; this demonstrates a direct connection between ecological structures of different dimensions.

Given that local and regional diversity covary, we may ask whether regional diversity expresses characteristics of the region that influence rates of species production, exchange of taxa between habitats, and the history of biotal exchange, or whether regional diversity is a collective property of the local habitats within the region. If local interactions regulated local diversity, the number of species within the region would merely represent the summation of local diversity over habitats within the region. Beta diversity would vary according to the local heterogeneity of habitats and other physical characteristics of the region, rather than reflecting the pressure of species production within the region as a whole. Thus, regional differences in diversity would arise from local- and landscape-scale considerations, rather than revealing differences between regions in the production and immigration of new forms. Because comparisons between continents usually do not consider area and habitat heterogeneity, these factors are difficult to assess.

The dissimilarities noted in this volume between the local diversities of species in matched habitats on different continents rule out any notion that regional diversity is merely a collective property of the diversities of ecologically saturated local habitats multiplied over the variety of habitats within the region. Variations in alpha and beta diversity may, however, reside at the landscape level rather than responding to processes at the regional level. For example, a more finely divided landscape having extensive borders between smaller areas of dissimilar habitat might exhibit higher local diversity as a result of enhanced exchange between habitats (the mass effect of Shmida and

Wilson [1985]). According to this scenario, however, beta diversity would decrease due to increased sharing of species between habitats, causing an inverse relationship between alpha and beta diversity. It should be obvious from the material presented in this volume that this prediction is falsified readily.

Species production is partly a function of geography and climate variation, which may result in intermittent isolation and reconnection of different areas within a region. The translation of increments in regional diversity into increments in local diversity depends on a different set of factors. The establishment of the local/regional diversity relationship depends upon a direct connection between the habitat breadth of a population (which is inversely related to beta diversity) and its local population density (which we presume is inversely related to alpha diversity). Understanding the nature of this connection is essential to a complete accounting of diversity phenomena.

Both McNaughton and Wolf (1970) and Brown (1984) suggested that abundant, widely distributed taxa specialize on resources or ecological conditions that are abundant both locally and over wide areas (Hanski, Kouki, and Halkka, chap. 10). Thus, the positive relationship, observed in many taxa, between geographical range and local abundance reflects the local and geographical arrangements of particular regions of the niche space that taxa occupy, rather than adaptations that influenced the niche breadth of the population. Brown (1984) elaborated this idea by envisioning a multitude of environmental factors whose favorableness for a particular population varies geographically and determines the local abundance of a population. As a consequence, each species exhibits a peak abundance where the greatest number of these environmental factors favors its growth, and population density declines with distance from this point as factors in the environment change geographically. Accordingly, the local population is in equilibrium with the local environment. Furthermore, both the local abundance and geographical range of a species increase in direct relation to ecological niche breadth.

Range-abundance relationships may also be addressed in the context of demographic models based on source-sink relationships. High productivity within a local area will likely support both a dense local population and a high rate of emigration to other habitats and localities. Thus, populations with high abundances locally may support sink populations in less favorable habitats (Pulliam 1988), thereby extending the species' geographical range. In this model, the equilibrium density of a local population depends on external factors (migration) as well as on the local environment (see Collins and Glen [1991] for an exploration of alternative views).

SPECIES DIVERSITY AND ECOLOGICAL DIVERSITY

Local diversity is sensitive to the production of species on a regional level. The mechanisms by which changes in diversity are accommodated within the local community are critical to understanding the regulation of local diversity as a balance between local interactions and regional processes. Species may invade communities either by crowding into the niche space occupied by residents or by taking up positions outside of this space. If species diversity and community niche space covaried, the larger niche space occupied by more diverse communities could represent intrinsic ecological properties of the habitats of diverse communities; alternatively, total niche space could expand in response to the pressure of regional processes that generate increased numbers of species. The second possibility presumably would depend upon the utilization of novel resources or evolution of novel methods of exploiting resources—in general, the evolutionary expansion of the community niche through diversification and adaptive radiation. Thus, the relationship between species diversity and ecological diversity may illuminate the development of community structure. Unfortunately, few studies have compared the ecological diversity of matched habitats in different regions that have marked differences in species diversity; most such studies rely on comparisons across habitat types or between latitudinal zones.

Food web analysis suggests that the small-scale structure of communities is conserved when species are added, but that the total community niche is enlarged. The number of links within the food web appears to increase in closer relation to the number of species than to the square of the number of species, which approximates the number of potential links in the food web (Cohen and Briand 1984). Thus, each species interacts with a similar number of other species regardless of the overall diversity of the community (Martinez 1992).

For example, in Beaver's (1985) analysis of food webs in the aquatic (inquiline) communities occupying *Nepenthes* pitcher plants (fig. 30.4), the total number of connections per species remained approximately constant over a threefold range of diversity of trophic types. Regardless of the diversity of the community, each species interacts (through predator-prey relationships) with between three and four other species. In addition, the variation in food web diversity in Beaver's study was unrelated to the resource base of the community; each food web was sustained by either recently drowned insects or organic debris, with live insects being consumed by only a single predator in several of the communities. Moreover, the lengths of food chains appeared to increase in direct relation to the total number of species (Cohen, Briand, and Newman 1986). In the *Nepenthes* example, very different communities have developed within similar habitats and on similar resource bases. Total community niche space varies in relation to the diversity of the local community independently of habitat; large community niches are associated with increased complexity—achieved by adding new ecological roles—rather than with an expanded resource base for the community.

Morphological similarity of organisms has frequently been used to assess ecological relationships (Hutchinson 1959; Hespenheide 1973; Karr and James 1975; Ricklefs and Travis 1980; Miles and Ricklefs 1984; Miles, Ricklefs, and Travis 1987). The value of morphological analysis arises from its ability to portray community organization with measurements taken independently of the habitat context. While this approach also has limitations, ecomorphological analyses of vertebrates have yielded an important generalization relevant to the question of spe-

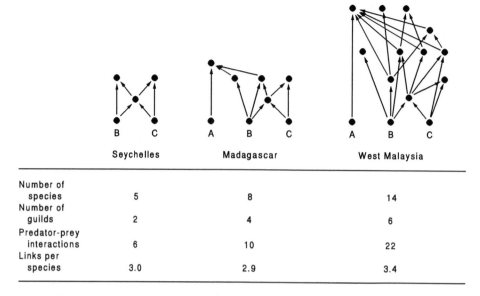

	Seychelles	Madagascar	West Malaysia
Number of species	5	8	14
Number of guilds	2	4	6
Predator-prey interactions	6	10	22
Links per species	3.0	2.9	3.4

Figure 30.4 Attributes of communities of invertebrates in *Nepenthes* pitcher plants in different regions. The food web diagrams show increasing ecological diversity (more guilds), but reduced connectance, with increasing species diversity. As a result, each species has a similar number of feeding links to others, on average, regardless of the overall species richness of the community. Sources of food: *A,* live insects; *B,* recently drowned insects; *C,* older organic debris. (Data from Beaver 1985.)

cies diversity: as the number of species in a community increases, the average distance between nearest neighbors in morphological space remains constant, while the total morphological volume occupied by the assemblage increases (Ricklefs and Miles 1993).

As we have seen in the case of food web structure, the fine-scale structure of vertebrate communities, assessed by packing in morphological space, appears invariant. Added species are associated with increased niche space, resulting in some cases from extension of the smaller axes of the space (Ricklefs and Travis 1980). This generalization summarizes comparisons of communities across ecological conditions within regions. Ecologists have not yet ascertained whether similar patterns will emerge from comparisons within similar habitat types between regions. These analyses nonetheless suggest that the core of the community niche space may resist invasion, and new species possibly must develop novel ways of exploiting resources to enter the community.

In other types of communities, patterns of niche occupancy are less clearly related to community diversity. For example, among herbivorous insects on bracken, there appears to be little relationship between total niche occupancy of the community or niche specialization by individual taxa and species diversity (Lawton, Lewinsohn, and Compton, chap. 16). Among mangrove plants, comparisons between the species-poor Western Hemisphere and the species-rich Indo–West Pacific revealed no relationship between microhabitat specialization and diversity; although the Indo–West Pacific species occupied a larger total niche volume, many more species were packed into each microhabitat as well (Ricklefs and Latham, chap. 20). Finally, Abrams (1981) determined that average competitive effects among eight species of Indo–West Pacific hermit crabs were about twice as strong as those among three Panamanian species, indicating tighter niche packing in the more diverse region. Thus, in some types of communities, the addition of species to the community is accompanied by increased niche packing.

THE TEMPORAL AND SPATIAL DIMENSIONS OF ECOLOGICAL, EVOLUTIONARY, AND HISTORICAL PROCESSES

The idea that regional and historical processes may leave a strong imprint on local diversity raises a major issue in community ecology concerning the relative rates of processes at different temporal and spatial dimensions (Ricklefs 1989b). How can processes such as species production, acting over evolutionary periods, shift local equilibria that are influenced by such local processes as competitive exclusion, acting over ecological periods? If interactions between species can reach equilibria within tens of generations (Miller 1967), the introduction of new species to the community at longer average intervals should have little effect on local diversity. The key to understanding the effects of regional processes and historical events on local diversity lies in reconciling the rates of each of these processes to within the same general order of magnitude. Because estimates of rates of species production within a region will not likely rise substantially, this will require a considerable downward revision of the time course of local extinction.

In laboratory systems, competitive exclusion runs its course within a few tens of generations when species compete exploitatively (Miller 1967). Laboratory observations impressed ecologists with the power of competition and other interactions to influence coexistence. However, laboratory environments are greatly simplified and force species into strong competition for a restricted variety of resources. When laboratory environments are made more complicated, species that would otherwise be excluded may persist as shifting arrays of subpopulations (metapopulations) occupying interconnected patches of habitat (Huffaker 1958; Pimentel et al. 1965).

In the special case of competition, theory tells us that the average time required for competitive exclusion increases as the competitive abilities of species become more nearly equal. To emphasize the importance of this relationship for understanding diversity patterns, Hubbell (1979) modeled tropical rainforest dynamics under the assumption of competitive equivalence. In such a system, species disappear slowly by chance fluctuations in their populations (death and replacement are randomly distributed across species in proportion to their abundance), in a process analogous to genetic drift. Eventually, the diversity of such a system will decline to a single species in the absence of invasions by new species. But the time course of this process is extremely long, requiring tens of thousands of generations or more, depending on the size of the system. Therefore, according to Hubbell's model, the addition of new species by species production or invasion from adjacent habitats may be sufficient to maintain an equilibrium diversity at a high level. Furthermore, the equilibrium number of species should be directly related to the rate of production/immigration.

Empirical data on tropical forest trees and, especially, various groups of tropical animals, indicate considerable specialization within diverse natural communities (e.g., Ashton 1969). Species are not ecologically equivalent, and each must predominate over some part of the total community niche. Thus, Hubbell's approach cannot fully explain diversity patterns. To be sure, some systems, such as the insects of bracken (Lawton, Lewinsohn, and Compton, chap. 16), appear to be noncompetitive in the sense that large portions of the potential niche volume for herbivores remain unoccupied. The diversity of these systems may depend largely on access to sources of colonists (Caswell 1976). In plants and many vertebrates, competition is readily demonstrated experimentally, and long-term coexistence of potential competitors may depend more on competition coefficients being nearly equal than on their being very small.

As a hypothesis, we suggest that within complex natural environments, large numbers of species can distribute themselves so as to achieve competitive equivalency. We assume that each taxon is specialized and is more productive than (competitively superior to) all others within some portion of the total niche space. Under pressure of interspecific competition, the niche space occupied by each species presumably contracts until, with an increasing specialization to utilize the most favorable portions of the niche space, the per capita growth rate of its population increases. The greater the number of species, the smaller the portion of niche space that each dominates. This is a conventional view of the relationship between specialization and diversity.

Specialization with respect to habitat may occur within ecological time when populations fail to maintain themselves in a variety of different habitat types, that is, when source populations cannot maintain sink populations. Presumably, similar specialization may also occur when individuals evolve to utilize narrower ranges of resources and conditions that they exploit more efficiently on average. Such specialization may require individuals to increase the area of habitat over which they forage to procure resources, thereby reducing the density of the population but permitting its continued existence in the face of strong interspecific competition.

How are species added to communities? Do communities become more resistant to invasion as diversity increases? Model systems in which species are added by drawing at random from a large pool exhibit increasing resistance to invasion as diversity increases (Case 1990; Drake 1990a; Haydon, Radtkey, and Pianka, chap. 11). The probability of a new species fitting into a suitable position diminishes as more species are added to the system and less niche space is available. However, such model systems lack several important attributes of natural systems that may affect these conclusions considerably. First, they are discrete and lack spatial structure. Thus, established species may not invade from adjacent habitat. Second, model species usually are given fixed niche relationships or interaction coefficients with respect to species already present; hence their niche space cannot shift or decrease to fit into the structure already established. Third, model systems have no historical component, in the sense that invading species are not related in any way to species already present in the system. In natural systems, the development of local diversity may include invasion by sister taxa (at some level) from other areas. Every invader comes from an established population existing in a similar type of community elsewhere, gradually encroaches upon the local community through intermediate localities, and may undergo evolutionary change during this period of expansion and invasion.

The fossil record of major evolutionary innovations illustrates how communities are invaded from the outside; these novelties may be used as tracers to follow the diversification and expansion of a clade through geographical and ecological space. In most cases, a major evolutionary novelty has required tens of millions of years to fully invade a biota (Steneck 1983; Vermeij 1987; Lidgard and Crane 1990; Jablonski and Bottjer 1991a). Consider the following scenario. A novelty conveying evolutionary advantage—perhaps increasing the efficiency of predation, or of escaping predation—arises at a certain point and quickly predominates within the local population, becoming a feature of the local species assemblage. Its spread by cladogenesis and ecological diversification occurs as new species that carry the novelty appear and invade new communities, either coexisting with or replacing sister taxa. The rate of spread would seem to be limited by cladogenesis and phyletic evolution; invasion of new ecological settings is enhanced by the ecological diversification of the carriers of the novelty. Thus, although advantageous novelties favor invading taxa, the taxa that bear them may require millions of years of evolution to expand their ecological distributions. We presume that this process describes the development of communities, requiring similarly long periods of evolutionary adjustment for invasion of ecological regions and accumulation of diversity within them.

ECOLOGICAL AND REGIONAL ASPECTS OF SPECIES PRODUCTION

Species production constitutes a major term in the species diversity equation, yet ecologists have largely ignored the consequences of speciation rate for the development of communities, as well as the ecological contexts that either promote or restrain speciation (see, however, Rosenzweig 1975). Speciation is an ecological as well as a genetic process, and it usually plays itself out over temporal and spatial dimensions coinciding with those of large regions. It seems impossible to consider speciation and species diversity separately. Our general understanding of speciation includes the importance of geographical isolation in most groups, the importance of ecological factors as well as geographical circumstances in determining isolation, and the role of intrinsic biological features of taxa in promoting isolation and genetic differentiation. Through their influence on the rate of species production, each of these factors contributes to diversity and may explain a part of global patterns in diversity.

The diversification of taxa on island archipelagoes demonstrates the importance of geographical isolation (Mayr 1963). The scenario involves the spread of a population to more than one island, independent evolution and divergence on each island, and subsequent reinvasion of one or more of the islands to increase local diversity. The degree of geographical isolation is critical to this process. Too much migration between islands prevents divergence; too little migration reduces the probability of subsequent reinvasion and biotal buildup. Thus, to promote diversification maximally, the geography of the archipelago must be properly tuned to the dispersal and other characteristics of the taxon. The Galápagos archipelago has fostered the production of thirteen species of geospizine finches with up to eleven species occupying each island (Grant 1986), whereas the four species of mockingbirds and one extremely variable species of giant tortoise are distributed one per island. In Hawaii, speciation of birds apparently requires isolation between islands, while speciation of picture-wing *Drosophila* flies may occur between forest fragments isolated by lava flows (Carson and Kaneshiro 1976). This suggests that particular geographical configurations of islands, or of ecological barriers to dispersal, might result in different rates of species production and different levels of regional and local diversity.

We may recognize a barrier to dispersal only when an animal or plant experiences difficulty crossing it. Difficulty depends on the physical and ecological nature of the barrier as well as the ability or willingness of the organism to cross it. Taxa with high vagility cross barriers by dispersing beyond them, and the distance between patches of suitable habitat determines the probability of colonization. Less vagile taxa spread by occupying continuous stretches of suitable habitat. Unsuitable habitat cannot be crossed, and thus their distribution reflects the history of the geographical extent of contiguous habitat. Fragmentation of habitat by climate change may create patches of isolated populations, which then may rejoin as the climate ameliorates and suitable habitat patches again spread back into contact. Clearly the effect of such ecological changes depends on the degree of physical heterogeneity in the environment and the degree of habitat specialization. Once more, diversification by speciation depends on the proper tuning of habitat specialization to the geographical and ecological characteristics of the region (Rosenzweig 1975).

Janzen (1967) suggested that climate zones may differ in the effectiveness of climatic barriers to dispersal, perhaps favoring species production at low latitudes. In the tropics, where temperatures vary little seasonally, species may become closely adapted to narrow temperature ranges. Under these circumstances, differences in temperature between localities, such as one experiences between different altitudes, may pose effective barriers to dispersal. At higher latitudes, plants and animals must tolerate wider ranges of temperature during the seasonal cycle. As a result, temperature differences between altitudes impose less stringent barriers. Thus, Janzen's suggestion that mountain passes are "higher" in the tropics is consistent with the hypothesis that new species are formed more readily in the tropics than in temperate regions.

Evolutionary divergence leading to species production depends as much on rates of evolutionary change as it does on barriers to dispersal. Gene substitution varies in relation to strength of selection and generation time. It is not clear that either of these differs between regions of low and high diversity, although neither has been well characterized for any group of organisms. Indeed, opportunities for directional selection may well diminish in more diverse assemblages in which tighter packing of niche space leaves fewer portions free of interspecific competition.

The importance of directional selection in rate of species production is suggested by the adaptive radiation of species that have entered new adaptive zones. The single taxon that invades an uninhabited archipelago does not feel the constraint of interspecific competition and may diverge rapidly from sister taxa, increasing the probability of coexistence and also the rate of evolutionary divergence (Schluter 1988a).

Farrell, Dussourd, and Mitter (1991) have shown that taxa of plants possessing canals for the accumulation and transport of resins and latex are more diverse than sister taxa that lack them. They suggest that the escalation of plant defenses may lead to greater diversification; escape from certain predators may have opened a new adaptive zone, relatively free of competitors, for these taxa. Mitter, Farrell, and Wiegmann (1988) have also suggested that shifts in host plant taxa have promoted the diversification of phytophagous insects in a similar manner of entry into new adaptive zones (see Farrell and Mitter, chap. 23).

Certain ecological attributes of species may promote reproductive isolation and species production. For example, Pierce (1989) has related the high diversity of lycaenid butterflies, which comprise approximately 40% of all butterfly species, to aspects of their close association with ants. Many taxa are attended and protected by ants, and females judge a plant's suitability for oviposition by their presence. When ants occur on unusual host plants, host switching, host race formation, and eventually speciation may occur (see Bush 1975). Pierce also suggested that lycaenid populations are finely subdivided, with little migration between them, because only individual plants with high nitrogen levels can support the growth of larvae and their production and secretion of nitrogen compounds used to attract ants. Thus, certain features of lycaenids and their mutualists would appear to create the conditions for high speciation rates.

Species flocks may provide important opportunities for studying the relationship between species production and diversity. The term "species flock" refers to an unusually large number of closely related species coexisting within a given region, among which there may be extensive sympatry of ecologically and morphologically similar species (Brooks 1950; table 30.1). Among the most frequently cited of such species flocks are the cichlid fishes of the African rift lakes (Fryer and Iles 1972; Greenwood 1974, 1984), the amphipod crustaceans of Lake Baikal (Kozhov 1963), and the *Drosophila* of the Hawaiian Islands (Carson and Kaneshiro 1976).

The 200 or so species of "haplochromine" cichlid fish of Lake Victoria are monophyletic and have evolved within the last 200,000 years, judging from mitochondrial DNA divergence (Meyer et al. 1990). If this diversity resulted from serial bifurcation of species into daughter species, speciation events would have to have occurred every 26,000 years on the average. Fryer (1977) suggested that allopatric speciation of rock-dwelling species may have occurred on the very small scale of rock outcroppings separated by sandy bottom, exaggerated by changes in lake level that alternately joined and separated suitable areas of habitat. However, McKaye and Gray (1984) experimentally demonstrated no relationship between the ease with which clades of rock-dwellers colonized new habitat and their diversity; in addition, in Lake Malawi, rock and sand dwellers are equally speciose, leaving open the mechanisms of species production by which these clades diversified.

Several authors have emphasized the idea that species flocks may form when a taxon enters a depauperate system, such as cichlids found in a newly formed lake or *Drosophila* encountered on a remote island archipelago (Strauss 1984). In most cases of species flocks, niche space is packed very densely with species, but the entire flock also exhibits a greater variety of morphologies and ecological roles than the parental taxon does (Dominey

1984). Thus, the haplochromines of Lake Victoria exhibit greater ecological diversity than cichlids in general, as do the Hawaiian *Drosophila* compared with other drosophiline species. Rapid speciation has been associated with conditions for establishing allopatry, reinforced by poor dispersal, habitat specialization, rapid genetic change in founder populations, and intense sexual selection (Strauss 1984). Clearly, the study of species flocks, especially in comparison to coexisting clades that have not undergone such spectacular radiation, may reveal much about the processes of species production and the development of ecological communities.

HISTORICAL DEVELOPMENT OF COMMUNITIES

The history of community development can provide information about the processes responsible for generating and maintaining diversity (e.g., Vermeij 1992). If diversity accumulates over time, then more diverse communities should contain relatively more old clades, and relatively more taxa per clade, than less diverse communities. If diversity represents an equilibrium between rates of species production and extinction, then at any given spatial dimension, shifts in diversity–specific extinction and production curves will produce diagnostic changes in rates of species turnover and the average age of taxa within communities (fig. 30.5). Specifically, if high diversity results from an increase in species production, then species should be younger on the whole than when high diversity results from a reduction in the rate of extinction (Rosenzweig 1975). Diversity also may occasionally be reduced by "mass extinctions," which may leave an imprint on diversity patterns for long periods (Kauffman and Fagerstrom, chap. 27). Information bearing upon these points comes from two sources, the fossil record and the systematics and biogeography of extant forms.

Historical (Fossil) Evidence

Fossil records provide direct evidence in some groups of the accumulation of diversity over time and of the roles of extinction, species production, faunal exchange, and provinciality in the generation of diversity. The value of fossil data can be enhanced by information on geographical ranges of taxa (local versus regional) and their ecological habitats (for example, inshore versus offshore in marine systems). Fossil evidence bearing upon community diversity has been presented in this volume for marine reefs (Kauffman and Fagerstrom, chap. 27), marine benthic fauna (Valentine and Jablonski, chap. 29), mammals (Van Valkenburgh and Janis, chap. 28), temperate trees (Latham and Ricklefs, chap. 26), and mangrove plants (Ricklefs and Latham, chap. 20). Fossil assemblages support the idea that increased regional diversity is accompanied by both increased local diversity and increased provincialism, or beta diversity (Bambach 1985; Sepkoski 1988; Van Valkenburgh and Janis, chap. 28). In many cases, diversity appears to have increased steadily over evolutionary time (e.g., angiosperms: Knoll 1986); but in many other cases, biotas have shown progressive, presumably environmentally induced, decreases in diversity. For example, the number of herbivorous mammal species

Table 30.1. Species Richness of Cichlid and Non-Cichlid Fishes in African Rift Valley Lakes and the River Systems from Which the Lake Faunas were Derived

| | Species richness | | Area | Age |
	Cichlids	Noncichlids	(km^2)	(my)[a]
Nile River	10	105		
Lake Turkana	7	32		
Lake Victoria	200	38	69,484	0.25–0.75
Lake Edward	40	17	2150	
Lake Albert	9	37	5346	
Zaire River	40	650		
Lake Tanganyika	136	111	32,893	2–4
Zambezi River	20	90		
Lake Malawi	200	42	28,490	1–2

Source: Greenwood 1984.
[a]Data from Meyer et al. 1990.

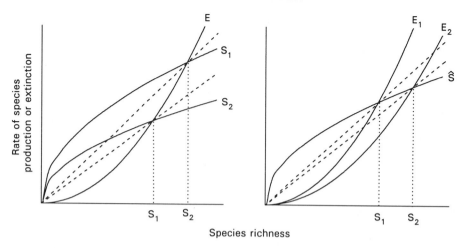

Figure 30.5 Speciation and extinction curves for continental areas illustrating variation in equilibrium diversity ($\hat{S}$) arising from variation in speciation rates (S_1 versus S_2, left) and extinction rates (E_1 versus E_2, right). The slope of the dashed lines represents the rate of species turnover, which is inversely related to the average age of taxa in the community. Variation in speciation rate causes turnover rate to increase (average age decrease) with increasing diversity; variation in extinction rate has the opposite effect. (After MacArthur 1969 and Rosenzweig 1975.)

in North America decreased steadily from the mid-Miocene onward, after a long period of diversification (Van Valkenburgh and Janis, chap. 28). The species richness of mangroves and temperate moist-forest trees appears to have remained steady since the mid-Tertiary, at least at the genus level (Ricklefs and Latham, chap. 20; Latham and Ricklefs, chap. 26).

Jablonski et al. (1983) and Jablonski and Bottjer (1991a) have demonstrated that higher taxa (orders) of marine invertebrates most frequently appear first in near-shore environments, and subsequently spread across the continental shelves, down the continental slopes, and into the deep-sea basins. Lower-level taxa (families and genera) seem to originate more evenly across this environmental gradient, suggesting differential origin of novel life forms with respect to environment, but not differential rates of species production. DiMichele and Aronson (1992) document similar habitat bias in the origins of higher taxa of terrestrial plants, which seem to arise more frequently in drier environments than in wetlands.

Another pattern that has been reported frequently from the fossil record is that regions of high diversity are associated with relatively short average taxonomic duration. This has been noted for foraminifera (Buzas 1972), mammals (Stehli, Douglas, and Newell 1969), and corals (Stehli and Wells 1971), among others. This generalization implies that differences in species richness between regions arise primarily from differences in rate of species production rather than rate of extinction (see fig. 30.5).

SYSTEMATIC AND BIOGEOGRAPHICAL EVIDENCE

The systematics of species provides data on origins and relationships of taxa. Taxa of higher rank are older than taxa of lower rank. Thus, the distribution of species within higher taxonomic groups provides information on the origins of differences in diversity between communities. For example, the higher diversity of trees and birds in the tropics, as compared with temperate latitudes, applies primarily to families and genera, rather than to species

within genera, suggesting a relatively great age for latitudinal gradients in diversity (Ricklefs 1989b). Phylogenetic trees may also provide information with which to evaluate the roles of differential rates of speciation and extinction in the development of diversity patterns (Hey 1992).

A comparison of local communities of songbirds (Passeriformes) in Panama and Illinois illustrates the potential value of phylogenetically based systematic analyses (table 30.2). The Panamanian localities had approximately twice as many species of residents as the Illinois locality, but the species were distributed heterogeneously between the regions with respect to taxonomic group. The moist-forest site in Panama was dominated by suboscine passerines of South American origin, whereas the Illinois forest was dominated by much younger tribes of the Fringillidae, which have northern roots but, in several cases, tropical centers of diversity in South America (e.g., Thraupini, Cardinalini, Icterini). The few tyrannid flycatchers that occur in Illinois represent a small subset of two subfamilies (Tyranninae, Fluvicoliinae), suggesting a young invasion of the temperate zone from the tropics.

Sibley and Alquist (1990) have provided DNA-DNA hybridization data on the origins of higher avian taxa. Genetic distances are proportional to differences in the melting points of hybridized DNA ($\Delta T_H 50$). Sibley and Alquist have suggested that a 1° C difference in melting point corresponds to approximately 2.3 million years since divergence in passerine taxa. The hybridization data allow one to date the origins of the clades that constitute a particular assemblage of species. For example, in the Illinois and Panama comparison, we see that the Panamanian clades are much older ($\Delta T_H 50$ averages 11.9° C) than the Illinoisan clades (4.9° C). The approximately twofold difference in average age of the taxa matches the twofold difference in diversity, although age and diversity are not well correlated among individual clades.

In a comparison of the Sonoran Desert of North America and the Monte Desert of northwestern Argentina, Blair, Hulse, and Mares (1976) noted high degrees of endemism within the deserts among reptiles and am-

Table 30.2. Phylogenetic Distribution of Species of Passerine Birds in Illinois and Panama Census Sites

| | Illinois | | Panama | | |
| | | | Forest | | Scrub |
	n	$\Delta T_H 50$	n	$\Delta T_H 50$	n
Furnariidae			7	12.1	1
Formicariidae			2	12.1	
Thamnophilidae			7	13.5	4
Tyrannidae	3	3.1	15	13.8	19
Vireonidae	2	8.8	2	8.8	2
Corvidae	1	6.2			
Sittidae	1	9.0			
Certhiidae	2	4.7	5	9.0	3
Paridae	2	10.6			
Fringillidae	10	2.1	3	5.4	17
Total	23		41		47
Average $\Delta T_H 50$	4.9		11.9		10.0

Source: Data from Karr 1968, 1971; phylogeny from Sibley and Alquist 1990.
Note: $\Delta T_H 50$, based on DNA-DNA hybridization studies of Sibley and Alquist, indicates the genetic divergence since the origin of the clade to which the local species belong. The $\Delta T_H 50$ data refer to the divergence of each family from its sister taxa. However, in the cases of several temperate-zone species, the $\Delta T_H 50$ values refer to a subgroup of species within a primarily tropical clade (*Empidonax*, *Contopus* in the Fluvicoliinae; *Myiarchus* in the Tyranninae; *Icterus* and *Agelaius* in the Icterini; *Troglodytes* and *Thryothorus* in the Certhiidae).

phibians, but not among mammals, most of which are wide-ranging taxa. In comparing the two regions, diversities are similar in reptiles and amphibians, but the Sonoran Desert has more than twice the species richness of mammals as the Monte Desert. Similarly, the arid Caatinga of eastern Brazil has a depauperate fauna containing only 31 nonflying mammal species, including only one endemic to that biome; 26 of the species also occur in evergreen mesophytic forest (Mares, Willig, and Larcher 1985). This low degree of endemism suggests a recent fauna with little adaptation to the stressful conditions of the Caatinga environment. Indeed, the water-concentrating ability of Caatinga mammals falls far below that of mammals in the Sonoran Desert of North America and deserts elsewhere in the world. The Caatinga does not lack endemic mammals because it is a young biome; the xeric-adapted flora exhibits a high degree of endemism. Why dry environments lack locally adapted mammals is an open question, but there is an evident relationship between low endemism and low diversity. Presumably, the stressful arid environment prevents many mammalian taxa from entering from more mesic environments, just as freezing conditions prevent plant taxa from entering temperate environments and high salt concentrations prevent most taxa from entering mangrove habitat.

These examples underscore the importance of treating species as unique entities that carry information about their own histories, and thus the histories of the assemblages to which they belong.

ECOLOGICAL AND GEOGRAPHICAL CENTERS OF ORIGIN

Many biologists have postulated that areas of high diversity for particular clades represent centers of origin of taxa within those clades (McCoy and Heck 1976; Croizat, Nelson, and Rosen 1974). Accordingly, diversity parallels species production, and differences in community

diversity between regions result from differences in the rate of production of new taxa. Regions distant from the center of origin receive new taxa primarily by migration from the center rather than by autochthonous production, although centers of species production may also spread out from a clade's origin with time. The center-of-origin principle may apply to habitats or biomes rather than areas within regions, in which case it refers to ecological rather than geographical patterns of diversity. Thus, the correlation between diversity and habitat may represent the historical origins of a clade in one type of habitat with subsequent spread through adaptive radiation into other habitats. The principle may also apply to major latitudinal gradients of diversity, in which case cladogenesis would proceed most rapidly in the tropics, and taxa would spread poleward from there, often having to cross major adaptive barriers along the way (Farrell, Mitter, and Futuyma, 1992; Latham and Ricklefs, 1993).

Several points have been raised against the center-of-origin concept. First, the present distributions of some taxa do not coincide with their fossil distributions. For example, the mangrove palm *Nypa* is known from fossil deposits in Eurasia and in North and South America, located entirely outside of its present geographical range (Ricklefs and Latham, chap. 20). Similarly, fossil evidence clearly places the origin of the Juglandaceae in North America; at present, however, except for the hickory genus *Carya*, the center of diversity of the family resides in eastern Asia. Second, the fossil record also reveals that regions of high diversity often have younger taxa on average than regions of low diversity (McCoy and Heck 1976). Stehli, Douglas, and Newell (1969) suggested that the average age of mammalian families (age since origin) is younger in the more diverse African region than it is in the less diverse European region, even though the oldest families were nonetheless African (fig. 30.6).

We are unimpressed with these objections for several reasons. First, detractors of the center-of-origin concept have presented anecdotal cases of major shifts in the biogeographical positions of taxa but have not produced more comprehensive analyses of the relationship between the locations of the oldest fossils and contemporary centers of diversity. Moreover, because of the dynamic nature of the earth's surface, one expects major taxa both to undergo wholesale shifts and to colonize new areas where they may proliferate. As taxa shift, they may carry their centers of origination, if not centers of origin, with them. Cladogenesis is an ongoing process that undoubtedly proceeds more rapidly in some areas than in others. Only when a center of origination remains in a given area for a long period will the center of diversity of a group correspond to the historical region of taxon production.

Second, the younger average age of taxa within centers of diversity does not contradict the center-of-origin concept. Centers of origination continuously produce new taxa, so that the center of origin should contain both young and old taxa, including the basal taxa of the clade. Peripheral areas will have received colonizing taxa, as well as autochthonously produced taxa, throughout the history of diversification of the clade. Thus, one might expect fewer very old taxa in peripheral areas, but otherwise a similar distribution of ages, provided that cladogenesis

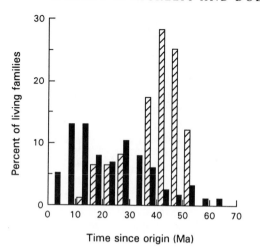

Figure 30.6 Distribution of ages of taxa in temperate (northern Eurasia: hatched bars) and tropical (central-western Africa: solid bars) mammalian faunas. (After Stehli, Douglas, and Newell 1969).

and extinction had proceeded more or less uniformly over time. We can also imagine a scenario in which a center of origination continuously produces new taxa, resulting in a more or less even distribution of bifurcation events, while a marginal region might have been colonized at some point in time and undergone a transient burst of diversification and adaptive radiation. If the latter could not be sustained in the marginal area, then one would perceive a relatively old and clumped distribution of bifurcation times, with a gradual decline in diversity due to extinction. Furthermore, if time has brought increasing provincialism, then the center of origination becomes progressively isolated from peripheral regions, leaving them with relictual old taxa while continuing to produce new taxa locally (Brown 1957). This almost certainly describes the situation in mangroves, in which the Indo–West Pacific seems particularly favored for the production of new taxa but has become isolated from other oceanic realms by latitudinal stratification of ocean temperature and the closure of the Tethys connection between the Indian and Atlantic Oceans (Ricklefs and Latham, chap. 20).

Third, assessment of the age of origin of taxa requires uniform application of taxonomic distinctions. If taxonomists overdistinguish taxa from more diverse areas, then those taxa will appear to have a younger age distribution for a given level of taxonomic rank. Thus, in Stehli, Douglas, and Newell's (1969) comparison of the ages of origin of mammalian families, the younger ages of the African taxa may reflect the finer subdivision of the more diverse tropical fauna. This has certainly happened in the case of birds, in which the tropical South American families of passerines (e.g., Tyrannidae, Formicariidae, Furnariidae) are far older by the metric of DNA-DNA hybridization ($\Delta T_H 50$ 11.4°–13.8° C) than taxa previously distinguished at the family level with representatives in north temperate regions (e.g., Parulidae, Thraupide, Icteridae: $\Delta T_H 50$ 5.5°–6.0° C). Sibley and Alquist (1990), in an attempt to make taxonomic distinctions correspond to genetic divergence, have reduced the latter taxa to tribes

within the subfamily Emberizinae (family Fringillidae: $\Delta T_H 50 = 10.0°$ c). This example does not bear on the center-of-origin issue, but illustrates the general problem of heterogeneous taxonomic distinctions and the need to evaluate the age and diversity of clades independently of taxonomic distinction, by direct fossil evidence where available, and otherwise by molecular methods.

The center-of-origin concept should not be used to support the hypothesis of high rates of species production in areas of high diversity. Rather, it emphasizes the need to marshal paleontological, biogeographical, and systematic evidence bearing on regional and historical factors in the generation and maintenance of patterns of biodiversity.

CONCLUSIONS

We have been able to touch only briefly on some of the major issues arising from global patterns of diversity. Biologists understand so little about the generation and maintenance of diversity that it is possible to entertain hypotheses as opposed as local environmental determinism and age-and-area scenarios based on centers of origin. Although these issues will not likely be resolved in the near future, patterns of diversity focus attention on the interconnectedness of natural phenomena and on the importance of understanding those connections. We predict that the present decade will see a resurgence of interest in comparative and historical analyses. In conclusion, we offer a number of suggestions concerning useful directions of inquiry that emphasize large-scale patterns and processes.

First, the phenomenon of diversity itself is poorly characterized geographically, taxonomically, and ecologically. We advocate systematic comparisons of diversity between continents or ocean basins within carefully matched habitats and with multiscale sampling, in order to assess regional and local components of diversity. In addition, information on taxonomy, guild structure, ecological relationships, and morphological diversity will greatly increase the value of such comparisons by allowing one to relate species richness to the occupation and generation of ecological (niche) space.

Second, alpha and beta diversity appear to vary in direct proportion to regional diversity. The generality of this relationship should be strengthened by the types of studies advocated above. In addition, however, ecologists should investigate the link between these two scales of diversity, and between all scales of diversity in general. The relationship between alpha and beta diversity would appear to be mediated by a direct relationship between the within-habitat utilization of resources, which expresses itself in abundance, and the occupation of habitats, which expresses itself in geographical distribution. This relationship arises either from adaptations of individual organisms, in which case the ability to exploit a variety of resources within a habitat correlates with ability to inhabit a variety of habitats; or from demography of populations, in which case locally productive populations can maintain themselves by migration in surrounding suboptimal habitats. Both these mechanisms should produce a positive correlation between local abundance and habitat distribution (and perhaps geographical distribution as well).

Third, ecologists should use historical, biogeographical, and systematic (including molecular) data to reconstruct the development of species assemblages. This will help to determine the degree to which age, area, and unique historical events contribute to contemporary diversity, and to establish the degree of historical relatedness between assemblages in different geographical areas, or between different habitats within areas. Ecological, geographical, and systematic data may help to sort out the relationships between regional and beta diversity, particularly if different habitats have different histories of community development.

Fourth, ecologists must take a more active role in investigating the processes of species production and extinction. In particular, it seems crucial to determine whether there are particular intrinsic qualities of species (genetic systems, developmental flexibility, mating systems, etc.), or particular ecological conditions and biogeographical configurations that promote speciation, and whether these qualities vary in direct relation to regional species diversity. Extinction has eluded ecological study beyond the determination of ecological correlates of small, localized populations that seem on the verge of extinction. Habitat alteration and other disturbances by humans are also changing the ecological context of extinction to the extent that what we can learn at present may have little bearing on the past. The study of extinction will require judicious choices of taxa at different levels of apparent vulnerability, and determination of how they differ in relation to other interacting populations, including (perhaps especially) disease organisms. As in the past, island systems will be particularly helpful in the study of extinction because they provide opportunities to study replicated populations whose relative ages can be estimated by molecular techniques.

Finally, ecologists should join with physiologists, functional morphologists, and evolutionists in studying adaptive radiation and entry into new adaptive zones. Diversity patterns indicate that major gradients in diversity may be established by adaptational barriers to entering habitats or regions different from those that supported the origin of a clade. Crossings of geographical barriers around depauperate island groups (Galápagos and Hawaiian Islands), of habitat barriers (terrestrial plants into the intertidal zone), of species-interaction barriers (herbivores to new host plants; host plants acquiring new defenses), and of climate zone barriers (tropical plants and animals invading frost zones) have all left their imprint on patterns of community diversity.

We join the authors of this book in the hope that the desire to explain diversity phenomena will provide the motivation for increased integration of ecology with the rest of the biological sciences and that it will bring about a new synthesis of biological knowledge.

References

Abbott, I. 1980. Theories dealing with the ecology of landbirds on islands. *Adv. Ecol. Res.* 11:329–71. (1)

Abele, L. G. 1984. Biogeography, colonization, and experimental community structure of coral-associated crustaceans. In *Ecological communities: Conceptual issues and the evidence*, ed. D. R. Strong, D. Simberloff, L. G. Abele, and A. B. Thistle, 123–37. Princeton, N.J.: Princeton University Press. (17, 22)

Aber, J. D., and J. M. Melillo. 1982. Nitrogen immobilization in decaying hardwood leaf litter as a function of initial nitrogen and lignin content. *Can. J. Bot.* 60:2263–69. (2)

Abrams, P. A. 1977. Density-independent mortality and interspecific competition: A test of Pianka's niche overlap hypothesis. *Am. Nat.* 111:539–52. (1)

Abrams, P. A. 1981. Competition in an Indo-Pacific hermit crab community. *Oecologia* 51:241–49. (30)

Abrams, P. A. 1988. Resource productivity-consumer species diversity: Simple models of competition in spatially heterogeneous environments. *Ecology* 69:1418–33. (2, 5, 6)

Abramsky, Z. N. 1978. Small mammal community ecology: Changes in species diversity in response to manipulated productivity. *Oecologia* 34:113–23. (5)

Abramsky, Z., S. Brand, and M. Rosenzweig. 1985. Geographical ecology of gerbilline rodents in sand dune habitats of Israel. *J. Biogeogr.* 12:363–72. (5, 14)

Abramsky, Z., and M. L. Rosenzweig. 1984. Tilman's predicted productivity-diversity relationship shown by desert rodents. *Nature* 309:150–51. (5, 6, 14)

Abramsky, Z. N., M. L. Rosenzweig, and B. Pinshow. 1991. The shape of a gerbil isocline: An experimental field study using principles of optimal habitat selection. *Ecology* 72:329–40. (5)

Abramsky, Z., M. L. Rosenzweig, B. Pinshow, J. S. Brown, B. Kotler, and W. A. Mitchell. 1990. Habitat selection: An experimental field test with two gerbil species. *Ecology* 71:2358–69. (5, 7)

Acciavatti, R. E., and D. L. Pearson. 1989. The tiger beetle genus *Cicindela* (Coleoptera, Insecta) from the Indian subcontinent. *Ann. Carnegie Mus.* 58:77–354. (18)

Adams, E. N. 1972. Consensus techniques and the comparison of taxonomic trees. *Syst. Zool.* 21:390–97. (24)

Adams, J. M., and F. I. Woodward. 1989. Patterns in tree species richness as a test of the glacial extinction hypothesis. *Nature* 339:699–701. (6, 26)

Adamson, R. S. 1927. The plant communities of Table Mountain. I. Preliminary account. *J. Ecol.* 15:278–309. (5)

Addicott, J. F., J. M. Aho, M. F. Antolin, D. K. Padilla, J. S. Richardson, and D. A. Soluk. 1987. Ecological neighborhoods: Scaling environmental patterns. *Oikos* 49:340–46. (8)

Adis, J., Y. D. Lubin, and G. G. Montgomery. 1984. Arthropods from the canopy of inundated and terra firme forests near Manaus, Brazil, with critical considerations of the Pyrethrum-fogging technique. *Stud. Neotropical Fauna Environ.* 19:223–36. (22)

Aho, J. M. 1990. Helminth communities of amphibians and reptiles: Comparative approaches to understanding patterns and processes. In *Parasite communities: Patterns and processes*, ed. G. W. Esch, A. O. Bush, and J. M. Aho, 157–95. London: Chapman and Hall. (17, 22)

Al-Mufti, M. M., C. L. Sydes, S. B. Furness, J. P. Grime, and S. R. Band. 1977. A quantitative analysis of shoot phenology and dominance in herbaceous vegetation. *J. Ecol.* 65:759–91. (2, 6)

Alcover, J. A., S. Moya-Sola, and J. Pons-Moya. 1981. *Les quimeres del passat.* Mallorca: Editorial Moll. (12)

Allen, T. F. H., and T. B. Starr. 1982. *Hierarchy: Perspectives for ecological complexity.* Chicago: University of Chicago Press. (1, 4)

Altaba, C. A. 1991. The importance of ecological and historical factors in the production of benzaldehyde by tiger beetles. *Syst. Zool.* 40:101–5. (18)

Alvarez, L. W., W. A. Alvarez, F. Asaro, and H. V. Michel. 1980. Extraterrestrial cause for the Cretaceous-Tertiary extinction. *Science* 208:1095–1108. (27)

Alvarez, W. A., E. G. Kauffman, F. Surlyk, L. Alvarez, F. Asaro, and H. V. Michel. 1984. The impact theory of mass extinctions and the marine invertebrate record across the Cretaceous-Tertiary boundary. *Science* 233:1135–41. (27)

Andersen, A. N. 1983. Species diversity and temporal distribution of ants in the semi-arid mallee region of northwestern Victoria. *Aust. J. Ecol.* 8:127–37. (14)

Andersen, A. N., and A. L. Yen. 1985. Immediate effects of fire on ants in the semi-arid mallee region of north-western Victoria. *Aust. J. Ecol.* 10:25–30. (14)

Andersen, T., S. Ligaard, T. Pedersen, and G. E. E. Soli. 1990. Pitfall catches of Carabidae and Staphylinidae (Coleoptera) in a temporarily protected forest area on the Eidanger peninsula, Telmark, SE Norway. *Fauna Norvegica*, ser. B 37:13–22. (10)

Andrewartha, H. G., and L. C. Birch. 1954. *The distribution and abundance of animals.* Chicago: University of Chicago Press. (10)

Antonovics, J. 1968. Evolution in closely adjacent plant populations. VI. Manifold effects of gene flow. *Heredity* 23:507–24. (7)

Arita, H. T., J. G. Robinson, and K. H. Redford. 1990. Rarity in Neotropical forest mammals and its ecological correlates. *Conserv. Biol.* 4:181–92. (10)

Armitage, P., B. West, and K. Steedman. 1984. New evidence of Black Rat in Roman London. *Lond. Archaeol.* 4:375–83. (12)

Armstrong, R. A. 1976. Fugitive species: Experiments with fungi and some theoretical considerations. *Ecology* 57:953–63. (2, 8)

Armstrong, R. A. 1989. Competition, seed predation, and species coexistence. *J. Theor. Biol.* 141:191–94. (22)

Armstrong, R. A., and R. McGehee. 1976. Coexistence of species competing for shared resources. *Theor. Popul. Biol.* 9:317–28. (2)

Armstrong, R. A., and R. McGehee. 1980. Competitive exclusion. *Am. Nat.* 115:151–70. (1, 2)

Arnold, S. J. 1972. Species densities of predators and their prey. *Am. Nat.* 106:220–36. (6)

Arnold, S. J. 1980. The microevolution of feeding behavior. In *Foraging behavior: Ecological, ethological and psychological approaches*, ed. A. Kamil and T. Sargent, 409–53. New York: Garland Press. (25)

Arnold, S. J. 1981. Behavioral variation in natural populations. I. Phenotypic, genetic, and environmental correlations between chemoreceptive responses to prey in the garter snake, *Thamnophis elegans*. *Evolution* 35:489–509. (25)

Aronson, J. A., and A. Shmida. In press. Plant species diversity along a Mediterranean-desert gradient and its correlation with interannual rainfall fluctuations. *J. Arid Environ.* (5)

Arthur, M. A., and S. O. Schlanger. 1979. Cretaceous "oceanic anoxic events" as causal factors in development of reef reservoired giant oil fields. *Bull. Am. Assoc. Petrol. Geol.* 63:870–85. (27)

Arthur, M. A., S. O. Schlanger, and H. C. Jenkyns. 1987. The Cenomanian-Turonian anoxic event. II. Paleoceanographic controls on organic matter production and preservation. In *Marine petroleum source rocks,* ed. J. Brooks and A. Fleet, 401–20. Geological Society, London, Special Publication no. 26. (27)

Ashton, P. S. 1969. Speciation among tropical forest trees: Some deductions in the light of recent research. *Biol. J. Linn. Soc.* 1:155–196. (30)

Askins, R. A., M. J. Philbrick, and D. S. Sugeno. 1987. Relationship between the regional abundance of forest and the composition of forest bird communities. *Biol. Conserv.* 39:129–52. (6)

Atkinson, I. 1989. Introduced animals and extinctions. Pages 54–75 In *Conservation for the twenty-first century,* ed. D. Western and M. C. Pearl, 54–75. New York: Oxford University Press. (29, 30)

Atkinson, I. A. E. 1985. The spread of commensal species of *Rattus* to oceanic islands and their effects on island avifaunas. Pages 35–81 In *Conservation of island birds,* ed. P. J. Moors, Technical Publication 3. Cambridge: International Council for Bird Preservation. (12)

Auerbach, M. J., and S. D. Hendrix. 1980. Insect-fern interactions: Macrolepidopteran utilization and species-area association. *Ecol. Entomol.* 5:99–104. (16)

Auerbach, M. J., and A. Shmida. 1987. Spatial scale and the determinants of plant species richness. *Trends Ecol. Evol.* 2:238–42. (1, 13)

Auffray, J. C., F. Vanlerberghe, and J. Britton-Davidian. 1990. The house mouse progression in Eurasia: A palaeontological and archaeozoological approach. *Biol. J. Linn. Soc.* 41:13–25. (12)

Ayre, D. J. 1982. Inter-genotype aggression in the solitary sea anemone *Actinia tenebrosa. Mar. Biol.* 68:199–205. (4)

Bailey, J. R. 1967. The synthetic approach to colubrid classification. *Herpetologica* 23:155–161. (25)

Bailey, L. H., and E. Z. Bailey. 1976. *Hortus Third.* New York: Macmillan. (26)

Bailey, R. C. 1988. Correlations between species richness and exposure: Freshwater molluscs and macrophytes. *Hydrobiologia* 162:183–91. (6)

Bakker, R. T. 1973. The deer flees, the wolf pursues: Incongruencies in predator-prey coevolution. In *Coevolution,* ed, D. J. Futuyma and M. Slatkin, 350–82. Sunderland, Mass: Sinauer Associates. (28)

Balouet, J. C. 1987. Extinction des vertébrés terrestres de Nouvelle Calédonie. *Mem. Soc. Geol. France,* n.s. 150:177–83. (12)

Bambach, R. K. 1977. Species richness in marine benthic habitats through the Phanerozoic. *Paleobiology* 3:152–67. (27)

Bambach, R. K. 1983. Ecospace utilization and guilds in marine communities through the Phanerozoic. Pages 719–746 In *Biotic interactions in Recent and fossil benthic communities,* ed. M. J. S. Tevesz and P. L. McCall, 719–46. New York: Plenum Press. (27)

Bambach, R. K. 1985. Classes and adaptive variety: The ecology of diversification in marine faunas through the Phanerozoic. In *Phanerozoic diversity patterns,* ed. J. W. Valentine, 191–254. Princeton, N.J.: Princeton University Press. (1, 28, 30)

Barbour, M., J. Burk, and W. Pitts. 1987. *Terrestrial plant ecology.* 2d ed. Menlo Park, Calif: Benjamin/Cummings. (2)

Barnes, H. 1955. The growth rate of *Balanus balanoides* (L.). *Oikos* 6:109–13. (4)

Barnett, M. A. 1983. Species structures and temporal stability of mesopelagic fish assemblages in the central gyre of the North and South Pacific Ocean. *Mar. Biol.* 74:245–56. (19)

Barnosky, A. D. 1989. The late Pleistocene event as a paradigm for widespread mammal extinction. In *Mass extinctions: Processes and evidence,* ed. S. K. Donovan, 235–54. London: Bellhaven Press. (29)

Barnosky, A. D., In press. Defining climate's role in ecosystem evolution: Clues from late Quaternary mammals. *Hist. Biol.* (29)

Barnosky, C. W., P. M. Anderson, and P. J. Bartlein. 1987. The northwestern U.S. during deglaciation: Vegetational history and paleoclimatic implications. In *North America and adjacent oceans during the last deglaciation,* ed. W. F. Ruddiman and H. E. Wright, Jr., 289–321. Boulder, Colo: Geological Society of America. (29)

Barron, E. J., and W. M. Washington. 1984. The role of geographic variables in explaining paleoclimates: Results from Cretaceous climate model sensitivity studies. *J. Geophys. Res.* 89:1267–79. (27)

Barry, J. C., M. E. Morgan, A. J. Winkler, L. J. Flynn, E. H. Lindsay, L. L. Jacobs, and D. Pilbeam. 1991. Faunal interchange and Miocene terrestrial vertebrates of southern Asia. *Paleobiology* 17:231–45. (30)

Barth, H. 1982. The biogeography of mangroves. In *Contributions to the ecology of halophytes,* ed. D. N. Sen and K. S. Rajpurohit, 35–60. The Hague: Junk. (20)

Bates, R. L., and J. A. Jackson, eds. 1980. *Glossary of geology,* 2nd ed. Falls Church, Va.: American Geological Institute. (27)

Baum, D. A., and A. Larson. 1991. Adaptation reviewed. A phylogenetic methodology for studying character macroevolution. *Syst. Zool.* 40:1–18. (25)

Baur, B., and J. Bengtsson. 1987. Colonizing ability in land snails on Baltic uplife archipelagoes. *J. Biogeog.* 14:249–86. (10)

Baverstock, P. 1984. Australia's living rodents: A restrained explosion. In *Vertebrate zoogeography and evolution in Australasia,* ed. M. Archer and G. Clayton, 913–19. Perth: Hesperian Press. (14)

Baverstock, P. R. 1982. Adaptations and evolution of the mammals of arid Australia. In *Evolution of the flora and fauna of arid Australia,* ed. W. R. Barker and P. J. M. Greenslade, 175–78. Adelaide: Peacock Publications. (14)

Baverstock, P. R., M. Archer, M. Adams, and B. J. Richardson. 1982. Genetic relationships among 32 species of Australian dasyurid marsupials, In *Carnivorous marsupials,* ed. M. Archer, 641–50. Sydney: Royal Zoological Society of New South Wales. (14)

Baverstock, P. R., C. H. S. Watts, M. Adams, and S. R. Cole. 1981. Genetical relationships among Australian rodents (Muridae). *Aust. J. Zool.* 29:289–303. (14)

Bazzaz, F. A., N. R. Chiariello, P. D. Coley, and L. F. Pitelka. 1987. Allocating resources to reproduction and defense. *BioScience* 37:58–67. (2)

Beadle, N. C. W. 1954. Soil phosphate and the delimitation of plant communities in eastern Australia. *Ecology* 35:370–75. (15)

Beadle, N. C. W. 1966. Soil phosphate and its role in molding segments of the Australian flora and vegetation with special reference to xeromorphy and sclerophylly. *Ecology* 47:991–1007. (2, 15)

Beard, J. S. 1981a. *Vegetation of central Australia.* In *Flora of Central Australia,* ed. J. Jessop, xxi–xxvi. Sydney: A. H. and A. W. Reed. (15)

Beard, J. 1981b. *Vegetation survey of Western Australia.* Nedlands, W.A.: University of Western Australia Press.

Beard, J. S. 1983. Ecological control of the vegetation of southwestern Australia: Moisture versus nutrients. In *Mediterranean-type ecosystems: The role of nutrients,* ed. F. J. Kruger, D. T. Mitchell, and J. U. M. Jarvis, 66–73. Berlin: Springer-Verlag. (15)

Beaver, R. A. 1985. Geographical variation in food web structure in *Nepenthes* pitcher plants. *Ecol. Entomol.* 10:241–48. (3, 30)

Beebe, W. 1946. Field notes on the snakes of Kartabo, British Guiana, and Caripito, Venezuela. *Zoologica* 31:11–52. (25)

Beehler, B. 1981. Ecological structuring of forest bird communities in New Guinea. *Monogr. Biol.* 42:837–61. (21)

Begon, M., J. L. Harper, and C. R. Townsend. 1990. *Ecology: Individuals, populations and communities.* 2d ed. Boston: Blackwell Scientific Publications. (5, 6)

Behrensmeyer, A. K., D. Western, and D. D. Boaz. 1979. New perspectives in vertebrate paleoecology from a Recent bone assemblage. *Paleobiology* 5:12–21. (28)

Bell, F. C. 1979. Precipitation. In *Arid-land ecosystems: Structure, functioning and management,* ed. D. W. Goodall and R. A. Perry, 51–72. Cambridge: Cambridge University Press. (14)

Belovsky, G. 1987. Extinction models and mammalian persistence. In *Viable Populations for Conservation,* ed. M. Soulé, 35–57. Cambridge: Cambridge University Press. (7)

Bennett, K. D. 1990. Milankovitch cycles and their effects on species in ecological and evolutionary time. *Paleobiology* 16:11–21. (29)

Benton, M. J. 1987a. Mass extinctions among families of non-marine tetrapods: The data. *Mem. Soc. Geol. France* 150:21–32. (29)

Benton, M. J. 1987b. Progress and competition in macroevolution. *Biol. Rev.* 62:305–38. (29)

Berenbaum, M. 1981. Patterns of furanocoumarin distribution and insect herbivory in the Umbelliferae: Plant chemistry and community structure. *Ecology* 62:1254–66. (23)

Berenbaum, M. 1983. Coumarins and caterpillars: A case for coevolution. *Evolution* 37:163–79. (23)

Berenbaum, M., and P. Feeny. 1981. Toxicity of angular furanocoumarins to swallowtail butterflies: Escalation in the coevolutionary arms race? *Science* 212:927–29. (23)

Berger, E. M. 1973. Gene-enzyme variation in three sympatric species of *Littorina. Biol. Bull. MBL, Woods Hole* 145:83–90. (4)

Berlocher, S. H., and G. L. Bush. 1982. An electrophoretic analysis of *Rhagoletis* (Diptera: Tephritidae) phylogeny. *Syst. Zool.* 31:136–55. (23)

Bernays, E. A., and R. Chapman. 1987. Evolution of plant deterrence to insects. In *Perspectives in chemoreception and behavior*, ed. E. A. Bernays and J. G. Stoffolano. New York: Springer-Verlag. (23)

Bernays, E. A., and M. Graham. 1988. On the evolution of host specificity in phytophagous arthropods. *Ecology* 69:886–92. (23)

Bernays, E. A., and D. H. Janzen. 1988. Saturniid and sphingid caterpillars: Two ways to eat leaves. *Ecology* 69:1153–60. (23)

Berner, R. A. 1991. A model for atmospheric CO_2 over Phanerozoic time. *Am. J. Sci.* 291:339–76. (27)

Berner, R. A., A. C. Lasaga, and R. M. Garrells. 1983. The carbonate-silicate geochemical cycle and its effect on atmospheric carbon dioxide over the past 100 million years. *Am. J. Sci.* 283:641–83. (27)

Bernstein, R. A. 1979. Relations between species diversity and diet in communities of ants. *Insectes Sociaux* 26:313–21. (14)

Bertness, M. 1984. Habitat and community modification by an introduced herbivorous snail. *Ecology* 65:370–81. (4)

Birks, H. J. B. 1980. British trees and insects: A test of the time hypothesis over the last 13,000 years. *Am. Nat.* 115:600–605. (6)

Bishop, Y. M. M., S. E. Fienberg, and P. W. Holland. 1975. Discrete multivariate analysis: Theory and practice. Cambridge, Mass: MIT Press. (18)

Blair, W. F., A. C. Hulse, and M. A. Mares. 1976. Origins and affinities of vertebrates of the North American Sonoran Desert and the Monte Desert of northwestern Argentina. *J. Biogeogr.* 3:1–18. (14, 30)

Blakers, M., S. J. J. F. Davies, and P. N. Reilly. 1984. *The atlas of Australian birds*. Carlton, Victoria: Melbourne University Press. (13)

Blatchley, W. S. 1910. Coleoptera or Beetles known to occur in Indiana. Indianapolis, Ind.: The Nature Publishing Co. (23)

Blondel, J. 1985. Habitat selection in island versus mainland birds. In *Habitat selection in birds*, ed. M. L. Cody, 477–516. New York: Academic Press. (12)

Blondel, J. 1986. Biogéographie Evolutive. Paris: Masson. (12)

Blondel, J. 1987. From biogeography to life history theory: A multithematic approach. *J. Biogeogr.* 14:405–22. (12)

Blondel, J. 1988. Biogéographie évolutive à différentes échelles: L'histoire des avifaunes méditerranéennes. In *Acta XIX Congressus Internationalis Ornithologici*, (Ottawa), ed. H. Ouellet, 155–88. (12)

Blondel, J. 1990a. Biogeography and history of forest bird faunas in the Mediterranean zone. In Biogeography and ecology of forest bird communities, ed. A. Keast, 95–107. The Hague: SPB Academic Publishing. (12)

Blondel, J. 1990b. Birds in biological isolates. In *Bird population studies: Their relevance to conservation and management*, ed. C. M. Perrins, J. D. Lebreton, and G. Hirons, 45–72. Oxford: Oxford University Press. (12)

Blondel, J., D. Chessel, and B. Frochot. 1988. Bird species impoverishment, niche expansion and density inflation in Mediterranean island habitats. *Ecology* 69:1899–1917. (12)

Blondel, J., and H. Farré. 1988. The convergent trajectories of bird communities in European forests. *Oecologia* 75:83–93. (12)

Blondel, J., P. Perret, and M. Maistre. 1990. On the genetical basis of the laying date in an island population of Blue Tit. *J. Evol. Biol.* 3:469–75. (12)

Blondel, J., P. Perret, M. Maistre, and P. C. Dias. 1992. Do harlequin Mediterranean environments function as source sink for Blue Tits (*Parus caeruleus* L.)? *Landscape Ecol.* 6:213–19. (12)

Blondel, J., R. Pradel, and J. D. Lebreton. 1992. Low fecundity insular Blue Tits do not survive better as adults than high fecundity mainland ones. *J. Anim. Ecol.* 61:205–13. (12)

Blondel, J., F. Vuilleumier, L. E. Marcus, and E. Terouanne. 1984. Is there ecomorphological convergence among mediterranean bird communities of Chile, California and France? *Evol. Biol.* 18:141–213. (21)

Boalch, G. T., and Potts, G. W. 1979. The first occurrence of *Sargassum muticum* (Yendo) Fensholt in the Plymouth area. *J. Mar. Biol. Assoc. UK* 57:29–32. (4)

Bochenski, Z. 1985. The development of western Palaearctic avifaunas from fossil evidence. *Acta XVIII Congr. Intern. Ornithol.* (Moscow), vol. 1, 338–347. (12)

Bock, C. E., and L. W. Lepthien. 1975. Patterns of bird species diversity revealed by Christmas counts versus breeding bird surveys. *West. Birds* 6:95–100. (6)

Bock, C. E., and R. E. Ricklefs. 1983. Range size and local abundance of some North American songbirds: a positive correlation. *Am. Nat.* 122:295–299. (10)

Boessneck, J., and J. Küver. 1970. Alluviale Tierknochenfunde aus der Ghar Dalam Hohle, Malta. *Senkenbergiana Biol.* 51:147–58. (12)

Boettcher, S. E., and P. J. Kalisz. 1990. Single-tree influence on soil properties in the mountains of eastern Kentucky. *Ecology* 71:1365–72. (8)

Bohnsack, J. A., and F. H. Talbot. 1980. Species packing by reef fishes on Australian and Caribbean reefs: An experimental approach. *Bull. Mar. Sci.* 30:710–23. (17, 22)

Bond, W. 1983. On alpha diversity and the richness of the Cape flora: A study in southern Cape fynbos. In *Mediterranean-type ecosystems: The role of nutrients*, ed. F. J. Kruger, D. T. Mitchell, and J. U. M. Jarvis, 337–56. Berlin: Springer-Verlag. (2, 6)

Boom, B. M. 1986. A forest inventory in Amazonian Bolivia. *Biotropica* 18:287–94. (23)

Bormann, F. H. 1990. Air pollution and temperate forest: Creeping degradation. In *The earth in transition: Patterns and processes of biotic impoverishment*, ed. G. M. Woodwell, 25–44. Cambridge: Cambridge University Press. (19)

Boucot, A. J. 1975a. *Evolution and extinction rate controls*. New York: Elsevier. (24)

Boucot, A. J. 1975b. Standing diversity of fossil groups in successive intervals of geologic time viewed in the light of changing levels of provincialism. *J. Paleontol.* 49:1105–11. (24)

Boucot, A. J. 1978. Community evolution and rates of cladogenesis. In *Evolutionary Biology*, vol. 11., ed. M. K. Hecht, W. C. Steere, and B. Wallace, 545–655. New York: Plenum. (24)

Boucot, A. J. 1981. *Principles of benthic marine paleoecology*. New York: Academic Press. (24)

Boucot, A. J. 1982. *Paleobiologic evidence of behavioral evolution and coevolution*. Corvallis, Ore.: A. J. Boucot. (24)

Boucot, A. J. 1983. Does evolution take place in an ecological vacuum? II. *J. Paleontol.* 57:1–30. (24, 27)

Boufford, D. E., and S. A. Spongberg. 1983. Eastern Asian-eastern North American phytogeographical relationships—a history from the time of Linnaeus to the twentieth century. *Ann. Mo. Bot. Gard.* 70:423–39. (26)

Bowers, M. A. 1988. Relationships between local distribution and geographic range of desert heteromyid rodents. *Oikos* 53:303–8. (14)

Bowers, M. D. 1988. Chemistry and coevolution: Iridoid glycosides, plants and herbivorous insects. In *Chemical mediation of coevolution*, ed. K. C. Spencer, 133–66. San Diego, Calif.: Academic Press. (23)

Bowler, J. M. 1982. Aridity in the late Tertiary and Quaternary of Australia. In *Evolution of the flora and fauna of arid Australia*, ed. W. R. Barker and P. J. M. Greenslade, 35–45. Frewville, South Australia: Peacock Publications. (14, 15)

Bowler, J. M., and R. J. Wasson. 1984. Glacial age environments of inland Australia. In *Late Cainozoic palaeoclimates of the south-*

ern hemisphere, ed. J. C. Vogel, 183–208. Rotterdam: A. A. Balkema. (14)

Box, E. O. 1981. Predicting physiognomic vegetation types with climate variables. *Vegetatio* 45:127–39. (15)

Bramlette, M. N. 1965. Massive extinctions in biota at the end of Mesozoic time. *Science* 148:1696–99. (5)

Branch, G. M. 1971. The ecology of *Patella* Linnaeus from the Cape Peninsula, South Africa. 1. Zonation, movements and feeding. *Zoologica Africana* 6:1–38. (4)

Branch, G. M. 1976. Intraspecific competition experienced by South African *Patella* species. *J. Anim. Ecol.* 45:507–30. (4)

Branch, G. M. 1984. Competition between marine organisms: Ecological and evolutionary implications. *Annu. Rev. Oceanogr. Mar. Biol.* 22:429–593. (4)

Braun, E. L. 1950. *Deciduous forests of eastern North America.* New York: Hafner. (26)

Breed, W. G. 1979. The reproductive rate of the hopping-mouse *Notomys alexis* and its ecological significance. *Aus. J. Zool.* 27:177–94. (14)

Breed, W. G. 1981. Unusual anatomy of the male reproductive tract in *Notomys alexis* (Muridae). *J. Mammal.* 62:373–75. (14)

Breed, W. G. 1986. Comparative morphology and evolution of the male reproductive tract in the Australian hydromyine rodents (Muridae). *J. Zool.* (Lond.) 209:607–29. (14)

Breed, W. G. 1990. Reproductive anatomy and sperm morphology of the long-tailed hopping-mouse, *Notomys longicaudatus* (Rodentia: Muridae). *Aust. Mammal.* 13:201–4. (14)

Brenchley, G. A., and J. T. Carlton. 1983. Competitive displacement of native mud snails by introduced periwinkles in the New England intertidal zone. *Biol. Bull.* 165:543–58. (29)

Brereton, J. L., and J. Kikkawa. 1963. Diversity of avian species. *Aust. J. Sci.* 26:12–14. (13)

Bretsky, P. W., and S. M. Klofak. 1986. "Rules of assembly" for two Late Ordovician communities. *Palaios* 1:462–77. (29)

Brewer, D. J. 1985. Herpetofaunas in the late Pleistocene: Extinctions and extralimital forms. In *Environments and extinctions: Man in late glacial North America,* ed. J. I. Mead and D. J. Meltzer, 31–52. Orono, Maine: Center for the Study of Early Man. (29)

Briand, F. 1983a. Biogeographic patterns in food web organization. In *Current trends in food web theory,* ed. D. L. DeAngelis, W. M. Post, and G. Sugihara, 37–40. Oak Ridge National Laboratory technical report ORNL-5983. (3)

Briand, F. 1983b. Environmental control of food web structure. *Ecology* 64:253–63. (1, 3)

Briand, F. 1985. Structural singularities of freshwater food webs. *Verh. Int. Verein. Limnol.* 22:3356–64. (3)

Briand, F., and J. E. Cohen. 1984. Community food webs have scale-invariant structure. *Nature* 307:264–67. (3)

Briand, F., and J. E. Cohen. 1987. Environmental correlates of food chain length. *Science* 238:956–60. (3, 7)

Briese, D. T. 1982. Relationship between the seed-harvesting ants and the plant community in a semi-arid environment. In *Ant-plant interactions in Australia,* ed. R. C. Buckley, 11–24. The Hague: Junk. (14)

Briese, D. T., and B. J. Macauley. 1977. Physical structure of an ant community in semi-arid Australia. *Aust. J. Ecol.* 2:107–20. (14)

Briggs, J. C. 1974. *Marine Zoogeography.* New York: McGraw-Hill. (20)

Brodkorb, P. 1971. Origin and evolution of birds. In *Avian Biology,* vol. I, ed. D. S. Farner and J. R. King, 19–55. New York: Academic Press. (12)

Brooks, D. R. 1981. Hennig's parasitological method: A proposed solution. *Syst. Zool.* 30:229–49. (24)

Brooks, D. R. 1985. Historical ecology: A new approach to studying the evolution of ecological associations. *Ann. Mo. Bot. Gard.* 72:660–80. (17, 18, 24, 25)

Brooks, D. R. 1988a. Macroevolutionary comparisons of host and parasite phylogenies. *Annu. Rev. Ecol. Syst.* 19:235–59. (17)

Brooks, D. R. 1988b. Scaling effects in historical biogeography: A new view of Space, Time and Form. *Syst. Zool.* 38:237–44. (24)

Brooks, D. R. 1990. Parsimony analysis in historical biogeography and coevolution: Methodological and theoretical update. *Syst. Zool.* 39:14–30. (24)

Brooks, D. R., and T. L. Deardorff. 1988. *Rhinebothrium devaneyi*

n. sp. (Eucestoda: Tetraphyllidea) and *Echinocephalus overstreeti* Deardorff and Ko, 1983 (Nematoda: Gnathostomidae) in a Thorny Back Ray, *Urogymnus asperrimus,* from Enewetak Atoll, with phylogenetic analysis of both species groups. *J. Parisitol.* 74:459–65. (24)

Brooks, D. R., and D. A. McLennan. 1991. *Phylogeny, ecology, and behavior: A research program in comparative biology.* Chicago: University of Chicago Press. (17, 24, 25, 30)

Brooks, D. R., T. B. Thorson, and M. A. Mayes. 1981. Freshwater stingrays (Potamotrygonidae) and their helminth parasites: Testing hypotheses of evolution and coevolution. In *Advances in Cladistics,* ed. V. A. Funk and D. R. Brooks, 147–75. New York: New York Botanical Garden. (24)

Brooks, J. L. 1950. Speciation in ancient lakes. *Q. Rev. Biol.* 25:30–60, 131–176. (30)

Brown, J. H. 1971. Mechanisms of competitive exclusion between two species of chipmunks. *Ecology* 52:305–11. (5)

Brown, J. H. 1973. Species diversity of seed-eating desert rodents in sand dune habitats. *Ecology* 54:775–87. (5, 6)

Brown, J. H. 1975. Geographical ecology of desert rodents. In *Ecology and Evolution of Communities,* ed. M. L. Cody and J. M. Diamond, 315–41. Cambridge, Mass: Belknap Press of Harvard University Press. (5)

Brown, J. H. 1981. Two decades of homage to Santa Rosalia: Toward a general theory of diversity. *Am. Zool.* 21:877–88. (6, 24)

Brown, J. H. 1984. On the relationship between abundance and distribution of species. *Am. Nat.* 124:253–79. (10, 24, 30)

Brown, J. H., and D. W. Davidson. 1977. Competition between seed-eating rodents and ants in desert ecosystems. *Science* 196:880–82. (5)

Brown, J. H., D. W. Davidson, and O. J. Reichman. 1979. An experimental study of competition between seed-eating desert rodents and ants. *Am. Zool.* 19:1129. (5)

Brown, J. H., and A. C. Gibson. 1983. *Biogeography.* St. Louis, Mo.: Mosby. (1, 6, 7, 17, 23, 28, 30)

Brown, J. H., and A. Kodric-Brown. 1977. Turnover rates in insular biogeography: Effect of immigration on extinction. *Ecology* 58:445–49. (7)

Brown, J. H., and M. A. Kurzius. 1987. Composition of desert rodent faunas: Combinations of coexisting species. *Ann. Zool. Fennici* 24:227–37. (10, 14)

Brown, J. H., and M. A. Kurzius. 1989. Spatial and temporal variation in guilds of North American granivorous desert rodents. In *Patterns in the structure of mammalian communities,* ed. D. W. Morris, Z. Abramsky, B. J. Fox, and M. R. Willig, 71–90. Special Publications, Museum, Texas Tech University, no. 28. Lubbock: Texas Tech University Press. (14)

Brown, J. H., and B. A. Maurer. 1987. Evolution of species assemblages: Effects of energetic constraints and species dynamics on the diversification of the North American avifauna. *Am. Nat.* 130:1–17. (24)

Brown, J. H., and B. A. Maurer. 1989. Macroecology: The division of food and space among species on continents. *Science* 243:1145–50. (24, 25)

Brown, J. H., and J. C. Munger. 1985. Experimental manipulation of a desert rodent community: Food addition and species removal. *Ecology* 66:1545–63. (5)

Brown, J. H., and R. A. Ojeda. 1987. Granivory: Patterns, processes, and consequences of seed consumption on two continents. *Revista Chilena de Historia Natural* 60:337–49. (14)

Brown, J. H., O. J. Reichman, and D. W. Davidson. 1979. Granivory in desert ecosystems. *Annu. Rev. Ecol. Syst.* 10:201–27. (5)

Brown, J. H., and Z. Zeng. 1989. Comparative population ecology of eleven species of rodents in the Chihuahuan Desert. *Ecology* 70:1507–25. (6, 14, 24)

Brown, J. S. 1986. Coexistence on a resource whose abundance varies: A test with desert rodents. Ph.D. diss., University of Arizona. (5)

Brown, J. S. 1989a. Coexistence on a seasonal resource. *Am. Nat.* 133:168–82. (5)

Brown, J. S. 1989b. Desert rodent community structure: A test of four mechanisms of coexistence. *Ecol. Monogr.* 59:1–20. (5)

Brown, J. S. 1990. Habitat selection as an evolutionary game. *Evolution* 44:732–46. (7)

Brown, J. S., and M. L. Rosenzweig. 1986. Habitat selection in slowly regenerating environments. *J. Theor. Biol.* 123:151–71. (5)

Brown, J. W. 1987. The peninsular effect in Baja California: An entomological assessment. *J. Biogeogr.* 14:359–365. (6)

Brown, K. S. 1981. The biology of *Heliconius* and related genera. *Annu. Rev. Entomol.* 26:427–56. (23)

Brown, V. K. 1990. Insect herbivory and its effect on plant succession. In *Pests, pathogens and plant communities*, ed. J. J. Burdon and S. R. Leather, 275–88. Oxford: Blackwell Scientific Publications. (23)

Brown, W. J. 1956. The new world species of *Chrysomela* L. *Can. Entomol.* 88 (Suppl. 3):1–54. (23)

Brown, W. L., Jr. 1957. Centrifugal speciation. *Q. Rev. Biol.* 32:247–77. (30)

Brown, W. L., and R. W. Taylor. 1970. Superfamily Formicoidea. In *The insects of Australia*, ed. CSIRO Division of Entomology, 951–59. Melbourne: Melbourne University Press. (14)

Bruns, D. A., A. B. Hale, and G. W. Minshall. 1987. Ecological correlates of species richness in three guilds of lotic macroinvertebrates. *J. Freshwater Ecol.* 4:163–76. (6)

Buckley, R., R. Wasson, and A. Gubb. 1987. Phosphorus and potassium status of arid dunefield soils in central Australia and southern Africa, and biogeographic implications. *J. Arid Environ.* 13:211–16. (14)

Bunt, J. S., W. T. Williams, and N. C. Duke. 1982. Mangrove distributions in northeast Australia. *J. Biogeogr.* 9:111–20. (20)

Bunt, J. S., W. T. Williams, J. F. Hunter, and H. J. Clay. 1991. Mangrove sequencing: Analysis of zonation in a complete river system. *Mar. Ecol. Prog. Ser.* 72:289–94. (20)

Burbidge, A. A., and N. L. McKenzie. 1989. Patterns in the modern decline of Western Australia's vertebrate fauna: Causes and conservation implications. *Biol. Conserv.* 50:143–98. (14)

Burrows, M. T., and R. N. Hughes. 1990. Variation in growth and consumption among individuals and populations of dogwhelks, *Nucella lapillus*: A link between foraging behaviour and fitness. *J. Anim. Ecol.* 59:723–42. (4)

Burton, R. W. 1983. Protein polymorphisms and genetic differentiation of marine invertebrate populations. *Mar. Biol. Lett.* 4:193–206. (4)

Bush, A. O. 1990. Helminth communities in avian hosts: Determinants of pattern. In *Parasite communities: Patterns and processes*, ed. G. W. Esch, A. O. Bush, and J. M. Aho. 197–232. London: Chapman and Hall. (17)

Bush, A. O., J. M. Aho, and C. R. Kennedy. 1990. A comparison of ecological and phylogenetic factors as determinants of helminth parasite community richness. *Evol. Ecol.* 4:1–20. (17)

Bush, A. O., and J. C. Holmes. 1986a. Intestinal helminths of lesser scaup ducks: Patterns of association. *Can. J. Zool.* 64:132–41. (10)

Bush, A. O., and J. C. Holmes. 1986b. Intestinal parasites of lesser scaup ducks: An interactive community. *Can. J. Zool.* 64:142–52. (17)

Bush, G. L. 1975. Modes of animal speciation. *Annu. Rev. Ecol. Syst.* 6:339–64. (22, 30)

Bush, G. L., S. M. Case, A. C. Wilson, and J. L. Patton. 1977. Rapid speciation and chomosomal evolution in mammals. *Proc. Natl. Acad. Sci., USA* 74:3942–46. (22, 30)

Buss, L. W., and P. O. Yund. 1988. A comparison of recent and historical populations of the colonial hydroid *Hydractinia*. *Ecology* 60:646–54. (24)

Butler, P. M. 1946. The evolution of carnassial dentitions in the Mammalia. *Proc. Zool. Soc. Lond.* 116:198–220. (28)

Butman, C. A. 1987. Larval settlement of soft-sediment invertebrates: The spatial scales of pattern explained by active habitat selection and the emerging role of hydrodynamical processes. *Annu. Rev. Oceanogr. Mar. Biol.* 25:113–65. (4)

Buzas, M. A. 1972. Patterns of species diversity and their explanation. *Taxon* 21:275–86. (30)

Cade, T. J. 1964. The evolution of torpidity in rodents. *Ann. Acad. Sci. Fennicae*, Ser. A IV 71(6):79–112. (14)

Cade, J. E. 1984a. Molecular systematics of neotropical xenodontine colubrid snakes: I. South American xenodontines. *Herpetologica* 40:8–20. (25)

Cadle, J. E. 1984b. Molecular systematics of neotropical xenodon-

tine colubrid snakes: II. Central American xenodontines. *Herpetologica* 40:21–30. (25)

Cadle, J. E. 1984c. Molecular systematics of noetropical xeondontine colubrid snakes: III. Overview of xenodontine phylogeny and the history of New World snakes. *Copeia* 1984:641–652. (25)

Cadle, J. E. 1985. The neotropical colubrid snake fauna (Serpentes: Colubridae): lineage components and biogeography. *Syst. Zool.* 34:1–20. (25)

Cadle, J. E. 1988. Phylogenetic relationships among advanced snakes: A molecular perspective. *Univ. Calif. Publ. Zool.* 119:1–70. (25)

Caffey, H. M. 1985. Spatial and temporal variation in settlement and recruitment of intertidal barnacles. *Ecol. Monogr.* 55:313–32. (4)

Calder, W. A. III. 1984. *Size, function, and life history.* Cambridge, Mass.: Harvard University Press. (25)

Calderon, E. 1983. Hallazgo de *Pelliciera rhizophorae* Tr. & Pl. en la costa del Atlántico con observaciones taxonómicas y biogeograficas preliminares. *Bol. Museo del Mar, Bogotá* 11:100–111. (20)

Camus, A. 1936–1938. *Les chênes: Monographie du genre quercus.* Paris: Paul Lechevalier. (26)

Cantino, P. D. 1982. Affinities of the Lamiales: A cladistic analysis. *Syst. Bot.* 7:237–48. (23)

Cantino, P. D., and R. W. Sanders. 1986. Subfamilial classification of Labiatae. *Syst. Bot.* 11:163–85. (23)

Carlton, J. T. 1979. Introduced invertebrates of San Francisco Bay. In *San Francisco Bay: The urbanized estuary*, ed. T. J. Conomos, 427–44. San Francisco: California Academy of Sciences. (29, 30)

Carlton, J. T. 1987. Patterns of transoceanic marine biological invasions in the Pacific Ocean. *Bull. Mar. Sci.* 41:452–65. (29)

Carlton, J. T. 1989. Man's role in changing the face of the ocean: Biological invasions and implications for conservation of nearshore environments. *Conserv. Biol.* 3:265–73. (29)

Carpenter, S. R., ed. 1988. *Complex interactions in lake communities.* Berlin: Springer-Verlag. (3)

Carson, H. L., and K. Y. Kaneshiro. 1976. *Drosophila* of Hawaii: Systematics and ecological genetics. *Annu. Rev. Ecol. Syst.* 7:311–45. (30)

Carson, H. L., and A. R. Templeton. 1984. Genetic revolutions in relation to speciation phenomena: The founding of new populations. *Annu. Rev. Ecol. Syst.* 15:97–131. (22)

Carson, W. P., and S. T. A. Pickett. 1990. Role of resources and disturbance in the organization of an old-field plant community. *Ecology* 71:226–38. (6)

Case, T. J. 1978. A general explanation for insular body size trends in terrestrial vertebrates. *Ecology* 59:1–18. (12)

Case, T. J. 1990. Invasion resistance arises in strongly interacting species-rich model competition communities. *Proc. Natl. Acad. Sci. USA* 87:9610–14. (1, 30)

Case, T. J., and D. T. Bolger. 1991. The role of introduced species in shaping the distribution and abundance of island reptiles. *Evol. Ecol.* 5:272–90. (30)

Case, T. J., D. T. Bolger, and A. D. Richman. 1992. Reptilian extinctions: The last ten thousand years. In *Conservation Biology*, ed. P. L. Fielder and S. K. Jain, 91–125. New York: Chapman and Hall. (30)

Case, T. J., and M. L. Cody. 1987. Testing theories of island biogeography. *Am. Sci.* 75:402–411. (7, 30)

Castenholz, R. W. 1961. The effect of grazing on marine littoral diatom populations. *Ecology* 42:783–94. (4)

Castenholz, R. W. 1963. An experimental study of the vertical distribution of littoral marine diatoms. *Limnol. Oceanogr.* 8:450–62. (4)

Caswell, H. 1976. Community structure: A neutral model analysis. *Ecol. Monogr.* 46:327–54. (10, 22, 30)

Caswell, H. 1978. Predator-mediated coexistence: A nonequilibrium model. *Am. Nat.* 112:127–54. (7, 9)

Caswell, H., and J. E. Cohen. 1991a. Communities in patchy environments: A model of disturbance, competition, and heterogeneity. In *Ecological heterogeneity*, ed. J. Kolasa, 97–122. New York: Springer-Verlag. (9)

Caswell, H., and J. E. Cohen. 1991b. Disturbance and diversity in metapopulations. *Biol. J. Linn. Soc.* 42:193–218. (9)

Caswell, H., and R. J. Etter. 1993. Ecological interactions in patchy environments: From patch-occupancy models to cellular automata. In *Patch dynamics*, ed. S. A. Levin, T. Powell and J. H. Steele. New York: Springer-Verlag. In press. (9)

Cazier, M. A. 1954. A review of the Mexican tiger beetles of the genus *Cicindela* (Coleoptera, Cicindelidae). *Bull. Am. Mus. Nat. Hist.* 103:227–310. (18)

Chaline, J. 1972. *Les Rongeurs du Pléistocène moyen et supérieur en France.* Cahiers de Paléontologie. Paris: CNRS. (12)

Chaline, J. 1974. Esquisse de l'évolution morphologique, biométrique et chromosomique du genre *Microtus* (Arvicolidae, Rodentia) dans le Pléistocène de l'hémisphère nord. *Bull. Soc. Geol. France,* 7ème série 16:440–50. (12)

Chandler, M. E. J. 1951. Note on the occurrence of mangrove in the London Clay. *Proc. Geol. Assoc.* 62:271–72. (20)

Chandler, M. E. J. 1961. *The Lower Tertiary floras of southern England I, Palaeocene floras: London clay flora* (Supplement). London: British Museum of Natural History. (20)

Chapin, F. S. III. 1980. The mineral nutrition of wild plants. *Annu. Rev. Ecol. Syst.* 11:233–60. (2, 15)

Chapman, M. G., and A. J. Underwood. 1992. Foraging behaviour of marine benthic grazers. In *Plant Animal Interactions in the Marine Benthos*, ed. D. M. John, S. J. Hawkins, and J. H. Price, 289–317. Oxford: Systematics Association and Clarendon Press. (4)

Chapman, V. J. 1970. Mangrove phytosociology. *Trop. Ecol.* 11:1–19. (20)

Chapman, V. J. 1975. Mangrove biogeography. In *Proceedings of the International Symposium on Biology and Management of Mangroves*, vol. 1, G. E. Walsh, S. C. Snedaker, and H. J. Teas, 3–22. (Eds.) Gainesville: University of Florida Press. (20)

Chapman, V. J. 1976. *Mangrove Vegetation.* Vaduz: J. Cramer. (20)

Cherepanov, A. I. 1988. *Cerambycidae of Northern Asia*, vol. 1. Acad. Sciences USSR; New Delhi: Amerind Publ. (23)

Cherry, J. F. 1990. The first colonization of the Mediterranean Islands: A review of recent research. *J. Mediterranean Archaeol.* 3:145–221. (12)

Chesson, P. L. 1983. Coexistence of competitors in a stochastic environment: The storage effect. In *Population biology*, ed. H. I. Freedman and C. Strobeck, 188–98. Berlin, New York: Springer-Verlag. (2)

Chesson, P. L. 1984. The storage effect in stochastic population models. In *Mathematical Ecology: Miramare-Trieste Proceedings*, ed. S. A. Levin and T. G. Hallam, 76–89. Berlin: Springer-Verlag. (2)

Chesson, P. L. 1985. Coexistence of competitors in spatially and temporally varying environments: A look at the combined effects of different sorts of variability. *Theor. Pop. Biol.* 28:263–87. (2, 8)

Chesson, P. L. 1986. Environmental variation and the coexistence of species. In *Community ecology*, ed. J. Diamond and T. Case, 240–56. New York: Harper and Row. (1, 2)

Chesson, P. L. 1991. A need for niches? *Trends Ecol. Evol.* 6:26–28. (1)

Chesson, P. L., and T. J. Case. 1986. Overview: Nonequilibrium community theories: Chance, variability, history, and coexistence. In *Community Ecology*, ed. J. Diamond and T. Case, 229–39. New York: Harper and Row. (1)

Chesson, P. L., and N. Huntly. 1988. Community consequences of life history traits in a variable environment. *Ann. Zool. Fennici* 25:5–16. (5)

Chesson, P. L., and N. Huntly. 1989. Short-term instabilities and long-term community dynamics. *Trends Ecol. Evol.* 4:293–98. (22)

Chesson, P. L., and R. R. Warner. 1981. Environmental variability promotes coexistence in lottery competitive systems. *Am. Nat.* 117:923–43. (1)

Chewning, W. C. 1975. Migratory effects in predator-prey models. *Math. Biosci.* 23:253–62. (8)

Cheylan, G. 1984. Les mammifères des îles de Provence et Méditerranée occidentale: Un exemple de peuplement insulaire nonéquilibré? *Revue d'Ecologie (Terre et Vie)* 39:37–54. (12)

Cheylan, G. 1991. Patterns of Pleistocene turnover, current distribution and speciation among Mediterranean mammals. In *Biogeog-*

raphy of Mediterranean invasions, ed. R. H. Groves and F. di Castri, 227–62. Cambridge: Cambridge University Press. (12)

Chi'ên, H. 1921. *Chinese economic trees.* Shanghai: Commercial Press. (26)

Choat, J. H. 1982. Fish feeding and the structure of benthic communities in temperate waters. *Annu. Rev. Ecol. Syst.* 13:423–49. (4)

Chow, P. L., and W. C. Tam. 1976. Periodic and traveling wave solutions to Lotka-Volterra equations with diffusion. *Bull. Math. Biol.* 38:643–58. (8)

Christensen, F. B., and T. M. Fenchel. 1977. *Theories of populations in biological communities.* Berlin: Springer-Verlag. (7)

Chubb, J. C. 1970. The parasite fauna of British freshwater fish. In *Aspects of fish parasitology*, ed. A. E. R. Taylor and R. Muller, 119–44. Symposia of the British Society for Parsitology, vol. 8. Oxford: Blackwell Scientific Publications. (17)

Churchill, D. M. 1973. The ecological significance of tropical mangroves in the early Tertiary floras of southern Australia. *Geol. Soc. Aust. Spec. Publ.* 4:79–86. (20)

Cisne, J. L., and B. D. Rabe. 1978. Coenocorrelation: Gradient analysis of fossil communities and its applications to stratigraphy. *Lethaia* 11:341–64. (29)

Claridge, M. F., and M. R. Wilson. 1981. Host plant associations, diversity and species-area relationships of mesophyll-feeding leafhoppers of trees and shrubs in Britain. *Ecol. Entomol.* 6:217–38. (23)

Clark, D. A., and D. B. Clark. 1984. Spacing dynamics of a tropical rainforest tree: Evaluation of the Janzen-Connell model. *Am. Nat.* 124:769–88. (1)

Clarke, J. A., and R. E. Johnson. 1990. Biogeography of white-tailed ptarmagin (*Lagopus leucurus*): Implications from an introduced population in the Sierra Nevada. *J. Biogeogr.* 17:649–56. (18)

Clarke, P. J., and P. J. Myerscough. 1991. Buoyancy of *Avicennia marina* propagules in south-eastern Australia. *Aust. J. Bot.* 39:77–83. (20)

Clarke, R. D. 1988. Chance and order in determining fish-species composition on small coral patches. *J. Exp. Mar. Biol. Ecol.* 115:197–212. (1)

Clements, F. E. 1936. Nature and structure of the climax. *J. Ecol.* 24:252–84. (4)(29)

Coaton, W. G. H. 1963. Survey of the termites (Isoptera) of the Kalahari thornveld and shrub bushveld of the R.S.A. *Koedoe* 6:38–50. (14)

Cocks, L. R. M., and R. B. Rickards. 1969. Five boreholes in Shropshire and the relationships of shelly and graptolitic facies in the Lower Silurian. *Q. J. Geol. Soc. Lond.* 124:213–38. (5)

Coddington, J. A. 1988. Cladistic tests of adaptational hypotheses. *Cladistics* 4:3–22. (18)

Cody, M. L. 1966a. The consistency of intra- and intercontinental grassland bird counts. *Am. Nat.* 100:371–76. (1, 17)

Cody, M. L. 1966b. A general theory of clutch-size. *Evolution* 20:174–84. (12)

Cody, M. L. 1974. *Competition and the structure of bird communities.* Princeton, N.J.: Princeton University Press. (1)

Cody, M. L. 1975. Toward a theory of continental species diversities: Bird distributions over Mediterranean habitat gradients. In *Ecology and evolution of communities*, ed. M. L. Cody and J. M. Diamond, 214–57. Cambridge, Mass.: Harvard University Press. (1, 13, 20, 21, 28, 30)

Cody, M. L. 1983. Continental diversity patterns and convergent evolution in bird communities. In *Mediterranean-type ecosystems*, ed. F. Kruger, D. T. Mitchell, and J. U. M. Jarvis, 347–402. Ecological Studies 43. Wein, Berlin: Springer-Verlag. (13)

Cody, M. L. 1986. Diversity, rarity, and conservation in Mediterranean-climate regions. In *Conservation biology: The science of scarcity and diversity*, ed. M. E. Soulé, 123–52. Sunderland, Mass.: Sinauer Associates. (13, 28)

Cody, M. L. 1991. Distributional notes on NE and SW Australian birds. *Sunbird* 21:1–9. (13)

Cody, M. L., and H. A. Mooney. 1978. Convergence versus nonconvergence in mediterranean-climate ecosystems. *Annu. Rev. Ecol. Syst.* 9:265–321. (1, 21)

Cogger, H. G., and H. Heatwole. 1981. The Australian reptiles: Origins, biogeography, distribution patterns and island evolution. In

Ecological biogeography of Australia, ed. A. Keast, 1331–1373. The Hague: Junk. (14)

Cohen, J. E. 1968. Alternative derivations of a species-abundance relation. *Am. Nat.* 102:165–72. (1)

Cohen, J. E. 1970. A Markov contingency-table model for replicated Lotka-Volterra systems near equilibrium. *Am. Nat.* 104:547–60. (9)

Cohen, J. E. 1978. *Food webs and niche space.* Princeton, N.J.: Princeton University Press. (3)

Cohen, J. E. 1989. Food webs and community structure. In *Perspectives in Ecological Theory,* ed. J. Roughgarden, R. M. May, and S. A. Levin, 181–202. Princeton, N.J.: Princeton University Press. (1)

Cohen, J. E., and F. Briand. 1984. Trophic links of community food webs. *Proc. Natl. Acad. Sci. USA* 81:4105–9. (3, 30)

Cohen, J. E., F. Briand, and C. M. Newman. 1986. A stochastic theory of community food webs. III. Predicted and observed lengths of food chains. *Proc. Roy. Soc.* (Lond.), ser. B, 228:317–53. (3, 30)

Cohen, J. E., F. Briand, and C. M. Newman. 1990. *Community food webs: Data and theory.* Berlin: Springer-Verlag. (3, 7)

Cohen, J. E., and C. M. Newman. 1985. A stochastic theory of community food webs. I. Models and aggregated data. *Proc. Roy. Soc.* (Lond.), ser. B, 224:421–48. (3)

Cohen, J. E., and C. M. Newman. 1988. Dynamic basis of food web organization. *Ecology* 69:1655–64. (3)

Cohen, J. E., C. M. Newman, and F. Briand. 1985. A stochastic theory of community food webs. II. Individual webs. *Proc. Roy. Soc.* (Lond.), ser. B, 224:449–61. (3)

Cohen, J. E., S. L. Pimm, P. Yodzis, and J. Saldaña. 1993a. Body sizes of animal predators and animal prey in food webs. *J. Anim. Ecol.* 62:67–78

Cohen, J. E., and 23 others. 1993b. Improving food webs. *Ecology* 74:252–8

Coleman, B. D., M. A. Mares, M. R. Willig, and Y. Hsieh. 1982. Randomness, area and species-richness. *Ecology* 63:1121–33. (5)

Coley, P. D., J. P. Bryant, and F. S. Chapin III. 1985. Resource availability and plant antiherbivore defense. *Science* 230:895–98. (2)

Colinvaux, P. 1987. Amazon diversity in light of the paleoecological record. *Quat. Sci. Rev.* 6:93–114. (29)

Collette, B. B. 1982. South American freshwater needlefishes of the genus *Potamorrhaphis* (Beloniformes: Belonidae). *Proc. Biol. Soc. Wash.* 95:714–47. (24)

Collins, S. L., and S. M. Glenn. 1991. Importance of spatial and temporal dynamics in species regional abundance and distribution. *Ecology* 72:654–64. (30)

Collinson, M. E. and P. R. Crane. 1978. *Rhododendron* seeds from the Palaeocene of southern England. *Bot. J. Linn. Soc.* 76:195–205. (26)

Colwell, R. K., and D. W. Winkler. 1984. A null model for null models in biogeography. In *Ecological communities: Conceptual issues and the evidence,* ed. D. R. Strong, D. Simberloff, L. G. Abele and A. B. Thistle, 344–59. Princeton, N.J.: Princeton University Press. (11)

Comins, H. N., and Blatt, D. W. E. 1974. Prey-predator models in spatially heterogeneous environments. *J. Theor. Biol.* 48:75–83. (8)

Comins, H. N., and M. P. Hassell. 1979. The dynamics of optimally foraging predators and parasitoids. *J. Anim. Ecol.* 48:335–51. (7)

Comins, H. N., and M. P. Hassell. 1987. The dynamics of predation and competition in patchy environments. *Theor. Popul. Biol.* 31:393–422. (7)

Comins, H. N., and I. R. Noble. 1985. Dispersal, variability, and transient niches: Species coexistence in a uniformly variable environment. *Am. Nat.* 126:706–23. (2)

Compton, S. G., J. H. Lawton, and V. K. Rashbrook. 1989. Regional diversity, local community structure, and vacant niches: The herbivorous insects of bracken in South Africa. *Ecol. Entomol.* 14:365–73. (7, 16, 22)

Connell, J. H. 1961b. The influence of interspecific competition and other factors on the distribution of the barnacle *Chthamalus stellatus. Ecology* 42:710–23. (1)

Connell, J. H. 1961a. The effects of competition, predation by *Thais*

lapillus, and other factors on natural populations of the barnacle, *Balanus balanoides. Ecol. Monogr.* 31:61–104. (1, 4)

Connell, J. H. 1970. A predator-prey system in the marine intertidal region. I. *Balanus glandula* and several predatory species of *Thais. Ecol. Monogr.* 40:49–78. (4)

Connell, J. H. 1971. On the role of natural enemies in preventing competitive exclusion in some marine animals and in rain forest trees. In *Dynamics of populations,* ed. P. J. den Boer and G. R. Gradwell, 298–313. Wageningen, The Netherlands: Centre for Agricultural Publishing and Documentation.

Connell, J. H. 1972. Community interactions on marine rocky intertidal shores. *Annu. Rev. Ecol. Syst.* 3:169–92. (4)

Connell, J. H. 1975. Some mechanisms producing structure in natural communities: A model and evidence from field experiments. In *Ecology and evolution of communities,* ed. M. L. Cody and J. Diamond, 460–90. Cambridge, Mass: Harvard University Press. (1, 4, 19)

Connell, J. H. 1978. Diversity in tropical rain forests and coral reefs. *Science* 199:1302–10. (1, 2, 5, 6, 19, 20, 21, 27, 30)

Connell, J. H. 1980. Diversity and the coevolution of competitors, or the ghost of competition past. *Oikos* 35:131–38. (21)

Connell, J. H. 1983. On the prevalence and relative importance of interspecific competition: Evidence from field experiments. *Am. Nat.* 122:661–96. (1, 2, 4, 22)

Connell, J. H. 1988. Long-term variation in coral populations at Heron Island. *Abs. VIth Int. Coral Reef Symp.* (Townsville, Aust.) 21. (27)

Connell, J. H., and Keough, M. J. 1985. Disturbance and patch dynamics of subtidal marine animals on hard substrata. In *Natural disturbance: The patch dynamics perspective,* ed. S. T. A. Pickett and P. S. White, 125–51. New York: Academic Press. (4)

Connell, J. H., and E. Orias. 1964. The ecological regulation of species diversity. *Am. Nat.* 98:399–414. (1, 5, 6, 26, 30)

Connor, E. F., S. H. Faeth, and D. S. Simberloff. 1983. Leafminers on oak: The role of immigration and in situ reproductive recruitment. *Ecology* 64:191–204. (10)

Connor, E. F., and E. D. McCoy. 1979. The statistics and biology of the species-area relationship. *Am. Nat.* 113:791–833. (1, 5, 7, 10)

Connor, E. F., and D. S. Simberloff. 1979. The assembly of species communities: Chance or competition? *Ecology* 60:1132–40. (1, 21)

Cook, R. E. 1969. Variation in species density in North American birds. *Syst. Zool.* 18:63–84. (1)

Coope, G. R. 1970. Interpretations of Quaternary insect fossils. *Annu. Rev. Entomol.* 15:97–120. (29)

Coope, G. R. 1979. Late Cenozoic fossil Coleoptera: Evolution, biogeography and ecology. *Annu. Rev. Ecol. Syst.* 10:247–67. (29)

Cooper, S. D., S. J. Walde, and B. L. Peckarsky. 1990. Prey exchange rates and the impact of predators on prey populations in streams. *Ecology* 71:1503–14. (7)

Cornell, H. V. 1985a. Local and regional richness of cynipine gall wasps on California Oaks. *Ecology* 66:1247–60. (6, 9, 16, 17, 22)

Cornell, H. V. 1985b. Species assemblages of cynipid gall wasps are not saturated. *Am. Nat.* 126:565–69. (1, 6, 9, 14, 16, 17, 22)

Cornell, H. V., and D. M. Kahn. 1989. Guild structure in the British arboreal arthropods: Is it stable and predictable? *J. Anim. Ecol.* 58:1003–20. (22, 23)

Cornell, H. V., and J. H. Lawton. 1992. Species interactions, local and regional processes, and limits to the richness of ecological communities: A theoretical perspective. *J. Anim. Ecol.* 61:1–12. (2, 7, 22, 30)

Cornell, H. V., and J. O. Washburn. 1979. Evolution of the richness-area correlation for cynipid gall wasps on oak trees: A comparison of two geographic areas. *Evolution* 33:257–74. (22, 23)

Courtney, S. P., and T. T. Kibota. 1990. Mother doesn't know best: Selection of hosts by ovipositing insects. In *Insect-plant interactions,* ed. E. A. Bernays, vol. 2. Boca Raton, Fla.: CRC Press. (23)

Cousins, S. H. 1991. Species diversity measurement: Choosing the right index. *Trends Ecol. Evol.* 6:190–92. (1)

Cowie, R. H. 1989. The zoogeographical composition and distribution of the Arabian termite fauna. *Biol. J. Linn. Soc.* 36:157–68. (14)

Cox, G. W., and R. E. Ricklefs. 1977. Species diversity, ecological release, and community structuring in Caribbean land bird faunas. *Oikos* 29:60–66. (1, 20)

Cracraft, J. 1973. Continental drift, paleoclimatology, and the evolution and biogeography of birds. *J. Zool.* (Lond.) 169:455–545. (14)

Cracraft, J. 1980. Avian phylogeny and intercontinental biogeographic patterns. In *Acta XVII Congr. Intern. Ornithol.* (Berlin), ed. R. Nohring, 1302–8. (12)

Cracraft, J. 1985. Biological diversification and its causes. *Ann. Mo. Bot. Gard.* 72:794–822. (1)

Cracraft, J. 1986. Origin and evolution of continental biotas: Speciation and historical congruence within the Australian avifauna. *Evolution* 40:977–96. (14)

Crame, J. A. 1986. Late Pleistocene molluscan assemblages from the coral reefs of the Kenya coast. *Coral Reefs* 4:183–96. (29)

Crane, P. R., and S. Lidgard. 1990. Angiosperm diversification and paleolatitudinal gradients in Cretaceous floristic diversity. *Science* 246:675–78. (26)

Crane, P. R., and S. Lidgard. 1991. Angiosperm diversification and Cretaceous patterns of palynological diversity. In *Major evolutionary radiations,* ed. P. D. Taylor and G. P. Larwood, 377–407. Oxford: Clarendon Press. (29)

Crawley, M. J. 1987. What makes a community invasible? In *Colonization, succession and stability,* ed. A. J. Gray, M. J. Crawley, and P. J. Edwards, 429–53. Oxford: Blackwell Scientific Publications. (30)

Creese, R. G., and Underwood, A. J. 1982. Analysis of inter- and intraspecific competition amongst limpets with different methods of feeding. *Oecologia* 53:337–46. (4)

Crepet, W. L. 1984. Advanced (constant) insect pollination mechanisms: Pattern of evolution and implications vis-à-vis angiosperm diversity. *Ann. Mo. Bot. Gard.* 71:607–30. (26)

Crepet, W. L., and K. C. Nixon. 1989. Earliest megafossil evidence of Fagaceae: Phylogenetic and biogeographic implications. *Am. J. Bot.* 76:842–55. (26)

Crisp, D. J. 1958. The spread of *Elminius modestus* Darwin in north-west Europe. *J. Mar. Biol. Assoc. UK* 37:483–520. (4)

Crisp, D. J. 1964. The effects of the severe winter of 1962–63 on marine life in Britain. *J. Anim. Ecol.* 33:165–210. (4)

Crisp, D. J., and A. J. Southward. 1959. The further spread of *Elminius modestus* in the British Isles to 1959. *J. Mar. Biol. Assoc. UK* 38:429–37. (4)

Critchley, A. T., W. F. Farnham, and S. L. Morrell. 1983. A chronology of new European sites of attachment for the invasive brown alga, *Sargassum muticum,* 1973–1981. *J. Mar. Biol. Assoc. UK* 63:799–812. (4)

Croizat, L. 1964. *Space, time and form.* Caracas: L. Croizat. (20)

Croizat, L. 1968. The biogeography of the tropical lands east of Suez-Madagascar: with particular reference to the dispersal and form-making of *Ficus* L., and other vegetal and animal groups. *Atti Ist. Bit. Univ. Crittogam Pavia,* ser. 6, 4:1–400. (20)

Croizat, L., G. Nelson, and D. E. Rosen. 1974. Centers of origin and related concepts. *Syst. Zool.* 23:265–87. (24, 30)

Cronquist, A. 1981. *An integrated system of classification of flowering plants.* New York: Columbia University Press. (20, 23, 26)

Crowell, K. L. 1983. Islands—insight or artifact? Population dynamics and habitat utilization in insular rodents. *Oikos* 41:442–54. (12)

Crowell, K. L., and S. I. Rothstein. 1981. Clutch-sizes and breeding strategies among Bermudan and North American passerines. *Ibis* 123:42–50. (12)

Crowley, P. H. 1979. Predator-mediated coexistence: An equilibrium interpretation. *J. Theor. Biol.* 80:129–44. (9)

Crowley, P. H. 1981. Dispersal and the stability of predator-prey interactions. *Am. Nat.* 118:673–701. (7, 8)

Crowson, R. A. 1981. *The biology of the Coleoptera.* London: Academic Press. (23)

Crozier, C. R., and R. E. J. Boerner. 1986. Stemflow induced soil nutrient heterogeneity in a mixed mesophytic forest. *Bartonia* 52:1–8. (2)

Cunha, O. R. da, and F. P. do Nascimento. 1978. Ofidios da Amazôniz X. As cobras da região leste do Pará. *Publicações Avulsas do Museu Goeldi* 31:1–218. (25)

Currie, D. J. 1991. Energy and large scale patterns of animal and plant species richness. *Am. Nat.* 137:27–49. (6, 30)

Currie, D. J., and V. Paquin. 1987. Large-scale biogeographical patterns of species richness of trees. *Nature* 329:326–27. (1, 5, 6, 26)

Czárán, T. 1989. Coexistence of competing populations along an environmental gradient: A simulation study. *Coenoses* 4:113–120. (8)

Daghlian, C. P. 1981. A review of the fossil record of monocotyledons. *Bot. Rev.* 47:517–55. (20)

Dahlgren, R. M. T. 1988. Rhizophoraceae and Anisophylleaceae: Summary statement, relationship. *Ann. Mo. Bot. Gard.* 75:1259–77. (20)

Dahlgren, R., and R. F. Thorne. 1984. The order Myrtales: Circumscription, variation, and relationships. *Ann. Mo. Bot. Gard.* 71:633–99. (20)

Dailey, P. J., R. C. Graves, and J. M. Kingsolver. 1978. Survey of Coleoptera collected on the common milkweed, *Asclepias syriaca,* at one site in Ohio. *Coleop. Bull.* 32:223–30. (23)

Daly, M., and S. Daly. 1975. Socio-ecology of Saharan gerbils, especially *Meriones libycus.* *Mammalia* 39:289–311. (14)

Danielson, B. J. 1991. Communities in a landscape: The influence of habitat heterogeneity on the interactions between species. *Am. Nat.* 138:1105–20. (7)

Darlington, P. J. 1957. *Zoogeography: The geographical distribution of animals.* New York: Wiley. (23)

Darwin, C. R. 1859. *The origin of species by means of natural selection, or the preservation of favoured races in the struggle for life.* London: John Murray. (4)

Darwin, C. R. 1872. *The origin of species.* 6th edition. London: John Murray. (24)

Davidson, D. W. 1977. Species diversity and community organization in desert seed-eating ants. *Ecology* 58:25–37. (6)

Davidson, D. W., R. S. Inouye, and J. H. Brown. 1984. Granivory in a desert ecosystem: Experimental evidence for indirect facilitation of ants by rodents. *Ecology* 65:1780–86. (5)

Davies, P. J. 1988. Evolution of the Great Barrier Reef—reductionist dream or expansionist vision? *Proc. VIth Int. Coral Reef Symp.* (Townesville, Aust.) 1:9–17. (27)

Davis, A. N., and R. T. Wilce. 1987. Algal diversity in relation to physical disturbance: A mosaic of successional stages in a subtidal cobble habitat. *Mar. Ecol. Prog. Ser.* 37:229–37. (6)

Davis, J. H. 1940. The ecology and geologic role of mangroves in Florida. *Carnegie Inst. Wash. Publ.,* no. 517, 303–412. (20)

Davis, M. B. 1986. Climatic instability, time lags, and community disequilibrium. In *Community Ecology,* ed. J. Diamond and T. J. Case, 269–84. New York: Harper and Row. (2, 7)

Davis, M. B. 1989. Insights from paleoecology on global change. *Bull. Ecol. Soc. Am.* 70:222–28. (15)

Davis, S. J. M. 1987. La faune. In *Le Néolithique Précéramique de Chypre,* ed. A. Le Brun, S. Cluzan, S. J. M. Davis, J. Hansen and J. Renault-Miskovsky. *L'Anthropologie* 91:305–311. (12)

Davis, S. J. M. 1989. Some animal remains from the Ceramic Neolithic of Cyprus. In *Fouilles récentes Khirokitia (Chypre), 1983–1986,* ed. A. Le Brun, 189–221. Paris: Recherches sur les Civilisations. (12)

Day, R. T., P. A. Keddy, J. McNeill, and T. Carleton. 1988. Fertility and disturbance gradients: A summary model for riverine marsh vegetation. *Ecology* 69:1044–54. (6)

Dayton, P. K. 1971. Competition, disturbance and community organization: The provision and subsequent utilization of space in a rocky intertidal community. *Ecol. Monogr.* 41:351–89. (4)

Dayton, P. K., V. Currie, T. Gerrodette, B. D. Keller, P. Rosenthal, and D. V. Tresca. 1984. Patch dynamics and stability of some California kelp communities. *Ecol. Monogr.* 54:253–89. (4)

Dayton, P. K., and M. J. Tegner. 1984. The importance of scale in community ecology: A kelp forest example with terrestrial analogs. In *A New Ecology: Novel Approaches to Interactive Systems,* ed. P. W. Price, C. N. Slobodchikoff, and W. S. Gaud, 457–81. London: John Wiley and Sons. (4)

Dean, R. L., and J. H. Connell. 1987a. Marine invertebrates in an algal succession. I. Variations in abundance and diversity with succession. *J. Exp. Mar. Biol. Ecol.* 109:195–216. (4)

Dean, R. L., and J. H. Connell. 1987b. Marine invertebrates in an algal succession. II. Tests of hypotheses to explain changes in diversity with succession. *J. Exp. Mar. Biol. Ecol.* 109:217–48. (4)

Dean, R. L., and J. H. Connell. 1987c. Marine invertebrates in an algal succession. III. Mechanisms linking habitat complexity with diversity. *J. Exp. Mar. Biol. Ecol.* 109:249–74. (4)

DeAngelis, D. L. 1975. Stability and connectance in food web models. *Ecology* 56:238–43. (3)

DeAngelis, D. L. 1980. Energy flow, nutrient cycling, and ecosystem resilience. *Ecology* 61:764–71. (3, 6)

DeAngelis, D. L. 1992. *Dynamics of nutrient cycling and food webs.* London: Chapman and Hall. (3)

Den Boer, P. J. 1990. The survival value of dispersal in terrestrial arthropods. *Biol. Conserv.* 54:175–92. (10)

Denley, E. J., and A. J. Underwood. 1979. Experiments on factors influencing settlement, survival and growth of two species of barnacles in New South Wales. *J. Exp. Mar. Biol. Ecol.* 36:269–293. (4)

Denno, R. F., K. L. Olmstead, and E. S. McCloud. 1989. Reproductive cost of flight capability: A comparison of life history traits in wing dimorphic planthoppers. *Ecol. Entomol.* 14:31–44. (23)

Denslow, J. S. 1980. Gap partitioning among tropical rainforest trees. *Biotropica* (Supplement) 12:47–55. (2)

DeSalle, R., and A. R. Templeton. 1988. Founder effects and the rate of mitochondrial DNA evolution in Hawaiian *Drosophila. Evolution* 42:1076–84. (23)

Diamond, J. M. 1972. Biogeographic kinetics: Estimation of relaxation times for avifaunas of Southwest Pacific Islands. *Proc. Natl. Acad. Sci. USA* 69:3199–3203. (30)

Diamond, J. M. 1975. Assembly of species communities. In *Ecology and Evolution of Communities,* ed. M. L. Cody and J. M. Diamond, 342–444. Cambridge, Mass.: Harvard University Press. (1, 7, 10, 25, 30)

Diamond, J. M. 1984. "Normal" extinctions of isolated populations. In *Extinctions,* ed. M. H. Nitecki, 191–246. Chicago: University of Chicago Press. (7)

Diamond, J. M. 1987. How do flightless mammals colonize oceanic islands. *Nature* 327:374. (11)

Diamond, J., and T. J. Case (eds.). 1986. *Community Ecology.* New York: Harper and Row. (1, 17)

Diamond, J., S. L. Pimm, M. E. Gilpin, and M. LeCroy. 1989. Rapid evolution of character displacement in myzomelid honeyeaters. *Am. Nat.* 134:675–708. (18)

Dickman, C. R. 1989. Patterns in the structure and diversity of marsupial carnivore communities. In *Patterns in the structure of mammalian communities,* ed. D. W. Morris, Z. Abramsky, B. J. Fox, and M. R. Willig, 241–51. Special Publications, Museum, Texas Tech University, No. 28. Lubbock: Texas Tech University Press. (14)

DiMichele, W. A., and R. B. Aronson. 1992. The Pennsylvanian-Permian vegetational transition: A terrestrial analogue to the onshore-offshore hypothesis. *Evolution* 46:807–24. (30)

DiMichele, W. A., and T. L. Phillips. 1990. Organization and dynamics of Pennsylvanian-age peat swamps (abstract). *IVth Int. Cong. Syst. Evol. Biol. Abst.:* 5. (29)

Ding Hou. 1958. Rhizophoraceae. *Flora Malesiana* I 5:429–93. (20)

Dix, R., and F. Smeins. 1967. The prairie, meadow and marsh vegetation of Nelson County, North Dakota. *Can. J. Bot.* 45:21–58. (2)

Dixon, A. F. G., P. Kindlmann, J. Leps, and J. Holman. 1987. Why there are so few species of aphids, especially in the tropics. *Am. Nat.* 129:580–92. (23)

Dixon, J. R., and P. Soini. 1977. The reptiles of the upper Amazon basin, Iquitos region, Peru. II. Crocodilians, turtles, and snakes. *Milwaukee Publ. Mus. Contr. Biol. Geol.* no. 12, 1–71. (25)

Dixon, J. R., R. A. Thomas, and H. W. Greene. 1976. Status of the neotropical snake *Rhabdosoma poeppigi* Jan, with notes on variation in *Atractus elaps* (Gunther). *Herpetologica* 32:221–27. (25)

Dobson, F. S. 1985. The use of phylogeny in behavior and ecology. *Evolution* 39:1384–88. (18)

Dobzhansky, T. 1951. *Genetics and the origin of species.* 3rd ed. New York: Columbia University Press. (24)

Dodd, J. R., and R. J. Stanton. 1981. *Paleoecology, concepts and applications.* New York: John Wiley and Sons. (27)

Dodge, Y. 1985. *Analysis of experiments with missing data.* New York: Wiley. (21)

Dodson, J. R. 1989. Late Pleistocene vegetation and environmental shifts in Australia and their bearing on faunal extinctions. *J. Archaeol. Sci.* 16:207–17. (14)

Dodson, J., and M. Westoby (eds.). 1985. Are Australian ecosystems different? *Proc. Ecol. Soc. Australia,* vol. 14. (15)

Dogiel, V. A. 1964. *General parasitology.* Edinburgh: Oliver and Boyd. (17)

Dominey, W. J. 1984. Effects of sexual selection and life history on speciation: Species flocks in African cichlids and Hawaiian *Drosophila.* In *Evolution of fish species flocks,* ed. A. A. Echelle and I. Kornfield, 231–49. Orono, Me.: University of Maine Press, (22, 30)

Donoghue, M. J. 1989. Phylogenies and the analysis of evolutionary sequences, with examples from seed plants. *Evolution* 43:1137–56. (18, 25)

Dorst, J., and F. Vuilleumier. 1986. Convergence in bird communities at high altitudes in the tropics (especially in the Andes and Africa) and at temperate latitudes (Tibet). In *High altitude tropical biogeography,* ed. F. Vuilleumier and M. Monasterio, 120–49. Oxford: Oxford University Press. (21)

Dott, R. H., and R. L. Batten. 1976. *Evolution of the earth.* 2d edition. New York: McGraw-Hill. (28)

Downs, F. L. 1967. Intrageneric relationships among colubrid snakes of the genus *Geophis* Wagler. *Misc. Publ. Mus. Zool., Univ. Michigan,* no. 131, 1–193. (25)

Drake, J. A. 1985. Some theoretical and empirical explorations of structure in food webs. Ph. D. diss., Purdue University. (3)

Drake, J. A. 1988. Models of community assembly and the structure of ecological landscapes. In *Second fall course in mathematical ecology* (Trieste, Italy), ed. T. Hallam, L. Gross, and S. A. Levin. Singapore: World Press. (3)

Drake, J. A. 1990b. The mechanics of community assembly and succession. *J. Theor. Biol.* 147:213–33. (1, 3, 7)

Drake, J. A. 1990a. Communities as assembled structures: Do rules govern pattern? *Trends Ecol. Evol.* 5:159–64. (24, 30)

Drake, J. A. 1991. Community assembly mechanics and the structure of an experimental species ensemble. *Am. Nat.* 137:1–26. (3)

Drake, J. A. 1993. The construction and assembly of an ecological landscape. *J. Anim. Ecol.* In press. (3)

Duellman, W. E. 1958. A monographic study of the colubrid snake genus *Leptodeira. Bull. Am. Mus. Nat. Hist.* 114:1–152. (25)

Duellman, W. E. 1963. Amphibians and reptiles of the rainforests of southern El Petén, Guatemala. *Publ. Mus. Nat. Hist., Univ. Kansas,* no. 15, 205–49. (25)

Duellman, W. E. 1978. The biology of an equatorial herpetofauna in Amazonian Ecuador. *Misc. Publ. Mus. Nat. Hist., Univ. Kansas,* no. 65, 1– 352. (25)

Duellman, W. E., and E. R. Pianka. 1990. Biogeography of nocturnal insectivores: Historical events and ecological filters. *Annu. Rev. Ecol. Syst.* 21:57–68. (14)

Duke, N. C. 1991a. *Nypa* in the mangroves of Central America: Introduced or relict? *Principes* 35:127–32. (20)

Duke, N. C. 1991b. A systematic revision of the mangrove genus *Avicennia* (Avicenniaceae) in Australasia. *Aust. Syst. Bot.* 4:299–324. (20)

Duke, N. C. 1993. Mangrove floristics and biogeography. In *Tropical Mangrove Ecosystems.* Springer-Verlag. In press. (20)

Duke, N. C., and B. R. Jackes. 1987. A systematic revision of the mangrove genus *Sonneratia* (Sonneratiaceae) in Australasia. *Blumea* 32:277–302. (20)

Dunbar, M. J. 1968. Ecological development in polar regions. Cliffs, N.J.: Prentice-Hall. (29)

Dunbar, S. R. 1983. Travelling wave solutions of diffusive Lotka-Volterra equations. *J. Math. Biol.* 17:11–32. (8)

Dungan, M. L. 1986. Three-way interactions: Barnacles, limpets, and algae in a Sonoran Desert rocky intertidal zone. *Am. Nat.* 127:292–316. (4)

Dunham, A. E., and D. B. Miles. 1985. Patterns of covariation in life history traits of squamate reptiles: The effects of size and phylogeny reconsidered. *Am. Nat.* 126:231–57. (18)

Dunn, E. R. 1949. Relative abundance of some Panamanian snakes. *Ecology* 30:39–57. (25)

Durham, J. W., and F. S. MacNeil. 1967. Cenozoic migration of marine invertebrates through the Bering Strait region. In *The Bering land bridge,* ed. D. M. Hopkins, 326–49. Stanford, Calif.: Stanford University Press. (29)

Dussourd, D. E., and R. F. Denno. 1991. Deactivation of plant defense: Correspondence between insect behavior and secretory canal architecture. *Ecology* 72:1383–96. (23)

Dussourd, D. E., and T. Eisner. 1987. Vein-cutting behavior: Insect counterploy to the latex defense of plants. *Science* 237:898–901. (23)

Eastop, V. F. 1973. Diversity of the *Sternorrhyncha* within major climatic zones. In *Diversity of insect faunas,* ed. L. A. Mound and N. Waloff, 71–88. London: Blackwell. (23)

Ehrlich, P. R., and P. H. Raven. 1964. Butterflies and plants: A study in coevolution. *Evolution* 18:586–608. (22)(23)

Ekman, S. 1953. *Zoogeography of the sea.* London: Sidgwick and Jackson. (27)

Eldredge, N. 1976. Differential evolutionary rates. *Paleobiology* 2:174–77. (22)

Elias, T. S. 1980. *The complete guide to North American trees.* New York: Van Nostrand Reinhold. (26)

Ellison, J. C. 1991. The Pacific paleogeography of *Rhizophora mangle* L. (Rhizophoraceae). *Biol. J. Linn. Soc.* 105:271–84. (20)

Eloranta, P. 1986. Phytoplankton structure in different lake types in central Finland. *Holarctic Ecol.* 9:214–24. (6)

Elseth, G. D., and K. D. Baumgardner. 1981. *Population biology.* New York: D. Van Nostrand. (5)

Elsol, J. A., and P. Saenger. 1983. A general account of the mangroves of Princess Charlotte Bay with particular reference to zonation of the open shoreline. In *Biology and ecology of mangroves,* ed. H. J. Teas, 37–46. The Hague: Junk. (20)

Elton, C. S. 1927. *Animal ecology.* New York: Macmillan. (1, 3)

Elton, C. S. 1958. *The ecology of invasions by animals and plants.* London: Methuen. (30)

Emerson, S., H. W. Greene, and E. R. Charnov. 1993. Allometric aspects of predator-prey interactions. In *Ecological morphology: Integrative organismal biology,* ed. P. C. Wainwright and S. M. Reilly. Chicago: The University of Chicago Press. In press. (25)

Emerson, W. K., and W. O. Addicott. 1958. Pleistocene invertebrates from Punta Baja, Baja California, Mexico. *American Museum Novitates,* no. 1909, 1–11. (29)

Emmons, L. H. 1984. Geographic variation in densities and diversities of non-flying mammals in Amazonia. *Biotropica* 16:210–222. (25)

Emmons, L. H., A. Gautier-Hion, and G. Dubost. 1983. Community structure of the frugivorous-folivorous forest mammals of Gabon. *J. Zool.* (Lond.) 199:209–22. (25)

Endean, R. 1982. *Australia's Great Barrier Reef.* St. Lucia, London, New York: University of Queensland Press. (27)

Endean, R., and A. M. Cameron. 1990. Trends and new perspectives in coral reef ecology. In *Coral reefs,* ed. Z. Dubinsky, 469–92. *Ecosystems of the world,* no. 25. Amsterdam: Elsevier. (27)

Endler, J. A. 1982. Problems in distinguishing historical from ecological factors in biogeography. *Am. Zool.* 22:441–52. (11, 18, 24)

Endler, J. A. 1989. Conceptual and other problems in speciation. In *Speciation and its consequences,* ed. D. Otte and J. A. Endler, 625–48. Sunderland, Mass.: Sinauer Associates. (24)

Eppley, R. W., and B. J. Peterson. 1979. Particulate organic matter flux and planktonic new production in the deep ocean. *Nature* 282(5740):677–80. (19)

Eppley, R. W., E. H. Renger, E. L. Venrick, and M. M. Mullin. 1973. A study of plankton dynamics and nutrient cycling in the central gyre of the North Pacific. *Limnol. Oceanogr.* 18:534–551. (19)

Erard, C. 1989. Bird community structure in two rainforests: Africa (Gabon) and South America (French Guiana)—a comparison. In *Vertebrates in complex tropical systems,* ed. M. L. Harmelin-Vivien and F. Bourliere, 89–122. New York: Springer. (21)

Ericsson, U. 1972. The invertebrate fauna of the Kilpisjärvi area, Finnish Lapland. *Acta Soc. Fauna Flora Fenn.* 80:37–39. (10)

Erwin, T. L. 1981. Taxon pulses, vicariance and dispersal: An evolutionary synthesis illustrated by carabid beetles. In *Vicariance bio-*

geography, a critique, ed. G. Nelson and D. E. Rosen. New York: Columbia University Press. (11)

Erwin, T. L. 1982. Tropical forests: Their richness in Coleoptera and other arthropod species. *Coleop. Bull.* 36:74–75. (15, 22)

Erwin, T. L. 1983. Beetles and other insects of tropical forest canopies at Manaus, Brazil, sampled by insecticidal fogging. In *Tropical rainforest ecology and management,* ed. S. L. Sutton, T. C. Whitmore, and A. C. Chadwick, 59–75. Oxford: Blackwell Scientific Publications. (22)

Esch, G. W. 1971. Impact of ecological succession on the parasite fauna in centrarchids from oligotrophic and eutrophic ecosystems. *Am. Midl. Nat.* 86:160–68. (17)

Esch, G. W., A. O. Bush, and J. M. Aho (eds.). 1990. *Parasite communities: Patterns and processes.* London: Chapman and Hall. (17)

Esch, G. W., C. R. Kennedy, A. O. Bush, and J. M. Aho. 1988. Patterns in helminth communities in freshwater fish in Great Britain: Alternative strategies for colonization. *Parasitology* 96:519–32. (17)

Estes, R., and A. M. Baez. 1985. Herpetofaunas of North and South America during the Late Cretaceous and Cenozoic: Evidence for interchange? In *The great American biotic interchange,* ed. F. G. Stehli and S. D. Webb, 139–97. New York: Plenum Press. (25)

Eurola, S., S. Hicks, and E. Kaakinen. 1984. Key to Finnish mire types. In *European mires,* ed. P. D. Moore, 11–111. London: Academic Press. (10)

Evans, E. W. 1989. Interspecific interactions among phytophagous insects of tallgrass prairie: An experimental test. *Ecology* 70:435–44. (22)

Excell, A. W. 1954. Combretaceae. *Flora Malesiana,* ser. 1, 4:533–589. (20)

Facelli, J. M., and S. T. A. Pickett. 1990. Markovian chains and the role of history in succession. *Trends Ecol. Evol.* 5:27–30. (24)

Fager, E. W., and J. A. McGowan. 1963. Zooplankton species groups in the North Pacific. *Science* 140:453–60. (19)

Fagerstrom, J. A. 1987. *The evolution of reef communities.* New York: John Wiley and Sons. (27)

Fagerström, T. 1988. Lotteries in communities of sessile organisms. *Trends Ecol. Evol.* 3:303–6. (1)

Fahrig, L., and G. Merriam. 1985. Habitat patch connectivity and population survival. *Ecology* 66:1762–68. (7, 10)

Fairweather, P. G. 1988. Predation creates haloes of bare space among prey on rocky shores in New South Wales. *Aust. J. Ecol.* 13:401–409. (4)

Fairweather, P. G., and A. J. Underwood. 1983. The apparent diet of predators and biases due to different handling times of their prey. *Oecologia* 56:169–79. (4)

Fairweather, P. G., A. J. Underwood, and M. J. Moran. 1984. Preliminary investigations of predation by the whelk *Morula marginalba.* *Mar. Ecol. Prog. Ser.* 17:143–56. (4)

Farrell, B. 1985. A Biosystematic and Evolutionary Study of *Phyllobrotica* (Coleoptera: Chrysomelidae). Master's thesis, University of Maryland. (23)

Farrell, B. D. 1991. Phylogenetics of insect/plant interactions: *Tetraopes* and *Asclepias.* Ph.D. diss., University of Maryland. (23)

Farrell, B. D., D. E. Dussourd, and C. Mitter. 1991. Escalation of plant defense: Do latex and resin canals spur plant diversification? *Am. Nat.* 138:881–900. (22, 23, 30)

Farrell, B. D., and T. L. Erwin. 1988. Leaf-beetle community structure in an Amazonian rainforest canopy. In *Biology of Chrysomelidae,* ed. P. Jolivet, E. Petitpierre and T. J. Hsiao, 73–90. Dordrecht: Kluwer. (23)

Farrell, B. D., and C. Mitter. 1990. Phylogenesis of insect/plant interactions: Have *Phyllobrotica* leaf beetles (Chrysomelidae) and the Lamiales diversified in parallel? *Evolution* 44:1389–1403. (23)

Farrell, B. D., C. Mitter, and D. J. Futuyma. 1992. Diversification at the insect-plant interface. *BioScience* 42:34–42. (1, 26, 30)

Feder, J. L., C. A. Chilcote, and G. L. Bush. 1988. Genetic differentiation between sympatric host races of the apple maggot fly *Rhagoletis pomonella.* *Nature* 336:61–64. (22)

Feder, J. L., C. A. Chilcote, and G. L. Bush. 1990. The geographic pattern of genetic differentiation between host associated populations of *Rhagoletis pomonella* (Diptera: Tephritidae) in the eastern United States and Canada. *Evolution* 44:570–94. (23)

Fedorov, A. A. 1966. The structure of the tropical rain forest and speciation in the humid tropics. *J. Ecol.* 54:1–11. (1)

Feeny, P. 1975. Biochemical coevolution between plants and their insect herbivores. In *Coevolution of insects and plants,* ed. L. E. Gilbert and P. R. Raven, 3–13. Austin: University of Texas Press. (23)

Feeny, P. 1977. Defensive ecology of the Cruciferae. *Ann. Mo. Bot. Gard.* 64:221–34. (23)

Feeny, P. P. 1987. The roles of plant chemistry in associations between swallowtail butterflies and their host plants. In *Insects—Plants,* ed. V. F. Labeyrie and D. Lachaise, 353–59. Dordrecht: Junk. (23)

Feinsinger, P., and Colwell, R. K. 1978. Community organization among neotropical nectar-feeding birds. *Am. Zool.* 19:779–95. (4)

Felsenstein, J. 1985. Phylogenies and the comparative method. *Am. Nat.* 125:1–15. (18)

Felten, H., and H. G. Storch. 1970. Kleinsäuger von den italienischen Mittelmeer-Inseln *Pantelleria* und *Lampedusa*. *Senkenbergiana Biol.* 51:159–73. (12)

Fernandes, G. W., and P. W. Price. 1988. Biogeographical gradients in galling species richness. *Oecologia* 76:161–67. (6)

Ferrar, P. 1982. Termites of a South African savanna. I. List of species and subhabitat preferences. *Oecologia* 52:125–32. (14)

Findley, J. S. 1969. Biogeography of southwestern boreal and desert mammals. *Univ. Kansas Publ., Mus. Nat. Hist.,* no. 51, 113–28. (14)

Fischer, A. G. 1960. Latitudinal variation in organic diversity. *Evolution* 14:64–81. (1, 6)

Fischer, A. G. 1961. Latitudinal variations in organic diversity. *Am. Sci.* 49:50–74. (1)

Fischer, A. G. 1984. The two Phanerozoic supercycles. In *Catastrophies in Earth history—the new uniformitarianism,* ed. W. A. Berggren and J. A. Van Couvering, 129–50. Princeton, N.J.: Princeton University Press. (27)

Fischer, A. G. 1986. Climatic rhythms recorded in strata. *Annu. Rev. Earth Planetary Sci.* 14:351–376. (29)

Fisher, R. A. 1932. *Statistical Methods for Research Workers.* 4th Edition. Edinburgh: Oliver and Boyd. (4)

Fisher, R. A. 1937. The wave of advance of advantageous genes. *Ann. Eugenics* 7:355–69. (8)

Flaherty, D. L. 1969. Ecosystem trophic complexity and Willamette mite, *Eotetranychus willamettei* Ewing (Acarina: Tetranychridae), densities. *Ecology* 50:911–15. (7)

Flessa, K. A. 1986. Causes and consequences of extinction. In *Patterns and processes in the history of life,* ed. D. M. Raup and D. Jablonski, 235–57. Berlin: Springer-Verlag. (1)

Fletcher, W. J. 1984. Intraspecific variation in the population dynamics and growth of the limpet, *Cellana tramoserica*. *Oecologia* 63:110–21. (4)

Florschutz, F., J. Menendez Amor, and T. A. Wijmstra. 1971. Palynology of a thick Quaternary succession in southern Spain. *Palaeogeogr. Palaeoclimatol. Palaeoecol.* 10:233–64. (12)

Flynn, L. J., R. H. Tedford, and Q. Zhanxiang. 1991. Enrichment and stability in the Pliocene mammalian fauna of North China. *Paleobiology* 17:246–65. (30)

Food and Agriculture Organization of the United Nations. 1972. *Atlas of the living resources of the sea.* Rome: FAO Department of Fisheries. (6)

Forcella, F. 1985. Final distribution is related to rate of spread in alien weeds. *Weed Res.* 25:181–91. (10)

Fosshagen, M.-S., P. Palmgren, and I. Valovirta. 1972. The invertebrate fauna of the Kilpisjärvi area, Finnish Lapland. 2. Terrestrial gastropods. *Acta Soc. Fauna Flora Fenn.* 80:37–39. (10)

Foster, D. R., P. K. Schoonmaker, and S. T. A. Pickett. 1990. Insights from paleoecology to community ecology. *Trends Ecol. Evol.* 5:119–22. (24)

Foster, J. B. 1964. Evolution of mammals on islands. *Nature* 202:234–35. (12)

Foster, M. S. 1982. Factors controlling the intertidal zonation of *Iridaea flaccida* (Rhodophyta). *J. Phycol.* 18:285–94. (4)

Foster, M. S. 1990. Organization of macroalgal assemblages in the Northeast Pacific: The assumption of homogeneity and the illusion of generality. *Hydrobiologia* 192:21–33. (4)

Foster, M. W. 1974. *Recent Antarctic and Subantarctic brachiopods.* Antarctic Research Series, vol. 21. Washington, D.C.: American Geophysical Union. (5)

Fox, B. J. 1987. Species assembly and the evolution of community structure. *Evol. Ecol.* 1:201–13. (25)

Fox, B. J., R. D. Quinn, and G. J. Breytenbach. 1985. A comparison of small-mammal succession following fire in shrublands of Australia, California, and South Africa. *Proc. Ecol. Soc. Aust.* 14:179–97. (21)

Frederiksen, N. O. 1984a. Sporomorph correlation and paleoecology, Piney Point and Old Church Formations, Pamunkey River, Virginia. In *Stratigraphy and paleontology of the outcropping Tertiary beds in the Pamunkey River region, central Virginia Coastal Plain—guidebook for Atlantic Coastal Plain Geological Association 1984 field trip,* ed. L. W. Ward and K. Krafft, 135–49. Atlantic Coastal Plain Geological Association. (26)

Frederiksen, N. O. 1984b. *Stratigraphic, paleoclimatic, and paleobiogeographic significance of Tertiary sporomorphs from Massachusetts.* U.S. Geological Survey Professional Paper no. 1308, 1–25. (26)

Freitag, R. 1979. Reclassification, phylogeny and zoogeography of the Australian species of *Cicindela* (Coleoptera: Cicindelidae). *Aust. J. Zool. suppl. ser.* 66:1–99. (18)

Freitag, R., and B. L. Barnes. 1989. Classification of Brazilian species of *Cicindela* and phylogeny and biogeography of subgenera *Brasiella, Gaymara* new subgenus, *Plectographa* and South American species of *Cylindera* (Coleoptera: Cicindelidae). *Quaestiones Entomologicae* 25:241–386. (18)

Fretter, V., and R. Manly. 1977. Settlement and early benthic life of *Littorina neritoides* (L.) at Wembury, South Devon. *J. Molluscan Stud.* 43:255–262. (4)

Fretwell, S. D. 1972. *Populations in a seasonal environment.* Princeton, N.J.: Princeton University Press. (7)

Fretwell, S. D. 1987. Food chain dynamics: the central theory of ecology? *Oikos* 50:291–301. (3)

Friis, E. M. 1979. The Damgaard flora: A new middle Miocene flora from Denmark. *Bull. Geol. Soc. Denmark* 27:117–42. (26)

Friis, E. M. 1985. Structure and function in late Cretaceous angiosperm flowers. *Biologiske Skrifter Danske Videnskabernes Selskab* 25:1–37. (26)

Fritz, R. S., C. F. Sacchi, and P. W. Price. 1986. Competition vs. host plant phenotype in species composition: Willow sawflies. *Ecology* 67:1608–18. (22)

Frost, S. H. 1977. Cenozoic reef systems of the Caribbean. In *Reefs and related carbonates—ecology and sedimentology,* ed. S. H. Frost et al., 92–110. American Association of Petroleum Geologists, Studies in Geology no. 4. (27)

Fry, W. G. 1979. Taxonomy, the individual, and the sponge. In *Biology and systematics of colonial organisms,* ed. G. Larwood and B. R. Rosen, 49–80. Special Publication 11 of the Systematics Association. London: Academic Press. (27)

Frye, R. J. 1983. Experimental field evidence of interspecific aggression between two species of kangaroo rats (*Dipodomys*). *Oecologia* 59:74–78. (5)

Fryer, G. 1977. Evolution of species flocks of cichlid fishes in African lakes. *Z. f. Zool. Syst. Evol.* 15:141–65. (30)

Fryer, G., and T. D. Iles. 1972. *The cichlid fishes of the Great Lakes of Africa.* Edinburgh: Oliver and Boyd. (22, 30)

Fuentes, E. R. 1981. Evolution of lizard niches in mediterranean habitats. In *Mediterranean-type shrublands,* ed. F. di Castri, D. W. Goodall, and R. L. Specht, 417–44. Amsterdam: Elsevier. (21)

Fulk, G. W. 1975. *Population ecology of rodents in the semiarid shrublands of Chile.* Occasional Papers of the Museum, Texas Tech University no. 33, 1–40. (14)

Fursich, F. T., and W. Werner. 1986. Benthic associations and their environmental significance in the Lusitanian Basin (Upper Jurassic, Portugal). *Neues Jahrb. Geol. Paläont.* 172:271–329. (29)

Futuyma, D. J. 1976. Food plant specialization and environmental predictability in Lepidoptera. *Am. Nat.* 110:285–92. (23)

Futuyma, D. J. 1983. Evolutionary interactions among herbivorous insects and plants. In *Coevolution,* ed. D. J. Futuyma and M. Slatkin, 207–31. Sunderland, Mass: Sinauer Associates. (23)

Futuyma, D. J. 1986. Evolution and coevolution in communities. In *Patterns and processes in the history of life,* ed. D. M. Raup and D. Jablonski, 369–82. Berlin: Springer-Verlag. (7, 23, 29)

Futuyma, D. J. 1991. Evolution of host specificity in herbivorous insects: Genetic, ecological, and phylogenetic aspects. In *Plant-animal interactions: Evolutionary ecology in tropical and temperate regions,* ed. P. W. Price, T. L. Lewinsohn, G. W. Fernandes, and W. W. Benson, 431–54. New York: John Wiley and Sons. (23)

Futuyma, D. J., and F. Gould. 1979. Associations of insects and plants in a deciduous forest. *Ecol. Monogr.* 49:33–50. (23)

Futuyma, D. J., and M. C. Keese. 1992. Evolution and coevolution of plants and phytophagous arthropods. In *Herbivores: Their interaction with secondary plant metabolites,* vol. 2, ed. G. A. Rosenthal and M. R. Berenbaum, 440–75. San Diego: Academic Press. (23)

Futuyma, D. J., and S. J. McCafferty. 1990. Phylogeny and the evolution of host associations in the leaf beetle genus *Ophraella* (Coleoptera, Chrysomelidae). *Evolution* 44:1885–1913. (23)

Futuyma, D. J., and G. Moreno. 1988. The evolution of ecological specialization. *Annu. Rev. Ecol. Syst.* 19:207–34. (23)

Gagne, W. C. 1979. Canopy-associated arthropods in *Acacia koa* and *Metrosideros* tree communities along an altitudinal transect on Hawaii Island. *Pacific Insects* 21:56–82. (22)

Gaillard, C. 1983. *Les bioherms a spongiaires et leur environment dans l'Oxfordien du Jura Meridional.* Document de Laboratoire de Géologie, Faculté de Science de Lyon no. 90. (27)

Gallardo, J. M. 1970. Estudio ecológico sobre los anfibios y reptiles del sudoeste de la Provincia de Buenos Aires, Argentina. *Rev. Mus. Argentino de Ciencias Nat. "Bernardino Rivadavia"* 10:27–63. (25)

Gallardo, J. M. 1986. La diversidad de la herpetofauna en la selva subtropical misionera. *An. Mus. Hist. Nat. Valparaiso* 17:153–59. (25)

Gallé, L. 1986. Habitat and niche analysis of grassland ants (Hymenoptera: Formicidae). *Entomol. Gener.* 11:197–211. (10)

Ganeshaiah, K. M., and V. V. Belavadi. 1986. Habitat segregation in four species of adult tiger beetles (Coleoptera: Cicindelidae). *Ecol. Entomol.* 11:147–154. (18)

Gans, C. 1974. *Biomechanics, an approach to vertebrate biology.* Philadelphia: J. B. Lippincott. (25)

Garrity, S. D., and S. C. Levings. 1981. A predator-prey interaction between two physically and biologically constrained tropical rocky shore gastropods: Direct, indirect and community effects. *Ecol. Monogr.* 51:267–86. (4)

Gaspar, C., and F. G. Werner. 1976. *The ants of Arizona: An ecological study of ants in the Sonoran Desert.* U.S./I.B.P. Desert Biome Research Memorandum 73-50. (14)

Gaston, K. J. 1990. Patterns in the geographical ranges of species. *Biol. Rev.* 65:105–29. (10)

Gaston, K. J., and J. H. Lawton. 1989. Insect herbivores on bracken do not support the core-satellite species hypothesis. *Am. Nat.* 134:761–77. (16, 22)

Gaston, K. J., and J. H. Lawton. 1990a. Effects of scale and habitat on the relationship between regional distribution and local abundance. *Oikos* 58:329–35. (10)

Gaston, K. J., and J. H. Lawton. 1990b. The population ecology of rare species. In *The biology and conservation of rare fish,* ed. A. Wheeler and D. Sutcliffe. *J. Fish. Biol.* 37 (suppl. A), 97–104. (10)

Gatz, A. J. 1979. Community organization in fishes as indicated by morphological factors. *Ecology* 60:711–18. (1)

Gauld, I. D. 1986. Latitudinal gradients in ichneumonid species-richness in Australia. *Ecol. Entomol.* 11:155–61. (23)

Gause, G. F. 1934. *The struggle for existence.* Baltimore: Williams and Wilkins. (1)

Geddes, D. S. 1985. Mesolithic domestic sheep in West Mediterranean Europe. *J. Archaeol. Sci.* 12:25–48. (12)

Gee, J. H. R., and P. S. Giller (Eds.). 1987. *Organization of communities past and present.* Oxford: Blackwell. (1)

Geiser, F. 1986. Thermoregulation and torpor in the kultarr, *Antechi-*

nomys laniger (Marsupialia: Dasyuridae). *J. Comp. Physiol.* B 156:751–57. (14)

Geiser, F., and R. V. Baudinette. 1987. Seasonality of torpor and thermoregulation in three dasyurid marsupials. *J. Comp. Physiol.* B 157:335–44. (14)

Geiser, F., and R. V. Baudinette. 1988. Daily torpor and thermoregulation in the small dasyurid marsupials *Planigale gilesi* and *Ningani yvonneae. Aust. J. Zool.* 36:473–81. (14)

Genoud, M. 1988. Energetic strategies of shrews: Ecological constraints and evolutionary implications. *Mammal Rev.* 18:173–93. (14)

Gentry, A. H. 1982. Phytogeographic patterns as evidence for a Choco refuge. In *Biological diversification in the tropics,* ed. G. T. Prance, 112–36. New York: Columbia University Press. (18, 20)

Gentry, A. H. 1988a. Changes in plant community diversity and floristic composition on environmental and geographical gradients. *Ann. Mo. Bot. Garden.* 75:1–34. (5, 26, 30)

Gentry, A. H. 1988b. Tree species richness of upper Amazonian forests. *Proc. Natl. Acad. Sci. USA* 85:156–59. (5)

Gentry, A. H., and C. Dodson. 1987. Contribution of nontrees to species richness of a tropical rain forest. *Biotropica* 19:149–56. (21)

Gentry, A. H., and L. H. Emmons. 1987. Geographical variation in fertility, phenology, and composition of the understory of Neotropical forests. *Biotropica* 19:216–27. (5)

Germeraad, J. H., C. A. Hopping, and J. Muller. 1968. Palynology of Tertiary sediments from tropical areas. *Rev. Palaeobot. Palynol.* 6:189–348. (20)

Gersper, P. L., and N. Holowaychuk. 1970. Effects of stemflow water of a Miami soil under a beech tree: II. Chemical properties. *Soil Sci. Soc. Am. Proc.* 34:786–94. (2)

Gersper, P. L., and N. Holowaychuk. 1971. Some effects of stemflow from forest canopy trees on chemical properties of soils. *Ecology* 52:691–702. (2)

Gertler, P. E., and J. Morales. 1980. Snake predation on *Marmosa noctivaga. J. Mammal.* 61:381. (25)

Gilbert, C. R. 1976. Composition and derivation of the North American freshwater fish fauna. *Fla. Scientist* 39:104–11. (24)

Gilpin, M. E., and J. M. Diamond. 1981. Immigration and extinction probabilities for individuals species: Relation to incidence functions and species colonization curves. *Proc. Natl. Acad. Sci. USA* 78:392–96. (7)

Gilpin, M. E., and I. Hanski. 1991. *Metapopulation dynamics: Empirical and theoretical investigations.* Cambridge: Cambridge University Press. (2, 7)

Gingerich, P. D. 1984a. Pleistocene extinctions in the context of origination-extinction equilibria. In *Quaternary extinctions: A prehistoric revolution,* ed. P. S. Martin and R. G. Klein, 211–22. Tucson: University of Arizona Press. (28, 29)

Gingerich, P. D. 1984b. Primate evolution. In *Mammals, notes for a short course,* ed. P. G. Gingerich and C. E. Badgley, 167–81. (28)

Gingerich, P. D. 1987. Evolution and fossil record: Patterns, rates and processes. *Can. J. Zool.* 65:1053–60. (28, 29)

Gingerich, P. D. 1989. New earliest Wasatchian mammalian fauna from the Eocene of northwestern Wyoming: Composition and diversity in a rarely sampled high-floodplain assemblage. *Univ. Michigan Pap. Paleont.* 28, 1–97. Ann Arbor: University of Michigan. (28)

Gittleman, J. L. 1985. Carnivore body size: Ecological and taxonomic correlates. *Oecologia* 67:540–54. (28)

Givnish, T. J. 1982. On the adaptive significance of leaf height in forest herbs. *Am. Nat.* 120:353–81. (2)

Gladfelter, W. D., J. C. Ogden, and E. H. Gladfelter. 1980. Similarity and diversity among coral reef fish and communities, a comparison between tropical western Atlantic (Virgin Islands) and tropical central Pacific (Marshall Islands) patch reefs. *Ecology* 69:1156–68. (4)

Glasser, J. W. 1982. On the causes of temporal change in communities: Modification of the biotic environment. *Am. Nat.* 119:375–90. (7)

Glazier, D. S. 1980. Ecological shifts and the evolution of geographically restricted species of North American *Peromyscus* (mice). *J. Biogeogr.* 7:63–83. (14)

Glazier, D. S. 1987. Energetics and taxonomic patterns of species diversity. *Syst. Zool.* 36:62–71. (24)

Gleason, H. A. 1926. The individualistic concept of the plant association. *Bull. Torrey Bot. Club* 53:7–26. (1, 29)

Goater, C. P., and A. O. Bush. 1988. Intestinal helminth communities in long-billed curlews: The importance of congeneric species. *Holarctic Ecol.* 11:140–45. (10, 17)

Godfray, H. C. J., and S. W. Pacala. 1992. Aggregation and the population dynamics of parasitoids and predators. *Am. Nat.* 140:30–40. (7)

Goldberg, D. E. 1987. Neighborhood competition in an old-field plant community. *Ecology* 68:1211–23. (2)

Goldberg, D. E., and T. E. Miller. 1990. Effects of different resource additions on species diversity in an annual plant community. *Ecology* 71:213–25. (5, 6)

Gooch, J. L. 1975. Mechanisms of evolution and population genetics. In *Marine ecology,* vol. 2, part 1, ed. O. Kinne, 349–409. Chichester: Wiley. (4)

Goodnight, C. J. 1990a. Experimental studies of community evolution. I: The response to selection at the community level. *Evolution* 44:1614–24. (29)

Goodnight, C. J. 1990b. Experimental studies of community evolution. II: The ecological basis of the response to community selection. *Evolution* 44:1625–36. (29)

Gopalsamy, K. 1977. Competition, dispersion and coexistence. *Math. Biosci.* 33:25–33. (8)

Goreau, T. F. 1959. The ecology of Jamaican coral reefs. I. Species composition and zonation. *Ecology* 40:67–90. (27)

Goreau, T. F., and N. I. Goreau. 1973. The ecology of Jamaican coral reefs. II. Geomorphology, zonation and sedimentary phases. *Bull. Mar. Sci.* 23:403–64. (27)

Gorham, E. 1979. Shoot height, weight and standing crop in relation to density of monospecific plant stands. *Nature* 279:148–50. (2)

Gorman, O. T. 1992. Evolutionary ecology vs. historical ecology: assembly, structure, and organization of stream fish communities. In *Systematics, historical ecology and North American freshwater fishes,* ed. R. L. Mayden. Palo Alto: Stanford University Press. (24)

Gosling, E. M., and D. McGrath. 1990. Genetic variability in exposed-shore mussels, *Mytilus* spp., along an environmental gradient. *Mar. Biol.* 104:413–18. (4)

Gosz, J. R., G. E. Likens, and F. H. Borman. 1973. Nutrient release from decomposing leaf and branch litter in the Hubbard Brook Forest, New Hampshire. *Ecol. Monogr.* 43:173–91. (2)

Gotelli, N. 1992. Metapopulation models: The rescue effect, the propagule rain, and the core-satellite hypothesis. *Am. Nat.* 138:768–76. (7)

Gottfried, B. M. 1979. Small mammal populations in woodlot islands. *Am. Midl. Nat.* 102:105–12. (10)

Gotto, R. V. 1969. *Marine animals: Partnership and other associations.* New York: American Elsevier. 1–96. (27)

Gould, F. 1991. Arthropod behaviour and the efficacy of plant protectants. *Annu. Rev. Entomol.* 36:305–30. (23)

Gould, S. J. 1985. The paradox of the first tier: An agenda for paleobiology. *Paleobiology* 11:2–12. (23)

Gould, S. J. 1989. *Wonderful life: The Burgess Shale and the nature of history.* New York: W. W. Norton. (11, 23)

Gould, S. J., and R. C. Lewontin. 1979. The spandrels of San Marco and the Panglossian paradigm: A critique of the adaptationist programme. *Proc. Roy. Soc.* (Lond.), ser. B 205:581–98. (14)

Gould, S. J., D. M. Raup, J. J. Sepkoski, Jr., T. J. M. Schopf, and D. S. Simberloff. 1977. The shape of evolution: A comparison of real and random clades. *Paleobiology* 3:23–40. (11)

Gould, S. J., and D. S. Woodruff. 1990. History as a cause of area effects: An illustration from *Cerion* on Great Inagua, Bahamas. *Biol. J. Linn. Soc.* 40:67–98. (18)

Gradstein, S. R., and T. Pocs. 1989. Bryophytes. In *Ecosystems of the world.* Vol. 14B, *Tropical rain forest ecosystems: Biogeographical and ecological studies,* ed. H. Lieth and M. J. A. Werger, 311–25. Amsterdam: Elsevier. (5)

Graham, A. (ed.). 1972. *Floristics and paleofloristics of Asia and eastern North America.* Amsterdam: Elsevier. (26)

Graham, A. 1977. New records of *Pelliceria* (Theaceae/Pelliceriaceae) in the Tertiary of the Caribbean. *Biotropica* 9:48–52. (20)

Graham, R. W. 1976. Late Wisconsin mammal faunas and paleoenvironmental gradients of the eastern United States. *Paleobiology* 2:343–50. (28)

Graham, R. W. 1985. Diversity and community structure of the late Pleistocene mammal fauna of North America. *Acta Zool. Fennici* 170:181–92. (29)

Graham, R. W. 1986. Response of mammalian communities to environmental changes during the late Quaternary. In *Community ecology,* ed. J. Diamond and T. J. Case, 300–313. New York: Harper and Row. (15, 25)

Graham, R. W., and E. L. Lundelius, Jr. 1984. Coevolutionary disequilibrium and Pleistocene extinctions. In *Quaternary extinctions: A prehistoric revolution,* ed. P. J. Martin and R. G. Klein, 223–49. Tucson: University of Arizona Press. (28, 29)

Grant, B. R., and P. R. Grant. 1989. *Evolutionary dynamics of a natural population: The large cactus finch of the Galápagos.* Chicago: University of Chicago Press. (12)

Grant, P. R. 1972. Convergent and divergent character displacement. *Biol. J. Linn. Soc.* 4:39–68. (1)

Grant, P. R. 1986. *Ecology and evolution of Darwin's Finches.* Princeton, N.J.: Princeton University Press. (1, 30)

Grant, P. R., and D. Schluter. 1984. Interspecific competition inferred from patterns of guild structure. In *Ecological communities: Conceptual issues and the evidence,* ed. D. R. Strong, D. S. Simberloff, L. G. Abele, and A. B. Thistle, 201–33. Princeton, N.J.: Princeton University Press. (1)

Grassle, J. F. 1973. Variety in coral reef communities. In *Biology and geology of coral reefs* 2, *Biology* 1, ed. O. A. Jones and R. Endean, 247–70. New York: Academic Press. (27)

Gray, A. 1878. Forest geography and archaeology. *Am. J. Sci.* 16:85–94, 183–196. (26)

Greene, H. W. 1975. Ecological observations on the red coffee snake, *Ninia sebae,* in southern Veracruz, Mexico. *Am. Midl. Nat.* 93:478–94. (25)

Greene, H. W. 1983. Dietary correlates of the origin and radiation of snakes. *Am. Zool.* 23:431–41. (25)

Greene, H. W. 1988. Species richness in tropical predators. In *Tropical rainforests: Diversity and conservation,* ed. F. Almeda and C. M. Pringle, 259–80. San Francisco, Calif.: California Academy of Sciences and Pacific Division, American Association for the Advancement of Science. (25)

Greene, H. W. 1989. Ecological, evolutionary, and conservation implications of feeding biology in Old World cat snakes, genus *Boiga* (Colubridae). *Proc. Calif. Acad. Sci.* 46:193–207. (25)

Greene, H. W., and G. M. Burghardt. 1978. Behavior and phylogeny: Constriction in ancient and modern snakes. *Science* 200:74–77. (25)

Greenslade, P. J. M. 1978. Ants. In *The physical and biological features of Kunoth Paddock in central Australia,* ed. W. A. Low, 109–13. CSIRO Division of Land Resources Management Technical Paper no. 4, Melbourne. (14)

Greenslade, P. J. M. 1979. *A guide to ants of South Australia.* Adelaide: South Australian Museum. (14)

Greenslade, P. J. M., and P. Greenslade. 1983. Ecology of soil invertebrates. In *Soils: An Australian viewpoint,* ed. CSIRO Division of Soils, 645–69. Melbourne: CSIRO; London: Academic Press. (14)

Greenslade, P. J. M., and P. Greenslade. 1984. Soil surface insects of the Australian arid zone. In *Arid Australia,* ed. H. G. Cogger and E. E. Cameron, 153–76. Sydney: Australian Museum. (14)

Greenslade, P. J. M., and R. B. Halliday. 1982. Distribution and speciation in meat ants, *Iridomyrmex purpureus* and related species (Hymenoptera: Formicidae). In *Evolution of the flora and fauna of arid Australia,* ed. W. R. Barker and P. J. M. Greenslade, 249–55. Adelaide: Peacock Publications. (14)

Greenwood, P. H. 1974. Cichlid fishes of Lake Victoria, East Africa: The biology and evolution of a species flock. *Bull. Brit. Mus. Nat. Hist.* (Zool.) suppl. 6:1–134. (30)

Greenwood, P. H. 1984. African cichlids and evolutionary theories. In *Evolution of fish species flocks,* ed. A. A. Echelle and I. Kornfield, 141–54. Orono, Me.: University of Maine Press. (30)

Gregor, H. J. 1982. *Die Juntertiären Floren Süddeutschlands: Pa-*

läokarpologie, Phytostratigraphie, Paläoökologie, Paläoklimatologie. Stuttgart: Ferdinand Enke Verlag. (26)

Gregory, R. D. 1990. Parasites and host geographic range as illustrated by waterfowl. *Funct. Ecol.* 4:645–54. (17)

Gregory, R. D., A. E. Keymer, and P. H. Harvey. 1991. Life history, ecology, and parasite community structure in Soviet birds. *Biol. J. Linn. Soc.* 43:249–62. (17)

Grigg, R. W. 1988. Paleoceanography of coral reefs in the Hawaiian-Emperor Chain. *Science* 240:1737–43. (29)

Grime, J. P. 1973. Control of species density in herbaceous vegetation. *J. Environ. Manage.* 1:151–67. (5)

Grime, J. P. 1979. *Plant strategies and vegetation.* Chichester: Wiley. (5)

Grinnell, J. 1917. The niche-relationships of the California thrasher. *Auk* 34:427–33. (1)

Grosberg, R. K. 1982. Intertidal zonation of barnacles: the influence of planktonic zonation of larvae on vertical distribution of adults. *Ecology* 63:894–99. (4)

Grover, J. P. 1988. Dynamics of competition in a variable environment: Experiments with two diatom species. *Ecology* 69:408–17. (2)

Grover, J. P. 1989. Effects of Si:P supply ratio, supply variability, and selective grazing in the plankton: An experiment with a natural algal and protistan assemblage. *Limnol. Oceanogr.* 34:349–67. (2)

Groves, R. H. (ed.) 1981. *Australian vegetation.* Cambridge, London, New York: Cambridge University Press. (13)

Guégan, J., and C. R. Kennedy. In press. Maximum local helminth parasite communities richness in British freshwater fish: A test of the colonisation time hypothesis. *Parasitology* 106:91–100. (17)

Gulmon, S. L., and H. A. Mooney. 1986. Costs of defense and their effects on plant productivity. In *On the economy of plant form and function,* ed. T. J. Givnish, 681–95. Cambridge: Cambridge University Press. (2)

Guo, S. 1990. A brief review on megafloral successions and climatic changes of the Cretaceous and early Tertiary in China. In *Proceedings of the Symposium on Paleofloristic and Paleoclimatic Changes in the Cretaceous and Tertiary,* (1989), ed. E. Knobloch and Z. Kvăek, 23–38. Prague: Geological Survey Publisher. (26)

Guthrie, R. D. 1984. Mosaics, allelochemics and nutrients: An ecological theory of late Pleistocene megafaunal extinctions. In *Quaternary Extinctions,* ed. P. S. Martin and R. G. Klein, 259–98. Tucson: University of Arizona Press. (28)

Guttman, S. I., T. K. Wood, and A. A. Karlin. 1981. Genetic differentiation along host plant lines in the sympatric *Enchenopa binotata* Say complex. *Evolution* 35:205–17. (22)

Guyer, C. In press. The reptile fauna of La Selva: Comparisons with other Neotropical sites. In *La Selva: Ecology and natural history of a Neotropical rainforest,* ed. L. A. McDade, K. S. Bawa, H. A. Hespenheide, and G. S. Hartshorn. Chicago: University of Chicago Press. (25)

Guyer, C., and M. A. Donnelly. 1990. Length-mass relationships among an assemblage of tropical snakes in Costa Rica. *J. Trop. Ecol.* 6:65–76. (25)

Guyer, C., and J. B. Slowinski. 1991. Comparisons of observed phylogenetic topologies with null expectations among three monophyletic lineages. *Evolution* 45:340–50. (11)

Gyllenberg, M., and I. Hanski. 1992. Single-species metapopulation dynamics: A structured model. *Theor. Pop. Biol.* 42:35–61. (10)

Hadač, E. 1976. Species diversity of mangrove and continental drift. *Folia Geobot. Phytotax.* (Praha) 11:213–16. (20)

Hadeler, K. P., an der Heiden, U., and Rothe, F. 1974. Nonhomogeneous spatial distributions of populations. *J. Math. Biol.* 1:165–76. (8)

Hadley, N. G., C. B. Knisley, T. D. Schultz, and D. L. Pearson. 1990. Water relations of tiger beetle larvae (*Cicindela marutha*): Correlations with habitat microclimate and burrowing activity. *J. Arid Environ.* 19:189–97. (18)

Hadlock, R. P. 1980. Alarm responses of the intertidal snail *Littorina littorea* (L.) to predation by the crab *Carcinus maenas* (L.). *Bull. MBL, Woods Hole* 159:269–79. (4)

Haedrich, R. L., G. T. Rowe, and P. T. Polloni. 1980. The megabenthic fauna in the deep sea south of New England, USA. *Mar. Biol.* 57:165–79. (5)

Hafner, M. S., and S. A. Nadler. 1988. Phylogenetic trees support the coevolution of parasites and their hosts. *Nature* 332:258–59. (23)

Haila, Y. 1983. Land birds on northern islands: A sampling metaphor for insular colonization. *Oikos* 41:334–51. (7)

Hairston, N. G., Sr. 1989. *Ecological experiments: Purpose, design, and execution.* Cambridge: Cambridge University Press. (22)

Hairston, N. G., F. E. Smith, and L. B. Slobodkin. 1960. Community structure, population control, and competition. *Am. Nat.* 44:421–25. (22, 24, 28)

Hall, B. P., and R. E. Moreau. 1970. *Atlas of speciation of African birds.* London: British Museum (Natural History). (12)

Hall, E. R. 1981. *The mammals of North America.* New York: John Wiley and Sons. (28)

Hall, J. B., and M. D. Swaine. 1976. Classification and ecology of closed-canopy forest in Ghana. *J. Ecol.* 64:913–51. (5)

Hall, S. J., and D. Raffaelli. 1991. Food web patterns: Lessons from a species-rich web. *J. Anim. Ecol.* 60:823–42. (3)

Hallam, A. 1987. Radiations and extinctions in relation to environmental change in the marine lower Jurassic of northwest Europe. *Paleobiology* 13:152–68. (29)

Hallam, A. 1989. The case for sealevel change as a dominant causal factor in mass extinction of marine invertebrates. *Phil. Trans. Roy. Soc.* (Lond.), ser. B 325:437–55. (27)

Hallock, P. 1987. Fluctuations in the trophic resource continuum: A factor in global diversity cycles. *Paleoceanography* 2:457–71. (27)

Hallock, P. 1988. The role of nutrient availability in bioerosion: Consequences to carbonate buildups. *Palaeogeogr., Palaeoceanogr. Palaeoecol.* 63:275–91. (27)

Hallock, P., A. C. Hine, G. A. Vargo, J. A. Elrod, and W. C. Jaap. 1988. Platforms of the Nicaragua Rise: Examples of the sensitivity of carbonate sedimentation to excess trophic resources. *Geology* 16:1104–7. (27)

Hallock, P., and W. Schlager. 1986. Nutrient excess and the demise of coral reefs and carbonate platforms. *Palaios* 1:389–98. (27)

Hämet-Ahti, L., J. Suominen, U. T. Ulvinen, P. Uotila, and S. Vuokko (eds.). 1986. *Retkeilykasvio.* 3d ed. Forssa: Suomen Luonnonsuojelun Tuki. (10)

Hamilton, T. H., I. Rubinoff, R. H. Barth, Jr., and G. L. Bush. 1963. Species abundance: Natural regulation of insular variation. *Science* 142:1575–77. (1)

Hancock, N. J., J. M. Hurst, and F. T. Fursich. 1974. The depths inhabited by Silurian brachiopod communities. *J. Geol. Soc. Lond.* 130:151–56. (5)

Handlirsch, A. 1908. *Die fossilen Insecten und die Phylogenie der rezenten Formen.* Leipzig: W. Engelmann. (23)

Hanski, I. 1981. Coexistence of competitors in patchy environment with and without predation. *Oikos* 37:306–12. (7)

Hanski, I. 1982. Dynamics of regional distribution: The core and satellite species hypothesis. *Oikos* 38:210–21. (3, 7, 10, 12, 22)

Hanski, I. 1983. Coexistence of competitors in a patchy environment. *Ecology* 64:493–500. (7, 8, 9)

Hanski, I. 1985. Single-species spatial dynamics may contribute to long-term rarity and commonness. *Ecology* 66:335–43. (10)

Hanski, I. 1991. Single-species metapopulation dynamics: Concepts, models, and observations. In *Metapopulation dynamics,* ed. M. E. Gilpin and I. Hanski, 17–38. London: Academic Press. (10)

Hanski, I., and Y. Cambefort (eds.). 1991. *Dung beetle ecology.* Princeton, N.J.: Princeton University Press. (2, 10)

Hanski, I., L. Hansson, and H. Henttonen. 1991. Specialist predators, generalist predators, and the microtine rodent cycle. *J. Anim. Ecol.* 60:353–67. (7)

Hansson, L., and H. Henttonen. 1988. Rodent dynamics as community processes. *Trends Ecol. Evol.* 3:195–200. (7)

Happold, M. 1976. Social behaviour of the Conilurine rodents (Muridae) of Australia. *Z. Tierpsychol.* 40:113–82. (14)

Haq, B. U. 1981. Paleogene paleoceanography: Early Cenozoic oceans revisited. *Oceanol. Acta, Colloque C4, Géologie des Océans,* 71–82. (20)

Haq, B. V., J. Hardenbol, and P. R. Vail. 1987. Chronology of fluctuating sealevels since the Triassic. *Science* 235:1159–67. (27)

Harland, W. B., R. L. Armstrong, A. V. Cox, L. E. Craig, A. G. Smith, and D. G. Smith. 1989. *A geologic time scale.* Cambridge: Cambridge University Press. (27)

Harmon, M. E. 1986. The ecology of woody debris in temperate ecosystems. *Adv. Ecol. Res.* 15:133–302. (2)

Harper, J. L. 1981. The meanings of rarity. *The biological aspects of rare plant conservation,* ed. H. Synge, 189–203. New York: John Wiley and Sons. (10)

Harrison, S. 1991. Local extinction in a metapopulation context: An empirical evaluation. In *Metapopulation dynamics: Empirical and theoretical investigations,* ed. M. E. Gilpin and I. Hanski. Cambridge: Cambridge University Press. (7)

Hart, D. D., and R. J. Horwitz. 1991. Habitat diversity and the species-area relationship: Alternative models and tests. In *Habitat structure,* ed. E. McCoy and G. Bell, 47–68. London: Chapman and Hall. (7)

Haseldonckx, P. 1972. The presence of *Nypa* in Europe: A solved problem. *Geol. Mijnbouw* 51:645–50. (20)

Hassell, M. P. 1978. *The dynamics of predator-prey systems.* Princeton, N.J.: Princeton University Press. (2)

Hassell, M. P., H. N. Comins, and R. M. May. 1991. Spatial structure and chaos in insect population dynamics. *Nature* 353:255–58. (7)

Hassell, M. P., and R. M. May. 1988. Spatial heterogeneity and the dynamics of parasitoid-host systems. *Ann. Zool. Fennici* 25:55–61. (8)

Hastings, A. 1977. Spatial heterogeneity and the stability of predator-prey systems. *Theor. Pop. Biol.* 12:37–48. (8, 9)

Hastings, A. 1978. Spatial heterogeneity and the stability of predator-prey systems: Predator-mediated coexistence. *Theor. Pop. Biol.* 14:380–95. (7, 9)

Hastings, A. 1980. Disturbance, coexistence, history, and competition for space. *Theor. Pop. Biol.* 18:363–73. (2)

Hastings, A., and C. L. Wolin. 1989. Within-patch dynamics in a metapopulation. *Ecology* 70:1261–66. (10)

Haukisalmi, V. 1986. Frequency distributions of helminths in microtine rodents in Finnish Lapland. *Ann. Zool. Fennici* 23:141–50. (10)

Haury, L. R., J. A. McGowan, and P. H. Wiebe. 1978. Patterns and processes in the time-space scales of plankton distribution. In *Spatial patterns in plankton communities,* ed. J. H. Steele, 277–327. New York: Plenum Press. (19)

Haverty, M. I., and W. L. Nutting. 1975. Density, dispersion, and composition of desert termite foraging populations and their relationship to superficial dead wood. *Environ. Entomol.* 4:480–86. (14)

Hawkins, B. A., and S. G. Compton. 1992. African fig wasp communities: Vacant niches and latitudinal gradients in species richness. *J. Anim. Ecol.* 61:361–72. (22)

Hawkins, B. A., and J. A. Lawton. 1987. Species richness for parasitoids of British phytophagous insects. *Nature* 326:788–90. (6)

Hay, M. 1985. Spatial patterns of herbivore impact and their importance in maintaining algal species richness. *Proc. Vth Int. Coral Reef Congr.* (Tahiti) 4:29–34. (5)

Hay, W. W. 1988. Paleoceanography. A review for the GSA Centennial. *Bull. Geol. Soc. Am.* 100:1934–56. (27)

Hays, J. D., J. Imbrie, and N. J. Shackleton. 1976. Variations in the earth's orbit: Pacemaker of the ice ages. *Science* 194:1121–32. (29)

Hayward, T. L. 1980. Spatial and temporal feeding patterns of copepods from the North Pacific central gyre. *Mar. Biol.* 58:295–309. (19)

Hayward, T. L., and J. A. McGowan. 1985. Spatial patterns of chlorophyll, primary production, macrozooplankton biomass, and physical structure in the central North Pacific ocean. *J. Plankton Res.* 7:147–67. (19)

Hayward, T. L., E. L. Venrick, and J. A. McGowan. 1983. Environmental heterogeneity and plankton community structure in the central North Pacific. *J. Mar. Res.* 41:711–29. (19)

Heads, P. A. 1986. Bracken, ants and extrafloral nectaries. IV. Do wood ants (*Formica lugubris*) protect the plant against insect herbivores? *J. Anim. Ecol.* 55:795–809. (16)

Heads, P. A., and J. H. Lawton. 1984. Bracken, ants and extrafloral nectaries. II. The effect of ants on the insect herbivores of bracken. *J. Anim. Ecol.* 53:1015–31. (16)

Heads, P. A., and J. H. Lawton. 1985. Bracken, ants and extrafloral nectaries. III. How insect herbivores avoid ant predation. *Ecol. Entomol.* 10:29–42. (16)

Heck, K. L. 1979. Some determinants of the composition and abundance of motile macroinvertebrate species in tropical and temperate turtle grass (*Thalassia testudinum*) meadows. *J. Biogeogr.* 6:183–97. (22)

Heckel, P. H. 1974. Carbonate buildups in the geologic record: A review. *Soc. Econ. Paleont. Mineral. Spec. Publ.* 18, 90–154. (27)

Hedgecock, D. 1986. Is gene flow from pelagic larval dispersal important in the adaptation and evolution of marine invertebrates? *Bull. Mar. Sci.* 39:550–564. (4)

Heer, O. 1865. *Die Urwelt der Schweiz.* Zürich: Schulthess. (23)

Heggberget, T. M. 1987. Number and proportion of southern bird species in Norway in relation to latitude, spring temperature and respiration equivalent. *Holarctic Ecol.* 10:81–89. (6)

Heim de Balsac, H. 1936. *Biogéographie des Mammifères et des Oiseaux de l'Afrique du Nord.* Paris: Presses Universitaires de France. (12)

Held, S. O. 1989. Colonization cycles on Cyprus. I: The Biogeographic and Paleontological Foundations of Early Prehistoric Settlement. *Report of the Department of Antiquities, Cyprus* (Nicosia), 7–28. (12)

Helvey, J. D., and J. H. Patric. 1965. Canopy and litter interception by hardwood of eastern United States. *Water Resources Res.* 1:193–206. (2)

Henderson, R. W. 1982. Trophic relationships and foraging strategies of some New World tree snakes (*Leptophis, Oxybelis, Uromacer*). *Amphibia-Reptilia* 3:71–80. (25)

Henderson, R. W. 1984. *Scaphiodontophis* (Serpentes: Colubridae): Natural history and test of a mimicry-related hypothesis. *Univ. Kansas Mus. Nat. Hist. Spec. Publ.* no. 10, 185–94. (25)

Henderson, R. W., and M. H. Binder. 1980. The ecology and behavior of vine snakes (*Ahaetulla, Oxybelis, Thelotornis, Uromacer*): A review. *Milwaukee Publ. Mus. Contrib. Biol. Geol.* no. 37, 1–38. (25)

Henderson, R. W., and B. I. Crother. 1989. Biogeographic patterns of predation in West Indian colubrid snakes. In *Biogeography of the West Indies: Past, present, and future,* ed. C. A. Woods, 479–518. Gainesville, Fla.: Sandhill Crane Press. (25)

Henderson, R. W., J. R. Dixon, and P. Soini. 1979. Resource partitioning in Amazonian snake communities. *Milwaukee Publ. Mus. Contrib. Biol. Geol.* no. 22, 1–11. (25)

Henderson, R. W., and L. G. Hoevers. 1977. The seasonal incidence of snakes at a locality in northern Belize. *Copeia* 1977:349–55. (25)

Herbold, B., and P. B. Moyle. 1986. Introduced species and vacant niches. *Am. Nat.* 128:751–60. (29)

Hespenheide, H. A. 1973. Ecological inferences from morphological data. *Annu. Rev. Ecol. Syst.* 4:213–29. (1, 30)

Hessler, R. R., and P. A. Jumars. 1974. Abyssal community analysis from replicate box cores in the central North Pacific. *Deep Sea Res.* 21:185–209. (5)

Hey, J. 1992. Using phylogenetic trees to study speciation and extinction. *Evolution* 46:627–40. (30)

Hickey, L. J., and R. W. Hodges. 1975. Lepidopteran leaf mine from the early Eocene Wind River formation of northwestern Wyoming. *Science* 189:718–20. (23)

Hilborn, R. 1975. The effect of spatial heterogeneity on the persistence of predator-prey interactions. *Theor. Pop. Biol.* 8:346–55. (8)

Hilborn, R., and S. C. Stearns. 1982. On inference in ecology and evolutionary biology: The problem of multiple causes. *Acta Biotheoretica* 31:145–64. (18)

Hilden, O. 1965. Habitat selection in birds. *Ann. Zool. Fennici* 2:53–75. (12)

Hill, J. E., and J. D. Smith. 1984. *Bats: A natural history.* London: British Museum (Natural History). (24)

Hochberg, M. E., and J. H. Lawton. 1990. Spatial heterogeneities in parasitism and population dynamics. *Oikos* 59:9–14. (7)

Hoffman, G. L. 1967. *Parasites of North American freshwater fishes.* Berkeley: University of California Press. (17)

Holdridge, L., W. Grenke, W. Hatheway, T. Liang, and J. Tosi, Jr. 1971. *Forest environments in tropical life zones: A pilot study.* Oxford: Pergamon Press. (2)

Hölldobler, B., and E. O. Wilson. 1990. *The ants.* Cambridge, Mass.: Belknap Press of Harvard University Press. (14)

Holliday, N. J. 1977. Population ecology of winter moth (*Operopht-*

era brumata) on apple in relation to larval dispersal and time of bud burst. *J. Appl. Ecol.* 4:803–14. (10)

Holling, C. S. 1973. Resilience and stability of ecological systems. *Annu. Rev. Ecol. Syst.* 4:1–24. (19)

Hollis, D. 1987. A review of the Malvales-feeding psyllid family Carsidaridae (Homoptera). *Bull. Brit. Mus. Nat. Hist.* (Entomology) 56:87–127. (23)

Holloway, J. D., and P. D. N. Hebert. 1979. Ecological and taxonomic trends in macrolepidopteran host selection. *Biol. J. Linn. Soc.* 11:229–51. (23)

Holman, E. W. 1989. Some evolutionary correlates of higher taxa. *Paleobiology* 15:357–63. (29)

Holmes, J. C. 1961. Effects of concurrent infections on *Hymenolepis diminuta* (Cestoda) and *Moniliformis dubius* (Acanthocephala). I. General effects and comparison with crowding. *J. Parasitol.* 47:209–16. (17)

Holmes, J. C. 1962. Effects of concurrent infections on *Hymenolepis diminuta* (Cestoda) and *Moniliformis dubius* (Acanthocephala). II. Effects on growth. *J. Parasitol.* 48:209–16. (17)

Holmes, J. C. 1973. Site selection by parasitic helminths: Interspecific interactions, site segregation, and their importance to the development of helminth communities. *Can. J. Zool.* 51:333–47. (17)

Holmes, J. C., and P. W. Price. 1986. Communities of parasites. In *Community ecology: Patterns and processes*, ed. J. Kikkawa and D. J. Anderson, 187–213. Melbourne: Blackwell Scientific Publications. (17)

Holt, R. D. 1977. Predation, apparent competition and the structure of prey communities. *Theor. Pop. Biol.* 12:197–229. (5)

Holt, R. D. 1984. Spatial heterogeneity, indirect interactions, and the coexistence of prey species. *Am. Nat.* 124:377–406. (5, 7, 8)

Holt, R. D. 1985. Population dynamics in two-patch environments: Some anomalous consequences of an optimal habitat distribution. *Theor. Pop. Biol.* 28:181–208. (7, 8)

Holt, R. D. 1987a. Population dynamics and evolutionary processes: The manifold roles of habitat selection. *Evol. Ecol.* 1:331–47. (7)

Holt, R. D. 1987b. Prey communities in patchy environments. *Oikos* 50:276–91. (7)

Holt, R. D. 1992. A neglected facet of island biogeography: The role of internal spatial dynamics in area effects. *Theor. Pop. Biol.* 41:354–71. (7)

Holt, R. D., and M. S. Gaines. 1992. The analysis of adaptation in heterogeneous landscapes: Implications for the evolution of fundamental niches. *Evol. Ecol.* 6:433–47. (7)

Holt, R. D., J. Grover, and D. Tilman, in press. Simple rules for interspecific dominance in systems with exploitative and apparent competition. *Am. Nat.*

Holt, R. D., and M. P. Hassell. 1993. Environmental heterogeneity and the stability of host-parasitoid interactions. *J. Anim. Ecol.* 62:89–100. (7)

Holt, R. D., and B. P. Kotler. 1987. Short-term apparent competition. *Am. Nat.* 130:412–30. (7)

Hoogmoed, M. S. 1980. Revision of the genus *Atractus* in Surinam, with the resurrection of two species (Colubridae, Reptilia). Notes on the herpetofauna of Surinam VII. *Zoologische Verhandelingen* 175:1–47. (25)

Hori, M. 1982. The biology and population dynamics of the tiger beetle, *Cicindela japonica* (Thunberg). *Physiol. Ecol. Japan* 19:77–212. (18)

Horn, H. S. 1971. *The adaptive geometry of trees.* Princeton, N.J.: Princeton University Press. (2)

Horn, H. S., and R. H. MacArthur. 1972. Competition among fugitive species in a harlequin environment. *Ecology* 53:749–52. (2, 8, 22)

Horrell, M. A. 1990. A global-scale reconstruction of late Cretaceous (Maestrichtian) climate. In *Paleographic atlas project, First annual meeting*, 28–33. University of Chicago. (20)

Horrell, M. A. 1991. Phytogeography and paleoclimatic interpretation of the Maestrichtian. *Palaeogeogr. Palaeoclimatol. Palaeoecol.* 86:87–138. (26)

Horton, D. R. 1972. Evolution in the genus *Egernia* (Lacertilia: Scincidae). *J. Herpetol.* 6:101–9. (14)

Hsu, S. B., S. P. Hubbell, and P. Waltman. 1977. A mathematical theory for single-nutrient competition in continuous cultures of microorganisms. *SIAM J. Appl. Math.* 32:366–83. (2)

Hu, H., and R. W. Chaney. 1940. *A Miocene flora from Shantung Province, China.* Carnegie Inst. Wash. Publ. no. 507, 1–147. (26)

Hubbell, S. P. 1979. Tree dispersion, abundance, and diversity in a tropical dry forest. *Science* 203:1299–1309. (1, 2, 19, 30)

Hubbell, S. P. 1980. Seed predation and the coexistence of tree species in tropical forests. *Oikos* 35:214–29. (2)

Hubbell, S. P., and R. B. Foster. 1983. Diversity of canopy trees in a neotropical forest and implications for conservation. In *Tropical rain forest: Ecology and management*, ed. S. Sutton, T. C. Whitmore, and A. Chadwick, 25–41. Oxford: Blackwell. (2)

Hubbell, S. P., and R. B. Foster. 1986. Biology, chance, and history and the structure of tropical rain forest tree communities. In *Community ecology*, ed. J. Diamond and T. J. Case, 314–29. New York: Harper and Row. (2, 22, 25, 26)

Huffaker, C. B. 1958. Experimental studies on predation: Dispersion factors and predator-prey oscillations. *Hilgardia* 27:343–83. (8, 10, 30)

Hughes, T. P. 1984. Population dynamics based on individual size rather than age: A general model with a reef coral example. *Am. Nat.* 123:778–95. (4)

Hughes, T. P. 1989. Community structure and diversity of coral reefs: The role of history. *Ecology* 70:275–79. (18, 24)

Hulse, A. C., R. D. Sage, and W. F. Blair. 1977. Reptiles. In *Convergent evolution in warm deserts*, ed. G. H. Orians and O. T. Solbrig, 133–37. Stroudsburg, Penn.: Dowden, Hutchinson and Ross. (14)

Humboldt, A., and A. Bonpland. 1807. *Essai sur la géographie des plantes accompagné d'un tableau physique des régions équinoxiales.* Paris: Schoell; reprint New York: Arno Press, 1977. (26)

Humphries, C. J., J. M. Cox, and E. S. Nielsen. 1986. *Nothofagus* and its parasites: A cladistic approach to coevolution. In *Coevolution and systematics*, ed. A. R. Stone and D. L. Hawksworth, 53–76. Systematics Association Special Vol. 32. Oxford: Clarendon Press. (23)

Humphries, C. J., and L. Parenti. 1986. *Cladistic biogeography.* London: Academic Press. (24)

Hunt, J. H. 1977. Ants. In *Convergent evolution in warm deserts*, ed. G. H. Orians and O. T. Solbrig, 147–52. Stroudsburg, Penn.: Dowden, Hutchinson and Ross. (14)

Huntley, B., and H. J. B. Birks. 1983. *An atlas of past and present pollen maps for Europe: 0–13,000 years ago.* Cambridge: Cambridge University Press. (12)

Huntley, B., and T. Webb III, eds. 1988. *Handbook of vegetation science.* Vol. 7, *Vegetation history.* Dordrecht: Kluwer Academic. (29)

Huntley, B., and T. Webb III. 1989. Migration: Species' response to climatic variations caused by changes in the Earth's orbit. *J. Biogeogr.* 16:5–19. (29)

Hurlbert, S. H. 1971. The nonconcept of species diversity: A critique and alternative parameters. *Ecology* 52:577–86. (1)

Hurlbert, S. H. 1984. Pseudoreplication and the design of ecological field experiments. *Ecol. Monog.* 54:187–211. (4)

Hurst, J. M. 1976. The depths inhabited by Silurian brachiopod communities: Comment and reply. *Geology* 4:709–10. (5)

Huston, M. 1979. A general hypothesis of species diversity. *Am. Nat.* 113:81–101. (1, 5, 6, 22, 26)

Huston, M. 1980. Soil nutrients and tree species richness in Costa Rican forests. *J. Biogeogr.* 7:147–57. (2)

Hutchings, P., and P. Saenger. 1987. *Ecology of mangroves*, St. Lucia, Australia: University of Queensland Press. (20)

Hutchinson, G. E. 1957. Concluding remarks. *Cold Spring Harbor Symp. Quant. Biol.* 22:415–27. (1)

Hutchinson, G. E. 1959. Homage to Santa Rosalia, or why are there so many kinds of animals? *Am. Nat.* 93:145–59. (1, 2, 3, 6, 22, 30)

Hutchinson, G. E. 1978. *An introduction to population ecology.* New Haven: Yale University Press. (27)

Illies, J. 1974. *Introduction to zoogeography.* London: Macmillan. (12)

Inouye, R. S., N. J. Huntly, D. Tilman, J. R. Tester, M. Stilwell, and K. C. Zinnel. 1987. Old-field succession on a Minnesota sand plain. *Ecology* 68:12–26. (5)

Issacs, J. D. 1972. Unstructured marine food webs and "pollutant analogues." *Fish. Bull.* 70:1053–59. (3)

Issacs, J. D. 1973. Potential trophic biomasses and trace-substance concentrations in unstructured marine food webs. *Mar. Biol.* 22:97–104. (3)

Issacs, J. D. 1976. Reproductive products in marine food webs. *Bull. South. Calif. Acad. Sci.* 75:220–23. (3)

Ives, A. R. 1991. Aggregation and coexistence in a carrion fly community. *Ecol. Monogr.* 61:75–94. (30)

Ives, A. R., and R. M. May. 1985. Competition within and between species in a patchy environment: Relations between microscopic and macroscopic models. *J. Theor. Biol.* 115:65–92. (8)

Iwasa, Y., and J. Roughgarden. 1986. Interspecific competition among metapopulations with space-limited subpopulations. *Theor. Pop. Biol.* 30:194–214. (8)

Jablonski, D. 1986. Background and mass extinctions: The alternation of macroevolutionary regimes. *Science* 231:129–33. (28)

Jablonski, D. 1987. Heritability at the species level: Analysis of geographic ranges of Cretaceous mollusks. *Science* 238:360–63. (1, 23)

Jablonski, D., and D. J. Bottjer. 1990a. The ecology of evolutionary innovation: The fossil record. In *Evolutionary innovations,* ed. M. H. Nitecki, 253–88. Chicago: University Press. (1, 29)

Jablonski, D., and D. J. Bottjer. 1990b. Onshore-offshore trends in marine invertebrate evolution. In *Causes of evolution: A paleontological perspective,* ed. R. M. Ross and W. D. Allmon, 21–75. Chicago: University of Chicago Press. (29)

Jablonski, D., and D. J. Bottjer. 1991b. The origin and diversification of major groups: Environmental patterns and macroevolutionary lags. In *Major evolutionary radiations,* ed. P. D. Taylor and G. P. Larwood, 17–57. Systematics Association Special Vol. 43. Oxford: Clarendon Press. (1, 29)

Jablonski, D., and D. J. Bottjer. 1991a. Environmental patterns in the origins of higher taxa: The post-Paleozoic fossil record. *Science* 252:1831–33. (30)

Jablonski, D., J. J. Sepkowski, Jr., D. J. Bottjer, and P. M. Sheehan. 1983. Onshore-offsore patterns in the evolution of Phanerozoic shelf communities. *Science* 222:1123–25. (30)

Jackson, D. A., and H. H. Harvey. 1989. Biogeographic associations in fish assemblages: Local vs. regional processes. *Ecology* 70:1472–84. (6, 18)

Jackson, D. J. 1956. The capacity of flight in certain water beetles and its bearing on their origin in the western Scottich isles. *Proc. Linn. Soc.* Lond. 167:76–96. (10)

Jackson, J. B. C. 1986. Modes of dispersal of clonal benthic invertebrates: Consequences for species' distributions and genetic structure of local populations. *Bull. Mar. Sci.* 39:588–606. (4)

Jackson, J. B. C., and F. K. McKinney. 1990. Ecological processes and progressive macroevolution of marine clonal benthos. In *Causes of evolution: A paleontological perspective,* ed. R. M. Ross and W. D. Allmon, 173–209. Chicago: University of Chicago Press. (29)

Jacobson, G. L., Jr., T. Webb III, and E. V. Grimm. 1987. Patterns and rates of vegetation change during the deglaciation of eastern North America. In *North America and adjacent oceans during the last deglaciation,* ed. W. F. Ruddiman and H. E. Wright, Jr., 227–88. Boulder, Colo.: Geological Society of America. (29)

Jaeger, J. -J. 1975. The mammalian faunas and hominid fossils of the Middle Pleistocene of the Maghreb. In *After the Australopithecines,* ed. Butzer and Isaac. The Hague: Mouton. (12)

Jaeger, J. -J, B. Coiffait, H. Tong, and C. Denys. 1987. Rodent extinctions following Messinian faunal exchanges between Western Europe and Northern Africa. *Mem. Soc. Geol. France,* n.s. 150:153–58. (12)

Jaenike, J. 1990. Host specialization in phytophagous insects. *Ann. Rev. Ecol. Syst.* 21:243–74. (23)

Jahn, G. 1991. Temperate deciduous forests of Europe. In *Ecosystems of the World 7: Temperate Deciduous Forests,* ed. E. Röhrig and B. Ulrich, 377–502. Amsterdam: Elsevier. (26)

Jaksic, F. M. 1981. Abuse and misuse of the term "guild" in ecological studies. *Oikos* 37:397–400. (4)

Jalas, J. (ed.). 1958. *Suuri Kasvikirja* 1. Helsinki: Otava. (10)

Jalas, J. (ed.). 1965. *Suuri Kasvikirja* 2. Helsinki: Otava. (10)

Jalas, J. (ed.). 1980. *Suuri Kasvikirja* 3. Helsinki: Otava. (10)

James, H. F., T. W. Stafford, D. W. Steadman, S. L. Olson, P. S. Martin, A. J. T. Jull, and P. C. McCoy. 1987. Radiocarbon dates on bones of extinct birds from Hawaii. *Proc. Natl. Acad. Sci. USA* 84:2350–54. (12)

Janis, C. M. 1982. Evolution of horns in ungulates: Ecology and paleoecology. Biol. Rev. 57:261–318. (28)

Janis, C. M. 1984. The significance of fossil ungulate communities as indicators of vegetation structure and climate. In *Fossils and climate,* ed. P. J. Brenchley, 85–104. New York: John Wiley and Sons. (28)

Janis, C. M. 1988. An estimation of tooth volume and hypsodonty indices in ungulate mammals and the correlation of these factors with dietary preferences. In *Teeth Revisited, Proceedings of the VIIIth International Symposium on Dental Morphology* (Paris, 1986), ed. D. E. Russell, J. -P Santoro, and D. Signeau-Russell, 367–87. Memoirs de Musée d'Histoire naturelle, Paris (sene C) 53. (28)

Janis, C. M. 1989. A climatic explanation for patterns of evolutionary diversity in ungulate mammals. *Palaeontology* 32:463–81. (28)

Janis, C. M. 1990. Correlation of cranial and dental variables with dietary preferences: A comparison of macropodoid and ungulate mammals. In *Proceedings of the De Vis Symposium,* ed. S. Turner, R. A. Thulborn, and R. E. Molnar, 349–66. Memoirs of the Queensland Museum 28. (28)

Janis, C. M. 1993. Do legs support the arms race hypothesis in mammalian predator/prey relationships? In *Vertebrate behavior derived from the fossil record,* ed. J. R. Horner and L. Ellis. New York: Columbia University Press. In press. (28)

Janis, C. M., and J. Damuth. 1990. Mammals. In *Evolutionary trends,* ed. K. J. McNamara, 301–45. London: Belhaven Press. (28)

Janzen, D. H. 1967. Why mountain passes are higher in the tropics. *Am. Nat.* 101:233–49. (5, 23, 30)

Janzen, D. H. 1970. Herbivores and the number of tree species in tropical forests. *Am. Nat.* 104:501–28. (1, 2, 22)

Janzen, D. H. 1977. Why are there so many species of insects? *Proc. XV Int. Cong. Entomol.* (Washington, D.C.), 84–94 (2)

Janzen, D. H. 1980. When is it coevolution? *Evolution* 34:611–12. (23)

Janzen, D. H. 1981. The peak in North American ichneumonid species richness lies between 38° and 42° N. *Ecology* 62:532–37. (1, 23)

Janzen, D. H. 1984. Two ways to be a tropical big moth: Santa Rosa saturniids and sphingids. *Ox. Sur. Evol. Biol.* 1:85–140. (23)

Jarman, M. R., and H. N. Jarman. 1968. The fauna and economy of Early Neolithic Knossos. *Annual of the British School at Athens.* 63:241–76. (12)

Jeffries, M. J., and J. H. Lawton. 1984. Enemy-free space and the structure of ecological communities. *Biol. J. Linn. Soc.* 23:269–86. (5)

Jenkyns, H. C. 1980. Cretaceous anoxic events—from continents to oceans. *J. Geol. Soc. Lond.* 137:171–88. (27)

Jermy, T. 1984. Evolution of insect/hostplant relationships. *Am. Nat.* 124:609–30. (23)

Jiménez, J. A. 1984. A hypothesis to explain the reduced distribution of the mangrove *Pelliciera rhizophorae* Tr. & Pl. *Biotropica* 16:304–8. (20)

Jitts, H. R., C. D. McAllister, K. Stephens, and J. D. Strickland. 1964. The cell division rates of some marine phytoplankters as a function of light and temperature. *J. Fish. Res. Board of Can.* 21:139–57. (2)

Johnson, C. C., and E. G. Kauffman. 1990. Originations, radiations and extinctions of Cretaceous rudistid bivalve species in the Caribbean Province. In *Extinction events in Earth history,* ed. E. G. Kauffman and O. H. Walliser, 307–24. Lecture Notes in Earth Sciences 30. Berlin: Springer-Verlag. (27)

Johnson, D. W. 1986. Desert buttes: Natural experiments for testing theories of island biogeography. *Natl. Geogr. Res.* 2:152–66. (6)

Johnson, J. G., and E. C. Potter. 1976. The depths inhabited by Silurian brachiopod communities: Comment and reply. *Geology* 4:189–91. (5)

Johnson, M. P., L. G. Mason, and P. H. Raven. 1968. Ecological parameters and plant species diversity. *Am. Nat.* 102:297–306. (6)

Jokiel, P., and F. J. Martinelli. 1992. The vortex model of coral reef biogeography. *J. Biogeogr.* 19:449–58. (20)

Jolivet, P. 1988. Food habits and food selection of Chrysomelidae:

Bionomic and evolutionary perspectives. In *Biology of Chryso-melidae,* ed. P. Jolivet, E. Petitpierre, and T. H. Hsiao, 1–20. Dordrecht: Kluwer. (23)

Jones, C. G., and R. D. Firn. 1978. The role of phytoecdysteroids in bracken fern, *Pteridium aquilinum* (L.) Kuhn, as a defense against phytophagous insect attack. *J. Chem. Ecol.* 4:117–38. (16)

Jones, C. G., and R. D. Firn. 1979a. Resistance of *Pteridium aquilinum* to attack by non-adapted phytophagous insects. *Biochem. Syst. Ecol.* 7:95–101. (16)

Jones, C. G., and R. D. Firn. 1979b. Some allelochemicals of *Pteridium aquilinum* and their involvement in resistance to *Pieris brassicae* (Lep., Pieridae). *Biochem. Syst. Ecol.* 7:197–92. (16)

Jones, C. G., and J. H. Lawton. 1991. Plant chemistry and insect species richness of British umbellifers. *J. Anim. Ecol.* 60:767–77. (16)

Jones, N. S., and H. L. Sanders. 1972. Distribution of Cumacea in the deep Atlantic. *Deep Sea Res.* 19:737–45. (5)

Jordan, C. F. 1983. Productivity of tropical forest ecosystems and the implications for their use as future wood and energy sources. In *Ecosystems of the world.* Vol. 14A, *Tropical rain forest ecosystems: Structure and function,* ed. F. B. Golley, 117–36. Amsterdam: Elsevier Scientific. (5)

Jordan, E. K. 1926. Molluscan fauna of the Pleistocene of San Quíntin Bay, Lower California. *Proc. Calif. Acad. Sci.,* ser. 4, 15:241–55. (29)

Judd, W. S., R. W. Sanders, and M. J. Donoghue. 1993. Angiosperm family pairs: Preliminary phylogenetic analyses. *Ann. Mo. Bot. Gard.* In press. (23)

Juncosa, A. M., and P. B. Tomlinson. 1988a. A historical and taxonomic synopsis of Rhizophoraceae and Anisophylleaceae. *Ann. Mo. Bot. Gard.* 75:1278–95. (20)

Juncosa, A. M., and P. B. Tomlinson. 1988b. Systematic comparison and some biological characteristics of Rhizophoraceae and Anisophylleaceae. *Ann. Mo. Bot. Gard.* 75:1296–1318. (20)

Kadmon, R., and A. Shmida. 1990. Spatiotemporal demographic processes in plant populations: An approach and a case study. *Am. Nat.* 135:382–97. (7)

Kappelman, J. 1988. Morphology and locomotor adaptations of the bovid femur in relation to habitat. *J. Morphol.* 198:119–30. (28)

Karban, R. 1989. Community organization of *Erigeron glaucus* folivores: Effects of competition, predation, and host plant. *Ecology* 70:1028–39. (22, 24)

Karban, R., and R. E. Ricklefs. 1983. Host characteristics, sampling intensity, and species richness of Lepidoptera larvae on broad-leaved trees in southern Ontario. *Ecology* 64:636–41. (1)

Kareiva, P. 1983. Influence of vegetation texture on herbivore populations: resource concentration and herbivore movement. In *Variable plants and herbivores in natural and managed systems,* ed. R. F. Denno and M. S. McClure, 259–86. New York: Academic Press. (16)

Karlin, S., and J. McGregor. 1972. Polymorphisms for genetic and ecological systems with weak coupling. *Theor. Pop. Biol.* 3:210–38. (8)

Karr, J. R. 1968. Habitat and avian diversity on strip-mined land in east-central Illinois. *Condor* 70:348–57. (30)

Karr, J. R. 1971. Structure of avian communities in selected Panama and Illinois habitats. *Ecol. Monogr.* 41:207–33. (30)

Karr, J. R. 1976. Within- and between-habitat avian diversity in African and neotropical lowland habitats. *Ecol. Monogr.* 46:457–81. (1, 21)

Karr, J. R., and F. C. James. 1975. Ecomorphological configurations and convergent evolution in species and communities. In *Ecology and evolution of communities,* ed. M. L. Cody and J. M. Diamond, 258–91. Cambridge, Mass.: Harvard University Press. (1, 25, 30)

Kartesz, J. R., and R. Kartesz. 1980. *A synonymized checklist of the vascular flora of the United States, Canada and Greenland.* Chapel Hill, N.C.: University of North Carolina Press. (26)

Kauffman, E. G. 1973. Cretaceous Bivalvia. In *Atlas of palaeobiogeography,* ed. A. Hallam, 353–83. Amsterdam: Elsevier. (27)

Kauffman, E. G. 1979. The ecology and biogeography of the Cretaceous-Tertiary extinction event. In *Cretaceous-Tertiary boundary events; a symposium.* II. *Proceedings,* ed. W. K. Christensen and T. Birkelund, 29–37. University of Copenhagen. (27)

Kauffman, E. G. 1984. The fabric of Cretaceous mass extinctions. In *Catastrophes and Earth history-the new uniformitarianism,* ed. W. A. Berggren and J. Van Couvering, 151–246. Princeton, N.J.: Princeton University Press. (27)

Kauffman, E. G. 1988. The dynamics of marine stepwise mass extinction. In *Paleontology and evolution: Extinction events—III,* ed. M. A. Lamolda, E. G. Kauffman, and O. H. Walliser, 57–71. Journadas de Paleontologia; 2nd International Conference on Global Bioevents. Revista Espanola Paleontologia, No. Extraordinario. (27)

Kauffman, E. G., and C. C. Johnson. 1988. The morphological and ecological evolution of middle and Upper Cretaceous reef-building rudistids. *Palaios* 3:194–216. (27)

Kauffman, E. G., and N. F. Sohl. 1974. Structure and evolution of Antillean Cretaceous rudist frameworks. *Verhandlungen Naturforschung Gesellschaft, Basel* 84 1:399–467. (27)

Kauffman, E. G., and N. F. Sohl. 1978. Rudists. In *Encyclopedia of paleontology,* ed. R. W. Fairbridge and D. Jablonski, 723–36. Stroudsburg, Penn.: Dowden, Hutchinson, and Ross. (27)

Keast, A. 1961. Bird speciation on the Australian continent. *Bull. Mus. Comp. Zool.* 123:305–495. (12, 13)

Keast, A. 1981a. Ecological biogeography of Australia. The Hague: Junk. (14)

Keast, A. 1981b. The evolutionary biogeography of Australian birds. In *Ecological biogeography of Australia,* ed. A. Keast, 1587–1635. The Hague: Junk. (13, 14)

Keast, A., H. F. Recher, H. Ford, and D. Saunders. 1985. *Birds of eucalypt forests and woodlands.* Chipping Norton, SA.: R.A.U.; Surrey Beatty and Sons. (13)

Keddy, P. A. 1981. Experimental demography of the sand-dune annual, *Cakile edentula,* growing along an environmental gradient in Nova Scotia. *Ecology* 69:615–30. (7)

Keddy, P. A. 1990. Competitive hierarchies and centrifugal organization in plant communities. In *Perspectives in Plant Competition,* ed. J. Grace and D. Tilman, 265–90. New York: Academic Press. (5)

Keeling, C. D., R. B. Bacastow, A. F. Carter, S. C. Piper, T. P. Worf, M. Heimann, W. G. Mook, and H. Roeloffzen. 1989. A three dimensional model of atmospheric CO_2 transport based on observed winds. I. Analysis of observational data. In *Aspects of climate variability in the Pacific and Western Americas,* ed. D. E. Peterson, 165–236. Geophysical Monograph no. 55. American Geophysical Union. (19)

Keigwin, L. D. 1978. Pliocene closing of the Isthmus of Panama, based on biostratigraphic evidence from nearby Pacific Ocean and Caribbean Sea cores. *Geology* 6:630–34. (20)

Keigwin, L. D. 1980. Palaeoceanographic change in the Pacific at the Eocene-Oligocene boundary. *Nature* 287:722–25. (20)

Kelly, J., C. Tripler, and S. Pacala. 1993. Spatial structure and the measurement of interspecific competition in plant communities. *Trends Ecol. Evol.* In Press. (2)

Kendall, D. G. 1948. On the generalized "birth-death" process. *Ann. Math. Stat.* 19:1–15. (11)

Kennedy, C. E. J., and T. R. E. Southwood. 1984. The number of species of insects associated with British trees: A reanalysis. *J. Anim. Ecol.* 53:455–78. (17, 22, 23)

Kennedy, C. R. 1978a. An analysis of the metazoan parasito-coenoses of brown trout *Salmo trutta* from British lakes. *J. Fish Biol.* 13:255–63. (17)

Kennedy, C. R. 1978b. The parasite fauna of resident charr *Salvelinus alpinus* from Arctic islands, with special reference to Bear Island. *J. Fish Biol.* 13:457–66. (17)

Kennedy, C. R., A. O. Bush, and J. M. Aho. 1986. Patterns in helminth communities: Why are birds and fish different? *Parasitology* 93:205–15. (17)

Kenny, D., and C. Loehle. 1991. Are food webs randomly connected? *Ecology* 72:1794–99. (3)

Keough, M. J. 1984. Effects of patch size on the abundance of sessile marine invertebrates. *Ecology* 65:423–437. (1)

Kerley, G. I. H. 1991. Seed removal by rodents, birds and ants in the semi-arid Karoo, South Africa. *J. Arid Environ.* 20:63–69. (14)

Kershaw, A. P. 1988. Australasia. In *Vegetation history,* ed. B. Huntley and T. Webb III, 237–306. Dordrecht: Kluwer Academic. (15)

Kiester, A. R., R. Lande, and D. W. Schemske 1984. Models of coevolution and speciation in plants and their pollinators. *Am. Nat.* 124:220–43. (29)

Kikkawa, J. 1968. Ecological association of birds and habitats in eastern Australia. *J. Anim. Ecol.* 37:143–65. (13)

Kikkawa, J. 1974. Comparison of avian communities between wet and semi-arid habitats of eastern Australia. *Aust. Wild. Res.* 1:107–16. (13)

Kikkawa, J. 1982. Ecological association of birds and vegetation structure in wet tropical forests of Australia. *Aust. J. Ecol.* 7:325–45. (13)

Kikkawa, J., and D. J. Anderson, eds. 1986. *Community ecology: Pattern and process.* Melbourne: Blackwell Scientific Publications. (1, 17)

Kikkawa, J., and K. Pearse. 1969. Geographical distribution of land birds in Australia—a numerical analysis. *Aust. J. Zool.* 17:821–40. (13)

Kikkawa, J., L. J. Webb, M. B. Dale, G. B. Monteith, J. G. Tracey, and W. T. Williams. 1981. Gradients and boundaries of monsoon forests in Australia. *Proc. Ecol. Soc. Aust.* 11:39–52. (13)

Kilpper, K. 1969. *Verzeichnis der im mittleren und unteren Rheinland gefundenen Grossreste von Tertiärpflanzen (von 1821–1968).* Essen: Ruhrland- und Heimatmuseum der Stadt Essen. (26)

Kiltie, R. A. 1984. Seasonality, gestation time, and large mammal extinctions. In *Quaternary extinctions,* ed. P. S. Martin and R. G. Klein, 299–314. Tucson: University of Arizona Press. (28)

King, P. B. 1977. *The evolution of North America.* Princeton, N.J.: Princeton University Press. (28)

Kingsford, M. J., and Choat, J. H. 1986. The fauna associated with drift algae captured with a plankton-mesh purse seine net. *Limnol. Oceanogr.* 30:618–30. (4)

Kingsland, S. E. 1985. *Modeling Nature: Episodes in the history of population ecology.* Chicago: University of Chicago Press. (1, 30)

Kirchner, T. 1977. The effects of resource enrichment on the diversity of plants and arthropods in a shortgrass prairie. *Ecology* 58:1334–44. (5)

Kirk, A. A. 1982. Insects associated with bracken fern *Pteridium aquilinum* (Polypodiaceae) in Papua New Guinea and their possible use in biological control. *Acta Oecologica/Oecologia Applicata* 3:343–59. (16)

Kitching, J. A., and J. Lockwood. 1974. Observations on shell form and its ecological significance in thaisid gastropods of the genus *Lepsiella* in New Zealand. *Mar. Biol.* 28:131–44. (4)

Kitching, J. A., L. Muntz, and F. J. Ebling. 1966. The ecology of Lough Line: XV. The ecological significance of shell and body form in *Nucella. J. Anim. Ecol.* 35:113–26. (4)

Kitching, R. L., and S. L. Pimm. 1985. The length of food chains: Phytotelmata in Australia and elsewhere. *Proc. Ecol. Soc. Aust.* 14:123–40. (21)

Klebs, R. 1910. Über Bernsteinenschlusse im Allgeimeinen und die Coleopteren meiner Bernsteinsammlung. *Schrift. Physik.-Ökonom. Ges. Konigsberg* 51:217–42. (23)

Knoll, A. H. 1984. Patterns of extinction in the fossil record of vascular plants. In *Extinctions,* ed. M. H. Nitecki, 21–68. Chicago: University of Chicago Press. (29)

Knoll, A. H. 1986. Patterns of change in plant communities through geological time. In *Community ecology,* ed. J. Diamond and T. J. Case, 126–44. New York: Harper and Row. (29, 30)

Koch, C. F. 1987. Prediction of sample size effects on the measured temporal and geographic distribution patterns of species. *Paleobiology* 13:100–107. (27)

Koch, C. F., and N. F. Sohl. 1983. Preservational effects in paleoecological studies: Cretaceous mollusc examples. *Paleobiology* 9:26–34. (27)

Kochmer, J. P., and S. N. Handel. 1986. Constraints and competition in the evolution of flowering phenology. *Ecol. Monogr.* 56:303–25. (23)

Koehn, P. K., F. J. Turano, and J. B. Mitton. 1973. Population genetics of marine pelecypods. II. Genetic differences in microhabitats of *Madiolus dermissus. Evolution* 27:100–105. (4)

Kohn, A. J. 1980. *Conus kahiko,* a new Pleistocene gastropod from Oahu, Hawaii. *J. Paleontol.* 54:534–41. (29)

Korpimaki, E., and K. Norrdahl. 1991a. Do breeding nomadic avian predators dampen population fluctuations of small mammals? *Oikos* 62:195–208. (7)

Korpimaki, E., and K. Norrdahl. 1991b. Numerical and functional responses of Kestrels, Short-eared Owls, and Long-eared Owls to vole densities. *Ecology* 72:814–26. (7)

Kotler, B. P. 1985. Microhabitat utilization in desert rodents: A comparison of two methods of measurement. *J. Mammal.* 66:374–78. (5)

Kotler, B. P., and J. S. Brown. 1988. Environmental heterogeneity and the coexistence of desert rodents. *Annu. Rev. Ecol. Syst.* 19:281–307. (5)

Kouki, J., and U. Häyrinen. 1991. On the relationship between distribution and abundance of birds breeding on Finnish mires: The effect of habitat specialization. *Ornis Fennica* 68:170–77. (10)

Kozhov, M. 1963. Lake Baikal and its life. Monographs in Biology 11, 1–344. (30)

Kräusel, R. 1939. Die fossilen Floren Ägyptens. *Abh. beyer. Akad. Wiss., math.-naturwiss.* 47:1–140. (20)

Krebs, C. J. 1972. *Ecology: The experimental analysis of distribution and abundance.* New York: Harper and Row. (6, 10)

Kremer, J. N., and S. W. Nixon. 1978. *A coastal marine ecosystem.* Berlin: Springer-Verlag. (3)

Krishna, K., and F. M. Weesner. 1970. *Biology of termites, vol. II.* New York: Academic Press. (14)

Krombein, K. V., P. D. Hurd, Jr., and D. R. Smith. 1979. *Catalog of the Hymenoptera of North America.* Washington, D.C.: Smithsonian Institution Press. (22)

Kruskal, J. B. 1964. Multidimensional scaling by optimizing goodness of fit to a nonmetric hypothesis. *Psychometrika* 29:1–27. (13)

Krüssman, G. 1979. *Die Baüme Europas.* Berlin: Parey. (26)

Krüssman, G. 1984. Manual of cultivated broad-leaved trees and shrubs. 3 vols. Beaverton, Ore.: Timber Press. (26)

Krutzsch, W. 1989. Paleogeography and historical phytogeography (phytochronology) in the *Neophyticum. Plant Syst. Ecol.* 162:5–61. (20)

Kuris, A. M., A. R. Blaustein, and J. J. Alio. 1980. Hosts as islands. *Am. Nat.* 116:570–86. (17)

Kurtén, B. 1957. Mammal migrations, Cenozoic stratigraphy, and the age of Peking man and the Australopithecines. *J. Paleontol.* 31:215–27. (28)

Kurtén, B. 1968. *Pleistocene mammals of Europe.* London: Weidenfeld and Nicholson. (12)

Kvaček, Z., and H. Walther. 1987. Revision der mitteleuropäischen tertiären Fagaceen nach blattepidermalen Characteristiken. *Feddes Repertorium* 98:637–52. (26)

Kvaček, Z., H. Walther, and C. Bužek. 1989. Paleogene floras of W. Bohemia (C.S.S.R.) and the Wiesselster Basin (G.D.R.) and their correlation. *Casopis pro Mineralogii a Geologii* 34:385–402. (26)

Lack, D. 1944. Ecological aspects of species-formation in passerine birds. *Ibis* 1944:260–86. (1)

Lack, D. 1947. *Darwin's finches.* Cambridge: Cambridge University Press. (1)

Lack, D. 1965. Evolutionary ecology. *J. Ecol.* 53:237–45. (12)

Lake, P. S., L. A. Barmuta, A. J. Boulton, I. C. Campbell, and R. M. St. Clair. 1985. Australian streams and Northern Hemisphere stream ecology: Comparisons and problems. *Proc. Ecol. Soc. Aust.* 14:61–82. (21)

Łańcucka-Środoniowa, M. 1975. Hydrangea L. (Saxifragaceae) and *Schefflera* Forst. (Araliaceae) in the Tertiary of Poland. *Acta Palaeobotanica* 16:103–12. (26)

Lande, R. 1987. Extinction thresholds in demographic models of territorial populations. *Am. Nat.* 130:624–35. (7)

Langenheim, J. H. 1973. Leguminous resin-producing trees in Africa and South America. In *Tropical forest ecosystems in Africa and South America: A comparative review,* ed. B. J. Meggers, E. S. Ayensu, and W. D. Duckworth, 89–104. Washington, D.C.: Smithsonian Institution Press. (23)

Larsson, S. G. 1978. *Baltic amber: A palaeobiological study.* Entomonograph, vol. 1. Klampenborg, Denmark: Scandinavian Science Press. (23)

Latham, R. E., and R. E. Ricklefs. 1993. Global patterns of tree species richness in moist forests: Energy-diversity theory does not account for variation in species richness. *Oikos* 67. In press. (1, 6, 26, 30)

Lawton, J. H. 1976. The structure of the arthropod community on bracken. *Bot. J. Linn. Soc. Lond.* 73:187–216. (16)

Lawton, J. H. 1978. Host-plant influences on insect diversity: The effects of space and time. *Symp. Roy. Entomol. Soc. Lond.* 9:105–25. (16)

Lawton, J. H. 1982. Vacant niches and unsaturated communities: A comparison of bracken herbivores at sites on two continents. *J. Anim. Ecol.* 51:573–95. (16, 22)

Lawton, J. H. 1984a. Herbivore community organization: General models and specific tests with phytophagous insects. In *A new ecology: Novel approaches to interactive systems*, ed. P. W. Price, C. N. Slobodchikoff, and W. S. Gauld, 329–52. New York: John Wiley. (16)

Lawton, J. H. 1984b. Non-competitive populations, non-convergent communities, and vacant niches: The herbivores of bracken. In *Ecological communities: Conceptual issues and the evidence*, ed. D. R. Strong, D. S. Simberloff, L. G. Abele, and A. B. Thistle, 67–100. Princeton, N.J.: Princeton University Press. (16, 17, 22, 23)

Lawton, J. H. 1986. The effect of parasitoids on phytophagous insect communities. In *Insect parasitoids*, ed. J. Waage and D. Greathead, 265–87. *Symp. Roy. Entomol. Soc. Lond.* 13: (16)

Lawton, J. H. 1988. Biological control of bracken in Britain: Constraints and opportunities. *Phil. Trans. Roy. Soc. (Lond.)*, ser. B 318:335–55. (16)

Lawton, J. H. 1989. Food webs. In *Ecological concepts*, ed. J. M. Cherrett. Oxford: Blackwell Scientific Publications. (3)

Lawton, J. H. 1990a. Developments in the UK biological control programme for bracken. In *Bracken biology and management*, ed. J. S. Thomson and R. T. Smith, 309–14. New South Wales, Australia: RIAS. (16)

Lawton, J. H. 1990b. Local and regional species-richness of bracken-feeding insects. In *Bracken biology and management*, ed. J. A. Thomson and R. T. Smith, 197–202. Sydney: Australian Institute of Agricultural Science. (22)

Lawton, J. H. 1990c. Species richness and population dynamics of animal assemblages: Patterns in body size, abundance, space. *Phil. Trans. Roy. Soc. (Lond.)*, ser. B. 330:283–91. (22)

Lawton, J. H., and K. J. Gaston. 1989. Temporal patterns in the herbivorous insects of bracken: A test of community predictability. *J. Anim. Ecol.* 58:1021–34. (16, 22)

Lawton, J. H., and P. A. Heads. 1984. Bracken, ants, and extrafloral nectaries. I. The components of the system. *J. Anim. Ecol.* 53:995–1014. (16)

Lawton, J. H., M. MacGarvin, and P. A. Heads. 1987. Effects of altitude on the abundance and species richness of insect herbivores on bracken. *J. Anim. Ecol.* 56:147–60. (16)

Lawton, J. H., and S. L. Pimm. 1978. Population dynamics and the length of food chains. *Nature* 272:189–90. (3)

Lawton, J. H., and P. W. Price. 1979. Species richness of parasites on hosts: Agromyzid flies on the British Umbelliferae. *J. Anim. Ecol.* 48:619–37. (17, 23)

Lawton, J. H., V. K. Rashbrook, and S. G. Compton. 1988. Biocontrol of British bracken: The potential of two moths from Southern Africa. *Ann. Appl. Biol.* 112:479–90. (16)

Lawton, J. H., and D. Schroder. 1977. Effects of plant type, size of geographical range, and taxonomic isolation on number of insect species associated with British plants. *Nature* 265:137–40. (17)

Lawton, J. H., and D. R. Strong. 1981. Community patterns and competition in folivorous insects. *Am. Nat.* 118:317–38. (17, 22)

Lawton, J. H., and P. H. Warren. 1989. Reply to P. Yodzis. *Trends Ecol. Evol.* 4:50. (3)

Laxton, J. H., and W. J. Stablum. 1974. Sample design for quantitative estimation of sedentary organisms of coral reefs. *Biol. J. Linn. Soc.* 6:1–18. (27)

Leather, S. R. 1986. Insect species richness of the British Rosaceae: The importance of host range, plant architecture, age of establishment, taxonomic isolation and species-area relationships. *J. Anim. Ecol.* 55:841–60. (17)

Lee, A. K., P. R. Baverstock, and C. H. S. Watts. 1981. Rodents—the late invaders. In *Ecological biogeography of Australia*, ed. A. Keast, 1521–53. The Hague: Junk. (14)

Lee, D. S., C. R. Gilbert, C. H. Hocutt, R. E. Jenkins, D. E. McAllister, and J. R. Stauffer, Jr. 1980. *Atlas of North American fishes*. Raleigh, N.C.: North Carolina State Museum of Natural History. (17)

Leigh, E. G., Jr. 1965. On the relationship between productivity, biomass, diversity and stability of a community. *Proc. Natl. Acad. Sci. USA* 53:777–83. (5)

Lemen, C. A., and M. L. Rosenzweig. 1978. Microhabitat selection in two species of heteromyid rodents. *Oecologia* 33:127–35. (5)

Leong, T. S., and J. C. Holmes. 1981. Communities of metazoan parasites in open water fishes of Cold Lake, Alberta. *J. Fish Biol.* 18:693–713. (17)

Leopold, E. B. 1967. Late-Cenozoic patterns of plant extinction. In *Pleistocene extinctions: The search for a cause*, ed. P. S. Martin and H. E. Wright, Jr. 203–46. New Haven: Yale University Press. (28)

Levin, B. R., F. M. Stewart, and L. Chao. 1977. Resource-limited growth, competition, and predation: A model and experimental studies with bacteria and bacteriophage. *Am. Nat.* 111:3–24. (2)

Levin, S. A. 1970. Community equilibria and stability, and an extension of the competitive exclusion principle. *Am. Nat.* 104:413–23. (2)

Levin, S. A. 1974. Dispersion and population interactions. *Am. Nat.* 108:207–28. (7, 8, 19, 30)

Levin, S. A. 1976a. Population dynamic models in heterogeneous environments. *Annu. Rev. Ecol. Syst.* 7:287–310. (8)

Levin, S. A. 1976b. Spatial patterning and the structure of ecological communities. *Lect. Math. Life Sci.* 8, 1–35. (7)

Levin, S. A. 1978. Population models and community structure in heterogeneous environments. In *Studies in mathematical biology. Part II, Populations and Communities*, ed. S. A. Levin, 439–76. Studies in Mathematics 16. Mathematical Association of America. (8)

Levin, S. A. 1979. Mechanisms for the generation and maintenance of diversity in ecological communities. In *The mathematical theory of the dynamics of populations*, 11, ed. R. W. Hiorms and D. Cooke, 173–94. London: Academic Press. (2)

Levin, S. A., and R. T. Paine. 1974. Disturbance, patch formation, and community structure. *Proc. Natl. Acad. Sci. USA* 71:2744–47. (2, 5)

Levin, S. A., and L. A. Segel. 1976. Hypothesis for origin of planktonic patchiness. *Nature* 259:659. (8)

Levins, R. 1969. Some demographic and genetic consequences of environmental heterogeneity for biological control. *Bull. Entomol. Soc. Am.* 15:237–40. (10)

Levins, R. 1970. *Extinction*. In *Some mathematical problems in biology*, ed. M. Gerstenhaber, 77–107. Providence, R.I.: American Mathematical Society. (9, 10)

Levins, R. 1979. Coexistence in a variable environment. *Am. Nat.* 114:765–83. (1, 2)

Levins, R., and D. Culver. 1971. Regional coexistence of species and competition between rare species. *Proc. Natl. Acad. Sci. USA* 6:1246–48. (2, 8)

Levinton, J. S. 1979. A theory of diversity equilibrium and morphological evolution. *Science*, 204:335–36. (2)

Levinton, J. S. 1985. Complex interactions of a deposit feeder with its resources: Roles of density, a competitor and detrital addition in the growth and survival of the mudsnail *Hydrobia totteni*. *Mar. Ecol. Prog. Ser.* 22:31–40. (4)

Levinton, J. S., and T. H. Suchanek. 1978. Geographic variation, niche breadth and genetic differentiation at different geographic scales in the mussels *Mytilus californianus* and *M. edulis*. *Mar. Biol.* 49:363–75. (4)

Lewin, R. 1986. Supply-side ecology. *Science* 234:25–27. (4, 17)

Lewinsohn, T. M. 1991. Insects in flower heads of Asteraceae in Southeast Brazil: A case study on tropical species richness. In *Plant-animal interactions: Evolutionary ecology in tropical and temperate regions*, ed. P. W. Price, T. M. Lewinsohn, G. W. Fernandes, and W. W. Benson, 525–59. New York: John Wiley. (7, 16, 22)

Lewis, H., and M. Lewis. 1955. The genus *Clarkia*. *Univ. Calif. Publ. Bot.* 20:241–392. (22)

Lewontin, R. C. 1977. Adaptation. *Sci. Am.* 239:212–30. (25)

Lewontin, R. C. 1987. The shape of optimality. *The latest on the best: Essays on evolution and optimality*, ed. J. Dupré, 151–59. Cambridge, Mass.: MIT Press. (25)

Li, H. L. 1952. Floristic relationships between eastern Asia and eastern North America. *Trans. Am. Phil. Soc.* 42, 371–429. (Reprint Philadelphia: Morris Arboretum, 1971.). (26)

Li, L. 1989. Global positive coexistence of a nonlinear elliptic biological interacting model. *Math. Biosci.* 97:1–15. (8)

Li, S. 1935. *Forest botany of China.* Shanghai: Commercial Press. (26)

Li, S. 1973. *Forest botany of China supplement.* Taipei: Chinese Foresty Association. (26)

Lidgard, S., and P. R. Crane. 1990. Angiosperm diversification and Cretaceous floristic trends: A comparison of palynofloras and leaf macrofloras. *Paleobiology* 16:77–93. (29, 30)

Lidgard, S., and J. B. C. Jackson. 1989. Growth in encrusting cheilostome bryozoans: I. Evolutionary trends. *Paleobiology* 15:255–82. (23)

Liem, K. F., and D. B. Wake. 1985. Morphology: Current approaches and concepts. In *Functional vertebrate morphology,* ed. M. Hildebrand, D. M. Bramble, K. F. Liem, and D. B. Wake, 366–77. Cambridge, Mass.: Belknap Press of Harvard University Press. (25)

Lieth, H. 1975. Modeling the primary productivity of the world. In *Primary productivity of the biosphere,* ed. H. Lieth and R. H. Whittaker, 237–63. New York: Springer-Verlag. (6)

Lillegraven, J. A. 1972. A biogeographical problem involving comparisons of later Eocene terrestrial vertebrate faunas of western North America. In *Historical biogeography, plate tectonics, and the changing environment,* ed. J. Gray and A. J. Boucot, 333–47. Corvallis: Oregon State University Press. (28)

Lindberg, D. R. 1991. Marine biotic interchange between the Northern and Southern hemispheres. *Paleobiology* 17:308–24. (29, 30)

Lindberg, H. 1948. Zur Kenntniss der Insektenfauna im Brackwasser des Baltischen Meeres. *Soc. Sci. Fenn. Comment. Biol.* 10:1–207. (10)

Lindeman, R. L. 1942. The trophic-dynamic aspect of ecology. *Ecology* 23:399–418. (3, 6, 24)

Lindroth, C. H. 1949. *Die Fennoscandischen Carabidae* III. Stockholm: Allegem. Teil. (10)

Lindroth, C. H. 1985. The Carabidae (Coleoptera) of Fennoscandia and Denmark. *Fauna Entomol. Scand.* 15:1–225. (10)

Lindroth, C. H. 1986. The Carabidae (Coleoptera) of Fennoscandia and Demark. *Fauna Entomol. Scand.* 15:233–497. (10)

Lindsay, A. M. 1985. Are Australian soils different? *Proc. Ecol. Soc. Aust.* 14:83–97. (14)

Linsley, E. G. 1942. A review of the fossil Cerambycidae of North America. *Proc. New Engl. Zool. Club* 21:17–42. (23)

Linsley, E. G. 1961. Cerambycidae of North America. Part I. *Univ. Calif. Publ. Entomol.* 18:1–135. (23)

Linsley, E. G. 1963. Bering arc relationships in Cerambycidae and their hostplants. In *Pacific Basin biogeography,* ed. J. L. Gressitt, 159–78. Honolulu: Bishop Museum Press. (23)

Lipps, J. H. 1970. Evolution in the pelagic realm. *Geol. Soc. Am. Abs. w. Programs,* 2:607–08. (5)

Littler, M. M., and D. S. Littler. 1984. Relationships between macroalgal functional form groups and substrata stability in a subtropical rocky-intertidal system. *J. Exp. Mar. Biol. Ecol.* 74:1–12. (4)

Livingstone, D. A. 1982. Quaternary geography of Africa and the refuge theory. In *Biological diversification in the tropics,* ed. G. T. Prance, 523–36. New York: Columbia University Press. (29)

Lockley, M. G. 1983. A review of brachiopod dominated palaeocommunities from the type Ordovician. *Palaeontology* 26:111–45. (5)

Loeb, V. J. 1979. Larval fishes in the zooplankton community of the North Pacific Central Gyre. *Mar. Biol.* 53:173–91. (19)

Loeb, V. J. 1980. Patterns of spatial and species abundance within the larval fish assemblage of the North Pacific central gyre during late summer. *Mar. Biol.* 60:189–200. (19)

Loehle, C., and J. H. K. Pechmann. 1988. Evolution: The missing ingredient in systems ecology. *Am. Nat.* 132:884–99. (24)

Losos, J. B. 1990. Ecomorphology, performance capability, and scaling of West Indian *Anolis* lizards: An evolutionary analysis. *Ecol. Monogr.* 60:369–88. (25)

Lotka, A. J. 1956. *Elements of mathematical biology.* New York: Dover. (6)

Lotz, J. M., and W. F. Font. 1985. Structure of enteric helminth communities in two populations of *Eptesicus fuscus* (Chiroptera). *Can. J. Zool.* 63:2969–78. (17)

Loveless, A. R. 1961. A nutritional interpretation of sclerophylly based on differences in the chemical composition of sclerophyllous and mesophytic leaves. *Ann. Bot.* n.s. 25:168–84. (15)

Low, B. S. 1978. Environmental uncertainty and the parental strategies of marsupials and placentals. *Am. Nat.* 112:197–213. (15)

Lubchenco, J. 1978. Plant species diversity in a marine intertidal community: Importance of herbivore food preference and algal competitive abilities. *Am. Nat.* 112:23–39. (2, 5)

Lynch, J. F., and D. F. Whigham. 1984. Effects of forest fragmentation on breeding bird communities in Maryland, USA. *Biol. Conserv.* 28:287–324. (6)

Lyons, H. L. 1976. Seasonality in central North Pacific chaetognaths. Ph.D. diss., University of California, San Diego. (19)

McAllister, D. E., S. R. Platania, F. W. Schueler, and D. S. Lee. 1986. Ichthyofaunal patterns on a geographic grid. In *The zoogeography of North American freshwater fishes,* ed. C. H. Hocutt and E. O. Wiley, 17–51. New York: Wiley. (6)

McArdle, B. H. 1990. When are rare species not there? *Oikos* 57:276–77. (10)

MacArthur, J. W. 1975. Environmental fluctuations and species diversity. In *Ecology and evolution of communities,* ed. M. L. Cody and J. M. Diamond, 74–80. Cambridge, Mass.: Harvard University Press. (6)

MacArthur, R. H. 1958. Population ecology of some warblers of northeastern coniferous forests. *Ecology* 39:599–619. (1)

MacArthur, R. H. 1964. Environmental factors affecting bird species diversity. *Am. Nat.* 98:387–97. (1, 21)

MacArthur, R. H. 1965. Patterns of species diversity. *Biol. Rev.* 40:510–33. (1, 6, 14, 30)

MacArthur, R. H. 1968. The theory of the niche. In *Population biology and evolution,* ed. R. C. Lewontin, 159–76. Syracuse, N.Y.: Syracuse University Press. (2)

MacArthur, R. H. 1969. Patterns of communities in the tropics. *Biol. J. Linn. Soc.* 1:19–30. (1, 30)

MacArthur, R. H. 1970. Species packing and competitive equilibrium for many species. *Theor. Pop. Biol.* 1:1–11. (2)

MacArthur, R. H. 1972. *Geographical ecology: Patterns in the distribution of species.* New York: Harper and Row. (1, 6, 17, 20, 21, 22, 23, 24, 26)

MacArthur, R. H., and R. Levins. 1964. Competition, habitat selection, and character displacement in a patchy environment. *Proc. Natl. Acad. Sci. USA* 51:1207–10. (2)

MacArthur, R. H., and R. Levins. 1967. The limiting similarity, convergence, and divergence of coexisting species. *Am. Nat.* 101:377–85. (1, 21)

MacArthur, R. H., and J. MacArthur. 1961. On bird species diversity. *Ecology* 42:594–98. (1, 13)

MacArthur, R. H., J. W. MacArthur, and J. Preer. 1962. On bird species diversity. II. Prediction of bird census from habitat measurements. *Am. Nat.* 96:167–174. (13)

MacArthur, R. H., and E. R. Pianka. 1966. On the optimal use of a patchy environment. *Am. Nat.* 100:603–9. (5)

MacArthur, R. H., H. Recher, and M. L. Cody. 1966. On the relation between habitat selection and species diversity. *Am. Nat.* 100:319–32. (5, 13, 20)

MacArthur, R. H., and E. O. Wilson. 1963. An equilibrium theory of insular zoogeography. *Evolution* 17:373–87. (1, 5, 24)

MacArthur, R. H., and E. O. Wilson. 1967. *The theory of island biogeography.* Princeton, N.J.: Princeton University Press. (1, 5, 7, 11, 12, 24)

McCartan, L., B. H. Tiffney, J. A. Wolfe, T. A. Ager, S. L. Wing, L. A. Sirkin, L. W. Ward, and J. Brooks. 1990. Late Tertiary floral assemblage from upland gravel deposits of the southern Maryland Coastal Plain. *Geology* 18:311–14. (26)

McCauley, D. E., and W. F. Eanes. 1987. Hierarchical population structure analysis of the milkweed beetle *Tetraopes tetropthalmus* (Forster). *Heredity* 58:193–201. (23)

McClaughtery, C. A., J. Pastor, J. D. Aber, and J. M. Melillo. 1985. Forest litter decomposition in relation to soil nitrogen dynamics and litter quality. *Ecology* 66:266–75. (2)

McCoy, E. D., and K. L. Heck, Jr. 1976. Biogeography of corals, seagrasses and mangroves: An alternative to the center of origin concept. *Syst. Zool.* 25:201–10. (20, 24, 30)

McCrea, K. D., and W. G. Abrahamson. 1987. Variation in herbivore infestation.: Historical vs. genetic factors. *Ecology* 68:822–27. (24)

Macdonald, I. A. W., F. J. Powrie, and W. R. Siegfried. 1986. The differential invasion of southern Africa's biomes and ecosystems by alien plants and animals. In *The ecology and management of biological invasions in Southern Africa,* ed. I. A. W. Macdonald, F. J. Kruger, and A. A. Ferrar, 209–25. (30)

MacGarvin, M., J. H. Lawton, and P. H. Heads. 1986. The herbivorous insect communities of open and woodland bracken: Experiments and habitat manipulations. *Oikos* 47:135–48. (16)

McGowan, J. A. 1971. Oceanic biogeography of the Pacific. In *The micropaleontology of oceans,* ed. B. M. Funnell and W. R. Riedel, 3–74. Cambridge: Cambridge University Press. (19)

McGowan, J. A. 1974. The nature of oceanic ecosystems. In *The biology of the Pacific Ocean,* ed. C. B. Miller, 9–28. Corvallis: Oregon State University Press. (19)

McGowan, J. A. 1986. The biogeography of pelagic ecosystems. In *Pelagic biogeography: Proceedings of an international conference* (The Netherlands, 25 May–5 June 1985), ed. A. C. Pierrot-Bults, S. van der Spoel, B. J. Zuhuranec, and R. K. Johnson, 191–200. UNESCO Technical Papers in Marine Science, no. 49. (19)

McGowan, J. A. 1990. Species dominance-diversity patterns in oceanic communities. In *The earth in transition: Patterns and processes of biotic impoverishment,* ed. G. M. Woodwell, 395–421. Cambridge: Cambridge University Press. (19)

McGowan, J. A., and T. L. Hayward. 1978. Mixing and oceanic productivity. *Deep Sea Res.* 25:771–93. (19)

McGowan, J. A., and C. B. Miller. 1980. Larval fish and zooplankton community structure. *Calif. Coop. Oceanic Fish. Invest. Rep.,* no. 21, 29–36. (19)

McGowan, J. A., and P. W. Walker. 1979. Structure in the copepod community of the North Pacific Central Gyre. *Ecol. Monogr.* 49:195–226. (19)

McGowan, J. A. and P. W. Walker. 1985. Dominance and diversity maintenance in an oceanic ecosystem. *Ecol. Monogr.* 55:103–18. (19)

McGuinness, K. A. 1984a. Equations and explanations in the study of species-area curves. *Biol. Rev.* 59:423–40. (5, 9)

McGuinness, K. A. 1984b. Species-area relations of communities on intertidal boulders: Testing the null hypothesis. *J. Biogeogr.* 11:439–56. (4)

McIntosh, R. P. 1985. *The background of ecology: Concept and theory.* Cambridge: Cambridge University Press. (24)

McIntosh, R. P. 1987. Pluralism in ecology. *Annu. Rev. Ecol. Syst.* 18:321–41. (24)

McKaye, K. R., and W. N. Gray. 1984. Extrinsic barriers to gene flow in rock-dwelling cichlids of Lake Malawi: Macrohabitat heterogeneity and reef colonization. In *Evolution of fish species flocks,* ed. A. A. Echelle and I. Kornfield, 169–83. Orono: University of Maine Press. (30)

McKillup, S. C. 1983. A behavioural polymorphism in the marine snail *Nassarius pauperatus:* Geographic variation correlated with food availability and differences in competitive ability between morphs. *Oecologia* 56:58–66. (4)

McKinney, M. L. 1990. Classifying and analysing evolutionary trends. In *Evolutionary trends,* ed. K. J. McNamara, 28–57. London: Belhaven Press. (28)

McLaren, D. J., and W. D. Goodfellow. 1990. Geological and biological consequences of giant impacts. *Annu. Rev. Earth Planetary Sci.* 18:123–71. (1)

McLaughlin, J. F., and J. Roughgarden. 1991a. Pattern and stability in predator-prey communities: How diffusion in spatially variable environments affects the Lotka-Volterra model. *Theor. Pop. Biol.* 40:148–72. (8)

McLaughlin, J. F., and J. Roughgarden. 1991b. Predation across spatial scales in heterogeneous environments. *Theor. Pop. Biol.* 41:277–99. (8)

MacLulich, J. H. 1986a. Colonization of bare rock surfaces by microflora in a rocky intertidal habitat. *Mar. Ecol. Prog. Ser.* 32:91–96. (4)

MacLulich, J. H. 1986b. Experimental evaluation of methods for sampling and assaying intertidal epilithic microalgae. *Mar. Ecol. Prog. Ser.* 34:275–80. (4)

McMurtrie, R. 1978. Persistence and stability of single-species and prey-predator systems in spatially heterogeneous environments. *Math. Biosci.* 39:11–51. (8)

McNab, B. K. 1982. The physiological ecology of South American mammals. In *Mammalian biology in South America,* ed. M. A. Mares and H. H. Genoways, 187–207. Special Publication Series, Pymatuning Laboratory of Ecology, vol. 6. Pittsburgh, Pa.: University of Pittsburgh. (14)

Macnae, W. 1966. Mangroves in eastern and southern Australia. *Aust. J. Bot.* 14:67–104. (20)

Macnae, W. 1968. A general account of the fauna and flora of mangrove swamps and forests in the Indo–West Pacific region. *Adv. Mar. Biol.* 6:73–270. (20)

McNaughton, S. J., and L. L. Wolf. 1970. Dominance and the niche in ecological systems. *Science* 167:131–39. (10, 30)

Madsen, R. W., and M. L. Moeschenberger. 1983. *Introducing statistics for business and economics.* New York: Prentice-Hall. (28)

Magurran, A. E. 1988. *Ecological diversity and its measurement.* Princeton, N.J.: Princeton University Press. (1)

Mahabale, T. S., and J. V. Deshpande. 1959. The genus *Sonneratia* and its fossil allies. *Paleobotanist* 6:51–64. (20)

Mai, D. H. 1960. Über neue Früchte und Samen aus dem deutschen Tertiär. *Paläontologische Zeitschrift* 34:73–90. (26)

Mai, D. H. 1971a. Fossile Funde von *Manglietia* Blume (Magnoliaceae). *Feddes Repertorium* 82:441–48. (26)

Mai, D. H. 1971b. Über fossile Lauraceae und Theaceae in Mitteleuropa. *Feddes Repertorium* 82:313–41. (26)

Mai, D. H. 1980. Zur Bedeutung von Relikten in der Florengeschichte. In *100 Jahre Arboretum (1879–1979),* 281–307. Berlin. (26)

Mai, D. H. 1981. Entwicklung und klimatische Differenzierung der Laubwaldflora Mitteleuropas im Tertiär. *Flora* 171:525–82. (26)

Mai, D. H. 1987a. Neue Arten nach Früchten und Samen aus dem Tertiär von Nordwestsachsen und der Lausitz. *Feddes Repertorium* 98:105–26. (26)

Mai, D. H. 1987b. Neue Früchte und Samen aus paläozänen Ablagerungen Mitteleuropas. *Feddes Repertorium* 98:197–229. (26)

Mai, D. H., and H. Walther. 1978. Die Floren der Haselbacher-Serien im Weißelster-Becken (Bezirk Leizig, D.D.R.). *Abh. Staat. Mus. Mineral. Geol. Dresden* 28:1–101. (26)

Mai, D. H., and H. Walther. 1988. Die pliozänen Floren von Thüringen, Deutsche Demokratische Republik. *Quartärpaläontologie Berlin* 7:55–297. (26)

Manchester, S. R. 1987. The fossil history of the Juglandaceae. *Ann. Mo. Bot. Gard. Monogr.* 21:1–137. (20, 26)

Manchester, S. R. 1989. Early history of the Juglandaceae. *Plant Syst. Evol.* 162:231–50. (20, 26)

Mares, M. A. 1976. Convergent evolution of desert rodents: Multivariate analysis and zoogeographic implications. *Paleobiology* 2:39–63. (1, 14)

Mares, M. A. 1980. Convergent evolution among desert rodents: A global perspective. *Bull. Carnegie Mus. Nat. Hist.* 16:1–51. (14, 21)

Mares, M. A. 1983. Desert rodent adaptation and community structure. *Great Basin Nat. Mem.* 7:30–43. (14)

Mares, M. A., and M. L. Rosenzweig. 1978. Granivory in North and South American deserts: Rodents, birds, and ants. *Ecology* 59:235–41. (14)

Mares, M. A., M. R. Willig, and T. E. Lacher, Jr. 1985. The Brazilian Caatinga in South American zoogeography: Tropical mammals in a dry region. *J. Biogeogr.* 12:57–69. (30)

Margalef, R. 1968. *Perspectives in ecological theory.* Chicago: University of Chicago Press. (6)

Margolis, L., and J. R. Arthur. 1979. *Synopsis of the parasites of fishes of Canada.* Bull. Fish. Res. Board of Canada, 199. Ottawa: Department of Fisheries and Oceans. (17)

Margulis, L., and K. V. Schwartz. 1988. *Five kingdoms.* 2d ed. New York: W. H. Freeman. (15)

Maristo, L. 1941. Die floristischen seetypen Finnlands. *Annales Botanici Societatis 'Vanamo'* 15:1–344. (10)

Marquet, P. A., S. A. Navarrete, and J. C. Castilla. 1990. Scaling population density to body size in rocky intertidal communities. *Science* 250:1125–27. (2)

Marsh, A. C. 1986. Ant species richness along a climatic gradient in the Namib Desert. *J. Arid Environ.* 11:235–41. (14)

Marshall, L. G., S. D. Webb., J. J. Sepkoski, and D. M. Raup. 1982. Mammalian evolution and the Great American Interchange. *Science* 215:1351–57. (28)

Martin, L. D. 1989. Fossil history of the terrestrial Carnivora. In *Carnivore behavior, ecology, and evolution*, ed. J. Gittleman, 536–68. Ithaca: Cornell University Press. (28)

Martin, P. S. 1958. A biogeography of reptiles and amphibians in the Gomez Farías region, Tamaulipas, Mexico. *Misc. Publ. Mus. Zool., Univ. Mich.*, 101:1–102. (25)

Martin, P. S. 1984. Prehistoric overkill: The global model. In *Quaternary extinctions*, ed. P. S. Martin and R. G. Klein, 354–403. Tucson: University of Arizona Press. (12)

Martinez, N. D. 1991. Artifacts or attributes? The effects of resolution on the Little Rock Lake food web. *Ecol. Monogr.* 61:367–92. (3)

Martinez, N. D. 1992. Constant connectance in community food webs. *Am. Nat.* 139:1208–18. (1, 3, 30)

Marx, H., and G. B. Rabb. 1972. Phyletic analysis of fifty characters of advanced snakes. *Fieldiana, Zool.* 63:1–321. (25)

Mather, P., and I. Bennett. 1984. *A coral reef handbook: A guide to the fauna, flora, and geology of Heron Island and adjacent reefs and cays*. 2d ed. Handbook Series 1. Brisbane: Australian Coral Reef Society. (27)

Maurer, B. A. 1990. The relationship between distribution and abundance in a patchy environment. *Oikos* 58:181–89. (6)

May, M. L., D. L. Pearson, and T. M. Casey. 1986. Oxygen consumption of active and inactive adult tiger beetles. *Physiol. Entomol.* 11:171–79. (18)

May, R. M. 1973. Stability and complexity in model ecosystems. Princeton, N.J.: Princeton University Press. (2, 3, 19, 21)

May, R. M. 1975a. Patterns of species abundance and diversity. In *Ecology and evolution of communities*, ed. M. L. Cody and J. M. Diamond, 81–120. Cambridge, Mass.: Harvard University Press. (1, 9)

May, R. M. 1975b. *Stability and complexity in model ecosystems*. Princeton, N.J.: Princeton University Press. (1)

May, R. M. 1977. Thresholds and breakpoints in ecosystems with a multiplicity of stable states. *Nature* 269:471–77. (2)

May, R. M. 1978. The dynamics and diversity of insect faunas. In *Diversity of insect faunas*, ed. L. A. Mound and N. Waloff, 188–204. Oxford: Blackwell. (2)

May, R. M. 1986. The search for patterns in the balance of nature: Advances and retreats. *Ecology* 67:1115–26. (2)

May, R. M. 1988. How many species are there on Earth? *Science* 241:1441–49. (15)

May, R. M., and R. H. MacArthur. 1972. Niche overlap as a function of environmental variability. *Proc. Natl. Acad. Sci. USA* 69:1109–13. (1, 2)

Mayden, R. L. 1987. Historical ecology and North American highland fishes: A research program in community ecology. In *Community and evolutionary ecology of North American stream fishes*, ed. W. J. Matthews and D. C. Heins, 210–22. Norman: University of Oklahoma Press. (24)

Mayden, R. L. 1988. Biogeography, parsimony, and evolution in North American freshwater fishes. *Syst. Zool.* 37:329–55. (24)

Mayden, R. L. 1989. Phylogenetic studies of North American minnows, with emphasis on the genus *Cyprinella* (Teleoster: Cypriniformes). *Univ. Kans. Mus. Nat. Hist. Misc. Publ.* 80:1–189. (24)

Mayer, H. 1984. *Wälder Europas*. Stuttgart: Gustav Fischer Verlag. (26)

Mayr, E. 1942. *Systematics and the origin of species from the viewpoint of the zoologist*. New York: Columbia University Press. (11)

Mayr, E. 1963. *Animal species and evolution*. Cambridge, Mass.: Belknap Press of Harvard University Press. (30)

Mayr, E. 1983. Introduction. In *Perspectives in ornithology*, ed. A. H. Brush and G. A. Clark, 1–21. Cambridge: Cambridge University Press. (12)

Mazer, S. J., and B. H. Tiffney. 1982. Fruits of *Wetherellia* and *Pa-*

leowetherellia (?Euphorbiaceae) from Eocene sediments in Virginia and Maryland. *Brittonia* 34:300–333. (20)

Mearns, A. J., D. R. Young, R. J. Olson, and H. A. Schafer. 1981. Trophic structure and the cesium-potassium ratio in pelagic ecosystems. *Calif. Coop. Oceanic Fish. Invest. Rep.*, no. 22, 99–110. (3)

Melillo, J. M., J. D. Aber, and J. F. Muratore. 1982. Nitrogen and lignin control of hardwood leaf litter decomposition dynamics. *Ecology* 63:621–26. (2)

Menge, B. A. 1972. Competition for food between two intertidal starfish species and its effect on body size and feeding. *Ecology* 53:635–44. (4)

Menge, B. A. 1976. Organization of the New England rocky intertidal community: Role of predation, competition and environmental heterogeneity. *Ecol. Monogr.* 46:335–93. (4)

Menge, B. A. 1978a. Predation intensity in a rocky intertidal community: Effect of an algal canopy, wave action and desiccation on predator feeding rates. *Oecologia* 34:17–35. (4)

Menge, B. A. 1978b. Predation intensity in a rocky intertidal community: Relation between predator foraging activity and environmental harshness. *Oecologia* 34:1–16. (4)

Menge, B. A. 1979. Coexistence between the seastars *Asterias vulgaris* and *A. forbesi* in a heterogeneous environment: A nonequilibrium explanation. *Oecologia* 41:245–72. (4)

Menge, B. A., L. R. Ashkenas, and A. Matson. 1983. Use of artificial holes in studying community development in cryptic marine habitats in a tropical rocky intertidal region. *Mar. Biol.* 77:129–42. (4)

Menge, B. A., and T. M. Farrell. 1989. Community structure and interaction webs in shallow marine hard-bottom communities: Tests of an environmental stress model. *Adv. Ecol. Res.* 19:189–262. (4)

Menge, B. A., and A. M. Olson. 1990. Role of scale and environmental factors in regulation of community structure. *Trends Ecol. Evol.* 5:52–56. (6, 24)

Menge, B. A., and J. P. Sutherland. 1976. Species diversity gradients: Synthesis of the roles of predation, competition and temporal heterogeneity. *Am. Nat.* 110:351–69. (4)

Menge, B. A., and J. P. Sutherland. 1987. Community regulation: Variation in disturbance, competition, and predation in relation to environmental stress and recruitment. *Am. Nat.* 130:730–57. (2, 4, 6)

Mengel, R. M. 1964. The probable history of species formation in some northern wood warblers (Parulidae). *Living Bird* 3:9–43. (12)

Menken, S. B. J. 1982. Biochemical genetics and systematics of small ermine moths. *Z. Zool. Syst. Evol.* 20:131–43. (23)

Mepham, R. H. 1983a. Mangrove floras of the southern continents. Part I. The geographical origin of Indo-Pacific mangrove genera and the development and present status of the Australia mangroves. *S. Afr. J. Bot.* 2:1–8. (20)

Mepham, R. H. 1983b. *Mangroves*. Cape Town and Rotterdam: Balkema. (20)

Mepham, R. H., and J. S. Mepham. 1985. The flora of tidal forests—a rationalization of the use of the term 'mangrove.' *S. Afr. J. Bot.* 51:77–99. (20)

Meserve, P. L., and W. E. Glanz. 1978. Geographical ecology of small mammals in the northern Chilean arid zone. *J. Biogeogr.* 5:135–48. (5, 14)

Meyen, S. V. 1987. *Fundamentals of paleobotany*. London: Chapman and Hall. (26)

Meyer, A. 1987. Phenotypic plasticity and heterochrony in *Cichlasoma managuense* (Pisces, Cichlidae) and their implications for speciation in cichlid fishes. *Evolution* 41:1357–69. (30)

Meyer, A., T. D. Kocher, P. Basasibwaki, and A. C. Wilson. 1990. Monophyletic origin of Lake Victoria cichlid fishes suggested by mitochondrial DNA sequences. *Nature* 347:550–53. (30)

Michaels, A. F., and M. W. Silver. 1988. Primary production, sinking fluxes and the microbial food web. *Deep Sea Res.* 35(4):473–90. (19)

Michaud, E. J., and J. R. Dixon. 1989. Prey items of 20 species of the neotropical colubrid snake genus *Liophis*. *Herpetol. Rev.* 20:39–41. (25)

Mikulic, D. G., and R. Watkins. 1981. Trilobite ecology in the Ludlow Series of the Welsh Borderland. In *Communities of the past,* ed. J. Gray, A. J. Boucot, and W. B. N. Berry, 101–17. Stroudsburg, Pa.: Hutchinson Ross. (5)

Miles, D. B., and R. E. Ricklefs. 1984. The correlation between ecology and morphology in deciduous forest passerine birds. *Ecology* 65:1629–40. (1, 30)

Miles, D. B., R. E. Ricklefs, and J. Travis. 1987. Concordance of eco-morphological relationships in three assemblages of passerine birds. *Am. Nat.* 129:347–64. (30)

Milewski, A. V. 1981a. A comparison of reptile communities in relation to soil fertility in the mediterranean and adjacent arid parts of Australia and southern Africa. *J. Biogeogr.* 8:493–503. (21)

Milewski, A. V. 1981b. A comparison of vegetation height in relation to the effectiveness of rainfall in the mediterranean and adjacent arid parts of Australia and South Africa. *J. Biogeogr.* 8:107–16. (15)

Milewski, A. V., and R. M. Cowling. 1985. Anomalies in the plant and animal communities in similar environments at The Barrens, Western Australia, and the Caledon Coast, South Africa. *Proc. Ecol. Soc. Aust.* 14:199–212. (15)

Miller, A. I. 1988. Spatio-temporal transitions in Paleozoic Bivalvia: An analysis of North American fossil assemblages. *Hist. Biol.* 1:251–73. (29)

Miller, A. I. 1989. Spatio-temporal transitions in Paleozoic Bivalvia: A field comparison of Upper Ordovician and Upper Paleozoic bivalve-dominated fossil assemblages. *Hist. Biol.* 2:227–60. (29)

Miller, J. S., and P. P. Feeny. 1989. Interspecific differences among swallowtail larvae (Lepidoptera: Papilionidae) in susceptibility to aristolochic acids and berberine. *Ecol. Entomol.* 14:287–96. (23)

Miller, R. S. 1967. Pattern and process in competition. *Adv. Ecol. Res.* 4:1–74. (1, 30)

Milliken, G. A., and D. E. Johnson. 1984. *Analysis of messy data.* Belmont, Calif.: Lifetime Learning Publications. (21)

Milne, B. T., and R. T. T. Forman. 1986. Peninsulas in Maine: Woody plant diversity, distance, and environmental patterns. *Ecology* 67:967–74. (6)

Mimura, M., Y. Kan-on, and Y. Nishiura. 1988. Oscillations in segregation of competing populations. In *Mathematical ecology,* ed. T. G. Hallam, L. J. Gross, and S. A. Levin, 717–33. Singapore: World Scientific. (8)

Mirov, N. T. 1967. *The genus* Pinus. New York: Ronald Press. (26)

Mitchell, A. 1974. *Trees of Britain and Northern Europe.* Boston: Houghton-Mifflin. (26)

Mithen, S. J., and J. H. Lawton. 1986. Food-web models that generate constant predator-prey ratios. *Oecologia* 69:542–50. (5)

Mitter, C., and B. D. Farrell. 1991. Macroevolutionary aspects of insect/plant interactions. In *Insect/plant Interactions,* vol. 3, ed. E. A. Bernays, 35–78. Boca Raton, Fla.: CRC Press. (23)

Mitter, C., B. Farrell, and B. Wiegmann. 1988. The phylogenetic study of adaptive zones: Has phytophagy promoted insect diversification? *Am. Nat.* 132:107–28. (22, 23, 30)

Mitter, C., and D. J. Futuyma. 1983. An evolutionary-genetic view of hostplant utilization by insects. In *Variable plants and herbivores in natural and managed systems,* ed. R. F. Denno and M. McClure, 427–59. New York: Academic Press. (23)

Miyamoto, M. M. 1985. Consensus cladograms and general classifications. *Cladistics* 1:186–89. (24)

Mole, R. R. 1924. The Trinidad snakes. *Proc. Zool. Soc. Lond.* 1924:235–78. (25)

Moll, E. J., and M. J. A. Werger. 1978. Mangrove communities. In *Biogeography and ecology of Southern Africa,* ed. M. J. A. Werger, 1231–38. The Hague: Junk. (20)

Monteith, G. B. 1990. Rainforest insects: Biodiversity, bioguesstimation, or just handwaving? *Myrmecia* 26:93–95. (15)

Mooi, R., P. F. Cannell, V. A. Funk, P. M. Mabee, R. T. O'Grady, and C. K. Starr. 1989. Historical perspectives, ecology, and tiger beetles: An alternative discussion. *Syst. Zool.* 38:191–95. (18)

Mooney, H. A., ed. 1977. *Convergent evolution in Chile and California: Mediterranean climate ecosystems.* Stroudsberg, Pa.: Dowden, Hutchinson, and Ross. (21)

Moore, D. R. J., and P. A. Keddy. 1989. The relationship between species richness and standing crop in wetlands: The importance of scale. *Vegetatio* 79:99–106. (60)

Moore, J., and D. Simberloff. 1990. Gastrointestinal helminth communities of bobwhite quail. *Ecology* 71:344–59. (17)

Moran, M. J. 1985. The timing and significance of sheltering and foraging behaviour of the predatory intertidal gastropod *Morula marginalba* Blainville (Muricidae). *J. Exp. Mar. Biol. Ecol.* 93:103–14. (4)

Moran, N. A. 1989. A 48-million-year-old aphid-host plant association and complex life cycle: Biogeographic evidence. *Science* 245:173–75. (23)

Moreau, R. E. 1954. The main vicissitudes of the European avifauna since the Pliocene. *Ibis* 96:411–31. (12)

Moreau, R. E. 1966. *The bird faunas of Africa and its islands.* New York: Academic Press. (12)

Morgan, A. V. 1987. Late Wisconsin and early Holocene paleoenvironments of east-central North America based on assemblages of fossil Coleoptera. In *North American and adjacent oceans during the last deglaciation,* ed. W. F. Ruddiman and H. E. Wright, Jr., 353–70. Boulder, Colo.: Geological Society of America. (29)

Morris, D. W. 1987. Ecological scale and habitat use. *Ecology* 68:362–69. (30)

Morris, D. W. 1988. Habitat-dependent population regulation and community structure. *Evol. Ecol.* 2:253–69. (30)

Morse, D. R., J. H. Lawton, M. M. Dodson, and M. H. Williamson. 1985. Fractal dimension of vegetation and the distribution of arthropod body lengths. *Nature* 314:731–32. (2)

Morse, D. R., N. E. Stork, and J. H. Lawton. 1988. Species number, species abundance, and body length relationships of arboreal beetles in Bornean lowland rain forest trees. *Ecol. Entomol.* 13:25–37. (22)

Morton, S. R. 1978. Torpor and nest-sharing in free-living *Sminthopsis crassicaudata* (Marsupialia) and *Mus musculus* (Rodentia). *J. Mammal.* 59:569–75. (14)

Morton, S. R. 1979. Diversity of desert-dwelling mammals: A comparison of Australia and North America. *J. Mammal.* 60:253–64. (14)

Morton, S. R. 1982. Dasyurid marsupials of the Australian arid zone: An ecological review. In *Carnivorous marsupials,* ed. M. Archer, 117–30. Sydney: Royal Zoological Society of New South Wales. (14)

Morton, S. R. 1985. Granivory in arid regions: Comparison of Australia with North and South America. *Ecology* 66:1859–66. (14)

Morton, S. R. 1990. The impact of European settlement on the vertebrate animals of arid Australia: A conceptual model. *Proc. Ecol. Soc. Aust.* 16:201–13. (14)

Morton, S. R., and A. Baynes. 1985. Small mammal assemblages in arid Australia: A reappraisal. *Aust. Mammal.* 8:159–69. (14)

Morton, S. R., and D. W. Davidson. 1988. Comparative structure of harvester ant communities in arid Australia and North America. *Ecol. Monogr.* 58:19–38. (14, 15, 21)

Morton, S. R., and P. H. Davies. 1983. Food of the zebra finch (*Poephila guttata*), and an examination of granivory in birds of the Australian arid zone. *Aust. J. Ecol.* 8:235–43. (14)

Morton, S. R., and C. D. James. 1988. The diversity and abundance of lizards in arid Australia: A new hypothesis. *Am. Nat.* 132:237–56. (14, 15, 20, 21)

Mourer-Chauviré, C. 1975. Les Oiseaux du Pléistocène Moyen et Supérieur de France. Thèse d'Etat, Universitaire Claude Bernard, Lyon. (12)

Muller, J. 1964. A palynological contribution to the history of the mangrove vegetation in Borneo. In *Ancient Pacific floras: The pollen story,* ed. L. J. Cranwell, 33–42. Honolulu: University of Hawaii Press. (20)

Muller, J. 1981. Fossil pollen record of extant angiosperms. *Bot. Rev.* 47:1–142. (20, 26)

Muller, J. 1984. Significance of fossil pollen for angiosperm history. *Ann. Mo. Bot. Gard.* 71:419–73. (23)

Murdoch, W. W., C. J. Briggs, R. M. Nisbet, W. S. C. Gurney, and A. Stewart-Oaten. 1992. Aggregation and stability in metapopulation models. *Am. Nat.* 140:41–58. (7)

Murdoch, W. W., and A. Oaten. 1975. Predation and population stability. *Adv. Ecol. Res.* 9:2–132. (7)

Murdoch, W. W., and A. Stewart-Oaten. 1989. Aggregation by parasitoids and predators: Effects on equilibrium and stability. *Am. Nat.* 134:288–310. (7)

Murphy, P. G. 1976. Electrophoretic evidence that selection reduces ecological overlap in marine limpets. *Nature* 261:228–30. (4)

Murray, J. D. 1975. Non-existence of wave solutions for the class of reaction-diffusion equations given by the Volterra interacting-population equations with diffusion. *J. Theor. Biol.* 52:459–69. (8)

Mushinsky, H. R. 1987. Foraging ecology. In *Snakes: Ecology and evolutionary biology*, ed. R. A. Seigel, J. T. Collins, and S. S. Novak, 302–34. New York: Macmillan. (25)

Myers, C. W. 1974. The systematics of *Rhadinaea* (Colubridae), a genus of New World snakes. *Bull. Am. Mus. Nat. Hist.* 153:1–262. (25)

Myers, C. W. 1982. Blunt-headed vine snakes (*Imantodes*) in Panama, including a new species and other revisionary notes. *American Museum Novitates*, no. 2738, 1–50. (25)

Myers, C. W. 1986. An enigmatic new snake from the Peruvian Andes, with notes on the Xenodontini (Colubridae, Xenodontinae). *American Museum Novitates*, no. 2853, 1–12. (25)

Nachmann, G. 1987. Systems analysis of acarine predator-prey interactions II. The role of spatial processes in system stability. *J. Anim. Ecol.* 56:267–81. (8)

Nadelhoffer, K. J., J. D. Aber, and J. M. Melillo. 1983. Leaf litter production and soil organic matter dynamics along a nitrogen availability gradient in southern Wisconsin (USA). *Can. J. For. Res.* 13:12–21. (2)

Nadelhoffer, K. J., J. D. Aber, and M. M. Melillo. 1984. Seasonal patterns of ammonium and nitrate uptake in nine temperate forest ecosystems. *Plant and Soil* 80:321–335. (2)

Nadelhoffer, K. J., A. E. Giblin, G. R. Shaver, and A. E. Linkins. 1990. Microbial processes and plant nutrient availability in arctic soils. In *Physiology of Arctic plants: Implications for climate change*, ed. Chapin et al. Academic Press. (2)

Nagy, K. A. 1987. Field metabolic rate and food requirement scaling in mammals and birds. *Ecol. Monogr.* 57:111–28. (6)

Namba, T. 1989. Competition for space in a heterogeneous environment. *J. Math. Biol.* 27:1–16. (8)

National Geographic Society. 1987. *Field Guide to the Birds of North America.* 2d. ed. Washington, D.C.: National Geographic Society. (13)

Nee, S., and R. M. May. 1992. Dynamics of metapopulations: Habitat destruction and competitive coexistence. *J. Anim. Ecol.* 61:37–40. (7)

Neill, W. E. 1974. The community matrix and the interdependence of the competition coefficients. *Am. Nat.* 108:399–408. (1)

Neimela, P., S. Hanhimaki, and R. Mannila. 1981. The relationship of adult size to noctuid moths (Lepidoptera, Noctuidae) to breadth of diet and growth form of host plants. *Ann. Entomol. Fennici* 47:17–20. (23)

Nel, J. A. J. 1975. Habitat heterogeneity and changes in small mammal community structure and resource utilization in the southern Kalahari. *Bull. Carnegie Mus. Nat. Hist.* 6:118–31. (14)

Nelsen, C. M. 1978. *Neptunea* (Gastropoda: Buccinacea) in the Neogene of the North Pacific and adjacent Bering Sea. *Veliger* 21:203–15. (29)

Nelson, C. J., J. N. Seiber, and L. P. Brower. 1981. Seasonal and intraplant variation of cardenolide content in the California milkweed *Asclepias eriocarpa*, and implications for plant defense. *J. Chem. Ecol.* 7:981–1010. (23)

Nelson, G. 1979. Cladistic analysis and synthesis: Principles and definitions, with a historical note on Adanson's *Familles des Plantes* (1763–1767). *Syst. Zool.* 28:1–21. (24)

Nelson, G. 1983. Reticulation in cladograms. In *Advances in cladistics: Proceedings of the second meeting of the Willi Hennig Society*, ed. N. I. Platnick and V. A. Funk, 105–11. New York: Columbia University Press. (24)

Nelson, G. 1984. Identity of the anchovy *Hildebrandichthys setiger* with notes on relationships and biogeography of the genera *Engraulis* and *Cetengraulis*. *Copeia* 1984:422–27. (24)

Nelson, G., and D. E. Rosen (eds.). 1981. *Vicariance biogeography: A critique.* New York: Columbia University Press. (26)

Nelson, G., and N. Platnick. 1981. *Systematics and biogeography: Cladistics and vicariance.* New York: Columbia University Press. (23, 24)

Nelson, H. F., C. W. Brown, and J. H. Brineman. 1962. Skeletal limestone classification. *Mem. Am. Assoc. Petroleum Geol.* 1:224–52. (27)

Neter, J., and W. Wasserman. 1974. Applied linear statistical models. Homewood, Ill.: Irwin. (21)

Neumann, A. C., and I. G. Macintyre. 1985. Reef response to sea-level rise: Keep-up, catch-up, or give-up? *Proc. Vth Int. Reef Cong.* (Tahiti) 3:105–10. (27)

Neuvonen, S., and P. Neimela. 1983. Species richness and faunal similarity of arboreal insect herbivores. *Oikos* 40:452–59. (23)

Newell, N. D. 1971. An outline history of tropical organic reefs. *American Museum Novitates*, no. 2465, 1–37. (27)

Newman, C. M., and J. E. Cohen. 1986. A stochastic theory of community food webs. IV. Theory of food chain lengths in large webs. *Proc. Roy. Soc.* (Lond.), ser. B 228:355–77. (3)

Newman, E. I. 1973. Competition and diversity in herbaceous vegetation. *Nature* 244:310–11. (5)

Newton, C. R. 1988. Significance of "Tethyan" fossils in the American Cordillera. *Science* 242:385–91. (20)

Nicholls, N. 1988. El Niño–Southern Oscillation and rainfall variability. *J. Clim.* 1:418–21. (15)

Nicholls, N. 1991. The El Niño–Southern Oscillation and Australian vegetation. *Vegetatio* 91:23–36. (15)

Nicholls, N., and K. Wong. 1990. Dependence of rainfall variability on mean rainfall, latitude, and the Southern Oscillation. *J. Clim.* 3:163–70. (14)

Nichols, F. H., J. K. Thompson, and L. E. Schemel. 1990. Remarkable invasions of San Francisco Bay (California, USA) by the Asian clam *Potamocorbula amurensis*. II. Displacement of a former community. *Mar. Ecol. Prog. Ser.* 68:95–101. (29)

Nicholson, A. J. 1954. An outline of the dynamics of animal populations. *Aust. J. Zool.* 2:9–65. (5)

Nicotri, M. E. 1977. Grazing effects of four marine intertidal herbivores on the microflora. *Ecology* 58:1020–32. (4)

Nielsen, B. E. 1971. Coumarin profiles in the Umbelliferae. In *The biology and chemistry of the Umbelliferae*, ed. V. Heywood, 25–336. London: Academic Press. (23)

Niemi, G. J., J. M. Hanowski, J. Kouki, and D. Rajasärkkä. 1983. Inter-continental comparisons of habitat structure as related to bird distributions in peatlands of eastern Finland and northern Minnesota. In *International symposium on peat utilization*, (Bemidji State University, Bemidji, Minn.), ed. C. H. Fuchsman and S. A. Spigarelli, 59–73. (21)

Niklas, K. J., B. H. Tiffney, and A. H. Knoll. 1985. Patterns in vascular land plant diversification: An analysis at the species level. In *Phanerozoic diversity patterns: Profiles in macroevolution*, ed. J. W. Valentine, 97–128. Princeton, N.J.: Princeton University Press. (26)

Nilsson, C. 1987. Distribution of a stream-edge vegetation along a gradient of current velocity. *J. Ecol.* 75:513–22. (6)

Nilsson, C., G. Grelsson, M. Johansson, and U. Sperens. 1989. Patterns of plant species richness along riverbanks. *Ecology* 70:77–84. (6)

Nilsson, C., and P. A. Keddy. 1988. Predictability of change in shoreline vegetation in a hydroelectric reservoir, northern Sweden. *Can. J. Fish. Aquat. Sci.* 45:1896–1904. (6)

Nilsson, S. G., and I. N. Nilsson. 1978. Breeding bird community densities and species richness in lakes. *Oikos* 31:214–21. (6)

Nilsson, T. 1983. *The Pleistocene.* Dordrecht: D. Reidel. (26)

Nishio, S., M. S. Blum, and S. Takahashi. 1983. Intraplant distribution of cardenolides in *Asclepias humistrata* (Asclepiadaceae), with additional notes on their fates in *Tetraopes melanurus* (Coleoptera: Cerambycidae) and *Rhyssomatus lineaticollis* (Coleoptera: Curculionidae). *Mem. Coll. Agric. Kyoto Univ.* 122:43–52. (23)

Noble, E. R. 1960. Fishes and their parasite-mix as objects for ecological studies. *Ecology* 41:593–96. (17)

Noon, B. R. 1981. The distribution of an avian guild along a temperate elevational gradient: The importance and expression of competition. *Ecol. Monogr.* 51:105–24. (1)

Odum, E. P. 1969. The strategy of ecosystem development. *Science* 164:262–70. (24)

Odum, E. P. 1971. *Fundamentals of Ecology.* 3d ed. Philadelphia: Saunders. (6)

Odum, H. T. 1970. Summary: An emerging view of the ecological system at El Verde. Chap. I-10 in *A Tropical Rainforest*, ed. H. T. Odum and R. F. Pigeon. I-191–I-289. Oak Ridge, Tenn.: U.S. Atomic Energy Commission. (6)

Ofer, J., A. Shulov, and I. Noy-Meir. 1978. Associations of ant species in Israel: A multivariate analysis. *Israel J. Zool.* 27:199–208. (14)

O'Grady, R. T. 1989. Parasite:host specificity. In *Parasitology: The biology of animal parasites,* 6th ed., ed. E. R. Noble and G. A. Noble, 495–511. Philadelphia: Lea and Febiger. (24)

Ohwi, J. 1965. *Flora of Japan* (in English). Washington, D.C.: Smithsonian Institution. (26)

Oksanen, L. 1988. Ecosystem organization: Mutualism and cybernetics or plain Darwinian struggle for existence? *Am. Nat.* 131:424–44. (24)

Oksanen, L., S. D. Fretwell, J. Arruda, and P. Niemela. 1981. Exploitation ecosystems in gradients of primary productivity. *Am. Nat.* 118:240–61. (5)

Oksanen, T. 1990. Exploitation ecosystems in heterogeneous habitat complexes. *Evol. Ecol.* 4:220–34. (7)

Okubo, A. 1980. *Diffusion and ecological problems: Mathematical models.* Biomathematics, 10. Berlin: Springer-Verlag. (7, 8)

Oliver, J. 1982. The geographic and environmental aspects of mangrove communities: Climate. In *Mangrove ecosystems in Australia,* ed. B. F. Clough, 19–30. Canberra: Australian National University Press. (20)

Oliver, J. A. 1948. The relationships and zoogeography of the genus *Thalerophis* Oliver. *Bull. Am. Mus. Nat. Hist.* 92:157–280. (25)

Olson, S. L., and H. F. James. 1982. Fossil birds from the Hawaiian Islands: Evidence for wholesale extinction by man before western contact. *Science* 217:633–35. (12)

O'Neill, R. V., D. L. DeAngelis, J. B. Waide, and T. F. H. Allen. 1986. *A hierarchical concept of ecosystems.* Princeton, N.J.: Princeton University Press. (1, 8)

Opler, P. A. 1973. Fossil lepidopterous leaf-mines demonstrate the age of some insect-plant relationships. *Science* 179:1321–23. (23)

Opler, P. A. 1974. Oaks as evolutionary islands for leaf-mining insects. *Am. Sci.* 62:67–73. (22)

Orians, G. H., and R. T. Paine. 1983. Convergent evolution at the community level. In *Coevolution,* ed. D. J. Futuyma and M. Slatkin, 431–58. Sunderland, Mass.: Sinauer Associates. (1, 14, 20, 21)

Orians, G. H., and O. T. Solbrig, eds. 1977. *Convergent evolution in warm deserts.* Stroudsberg, Pa.: Dowden, Hutchinson and Ross. (21)

Ortega, S. 1985. Competitive interactions among tropical intertidal limpets. *J. Exp. Mar. Biol. Ecol.* 90:21–25. (4)

Otte, D., and J. A. Endler, eds. 1989. *Speciation and its consequences.* Sunderland, Mass.: Sinauer Associates. (1)

Owen, J. G. 1988. On productivity as a predictor of rodent and carnivore diversity. *Ecology* 69:1161–65. (5, 6)

Pacala, S. W. 1986a. Neighborhood models of plant population dynamics. 2. Multispecies models of annuals. *Theor. Pop. Biol.* 29:262–92. (2)

Pacala, S. W. 1986b. Neighborhood models of plant population dynamics. 4. Single-species and multispecies models of annuals with dormant seeds. *Am. Nat.* 128:859–78. (2)

Pacala, S. W. 1987. Neighborhood models of plant population dynamics. 3. Models with spatial heterogeneity in the physical environment. *Theor. Pop. Biol.* 31:359–92. (8)

Pacala, S. W., and J. Roughgarden. 1982. Spatial heterogeneity and interspecific competition. *Theor. Pop. Biol.* 121:92–113. (8)

Pacala, S. W., and J. A. Silander, Jr. 1985. Neighborhood models of plant population dynamics. I. Single-species models of annuals. *Am. Nat.* 125:385–411. (2)

Pace, M. L., G. A. Knauer, D. M. Karl, and J. H. Martin. 1987. Primary production, new production and vertical flux in the eastern Pacific Ocean. *Nature* 325:803–04. (6)

Page, C. N. 1976. The taxonomy and phytogeography of bracken—a review. *Bot. J. Linn. Soc. Lond.* 73:1–34. (16)

Page, C. N. 1990. Taxonomic evaluation of the fern genus *Pteridium* and its active evolutionary state. In *Bracken biology and management,* ed. J. A. Thomson and R. T. Smith, 23–34. New South Wales, Australia: RIAS. (16)

Page, R. D. M. 1987. Graphs and generalized tracks: Quantifying Croizat's panbiogeography. *Syst. Zool.* 36:1–17. (24)

Page, R. D. M. 1988. Quantitative cladistic biogeography: Constructing and comparing area cladograms. *Syst. Zool.* 37:254–70. (23, 24)

Page, R. D. M. 1990. Temporal congruence and cladistic analysis of biogeography and cospeciation. *Syst. Zool.* 39:205–26. (23)

Paine, R. T. 1966. Food web complexity and species diversity. *Am. Nat.* 100:65–75. (1, 2, 5, 11, 24)

Paine, R. T. 1974. Intertidal community structure: Experimental studies on the relationship between a dominant competitor and its principal predator. *Oecologia* 15:93–120. (1, 4)

Paine, R. T. 1980. Food webs: Linkage, interaction strength and community infrastructure. *J. Anim. Ecol.* 49:667–85. (3)

Paine, R. T., and S. A. Levin. 1981. Intertidal landscapes: Disturbance and the dynamics of patterns. *Ecol. Monogr.* 51:145–78. (5, 8, 19)

Paine, R. T., and A. R. Palmer. 1978. *Sicyases sanguineus:* A unique trophic generalist from the Chilean intertidal zone. *Copeia* 1:75–81. (4)

Palmer, A. R. 1990. Effect of crab effluent and scent of damaged conspecifics on feeding, growth, and shell morphology of the Atlantic dogwhelk *Nucella lapillus* (L.). *Hydrobiologia* 193:155–82. (4)

Palmer, E., and N. Pitman. 1972. *Trees of Southern Africa,* vol. III. Cape Town: A. A. Balkema. (20)

Palombo, M. R. 1985. I grandi mammiferi pleistocenici delle isole del Mediterraneo: Tempi e vie di migrazione. *Bolletina della Societa Paleontologica Italiana* 24:201–224. (12)

Parker, G. G. 1983. Throughfall and stemflow in the forest nutrient cycle. *Adv. Ecol. Res.* 13:57–133. (2)

Parmenter, R. R., and J. A. MacMahon. 1988. Factors influencing species composition and population sizes in a ground beetle community (Carabidae): Predation by rodents. *Oikos* 52:350–56. (18)

Parrish, J. D., and S. B. Saila. 1970. Interspecific competition, predation and species diversity. *J. Theor. Biol.* 27:207–20. (24)

Parish, J. T. 1982. Upwelling and petroleum source beds, with reference to the Paleozoic. *Bull. Am. Assoc. Petroleum Geol.* 66:750–74. (27)

Pastor, J., J. D. Aber, C. A. McClaugherty, and J. M. Melillo. 1984. Aboveground production and N and P cycling along a nitrogen mineralization gradient on Blackhawk Island, Wisconsin. *Ecology* 65:256–68. (2)

Patrick, R. A. 1963. The structure of diatom communities under varying ecological conditions. *Trans. N.Y. Acad. Sci.* 108:359–65. (5)

Patten, B. C., and E. P. Odum. 1981. The cybernetic nature of ecosystems. *Am. Nat.* 118:886–95. (24)

Paulay, G. 1991a. Effects of late Cenozoic sea-level fluctuations on the bivalve faunas of tropical oceanic islands. *Paleobiology* 16:415–34. (29)

Paulay, G. 1991b. Late Cenozoic sea level fluctuations and the diversity and species composition of insular shallow water marine faunas. In *The unity of evolutionary biology,* ed. E. C. Dudley, 184–93. Portland, Ore.: Dioscorides Press. (29)

Paulay, G., and T. Spencer. 1989. Geomorphology, paleoenvironments and faunal turnover, Henderson Island, S. E. Polynesia. *Proc. VIth Int. Coral Reef Symp.* 3:461–66. (29)

Pearl, R., and L. J. Reed. 1920. On the growth rate of the population of the United States since 1790 and its mathematical representation. *Proc. Natl. Acad. Sci. USA* 6:275–88. (1)

Pearson, D. L. 1977. A pantropical comparison of bird community structure on six lowland forest sites. *Condor* 79:232–44. (1, 18)

Pearson, D. L. 1980. Patterns of limiting similarity in tropical forest tiger beetles (Coleoptera: Cicindelidae). *Biotropica* 12:195–204. (18)

Pearson, D. L. 1982. Historical factors and bird species richness. In *Biological diversification in the tropics,* ed. G. T. Prance, 441–52. New York: Columbia University Press. (18)

Pearson, D. L. 1985. The function of multiple anti-predator mechanisms in adult tiger beetles (Coleoptera: Cicindelidae). *Ecol. Entomol.* 10:65–72. (18)

Pearson, D. L. 1986. Community structure and species co-occurrence: A basis for developing broader generalizations. *Oikos* 46:419–23. (18)

Pearson, D. L. 1988. Biology of tiger beetles. *Annu. Rev. Entomol.* 33:123–147. (18)

Pearson, D. L., M. S. Blum, T. H. Jones, H. M. Fales, E. Gonda, and

B. R. Witte. 1988. Historical perspective and the interpretation of ecological patterns: Defensive compounds of tiger beetles (Coleoptera: Cicindelidae). *Am. Nat.* 132:404–16. (18)

Pearson, D. L., and F. Cassola. 1992. Worldwide species richness patterns of tiger beetles (Coleoptera: Cicindelidae): Indicator taxon for biodiversity and conservation. *Conserv. Biol.* 6:376–91. (18)

Pearson, D. L., and K. Ghorpade. 1989. Geographical distribution and ecological history of tiger beetles (Coleoptera: Cicindelidae) of the Indian subcontinent. *J. Biogeogr.* 16:333–34. (18)

Pearson, D. L., and S. A. Juliano. 1991. Mandible length ratios as a mechanism for co-occurence: Evidence from a world-wide comparison of tiger beetle assemblages (Cicindelidae). *Oikos* 60:223–33. (18)

Pearson, D. L., and C. B. Knisley. 1985. Evidence for food as a limiting resource in the life cycle of tiger beetles (Coleoptera: Cicindelidae). *Oikos* 45:161–68. (18)

Pearson, D. L., and R. C. Lederhouse. 1987. Thermal ecology and the structure of an assemblage of adult tiger beetle species (Cicindelidae). *Oikos* 50:247–55. (18)

Pearson, D. L., and E. J. Mury. 1979. Character divergence and convergence among tiger beetles (Coleoptera: Cicindelidae). *Ecology* 60:557–66. (18)

Pease, C. M., R. Lande, and J. J. Bull. 1989. A model of population growth, dispersal, and evolution in a changing environment. *Ecology* 70:1657–64. (7)

Pérez Higareda, G. 1978. Reptiles and amphibians from the Estación Biología Tropical "Los Tuxtlas" (U.N.A.M.), Veracruz, México. *Bull. Md. Herpetol. Soc.* 14:67–74. (25)

Perrins, C. M. 1965. Population fluctuations and clutch-size in the Great Tit *Parus major. J. Anim. Ecol.* 34:601–47. (12)

Perry, D. A., M. T. Amaranthus, J. G. Borchers, and F. L. Borchers. 1990. Species migration and ecosystem stability during climate change: The belowground connection. *Conserv. Biol.* 4:266–74. (29)

Persson, L., S. Diehl, L. Johansson, G. Anderson, and S. F. Hamrin. 1992. Trophic interactions in temperate lake ecosystems: A test of food chain theory. *Am. Nat.* 140:59–84. (3)

Peters, J. A. 1960. The snakes of the subfamily Dipsadinae. *Misc. Publ. Mus. Zool., Univ. Mich.* 114:1–224. (25)

Peters, R. H. 1983. *The ecological implications of body size.* New York: Cambridge University Press. (2)

Peterson, C. H., and R. Black. 1988. Density-dependent mortality caused by physical stress interacting with biotic history. *Am. Nat.* 131:257–70. (24)

Petraitis, P. S. 1991. Requitment of the mussel *Mytilus edulis* L. on sheltered and exposed shores in Maine, U.S.A. *J. Exp. Mar. Biol. Ecol.* 147:65–80. (4)

Petraitis, P. S., R. E. Latham, and R. A. Niesenbaum. 1989. The maintenance of species diversity by disturbance. *Q. Rev. Biol.* 64:393–418. (2, 5, 26)

Petrides, G. A. 1972. *A field guide to trees and shrubs.* Boston: Houghton-Mifflin. (26)

Phipps, J. B. 1983. Biogeographic, taxonomic, and cladistic relationships between east Asiatic and North American *Crataegus. Ann. Mo. Bot. Gard.* 70:667–700. (26)

Phipps, J. B., K. R. Robertson, P. G. Smith, and J. R. Rohrer. 1990. A checklist of the subfamily Maloideae (Rosaceae). *Can. J. Bot.* 68:2209–69. (26)

Pianka, E. R. 1966. Latitudinal gradients in species diversity: A review of concepts. *Am. Nat.* 100:33–46. (1, 5, 20, 26, 30)

Pianka, E. R. 1967. On lizard species diversity: North American flatland deserts. *Ecology* 48:333–51. (1)(6)

Pianka, E. R. 1972. Zoogeography and speciation of Australian desert lizards: An ecological perspective. *Copeia* 1972:127–45. (14)

Pianka, E. R. 1973. The structure of lizard communities. *Annu. Rev. Ecol. Syst.* 4:53–74. (1, 14, 21)

Pianka, E. R. 1975. Niche relations of desert lizards. In *Ecology and Evolution of Communities*, ed., M. L. Cody and J. M. Diamond, 292–314. Harvard University Press, Cambridge, Massachusetts. (1)

Pianka, E. R. 1978. *Evolutionary Ecology.* New York: Harper and Row. (4, 23)

Pianka, E. R. 1986. *Ecology and natural history of desert lizards.* Princeton, N.J.: Princeton University Press. (14)(15)(21)(25)

Pianka, E. R. 1989a. Desert lizard diversity: Additional comments and some data. *Am. Nat.* 134:344–64. (14, 18, 21)

Pianka, E. R. 1989b. Latitudinal gradients in species diversity. *Trends Ecol. Evol.* 4:223. (25, 26)

Pianka, E. R., and R. B. Huey. 1971. Bird species density in the Kalahari and the Australian deserts. *Koedoe* 14:123–29. (6, 14)

Pianka, E. R., and R. B. Huey. 1978. Comparative ecology, niche segregation, and resource utilization among gekkonid lizards in the southern Kalahari. *Copeia* 1978:691–701. (25)

Pichon, M. 1985. Scleractinia. In *Faune et flore: Premier abrege des organismes marins de Polynesie Française,* ed. G. Richard, 399–406. Vth International Congress on Coral Reefs (Tahiti), no. 1. (27)

Pickett, S. T. A. 1976. Succession: An evolutionary interpretation. *Am. Nat.* 110:107–19. (7)

Pickett, S. T. A., and P. S. White, eds. 1985. *The ecology of natural disturbance and patch dynamics.* New York: Academic Press. (2, 12)

Pielou, E. C. 1969. *An introduction to mathematical ecology.* New York: Wiley. (1, 11)

Pielou, E. C. 1977. *Mathematical ecology.* New York: Wiley. (1)

Pielou, E. C. 1979. *Biogeography.* New York: John Wiley and Sons. (13)

Pierce, N. E. 1989. Amplified species diversity: A case study of an Australian lycaenid butterfly and its attendent ants. In *The biology of butterflies*, ed. R. I. Vane-Wright and P. R. Ackery, 197–200. Princeton, N.J.: Princeton University Press. (30)

Pimentel, D., E. H. Feinberg, P. W. Wood, and J. T. Hayes. 1965. Selection, spatial distribution, and the coexistence of competing fly species. *Am. Nat.* 99:97–109. (30)

Pimm, S. L. 1982. *Food webs.* London: Chapman and Hall. (1, 3, 5)

Pimm, S. L. 1991. *The Balance of Nature? Ecological Issues in the Conservation of Species and Communities.* Chicago: University of Chicago Press. (7, 29)

Pimm, S. L., J. H. Lawton, and J. E. Cohen. 1991. Food web patterns and their consequences. *Nature* 350:669–74. (5, 7)

Pimm, S. L., and R. L. Kitching. 1987. The determinants of food chain lengths. *Oikos* 50:302–07. (3)

Pimm, S. L., and J. H. Lawton. 1977. The number of trophic levels in ecological communities. *Nature* 268:329–31. (3)

Pimm, S. L., and J. H. Lawton. 1978. On feeding on more than one trophic level. *Nature* 275:542–44. (3)

Pimm, S. L., and J. H. Lawton. 1980. Are food webs divided into compartments? *J. Anim. Ecol.* 49:879–98. (1)

Pimm, S. L., and J. C. Rice. 1987. The dynamics of multispecies, multi-life-stage models of aquatic food webs. *Theor. Pop. Biol.* 32:303–25. (3)

Pimm, S. L., and M. L. Rosenzweig. 1981. Competitors and habitat use. *Oikos* 37:1–6. (5)

Pimm, S. L., M. L. Rosenzweig, and W. A. Mitchell. 1985. Competition and food selection: Field tests of a theory. *Ecology* 66:798–807. (5)

Pizzey, G. 1980. *A Field Guide to the Birds of Australia.* Princeton, N.J.: Princeton University Press. (13)

Platnick, N. I., and G. Nelson. 1978. A method of analysis for historical biogeography. *Syst. Zool.* 27:1–16. (24)

Platt, T., and K. Denman. 1977. Organization in the pelagic ecosystem. *Helgolander wiss. Meer.* 30:575–81. (3)

Platt, T., and K. Denman. 1978. The structure of pelagic marine ecosystems. *Rapp. P.-V. Reun. Cons. Int. Explor. Mer* 173:60–65. (3)

Platt, W., and I. Weis. 1977. Resource partitioning and competition within a guild of fugitive prairie plants. *Am. Nat.* 111:479–513. (2)

Polis, G. A. 1991. Complex desert food webs: An empirical critique of food web theory. *Am. Nat.* 138:123–55. (3)

Pollard, E., M. L. Hall, and T. J. Bibby. 1986. *Monitoring the abundance of butterflies, 1976–1985.* Publication no. 2, Nature Conservation Council, Britain. (10)

Pons, A. 1981. The history of the Mediterranean shrublands. In *Maquis and chaparrals,* ed. F. di Castri, D. W. Goodall, and

R. L. Specht, 131–38. Ecosystems of the world. Amsterdam: Elsevier. (12)

Pons, A. 1984. Les changements de la végétation de la région méditerrané durant le Pliocène et le Quaternaire en relation avec l'histoire du climat et de l'action de l'homme. *Webbia* 38:427–39. (12)

Pons, A., and M. Reille. 1986. Nouvelles recherches pollenanalytiques à Padul (Granada), à la fin du dernier glaciaire et l'Holocène. *Proc. Symp. Quat. Clim. West. Med.* (Madrid), 405–20. (12)

Popescu, C. P., J. P. Quere, and P. Franceschi. 1980. Observations chromosomiques sur le sanglier français (*Sus scrofa scrofa*). *Ann. Gén. Sél. anim.* 12:395–400. (12)

Poplin, F. 1979. Origine du Mouflon de Corse dans une nouvelle perspective paléontologique: par marronnage. *Ann. Gén. Sél. anim.* 11:133–43. (12)

Por, F. D. 1978. *Lessepsian migration: The influx of Red Sea biota into the Mediterranean by way of the Suez Canal.* Berlin: Springer-Verlag. (29, 30)

Post, W. M., and S. L. Pimm. 1983. Community assembly and food web stability. *Math. Biosci.* 64:169–92. (3)

Potter, F. W., and D. L. Dilcher. 1980. Biostratigraphic analysis of Eocene clay deposits in Henry County, Tennessee. In *Biostratigraphy of fossil plants: Successional and paleoecological analyses,* ed. D. L. Dilcher and T. N. Taylor, 211–25. Stroudsburg, Pa.: Dowden, Hutchinson and Ross. (26)

Pough, F. H., and J. D. Groves. 1983. Specializations of the body form and food habits of snakes. *Am. Zool.* 23:443–54. (25)

Powell, T. M. 1989. Physical and biological scales of variability in lakes, estuaries, and the coastal ocean. In *Perspectives in ecological theory,* ed. J. Roughgarden, R. M. May, and S. A. Levin, 157–76. Princeton, N.J.: Princeton University Press. (8)

Prakash, U. 1960. A survey of the Deccan Intertrappean flora of India. *J. Palaeontol.* 34:1027–40. (20)

Prance, G. T., ed. 1982. *The biological model of diversification in the tropics.* New York: Columbia University Press. (20, 23, 25)

Prance, G. T., W. A. Rodriguez, and M. F. DaSilva. 1976. Inventario florestal de um hectare de mata de terra firme km 30 da estrada Manaus-Itacoatiara. *Acta Amazonica* 6:9–35. (23)

Preston, F. W. 1962. The canonical distribution of commonness and rarity. *Ecology* 43:185–215; 410–32. (5, 7)

Price, P. W. 1977. General concepts in the evolutionary biology of parasites. *Evolution* 31:405–20. (22)

Price, P. W. 1980. *Evolutionary biology of parasites.* Princeton, N.J.: Princeton University Press. (17, 23)

Price, P. W. 1983. Hypotheses on organization and evolution in herbivorous insect communities. In *Variable plants and herbivores in natural and managed systems,* ed. R. F. Denno and M. S. McClure, 559–96. New York: Academic Press. (22, 23)

Price, P. W., and K. M. Clancy. 1983. Patterns in number of helminth parasite species in freshwater fishes. *J. Parasitol.* 69:449–54. (17)

Price, P. W., and H. Pschorn-Walcher. 1988. Are galling insects better protected against parasitoids than exposed feeders?: A test using tenthredinid sawflies. *Ecol. Entomol.* 13:195–205. (23)

Price, P. W., C. N. Slobodchikoff, and W. S. Gaud, eds. 1984. *A new ecology: Novel approaches to interactive systems.* New York: Wiley Interscience. (17)

Prodon, R., R. Fons, and F. Athias-Binche. 1987. The impact of fire on animal communities in mediterranean area. In *The role of fire in ecological systems,* ed. L. Trabaud, 121–57. The Hague: SPB Academic Publishing. (12)

Prodon, R., and J. D. Lebreton. 1981. Breeding avifauna of a Mediterranean succession: The Holm Oak and Cork Oak series in the eastern Pyrénées. I. Analysis and modelling of the structure gradient. *Oikos,* 37:21–38. (12)

Puerto, A., M. Rico, M. D. Matias, and J. A. Garcia. 1990. Variation in structure and diversity in Mediterranean grasslands related to trophic status and grazing intensity. *J. Vegetation Sci.* 1:445–452. (2)

Pulliam, H. R. 1988. Sources, sinks, and population regulation. *Am. Nat.* 132:652–61. (1, 7, 12, 22, 30)

Quézel, P. 1985. Definition of the Mediterranean region and origin of its flora. In *Plant conservation in the Mediterranean area,* ed. C. Gomez-Campo, 9–24. Dordrecht: Junk. (12)

Quinn, J. F., and A. E. Dunham. 1983. On hypothesis testing in ecology and evolution. *Am. Nat.* 122:602–17. (18)

Quinn, J. F., and S. P. Harrison. 1988. Effect of habitat fragmentation and isolation on species richness: Evidence from biogeographic patterns. *Oecologia* 75:132–40. (6)

Rabinovich, J. E., and E. H. Rapoport. 1975. Geographical variation of diversity in Argentine passerine birds. *J. Biogeogr.* 2:141–57. (6)

Rabinowitz, D. 1978. Dispersal properties of mangrove propagules. *Biotropica* 10:47–57. (20)

Rachele, L. D. 1976. Palynology of the Lagler lignite: A deposit in the Tertiary Cohansey Formation of New Jersey, USA. *Rev. Palaeobot. Palynol.* 22:225–52. (26)

Rage, J.-C. 1972. Les amphibiens et les reptiles du gisement des abîmes de la Fage. *Nouv. Arch. Mus. d'Hist. Nat. Lyon* 10:79–90. (12)

Ragni, B. 1981. Gatto Selvatico *Felis silvestris* Schreber 1777. In *Distribuzione e Biologie di 22 specie di Mammiferi in Italia,* 105–113. Roma: Consilio nazionale delle Ricerche. (12)

Ralph, C. J. 1985. Habitat association patterns of forest and steppe birds of northern Patagonia, Argentina. *Condor* 87:471–83. (21)

Rand, A. S., and C. W. Myers. 1990. The herpetofauna of Barro Colorado Island, Panama: An ecological summary. In *Four Neotropical rainforests,* ed. A. Gentry, 386–409. New Haven: Yale University Press. (25)

Rand McNally. 1969. *International Atlas.* Chicago: Rand McNally (26)

Rashbrook, V. K., S. G. Compton, and J. H. Lawton. 1992. Antherbivore interactions: reasons for the absence of benefits to a fern with foliar nectaries. *Ecology* 73:2167–74. (16)

Rau, G. H., A. J. Mearns, D. R. Young, R. J. Olson, H. A. Schafer, and I. R. Kaplan. 1983. Animal $^{13}C/^{12}C$ correlates with trophic level in pelagic food webs. *Ecology* 64:1314–18. (3)

Raup, D. M. 1972. Taxonomic diversity during the Phanerozoic. *Science* 177:1065–71. (27)

Raup, D. M. 1975. Taxonomic diversity estimation using rarefaction. *Paleobiology* 1:333–42. (27)

Raup, D. M., and S. J. Gould. 1974. Stochastic simulation and evolution of morphology—towards a nomothetic paleontology. *Syst. Zool.* 23:305–22. (11)

Raup, D. M., S. J. Gould, T. J. M. Schopf, and D. S. Simberloff. 1973. Stochastic models of phylogeny and the evolution of diversity. *J. Geol.* 81:525–42. (11, 23)

Raup, D. M., and J. J. Sepkoski. 1984. Periodicity of extinctions in the geologic past. *Proc. Natl. Acad. Sci. USA* 81:801–05. (27)

Raup, D. M., and J. J. Sepkoski. 1986. Periodic extinctions of families and genera. *Science* 231:833–36. (27)

Recher, H. F. 1969. Bird species diversity and habitat diversity in Australia and North America. *Am. Nat.* 103:75–80. (1, 6, 13, 15, 20, 21)

Reeve, J. D. 1988. Environmental variability, migration, and persistence in host-parasitoid systems. *Am. Nat.* 132:810–36. (7, 8)

Reeve, J. D. 1990. Stability, variability, and persistence in host-parasitoid systems. *Ecology* 71:422–26. (8)

Rehder, A. 1940. *Manual of cultivated trees and shrubs.* 2d ed. New York: Macmillan; Reprint Portland Ore.: Dioscorides Press, 1986. (26)

Reichman, O. J. 1983. Behaviour of desert heteromyids. In *Biology of desert rodents,* ed. O. J. Reichman and J. H. Brown, 77–90. Great Basin Naturalist Memoirs, no. 7. (14)

Reichman, O. J., and J. H. Brown, eds. 1983. *Biology of desert rodents.* Great Basin Naturalist Memoirs, no. 7. (14)

Reid, D. G. 1990. Trans-Arctic migration and speciation induced by climatic change: The biogeography of *Littorina* (Mollusca: Gastropoda). *Bull. Mar. Sci.* 235:64–66. (29)

Reid, E. M., and M. E. J. Chandler. 1933. *The flora of the London Clay.* London: British Museum of Natural History. (20, 26)

Reid, J. L., E. Brinton, A. Fleminger, E. L. Venrick, and J. A. McGowan. 1978. Ocean circulation and marine life. In ed. H. Carnock and G. Deacon, *Advances in oceanography: Proceedings of the General Symposium of the Joint Oceanographic Assembly* (13–24 September, 1976, Edinburgh, Scotland), 66–130. New York: Plenum Press.

Reille, M. 1984. Origine de la végétation actuelle de la Corse sudorientale; analyse pollinique de cinq marais côtiers. *Pollen et Spores* 26:43–60. (12)

Rejmánek, M., and P. Stary. 1979. Connectance in real biotic communities and critical values for stability of model ecosystems. *Nature* 280:311–13. (3)

Reumer, J. W. F., and S. Payne. 1986. Notes on the Soricidae (Insectivora, Mammalia) from Crete. II. The shrew remains from Minoan and Classical Kommos. *Bonn zoologischer Beiträge* 3:173–82. (12)

Rex, M. A. 1981. Community structure in the deep-sea benthos. *Annu. Rev. Ecol. Syst.* 12:331–53. (5, 6)

Rey, J. R. 1981. Ecological biogeography of arthropods on *Spartina* islands in northwest Florida. *Ecol. Monogr.* 51:237–65. (10)

Rey, J. R., and D. R. J. Strong. 1983. Immigration and extinction of salt marsh arthropods on islands: An experimental study. *Oikos* 41:396–401. (10)

Reyment, R. A. 1983. Palaeontological aspects of island biogeography: Colonization and evolution of mammals on Mediterranean islands. *Oikos* 41:299–306. (12)

Reynolds, R. P., and N. J. Scott. 1982. Use of a mammalian resource by a Chihuahuan snake community. *U.S. Dept. Interior, Fish Wild. Ser., Wild. Res. Rep.* 13:99–118. (25)

Rice, B. 1985. No evidence for divergence between Australia and elsewhere in plant species richness at tenth-hectare scale. *Proc. Ecol. Soc. Aust.* 14:99–101. (15)

Rice, B., and M. Westoby. 1983. Plant species richness at the 0.1 hectare scale in Australian vegetation compared to other continents. *Vegetatio* 52:129–40. (14, 15)

Rice, B., and M. Westoby. 1985. Plant species richness at the 0.1 hectare scale in Australian vegetation compared to other continents. In *Plant community ecology: Papers in honor of Robert H. Whittaker,* ed. R. K. Peet, 237–48. Dordrecht: Junk. (21)

Richard, G., ed. 1985. Faune et flore: Premier abrege des organismes marins de Polynesie Française. *Proc. Vth Int. Cong. Coral Reefs* (Tahiti) 1:379–518. (27)

Richerson, P., R. Armstrong, and C. R. Goldman. 1970. Contemporaneous disequilibrium: A new hypothesis to explain the "paradox of the plankton." *Proc. Natl. Acad. Sci. USA* 67:1710–14. (19)

Richerson, P. J., and K.-L. Lum. 1980. Patterns of plant species diversity in California: Relation to weather and topography. *Am. Nat.* 116:504–36. (1, 6, 26)

Richman, A., T. J. Case, and T. D. Schwaner. 1988. Natural and unnatural extinction rates of reptiles on islands. *Am. Nat.* 131:611–30. (7, 30)

Ricklefs, R. E. 1977. Environmental heterogeneity and plant species diversity: An hypothesis. *Am. Nat.* 111:376–81. (1, 26, 30)

Ricklefs, R. E. 1987. Community diversity: Relative roles of local and regional processes. *Science* 235:167–71. (1, 6, 7, 9, 11, 12, 14, 16, 18, 19, 20, 21, 22, 24, 26, 28, 30)

Ricklefs, R. E. 1989a. Spatial and temporal patterns and processes in communities of forest birds. *Ostrich,* suppl. 14:85–95. (12)

Ricklefs, R. E. 1989b. Speciation and diversity: Integration of local and regional processes. In *Speciation and its consequences,* ed. D. Otte and J. Endler, 599–622. Sunderland, Mass.: Sinauer Associates. (7, 11, 17, 23, 26, 30)

Ricklefs, R. E. 1990. Ecology. 3d ed. New York: W. H. Freeman. (1, 24)

Ricklefs, R. E., and G. W. Cox. 1972. Taxon cycles in the West Indian avifauna. *Am. Nat.* 106:195–219. (1, 20, 30)

Ricklefs, R. E., and R. E. Latham. 1992. Intercontinental correlation of geographical ranges suggests stasis in ecological traits of relict genera of temperate perennial herbs. *Am. Nat.* 139:1305–21. (1)

Ricklefs, R. E., and D. B. Miles. 1993. Ecological and evolutionary inferences from morphology: an ecological perspective. In *Ecological morphology: Integrative organismal biology,* ed. P. C. Wainright and S. Reilly. Chicago: University of Chicago Press. In press. (1, 30)

Ricklefs, R. E., and J. Travis. 1980. A morphological approach to the study of avian community organization. *Auk* 97:321–38. (1, 30)

Riddle, B. R., and R. L. Honeycutt. 1990. Historical biogeography in North American arid regions: An approach using mitochon-drial-DNA phylogeny in grasshopper mice (genus *Onychomys*). *Evolution* 44:1–15. (14)

Riebesell, J. F. 1974. Paradox of enrichment in competitive systems. *Ecology* 55:183–87. (5, 6)

Rigby, C., and J. H. Lawton. 1981. Species-area relationships of arthropods on host plants: Herbivores on bracken. *J. Biogeogr.* 8:125–33. (16)

Riley, E. G., and W. R. Enns. 1979. An annotated checklist of Missouri leaf beetles (Coleoptera: Chrysomelidae). *Trans. Mo. Acad. Sci.* 13:53–82. (23)

Riordan, J. 1958. An introduction to combinatorial analysis. New York: Wiley. (9)

Rivalier, E. 1954. Démembrement du genre *Cicindela* L. Faune américaine. *Rev. fr. d'Entomol.* 21:249–68. (18)

Rivalier, E. 1963. Démembrement du genre *Cicindela* L. (fin). V. Faune australienne. (Et liste récapitulative des genres et sousgenres proposés pour la faune mondiale). *Rev. fr. d'Entomol.* 30:30–48. (18)

Rivalier, E. 1969. Démembrement du genre *Odontochila* [Col. Cicindelidae] et révision des principales espèces. *Ann. Soc. Entomol. France* 5:195–237. (18)

Robinson, G. R., R. D. Holt, M. S. Gaines, S. P. Hamburg, E. A. Martinko, and H. S. Fitch. 1992. Diverse and contrasting effects of habitat fragmentation. *Science.* 257:524–26. (7)

Robinson, J. V., and J. E. Dickerson, Jr. 1987. Does invasion sequence affect community structure? *Ecology* 68:587–95. (24)

Roer, H. 1984. Zum Vorkommen und Beutefangverhalten des Sandlaufäkfers *Mantichora latipennis* Waterh. (Col.: Cicindelidae) in Südwestafrika/Namibia. *SWA Wissenschaftliche Gesellschaft* (Windhoek) 38:87–93. (18)

Roeske, C. N., J. N. Seiber, L. P. Brower, and C. M. Moffitt. 1976. Milkweed cardenolides and their comparative processing by monarch butterflies (*Danaus plexippus*). *Recent Adv. Phytochem.* 10:93–159. (23)

Rogovin, K. A., and A. V. Surov. 1990. Morpho-ecological structure of desert rodent communities in central Asia and southwestern North America: A multivariate approach. *Acta Theriologica* 35:225–39. (14)

Rohde, K. 1989. Simple ecological systems, simple solutions to complex problems? *Evol. Theory* 8:305–50. (17)

Romer, A. S. 1966. *Vertebrate Paleontology.* 3d ed. Chicago: University of Chicago Press. (28)

Root, R. B. 1967. The niche exploitation pattern of the blue-gray gnatcatcher. *Ecol. Monogr.* 37:317–50. (1, 3, 4)

Root, R. B. 1973. Organization of the plant-arthropod association in simple and diverse habitats: The fauna of collards (*Brassica oleracea*). *Ecol. Monogr.* 43:95–124. (17)

Rose, K. D. 1981. Composition and species diversity in Paleocene and Eocene assemblages: An empirical study. *J. Vert. Paleontol.* 1:367–88. (28)

Rose, K. D. 1990. Postcranial skeletal remains and adaptations in Early Eocene mammals from the Willwood Formation, Bighorn Basin, Wyoming. *Geol. Soc. Am. Spec. Paper,* no. 243, 107–33. (28)

Rosen, B. R. 1977. The depth distribution of Recent hermatypic corals and its palaeontological significance. *Mem. Bur. d'Rech. Geol. Mineral.* 89:507–17. (27)

Rosen, B. R. 1988. Progress, problems and patterns in the biogeography of reef corals and other tropical marine organisms. *Helgolander Meeresunters.* 42:269–301. (20)

Rosen, D. E. 1975. A vicariance model of Caribbean biogeography. *Syst. Zool.* 24:431–64. (24)

Rosen, D. E. 1985. Geological hierarchies and biogeographic congruence in the Caribbean. *Ann. Mo. Bot. Gard.* 72:636–59. (24)

Rosenzweig, M. L. 1966. Community structure in sympatric carnivores. *J. Mammal.* 47:602–20. (28)

Rosenzweig, M. L. 1968. Net primary productivity of terrestrial environments: Predictions from climatological data. *Am. Nat.* 102:67–84. (5, 6)

Rosenzweig, M. L. 1971. Paradox of enrichment: Destabilization of exploitation ecosystems in ecological time. *Science* 171:385–87. (5, 6)

Rosenzweig, M. L. 1972. Stability of enriched aquatic ecosystems. *Science* 175:564–65. (5)

Rosenzweig, M. L. 1973. Habitat selection experiments with a pair of coexisting heteromyid rodent species. *Ecology* 54:111–17. (5)

Rosenzweig, M. L. 1975. On continental steady states of species diversity. In *Ecology and evolution of communities,* ed. M. L. Cody and J. M. Diamond, 121–40. Cambridge, Mass.: Belknap Press of Harvard University Press. (1, 5, 30)

Rosenzweig, M. L. 1977a. Aspects of biological exploitation. *Q. Rev. Biol.* 52:371–80. (5)

Rosenzweig, M. L. 1977b. Coexistence and diversity in heteromyid rodents. In *Evolutionary ecology,* ed. B. Stonehouse and C. Perrins, 89–99. London: Macmillan. (5)

Rosenzweig, M. L. 1977c. Geographical speciation: On range size and the probability of isolate formation. In *Proceedings of the Washington State University Conference on Biomathematics and Biostatistics* (Pullman, Wash.) ed. D. Wollkind, 172–94. (5)

Rosenzweig, M. L. 1979. Optimal habitat selection in two-species competitive systems. In *Population ecology,* ed. U. Halbach and J. Jacobs, 283–93. Fortschritte der Zoologie 25. Stuttgart: Gustav Fischer. (5)

Rosenzweig, M. L. 1981. A theory of habitat selection. *Ecology* 62:327–35. (5, 30)

Rosenzweig, M. L. 1985. Some theoretical aspects of habitat selection. In ed. M. L. Cody, 517–40. *Habitat selection in birds,* New York: Academic Press. (30)

Rosenzweig, M. L. 1986. Hummingbird isolegs in an experimental system. *Behav. Ecol. Sociobiol.* 19:313–22. (5)

Rosenzweig, M. L. 1987a. Community organization from the point of view of habitat selectors. In *Organization of communities past and present,* ed. J. H. R. Gee and B. J. Giller, 469–90. Oxford: Blackwell Scientific Publications. (5, 7)

Rosenzweig, M. L. 1987b. Habitat selection as source of biological diversity. *Evol. Ecol.* 1:315–30. (5)

Rosenzweig, M. L. 1991. Habitat selection and population interactions: The search for mechanisms. *Am. Nat.,* suppl., 137:S5–S28. (5)

Rosenzweig, M. L. 1992. Species diversity gradients: We know more and less than we thought. *J. Mammal.* 73:715:30. (5)

Rosenzweig, M. L., and Z. N. Abramsky. 1980. Microtine cycles: The role of habitat variability. *Oikos* 34:141–46. (5)

Rosenzweig, M. L., and Z. N. Abramsky. 1985. Detecting density-dependent habitat selection. *Am. Nat.* 126:405–17. (5, 30)

Rosenzweig, M. L., and Z. N. Abramsky. 1986. Centrifugal community organization. *Oikos* 47:339–48. (5)

Rosenzweig, M. L., Z. N. Abramsky, and S. Brand. 1984. Estimating species interactions in heterogeneous environments. *Oikos* 43:329–40. (5)

Rosenzweig, M. L., and W. M. Schaffer. 1978. Homage to the Red Queen II: Coevolutionary response to enrichment of exploitation ecosystems. *Theor. Pop. Biol.* 9:158–63. (5)

Ross, C. A., and J. R. P. Ross. 1988. Late Paleozoic transgressive-regressive deposition. In *Sea-level changes: An integrated approach,* ed. C. K. Wilgus et al., 227–247. SEPM Special Publication 42. Society of Economic Paleontologists and Mineralogists. (27)

Ross, H. H. 1972a. The origin of species diversity in ecological communities. *Taxon* 21:253–59. (1, 24)

Ross, H. H. 1972b. An uncertainty principle in ecological evolution. In *A symposium on ecosystematics,* ed. R. T. Allen and F. C. James, 133–57. Occasional Paper no. 4. Fayetteville: University of Arkansas Museum. (24)

Ross, S. T. 1986. Resource partitioning in fish assemblages: A review of field studies. *Copeia* 1986:352–88. (24)

Roth, L. C., and A. Grijalva. 1991. New record of the mangrove *Pelliciera rhizophorae* (Theaceae) on the Caribbean Coast of Nicaragua. *Rhodora* 93:183–86. (20)

Rothe, F. 1976. Convergence to the equilibrium state in the Volterra-Lotka diffusion equations. *J. Math. Biol.* 3:319–24. (8)

Rothe, F. 1983. Asymptotic behavior of a nonhomogeneous predator-prey system with one diffusing and one sedentary species. *Math. Methods Appl. Sci.* 5:40–67. (8)

Rothe, F. 1984. *Global solutions of reaction-diffusion systems.* Lecture Notes in Mathematics 1072. Berlin: Springer-Verlag. (8)

Roughgarden, J. 1974. Population dynamics in a spatially varying environment: How population size tracks spatial variation in carrying capacity. *Am. Nat.* 108:649–64. (8)

Roughgarden, J. 1978. Influence of competition on patchiness in a random environment. *Theor. Pop. Biol.* 14:185–203. (8)

Roughgarden, J., S. D. Gaines, and Y. Iwasa. 1984. Dynamics and evolution of marine populations with pelagic larval dispersal. In *Exploitation of marine communities,* ed. R. M. May. Dahlem Konferenzen. Berlin: Springer-Verlag. (4)

Roughgarden, J., S. Gaines, and S. Pacala. 1987. Supply side ecology: The role of physical transport processes. In *Organization of communities past and present,* ed. J. H. R. Gee and P. S. Giller, 491–518. Oxford: Blackwell Scientific Publications. (1, 2, 4, 11, 30)

Roughgarden, J., S. Gaines, and H. Possingham. 1988. Recruitment dynamics in complex life cycles. *Science* 241:1460–66. (2, 7)

Roughgarden, J., D. Heckel, and E. R. Fuentes. 1983. Coevolutionary theory and the biogeography and community structure of *Anolis.* In *Lizard ecology: Studies of a model organism,* ed. R. B. Huey, E. R. Pianka, and T. W. Schoener, 371–410. Cambridge, Mass.: Harvard University Press. (25)

Rubner, K., and F. Reinhold. 1953. *Das Natürliche Waldbild Europas.* Hamburg: Paul Parey. (26)

Russell, M. P. 1990. Quantitative and experimental applications in paleoecology: Integrating neontological and paleontological evidence. Ph.D. diss., University of California, Berkeley. (29)

Rutzler, K., and I. G. MacIntyre. 1982. *The Atlantic barrier reef ecosystem at Carrie Bow Cay, Belize. I. Structure and communities.* Smithson. Contrib. Mar. Sci. 12. (27)

Rydin, H., and S. -O. Borgegård. 1991. Plant characteristics over a century of primary succession on islands: Lake Hjälmarem. *Ecology* 72:1089–1101. (6)

Ryther, J. H. 1969. Photosynthesis and fish production in the sea. *Science* 166:72–76.

Ryti, R. T., and T. J. Case. 1992. The role of neighborhood competition in the spacing and diversity of ant communities. *Am. Nat.* 139:355–74. (14)

Sabelis, M. W., and O. Diekmann. 1988. Overall population stability despite local extinction: The stabilizing influence of prey dispersal from predator-invaded patches. *Theor. Pop. Biol.* 34:169–76. (7, 8)

Saenger, P. 1982. Morphological, anatomical and reproductive adaptations of Australian mangroves. In *Mangrove ecosytems in Australia,* ed. B. F. Clough, 153–91. Canberra: Australian National University Press. (20)

Saenger, P., E. J. Hegerl, and J. D. S. Davie. 1983. Global status of mangrove ecosystems. *Environmentalist* 3, suppl. 3:1–88. (20)

Saigusa, T., A. Nakanishi, H. Shima, and O. Yata. 1982. Phylogeny and geographical distribution of the swallowtail subgenus *Graphium* (Lepidoptera: Papilionidae). *Entomol. Genet.* 8:59–69. (23)

Saito, T. 1976. Geologic significance of cooling direction in planktonic foraminifera *Pulleniatina. Geology* 4:305–09. (20)

Sakai, A., and W. Larcher. 1987. *Frost survival of plants.* Berlin: Springer-Verlag. (26)

Sale, P. F. 1977. Maintenance of high diversity in coral reef fish communities. *Am. Nat.* 111:337–59. (1, 22)

Sale, P. F. 1978. Coexistence of coral reef fishes—a lottery for living space. *Environ. Biol. Fish.* 3:85–102. (1)

Sale, P. F., and R. Dybdahl. 1975. Determinants of community structure for coral reef fishes in an experimental habitat. *Ecology* 56:1343–55. (4)

Sandberg, P. A. 1983. An oscillating trend in Phanerozoic nonskeletal carbonate mineralogy. *Nature* 305:19–22. (27)

Sanders, H. L. 1968. Benthic marine diversity: A comparative study. *Am. Nat.* 102:243–82. (5)

Sanders, H. L. 1969. Benthic marine diversity and the time-stability hypothesis. In *Diversity and stability in ecological systems,* ed. G. M. Woodwell and H. H. Smith, 71–81. Brookhaven Symposium on Biology 22. (27)

SAS Institute, Inc. 1987. *SAS/STAT guide for personal computers,* Version 6. Cory, N.C.: SAS Institute. (18)

Sauer, J. D. 1988. *Plant migration: The dynamics of geographic patterning in seed plant species.* Berkeley: University of California Press. (26)

Savage, A. A., and G. M. Gazey. 1987. Relationships of physical and chemical conditions to species diversity and density of gastropods in English lakes. *Biol. Conserv.* 42:95–113. (6)

Savage, D. E., and D. E. Russell. 1983. *Mammalian paleofaunas of the World.* Reading, Mass.: Addison-Wesley. (28)

Savage, J. M. 1982. The enigma of the Central American herpetofauna: Dispersals or vicariance? *Ann. Mo. Bot. Gard.* 69:464–547. (25)

Savage, R. G. J. 1977. Evolution in carnivorous mammals. *Paleontology* 20:237–71. (28)

Savin, S. M. 1977. The history of the earth's surface temperature during the past 100 million years. *Annu. Rev. Earth Planetary Sci.* 5:319–55. (20)

Savitzky, A. H. 1983. Coadapted character complexes among snakes: Fossoriality, piscivory, and durophagy. *Am. Zool.* 23:397–409. (25)

Sawyer, C. N. 1966. Basic concepts of eutrophication. *J. Water Pollution Fed.* 38:737–44. (5)

Sazima, I. 1986. Similarities in feeding behaviour between some marine and freshwater fishes in two tropical communities. *J. Fish. Biol.* 29:53–65. (25)

Sazima, I. 1989. Feeding behavior of the snail-eating snake, *Dipsas indica. J. Herpetol.* 23:464–68. (25)

Schad, G. A. 1963. Niche diversification in a parasitic species flock. *Nature* 198:404–6. (17)

Schaefer, S. A., and G. V. Lauder. 1986. Historical transformation of functional design: Evolutionary morphology of feeding mechanisms in loricarioid catfishes. *Syst. Zool.* 35:489–508. (25)

Schaffer, W. M., D. B. Jensen, D. E. Hobbs, J. Gurevitch, J. R. Todd, and M. V. Schaffer. 1979. Competition, foraging energetics and the cost of sociality in three species of bees. *Ecology* 60:976–87. (5)

Schall, J. J., and E. R. Pianka. 1978. Geographical trends in the numbers of species. *Science* 201:679–86. (6, 15)

Schaller, G. B. 1972. *The Serengeti lion.* Chicago: University of Chicago Press. (28)

Scheibe, J. S. 1987. Climate, competition, and the structure of temperate zone lizard communities. *Ecology* 68:1424–36. (6)

Scheltema, R. S. 1971. Larval dispersal as a means of genetic exchange between geographically separated populations of shallow-water benthic marine gastropods. *Biol. Bull. MBL, Woods Hole* 140:284–322. (4)

Schimper, A. F. W. 1903. *Plant geography upon a physiological basis.* Oxford: Oxford University Press. (15)

Schindler, D. W. 1990. Experimental perturbations of whole lakes as tests of hypotheses concerning ecosystem structure and function. *Oikos* 57:25–41. (5, 6)

Schlager, W. 1981. The paradox of drowned reefs and carbonate platforms. *Bull. Geol. Soc. Am.* 92:197–211. (27)

Schlosser, I. J. 1982. Trophic structure, reproductive success, and growth rate of fishes in a natural and modified head water stream. *Can. J. Fish. Aquat. Sci.* 39:968–78. (17)

Schluter, D. 1982. Distributions of Galápagos ground finches along an altitudinal gradient: The importance of food supply. *Ecology* 63:1504–17. (1)

Schluter, D. 1986. Tests for similarity and convergence of finch communities. *Ecology* 67:1073–85. (1, 16, 21, 30)

Schluter, D. 1988a. Character displacement and the adaptive divergence of finches on islands and continents. *Am. Nat.* 131:799–824. (30)

Schluter, D. 1988b. The evolution of finch communities on islands and continents: Kenya vs. Galápagos. *Ecol. Monogr.* 58:229–49. (25)

Schluter, D., and P. R. Grant. 1982. The distribution of *Geospiza difficilis* in relation to *G. fuliginosa* in the Galápagos Islands: Tests of three hypotheses. *Evolution* 36:1213–26. (1)

Schluter, D., and P. R. Grant. 1984. Determinants of morphological patterns in communities of Darwin's finches. *Am. Nat.* 123:175–96. (1)

Schluter, D., T. D. Price, and P. R. Grant. 1985. Ecological character displacement in Darwin's finches. *Science* 227:1056–59. (1)

Schmitt, R. J. 1987. Indirect interactions between prey: Apparent competition, predator aggregation, and habitat segregation. *Ecology* 68:1887–97. (7)

Schodde, R. 1982. Origin, adaptation and evolution of birds in arid Australia. In *Evolution of the flora and fauna of arid Australia,* ed. W. R. Barker and P. J. M. Greenslade, 191–224. Adelaide: Peacock Publications. (13, 14)

Schoener, T. W. 1974. Resource partitioning in ecological communities. *Science* 185:27–39. (1, 28)

Schoener, T. W. 1983. Field experiments on interspecific competition. *Am. Nat.* 122:240–85. (1, 4, 22)

Schoener, T. W. 1984. Size differences among sympatric, bird-eating hawks: A worldwide survey. In *Ecological communities: Conceptual issues and the evidence,* ed. D. R. Strong, Jr., D. Simberloff, L. G. Abele, and A. B. Thistle, 254–81. Princeton, N.J.: Princeton University Press. (1)

Schoener, T. W. 1986. Overview: Kinds of ecological communities—ecology becomes pluralistic. In *Community ecology,* ed. J. Diamond and T. J. Case, 467–79. New York: Harper and Row. (22)

Schoener, T. W. 1989. Food webs from the small to the large. *Ecology* 70:1559–89. (3, 7)

Schoener, T. W. 1991. Extinction and the nature of the metapopulation: A case study. *Acta Oecologia* 12:53–75. (7)

Schoener, T. W., and D. A. Spiller. 1987. High population persistence in a system with high turnover. *Nature* 330:474–77. (7)

Schoenly, K., and J. E. Cohen. 1991. Temporal variation in food web structure: 16 empirical cases. *Ecol. Monogr.* 61:267–98. (3)

Schultz, G. E. 1977. *Guidebook: Field conference on late Cenozoic biostratigraphy of the Texas Panhandle and adjacent Oklahoma.* Special Publication of the Kilgore Research Center, Department of Geology and Anthropology, West Texas State University, Canyon, no. 1. (28)

Schultz, T. D., and N. F. Hadley. 1987. Microhabitat segregation and physiological differences in co-occurring tiger beetle species, *Cicindela oregona* and *Cicindela tranquebarica. Oecologia* 73:363–70. (18)

Schultz, T. D., M. C. Quinlan, and N. F. Hadley. 1992. Preferred body temperature, metabolic physiology, and water balance of adult *Cicindela longilabris:* A comparison of populations from boreal habitats and climatic refugia. *Physiol. Zool.* 65:226–42. (18)

Schulze, R. E., and O. S. McGee. 1978. Climatic indices and classifications in relation to the geography of southern Africa. In *Biogeography and ecology of Southern Africa,* ed. M. J. A. Werger, 19–52. The Hague: Junk. (20)

Schwarzwalder, R. N. 1986. Systematics and early evolution of the Plantanaceae. Ph.D. diss., Indiana University, Bloomington. (26)

Schwinning, S., and M. L. Rosenzweig. 1990. Periodic oscillations in an ideal-free predator-prey distribution. *Oikos* 59:85–91. (7)

Scotese, C. R., and W. S. McKerrow. 1990. Palaeozoic paleogeography and biogeography. *Mem. Geol. Soc. Lond.* 12:1–21. (27)

Scotese, C. R., and W. W. Sager. 1988. Mesozoic and Cenozoic plate reconstructions. *Tectonophysics* 155:317–30. (27)

Scott, K. M. 1985. Allometric trends and locomotor adaptations in the Bovidae. *Bull. Am. Mus. Nat. Hist.* 179:197–288. (28)

Scott, N. J. 1969. A zoogeographic analysis of the snakes of Costa Rica. Ph.D. diss., University of Southern California, Los Angeles. (25)

Scott, N. J. 1976. The abundance and diversity of the herpetofauna of tropical forest litter. *Biotropica* 8:41–58. (5)

Scott, R. W. 1984a. Evolution of Early Cretaceous reefs in the Gulf of Mexico. *Palaeontographica Americana* 54:406–12. (27)

Scott, R. W. 1984b. Mesozoic biotas and depositional systems of the Gulf of Mexico-Caribbean region. In *Jurassic-Cretaceous biochronology and biogeography of North America,* ed. G. E. G. Westermann, 49–64. Geological Association of Canada Special Paper 27. (27)

Scott, R. W., P. A. Fernandez-Mendiola, E. Gili, and A. Simo. 1990. Persistence of coral-rudist reefs into the Late Cretaceous. *Palaios* 5:98–110. (27)

Seaward, M. R. D. 1976. Observations on the bracken component of the pre-Hadrianic deposits at Vindolanda, Northumberland. *Bot. J. Linn. Soc.* 73:177–85. (16)

Segel, L. A., and S. A. Levin. 1976. Application of nonlinear stability theory to the study of the effects of diffusion on predator-prey interactions. In *Topics in statistical mechanics and biophysics: A memorial to Julius J. Jackson,* ed. R. A. Piccirelli, 123–52. American Institute of Physics Conference Proceedings, 27. (8)

Seib, R. L. 1984. Prey use in three syntopic neotropical racers. *J. Herpetol.* 18:412–20. (25)

Seib, R. L. 1985a. Euryphagy in a tropical snake, *Coniophanes fissidens. Biotropica* 17:57–64. (25)

Seib, R. L. 1985b. Feeding ecology and organization of Neotropical snake faunas. Ph.D. diss., University of California, Berkeley. (25)

Sepkoski, J. J., Jr. 1981. A factor analytic description of the Phanerozoic marine fossil record. *Paleobiology* 7:36–53. (29)(30)

Sepkoski, J. J., Jr. 1984. A kinetic model of Phanerozoic taxonomic diversity. III. Post-Paleozoic families and mass extinctions. *Paleobiology* 10:246–67. (29)

Sepkoski, J. J., Jr. 1987. Environmental trends in extinction during the Paleozoic. *Science* 235:64–66. (29)

Sepkoski, J. J., Jr. 1988. Alpha, beta, or gamma: Where does all the diversity go? *Paleobiology* 14:221–34. (28, 30)

Sepkoski, J. J., Jr., and A. I. Miller. 1985. Evolutionary faunas and the distribution of Paleozoic benthic communities in space and time. In *Phanerozoic diversity patterns: Profiles in macroevolution,* ed. J. W. Valentine, 153–90. Princeton, N.J.: Princeton University Press. (29)

Sepkoski, J. J. Jr., and P. M. Sheehan. 1983. Diversification, faunal change, and community replacement during the Ordovician radiations. In *Biotic interactions in recent and fossil benthic communities,* ed. M. J. S. Tevesz and P. L. McCall, 673–717. New York: Plenum Press. (29)

Serie, P. 1920. Notas sobre la alimentación de algunos ofidios. *Revista del Jardin Zoológico,* Buenos Aires (December 1919):1–24. (25)

Settle, W. H., and L. T. Wilson. 1990. Invasion by the variegated leafhopper and biotic interactions: Parasitism, competition, and apparent competition. *Ecology* 71:1461–70. (7)

Sexton, O. J., and H. Heatwole. 1965. Life history notes on some Panamanian snakes. *Caribbean J. Sci.* 5:39–43. (25)

Shabica, S. V., and A. J. Boucot. 1976. The depths inhabited by Silurian brachiopod communities: Comment and reply. *Geology* 4:132, 187–89, 710–12. (5)

Shackleton, N. J., J. Imbrie, and N. G. Pisias. 1988. The evolution of oceanic oxygen-isotope variability in the North Atlantic over the past three million years. *Phil. Trans. Roy. Soc.* (Lond.), ser B 318:679–88. (29)

Shackleton, N. J., and N. D. Opdyke. 1973. Oxygen isotope and paleomagnetic stratigraphy of equatorial Pacific core V28-238: Oxygen isotope temperatures and ice volumes on a 10^5-year and 10^6-year scale. *Quat. Res.* 3:39–55. (29)

Sharp, A. J. 1951. The relation of the Eocene Wilcox flora to some modern floras. *Evolution* 5:1–5. (26)

Sheehan, P. M. 1985. Reefs are not so different—they follow the evolutionary pattern of level bottom communities. *Geology* 13:46–49. (27)

Sheehan, P. M. 1988. Late Ordovician events and the terminal Ordovician extinction. *New Mexico Bureau of Mines and Mineral Resources Memoir* 44:405–15. (29)

Shenbrot, G. I., K. A. Rogovin, and A. V. Surov. 1991. Comparative analysis of spatial organization of desert lizard communities in Middle Asia and Mexico. *Oikos* 61:157–68. (14)

Shigesada, N. 1984. Spatial distribution of rapidly dispersing animals in heterogeneous environments. In *Mathematical Ecology,* ed. S. A. Levin and T. G. Hallam, 478–91. Lecture Notes in Biomathematics 54. Berlin: Springer-Verlag. (8)

Shigesada, N., K. Kawasaki, and E. Teramoto. 1979. Spatial segregation of interacting species. *J. Theor. Biol.* 79:83–99. (8)

Shigesada, N., and J. Roughgarden. 1982. The role of rapid dispersal in the population dynamics of competition. *Theor. Pop. Biol.* 21:353–72. (8)

Shimwell, D. W. 1971. *Description and classification of vegetation.* Seattle: University of Washington Press. (1)

Shmida, A. 1981. Mediterranean vegetation in California and Israel: Similarities and differences. *Israel J. Bot.* 30:105–23. (1)

Shmida, A., and S. Ellner. 1984. Coexistence of plant species with similar niches. *Vegetatio* 58:29–55. (2, 7)

Shmida, A., M. Evenari, and I. Noy-Meir. 1986. Hot desert ecosystems: An integrated view. In *Ecosystems of the world.* Vol. 12B, *Hot deserts and arid shrublands,* ed. M. Evenari, I. Noy-Meir, and D. W. Goodall, 379–87. Amsterdam: Elsevier Scientific. (5)

Shmida, A., and M. V. Wilson. 1985. Biological determinants of species diversity. *J. Biogeogr.* 12:1–20. (1, 7, 13, 30)

Shorrocks, B., W. Atkinson, and P. Charlesworth. 1979. Competition on a divided and ephemeral resource. *J. Anim. Ecol.* 48:899–908. (8)

Shorrocks, B., and J. Rosewell. 1986. Guild size in drosophilids: A simulation model. *J. Anim. Ecol.* 55:527–41. (22, 30)

Shorrocks, B., and I. R. Swingland. 1990. *Living in a patchy environment.* Oxford: Oxford University Press. (7)

Shulenberger, E. 1977. Hyperiid amphipods from the zooplankton community of the North Pacific central gyre. *Mar. Biol.* 42:375–85. (19)

Shuter, E. 1990. Herbivorous arthropods on bracken fern (*Pteridium*) in Australia. M.Sc. thesis, Macquarie University, Sydney. (16)

Shuter, E., and M. Westoby. 1992. Herbivorous arthropods on bracken (*Pteridium aquilinum* (L.) Khun) in Australia compared with elsewhere. *Aust. J. Ecol.* 17:329–39. (15)

Sibley, C. G., and J. E. Alquist. 1990. *Phylogeny and classification of the birds of the world.* New Haven: Yale University Press. (30)

Siegfried, W. R., and T. M. Crowe. 1983. Distribution and species diversity of birds and plants in fynbos vegetation of Mediterranean-climate zone, South Africa. In *Mediterranean-type ecosystems: The role of nutrients,* ed. F. J. Kruger, D. T. Mitchell, and J. U. M. Jarvis, 403–16. Berlin: Springer-Verlag. (6)

Signor, P. W. III. 1985. Real and apparent trends in species richness through time. In *Phanerozoic diversity patterns: Profiles in macroevolution,* ed. J. W. Valentine, 129–50. Princeton, N.J.: Princeton University Press. (27)

Sih, A. 1987. Prey refuges and predator-prey stability. *Theor. Pop. Biol.* 31:1–13. (7)

Sih, A., P. Crowley, M. McPeek, J. Petranka, and K. Strohmeier. 1985. Predation, competition, and prey communities: A review of field experiments. *Annu. Rev. Ecol. Syst.* 16:269–311. (1, 2, 4)

Silvert, W., and T. Platt. 1978. Energy flux in the pelagic ecosystem: A time-dependent equation. *Limnol. Oceanogr.* 23:813–16. (3)

Silvert, W., and T. Platt. 1980. Dynamic energy flow model of the particle size distribution in pelagic ecosystems. In *Evolution and ecology of zooplankton communities,* ed. W. C. Kerfoot. Hanover, N.H.: University Press of New England. (3)

Silvertown, J. 1980. The dynamics of a grassland ecosystem: Botanical equilibrium in the park grass experiment. *J. Appl. Ecol.* 17:491–504. (5)

Silvertown, J. 1985. History of a latitudinal diversity gradient: Woody plants in Europe 13,000–1000 years B.P. *J. Biogeogr.* 12:519–25. (6)

Simberloff, D. S. 1979. Rarefaction as a distribution-free method of expressing and estimating diversity. In *Ecological diversity in theory and practice,* J. F. Grassle, G. P. Patil, W. K. Smith, and C. Taillie, 159–76. Fairland, Md.: International Cooperative Publishing House. (1, 26)

Simberloff, D. S. 1980. A succession of paradigms in ecology: Essentialism, materialism and probabilism. In *Conceptual issues in ecology,* ed. E. Saarinen, 63–99. Dordrecht: Reidel. (4)

Simberloff, D. S. 1981. Community effects of introduced species. In *Biotic crises in ecological and evolutionary time,* ed. M. H. Nitecki, 53–81. New York: Academic Press. (29, 30)

Simberloff, D. S. 1986. Introduced insects: A biogeographic and systematic perspective. In *Ecology of biological invasions of North America and Hawaii,* ed. H. A. Mooney and J. A. Drake, 3–26. Berlin: Springer-Verlag. (29, 30)

Simberloff, D. S. 1987. Calculating probabilities that cladograms match: A method of biogeographical inference. *Syst. Zool.* 36:175–95. (24)

Simberloff, D. S. 1988. Effects of drift and selection on detecting similarities between large cladograms. *Syst. Zool.* 37:56–59. (24)

Simberloff, D. S., and W. Boecklen. 1981. Santa Rosalia reconsidered: Size ratios and competition. *Evolution* 35:1206–28. (1)

Simberloff, D. S., and E. O. Wilson. 1969a. Experimental zoogeography of islands: The colonization of empty islands. *Ecology* 50:278–96. (7)

Simberloff, D. S., and E. O. Wilson. 1969b. Experimental zoogeography of islands: Defaunation and monitoring techniques. *Ecology* 50:267–78. (5)

Simbotwe, M. P., and G. R. Friend. 1985. Comparison of the herpetofaunas of tropical wetland habitats from Lochinvar National Park, Zambia and Kakadu National Park, Australia. *Proc. Ecol. Soc. Aust.* 14:141–51. (21)

Simmons, A. H. 1988. Extinct pygmy hippopotamus and early man in Cyprus. *Nature* 333:554–57. (12)

Simpson, B. B., and J. Haffer. 1978. Speciation patterns in the Amazonian forest biota. *Annu. Rev. Ecol. Syst.* 9:497–518. (12)

Simpson, G. G. 1944. *Tempo and mode in evolution.* New York: Columbia University Press. (29)

Simpson, G. G. 1953. *The major features of evolution.* New York: Columbia University Press. (22, 23, 29)

Simpson, G. G. 1964a. Species densities of North American mammals. *Syst. Zool.* 13:361–89. (6)

Simpson, G. G. 1964b. Species density of North American recent mammals. *Syst. Zool.* 13:57–73. (1)

Simpson, G. G. 1970. The Argyrolagidae, extinct South American marsupials. *Bull. Mus. Comp. Zool.* (Harvard) 139:1–86. (14)

Singh, G. 1988. History of aridland vegetation and climate: A global perspective. *Biol. Rev.* 63:159–95. (28)

Singh, G., and J. Luly. 1989. Changes in vegetation and seasonal climate since the last full glacial at Lake Frome, South Australia. *Palaeogeogr. Palaeoclimatol. Palaeoecol.* 84:75–86. (14)

Skellam, J. G. 1951. Random dispersal in theoretical populations. *Biometrika* 38:196–218. (2, 8, 22)

Slatkin, M. 1974. Competition and regional coexistence. *Ecology* 55:128–34. (8, 9)

Slatkin, M. 1987. Gene flow and the geographic structure of natural populations. *Science* 236:787–92. (7)

Slobodkin, L. B., F. E. Smith, and N. G. Hairston. 1967. Regulation in terrestrial ecosystems, and the implied balance of nature. *Am. Nat.* 101:109–24. (22)

Slowinski, J. B., and C. Guyer. 1989. Testing the stochasticity of patterns of organismal diversity: An improved null model. *Am. Nat.* 134:907–21. (11, 14, 22)

Smith, A. G., and J. C. Briden. 1977. *Mesozoic and Cenozoic paleocontinental maps.* Cambridge: Cambridge University Press. (27)

Smith, A. G., J. C. Briden, and G. E. Drewry. 1973. Phanerozoic world maps. In *Organisms and continents through time,* ed. N. F. Hughes, 1–42. Palaeontological Association of London, Special Papers in Palaeontology 12. (27)

Smith, A. G., A. M. Hurley, and J. C. Briden. 1981. *Phanerozoic paleocontinental world maps.* Cambridge: Cambridge University Press. (27)

Smith, A. T. 1974. The distribution and dispersal of pikas: Consequences of insular population structure. *Ecology* 55:1112–19. (10)

Smith, F. E. 1972. Spatial heterogeneity, stability, and diversity in ecosystems. In *Growth by intussesception: Ecological essays in honor of Evelyn Hutchinson,* ed. E. S. Deevey. Transactions of the Connecticut Academy of Arts and Sciences 44. (7)

Smith, K. L., Jr. 1978. Benthic community respiration in the N.W. Atlantic Ocean. *Mar. Biol.* 47:337–47. (5)

Smith, M. A. 1989. The case for a resident human population in the central Australian ranges during full glacial aridity. *Archaeol. Oceania* 24:93–105. (14)

Smith, S. V., W. J. Kimmerer, E. A. Laws, R. E. Brock, and P. W. Walsh. 1981. Kaneohe Bay sewage diversion experiment: Perspectives on ecosystem response to nutritional perturbation. *Pacific Sci.* 35:279–402. (27)

Smith, T. J., and N. C. Duke. 1987. Physical determinants of interestuary variation in mangrove species richness around the tropical coastline of Australia. *J. Biogeogr.* 14:9–19. (6)

Snedaker, S. C. 1982. Mangrove species zonation: Why? In *Contributions to the ecology of halophytes,* ed. D. N. Sen and J. S. Rajjpurohit, 111–26. The Hague: Junk. (20)

Snow, D. W. 1978. *An atlas of speciation in African non-passerine birds,* British Museum (Natural History), Tring. (12)

Snyder, T. E. 1954. *Order Isoptera: The termites of the United States and Canada.* New York: National Pest Control Association. (14)

Söderström, L. 1989. Regional distribution patterns of bryophyte species on spruce logs in northern Sweden. *Bryologist* 92:349–55. (10)

Sokal, R. R., and F. J. Rohlf. 1981. *Biometry.* 2d ed. New York: W. H. Freeman and Co. (17, 26)

Solounias, N., and B. Dawson-Saunders. 1988. Dietary adaptations and paleoecology of the late Miocene ruminants from Pikermi and Samos in Greece. *Palaeogeogr. Palaeoclimatol. Palaeoecol.* 65:149–72. (28)

Sommer, U. 1984. The paradox of the plankton: Fluctuations of phosphorus availability maintain diversity of phytoplankton in flow-through cultures. *Limnol. Oceanogr.* 29:633–36. (2)

Sommer, U. 1985. Comparison between steady state and non-steady state competition: Experiments with natural phytoplankton. *Limnol. Oceanogr.* 30:335–46. (2)

Sondaar, P. Y. 1977. Insularity and its effect on mammal evolution. In *Major patterns of vertebrate evolution,* ed. P. C. Goody and B. M. Heckt, 671–707. New York: Plenum Publishing Corp. (12)

Sondaar, P. Y., M. Sanges, T. Kotsakis, and P. L. De Boer. 1986. The Pleistocene deer hunters of Sardinia. *Geobios* 19:17–25. (12)

Song, Z., H. Li, and G. Liu. 1981. Miocene floristic regions of China. *Geol. Soc. Am. Spec. Pap.* no. 187, 249–54. (26)

Song, Z., W. Li, and C. He. 1983. Cretaceous and Palaeogene palynofloras and distribution of organic rocks in China. *Scientia Sinica* 26:538–49. (26)

Soulé, M. E., A. C. Alberts, and B. T. Bolger. 1992. The effects of habitat fragmentation in chaparral plants and vertebrates. *Oikos* 63:39–47. (7)

Sousa, W. P. 1979a. Disturbance in marine intertidal boulder fields: The nonequilibrium maintenance of species diversity. *Ecology* 60:1225–39. (4, 5)

Sousa, W. P. 1979b. Experimental investigations of disturbance and ecological succession in a rocky intertidal algal community. *Ecol. Monogr.* 49:227–54. (4)

Sousa, W. P. 1980. The responses of a community to disturbance—the importance of successional age and species life histories. *Oecologia* 45:72–81. (4)

Sousa, W. P. 1984. The role of disturbance in natural communities. *Annu. Rev. Ecol. Syst.* 15:353–91. (12)

Southwood, T. R. E. 1961. The number of species of insect associated with various trees. *J. Anim. Ecol.* 30:1–8. (1)

Southwood, T. R. E. 1973. The insect/plant relationship—an evolutionary perspective. In *Insect-plant relationships,* ed. H. F. van Emden, 3–30. London: Blackwell Scientific Publications. (23)

Southwood, T. R. E., V. C. Moran, and C. E. J. Kennedy. 1982. The richness, abundance and biomass of the arthropod communities on trees. *J. Anim. Ecol.* 51:635–49. (1)

Specht, R. L. 1970. *The vegetation of South Australia.* 2d ed. Adelaide: Government Printer. (15)

Specht, R. L. 1981a. Biogeography of halophytic angiosperms (salt marsh, mangrove and seagrass). In *Ecological biogeography of Australia,* ed. A. Keast, 577–89. The Hague: Junk. (20)

Specht, R. L. 1981b. *Major vegetation formations in Australia.* In *Ecological biogeography of Australia,* ed. A. Keast, 165–297. The Hague: Junk. (13)

Springer, D. A. 1985. Gradient versus cluster analysis of fossil assemblages: A comparison from the Ordovician of southwestern Virginia. *Lethaia* 18:181–98. (29)

Springer, D. A., and A. I. Miller 1990. Levels of spatial variability: The "community" problem. In *Paleocommunity temporal dynamics: The long-term development of multispecies assemblies,* ed. W. Miller III, 13–30. Paleontological Society Special Publication 5. (29)

Stafford Smith, D. M., and S. R. Morton. 1990. A framework for the ecology of arid Australia. *J. Arid Environ.* 18:255–78. (14, 15)

Staiger, H. 1957. Genetical and morphological variation in *Purpura lapillus* with respect to local and regional differentiation of population groups. *Année Biologique* 33:251–58. (4)

Stanley, G. D., Jr. 1981. Early history of scleractinian corals and its geological consequences. *Geology* 9:507–11. (27)

Stanley, G. D., Jr. 1988. The history of Early Mesozoic reef communities: A three-step process. *Palaios* 3:170–83. (27)

Stanley, S. M. 1979. *Macroevolution: Pattern and process.* San Francisco: W. H. Freeman. (23, 29)

Stanley, S. M. 1984. Marine mass extinctions: A dominant role for temperature. In *Extinctions,* ed. M. H. Nitecki, 69–117. Chicago: University of Chicago Press. (27)

Stark, N. M., and C. F. Jordan. 1978. Nutrient retention by the root mat of an Amazonian rain forest. *Ecology* 59:434–37. (5)

Statz, G. 1938. Funf neue fossile Cermabyciden-arten aus den mitteloligocenen Ablagerungen von Rott am Siebengebirge. *Entomologische Bletter* 34:173–79. (23)

Stehli, F. G., R. G. Douglas, and N. D. Newell. 1969. Generation and maintenance of gradients in taxonomic diversity. *Science* 164:947–49. (1, 30)

Stehli, F. G., and S. D. Webb. 1985. *The great American biotic interchange.* New York: Plenum Press. (1)

Stehli, F. G., and J. W. Wells. 1971. Diversity and age patterns in hermatypic corals. *Syst. Zool.* 20:115–26. (27, 30)

Steneck, R. S. 1983. Escalating herbivory and resulting adaptive trends in calcareous algal crusts. *Paleobiology* 9:44–61. (30)

Stevens, G. C. 1986. Dissection of the species-area relationship among wood-boring insects and their host plants. *Am. Nat.* 128:35–46. (22)

Stevens, G. C. 1989. The latitudinal gradient in geographical range: How so many species coexist in the tropics. *Am. Nat.* 133:240–56. (1, 23, 26)

Stiling, P. D., and D. R. Strong. 1984. Experimental density manipulation of stem-boring insects: Some evidence for interspecific competition. *Ecology* 65:1683–85. (22)

Stimson, J. 1970. Territorial behavior of the owl limpet *Lottia gigantea. Ecology* 51:113–18. (4)

Stimson, J. 1973. The role of the territory in the ecology of the intertidal limpet *Lottia gigantea* (Gray). *Ecology* 54:1020–30. (4)

Stock, T. M., and J. C. Holmes. 1988. Functional relationships and microhabitat distributions of enteric helminths of grebes (Podicipedidae): The evidence for interactive communities. *J. Parasitol.* 74:214–27. (17)

Stocker, G. C., and G. L. Unwin. 1989. The rain forests of northeastern Australia—their environment, evolutionary history and dynamics. In *Ecosystems of the world.* Vol. 14B, *Tropical rain forest ecosystems: Biogeographical and ecological studies,* ed. H. Leith and M. J. A. Werger, 241–59. Amsterdam: Elsevier Scientific. (5)

Storch, G. 1970. Holozäne Kleinsäugerfunde aus der Ghar Dalam Höhle Malta (Mammalia: Insectivora, Chiroptera, Rodentia). *Senkenbergiana Biol.* 51:135–45. (12)

Stork, N. E. 1988. Insect diversity: Facts, fiction and speculation. *Biol. J. Linn. Soc.* 35:321–37. (15)

Stradling, D. J. 1978. The influence of size on foraging in the ant, *Atta cephalotes,* and the effect of some plant defense mechanisms. *J. Anim. Ecol.* 47:173–88. (23)

Strauss, R. E. 1984. Allometry and functional feeding morphology in haplochromine cichlids. In *Evolution of fish species flocks,* ed. A. A. Echelle and I. Kornfield, 217–30. Orono: University of Maine Press. (30)

Strong, D. R. 1979. Biogeographical dynamics of insect-host plant communities. *Annu. Rev. Entomol.* 24:89–119. (17)

Strong, D. R., J. H. Lawton, and T. R. E. Southwood. 1984. *Insects on plants: Community patterns and mechanisms.* Oxford: Blackwell. (16, 17, 22, 23)

Strong, D. R., E. D. McCoy, and J. R. Rey. 1977. Time and the number of herbivore species: The pests of sugarcane. *Ecology* 58:167–75. (6)

Strong, D. R., D. Simberloff, L. G. Abele, and A. B. Thistle (Eds.). 1984. *Ecological communities: Conceptual issues and the evidence.* Princeton, N.J.: Princeton University Press. (1, 17, 24, 30)

Strong, D. R., L. A. Szyska, and D. S. Simberloff. 1979. Tests of community-wide character displacement against null hypotheses. *Evolution* 33:897–913. (1)

Stucky, R. K. 1990. Evolution of land mammal diversity in North America during the Cenozoic. In *Current mammalogy,* ed. H. H. Genoways, 375–432. New York: Plenum Press. (1, 28)

Suc, J. P. 1978. Analyse pollinique de dépôts plio-pléistocènes du sud du massif basaltique de l'Escandorgue. *Pollen et Spores* 20:497–512. (12)

Sugihara, G. 1981. $S = cA^z$, $z = \frac{1}{4}$: A reply to Connor and McCoy. *Am. Nat.* 117:790–93. (7, 17)

Sugihara, G. 1982. Niche hierarchy: Structure, organization and assembly in natural communities. Ph.D. diss., Princeton University. (3)

Sugihara, G. 1984. Graph theory, homology, and food webs. *Proc. Symp. Appl. Math.* 30:83–101. (3)

Sugihara, G., K. Schoenly, and A. Trombla. 1989. Scale invariance in food web properties. *Science* 245:48–52. (1, 3)

Sumlin, W. D. III. 1984. Studies on the Australian Cicindelidae. III. Observations on the Australian members of the genus *Cicindela* L. (Coleoptera). *Entomol. News* 95:189–99. (18)

Summerhayes, V. S. 1941. The effect of voles (*Microtus agrestis*) on vegetation. *J. Ecol.* 29:14–48. (5)

Svensson, B. W. 1992. Changes in occupancy, niche breadth and abundance of three *Gyrinus* species as their respective range limits are approached. *Oikos* 63:147–56. (10)

Swanson, P. 1945. Herpetological notes from Panama. *Copeia* 1945:210–16. (25)

Swift, L. W., W. T. Swank, J. B. Mankin, R. J. Luxmoore, and R. A. Goldstein. 1975. Simulation of evapotranspiration and drainage from mature and clearcut deciduous forest and young pine plantation. *Water Resources Res.* 11:667–73. (2)

Swingle, H. S. 1946. Experiments with combinations of largemouth black bass, bluegills and minnows in ponds. *Trans. Am. Fish. Soc.* 76:46–62. (5)

Swisher, C. C., and D. R. Prothero. 1990. Single-crystal[40] AR/[39] AR dating of the Eocene-Oligocene transition in North America. *Science* 249:760–62. (28)

Takhtajan, A., ed. 1974. *Magnoliophyta Fossilia U.S.S.R.* Vol. 1, *Magnoliaceae-Eucommiaceae.* Leningrad: Nauka. (26)

Talent, J. A. 1988. Organic reef-building: Episodes of extinction and symbiosis. *Senckenbergiana Lethaea* 69:315–68. (27)

Tanai, T. 1972. Tertiary history of vegetation in Japan. In *Floristics and paleofloristics of Asia and Eastern North America.* ed. A. Graham, 235–55. Amsterdam: Elsevier. (26)

Tappan, H. 1968. Primary production: Isotopes, extinctions and the atmosphere. *Palaeogeogr. Palaeoclimatol. Palaeoecol.* 4:187–210. (5)

Tappan, H., and A. R. Loeblich. 1973. Evolution of the oceanic plankton. *Earth Sci. Rev.* 9:207–40. (5)

Tauber, M. J., C. A. Tauber, and S. Masaki. 1986. *Seasonal adaptations of insects.* New York: Oxford University Press. (23)

Taylor, A. D. 1988. Large scale spatial structure and population dynamics in arthropod predator-prey systems. *Ann. Zool. Fennici* 25:63–74. (8)

Taylor, A. D. 1990. Metapopulations, dispersal, and predator-prey dynamics: An overview. *Ecology* 71:429–33. (7, 8)

Taylor, J. D. 1978. Faunal response to the instability of reef habitats: Pleistocene molluscan assemblages of Aldabra Atoll. *Palaeontology* 21:1–30. (29)

Taylor, L. R., I. P. Woiwod, and J. N. Perry. 1978. The density-dependence of spatial behaviour and the rarity of randomness. *J. Anim. Ecol.* 47:383–406. (10)

Taylor, R. W. 1983. Descriptive taxonomy: Past, present and future. In *Australian systematic entomology: A bicentenary perspective,* ed. E. Highley and R. W. Taylor, 93–134. Melbourne: CSIRO. (15)

Taylor, W. D. 1979. Sampling data on the bactivorous ciliates of a small pond compared to neutral models of community structure. *Ecology* 60:876–83. (22)

Tchernov, E. 1984. Faunal turnover and extinction rate in the Levant. In *Quaternary extinctions,* ed. P. S. Martin and R. G. Klein, 528–52. Tucson: University of Arizona Press. (12)

Tedford, R. H., M. F. Skinner, R. W. Fields, J. M. Rensberger, D. P. Whistler, T. Galusha, B. E. Taylor, J. R. Macdonald, and S. D. Webb. 1987. Faunal succession and biochronology of the Arikareean through Hemphillian interval (late Oligocene through earliest Pliocene epochs) in North America. In *Cenozoic mammals of North America,* ed. M. O. Woodburne. Berkeley: University of California Press. (28)

Teichert, C. 1990. The Permian-Triassic boundary revisited. In *Extinction events in earth history,* ed. E. G. Kauffman and O. H. Walliser, 199–238. Lecture notes in Earth Sciences 30. Berlin: Springer-Verlag. (27)

Templeton, A. R. 1989. The meaning of species and speciation: A genetic perspective. In *Speciation and its consequences,* ed. D. Otte and J. Endler, 3–27. Sunderland, Mass.: Sinauer Associates. (24)

Terborgh, J. 1971. Distribution on environmental gradients: Theory and a preliminary interpretation of distributional patterns in the avifauna of the Cordillera Vilcabamba, Peru. *Ecology* 52:23–40. (1)

Terborgh, J. 1973. On the notion of favorableness in plant ecology. *Am. Nat.* 107:481–501. (1, 5, 6)

Terborgh, J. 1974. Preservation of natural diversity: The problem of species extinction. *BioScience* 24:715–22. (30)

Terborgh, J. 1985. The role of ecotones in the distribution of Andean birds. *Ecology* 66:1237–46. (1)

Terborgh, J. 1990. *Where have all the birds gone?* Princeton, N.J.: Princeton University Press. (7)

Terborgh, J. W., and J. Faaborg. 1980. Saturation of bird communities in the West Indies. *Am. Nat.* 116:178–95. (1, 7, 9, 16, 17, 18, 22)

Terborgh, J., S. K. Robinson, T. A. Parker III, C. A. Munn, and N. Pierpont. 1990. Structure and organization of an Amazonian forest bird community. *Ecol. Monogr.* 60:213–38. (5)

Test, F. H., O. J. Sexton, and H. Heatwole. 1966. Reptiles of Rancho Grande and vicinity, Estado Aragua, Venezuela. *Misc. Publ. Mus. Zool. Univ. Mich.* 128:1–63. (25)

Thayer, C. W. 1983. Sediment-mediated biological disturbance and the evolution of marine benthos. In *Biotic interactions in recent and fossil benthic communities,* ed. M. J. S. Tevesz and P. L. McCall, 400–595. New York: Plenum. (29)

Thiery, R. G. 1982. Environmental instability and community diversity. *Biol. Rev.* 57:671–710. (1)

Thom, B. G. 1982. Mangrove ecology—a geomorphological perspective. In *Mangrove ecosystems in Australia,* ed. B. F. Clough, 3–17. Canberra: Australian National University Press. (20)

Thomas, C. D. 1990. Fewer species. *Nature* 347:237. (15)

Thomas, C. D., and H. C. Mallorie. 1985. Rarity, species richness and conservation: Butterflies of the Atlas Mountains in Morocco. *Biol. Conser.* 33:95–117. (10)

Thomas, J. A. 1984. The conservation of butterflies in temperate countries: Past efforts and lessons for the future. In *The biology of butterflies,* ed. Vane-Wright, R. I., and P. R. Ackery, 333–53. Symposium of the Royal Entomological Society of London, 11. (10)

Thomas, J. A. 1986. *Butterflies of the British Isles.* Country Life Books. (10)

Thomas, R. A. 1976. A revision of the South American colubrid snake genus *Philodryas* Wagler, 1830. Ph.D. diss., Texas A&M University, College Station. (25)

Thompson, J. N. 1982. *Interaction and coevolution.* New York: John Wiley and Sons. (10)

Thompson, J. N. 1989. Concepts of coevolution. *Trends Ecol. Evol.* 4:179–83. (23)

Thompson, J. N., and M. E. Moody. 1985. Assessing probability of interaction in size-structured populations: *Depressaria* attack on *Lomatium. Ecology* 66:1597–1607. (10)

Thomson, J. A. 1990. Bracken morphology and life cycle: Preferred terminology. In *Bracken biology and management,* ed. J. A. Thomson and R. T. Smith, 333–39. New South Wales, Australia: RIAS. (16)

Thresher, R. E. 1991. Geographic variability in the ecology of coral reef fishes: Evidence, evolution, and possible implications. In *The ecology of fishes on coral reefs,* ed. P. F. Sale, 401–36. San Diego: Academic Press. (30)

Tiffney, B. H. 1985a. The Eocene North Atlantic land bridge: Its importance in Tertiary and modern phytogeography of the Northern Hemisphere. *J. Arnold Arboretum* 66:243–73. (26)

Tiffney, B. H. 1985b. Perspectives on the origin of the floristic similarity between eastern Asia and eastern North America. *J. Arnold Arboretum* 66:73–94. (23)

Tilman, D. 1982. *Resource competition and community structure.* Princeton, N.J.: Princeton University Press. (1, 2, 5, 6)

Tilman, D. 1985. The resource ratio hypothesis of succession. *Am. Nat.* 125:827–52. (22)

Tilman, D. 1987. Secondary succession and the pattern of plant dominance along experimental nitrogen gradients. *Ecol. Monogr.* 57:189–214. (5, 6)

Tilman, D. 1988. Plant strategies and the dynamics and structure of plant communities. Princeton, N.J.: Princeton University Press. (1, 2)

Tilman, D. 1990. Constraints and tradeoffs: Toward a predictive theory of competition and succession. *Oikos* 58:3–15. (2)

Tilman, D. 1993. Competition, coexistence, and biodiversity in spatially subdivided habitats. *Ecology.* In press. (2)

Tilman, D., and A. El Haddi. 1992. Climatic variation and biodiversity in grasslands. *Oecologia* 89:257–64. (2)

Tilman, D., M. Mattson, and S. Langer. 1981. Competition and nutrient kinetics along a temperature gradient: An experimental test

of a mechanistic approach to niche theory. *Limnol. Oceanogr.* 26:1020–33. (2)

Tilman, D., and D. Wedin. 1991a. Dynamics of nitrogen competition between successional grasses. *Ecology* 72:1038–49. (2)

Tilman, D., and D. Wedin. 1991b. Plant traits and resource reduction for five grasses growing on a nitrogen gradient. *Ecology* 72:685–700. (2)

Timms, B. V. 1985. The structure of macrobenthic communities of Australian lakes. *Proc. Ecol. Soc. Aust.* 14:51–59. (21)

Tipper, J. C. 1979. Rarefaction and rarefiction—the use and abuse of a method in paleoecology. *Paleobiology* 5:423–434. (27)

Tobe, H., and P. H. Raven. 1988. Seed morphology and anatomy of Rhizophoraceae, inter- and intrafamilial relationships. *Ann. Mo. Bot. Gard.* 75:1319–42. (20)

Todd, C. D., J. N. Havenhand, and J. P Thorpe. 1988. Genetic differentiation, pelagic larval transport and gene flow between local populations of the intertidal marine mollusc *Adalaria proxima* (Alder and Hancock). *Func. Ecol.* 2:441–51. (4)

Toland, J. 1976. *Adolf Hitler.* New York: Ballantine Books. (11)

Tomialojc, L., T. Wesolowski, and W. Walankiewicz. 1984. Breeding bird community of a primaeval temperate forest (Bialowieza National Park, Poland). *Acta Ornithologica* 20:241–310. (12, 21)

Tomlinson, P. B. 1986. *The botany of mangroves.* Cambridge: Cambridge University Press. (20, 26)

Tomlinson, P. B., R. B. Primack, and J. S. Bunt. 1979. Preliminary observations on floral biology in mangrove Rhizophoraceae. *Biotropica* 11:256–77. (20)

Tonn, W. M., J. J. Magnuson, M. Rask, and J. Toivonen. 1990. Intercontinental comparison of small-lake fish assemblages: The balance between local and regional processes. *Am. Nat.* 136:345–75. (16, 21, 22)

Toomey, D. F. 1981. Organic buildup constructional capability in Lower Ordovician and Late Paleozoic mounds. In *Communities in the past,* ed. J. Gray and A. J. Boucot, 35–68. Stroudsburg, Pa.: Hutchinson and Ross. (27)

Tralau, H. 1964. The genus *Nypa* van Wurmb. *Kongl. Svenska Vetensk. Acad. Handl.* 10:1–29. (20)

Traverse, A. 1955. *Pollen analysis of the Brandon lignite of Vermont.* U.S. Bureau of Mines Report of Investigations, no. 5151, 1–107. (26)

Tryon, R. 1989. Pteridophytes. In *Ecosystems of the world.* Vol. 14B, *Tropical rain forest ecosystems: Biogeographical and ecological studies,* ed. H. Lieth and M. J. A. Werger, 327–38. Amsterdam: Elsevier Scientific. (5)

Turner, J. R. G., C. M. Gatehouse, and C. A. Corey. 1987. Does solar energy control organic diversity? Butterflies, moths, and the British climate. *Oikos* 48:195–205. (5, 6)

Turner, J. R. G., J. J. Lennon, and J. A. Lawrenson. 1988. British bird species distributions and the energy theory. *Nature* 335:539–541. (5, 6, 26)

Turner, M. G. 1989. Landscape ecology: The effect of pattern on process. *Annu. Rev. Ecol. Syst.* 20:171–97. (7, 30)

Tyndale-Biscoe, H. 1973. *Life of marsupials.* London: Edward Arnold. (15)

Uerpmann, H. P. 1971. Die Tierknochenfunde aus der Talayot Siedlung von S'Illot (San Lorenzo, Mallorca). *Studien über frühe Tierknochenfunde von iberischen Halbinsel* 2:1–111. (12)

Uerpmann, H. P. 1987. The origin and relations of Neolithic sheep and goats in the Western Mediterranean. In *Premières communautés paysannes en Méditerranée occidentale,* ed. J. Guilaine, J. Courtin, and J. L. Vernet, 175–79. Paris: Centre National de la Recherche Scientifique. (12)

Underwood, A. J. 1976. Food competition between age-classes in the intertidal neritacean *Nerita atramentosa* Reeve (Gastropoda: Prosobranchia). *J. Exp. Mar. Biol. Ecol.* 23:145–54. (4)

Underwood, A. J. 1978. An experimental evaluation of competition between three species of intertidal prosobranch gastropods. *Oecologia* 33:185–208. (4)

Underwood, A. J. 1981. Techniques of analysis of variance in experimental marine biology and ecology. *Annu. Rev. Oceanogr. Mar. Biol.* 19:513–605. (4)

Underwood, A. J. 1984. Vertical and seasonal patterns in competition for microalgae between intertidal gastropods. *Oecologia* 64:211–22. (4)

Underwood, A. J. 1985. Physical factors and biological interactions: The necessity and nature of ecological experiments. In *The ecology of rocky coasts,* ed. P. G. Moore and R. Seed, 371–90. London: Hodder and Stoughton. (4)

Underwood, A. J. 1986a. The analysis of competition by field experiments. In *Community ecology: Pattern and process,* ed. J. Kikkawa and D. J. Anderson, 240–58. Melbourne: Blackwell. (4)

Underwood, A. J. 1986b. What is a community? In *Patterns and processes in the history of life,* ed. D. M. Raup and D. Jablonski, 351–68. Berlin: Springer-Verlag. (4, 7)

Underwood, A. J. 1992. Competition in marine plant/animal interactions. In *Plant-animal interactions in the marine benthos,* ed. D. M. John, S. J. Hawkins and J. H. Price, 443–75. Oxford: Systematics Association and Clarendon Press. (4)

Underwood, A. J., and E. J. Denley. 1984. Paradigms, explanations and generalizations in models for the structure of intertidal communities on rocky shores. In *Ecological communities: Conceptual issues and the evidence,* ed. D. R. Strong, D. Simberloff, L. G. Abele and A. Thistle, 151–80. Princeton, N.J.: Princeton University Press. (4)

Underwood, A. J., E. J. Denley, and M. J. Moran. 1983. Experimental analyses of the structure and dynamics of mid-shore rocky intertidal communities in New South Wales. *Oecologia* 56:202–19. (4)

Underwood, A. J., and P. G. Fairweather. 1986. Intertidal communities: Do they have different ecologies or different ecologists? *Proc. Ecol. Soc. Aust.* 14:7–16. (4)

Underwood, A. J., and P. G. Fairweather. 1989. Supply-side ecology and benthic marine assemblages. *Trends Ecol. Evol.* 4:16–20. (4)

Underwood, A. J., and P. Jernakoff. 1981. Interactions between algae and grazing gastropods in the structure of a low-shore algal community. *Oecologia* 48:221–33. (4)

Underwood, A. J., and P. H. Verstegen. 1988. Experiments on the association between the intertidal amphipod *Hyale media* Dana and the limpet *Cellana tramoserica* (Sowerby). *J. Exp. Mar. Biol. Ecol.* 119:83–98. (4)

Uphof, J. C. T. 1968. *Dictionary of economic plants.* Lehre, Germany: Verlag von Cramer. (26)

Vail, P. R., R. M. Mitchum, Jr., R. G. Todd, J. M. Widmier, S. Thompson III, J. B. Sangree, J. N. Bubb, and W. G. Hatlelid. 1977. Seismic stratigraphy and global changes in sea level. *Am. Assoc. Petroleum Geol. Mem.* 26:49–212. (27)

Valentine, J. W. 1961. Paleoecologic molluscan geography of the Californian Pleistocene. *Univ. Calif. Publ. Geol. Sci.* 34:309–442. (29)

Valentine, J. W. 1966. Numerical analysis of marine molluscan ranges on the extratropical northeastern Pacific shelf. *Limnol. Oceanogr.* 11:198–211. (29)

Valentine, J. W. 1968. Climatic regulation of species diversification and extinction. *Bull. Geol. Soc. Am.* 79:273–75. (27)

Valentine, J. W. 1970. How many fossil species? A new approximation. *J. Paleontol.* 44:410–15. (27)

Valentine, J. W. 1971. Resource supply and species diversity patterns. *Lethaia* 4:51–61. (27)

Valentine, J. W. 1973. *Evolutionary ecology of the marine biosphere.* Englewood Cliffs, N.J.: Prentice-Hall. 1–511. (27)

Valentine, J. W. 1976. Genetic strategies of adaptation. In *Molecular Evolution,* ed. F. J. Ayala, 78–94. Sunderland, Mass.: Sinauer Associates. (5)

Valentine, J. W. 1989a. How good was the fossil record? Clues from the Californian Pleistocene. *Paleobiology* 15:83–94. (29)

Valentine, J. W. 1989b. Phanerozoic marine faunas and the stability of the Earth system. *Global and Planetary Change* 1:137–155. (29)

Valentine, J. W., and D. Jablonski. 1991. Biotic effects of sea level change: The Pleistocene test. *J. Geophys. Res.* 96:6873–78. (29)

Valentine, J. W., and B. Mallory 1965. Recurrent groups of bonded species in mixed death assemblages. *J. Geol.* 73:683–701. (29)

Valentine, J. W., and R. F. Meade. 1961. Californian Pleistocene paleotemperatures. *Univ. Calif. Publ. Geol. Sci.* 40:1–46. (29)

Valentine, J. W., B. H. Tiffney, and J. J. Sepkoski, Jr. 1991. Evolutionary dynamics of plants and animals: A comparative approach. *Palaios* 5:81–88. (29)

Valladares, G., and J. H. Lawton. 1991. Host-plant selection in the holly leaf-miner: Does mother know best? *J. Anim. Ecol.* 60:227–40. (7)

Van Balen, J. H. 1973. A comparative study of the Great Tit *Parus major* in different habitats. *Ardea* 61:1–93. (12)

Vance, R. R. 1980. The effect of dispersal on population size in a temporally varying environment. *Theor. Pop. Biol.* 18:342–62. (7)

van der Hammen, T., T. A. Wijmstra, and W. H. Zagwijn. 1971. The floral record of the late Cenozoic of Europe. In *The late Cenozoic glacial ages,* ed. K. K. Turekian, 391–424. New Haven: Yale University Press. (26)

Vandermeer, J. H. 1969. The competitive structure of communities: an experimental approach with protozoa. *Ecology* 50:362–71. (1)

Vandermeer, J. H. 1972. Niche theory. *Annu. Rev. Ecol. Syst.* 3:107–32. (1)

Vandermeer, J. H. 1973. On the regional stabilization of locally unstable predator-prey relationships. *J. Theor. Biol.* 41:161–70. (8)

Van Noordwijk, A. J., J. H. van Balen, and W. Scharloo. 1980. Heritability of ecologically important traits in the Great Tit. *Ardea* 68:193–203. (12)

Van Riper, C. III., S. G. Van Riper, M. L. Goff, and M. Laird. 1986. The epizootiology and ecological significance of malaria in Hawaiian land birds. *Ecol. Monogr.* 56:327–44. (30)

Van Steenis, C. G. G. J. 1949. Plumbaginaceae. *Flora Malesiana,* ser. 1, 4(2):107–12. (20)

Van Valen, J. 1973a. Body size and numbers of plants and animals. *Evolution* 27:27–35. (2)

Van Valen, L. 1973b. Pattern and the balance of nature. *Evol. Theory* 1:31–49. (12)

Van Valen, L. M. 1985. A theory of origination and extinction. *Evol. Theory* 7:133–42. (29)

Van Valkenburgh, B. 1985. Locomotor diversity within past and present guilds of large predatory mammals. *Paleobiology* 11:406–28. (28)

Van Valkenburgh, B. 1987. Skeletal indicators of locomotor behavior in living and extinct carnivores. *J. Vertebrate Paleontol.* 7:162–182. (28)

Van Valkenburgh, B. 1988. Trophic diversity within past and present guilds of large predatory mammals. *Paleobiology* 14:156–73. (25, 28)

Van Valkenburgh, B. 1989. Carnivore dental adaptations and diet: A study of trophic diversity within guilds. In *Carnivore behavior, ecology and evolution,* ed. J. L. Gittleman, 410–36. Ithaca: Cornell University Press. (28)

Van Valkenburgh, B. 1991. Iterative evolution of hypercarnivory in canids (Mammalia: Carnivora): Evolutionary interactions among sympatric predators. *Paleobiology* 17:240–62. (28)

Vanzolini, P. E., and W. R. Heyer. 1985. The American herpetofauna and the interchange. In *The great American biotic interchange,* ed. F. G. Stehli and S. D. Webb, 475–87. New York: Plenum Press. (25)

Vanzolini, P. E., A. M. Ramos-Costa, and L. J. Vitt. 1980. *Repteis das Caatingas.* Academia Brasileira de Ciências. (25)

Venrick, E. L. 1982. Phytoplankton in an oligotrophic ocean: Observations and questions. *Ecol. Monogr.* 52:129–54. (19)

Venrick, E. L. 1990. Phytoplankton in an oligotrophic ocean: Species structure and interannual variability. *Ecology* 71:1547–63. (19)

Venrick, E. L., J. A. McGowan, D. R. Cayan, and T. L. Hayward. 1987. Climate and chlorophyll a: Long-term trends in the central North Pacific ocean. *Science* 238:70–72. (19)

Vermeij, G. J. 1987. *Evolution and escalation: An ecological history of life.* Princeton, N.J.: Princeton University Press. (1, 23, 29, 30)

Vermeij, G. J. 1989. Invasion and extinction: The last three million years of North Sea pelecypod history. *Conserv. Biol.* 3:274–81. (29)

Vermeij, G. J. 1991a. Anatomy of an invasion: The trans-Arctic interchange. *Paleobiology* 17:281–307. (30)

Vermeij, G. J. 1991b. When biotas meet: Understanding biotic interchange. *Science* 253:1099–1104. (1, 30)

Vermeij, G. J. 1992. Time of origin and biogeographical history of specialized relationships between northern marine plants and herbivorous molluscs. *Evolution* 46:657–64. (30)

Vernet, J. L. 1973. Étude sur l'histoire de la végétation du sud-est de la France au Quaternaire d'après les charbons de bois principalement. *Paléobiologie continentale* 4:1–90. (12)

Vezey, E. L., V. P. Shah, J. J. Skvarla, and P. H. Raven. 1988. Morphology and phenetics of Rhizophoraceae pollen. *Ann. Mo. Bot. Gard.* 75:1369–86. (20)

Vigne, J. D. 1988a. Apports de la biogéographie insulaire à la connaissance de la place des mammifères sauvages dans les sociétés néolithiques méditerranéennes. *Anthropozoologica* 8:31–52. (12)

Vigne, J. D. 1988b. Données préliminaires sur l'histoire du peuplement mammalien de l'îlot de Zembra (Tunisie). *Mammalia* 52:567–74. (12)

Vigne, J. D. 1988c. *Les Mammifères post-glaciaires de Corse, étude archéozoologique.* XXVIe suppl., Gallia Préhistoire. Paris: Centre National de la Recherche Scientifique. (12)

Vigne, J. D. 1990. Biogeographical history of the mammals on Corsica (and Sardinia) since the Final Pleistocene. *Academia Nazionale dei Lincei* (Roma) 85:370–92. (12)

Vigne, J. D., and M. C. Marinval-Vigne. 1983. Méthode pour la mise en évidence de la consommation du petit gibier. *Brit. Archaeol. Rep., Int. Ser.* 163:239–42. (12)

Vitousek, P. 1982. Nutrient cycling and nutrient use efficiency. *Am. Nat.* 119:553–72. (2)

Vitousek, P. 1986. Biological invasions and ecosystem properties: Can species make a difference? In *Ecology of biological invasions of North America and Hawaii,* ed. H. A. Mooney and J. A. Drake, 163–76. Berlin: Springer-Verlag. (29, 30)

Vitt, L. J. 1980. Ecological observations on sympatric *Philodryas* (Colubridae) in northeastern Brazil. *Papeis Avulsos de Zoología* (São Paulo) 34:87–98. (25)

Vitt, L. J. 1983. Ecology of an anuran-eating guild of terrestrial tropical snakes. *Herpetologica* 39:52–66. (25)

Vitt, L. J. 1987. Communities. In *Snakes, ecology and evolutionary biology,* ed. R. A. Seigel, J. T. Collins, and S. S. Novak, 335–65. New York: Macmillan. (25)

Vitt, L. J., and L. D. Vangilder. 1983. Ecology of a snake community in northeastern Brazil. *Amphibia-Reptilia* 4:273–96. (25)

Vogel, P., T. Maddalena, and P. J. Schembri. 1990. Cytataxonomy of shrews of the genus *Crocidura* from Mediterranean islands. *Vie et Milieu* 40:124–29. (12)

Vollenweider, R. A. 1975. Input-output models with special reference to the phosphorus loading concept in limnology. *Schweizerische Zeitschrift für Hydrologie* 37:53–84. (6)

Voous, K. H. 1960. *Atlas of European birds.* Edinburgh: Nelson. (12)

Vuilleumier, F. 1972. Bird species diversity in Patagonia (temperate South America). *Am. Nat.* 106:266–71. (21)

Vuilleumier, F., and D. Simberloff. 1980. Ecology versus history as determinants of patchy and insular distribution in high Andean birds. *Evol. Biol.* 12:235–379. (21)

Waldren, W. H. 1982. Balearic prehistoric ecology and culture. *Brit. Archaeol. Rep., Int. Ser* 149 (i, ii, iii). (12)

Walker, D., and Y. Chen. 1987. Palynological light on tropical forest dynamics. *Quat. Sci. Rev.* 6:77–92. (29)

Walker, S. E. 1989. Hermit crabs as taphonomic agents. *Palaios* 4:439–52. (29)

Walker, T. D., and J. W. Valentine. 1984. Equilibrium models of evolutionary species diversity and the number of empty niches. *Am. Nat.* 124:887–99. (29)

Wallace, A. R. 1876. *The geographical distribution of animals,* Vols. 1 and 2. Reprint, New York: Hafner 1962. (1)

Wallace, A. R. 1878. *Tropical nature and other Essays.* New York: Macmillan. (2, 6)

Wallis, J. B. 1961. *The Cicindelidae of Canada.* Toronto: University of Toronto Press. (18)

Wallis, R. L. 1982. Adaptation to low environmental temperatures in the carnivorous marsupials. In *Carnivorous marsupials,* ed. M. Archer, 285–91. Sydney: Royal Zoological Society of New South Wales. (14)

Walter, H. 1979. *Vegetation of the earth.* New York: Springer-Verlag. (26)

Walter, H. 1985. *Vegetation of the earth and ecological systems of the geo-biosphere.* 3d ed. Trans. from the 5th, revised German edition by Owen Muise. Berlin: Springer-Verlag. (20)

Wang, C. 1961. *The Forests of China.* Maria Moors Cabot Foundation Publication no. 5. Cambridge, Mass.: Harvard University. (26)

Wanntorp, H. E., D. R. Brooks, T. Nilsson, S. Nylin, F. Ronquist, S. Stearns, and N. Wedell. 1990. Phylogenetic approaches in ecology. *Oikos* 57:119–32. (17)

Ward, C. J., and T. D. Steinke. 1982. A note on the distribution and approximate areas of mangroves in South Africa. *S. Afr. J. Bot.* 1:51–53. (20)

Ward, D. B. 1979. Plants. In *Rare and endangered biota of Florida,* ed. P. C. H. Pritchard, vol. 5. Gainesville: University Presses of Florida. (23)

Warner, R. E. 1968. The role of introduced diseases in the extinction of the endemic Hawaiian avifauna. *Condor* 70:101–20 (30)

Warner, R. R., and P. L. Chesson. 1985. Coexistence mediated by recruitment fluctuations: A field guide to the storage effect. *Am. Nat.* 125:769–87. (2)

Warren, P. H., and J. H. Lawton. 1987. Invertebrate predator-prey body size relationships: An explanation for upper triangular food webs and patterns in food web structure? *Oecologia* 74:231–35. (3)

Washburn, A. L. 1980. *Geocryology.* New York: Halsted Press. (18)

Wasserman, S. S., and C. Mitter. 1978. The relationship of body size to breadth of diet in some Lepidoptera. *Ecol. Entomol.* 3:155–60. (23)

Wasson, R. J. 1982. Landform development in Australia. In *Evolution of the flora and fauna of arid Australia,* ed. W. R. Barker and P. J. M. Greenslade, 23–33. Adelaide: Peacock Publications. (14)

Wasson, R. J. 1984. Late Quaternary palaeoenvironments in the desert dunefields of Australia. In *Late Cainozoic palaeoclimates of the Southern Hemisphere,* ed. J. C. Vogel, 419–432. Rotterdam: A. A. Balkema. (14)

Watkins, R. 1979. Benthic community organization in the Ludlow Series of the Welsh Borderland. *Bull. Brit. Mus. (Nat. Hist.) Geol.* 31:175–280. (5)

Watson, J. A. L. 1982. Distribution, biology and speciation in the Australian harvester termites, *Drepanotermes* (Isoptera: Termitinae). In *Evolution of the flora and fauna of arid Australia,* ed. W. R. Barker and P. J. M. Greenslade, 263–265. Adelaide: Peacock Publications. (14)

Watson, J. A. L., R. A. Barrett, and C. Lendon. 1978. Termites. In *The physical and biological features of Kunoth Paddock in central Australia,* ed. W. A. Low, 101–8. CSIRO Division of Land Resources Management, Technical Paper no. 4. Melbourne. (14)

Watson, J. A. L., and F. J. Gay. 1983. Taxonomy and applied entomology of Australian termites: A small order in perspective. In *Australian systematic entomology: A bicentenary perspective,* ed. E. Highley and R. W. Taylor, 34–49. Melbourne: CSIRO. (14)

Watson, J. G. 1928. Mangrove forests of the Malay Peninsula. *Malayan Forest Records* 6:1–275. (20)

Watt, A. S. 1947. Pattern and process in the plant community. *J. Ecol.* 35:1–22. (1)

Watt, A. S. 1955. Bracken versus heather: A study in plant sociology. *J. Ecol.* 43:490–506. (4)

Webb, L. J., and J. G. Tracey. 1981. The rainforests of northern Australia. In *Australian vegetation,* ed. R. H. Groves, 67–101. Cambridge: Cambridge University Press. (13)

Webb, S. D. 1969a. The Burge and Minnechaduza Clarendonian mammalian faunas of north-central Nebraska. *Univ. Calif. Publ. Geol. Sci.* 78:1–191. (28)

Webb, S. D. 1969b. Extinction-origination equilibrium in late Cenozoic land mammals of North America. *Evolution* 23:688–702. (28)

Webb, S. D. 1977. A history of savannah vertebrates in the New World. Part I: North America. *Annu. Rev. Ecol. Syst.* 8:355–80. (28)

Webb, S. D. 1983. The rise and fall of the Late Miocene ungulate fauna in North America. In *Coevolution,* ed. M. H. Nitecki, 267–306. Chicago: University of Chicago Press. (28)

Webb, S. D. 1984. Ten million years of mammal extinction in North America. In *Quaternary Extinctions,* ed. P. S. Martin and R. G. Klein, 189–210. Tucson: University of Arizona Press. (28)

Webb, S. D. 1989. The fourth dimension in North American terrestrial mammal communities. In *Patterns in the structure of mammalian communities,* ed. D. W. Morris, Z. Abramsky, B. J. Fox, and M. R. Willig, 181–203. Special Publications, Museum, Texas Tech University, 28. Lubbock: Texas Tech University Press. (28)

Webb, S. D. 1991. Ecogeography and the Great American Interchange. *Paleobiology* 17:266–80. (30)

Webb, S. D., and A. D. Barnosky. 1989. Faunal dynamics of Pleistocene mammals. *Ann. Rev. Earth Planetary Sci.* 17:413–38. (29)

Webb, T. III. 1987. The appearance and disappearance of major vegetational assemblages: Long-term vegetational dynamics in eastern North America. *Vegetatio* 69:177–87. (15)

Wedin, D., and D. Tilman. 1990. Species effects on nitrogen cycling: A test with perennial grasses. *Oecologia* 84:433–41. (2)

Wehmiller, J. F., K. R. Lajoie, K. A. Kvenvolden, E. Peterson, D. F. Belknap, G. L. Kennedy, W. O. Addicott, J. G. Vedder, and R. W. Wright. 1977. Correlation and chronology of Pacific coast marine terrace deposits of continental United States by fossil amino acid stereochemistry—technique evaluation, relative ages, kinetic model ages, and geologic implications. *U.S. Geological Survey, Open File Report,* 77–680. (29)

Welden, C. W., and W. L. Slauson. 1986. The intensity of competition versus its importance: An overlooked distinction and some implications. *Q. Rev. Biol.* 61:23–44. (4)

Wells, A. G. 1983. Distribution of mangrove species in Australia. In *Biology and ecology of mangroves,* ed. H. J. Teas, 57–76. The Hague: Junk. (20)

Wells, J. W. 1955. A survey of the distribution of coral genera in the Great Barrier Reef region. *Reports of the Great Barrier Reef Committee* 6:21–29. (27)

Wells, J. W. 1957. Coral reefs. *Geol. Soc. Am. Mem.* 67; 1:609–31. (27)

Werner, P. A., and W. J. Platt. 1976. Ecological relationships of co-occurring goldenrods (*Solidago*: Compositae). *Am. Nat.* 110:959–71. (2)

West, N. E., and J. O. Klemmedson. 1978. Structural distribution of nitrogen in desert ecosystems. In *Nitrogen in desert ecosystems,* ed. N. E. West and J. J. Skujins, 1–16. Stroudsburg, Pa.: Dowden, Hutchinson and Ross. (14)

Wester, L. 1981. Introduction and spread of mangroves in the Hawaiian Islands. *Assoc. Pacific Coast Geogr. Yrbk.* 43:125–37. (20)

Westgate, J. W., and C. T. Gee. 1990. Paleoecology of a Middle Eocene mangrove biota (vertebrates, plants, and invertebrates) from southwest Texas. *Palaeogeogr. Palaeoclim. Palaeoecol.* 787:163–77. (20)

Westman, W. E. 1990. Park management of exotic plant species: Problems and issues. *Conserv. Biol.* 4:251–60. (29)

Westoby, M. 1985. Two main relationships among the components of species richness. *Proc. Ecol. Soc. Aust.* 14:103–7. (15, 16, 22)

Westoby, M. 1988. Comparing Australian ecosytems to those elsewhere. *BioScience* 38:549–56. (14, 15)

Westoby, M., J. M. Cousins, and A. C. Grice. 1982. Rate of decline of some soil seed populations during drought in western New South Wales. In *Ant-plant interactions in Australia,* ed. R. C. Buckley, 7–10. The Hague: Junk. (14)

Westoby, M., B. Rice, and J. Howell. 1990. Seed size and plant growth form as factors in dispersal spectra. *Ecology* 71:1307–15. (15)

Wethey, D. S. 1984. Spatial pattern in barnacle settlement—day to day changes during the settlement season. *J. Mar. Biol. Assoc. UK* 64:687–98. (4)

Wethey, D. S. 1985. Catastrophe, extinction, and species diversity: A rocky intertidal example. *Ecology* 66:445–56. (29)

Whiteside, M. C., and R. V. Harmsworth. 1967. Species diversity in chydorid (Cladocera) communities. *Ecology* 48:664–67. (5)

Whitford, W. G. 1978. Structure and seasonal activity of Chihuahua Desert ant communities. *Insectes Sociaux* 25:79–88. (14)

Whitford, W. G., S. Dick-Peddie, D. Walters, and J. Ludwig. 1978. Effects of shrub defoliation on grass cover and rodent species in a Chihuahuan desert ecosystem. *J. Arid Environ.* 1:237–42. (5)

Whitford, W. G., J. A. Ludwig, and J. C. Noble. 1992. The importance of subterranean termites in semi-arid ecosystems in southeastern Australia. *J. Arid Environ.* 22:87–91. (14)

Whittaker, R. H. 1953. A consideration of climax theory: The climax as a population and pattern. *Ecol. Monogr.* 23:41–78. (1)

Whittaker, R. H. 1967. Gradient analysis of vegetation. *Biol. Rev.* 42:207–64. (1)

Whittaker, R. H. 1972. Evolution and measurement of species diversity. *Taxon* 21:213–51. (1, 11, 16)

Whittaker, R. H. 1975. *Communities and ecosystems.* 2d ed. New York: Macmillan. (2, 5, 19)

Whittaker, R. H. 1977. Evolution of species diversity in land communities. *Evol. Biol.* 10:1–67. (15)

Whittaker, R. H., and W. A. Niering. 1975. Vegetation of the Santa Catalina Mountains, Arizona. V. Biomass, production, and diversity along an elevational gradient. *Ecology* 56:771–90. (1, 5)

Wiegmann, B. M., C. Mitter, and B. D. Farrell. 1993. Diversification of carnivorous parasitic insects: Extraordinary radiation, or specialized dead end? *Am. Nat.* In press.

Wiens, J. A. 1981. Scale problems in avian censusing. *Stud. Avian Biol.* 6:513–21. (12)

Wiens, J. A. 1986. Spatial scale and temporal variation in studies of shrubsteppe birds. In *Community ecology,* ed. J. Diamond and T. J. Case, 154–72. New York: Harper and Row. (8, 12)

Wiens, J. A. 1989a. *The ecology of bird communities.* Vols. 1 and 2. Cambridge: Cambridge University Press. (1, 12, 16, 22)

Wiens, J. A. 1989b. Spatial scaling in ecology. *Funct. Ecol.* 3:385–97. (7, 8)

Wiens, J. A. 1991. Ecological similarity of shrub-desert avifaunas of Australia and North America. *Ecology* 72:479–95. (14, 15, 21)

Wiens, J. A., and J. T. Rotenberry. 1985. Response of breeding passerine birds to rangeland alteration in a North American shrubsteppe locality. *J. Appl. Ecol.* 22:655–68. (12)

Wiesner, J. 1988. Die Gattung *Therates* Latr. und ihre Arten. 15. Beitrag zur Kenntnis der Cicindelidae (Coleoptera). *Mitteilungen münchener entomologische Gesellschaft* 78:5–107. (18)

Wilbur, H. M. 1972. Competition, predation, and the structure of the *Ambystoma-Rana sylvatica* community. *Ecology* 53:3–21. (1)

Wilcove, D. S., C. H. McLellan, and A. P. Dobson. 1986. Habitat fragmentation in the temperate zone. In *Conservation biology: The science of scarcity and diversity,* ed. M. E. Soulé, 237–256. Sunderland, Mass.: Sinauer Associates. (12)

Wilcox, B. A. 1978. Supersaturated island faunas: A species-area relationship for lizards on post-Pleistocene land-bridge islands. *Science* 199:996–98. (30)

Wiley, E. O. 1978. The evolutionary species concept revisited. *Syst. Zool.* 27:17–26. (24)

Wiley, E. O. 1980. Is the evolutionary species concept fiction? A consideration of classes, individuals, and historical entities. *Syst. Zool.* 29:76–80. (24)

Wiley, E. O. 1981. Phylogenetics: The theory and practice of phylogenetic systematics. New York: Wiley. (21, 24)

Wiley, E. O. 1998a. Parsimony analysis and vicariance biogeography. *Syst. Zool.* 37:271–90. (24)

Wiley, E. O. 1988b. Vicariance biogeography. *Annu. Rev. Ecol. Syst.* 19:513–42. (24)

Wiley, E. O., and R. L. Mayden. 1985. Species and speciation in phylogenetic systematics, with examples from the North American fish fauna. *Ann. Mo. Bot. Gard.* 72:596–635. (24)

Wiley, E. O., D. J. Siegel-Causey, D. R. Brooks, and V. A. Funk. 1990. *The compleat cladist: A primer of phylogenetic procedures.* Special Publications of the Museum of Natural History, University of Kansas, Lawrence, Kansas. (24)

Wilkinson, H. P. 1981. The anatomy of the hypocotyls of *Ceriops* Arnott (Rhizophoraceae), recent and fossil. *Biol. J. Linn. Soc.* 82:139–64. (20)

Wilkinson, H. P. 1983. Starch grain casts and moulds in Eocene (Tertiary) fossil mangrove hypocotyls. *Ann. Bot.* 51:39–45. (20)

Wilkinson, L. 1988. *SYSTAT: The system for statistics.* Evanston: SYSTAT, Inc. (24)

Williams, C. B. 1964. *Patterns in the balance of nature.* London: Academic Press. (24)

Williams, C. H., and M. Raupach. 1983. Plant nutrients in Australian soils. In *Soils: An Australian viewpoint,* ed. CSIRO Division of Soils, 777–94. Melbourne: CSIRO/Academic Press. (15)

Williams, E. E. 1972. The origin of faunas: Evolution of lizard congeners in a complex island fauna—a trial analysis. *Evol. Biol.* 6:47–89. (25)

Williams, E. E. 1983. Ecomorphs, faunas, island size, and diverse end points in island radiations of *Anolis*. In *Lizard ecology: Studies of a model organism,* ed. R. B. Huey, E. R. Pianka, and

T. W. Schoener, 326–70. Cambridge, Mass.: Harvard University Press. (25)

Williams, O. B., and J. H. Calaby. 1985. The hot deserts of Australia. In *Hot deserts and arid shrublands,* ed. M. Evenari, I. Noy-Meir, and D. W. Goodall, 269–312. Amsterdam: American Elsevier. (14)

Williams, P. J., and M. W. Smith. 1989. *The frozen earth: Fundamentals of geocryology.* Oxford: Alden Press. (18)

Williams, R. J., F. B. Griffiths, E. J. Van der Wel, and J. Kelly. 1988. Cargo vessel ballast water as a vector for the transport of nonindigenous marine species. *Estuarine Coastal Shelf Sci.* 26:409–21. (29)

Williamson, M. 1981. *Island populations.* Oxford: Oxford University Press. (7, 12)

Willis, H. L. 1972. Species density of North American *Cicindela* (Coleoptera, Cicindelidae). *Cicindela* 4:29–43. (18)

Willis, J. C. 1922. *Age and area: A study in geographical distribution and origin in species.* Cambridge: Cambridge University Press. (1, 30)

Willis, J. C. 1966. *A Dictionary of the flowering plants and ferns.* 7th ed., rev. by H. K. Airy Shaw. Cambridge: Cambridge University Press. (20)

Willson, M. F., B. L. Rice, and M. Westoby. 1990. Seed dispersal spectra: A comparison of temperate plant communities. *J. Vegetation Sci.* 1:547–62. (15)

Wilson, E. O. 1961. The nature of the taxon cycle in the Melanesian ant fauna. *Am. Nat.* 95:169–93. (20)

Wilson, E. O. 1969. The species equilibrium. In *Diversity and stability in ecological systems,* ed. G. M. Woodwell and H. H. Smith, 38–47. Brookhaven, N.Y.: Brookhaven National Laboratory. (22)

Wilson, E. O. 1987. Causes of ecological success: The case of the ants. *J. Anim. Ecol.* 56:1–9. (14)

Wilson, L. D. 1987. A resume of the colubrid snakes of the genus *Tantilla* of South America. *Milwaukee Public Museum Contributions in Biology and Geology,* no. 68, 1–35. (25)

Wilson, S. D., and P. A. Keddy. 1988. Species richness, survivorship, and biomass accumulation along an environmental gradient. *Oikos* 53:375–80. (6)

Winemiller, K. O. 1989. Must connectance decrease with species richness? *Am. Nat.* 134:960–968. (3)

Winemiller, K. O. 1990. Spatial and temporal variation in tropical fish trophic networks. *Ecol. Monogr.* 60:331–67. (3)

Winemiller, K. O., and E. R. Pianka. 1990. Organization in natural assemblages of desert lizards and tropical fishes. *Ecol. Monogr.* 60:27–55. (25)

Winer, B. J. 1971. *Statistical principles in experimental design,* 2nd ed. Tokyo: McGraw-Hill Kogakusha. (4)

Wing, S. L., and M. B. Farley. 1990. Stability in the composition of Paleogene forests (Abstract). *IVth Int. Cong. Syst. Evol. Biol.,* Abstracts, 7. (29)

Wing, S. L., and L. J. Hickey. 1984. The *Platycarya* perplex and the evolution of the Juglandaceae. *Am. J. Bot.* 71:388–411. (26)

Winograd, M. 1983. Observaciones sobre el hallazgo de *Pelliciera rhizophorae* (Theaceae) en el Caribe Colombiano. *Biotropica* 15:297–98. (20)

Wise, K. P., and T. J. M. Schopf. 1981. Was marine faunal diversity in the Pleistocene affected by changes in sea level? *Paleobiology* 7:394–99. (27)

Wisniewski, W. L. 1958. Characterization of the parasitofauna of an eutrophic lake. *Acta Parasitologica Polonica* 6:1–64. (17)

Witman, J. D. 1987. Subtidal coexistence: Storms, grazing, mutualism, and the zonation of kelps and mussels. *Ecol. Monogr.* 57:167–87. (1)

Woinarski, J. C. Z., and J. M. Cullen. 1985. Breeding biology and life history of small insectivorous birds in Australian forests: Response to a stable environment? *Proc. Ecol. Soc. Aust.* 14:159–68. (15)

Wolfe, J. A. 1978. A paleobotanical interpretation of Tertiary climates in the Northern Hemisphere. *Am. Sci.* 66:694–703. (23, 28)

Wolfe, J. A. 1979. *Temperature parameters of humid to mesic forests of eastern Asia and relation to forests of other regions of the Northern Hemisphere and Australia.* U.S. Geological Survey Survey Professional Paper, no. 1106. (15, 26)

Wolfe, J. A. 1981. Vicariance biogeography of angiosperms in relation to paleobotanical data. In *Vicariance biogeography: A critique,* ed. G. Nelson and D. E. Rosen, 413–27. New York: Columbia University Press. (26)

Wolfe, J. A. 1985. Distribution of major vegetation types during the Tertiary. In *The carbon cycle and atmospheric CO_2,* ed. E. T. Sundquist and W. S. Broecker, 357–76. Geophysical Monograph 32. Washington, D.C.: American Geophysical Union. (28)

Wollkind, D. J. 1976. Exploitation in three trophic levels: An extension allowing intraspecies carnivore interaction. *Am. Nat.* 110:431–47. (5)

Wood, S. L. 1982. *The bark and ambrosia beetles (Coleoptera: Scolytidae) of North and Central America, a taxonomic monograph.* Great Basin Naturalist Memoirs, no. 6. (23)

Wood, T. K., K. L. Olmstead, and S. I. Guttman. 1990. Insect phenology mediated by host-plant water relations. *Evolution* 44:629–36. (23)

Wood, T. K., and S. I. Guttman. 1981. The role of host plants in the speciation of treehoppers: An example from the *Enchenopa binotata* complex. In *Insect life history patterns: Habitat and geographic variation,* ed. R. F. Denno and H. Dingle, 39–54. New York: Springer-Verlag. (22)

Woodburne, M. (Ed.). 1987. *Cenozoic mammals of North America.* Berkeley: University of California Press. (28)

Woodroffe, C. D., and J. Grindrod. 1991. Mangrove biogeography: The role of Quaternary environmental and sea-level change. *J. Biogeogr.* 18:479–92. (20)

Woodson, R. E. 1954. The North American species of *Asclepias* L. *Ann. Mo. Bot. Gard.* 41:1–211. (23)

Wright, D. H. 1983. Species-energy theory: An extension of species-area theory. *Oikos* 41:496–506. (2, 5, 6, 24, 26)

Wright, D. H. 1985. Patch dynamics of a foraging assemblage of bees. *Oecologia* 65:558–65. (6)

Wright, D. H. 1987. Estimating human effects on global extinction. *Int. J. Biometeorology* 31:293–99. (6)

Wright, D. H. 1990. Human impacts on energy flow through natural ecosystems, and implications for species endangerment. *Ambio* 19:189–94. (6)

Wright, D. H. 1991. Correlations between incidence and abundance are expected by chance. *J. Biogeogr.* 18:463–66, and erratum following author index for volume. (6, 10)

Wright, S. J. 1992. Seasonal drought, soil fertility, and the species diversity of tropical forest plant communities. *Trends Ecol. Evol.* 7:260–63. (30)

Wyndham, E. 1983. Movements and breeding seasons of the budgerigar. *Emu* 82:276–82. (14)

Yoda, K., T. Kira, H. Ogawa, and H. Hozumi. 1963. Self-thinning in overcrowded pure stands under cultivated and natural conditions. *J. Inst. Polytech.* (Osaka City University), ser. D 14:107–29. (2)

Yodzis, P. 1978. *Competition for space and the structure of ecological communities.* New York: Springer-Verlag. (7, 8, 19, 30)

Yodzis, P. 1980. The connectance of real ecosystems. *Nature.* 284:544–45. (3)

Yodzis, P. 1981. The structure of assembled communities. *J. Theor. Biol.* 92:103–17. (3)

Yodzis, P. 1982. The compartmentation of real and assembled ecosystems. *Am. Nat.* 120:551–70. (1, 3)

Yodzis, P. 1984a. Energy flow and the vertical structure of real ecosystems. *Oecologia* 65:86–88. (3)

Yodzis, P. 1984b. The structure of assembled communities. II. *J. Theor. Biol.* 107:115–26. (3)

Yodzis, P. 1986. Competition, mortality, and community structure. In *Community ecology,* ed. J. Diamond and T. J. Case. New York: Harper and Row. (3)

Yodzis, P. 1988. The indeterminancy of ecological interactions, as perceived through perturbation experiments. *Ecology* 69:508–15. (3, 7)

Yodzis, P. 1989a. *Introduction to theoretical ecology.* New York: Harper and Row. (3)

Yodzis, P. 1989b. Patterns in food webs. *Trends Ecol. Evol.* 4:49–50. (3)

Yom-Tov, Y. 1989. Environmental predictability and life history traits in arid environments. *J. Arid Environ.* 17:193–97. (14)

Young, C. M. 1987. Novelty of "Supply-side ecology." *Science* 235:415–16. (4)

Yount, J. L. 1956. Factors that control species numbers in Silver Springs, Florida. *Limnol. Oceanogr.* 1:286–95. (5)

Zandee, M., and M. C. Roos. 1987. Component-compatibility in historical biogeography. *Cladistics* 3:305–32. (24)

Zeigler, B. P. 1977. Persistence and patchiness of predator-prey systems induced by discrete event population exchange mechanisms. *J. Theor. Biol.* 67:687–713. (8)

Zheng W. (chief editor). 1983, 1985. *Zhongguo Shu Mu Zhi: Woody Plants of China.* Beijing: Chu Ban She. (26)

Ziegler, A. M. 1965. Silurian marine communities and their environmental significance. *Nature* 207:270–72. (5)

Ziegler, A. M. 1990. Paleogeographic Atlas Project: mapping the Mesozoic and Cenozoic at the University of Chicago. *Geotimes* 35:22–24. (27)

Ziegler, A. M., L. R. M. Cocks, and R. K. Bambach. 1968. The composition and structure of Lower Silurian marine communities. *Lethaia* 1:1–27. (5)

Ziegler, A. M., C. R. Scotese, and S. F. Barrett. 1982. Mesozoic and Cenozoic paleogeographic maps. In *Tidal friction and the Earth's rotation II; A workshop,* ed. P. Brosche and J. Suendermann, 240–52. Berlin: Springer-Verlag. (27)

Zimmerman, B. L., and M. T. Rodrigues. 1990. Frogs, snakes, and lizards of the INPA-WWF Reserves near Manaus, Brazil. In *Four Neotropical rainforests,* ed. A. Gentry, 426–54. New Haven: Yale University Press. (25)

Zinke, P. J. 1962. The pattern of influence of individual forest trees on soil properties. *Ecology* 43:130–33. (2)

Zwölfer, H. 1978. Mechanismen und Ergebnisse der Co-evolution von phytophagen und entomophagen Insekten und Hoheren Pflanzen. *Sonderbd. naturwiss. Ver. Hamburg* 2:7–50. (23)

Zwölfer, H. 1987. Species richness, species packing, and evolution in insect-plant systems. *Ecol. Stud.* 61:301–19. (16, 22, 23)

Zwölfer, H. 1988. Evolutionary and ecological relationships of the insect fauna of thistles. *Annu. Rev. Entomol.* 33:103–229. (23)

Zwölfer, H., and J. Herbst. 1988. Preadaptation, Wirtskreiserweiterung und Parallel-Cladogenese in der Evolution von phytophagen Insekten. *Zeitschrift für Zoologische Systematik und Evolutionsforschung* 26:320–40. (23)

Contributors

Zvika Abramsky
Department of Biology
Ben-Gurion University of the Negev
Beersheba
Israel

John M. Aho
Department of Biology
Auburn University
7300 University Drive
Montgomery, AL 36117
United States of America

Jacques Blondel
Centre National de la Recherche Scientifique
Centre d'Ecologie Fonctionelle et Evolutive
B.P. 5051
34033 Montpellier Cedex
France

Daniel R. Brooks
Department of Zoology
University of Toronto
Toronto, Ontario M5S 1A1
Canada

Albert O. Bush
Department of Zoology
Brandon University
Brandon, Manitoba R7A 6A9
Canada

John E. Cadle
Museum of Comparative Zoology
Harvard University
Cambridge, MA 02138
United States of America

Hal Caswell
Biology Department
Woods Hole Oceanographic Institution
Woods Hole, MA 02543
United States of America

Martin L. Cody
Department of Biology
University of California
Los Angeles, CA 90024
United States of America

Joel E. Cohen
The Rockefeller University
1230 York Avenue
New York, NY 10021
United States of America

Stephen G. Compton
Department of Zoology and Entomology
Rhodes University
Grahamstown
South Africa

Howard V. Cornell
School of Life and Health Sciences
University of Delaware
Newark, DE 19716
United States of America

David J. Currie
Biologie/Biology
University of Ottawa
Ottawa, Ontario K1N 6N5
Canada

J. A. Fagerstrom
Department of Geological Sciences
University of Colorado
Boulder, CO 80309
United States of America, and

Centre de l'Environment
École Pratique des Hautes Études
Centre National de la Récherche Scientifique
1453 Récifs Coralliens
Moorea
French Polynesia

Brian D. Farrell
Department of Environmental, Population, and
    Organismic Biology
University of Colorado
Boulder, CO 80309
United States of America

Harry W. Greene
Museum of Vertebrate Zoology
University of California
Berkeley, CA 94720
United States of America

Antti Halkka
Department of Zoology
Division of Ecology
P.O. Box 17 (P. Rautatiekatu 13)
SF-00014 University of Helsinki
Finland

Ilkka Hanski
Department of Zoology
Division of Ecology
P.O. Box 17 (P. Rautatiekatu 13)
SF-00014 University of Helsinki
Finland

Daniel Haydon
Department of Zoology
Oxford University
Oxford OX1 3PS
England

Robert D. Holt
Museum of Natural History
Department of Systematics and Ecology
The University of Kansas
Lawrence, KA 66045
United States of America

David Jablonski
Department of Geophysical Sciences
University of Chicago
Chicago, IL 60637
United States of America

Christine M. Janis
Program in Ecology and Evolutionary Biology
Division of Biology and Medicine
Brown University
Providence, RI 02912
United States of America

Steven A. Juliano
Department of Biological Sciences
Illinois State University

Normal, IL 61761
United States of America

Erle G. Kauffman
Department of Geological Sciences
University of Colorado
Boulder, CO 80309
United States of America

Jari Kouki
Department of Zoology
Division of Ecology
P.O. Box 17 (P. Rautatiekatu 13)
SF-00014 University of Helsinki
Finland

Roger Earl Latham
Department of Geology
University of Pennsylvania
Philadelphia, PA 19104–6316
United States of America

John H. Lawton
NERC Centre for Population Biology
Imperial College
Ascot, Berkshire SL5 7PY
United Kingdom

Thomas M. Lewinsohn
Laboratório de Interações Insectos/Plantas
Departimento Zoologia
Unicamp
13081 Campinas
Sao Paulo
Brazil

John A. McGowan
Scripps Institution of Oceanography
9500 Gilman Drive
La Jolla, CA 92093–0227

John F. McLaughlin
Center for Conservation Biology
Department of Biological Sciences
Stanford University
Stanford, CA 94305
United States of America

Deborah A. McLennan
Department of Zoology
University of Toronto
Toronto, Ontario M5S 1A1
Canada

Brian A. Maurer
Department of Zoology

Brigham Young University
Provo, UT 84602
United States of America

Charles Mitter
Department of Entomology
University of Maryland
College Park, MD 20742
United States of America

Stephen R. Morton
CSIRO Wildlife
P.O. Box 2111
Alice Springs, NT 0871
Australia

Stephen Pacala
Department of Ecology and Evolutionary Biology
Princeton University
Princeton, NJ 08544
United States of America

David L. Pearson
Department of Zoology
Arizona State University
Tempe, AZ 85287
United States of America

Peter S. Petraitis
Department of Biology
University of Pennsylvania
Philadelphia, PA 19104–6018
United States of America

Eric R. Pianka
Department of Zoology
University of Texas
Austin, TX 78712
United States of America

Ray R. Radtkey
Department of Biology
University of California
La Jolla, CA 92093
United States of America

Robert E. Ricklefs
Department of Biology
University of Pennsylvania
Philadelphia, PA 19104–6018
United States of America

Michael L. Rosenzweig
Department of Ecology and Evolutionary Biology
University of Arizona
Tucson, AZ 85721
United States of America

Jonathan Roughgarden
Department of Biological Sciences
Stanford University
Stanford, CA 94305
United States of America

Dolph Schluter
Department of Zoology
University of British Columbia
Vancouver, B.C. V6T 2A9
Canada

David Tilman
Department of Ecology, Evolution, and Behavior
University of Minnesota
Minneapolis, MN 55455
United States of America

A. J. Underwood
Institute of Marine Ecology
University of Sydney
New South Wales 2006
Australia

James W. Valentine
Department of Integrative Biology
Museum of Paleontology
University of California
Berkeley, CA 94720
United States of America

Blaire Van Valkenburgh
Department of Biology
University of California
Los Angeles, CA 90024–1606
United States of America

Jean-Denis Vigne
Centre National de la Recherche Scientifique
Laboratoire d'Anatomie Comparée
Muséum National d'Histoire Naturelle
55, rue Buffon
75005 Paris
France

Patricia W. Walker
Scripps Institution of Oceanography
9500 Gilman Drive
La Jolla, CA 92093–0227
United States of America

Mark Westoby
School of Biological Sciences
Mac Quarie University
Sydney, NSW 2109
Australia

David H. Wright
Department of Entomology
University of Georgia
Athens, GA 30602
United States of America

Peter Yodzis
Department of Zoology
University of Guelph
Guelph, Ontario N1G 2W1
Canada

Index